STUDY GUIDE

to accompany

CURTIS AND BARNES

Biology FIFTH EDITION

David J. Fox
University of Tennessee

Adele C. Monroe

WORTH PUBLISHERS, INC.

Study Guide
to accompany
Curtis and Barnes: Biology, Fifth Edition
By David J. Fox and Adele C. Monroe

Printed in the United States of America

ISBN: 0-87901-395-8
Printing: 5 4 3 2 1   Year: 93 92 91 90 89

Worth Publishers, Inc.
33 Irving Place
New York, New York 10003

Preface

This Study Guide is designed to accompany the fifth edition of BIOLOGY by Helena Curtis and N. Sue Barnes. Its unique format and content were developed from years of classroom testing.

We assume that you, the user of this Guide, have a serious interest in learning about biology and in doing well on examinations. We have written this Study Guide to help you achieve these goals.

Each Study Guide chapter opens with an outline of the organization of the text chapter. Next, the *Major Concepts* of the chapter are presented in several paragraphs. With these in mind, you should read the chapter quickly; do not slow down to concentrate on details. Next, read the chapter summary to solidify your understanding of the major concepts.

A second close reading of the chapter will constitute your major learning experience. As you read the chapter this time, stop to answer the questions in the *Guided Study of the Chapter* section of this Study Guide. These questions will help you to extract the most important information from each paragraph of the text. Your written answers will be a valuable study aid prior to examinations.

Now you are ready to assess your understanding of the course material by answering sample exam questions. The *Testing Your Understanding* questions are similar to the types of questions you are likely to encounter on quizzes and exams. Each chapter ends with a section, *Performance Analysis,* that also provides a valuable learning experience. In it we discuss why the given response is correct and why the other responses are incorrect. This pedagogical technique further reinforces the correct answer in your mind and helps clarify any misconceptions that may have led you to an incorrect response.

You may find your own ways of studying from the text and this Study Guide. We would like to suggest, however, that you read the essay, *How to Manage Your Time Efficiently and Study More Effectively.* This Study Guide was written with these learning principles in mind; your use of the essay's many helpful suggestions will undoubtedly enhance your success in this course.

We are enthusiastic about this approach to mastering the material presented in the text and hope that you will find these methods effective. If you have suggestions, comments, or criticisms, please do not hesitate to write us, in care of Worth Publishers.

May 1989

David J. Fox
Adele C. Monroe

Contents

How to Manage Your Time Efficiently and Study More Effectively

Richard O. Straub, The University of Michigan at Dearborn

How effectively do you study? Good study habits make the job of being a college student much easier. Many students, who *could* succeed in college, fail or drop out because they have never learned to manage their time efficiently. Even the best students can usually benefit from an in-depth evaluation of their current study habits.

There are many ways to achieve academic success, of course, but your approach may not be the most effective or efficient. Are you sacrificing your social life or your physical or mental health in order to get A's on your exams? Good study habits result in better grades *and* more time for other activities.

EVALUATE YOUR CURRENT STUDY HABITS

To improve your study habits, you must first have an accurate picture of how you currently spend your time. Begin by putting together a profile of your present living and studying habits. Answer the following questions by writing *yes* or *no* on each line.

_____ 1. Do you usually set a schedule to budget your time for studying, recreation, and other activities?

_____ 2. Do you often put off studying until time pressures force you to cram?

_____ 3. Do other students seem to study less than you do, but get better grades?

_____ 4. Do you usually spend hours at a time studying one subject, rather than dividing that time between several subjects?

_____ 5. Do you often have trouble remembering what you have just read in a textbook?

_____ 6. Before reading a chapter in a textbook, do you skim through it and read the section headings?

_____ 7. Do you try to predict exam questions from your lecture notes and reading?

_____ 8. Do you usually attempt to paraphrase or summarize what you have just finished reading?

_____ 9. Do you find it difficult to concentrate very long when you study?

_____ 10. Do you often feel that you studied the wrong material for an exam?

Thousands of college students have participated in similar surveys. Students who are fully realizing their academic potential usually respond as follows: (1) yes, (2) no, (3) no, (4) no, (5) no, (6) yes, (7) yes, (8) yes, (9) no, (10) no.

Compare your responses to those of successful students. The greater the discrepancy, the more you could benefit from a program to improve your study habits. The questions are designed to identify areas of weakness. Once you have identified your weaknesses, you will be able to set specific goals for improvement and implement a program for reaching them.

MANAGE YOUR TIME

Do you often feel frustrated because there isn't enough time to do all the things you must and want to do? Take heart. Even the most productive and successful people feel this way at times. But they establish priorities for their activities and they learn to budget time for each of them. There's much in the saying "If you want something done, ask a busy person to do it." A busy person knows how to get things done.

If you don't now have a system for budgeting your time, develop one. Not only will your academic accomplishments increase, but you will actually find more time in your schedule for other activities. And you won't have to feel guilty about "taking time off," because all your obligations will be covered.

Establish a Baseline

As a first step in preparing to budget your time, keep a diary for a few days to establish a summary, or baseline, of the time you spend in studying, socializing, working, and so on. If you are like many students, much of your "study" time is nonproductive; you may sit at your desk and leaf through a book, but the time is actually wasted. Or you may procrastinate. You are always getting ready to study, but you rarely do.

Besides revealing where you waste time, your diary will give you a realistic picture of how much time you need to allot for meals, commuting, and other fixed activities. In addition, careful records should indicate the times of the day when you are consistently most productive. A sample time-management diary is shown in Table 1.

Table 1   Sample Time-Management Diary: Monday

BEHAVIOR	TIME COMPLETED	DURATION HOURS:MINUTES
Sleep	7:00	7:30
Dress	7:25	:25
Breakfast	7:45	:20
Commute	8:20	:35
Coffee	9:00	:40
French	10:00	1:00
Socialize	10:15	:15
Videogame	10:35	:20
Review Biology	11:00	:25
Biology	12:00	1:00
Lunch	12:25	:25
Study Lab	1:00	:35
Biology Lab	4:00	3:00
Work	5:30	1:30
Commute	6:10	:40
Dinner	6:45	:35
TV	7:30	:45
Study Biology	10:00	2:30
Socialize	11:30	1:30
Sleep		

Prepare a similar chart for each day of the week. When you finish an activity, note it on the chart and write down the time it was completed. Then determine its duration by subtracting the time the previous activity was finished from the newly entered time.

Plan the Term

Having established and evaluated your baseline, you are ready to devise a more efficient schedule. Buy a calendar that covers the entire school term and has ample space for each day. Using the course outlines provided by your instructors, enter the dates of all exams, term paper deadlines, and other important academic obligations. If you have any long-range personal plans (concerts, weekend trips, etc.), enter the dates on the calendar as well. Keep your calendar up to date and refer to it often. I recommend carrying it with you at all times.

Develop a Weekly Calendar

Now that you have a general picture of the school term, develop a weekly schedule that includes all of your activities. Aim for a schedule that you can live with for the entire school term. A sample weekly schedule, incorporating the following guidelines, is shown in Table 2.

1. Enter your class times, work hours, and any other fixed obligations first. *Be thorough.* Using information from your time management diary, allow plenty of time for such things as commuting, meals, laundry, and the like.

2. Set up a study schedule for each of your courses. The study habits survey and your time management diary will direct you. The following guidelines should also be useful.

(a) Establish regular study times for each course. The 4 hours needed to study one subject, for example, are most profitable when divided into shorter periods spaced over several days. If you cram your studying into one 4-hour block, what you attempt to learn in the third or fourth hour will interfere with what you studied in the first 2 hours. Newly acquired knowledge is like wet cement. It needs some time to "harden" to become memory.

(b) Alternate subjects. The type of interference just mentioned is greatest between similar topics. Set up a schedule in which you spend time on several *different* courses during each study session. Besides reducing the potential for interference, alternating subjects will help to prevent mental fatigue with one topic.

(c) Set weekly goals to determine the amount of study time you need to do well in each course. This will depend on, among other things, the difficulty of your courses and the effectiveness of your methods. Many professors recommend studying at least 1 to 2 hours for each hour in class. If your time diary indicates that you presently study less time than that, do not plan to jump immediately to a much higher level. Increase study time from your baseline by setting weekly goals (see **4.**) that will gradually bring you up to the desired level. As an initial schedule, for example, you might set aside an amount of study time for each course that matches class time.

(d) Schedule for maximum effectiveness. Tailor your schedule to meet the demands of each course. For the course that emphasizes lecture notes, schedule time for a daily review soon after the class. This will give you a chance to revise your notes and clean up any hard-to-decipher shorthand while the material is still fresh in your mind. If you are evaluated for class participation (for example, in a language course), allow time for a review just *before* the class meets. Schedule study time for your most difficult (or least motivating) courses during times when you are the most alert and distractions are fewest.

(e) Schedule open study time. Emergencies or additional obligations could throw off your schedule. And you may simply need some extra time periodically for a project or for review in one of your courses. Schedule several hours each week for such purposes.

3. After you have budgeted time for studying, fill in slots for recreation, hobbies, relaxation, household errands, and the like.

4. Set specific goals. Before each study session, make a list of specific goals. The simple note "7–8 PM: study biology" is too broad to ensure the most effective use of the time. Formulate your daily goals according to what you know you must

Table 2   Sample Weekly Schedule

TIME	MONDAY	TUESDAY	WEDNESDAY	THURSDAY	FRIDAY	SATURDAY	SUNDAY
7–8	Dress, eat	Dress, eat	Dress, eat	Dress, eat	Dress, eat		
8–9	Psychology	Study Psych.	Psychology	Study Psych.	Psychology	Dress, eat	
9–10	English	Study English	English	Study English	English	Study English	
10–11	Study French	Free	Study French	Open study	Study French	Study Biology	
11–12	French	Study Biology	French	Open study	French	Study Stats.	
12–1	Lunch	Lunch	Lunch	Lunch	Lunch	Lunch	
1–2	Statistics	Biology Lab	Statistics	Study or free	Statistics	Free or errands	Study Psych.
2–3	Biology	Biology Lab	Biology	Free	Biology	Free or errands	Study French
3–4	Free	Biology Lab	Free	Free	Study Psych.	Free or errands	Study Stats.
4–5	Job	Job	Job	Job	Job	Free	Free
5–6	Job	Job	Job	Job	Job	Free	Free
6–7	Dinner	Dinner	Dinner	Dinner	Dinner	Dinner	Dinner
7–8	Study Biology	Study Biology	Study Biology	Study Biology	Free	Free	Study Biology
8–9	Study English	Study Stats.	Study Psych.	Study Stats.	Free	Free	Study Stats.
9–10	Study Stats.	Study English	Study Stats.	Study French	Free	Free	Study English
10–11	Study Psych.	Study French	Study English	Open study	Free	Free	Open study

This is a sample schedule for a student with a 16-credit load and a 10-hour-per-week part-time job. Using this chart as an illustration, make up a weekly schedule, following the guidelines outlined here.

accomplish during the term. If you have course outlines with advance assignments, set systematic daily goals that will allow you, for example, to cover fifteen chapters before the exam. And be realistic: can you actually expect to cover several chapters in one session? Divide large tasks into smaller units; stop at the most logical resting points. When you complete a specific goal, take a 5- or 10-minute break before tackling the next goal.

5. Evaluate how successful or unsuccessful your studying has been on a daily or weekly basis. Did you reach most of your goals? If so, reward yourself immediately. You might even make a list of five to ten rewards to choose from. If you have trouble studying regularly, you may be able to motivate yourself by making such rewards contingent on completing specific goals.

6. Finally, until you have lived with your schedule for several weeks, don't hesitate to revise it. You may need to allow more time for chemistry, for example, and less for some other course. If you are trying to study regularly for the first time and are feeling burned out, you probably have set your initial goals too high. Don't let failure cause you to despair and abandon the program. Accept your limitations and revise your schedule so that you are studying only 15 to 20 minutes more each evening than you are used to. The point is to *identify a regular schedule with which you can achieve some success.* Time management, like any skill, must be practiced to become effective.

TECHNIQUES FOR EFFECTIVE STUDY

Knowing how to put study time to best use is, of course, as important as finding a place for it in your schedule. Here are some suggestions that should enable you to increase your reading comprehension and improve your notetaking. A few study tips are included as well.

Using SQ3R to Increase Reading Comprehension

How do you study from a textbook? If you are like many students, you simply read and reread in a *passive* manner. Studies have shown, however, that most students who simply read a textbook cannot remember more than half the material ten minutes after they have finished. Often, what is retained is the unessential material rather than the important points upon which exam questions will be based.

This Study Guide employs a program known as SQ3R (Survey, Question, Read, Recite, and Review) to facilitate, and allow you to assess, your comprehension of the important facts and concepts in BIOLOGY, Fifth Edition, by Helena Curtis and N. Sue Barnes.

Research has shown that students using SQ3R achieve significantly greater comprehension of textbooks than students reading in the more traditional passive manner. Once you have learned this program, you can improve your comprehension of any textbook.

Survey Before reading a chapter, determine whether the text or the study guide has an outline or list of objectives. Read this material and the summary at the end of the chapter. Next, read the textbook chapter fairly quickly, paying special attention to the major headings and subheadings. This survey will give you an idea of the chapter's contents and organization. You will then be able to divide the chapter into logical sections in order to formulate specific goals for a more careful reading of the chapter. In this Study Guide, the *Major Concepts* summarizes the major topics of the textbook chapter.

Question You will retain material longer when you have a use for it. If you look up a word's definition in order to solve a crossword puzzle, for example, you will remember it longer than if you merely fill in the letters as a result of putting other words in. Surveying the chapter will allow you to generate important questions that the chapter will proceed to answer. These questions correspond to "mental files" into which knowledge will be sorted for easy access.

As you survey, jot down several questions for each chapter section. One simple technique is to generate questions by rephrasing a section heading. For example, "The Spindle" head could be turned into "What is the spindle?" Good questions will allow you to focus on the important points in the text. Examples of good questions are those that begin as follows: "List two examples of . . ." "What is the function of . . . ?" "What is the significance of . . . ?" Such questions give a purpose to your reading. Similarly, you can formulate questions based on the chapter outline.

The *Guided Study of the Chapter* section of this Study Guide provides the types of questions you might formulate while surveying each chapter. This section is a detailed set of questions covering the points made in the text.

Read When you have established "files" for each section of the chapter, review your first question, begin reading, and continue until you have discovered its answer. If you come to material that seems to answer an important question you don't have a file for, stop and write down the question.

Using this Study Guide, read the chapter one section at a time. First, preview the section by skimming it, noting headings and boldface items. Next, study the appropriate section questions in the *Guided Study of the Chapter*. Then, as you read the chapter section, search for the answer to each question.

Be sure to read everything. Don't skip photo or art captions, graphs, or footnotes. In some cases, what may seem vague in reading will be made clear by a simple graph. Keep in mind that test questions are sometimes drawn from illustrations and charts.

Recite When you have found the answer to a question, close your eyes and mentally recite the question and its answer. Then *write* the answer next to the question. It is important that you recite an answer in your own words rather than the authors'. Don't rely on your short-term memory to repeat the authors' words verbatim.

In responding to the objectives, pay close attention to what is called for. If you are asked to identify or list, do just that. If asked to compare, contrast, or do both, you should focus on the similarities (compare) and differences (contrast) between the concepts or theories. Answering the questions carefully will not only help you to focus your attention on the important concepts of the text, but it will also provide excellent practice for essay exams.

Recitation is an extremely effective study technique, recommended by many learning experts. In addition to increasing reading comprehension, it is useful for review. Trying to explain something in your own words clarifies your knowledge, often by revealing aspects of your answer that are vague or incomplete. If you repeatedly rely upon "I know" in recitation, you really *may not know*.

Recitation has the additional advantage of simulating an exam, especially an essay exam; the same skills are required in both cases. Too often students study without ever putting the book and notes aside, which makes it easy for them to develop false confidence in their knowledge. When the material is in front of you, you may be able to *recognize* an answer, but will you be able to *recall* it later, when you take an exam that does not provide these retrieval cues?

After you have recited and written your answer, continue with your next question in the same way.

Review When you have answered the last question on the material you have designated as a study goal, go back and review. Read over each question and your written answer to it. Your review might also include a brief written summary that integrates all of your questions and answers. This review need not take longer than a few minutes, but it is important. It will help you retain the material longer and will greatly facilitate a final review of each chapter before the exam. (An excellent way to review your understanding of the chapters of BIOLOGY, Fifth Edition, is to complete the questions at the end of each chapter and those in the *Testing Your Understanding* section of this Study Guide. Then go through the *Performance Analysis* section, which explains why the correct answers are correct. You may discover that you don't know the chapter as well as you thought you did!)

One final suggestion: Incorporate SQ3R into your time-management calendar. Set specific goals for completing SQ3R with each assigned chapter. Keep a record of chapters completed, and reward yourself for being conscientious. Initially, it takes more time and effort to "read" using SQ3R, but with practice, the steps will become automatic. More important, you will comprehend significantly more material and retain knowledge longer than passive readers do.

Taking Lecture Notes

Are your class notes as useful as they might be? One way to determine their worth is to compare them with those taken by other good students. Are yours as thorough? Do they provide you with a comprehensible outline of each lecture? If not, then the following suggestions might increase the effectiveness of your notetaking.

1. Keep a separate notebook for each course. Use 8 1/2 × 11-inch pages. Consider using a ring binder, which would allow you to revise and insert notes while still preserving lecture order.

2. Take notes in the format of a lecture outline. Use roman numerals for major points, letters for supporting arguments, and so on. Some instructors will make this easy by delivering organized lectures and, in some cases, by outlining their lectures on the board. If a lecture is disorganized, you will probably want to reorganize your notes soon after the class.

3. As you take notes in class, leave a wide margin on one side of each page. After the lecture, expand or clarify any shorthand notes while the material is fresh in your mind. Use this time to write important questions in the margin next to notes that answer them. This will facilitate later review and will allow you to anticipate similar exam questions.

EVALUATE YOUR EXAM PERFORMANCE

How often have you received a grade on an exam that did not do justice to the effort you spent preparing for the exam? This is a common experience that can leave one feeling bewildered and abused. "What do I have to do to get an A?" "The test was unfair!" "I studied the wrong material!"

The chances of this happening are greatly reduced if you have an effective time-management schedule and use the study techniques described here. But it can happen to the best-prepared student and is most likely to occur on your first exam with a new professor.

Remember that there are two main reasons for studying. One is to learn for your own general academic development. Many people believe that such knowledge is all that really matters. Of course, it is possible, though unlikely, to be an expert on a topic without achieving commensurate grades, just as one can, occasionally, earn an excellent grade without truly mastering the course material. During a job interview or in the workplace, however, your A in Fortran won't mean much if you can't actually program a computer.

In order to keep career options open after you graduate, you must know the material *and* maintain competitive grades. In the short run, this means performing well on exams, which is the second main objective in studying.

Probably the single best piece of advice to keep in mind when studying for exams is to *try to predict exam questions*. This means ignoring the trivia and focusing on the important questions and their answers (with your instructor's emphasis in mind).

A second point is obvious. How well you do on exams is determined by your mastery of *both* lecture and textbook material. Many students (partly because of poor time manage-

ment) concentrate too much on one at the expense of the other.

To evaluate how well you are learning lecture and textbook material, analyze the questions you missed on the first exam. If your instructor does not review exams during class, you can easily do it yourself. Divide the questions into two categories: those drawn primarily from lectures and those drawn primarily from the textbook. Determine the percentage of questions you missed in each category. If your errors are evenly distributed and you are satisfied with your grade, you have no problem. If you are weaker in one area, you will need to set future goals for increasing and/or improving your study of that area.

Similarly, note the percentage of test questions drawn from each category. Although most courses involve exams that cover *both* lecture notes and the textbook, the relative emphasis of each may vary from instructor to instructor. While your instructors may not be entirely consistent in making up future exams, you may be able to tailor your studying for each course by placing *additional* emphasis on the appropriate area.

Exam evaluation will also point out the types of questions your instructor prefers. Does the exam consist primarily of multiple-choice, true-false, or essay questions? You may also discover that an instructor is fond of wording questions in certain ways. For example, an instructor may rely heavily on questions that require you to draw an analogy between a theory or concept and a real-world example. Evaluate both your instructor's style and how well you do with each format. Use this information to guide your future exam preparation.

The Testing Your Understanding sections and Review Tests of this Study Guide will provide you with an important aid in studying for exams. Although the practice exams will help you determine how well prepared you are, they do not cover all the important topics in the chapter. If these tests don't include all of the types of questions your instructor typically writes, make up your own practice exam questions. Spend extra time testing yourself with the question formats that are most difficult for you. There is no better way to evaluate your preparation for an upcoming exam than by testing yourself under the conditions most likely to be in effect during the actual exam.

A FEW PRACTICAL TIPS

Even the best intentions for studying sometimes fail. Some of these failures occur because students attempt to work under conditions that are simply not conducive to concentrated study. To help ensure the success of your self-management program, here are a few suggestions that should assist you in reducing the possibility of procrastination or distraction.

1. If you have set up a schedule for studying, make your roommate, family, and friends aware of this commitment, and ask them to honor your quiet study time. Close your door and post a "Do Not Disturb" sign.

2. Set up a place to study that minimizes potential distractions. Use a desk or table, not your bed or an extremely comfortable chair. Keep your desk and the walls around it free from clutter. If you need a place other than your room, find one that meets as many of the above requirements as possible —for example, in the library stacks.

3. Do nothing but study in this place. It should become associated with studying so that it "triggers" this activity, just as a mouth-watering aroma elicits an appetite.

4. Never study with the television on or with other distracting noises present. If you must have music in the background in order to mask outside noises, for example, play soft instrumental music. Don't pick vocal selections; your mind will be drawn to the lyrics.

5. Study by yourself. Other students can be distracting or can break the pace at which *your* learning is most efficient. In addition, there is always the possibility that group studying will become a social gathering. Reserve that for its own place in your time schedule.

If you continue to have difficulty concentrating for very long, try the following suggestions.

6. Study your most difficult or most challenging subjects first, when you are most alert.

7. Start with relatively short periods of concentrated study, with breaks in between. If your attention starts to wander, get up immediately and take a break. It is better to study effectively for 15 minutes and then take a break than to fritter away 45 minutes out of an hour. Gradually increase the length of study periods, using your attention span as an indicator of successful pacing.

SOME CLOSING THOUGHTS

I hope that these suggestions not only help make you more successful academically, but also enhance the quality of your college life in general. Having the necessary skills makes any job a lot easier and more pleasant. Let me repeat my warning not to attempt to make too drastic a change in your lifestyle immediately. Start by establishing a few realistic goals; then gradually shape your performance to the desired level. Good habits require time and self-discipline to develop. Once established they can last a lifetime.

Introduction

MAJOR CONCEPTS

The theory of evolution is today the foremost unifying idea in all of biology. But it was not always so. Even though certain early Greek philosophers had conceived of a form of evolution, and Aristotle believed in a natural hierarchy of living things, most biologists believed in some form of special creation well into the nineteenth century. Carolus Linnaeus (eighteenth century) believed that all species living were created by God in their present form and have remained unchanged since creation. Although Georges-Louis Leclerc de Buffon (nineteenth century) believed in divine creation, he was the first to propose that species might undergo change over time.

The age of the earth was estimated by Christian theologians to be six thousand years and was widely accepted until geologists of the eighteenth and nineteenth centuries challenged this "fact." This was considered to be too short a time for evolutionary processes to yield the diversity of life forms in evidence on the earth. The uniformitarianism theory of James Hutton proposed that the earth had been shaped by gradual processes, which implied that the earth has a long history and that change is the *normal* course of events. These implications were supported by the fossil strata studies of William Smith. Uniformitarianism further implied that there may be alternatives to a literal interpretation of the Bible.

In an effort to explain his findings in the fossil record, Georges Cuvier proposed the theory of catastrophism, which stated that species became extinct due to series of catastrophes. Louis Agassiz expounded on this concept and contended that each catastrophe was followed by a new separate creation.

Jean Baptiste Lamarck was the first modern scientist to propose a systematic concept of evolution. He believed in the inheritance of acquired characteristics and that all creatures are motivated by an unconscious striving upward toward greater complexity. Both of these concepts are inconsistent with current knowledge of genetics and evolution.

During his voyage on the *Beagle*, Charles Darwin was struck with the great diversity of animal and plant life. While in the Galapagos islands he imagined that the diversity exhibited by the birds, tortoises, and other animals could be accounted for by modifications undergone by a single species. An essay by Thomas Malthus on the factors that limit the growth of populations stimulated Darwin to consider natural selection as a mechanism of evolution. He considered the variations among individuals in a population to have occurred by chance and that these variations are the raw materials upon which natural selection acts.

There are three fundamental principles of modern biology. One: The cell theory states that all living things are made of cells and that all cells come from preexisting cells. Two: All

1

living organisms are subject to the same chemical and physical laws as nonliving things. Three: Living organisms are subject to the laws of thermodynamics which state that energy cannot be created or destroyed, only changed from one form to another, and that natural events tend to proceed in such a way that energy dissipates or becomes random.

The numerous forms of life are classified into five kingdoms. The kingdom Monera includes the simplest single-celled organisms that lack an organized cell nucleus. Most members of the kingdom Protista are also one-celled, but the cells of these organisms have a true nucleus. The kingdom Fungi includes yeasts, molds, and mushrooms. The final two kingdoms, Plantae and Animalia, contain more complex life forms that reflect a higher level of evolutionary development than the other three kingdoms.

The nature of science is such that it deals only with concepts that can be tested. Once an idea or "hunch" is formulated in such a way that its validity can be tested, it is called a hypothesis. When substantial evidence accumulates to support the hypothesis, it becomes a theory. In order for a theory to be accepted as a law or principle, it must withstand repeated testing over time.

Science is an ongoing process that is constantly changing our understanding of the natural world. Through new technologies, science may provide alternative techniques that may be applied to solve problems, but it cannot determine whether or not (or how) these new advances should be used.

HOW TO STUDY THE CHAPTER

Read the entire chapter through quickly, focusing on the major concepts.

Use the GUIDED STUDY OF THE CHAPTER to help you identify the important details as you **reread** the chapter. Writing out the answers to these questions will help fix them in your mind as well as provide you with a valuable study aid.

Answer the questions in TESTING YOUR UNDERSTANDING without the aid of your text. Check your answers against those in PERFORMANCE ANALYSIS. Analyzing your answers will give you valuable feedback on your level of understanding and preparedness for classroom testing.

GUIDED STUDY OF THE CHAPTER

I. Introduction (page 1)

1. In what activities did Darwin engage as a young man that prepared him for his role as the *Beagle's* naturalist?

2. Upon Darwin's return, what circumstance allowed him to pursue a lifetime of independent work and study?

II. The Road to Evolutionary Theory (pages 1–5)

A. General Remarks (pages 1–2)

Focus: **Although Darwin was not the first to propose that organisms evolve, he did amass a large body of supporting evidence and also was the first to propose a mechanism by which evolution might occur.**

3. Describe the intellectual climate during Darwin's time by (a) contrasting the beliefs of Aristotle and the later Occidental biologists, and (b) summarizing the belief of Carolus Linnaeus.

   a.

   b.

B. Evolution before Darwin (page 2)

4. What two theories that were similar to modern theories were developed by the school of Greek philosophy to which Anaximander and Lucretius belonged?

5. The ideas of Georges-Louis Leclerc de Buffon seem to be intermediate between the creationist and evolutionist positions. Explain.

6. Summarize the evolutionary views of Charles Darwin's grandfather, Erasmus.

C. **The Age of the Earth** *(pages 2–3)*

7. a. What is the basic premise of the theory of uniformitarianism?

   b. What three important implications did Hutton's theory of uniformitarianism have for the development of evolutionary theory?

D. **The Fossil Record** *(page 3)*

8. William Smith's study of the distribution of fossils within geological strata eventually established that:

9. How did the revolution in geological thinking set the stage for a reevaluation of biological thinking with respect to the history of living organisms?

E. **Catastrophism** *(pages 3–4)*

10. What were Cuvier's contributions to paleontology?

11. How did the explanations of Cuvier and Agassiz differ with respect to repopulation after extinctions?

F. **The Concepts of Lamarck** *(pages 4–5)*

> *Focus:* Lamarck's concept of evolution was undoubtedly influenced by his work on one-celled organisms and invertebrates.

12. How did Lamarck's interpretation of the fossil record differ from that of Cuvier?

13. What two concepts did Lamarck propose to explain the increasing complexity of organisms?

III. **Development of Darwin's Theory** *(pages 5–9)*

A. **The Earth Has a History** *(page 5)*

14. What conclusion of Charles Lyell had the greatest impact on the development of Charles Darwin's theory?

B. **The Voyage of the *Beagle*** *(pages 5–6)*

15. What impressed Darwin the most as he collected specimens in the interior and on the two coasts of South America?

16. Why did Darwin find the Galapagos islands the most interesting of all his study sites?

17. What observations led Darwin to suspect that all the birds on the Galapagos islands originally arose from one species?

C. **The Darwinian Theory** *(pages 7–9)*

18. a. What was the basic conclusion of the essay by Thomas Malthus?

    b. How did it influence Darwin's thinking?

19. How did Darwin view natural selection?

20. a. According to Darwin, inherited variations arise by _____ and their value to an organism is measured by the organism's _____ and _____.

    b. What role does variation play in Darwin's scheme of evolution?

21. How did Darwin's explanation of the giraffe's long neck differ from that of Lamarck?

22. Why did Darwin call *The Origin of Species* "one long argument"?

> *Focus:* "The theory of evolution is quite rightly called the greatest unifying theory in biology."

D. *Essay:* **Darwin's Long Delay** *(page 8)*

23. What events prompted Darwin to begin working on his manuscript after a delay of 20 years, and culminated in its publication?

24. Cite a possible reason for Darwin's long delay in publishing his theory of evolution.

E. **Challenges to Evolutionary Theory** *(page 9)*

25. List six areas of biology from which evidence consistent with evolutionary theory has been drawn.

IV. **Unifying Principles of Modern Biology** *(pages 10–12)*

A. **All Organisms Are Made Up of Cells** *(page 10)*

26. What did Schleiden, Schwann, and Virchow each contribute to the development of the modern cell theory?

> *Focus:* There is an unbroken continuity from the first primitive cells that evolved to the organisms living on earth today.

B. **All Organisms Obey the Laws of Physics and Chemistry** *(pages 10–11)*

27. Summarize the basic tenets of the vitalists and mechanists.

28. Cite the evidence supporting the viewpoints of the reductionists and the vitalists.

29. How did the experiments of Eduard and Hans Büchner lay to rest the vitalists' position, to which Louis Pasteur ascribed?

30. The capacity of an organism to reproduce resides in the _____ molecule.

C. All Organisms Require Energy (page 12)

31. State in your own words the two thermodynamic laws that apply to biology.

Focus: Cells can be best understood as a complex of systems for transforming energy. Evolution may be viewed as a competition among organisms for the most efficient use of energy resources.

V. The Forms of Life (pages 12–14)

Focus: Diversity among living organisms (there are some 5 million *different* species) is the result of evolutionary competition.

---

32. Our current system for classifying living organisms reveals:

   a.

   b.

   c.

33. Summarize the characteristics of the organisms that belong to each of the five kingdoms.

   Monera

   Protista

   Fungi

Plantae

Animalia

VI. The Nature of Science (pages 14–18)

A. General Remarks (pages 14–15)

34. Name at least four ways of seeking principles of order.

35. Identify three ways in which scientists accumulate data about a hypothesis.

36. In order to qualify as science, what principles must be observed by a discipline of study?

37. a. Arrange the following terms that apply to scientific ideas in order of increasing validity: hypothesis, law, theory, hunch.

   b. Why is it not possible to test some hypotheses immediately?

38. a. What steps are involved in conducting a controlled experiment?

   b. Of what value is a controlled experiment?

39. In what two ways can hypotheses be repeatedly and successfully tested?

40. By what means and for what reasons do scientists report their experimental methods, results, and conclusions to the rest of the scientific community?

41. a. How does a hypothesis become a theory in science?

    b. How does the meaning of the word theory as used in science differ from its meaning in common usage?

42. List some of the methods and approaches used in science.

B. *Essay:* **Some Comments on Science and Scientists** (*page 16*)

43. *Science and Theory:* What do the two quotes in this section have in common?

44. *The Scientific Method:* What does this short essay have to do with the process of science?

45. What overall impression of scientists do you get from reading the selections in this section?

> *Focus:* It is the nature of science and scientists to continually question the details as well as the basic tenets of their hypotheses. Only by a continual process of questioning and experimentation can scientists hope to approach the truth.

C. **Science and Human Values** (*pages 17–18*)

46. How does science differ from art, religion, and philosophy in its approach to natural phenomena?

47. Explain why science is unable to make value judgments.

> *Focus:* Scientists are limited to studying only those hypotheses and phenomena that can be investigated with objective techniques.

48. Identify at least three examples of situations in which science has provided new technology but could not indicate how that tool should be used.

D. **Science as Process** (*page 18*)

49. a. Why do text books stress only what is known at the present time?

    b. What is the potential danger of this approach?

> *Focus:* Science is not information contained in textbooks; it is a dynamic process taking place in the minds of living scientists. Study biology because it is an adventure for the mind and nourishment for the spirit.

TESTING YOUR UNDERSTANDING

After you have completed this examination, compare your answers with those in the section that follows.

1. After his voyage, Charles Darwin:
   a. gave up the study of natural history.
   b. took up the typical life of an English country gentleman.
   c. lived on his mariner's pension.
   d. became an amateur botanist.
   e. devoted the rest of his life to work and study.

2. Which of the following scientists believed that all species now in existence were created by the sixth day of God's labor?
   a. Erasmus Darwin
   b. Georges-Louis Leclerc de Buffon
   c. Aristotle
   d. Lucretius
   e. Carolus Linnaeus

3. Who said "there are lesser families conceived by Nature and produced by Time"?
   a. Aristotle
   b. James Hutton
   c. Georges-Louis Leclerc de Buffon
   d. William Smith
   e. Carolus Linnaeus

4. Erasmus Darwin believed that:
   a. changes in organisms take place by a series of degenerations.
   b. changes in organisms take place in response to their environments.
   c. all organisms were created by God.
   d. organisms exist in a natural hierarchy, with man on top.
   e. the evidence before one's own eyes confirms the concept of special creation.

5. To whom can the following concept be attributed? Change is itself the normal course of events, as opposed to a static system interrupted by an occasional unusual event.
   a. James Hutton
   b. Charles Darwin
   c. William Smith
   d. Carolus Linnaeus
   e. Georges Cuvier

6. Whose work on the distribution of fossils led to the unavoidable conclusion that the earth had been formed layer by layer over the course of time?
   a. Charles Darwin
   b. Charles Lyell
   c. Thomas Malthus
   d. William Smith
   e. James Hutton

7. Which concept does NOT require the earth to have a long history?
   a. uniformitarianism
   b. catastrophism
   c. special creation
   d. evolution
   e. natural selection

8. In order for Charles Darwin's ideas to be accepted as valid, the concept of _____ had to be revised, and the work of _____ was a great contribution in this area.
   a. the age of the earth; Louis Agassiz
   b. geographic space; William Smith
   c. the age of the earth; Charles Lyell
   d. geographic space; Ernst Mayr
   e. the diversity of species; Aristotle

9. The number of elephants descended from a single pair does not increase to 19 million in 750 years. Charles Darwin proposed this is the result of:
   a. artificial selection.
   b. natural selection.
   c. insufficient hereditary variability.
   d. the inheritance of acquired characteristics.
   e. reproductive failure resulting from the breeding of closely related individuals.

10. Charles Darwin's ideas about natural selection were stimulated by:
    a. Thomas Malthus's essay on limits to population growth.
    b. James Hutton's theory of uniformitarianism.
    c. Charles Lyell's two-volume work on geology.
    d. Erasmus Darwin's notes and poems.
    e. Jean Baptiste Lamarck's hypothesis of acquired characteristics.

11. Which of these scientists proposed that cells can arise only from preexisting cells?
    a. Robert Hooke
    b. Matthias Schleiden
    c. Theodor Schwann
    d. Rudolf Virchow
    e. Charles Darwin

12. One feature of scientific investigation that is particularly important to the indirect method of hypothesis testing is:
    a. gathering background data on the organisms being studied.
    b. reporting one's findings in journals or at scientific meetings.
    c. the use of well-established and respected techniques.
    d. collecting data of many different types to provide balance to the observations.
    e. the generation of logical deductions that predict what will happen if the hypothesis is correct.

13. Ninety-five percent of the organisms in kingdom Animalia are:
    a. invertebrates.
    b. mammals.
    c. insects.
    d. microscopic multicellular organisms.
    e. single-celled organisms that prey on other single-celled organisms.

14. Yeasts belong to kingdom:
    a. Plantae.
    b. Animalia.
    c. Protista.
    d. Fungi.
    e. Monera.

15. Which of these men first showed that an organic molecule could be synthesized from an inorganic one?
    a. René Descartes
    b. Friedrich Wöhler
    c. Eduard Büchner
    d. Hans Büchner
    e. Louis Pasteur

16. T or F   The only experiments that produce valid results are those performed in accordance with the scientific method.

17. T or F   Eduard and Hans Büchner claimed that the transformation of fruit juice into wine was a vital reaction, while Louis Pasteur demonstrated that it was due to enzymatic activity.

18. T or F   More knowledge about the diversity of organisms has been gained in the last 100+ years than in all the prior history of science.

19. T or F   A Greek school of philosophy developed a theory of evolution that was unknown in Europe when the science of biology began to take shape.

20. T or F   None of Jean Baptiste Lamarck's contemporary scientists believed his ideas about acquired characteristics and an unconscious upward striving of every living creature.

21. Using the giraffe as an example, explain how Jean Baptiste Lamarck's principle of acquired characteristics was supposed to have worked.

22. What was peculiar about the tortoises on the Galapagos islands? What was it about the Galapagos birds that haunted Charles Darwin?

23. Name five of the six fields of biology that yield data consistent with the theory of evolution.

24. "A modern science is not a static accumulation of facts . . ." Explain what the authors mean by this statement.

PERFORMANCE ANALYSIS

1. e   After Charles Darwin returned from his voyage on the *Beagle,* he spent the rest of his life reflecting on his observations, studying the work of his contemporaries, and formulating his theory of evolution which he published in *The Origin of Species.* (page 1)

2. e   Carolus Linnaeus joined many of his contemporaries in the belief that all species in existence at that time were created by the sixth day of God's labor. He maintained this belief throughout his life in spite of the fact that his *Species Plantarum* underwent multiple revisions to accommodate new species as they were discovered and described. (page 2)

3. c   Georges-Louis Leclerc de Buffon made this statement in an attempt to explain the overwhelming variety of living organisms. He believed that the "lesser families conceived by Nature and produced by Time" were species created in addition to those created divinely. He believed changes in organisms over time reflected a degeneration process. (page 2)

4. b   Erasmus Darwin was among the first to suggest that species change over time. He proposed that species are historically connected to one another, that organisms change in response to their environments, and that these changes may be passed on to the offspring. (page 2)

5. a   In developing the theory of uniformitarianism, the geologist James Hutton proposed that the earth's surface was molded by the physical forces of nature over long periods of time and that change was a continual and gradual process, rather than sudden and catastrophic. This theory eventually led to the conclusion that the earth is much older than had been previously believed. (page 2)

6. d   William Smith was an English surveyor who recognized that the different geological strata have fossils characteristic of each layer regardless of the region of England in which they were found. He did not interpret these findings, but others concluded that the surface of the earth had been formed one layer at a time over a long period. (page 3)

7. c   The concept of special creation is the only one that does not require earth to have a long history. Even catastrophism, a concept based on the extinction of species by a series of catastrophes, requires a fairly long time period to explain why so many species no longer exist. (page 3)

8. c   Charles Darwin's concept of the evolution of species required that the earth be much older than the estimate of 6000 years which was derived by Christian theologians. The theory of uniformitarianism proposed by James Hutton and later work by Charles Lyell which produced evidence supporting this theory demonstrated that the earth's age was far older than previous investigators had imagined. (page 5)

9. b   Charles Darwin proposed that a process called natural selection limited the survival and reproduction of organisms in natural environments. He argued that those individuals that were the best adapted to their environment (i.e., had characteristics promoting their survival and

reproduction under the conditions present in the environment) were able to reproduce and pass those traits on to their offspring. (page 7)

10. **a** Thomas Malthus's essay warned that if the human population continued to grow, it would outgrow its food supply. Darwin applied this basic concept (that food supply and other factors limit the growth of populations) to other species. Darwin also recognized the very important role of genetic variation in providing a variety of characteristics upon which the forces of natural selection can act. (page 7)

11. **d** Rudolf Virchow contributed to the modern cell theory by proposing that all cells arise from preexisting cells. Schleiden and Schwann concluded that all plants and animals (respectively) are composed of cells. Robert Hooke first applied the word "cell" to living tissues. (page 10)

12. **e** Although all of the procedures listed are important in attempting to validate a hypothesis, *indirect* methods of testing hypotheses are strongly dependent upon the generation of logical conclusions predicting possible events if the hypothesis is correct. Indirect methods are often employed to study situations that are not amenable to direct tests. (page 15)

13. **a** Ninety-five percent of the 1.5 million known animal species are invertebrates, animals without a backbone. One million of these are insect species. (page 14)

14. **d** Yeasts are members of the kingdom Fungi. (page 13)

15. **b** Friedrich Wöhler synthesized urea (an organic molecule) from ammonium cyanate (an inorganic compound) in the laboratory. René Descartes was a leading proponent of the mechanistic view of living organisms. Eduard and Hans Büchner dealt a fatal blow to the concept of vitalism when they demonstrated that fermentation is an enzymatic process that can occur outside a living cell. Louis Pasteur was a leading supporter of vitalism. (pages 10, 11)

16. **False** The authors report that there are many methods of investigating hypotheses (some of which are indirect) that can yield important information concerning the accuracy of a hypothesis. The method of investigation is determined by the nature of the problem. (page 15)

17. **False** See the answer to question 15.

18. **True** The information obtained about the diversity and functioning of living organisms within the last 100+ years far outweighs all the information accumulated prior to that time. (page 14)

19. **True** A Greek school of philosophy that was founded by Anaximander developed an atomic theory and an evolutionary theory that bear striking similarities to modern concepts. However, this school of thought was unknown to Europeans who initiated the development of modern biology. (page 2)

20. **False** Most of Lamarck's contemporaries did not object to his thoughts on the inheritance of acquired characteristics and an unconscious upward striving of organisms. However, Georges Cuvier did object very strenuously to these ideas. (page 5)

21. Lamarck believed that organs in animals become stronger or weaker through use or disuse. However, he also believed that these changes could be passed on to the offspring. His explanation of the long necks in giraffes was that by stretching their necks, giraffe's necks grew longer and their offspring inherited that length. Then the offspring stretched their necks, which grew even longer than those of the parents. (Contrast this with Darwin's explanation which would be that the giraffes with necks long enough to reach certain food items would leave more offspring than giraffes with shorter necks, therefore longer necks would become more common.) (page 4)

22. Each of the Galapagos islands had a type of tortoise unique to that island that was not found on any of the other islands. Among the birds, Darwin observed a group of 13 finchlike species that differed with respect to the sizes and shapes of their beaks and bodies and most especially with respect to the types of foods they ate. Many of their physical characteristics were found only on completely dissimilar birds on the mainland. Darwin later speculated that one bird species may have "been taken and modified for different ends." (page 6)

23. The six fields of biology that have produced data consistent with the theory of evolution are anatomy, physiology, biochemistry, embryology, behavior, and genetics. (page 9)

24. Modern science is an ongoing process. Our body of scientific knowledge is constantly being revised and updated. The hypotheses being proposed and tested this year will be substantiated or disproven over time. New questions are constantly being asked and new techniques are being developed to answer those questions. Just as the concept of vitalism (once held as truth by some of the most respected scientists of the time) gave way in the face of overwhelming evidence to the contrary, so will some of today's ideas fall by the wayside in the future. (page 18)

PART **1**

Biology of Cells

SECTION **1**

The Unity of Life

Atoms and Molecules

CHAPTER ORGANIZATION

MAJOR CONCEPTS

Our universe is currently thought to have arisen subsequent to an explosion that sent matter and energy hurtling through space. In the aftermath of this explosion particles of matter combined to form the elements that are the building blocks of all living and nonliving systems.

Seven characteristics that distinguish living from nonliving objects are (1) structural and functional organization and complexity, (2) homeostasis, (3) reproduction, (4) growth and development, (5) energy conversion, (6) responsiveness to stimuli, and (7) adaptation to the environment.

An atom consists of protons and neutrons in a nucleus with electrons in orbitals and energy levels around the nucleus. The electrons in the energy level closest to the nucleus contain the least energy and those in the outer energy levels contain the most energy. The number and arrangement of electrons determines the chemical activity of an atom. The most stable electron configuration is for the outer energy level of an atom to be filled.

The number of protons in an atom determines its identity as an element. Atoms with the same number of protons but with differing numbers of neutrons are isotopes of the same element. The isotopes of an element exhibit the same chemical activity.

Atoms of the same or different elements can bond together to form molecules. The bonds may result from a transfer of electrons (ionic) or from sharing electrons (covalent). Covalent bonds may be single, double, or triple, depending on the number of pairs of electrons being shared. Covalent bonds may be polar or nonpolar in character. Chemical reactions between molecules involve exchanges of electrons among the atoms involved.

Of all the elements known to humans, six elements (carbon, hydrogen, nitrogen, oxygen, phosphorus, and sulfur) constitute 99 percent of all living matter. These elements are biologically significant because the atoms are small, they tend to form stable covalent bonds, and all (except hydrogen) can form bonds with two or more atoms.

Within a cell, the levels of biological organization from least to most complex are subatomic particles, atom, molecule, and cell. In multicellular organisms, cells are organized into tissues, which may be grouped into organs. Organs that participate in a common function are grouped into an organ system. Each multicellular organism is a collection of integrated organ systems.

HOW TO STUDY THE CHAPTER

Read the entire chapter through quickly, focusing on the major concepts.

Use the GUIDED STUDY OF THE CHAPTER to help you identify the important details as you **reread** the chapter. Writing out the answers to these questions will help fix them in your mind as well as provide you with a valuable study aid.

Answer the questions in TESTING YOUR UNDERSTANDING without the aid of your text. Check your answers against those in PERFORMANCE ANALYSIS. Analyzing your answers will give you valuable feedback on your level of understanding and preparedness for classroom testing.

GUIDED STUDY OF THE CHAPTER

I. Introduction *(page 23)*

> *Focus:* **Our universe is believed to have originated following an explosion that scattered energy and matter into space.**

1. Describe how atoms are thought to have formed in the aftermath of the "big bang."

II. Atoms *(pages 23–28)*

A. General Remarks *(pages 23–24)*

2. *Vocabulary*: Distinguish between atoms and elements.

3. a. How do protons, neutrons, and electrons differ with respect to electrical charge?

   b. Describe the arrangement of protons, neutrons, and electrons in an atom.

c. Which subatomic particle determines the chemical properties of an atom?

4. *Vocabulary*: Distinguish between the terms atomic weight and atomic number.

B. Isotopes *(pages 24–25)*

5. *Vocabulary*: What are isotopes?

> *Focus:* **Since isotopes of an element have the same number of electrons, they behave the same chemically.**

6. a. What is meant when an isotope is said to be "radioactive"?

   b. Identify two forms in which energy may be emitted from a radioactive isotope.

   c. Name two devices that may be used to detect radioactive isotopes.

   d. Describe four uses of radioactive isotopes in biological research or medicine.

C. Models of Atomic Structure *(page 28)*

> *Focus:* **Various models of atomic structure have been constructed throughout history. As our understanding changes, the models change also.**

7. Identify the characteristics of atomic structure that are emphasized by the (a) plum-pudding model, (b) billiard-ball model, (c) planetary model, (d) Bohr model, and (e) orbital model.

a.

b.

c.

d.

e.

III. *Essay:* The Signs of Life *(pages 26–27)*

8. Identify seven characteristics of living things.

IV. Electrons and Energy *(pages 28–30)*

A. General Remarks *(pages 28–29)*

9. *Vocabulary:* Define potential energy as it applies to electrons.

10. a. *Vocabulary:* What is an electron energy level?

b. What is the relationship between the amount of energy possessed by an electron and its location relative to the nucleus?

11. *Vocabulary:* Define the term quantum.

Focus: In order for an electron to be raised to the next higher energy level, it must absorb a specified quantity of energy (a quantum). That same amount of energy is released when an electron moves to a lower energy level.

---

12. a. What effect does sunlight have on electrons in chlorophyll molecules in plant cells?

b. What is the consequence of this reaction?

B. The Arrangement of Electrons *(pages 29–30)*

13. *Vocabulary:* What is an electron's orbital?

14. In the first two energy levels of an atom, how many electrons may occupy one *orbital?*

15. What is the maximum number of electrons that may occupy (a) the first energy level and (b) the second energy level for atoms with an atomic number less than or equal to 20?

16. a. Describe the arrangement of electrons that is consistent with maximum stability for an atom.

b. The atoms of which elements exhibit this arrangement?

c. Comment on the reactivity of the elements mentioned in (b) and relate this to their subatomic structure.

17. If an atom has a partially filled energy level, what are three means by which that atom can obtain a filled outer energy level?

V. Bonds and Molecules (pages 30–34)

A. General Remarks (page 30)

18. *Vocabulary:* What are molecules?

19. *Vocabulary:* What are bonds?

B. Ionic Bonds (page 31)

20. a. *Vocabulary:* What are ions?

    b. How are ions created?

21. What force maintains an ionic bond?

22. For each of the following ions, identify one role in living organisms.

    Potassium (K^+)

    Calcium (Ca^{2+})

    Sodium (Na^+)

    Magnesium (Mg^{2+})

C. Covalent Bonds (pages 32–34)

23. *Vocabulary:* Define or describe a covalent bond.

24. *Vocabulary:* How is a molecular orbital similar to and different from an atomic orbital?

25. Of what significance is the fact that carbon has four electrons in its outer energy level?

> *Focus:* An atom can acquire a filled outer energy level by gaining or losing electrons (thereby becoming an ion) or by sharing electrons in a covalent bond with another atom.

Polar Covalent Bonds (pages 32–33)

26. a. *Vocabulary:* What is a polar covalent bond?

    b. How does a polar covalent bond differ from an ionic bond?

27. a. *Vocabulary:* Characterize a polar molecule.

    b. What element is often present in polar molecules in living organisms?

28. Under what conditions is a covalent bond wholly nonpolar?

Double and Triple Bonds (pages 33–34)

29. Account for the bonding properties of oxygen based on the arrangement of its electrons.

30. *Vocabulary:* Distinguish double and triple bonds from single covalent bonds.

31. Describe the changes that occur in the electron orbitals of a carbon atom when (a) single and (b) double bonds are formed.

32. a. Describe the bonding between carbon and oxygen in carbon dioxide.

    b. What is the resulting shape of the carbon dioxide molecule and what consequences does this shape have for the character of the molecule?

33. How may the shape of a molecule that contains polar covalent bonds influence the overall polarity of the molecule? Cite two examples.

Focus: The inability of atoms to rotate around double and triple bonds influences the physical properties of compounds in which they are present.

34. Correlate the bonding patterns in fats and oils with their physical state at room temperature.

35. *Review Question:* For the following Bohr diagrams, indicate the most likely type of bond the atom would form in order to achieve a filled outer electron energy level. Include in your answer the number of electrons involved and what happens to those electrons (gained, lost, shared).

    a. Carbon

    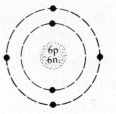

    b. Nitrogen

    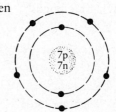

    c. Sodium

    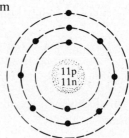

    d. Sulfur

    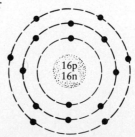

e. Chlorine

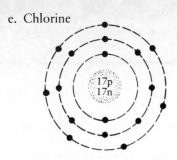

f. Calcium

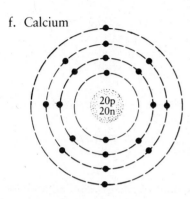

VI. Chemical Reactions *(pages 34–35)*

A. General Remarks *(page 34)*

36. *Vocabulary:* Define chemical reactions.

Focus: In a correctly constructed chemical equation, the number of atoms of each element on the left side equals the number on the right side.

---

37. What information does an equation for a chemical reaction provide?

38. a. *Vocabulary:* Define a chemical compound.

    b. Cite two examples of chemical compounds.

B. Types of Reactions *(pages 34–35)*

39. *Vocabulary:* Distinguish between combination, dissociation, and exchange reactions.

40. Identify the following chemical reactions as combination, dissociation, or exchange reactions.

    a. HCl + KOH $\longrightarrow$ KCl + H_2O
       Hydrochloric   Potassium   Potassium   Water
       acid   hydroxide   chloride

    b. H_2CO_3 $\longrightarrow$ H^+ + HCO_3^-
       Carbonic   Hydrogen   Bicarbonate
       acid   ion   ion

    c. K^+ + Cl^- $\longrightarrow$ KCl
       Potassium   Chloride   Potassium
       ion   ion   chloride

VII. The Biologically Important Elements *(page 35)*

41. Name the six elements that constitute 99 percent of all living matter.

42. Identify three characteristics of these elements that are significant to the role they play in living systems.

VIII. Levels of Biological Organization *(pages 35–37)*

Focus: Living systems and nonliving matter are composed of the same chemical and physical components.

---

43. List the levels of biological organization from subatomic particles up to the level at which the characteristics of living organisms appear.

44. a. Most of the weight of an *E. coli* cell consists of what compound?

    b. How many different kinds of macromolecules may be present in a single *E. coli* cell?

    c. Name three types of functions these macromolecules perform.

45. a. What is the preferred residence of *E. coli*?

    b. How are the cells of the human intestinal tract similar to and different from *E. coli* cells?

46. Describe one situation that illustrates the delicate balance between the various microorganisms in the human intestinal tract.

47. *Vocabulary*: Define the term biosphere.

IX. **Summary** *(page 38)*: Read the summary. If you are familiar with the essential features of the material presented there, you are ready to complete the section TESTING YOUR UNDERSTANDING.

TESTING YOUR UNDERSTANDING

After you have completed this examination, compare your answers with those in the section that follows.

1. Several hundred thousand years after the "big bang":
   a. the temperature had dropped to 10^{11} degrees Celsius.
   b. protons and neutrons began to assemble.
   c. atoms split to form protons and neutrons.
   d. molecules began to assemble from atoms.
   e. the first forms of life began to assemble from molecules.

2. The atomic number of an element describes the number of:
   a. neutrons in its outer energy level.
   b. neutrons in the atomic nucleus.
   c. electrons in the atomic nucleus.
   d. protons in the atomic nucleus.
   e. electrons in its outer energy level.

3. There are _____ naturally occurring elements.
   a. 88
   b. 92
   c. 96
   d. 100
   e. 108

4. Which statement is NOT true of isotopes?
   a. Isotopes of silicon are frequently used in biological research.
   b. Isotopes can be used to determine the age of rocks and fossils.
   c. Isotopes of carbon are frequently used in biological research.
   d. Certain isotopes are useful in treating various forms of cancer.
   e. Radioactive isotopes exhibit the same chemical properties as do their corresponding nonradioactive isotopes.

5. The concept of the atom as the indivisible unit of elements is almost _____ years old.
   a. 50   d. 200
   b. 100   e. 250
   c. 150

6. Which model of atomic structure is the most accurate and most current?
   a. Bohr   d. planetary
   b. plum pudding   e. billiard ball
   c. orbital

7. The discrete amount of energy required to move an electron from one energy level to another is known as a(n):
   a. quark.   d. ionophore.
   b. isotope.   e. quantum.
   c. jump.

8. The lowest electron energy level of an atom consists of a single _____ orbital that can contain a maximum of _____ electrons.
   a. helical; 2   d. tetrahedral; 2
   b. spherical; 1   e. spherical; 2
   c. tetrahedral; 1

9. The elements hydrogen and helium both:
   a. are noble gases.
   b. contain two electrons in their outer energy level.
   c. have a single spherical orbital.
   d. possess one neutron in the nucleus.
   e. have a completely filled outer energy level.

10. How many orbitals are present in the first (_____) and second (_____) energy levels?
    a. 1; 2
    b. 1; 4
    c. 2; 2
    d. 2; 4
    e. 2; 8

11. An important characteristic of ionic bonds is that they:
    a. are very weak.
    b. break apart easily in water.
    c. can be formed between any two atoms.
    d. involve the sharing of electrons between two atoms.
    e. involve the transfer of protons from one atom to another resulting in the formation of charged particles.

12. Which ion(s) is (are) required in the production and propagation of nerve impulses?
   a. calcium
   b. potassium
   c. sodium
   d. potassium and sodium
   e. calcium, potassium, and sodium

13. The bonds joining the two identical atoms of a hydrogen molecule (H_2) and an oxygen molecule (O_2) are:
   a. ionic.
   b. noble.
   c. nonpolar covalent.
   d. polar covalent.
   e. triple.

14. In carbon dioxide (CO_2), oxygen forms _____ bonds with carbon.
   a. ionic
   b. covalent
   c. double ionic
   d. double covalent
   e. triple covalent

15. When carbon forms four single bonds with other atoms, its bonds:
   a. all lie in a straight line.
   b. form the shape of a "Y".
   c. point toward the corners of a cube.
   d. point toward the corners of a tetrahedron.
   e. point toward the corners of a square.

16. Carbonic acid may be formed by bubbling carbon dioxide through water. Balance this equation by determining the numbers that go in each blank.
   _____ CO_2 + _____ H_2O ⟶ _____ H_2CO_3
   a. 1; 1; 1
   b. 1; 2; 1
   c. 2; 1; 2
   d. 2; 2; 1
   e. 1; 2; 2

17. Which molecule is NOT a compound?
   a. H_2
   b. NaCl
   c. H_2CO_3
   d. CH_4
   e. CO_2

18. The binomial *Escherichia coli* means that:
   a. all species with the name *coli* belong to the genus *Escherichia*.
   b. all species with the name *Escherichia* belong to the genus *coli*.
   c. within the genus *Escherichia*, there is a species called *E. coli*.
   d. within the genus *coli*, there is a species called *Escherichia*.
   e. the only species within the genus *Escherichia* is called *E. coli*.

19. T or F   All isotopes are radioactive, which means that their atomic nuclei are unstable and emit rapidly moving subatomic particles or electromagnetic radiation.

20. T or F   The greater the potential energy of an electron, the closer it will be found to its atomic nucleus.

21. T or F   In plants, the radiant energy of sunlight is transformed into the chemical energy on which life depends as electrons are passed from higher to lower energy levels.

22. T or F   Atoms with partially filled outer energy levels tend to interact in such a way that after the reaction both atoms have completely filled outer energy levels.

23. Distinguish between the atomic weight and the atomic number of an element.

24. Using sodium and chlorine as examples, describe the process by which two interacting atoms become ions.

25. At room temperature, fats are solids whereas oils are liquids. Account for this fact in terms of the bonding structures of fats and oils.

PERFORMANCE ANALYSIS

1. **b**   According to current information, protons and neutrons began to assemble several hundred thousand years after the "big bang" as the universe cooled. The temperature had dropped to 2500°C by this time. (page 23)

2. **d**   By definition, the atomic number of an element is the number of protons in the nucleus. (page 24)

3. **b**   There are 92 naturally occurring elements. (Several more have been created under laboratory conditions by researchers.) (page 23)

4. **a**   The rate of decay of isotopes occurring in geological strata is used to determine the age of rocks and fossils. Isotopes of carbon are sometimes used in biological research to trace the flow of certain atoms or molecules through an organism. Some radioactive isotopes are helpful in combating certain types of cancer. Since isotopes of the same element have the same electron configuration, radioactive isotopes behave chemically and physically the same as nonradioactive isotopes of the same element. Isotopes of silicon would not be very useful in biological research since silicon is unreactive in living systems and tends to form stable compounds. (pages 24, 25)

5. **d**   The concept of the atom as the smallest particle identifying an element emerged nearly 200 years ago. (page 28)

6. **c**   The orbital model of atomic structure is the most accurate representation of the relationship between electrons and the nucleus of an atom. It replaced the Bohr model, which emphasizes the relationship between electron energy level and distance from the nucleus. The planetary model pictured electrons orbiting the nucleus in a fashion analogous to the orbit of the planets around the sun. The plum pudding model proposed that the atom is a spherical mass with electrons embedded in the surface. The oldest model of atomic structure presented the atom as a solid sphere resembling a billiard ball. (page 28)

7. **e**   A quantum is defined as the discrete amount of energy needed to move an electron from one energy level to a higher energy level. (page 29)

8. **e**   The first (and lowest) energy level of an atom contains one orbital that is spherical in shape and contains at most two electrons. The second energy level contains one

spherical orbital and three dumbbell-shaped orbitals; each of the four orbitals can hold up to two electrons. (page 29)

9. **c** Helium has only two electrons, both of which occupy the inner energy level which has a single spherical orbital. The term "noble gas" refers to the fact that atoms of elements in which the outer energy level is completely filled tend not to interact with other atoms. Hydrogen has only one electron, and it occupies the spherical orbital of the lowest energy level. The nucleus of helium consists of two protons and two neutrons; hydrogen has a single proton as its nucleus. (page 30 and Table 1–3)

10. **b** The first energy level consists of one spherical orbital that can hold two electrons. The second energy level consists of one spherical and three dumbbell-shaped orbitals that can hold two electrons each. (page 29 and Figure 1–6)

11. **b** Ionic bonds form between oppositely charged ions. An ion is an atom that has acquired a positive or negative charge due to the loss or gain of one or more electrons. Although not as strong as covalent bonds, ionic bonds are stronger than other types of attractive forces between atoms including hydrogen bonds. The fact that ions dissociate readily in water is very important to their roles in living organisms. (page 31)

12. **e** Calcium, potassium, and sodium are all essential to the production and propagation of a nerve impulse. Their roles will be discussed in more detail in Chapter 40. (page 31)

13. **c** In molecular hydrogen (H_2) and molecular oxygen (O_2), electrons are shared between the two atoms. In each case, the atoms are of the same element (their attraction for electrons is identical) and the resulting covalent bond is nonpolar. (page 33)

14. **d** Carbon shares electrons via two double bonds with two oxygen atoms in the carbon dioxide molecule. Each bond is polar since the oxygen atom attracts electrons more strongly than the carbon atom. However, the molecule itself is nonpolar because it is linear with the slightly negative oxygen atoms at the ends of the molecule. (page 33)

15. **d** When carbon makes four single covalent bonds, new electron orbitals are created. These orbitals, and the bonds involved, point toward the corners of a tetrahedron. (It will only be a regular tetrahedron if atoms of the same element are involved in all four bonds.) (page 33)

16. **a** An equation is balanced when the total number of atoms of an element on the left side equals the total number on the right side. This equation is balanced as it is written, with one molecule of each reactant combining to yield one molecule of product. (page 34)

17. **a** A compound is a substance made of molecules containing atoms of two or more different elements. Molecu-

lar hydrogen (H_2) is the only substance listed that is not a compound. (page 34)

18. **c** A species of organism consists of individuals that are reproductively isolated from all other types of organisms. Matings within a species produce fertile offspring. Matings of individuals of two different species may or may not result in offspring, but any offspring produced will be sterile. A genus is a collection of species that are similar and are probably evolutionarily related. In the binomial system of naming organisms, a species is identified by its genus name (or the letter abbreviation of its genus name) and by its species name (or more accurately, its specific epithet). Within the genus *Escherichia*, there is a species called *Eschericia coli*, or abbreviated *E. coli*. (footnote, page 35)

19. **False** All isotopes are not radioactive. However, those that are radioactive do emit particles and/or electromagnetic radiation from their nuclei. (page 25)

20. **False** Electrons that are closest to the nucleus have the lowest potential energy. The farther away an electron is from the nucleus, the greater its potential energy. (page 28)

21. **True** During the process of photosynthesis, plants convert the radiant energy of the sun into chemical energy. As you will learn in detail in Chapter 10, this conversion process involves the transfer of electrons from lower to higher and back to lower energy levels. (page 29)

22. **True** The most stable electron configuration for an atom is one in which the outer energy level is filled. Atoms tend to interact by either transferring or sharing electrons until the outer energy level is filled. (page 30)

23. The atomic weight of a element refers to the sum of the number of protons and neutrons in the nucleus. The atomic number refers only to the number of protons. (page 24)

24. Sodium has one electron in its outer energy level and would have a stable configuration (and a single positive charge) if it lost that electron. Chlorine has seven electrons in its outer energy level and would have a stable configuration (and a single negative charge) if it gained one more electron. When sodium and chlorine atoms interact, sodium loses its one outer electron to chlorine. The resulting oppositely charged ions are attracted to each other and form ionic bonds. (page 31)

25. The fatty acid chains in oils have double bonds between some of the carbon atoms whereas the fatty acid chains in fats do not have double bonds in the carbon backbone. Molecules can rotate freely around single bonds but cannot rotate around double bonds. The double bonds in oils increase rigidity of individual molecules, preventing them from being closely packed. Fats do not have this restriction of movement and can therefore be packed into a solid form. (page 34)

Water

CHAPTER ORGANIZATION

MAJOR CONCEPTS

A water molecule consists of two hydrogen atoms bonded to one oxygen atom in polar covalent bonds. The resulting molecule has polar character. Hydrogen bonds form when the slightly positive hydrogen atoms of one water molecule are attracted to the slightly negative oxygen atom of another water molecule. Although not as strong as ionic or covalent bonds, these hydrogen bonds are responsible for the following important characteristics of water: surface tension, capillary action, resistance to temperature change, high heat of vaporization (responsible for the phenomenon of evaporative cooling), and the lower density of ice relative to liquid water.

These properties of the water molecule have important consequences for living organisms.

The polar nature of water makes it well suited as a solvent for ionic compounds and other polar molecules. Nonpolar molecules do not dissolve in water.

Water has a mild tendency to ionize into hydronium ions (H_3O^+) and hydroxyl ions (OH^-). By convention, hydronium ions are referred to as hydrogen ions (H^+). Acids are substances that increase the hydrogen ion concentration of a solution and bases increase the hydroxyl ion concentration. The acidity of a solution is measured using a pH scale where the pH equals the negative log of the hydrogen ion concentration. Strong acids and bases are those that ionize nearly completely. Weak acids and bases ionize only slightly. In living systems, buffers serve the important function of resisting changes in pH.

All the water in the biosphere is continuously being cycled between the living and nonliving components of the biosphere.

HOW TO STUDY THE CHAPTER

Read the entire chapter through quickly, focusing on the major concepts.

Use the GUIDED STUDY OF THE CHAPTER to help you identify the important details as you **reread** the chapter. Writing out the answers to these questions will help fix them in your mind as well as provide you with a valuable study aid.

Answer the questions in TESTING YOUR UNDERSTANDING without the aid of your text. Check your answers against those in PERFORMANCE ANALYSIS. Analyzing your answers will give you valuable feedback on your level of understanding and preparedness for classroom testing.

GUIDED STUDY OF THE CHAPTER

I. Introduction (page 40)

> *Focus:* Water is essential for the survival of all living organisms. The unique properties of water have several biologically important consequences.

II. The Structure of Water (page 41)

1. a. Diagram a water molecule.

   b. What type of bond is present between the atoms of a water molecule?

   c. Describe the polar nature of the water molecule and illustrate it on your diagram in (a).

2. a. *Vocabulary:* Define a hydrogen bond.

   b. Under what conditions do hydrogen bonds occur?

   c. How do the strength and permanence of hydrogen bonds compare to those of covalent and ionic bonds?

III. Consequences of the Hydrogen Bond (pages 42–45)

A. Surface Tension (page 42)

3. a. *Vocabulary:* Define surface tension.

   b. Surface tension is a consequence of what property of water molecules?

4. *Vocabulary:* Distinguish between cohesion and adhesion.

> *Focus:* The polarity of the water molecule is responsible for the ability of water to adhere strongly to charged or polar molecules.

B. Capillary Action and Imbibition (page 42)

5. a. *Vocabulary:* Distinguish between capillary action and imbibition.

   b. Cite one example illustrating the importance of each force in the life of a plant.

C. Resistance to Temperature Change (page 43)

6. *Vocabulary:* What is meant by the specific heat of a substance?

7. *Vocabulary:* What is a calorie?

8. Kinetic energy refers to:

9. *Vocabulary:* Distinguish between heat and temperature.

10. a. Describe how hydrogen bonding between water molecules determines the specific heat of water.

b. Name two consequences of the high specific heat of water that are important to living systems.

D. Vaporization (pages 43–44)

11. *Vocabulary:* Define vaporization and heat of vaporization.

12. Explain why perspiration is an effective mechanism for cooling the body.

E. Freezing (page 44)

13. *Vocabulary:* What is the density of a substance?

14. a. Account for the fact that ice occupies more space than an equivalent weight of water.

b. Of what importance is this fact to aquatic life forms?

15. *Vocabulary:* Define the heat of fusion of a substance.

16. Describe how ice and snow act as temperature stabilizers in the environment.

17. a. What influence does a dissolved substance have on the freezing point of water?

b. Cite three examples in which the adaptation of living organisms to freezing temperatures involves this principle.

IV. Water as a Solvent (pages 45–47)

18. a. *Vocabulary:* Define the term solution.

b. *Vocabulary:* Distinguish between solvent and solute.

19. What two types of molecules (or compounds) are soluble in water?

20. *Vocabulary:* Distinguish between hydrophilic and hydrophobic molecules and cite examples of each type of molecule.

V. *Essay:* The Seasonal Cycle of a Lake (page 46)

21. a. Describe the temperature differences in the epilimnion, thermocline, and hypolimnion in the spring, summer, fall, and winter.

b. Discuss the consequences of these seasonal variations to inhabitants of a lake.

VI. Ionization of Water: Acids and Bases *(pages 47–52)*

A. General Remarks *(page 47)*

22. *Vocabulary:* What is meant by the term dynamic equilibrium?

23. *Vocabulary:* Distinguish between a hydronium ion and hydroxide ion.

24. *Vocabulary:* Distinguish between an acid and a base.

B. Strong and Weak Acids and Bases *(page 48)*

25. What characteristic determines whether an acid (or base) is classified as strong or weak?

26. What happens to the concentration of H^+ and OH^- ions when equal amounts of an acid and a base of comparable strength are mixed?

27. Discuss the action of (a) carboxyl groups and (b) amino groups in affecting hydrogen ion and hydroxyl ion concentrations in living systems.

C. The pH Scale *(pages 48–49)*

28. a. What is the use of the pH scale?

    b. What does the term pH represent?

29. *Vocabulary:* Define molecular weight.

> *Focus:* The term "mole" is a quantitative term in the same sense that "dozen" is a quantitative term. A mole is 6.02×10^{23} atoms, molecules, etc. The weight of a mole of a compound is equal to the molecular weight expressed in grams. Example: A mole (6.02×10^{23}) of water molecules weighs 18 grams (the molecular weight of water is 18).

30. A solution with a pH of 5 has _____ times the number of hydrogen ions as a solution with a pH of 6.

31. *Vocabulary:* Redefine acids and bases in terms of pH.

32. Fill in the following table to indicate what happens when an acid or base is added to a solution.

Table for Question 32

SUBSTANCE ADDED TO THE SOLUTION:	ACID	BASE
Which ion increases in concentration?		
Which ion decreases in concentration?		
Which ion may be donated to the solution?		
Which ion may be removed from the solution?		

D. *Essay: Acid Rain* (pages 50–51)

33. What is the pH of unpolluted rainfall?

34. a. Name the two main acids responsible for acid rain.

    b. What are the major sources of each of these two acids?

35. Identify two direct consequences of acid rain pollution to animal life.

36. a. What are four effects of acid rain on plant life?

    b. What factors influence the degree to which acid rain will affect plants in a particular region?

E. **Buffers** *(page 52)*

37. a. Why is it important for living systems to maintain a fairly constant pH?

    b. What is the *most common* pH range for the chemical processes that occur in living organisms?

38. a. *Vocabulary:* What is a buffer?

    b. By what mechanism do buffers resist changes in pH?

c. Under what condition is a buffer system most effective?

> *Focus:* One component of the blood buffer system (carbonic acid) is in equilibrium with dissolved carbon dioxide in the blood, which is in equilibrium with gaseous carbon dioxide in the lungs. This means that the process of respiration is directly associated with blood acid-base balance.

39. The following equation represents the relationship between the carbonic acid buffer system in the blood and CO_2 in the lungs. What would you expect to happen to a animal's respiratory rate if there was and increase in the H^+ ion concentration in the blood? Explain your answer.

$$H^+ + \underset{\substack{\text{Bicarbonate} \\ \text{ion}}}{HCO_3^-} \longleftrightarrow \underset{\substack{\text{Carbonic} \\ \text{acid}}}{H_2CO_3} \longleftrightarrow H_2O + \underset{\substack{CO_2 \text{ (gas dissolved} \\ \text{in the blood)}}}{CO_2} \overset{\substack{CO_2 \text{ (gas in the} \\ \text{lungs)}}}{\updownarrow}$$

VII. **The Water Cycle** *(page 53)*

40. Describe the possible fates of a water molecule that has evaporated from the Atlantic Ocean.

41. *Vocabulary:* Distinguish between zone of saturation and water table.

> *Focus:* All the water in the biosphere is continuously being recycled through living and nonliving components.

VIII. **Summary** *(page 54):* Read the summary. If you are familiar with the essential features of the material presented there, you are ready to complete the section TESTING YOUR UNDERSTANDING.

TESTING YOUR UNDERSTANDING

After you have completed this examination, compare your answers with those in the section that follows.

1. Water is drawn into a paper towel by:
   a. surface tension.
   b. specific heat.
   c. ionization.
   d. buffers in the towel.
   e. adhesion and cohesion.

2. Water is at its maximum density at:
   a. 100°C.
   b. 32°C.
   c. 20°C.
   d. 4°C.
   e. 0°C.

3. Suppose some water at 20°C is placed in a freezer which is at −10°C. The temperature of the water would drop steadily, and then would remain constant for a while until the water turned into ice at the same temperature. The water stays at the same temperature until it loses an amount of energy equal to its:
   a. heat of vaporization.
   b. heat of fusion.
   c. specific heat.
   d. heat of ionization.
   e. heat of solidification.

4. Substances that readily dissolve in water are _____ whereas substances that do not readily dissolve in water are _____.
   a. nonpolar; polar
   b. oxidized; reduced
   c. acids; bases
   d. hydrophilic; hydrophobic
   e. bases; acids

5. Many organic molecules contain a carboxyl group (−COOH) which ionizes slightly in water to form −COO$^-$ and H$^+$. These molecules are examples of:
   a. strong acids.
   b. weak acids.
   c. strong bases.
   d. weak bases.
   e. neither acids nor bases.

6. A solution with a pH of 7:
   a. is basic.
   b. is acidic.
   c. has more hydrogen ions than hydroxyl ions.
   d. has more hydroxyl ions than hydrogen ions.
   e. has a hydrogen ion concentration of 10^{-7} moles per liter.

7. Combinations of weak acids and weak bases will resist changes in the pH of a solution and are:
   a. hydrophobic.
   b. acidophobic.
   c. always at a pH of 7.
   d. buffers.
   e. able to absorb unlimited amounts of hydrogen ions.

8. Nutrients found on the bottom of a deep lake are carried throughout the lake:
   a. in the fall.
   b. during the winter.
   c. during the summer.
   d. when the lake is stratified.
   e. by the epilimnion.

9. Which of the following is NOT a confirmed or potential problem associated with acid rain?
   a. decreased need for fertilizers
   b. acidification of mountain lakes
   c. loss of vegetation
   d. decreased nutrient availability
   e. death of fish in the Adirondacks

10. Each water molecule is made up of two:
    a. hydrogen atoms covalently linked to one oxygen atom.
    b. oxygen atoms covalently linked to one hydrogen atom.
    c. oxygen atoms ionically linked to two hydrogen atoms.
    d. hydrogen atoms ionically linked to one oxygen atom.
    e. oxygen atoms ionically linked to one oxygen atom.

11. Which of the following is NOT a property of water?
    a. Its greatest density is at 4°C.
    b. Dissolved substances increase the temperature at which water freezes.
    c. It has a highly polar molecular structure.
    d. Its melting and freezing points are the same.
    e. Its most stable molecular structure, an open lattice-work, is created by ice crystal formation.

12. Which statement is NOT true of hydrogen bonds?
    a. They are very short lived.
    b. They are stronger than either covalent or ionic bonds.
    c. They usually involve hydrogen atoms covalently bonded to an oxygen or a nitrogen atom.
    d. They occur in large molecules to help maintain structural stability.
    e. Every water molecule can form hydrogen bonds with four other water molecules.

13. The capillary movement of water molecules into a material that subsequently swells is called:
    a. adhesion.
    b. cohesion.
    c. imbibition.
    d. buffering capacity.
    e. surface tension.

14. When making a solution, the substance (usually a liquid) that is present in greatest amount is known as the:
    a. solvent.   d. solute.
    b. buffer.   e. base.
    c. acid.

15. Identify the phenomenon for which the surface tension of water is NOT at least partially responsible.
    a. Water forms spherical droplets as it falls from a leaky faucet.
    b. Water striders are able to walk on water.
    c. A needle may be floated on the surface of a glass of water.
    d. Water is capable of wetting things that are polar or charged.
    e. Water beads up on the surface of a waxed car.

16. The energy it takes to change 1 gram of a liquid to a gas is called the:
    a. heat of fusion.
    b. energy of ionization.
    c. equilibrium constant.
    d. specific heat.
    e. heat of vaporization.

17. Surface tension is the result of:
    a. the adhesion of water molecules to one another.
    b. the fusion of water molecules to one another.
    c. the cohesion of water molecules to one another.
    d. imbibition.
    e. ionization.

18. The amount of heat that is needed to raise the temperature of 1 gram of water 1°C is known as:
    a. the heat of fusion.
    b. kinetic energy.
    c. a calorie.
    d. the heat of vaporization.
    e. the energy of ionization.

19. The water molecule as a whole is neutral in charge, having equal numbers of protons and electrons. The electrons:
    a. are evenly dispersed between the hydrogen and oxygen nuclei so that all regions of the water molecule are uncharged.
    b. spend more time around the hydrogen nuclei so that the hydrogens are slightly negatively charged and the oxygen is slightly positively charged.
    c. spend more time around the oxygen nucleus so that the oxygen is slightly negatively charged and the hydrogens are slightly positively charged.
    d. spend more time around the hydrogen nuclei so that the oxygen is slightly negatively charged and the hydrogens are slightly positively charged.
    e. spend more time around the oxygen nucleus so that the hydrogens are slightly negatively charged and the oxygen is slightly positively charged.

20. **T or F** The climate of coastal regions is more moderate than that of comparable inland regions due to the high specific heat of the surrounding water.

21. **T or F** In the water cycle, which is powered by geothermal energy, recirculated water is made available to supply the needs of terrestrial plants and animals.

22. **T or F** A mole of any substance, regardless of its molecular weight, has the same number of particles in it as a mole of any other substance.

23. Describe the water cycle. Show how it provides a continual source of water for land plants and animals.

24. Why is it important, biologically, that ice floats?

PERFORMANCE ANALYSIS

1. **e** Water is drawn into a paper towel by capillary action, which involves the adhesion of water molecules to the paper towel fibers and the cohesion of water molecules to one another. (page 42)

2. **d** The maximum density of water occurs at 4°C, a temperature at which molecular motion is slow. However, as water cools to the freezing point of 0°C, the motion of molecules decreases to the point where each molecule can make hydrogen bonds with four other water molecules. In order to make these bonds, the molecules must move apart slightly (decreasing the density) and assume a regular pattern. (page 44)

3. **b** By definition, the heat of fusion is the amount of energy required to convert 1 gram of a substance from a solid to a liquid. The same amount of energy is lost when the substance changes from the liquid to the solid phase. The heat of vaporization is the amount of energy required to change 1 gram of a substance from a liquid to gas. The specific heat is the quantity of energy needed to raise the temperature of 1 gram of a substance by 1°C. All three quantities are measured in calories. (pages 43–45)

4. **d** Hydrophilic molecules are those that dissolve readily in water. They are characterized by polar or charged regions. Hydrophobic molecules do not have any polar regions and therefore do not dissolve in water; they are soluble in nonpolar solvents. (pages 46, 47)

5. **b** Acids are substances that increase the relative concentration of H^+ in a solution by either donating H^+ to or removing OH^- from the solution. Bases are substances that raise the relative concentration of OH^- in a solution either by donating OH^- to or removing H^+ from the solution. Strong acids and bases dissociate completely in solution whereas weak acids and bases ionize only slightly. (page 48)

6. **e** The pH of a solution is the negative logarithm of the hydrogen ion concentration. A solution with a pH of 7 is neutral and has equal numbers of hydrogen and hydroxyl ions; the concentration of both ions is 10^{-7} moles per liter. (page 49)

7. **d** A buffer is a substance that resists a change in the pH of a solution in which it is present. Buffers typically consist of the acid-base pair (H^+-donor and H^+-acceptor forms) of a weak acid. (page 52)

8. **a** In the seasonal cycle of a lake, sediment nutrients are mixed throughout the lake during the fall overturn. The fall overturn occurs as surface water is cooled, sinks to the bottom, and displaces the warmer water which rises to the surface. A similar mixing occurs in the spring. (essay, page 50)

9. **a** Acid rain has been connected with the acidification of mountain lakes, with subsequent death of fish and other aquatic life; loss of vegetation; and a decreased nutrient content and availability in soils. (essay, page 50)

10. **a** A water molecule consists of two hydrogen atoms covalently bonded to one oxygen atom. The bonds are polar covalent bonds and the molecule as a whole is polar. (page 41)

11. **b** Dissolved substances typically decrease the temperature at which water freezes because they interfere with the formation of the regular latticework that results from hydrogen bonding between water molecules. (page 45)

12. **b** Hydrogen bonds are weaker than either covalent or ionic bonds but are still strong enough to have a significant influence on the behavior of molecules. (page 41)

13. **c** Imbibition is the capillary movement of water molecules into substances that swell as a result of the water movement. Capillary action is due to cohesive and adhesive forces. (page 42)

14. **a** In a solution, the substance present in the largest quantity is the solvent. Any component present in a smaller amount is a solute. (page 45)

15. **d** The wetting capacity of water is due to the adhesion of water molecules to charged or polar molecules on the surface of an object. Surface tension results from the cohesion of water molecules to each other. Both cohesion and adhesion are consequences of the polar character of the water molecule. (page 42)

16. **e** See the answer to question 3.

17. **c** See the answer to question 15.

18. **c** A calorie is defined as the amount of heat needed to raise the temperature of 1 gram of water by 1°C. In nutritional literature, the term "calorie" refers to one kilocalorie, or 1,000 calories. (page 43)

19. **c** The electrons of a water molecule spend more time close to the oxygen atom because oxygen attracts electrons much more strongly than does hydrogen. The result is a polar molecule, with the oxygen being slightly negative in character and the hydrogens being slightly positive. (page 41)

20. **True** Oceans resist cooling during the winter and heating during the summer as a consequence of the high specific heat of water. This serves to moderate the climates of coastal regions. (page 43)

21. **False** The water cycle is powered by solar, not geothermal energy. (essay, page 53)

22. **True** A mole is a quantitative term in the same way that the term "dozen" is a quantitative term. There are 12 donuts in a dozen, just as there are 12 apples in a dozen. One mole of water (H_2O) contains 6.02×10^{23} molecules and weighs 18 grams. One mole of methane (CH_4) contains 6.02×10^{23} molecules and weighs 16 grams. (page 48)

23. The water cycle is powered by solar energy that causes water evaporation from bodies of water and living organisms. Evaporated water condenses to form clouds, from which water falls in various forms of precipitation. The precipitated water strikes terrestrial bodies of water and land masses. Plants take up water from soil and surface water. Animals obtain the water they need from surface sources and by ingestion of plant and animal matter. Water eventually finds its way to the oceans by various surface and underground routes. (essay, page 53)

24. If ice was more dense than liquid water and sank to the bottom of a body of water, all bodies of water would eventually be frozen solid. The surface might thaw when exposed to direct sunlight, but there would not be enough depth or volume to support the vast quantities of aquatic life that currently exist on the planet. Since ice floats, it is able to act as an insulator, ensuring that the water below the ice never freezes and maintaining a year-round habitat for aquatic organisms. (page 44)

Organic Molecules

CHAPTER ORGANIZATION

MAJOR CONCEPTS

Organic molecules are those that contain carbon. Carbon-containing molecules are well-suited as the building blocks of living organisms because the carbon atom is small, it can make four strong covalent bonds with other atoms, and it can make double bonds. Hydrocarbons are compounds that contain only carbon and hydrogen. The four main types of organic molecules present in living systems are carbohydrates, lipids, proteins, and nucleic acids.

The chemical activity of an organic molecule is determined by the number and type of functional groups it contains. Functional groups important in biological systems include aldehyde, amino, carboxyl, hydroxyl, ketone, methyl, and phosphate groups. Polar or charged functional groups make an organic molecule soluble in polar solvents, such as water. Hydrophobic functional groups make a molecule insoluble in water.

Covalent bonds, which are strong and stable, are typically found in organic molecules. The strength of a covalent bond is determined by the configuration of the electron orbitals. Any chemical reaction resulting in new combinations of atoms involves changes in electron configuration and consequently in bond strengths. During a reaction, energy will either be released from the system or taken into the system from the environment depending on the relative bond strengths of the products and the reactants. The energy change ($\triangle H°$) of a reaction is expressed in kilocalories per mole of reactant. A minus sign indicates energy has been released; a plus sign means that energy was taken into the system. In living systems, enzymes act as catalysts that lower the energy needed to start a chemical reaction and minimize the proportion of energy released as heat.

Carbohydrates are divided into three main categories based on the number of sugar units each molecule contains. Monosaccharides have one sugar unit, disaccharides have two, and polysaccharides are composed of long chains of sugar units. Carbohydrates function in energy storage, supply, and transport and as structural components of cells.

Monosaccharides contain carbon, hydrogen, and oxygen and can be described by the formula $(CH_2O)_n$. The basic structure is that of a carbon backbone with attached hydroxyl groups and an aldehyde or ketone group. In a solution, monosaccharides typically form a ring structure in which the aldehyde or ketone group reacts with one of the hydroxyl groups. The principal energy source for vertebrate cells is the six-carbon sugar glucose.

Disaccharides are energy transport forms in plants and many insects. The two sugar units are joined together by a condensation reaction in which one molecule of water is removed. In the reverse reaction, hydrolysis, the bond between the two sugar units is broken with the addition of one molecule of water.

Polysaccharides function in energy storage and as structural components of cells. The most common storage polysaccharide of plants (starch) and the most common storage polysaccharide of higher animals (glycogen) are both constructed of alpha-glucose units. Cellulose, the major structural polysaccharide of plants, is a polymer of beta-glucose units.

Lipids are nonpolar compounds that are not soluble in water. Lipids include fats and oils, phospholipids, glycolipids, and steroids. Lipids function in energy storage, insulation, protection (cushioning), cellular structure, and as components of hormones.

A fat molecule is constructed of one glycerol molecule bonded to three fatty acid molecules. Fats and oils contain more than twice the energy of carbohydrates or proteins on a per gram basis. If the hydrophobic nature of fats and oils and the hydrophilic nature of carbohydrates is taken into account, fats store six times as much energy as glycogen. Fats, which are solids at room temperature, contain saturated fatty acids. Oils, which are liquids at room temperature, contain unsaturated fatty acids.

Phospholipids are similar to fats but have a phosphate group substituted for a fatty acid at one terminal glycerol carbon. In glycolipids, a short carbohydrate chain is attached to the third carbon of the glycerol molecule. The phosphate group of a phospholipid and the carbohydrate group of a glycolipid are both hydrophilic. In an aqueous solution, the hydrophobic lipid tails cluster together and the hydrophilic heads are exposed. Phospholipids and glycolipids are important components of cell membranes because of this behavior.

The basic structure of steroids is that of four carbon rings joined together. Important steroids are cholesterol, sex hormones, and the hormones of the adrenal cortex.

Proteins are long chains of amino acids that can fold into various three-dimensional shapes. The same 20 amino acids are present in the proteins of all living organisms. In the synthesis of a polypeptide, each amino acid is added by a condensation reaction. Conversely, the peptide bonds between adjacent amino acids are broken by hydrolysis.

A particular protein may consist of one or more polypeptide chains. There are four levels of protein structure. The primary structure, the sequence of amino acids, determines the final shape of a protein. The secondary structure is the regular configuration caused by hydrogen bonding between atoms of the polypeptide backbone. The tertiary structure is the three-dimensional structure resulting from interactions among R groups. The quaternary structure is determined by the interaction of two or more polypeptides in a protein.

Proteins play important roles in cellular structure and in the regulation of biochemical processes. Structural proteins may be either fibrous or globular in shape. The enzymes that catalyze chemical reactions are typically globular proteins.

Nucleic acids are composed of nucleotide units, each of which consists of a sugar, a phosphate group, and a nitrogenous base. The molecules that transmit genetic information between cells and that translate the genetic message into proteins are nucleic acids.

HOW TO STUDY THE CHAPTER

Read the entire chapter through quickly, focusing on the major concepts.

Use the GUIDED STUDY OF THE CHAPTER to help you identify the important details as you **reread** the chapter. Writing out the answers to these questions will help fix them in your mind as well as provide you with a valuable study aid.

Answer the questions in TESTING YOUR UNDERSTANDING without the aid of your text. Check your answers against those in PERFORMANCE ANALYSIS. Analyzing your answers will give you valuable feedback on your level of understanding and preparedness for classroom testing.

GUIDED STUDY OF THE CHAPTER

I. **Introduction** (*page 55*)

   1. *Vocabulary:* What are organic molecules?

   > *Focus:* Living organisms are 50 to 95 percent water and less than one percent small ions. Organic molecules account for nearly all the remaining material.

   2. a. List the four types of organic molecules that are found in large numbers in living organisms.

b. What three elements are common to all four types of molecules?

c. What other three elements are also commonly found in biologically important organic molecules?

II. The Central Role of Carbon (pages 55–59)

A. The Carbon Backbone (pages 55–56)

3. What characteristic of an organic molecule determines it shape?

4. Of what significance is the overall shape of an organic molecule?

5. *Vocabulary:* What are hydrocarbons?

6. a. Name three hydrocarbon fuels of economic importance.

b. From what are these fuels derived?

B. Functional Groups (pages 56–57)

7. Of what significance are functional groups to organic molecules?

8. *Vocabulary:* Define an alcohol in terms of chemical structure.

9. Name and draw the structural formulas of seven types of functional groups of importance in living systems.

10. What properties are conferred on organic molecules by the presence of a polar functional group?

C. The Energy Factor (pages 57–59)

11. How is the strength of a chemical bond measured?

Focus: When a covalent bond is broken, atoms with unfilled outer energy levels are formed. These atoms are unstable and readily form covalent bonds with available atoms.

12. What factors determine whether the bonds formed are the same as or different from those that were broken?

13. What factor determines whether energy will be released to or absorbed from the environment during a chemical reaction?

14. Describe how a calorimeter can be used to measure energy stored in an organic compound. (Figure 3–4)

15. a. *Vocabulary:* What does the term $\triangle H°$ represent?

b. Explain the significance of a positive and a negative value for $\triangle H°$.

16. a. What two aspects of chemical reactions are minimized in those chemical reactions that occur in living organisms?

    b. What type of molecule assists in accomplishing these feats?

17. *Vocabulary:* State the biological definition of the term "strategy" in your own words.

D. *Essay:* **Why Not Silicon?** *(page 58)*

    18. a. What characteristics of silicon's atomic structure and bonding properties are responsible for the fact that silicon is not very common in living organisms?

        b. How do these properties compare with those of carbon?

III. **Carbohydrates: Sugars and Polymers of Sugars** *(pages 59–66)*

A. **General Remarks** *(page 59)*

    19. Name two functions of carbohydrates in living systems.

20. Distinguish among monosaccharides, disaccharides, and polysaccharides and name one example of each type of carbohydrate.

21. *Vocabulary:* Distinguish between the terms polymer and monomer.

B. **Monosaccharides: Ready Energy for Living Systems** *(pages 60–61)*

    22. The formula for a monosaccharide that emphasizes the proportions of the constituent elements is:

    23. *Vocabulary:* Distinguish between aldoses and ketoses.

    24. What happens to the molecular structure of a ketose or an aldose when it is placed in water?

    25. Describe the structural difference between alpha-glucose and beta-glucose.

> *Focus:* Carbohydrates can be oxidized by living organisms to release the energy contained in their molecular bonds for use by the organism. The amount of energy released equals the amount released if the carbohydrate is burned in a calorimeter.

26. Which monosaccharide is a principal energy source for vertebrates?

27. a. When 1 mole of glucose is oxidized, how many moles of oxygen are used?

   b. What are the chemical products?

   c. How much energy is released?

C. Disaccharides: Transport Forms *(pages 61–64)*

Focus: **In plants and many insects, the transport carbohydrate is a disaccharide.**

28. Name three common disaccharides and indicate their constituent sugars.

29. *Vocabulary:* Distinguish between the chemical processes of condensation and hydrolysis. (Include energy input versus energy released in your distinction.)

30. Match the following disaccharides with their component monosaccharides. An answer may be used more than once. A blank may have more than one answer as indicated in parentheses.
   1. sucrose
   2. lactose
   3. trehalose
   4. maltose

   a. _____ two glucose molecules (2 answers)
   b. _____ glucose and fructose
   c. _____ glucose and galactose

D. *Essay:* Representations of Molecules *(pages 62–63)*

31. Name three types of molecular models and discuss the advantages and disadvantages of each type.

32. a. *Vocabulary:* Distinguish between molecular and structural formulas. (Use examples in your answer.)

   b. In what situations are molecular and structural formulas more practical than molecular models?

Focus: **Glucose and fructose provide an example of two compounds that have the same molecular formula but different structural formulas.**

33. Match the following representations of atomic structure with the applicable description. Each answer may be used only once. A blank will have only one answer.
   1. orbital model
   2. ball-and-stick model
   3. structural formula
   4. space-filling model
   5. molecular formula

   a. _____ emphasizes the geometry of the molecule, especially the bonds between atoms
   b. _____ shows only the number of each type of atom within a molecule
   c. _____ approximates overall molecular shape
   d. _____ most accurate two-dimensional representation, but impractical for molecules with more than a few atoms
   e. _____ the most simple representation of how atoms are bonded together

E. **Storage Polysaccharides** *(page 64)*

34. *Vocabulary:* Compare and contrast starch and glycogen with respect to monosaccharide component and type of living organism by which they are used.

35. *Vocabulary:* Distinguish between amylose and amylopectin.

36. What are the two main storage sites of glycogen in vertebrates?

37. Briefly describe the role of glucagon in regulating blood glucose levels.

38. The energy stored in polysaccharides is released to be used by a living organism during what two general processes?

F. **Structural Polysaccharides** *(pages 65–66)*

> *Focus:* In addition to their other functions, polysaccharides are structural components of living cells and tissues.

39. Cellulose is found in what part of a plant cell?

40. How does cellulose differ from starch and glycogen (a) as an energy source for living organisms and (b) with respect to type of glucose units involved?

41. a. How does the monosaccharide unit of chitin differ from that of starch and glycogen?

    b. Where is chitin commonly found in nature?

IV. **Lipids** *(pages 67–70)*

A. **General Remarks** *(page 67)*

42. a. *Vocabulary:* What are lipids?

    b. Identify three functions of lipids in living organisms.

B. **Fats and Oils: Energy in Storage** *(page 67)*

43. a. What feature of fats and oils enables them to store more chemical energy than carbohydrates?

    b. How much energy do fats store on a dry weight basis relative to carbohydrates?

44. a. Why is less water associated with stored fat than with stored glycogen?

    b. What might be one evolutionary consequence of this fact?

45. a. Describe the structure of a fat molecule.

b. *Vocabulary:* Describe glycerol and a fatty acid.

c. *Vocabulary:* What is a triglyceride?

46. What feature of fat molecules is responsible for their hydrophobic nature?

47. *Vocabulary:* Distinguish between saturated and unsaturated fatty acids with respect to structure, sources, and melting temperature.

Sugars, Fats, and Calories (page 68)

48. Under what circumstances are fats (a) produced by the body and (b) broken down to supply energy for the body?

Insulators and Cushions (page 68)

49. Describe examples of two functions of body fat other than energy storage.

C. **Phospholipids and Glycolipids** *(page 69)*

50. a. How does the structure of (a) phospholipids and (b) glycolipids differ from that of triglycerides?

b. What consequence do these structural characteristics have for the physical behavior of phospholipids and glycolipids?

Focus: **Phospholipids and glycolipids are major structural components of cellular membranes because of their dual hydrophobic/hydrophilic nature.**

D. **Waxes** *(page 70)*

51. List six biological functions of waxes.

E. **Cholesterol and Other Steroids** *(page 70)*

52. a. What structural feature is common to all steroids?

b. Name two structural variations also found in steroids.

53. a. Cholesterol is a component of what two structures?

b. Where is cholesterol synthesized?

54. Name three dietary sources of cholesterol.

55. What is one consequence of high blood levels of cholesterol?

56. Name two categories of hormones that are formed from steroids.

57. *Vocabulary:* What are prostaglandins?

F. *Essay:* **Regulation of Blood Cholesterol** *(page 71)*

58. Which organ regulates blood cholesterol levels?

59. *Vocabulary:* Define and distinguish between LDLs and HDLs. (Include their functions.)

60. Identify three factors that may cause elevated blood cholesterol levels.

61. What happens when the supply of blood LDLs is greater than the amount that can be managed by the liver and hormone-synthesizing organs?

62. Describe two situations in which heredity influences cholesterol levels.

63. Identify three behavioral factors that influence blood cholesterol levels.

V. **Proteins** *(pages 70–80)*

A. **General Remarks** *(pages 70–72)*

64. Name seven functional categories of proteins.

65. Describe the general structure of a protein.

66. Approximately how many different types of protein molecules are found (a) in an *E. coli* cell and (b) in more complex organisms?

B. **Amino Acids: The Building Blocks of Proteins** *(pages 72–74)*

67. Diagram the general structure of an amino acid.

68. Describe three ways in which the pH of a solution may affect a free amino acid.

Focus: **The same 20 amino acids are used to build all the proteins found in living systems.**

69. a. The variation among amino acids occurs due to what structural component?

b. Describe the four general categories of amino acids.

Secondary

Tertiary

Quaternary

c. Distinguish between the two possible secondary structures named by Pauling and Corey.

Focus: **Amino acids are linked together to form proteins during condensation reactions similar to those that occur during the formation of polysaccharides from monosaccharides.**

---

70. *Vocabulary:* Define peptide bond and polypeptide.

72. *Vocabulary:* What are fibrous proteins?

73. a. What are disulfide bridges and what role do they play in protein structure?

C. **The Levels of Protein Organization** *(pages 74–76)*

71. a. *Vocabulary:* Distinguish between primary, secondary, tertiary, and quaternary protein structure. (You will need to read through page 76 to answer this question completely.)

b. Identify four other features of R groups that may affect the three-dimensional structure of a protein.

74. Name two important types of proteins that are globular.

75. *Vocabulary:* Distinguish between multimeric, dimer, trimer, and tetramer.

b. For each level of protein organization, indicate the type(s) of force(s) maintaining that structure.

Primary

> *Focus:* The secondary, tertiary, and quaternary structures of a protein are all determined by the primary structure.

D. *Essay:* Amino Acids and Nitrogen *(page 76)*

> *Focus:* Certain microscopic organisms can convert atmospheric nitrogen into ammonia, nitrites, and nitrates. Plants incorporate the nitrogen in these compounds into amino acids that can be used by animals directly or used to make other amino acids.

> *Focus:* Some amino acids cannot be manufactured by animals and must be obtained in the diet. Obtaining a balanced supply of amino acids is as important as acquiring a minimum amount of total protein.

76. A meal of beans and rice provides protein equivalent to that of a steak. Explain this phenomenon.

E. Structural Uses of Proteins *(pages 77–78)*

Fibrous Proteins (page 77)

77. What features of collagen determine its classification as a fibrous protein?

> *Focus:* Collagen strands are arranged differently in tissues that have different functions.

78. Name three other fibrous proteins.

Globular Proteins (page 78)

> *Focus:* Globular proteins play important structural roles in cells.

79. a. Name two functions of microtubules.

b. Describe the general structure of microtubules.

F. Hemoglobin: An Example of Specificity *(pages 78–80)*

80. a. Describe the heme group of hemoglobin.

b. What is the function of the heme group?

81. Describe the relationship of the heme groups to the protein chains in hemoglobin.

82. a. What feature of sickle cell hemoglobin is responsible for the altered structure of the hemoglobin molecule?

b. What effect does this change in the primary structure have on the behavior of the molecule and what are the consequences for an individual with sickle cell anemia?

83. *Review Question:* Match the following proteins with the correct function or functional category. An answer may be used more than once. Each blank will have only one answer. (Table 3–2)

1. structural
2. regulatory
3. contractile
4. storage
5. protection (immune function)
6. enzyme
7. transport

a. _____ myosin

b. _____ hemoglobin

c. _____ seed protein

d. _____ pepsin

e. _____ collagen

f. _____ growth hormone

g. _____ egg white

h. _____ silk

i. _____ antibodies

VI. Nucleotides *(page 80)*

84. *Vocabulary:* What are nucleic acids?

85. Name the three subunits of a nucleotide.

86. Identify the sugar subunit present in (a) RNA and (b) DNA.

87. a. Distinguish between purines and pyrimidines.

b. Classify each of the following as a purine or pyrimidine:

adenine   thymine

cytosine   uracil

guanine

c. Which of these nitrogenous bases are found in DNA?

d. Which of these nitrogenous bases are found in RNA?

> *Focus:* DNA is the molecule that transmits genetic information from one generation to the next. RNA is the molecule that translates the genetic message of DNA into functional proteins.

88. *Review Question:* Fill in the following table indicating which elements are components of each type of molecule.

Table for Question 88

	CARBO-HYDRATES	LIPIDS	PROTEINS	NUCLEIC ACIDS
Carbon (C)				
Hydrogen (H)				
Oxygen (O)				
Nitrogen (N)				
Phosphorus (P)				
Sulfur (S)				

VII. Summary *(pages 81–82)*: Read the summary. If you are familiar with the essential features of the material presented there, you are ready to complete the section TESTING YOUR UNDERSTANDING.

TESTING YOUR UNDERSTANDING

After you have completed this examination, compare your answers with those in the section that follows.

1. Which of the following are NOT among the major types of organic molecules found in living organisms?
   a. carbohydrates
   b. hydrocarbons
   c. lipids
   d. proteins
   e. nucleotides

2. Which of these functional groups tends to make an organic molecule insoluble in water?
   a. hydroxyl
   b. amino
   c. aldehyde
   d. methyl
   e. carboxyl

3. The principal energy-storage molecules in most living things are:
   a. proteins.
   b. lipids.
   c. carbohydrates.
   d. hydrocarbons.
   e. nucleotides.

4. The most abundant organic compound in the biosphere is:
   a. DNA.
   b. hemoglobin.
   c. glucose.
   d. cellulose.
   e. ethyl alcohol.

5. When a monosaccharide forms a ring structure in aqueous solution, a bond is formed between:
   a. the aldehyde group and the ketone group.
   b. a hydroxyl group and either an aldehyde or a ketone group.
   c. two hydroxyl groups.
   d. the group on carbon 1 and the group on carbon 6.
   e. the ketone group and either a hydroxyl or an aldehyde group.

6. How much energy does a mole of glucose release upon being oxidized?
   a. 73 kilocalories
   b. 300 kilocalories
   c. 373 kilocalories
   d. 600 kilocalories
   e. 673 kilocalories

7. The disaccharide sucrose is the major transport sugar in:
   a. plants.
   b. insects.
   c. vertebrates.
   d. plants and insects.
   e. insects and animals.

8. The hydrolysis of sucrose to yield glucose and fructose requires the addition of:
   a. water.
   b. hydrogen.
   c. 673 kilocalories of energy.
   d. a monosaccharide.
   e. a carboxyl functional group.

9. The presence of glucagon in the bloodstream causes:
   a. hydrolysis of glycogen in the liver.
   b. decrease of glucose levels in the bloodstream.
   c. synthesis of glycogen in the muscles.
   d. storage of lipid in the fatty tissues.
   e. storage of glucose in muscle.

10. The storage polysaccharides starch and glycogen are made of _____-glucose, whereas the structural polysaccharide cellulose is made of _____-glucose.
    a. alpha; alpha
    b. alpha; beta
    c. beta; alpha
    d. beta; beta
    e. All polysaccharides are made of alternating alpha- and beta-glucose units.

11. Taking the water factor into account, fats store almost _____ times more energy per gram than glycogen.
    a. two
    b. three
    c. four
    d. five
    e. six

12. Some fatty acids, such as stearic acid, are saturated. This means that:
    a. each carbon atom is bonded to a hydroxyl group.
    b. the fatty acid contains one or more double bonds.
    c. the fatty acid contains at least one triple bond.
    d. each carbon forms bonds with four other atoms.
    e. the fatty acid is bonded to a glycerol molecule.

13. Which of the following is NOT a function of triglycerides?
    a. thermal insulation
    b. sustained energy for flight
    c. an efficient form of energy storage
    d. a cushion against physical shock
    e. building blocks for steroid hormones

14. Cholesterol and the sex hormones are examples of:
    a. phospholipids.
    b. steroids.
    c. triglycerides.
    d. fatty acids.
    e. oils.

15. Proteins are polymers composed of a linear sequence of _____ monomers.
    a. alpha-glucose
    b. glycerol
    c. amino acid
    d. nucleotide
    e. triglyceride

16. Side groups of amino acids that have regions of slightly positive charge and regions of slightly negative charge (but no overall charge) are:
    a. acidic.
    b. basic.
    c. neutral.
    d. polar.
    e. nonpolar.

17. The alpha helix and beta pleated sheet structures of proteins were discovered by:
    a. Linus Pauling and Robert Hooke.
    b. Robert Hooke and Robert Corey.
    c. Linus Pauling and Rudolf Virchow.
    d. Rudolf Virchow and Robert Corey.
    e. Robert Corey and Linus Pauling.

18. The three-dimensional configuration of a protein is known as its _____ structure.
    a. primary
    b. secondary
    c. tertiary
    d. quaternary
    e. pentanary

19. Collagen, silk, keratin, and elastin are examples of proteins with regular, repetitive sequences of amino acids and are known as _____ proteins.
    a. fibrous
    b. globular
    c. primary
    d. alpha
    e. beta

20. In sickle cell hemoglobin, the defective polypeptide differs from the normal polypeptide in that one of the amino acids:
    a. forms a double bond to the adjacent amino acid.
    b. has an aldehyde group instead of a carboxyl group.
    c. has two R groups instead of one.
    d. has a different R group.
    e. has lost an amino group.

21. **T or F** Cows, other ruminants, termites, and cockroaches are some of the few organisms that can actually digest cellulose.

22. **T or F** There are approximately 25 different naturally occurring fatty acids, most of which have an odd number of carbons.

23. **T or F** Extra fat in a pregnant or nursing mother can prevent damage to the child when food is in short supply.

24. **T or F** Silicon dioxide and carbon dioxide are both gases that can dissolve readily in water.

25. **T or F** All of the biochemistry that goes on in a cell is based on the ability of one class of molecules to "recognize" other molecules.

26. **T or F** The ideal levels of HDL and LDL in the bloodstream are high HDL and low LDL.

27. **T or F** The pyrimidines present in DNA are cytosine, adenine, and thymine.

28. How does the word strategy as used by biologists differ from its conventional usage?

29. A glycerol molecule is polar because of its three hydroxyl groups. A fatty acid has a long, nonpolar hydrocarbon chain, but its end is polar or negatively charged because of its carboxyl group. Explain how the triglyceride that is formed from three fatty acids and one glycerol molecule can be neutral and nonpolar.

30. What is the relationship between the position of hydroxyl groups in the straight-chain representation of glucose and fructose and their positions above or below the plane in the ring representation?

31. Explain why vegetarians must be careful to achieve adequate protein intake in their diets.

PERFORMANCE ANALYSIS

1. **b** Carbohydrates, lipids, proteins, and nucleotides are all types of organic molecules important to living systems. Hydrocarbons, which contain only hydrogen and carbon, are relatively unimportant in living systems. (pages 55, 56)

2. **d** Hydroxyl, amino, aldehyde, and carboxyl functional groups all possess polar character and impart this polar character (and water solubility) to any compound of which they are a part. Further, amino and carboxyl groups may become charged, which also imparts water solubility to a compound. Methyl groups are nonpolar and therefore insoluble in water. (page 56 and Table 3–1)

3. **c** Carbohydrates are the principal energy storage form for living organisms. However, vertebrates use mainly lipids, which are considerably more efficient, for energy storage. (pages 59, 67)

4. **d** Cellulose, which constitutes 40 percent of the cell wall of young plant cells, is the most abundant organic compound in the biosphere. (page 59)

5. **b** The ring structure that forms when a monosaccharide is placed in water results from a bond between the aldehyde or ketone group and one of the hydroxyl groups. (page 60)

6. **e** When one mole of glucose (remember a mole is 6.02×10^{23} molecules) is oxidized either by a living organism or by being burned in the presence of oxygen, 673 kilocalories are released. (page 61)

7. **a** Sucrose (which is table sugar) is the major transport carbohydrate in plants. Glucose is the main transport form in vertebrates and trehalose is a major transport sugar in insects. (page 61)

8. **a** Hydrolysis can be thought of as a process that splits two molecules apart by the addition of water. It is the reverse of a condensation reaction. Condensation and hydrolysis reactions are the means by which monomer units of polymers are joined together and broken apart. (page 64)

9. **a** Glucagon is a hormone produced by the pancreas that is released into the bloodstream when the blood concentration of glucose falls to a certain level. Glucagon stimulates the liver to hydrolyze glycogen into glucose, thereby raising blood glucose levels. (page 64)

10. **b** The storage polysaccharides glycogen and starch are composed of alpha-glucose units. The structural polysaccharide cellulose is constructed of beta units. (pages 65, 66)

11. **e** Fats are a more concentrated form of energy storage than carbohydrates, in part because polar carbohydrates have water molecules associated with them. With the weight of water excluded from the calculation, fats contain six times as much energy as glycogen on a per weight basis. (page 67)

12. **d** By definition, a saturated fatty acid is one in which each carbon atom forms bonds with four other atoms; no double bonds are present. An unsaturated fatty acid has one or more carbon-carbon double bonds in its carbon backbone. Saturated fats tend to be solids at room temperature. Unsaturated fats tend to be liquids (oils) at room temperature because the double bonds lend rigidity to the carbon chains and as a consequence they do not pack as tightly as chains with no double bonds. (pages 34, 67)

13. **e** Triglycerides (or neutral fats) function in thermal insulation, energy storage (especially long-term storage), and protection from physical shock. The building block of steroid hormones is cholesterol, a lipid that has the ster-

oid skeleton, which is composed of four linked carbon rings. (pages 67, 68, 70)

14. **b** Cholesterol and sex hormones are examples of steroids, which have a skeleton of four linked carbon rings. (page 70)

15. **c** Proteins are polymers composed of amino acid monomers joined together by condensation reactions. The linear sequence of amino acids in a protein is its primary structure. (page 70)

16. **d** Amino acids that have R groups with slight positive and negative charges, but no overall charge, are polar. If the amino acid is put into an acidic or basic solution, these R groups may become charged. (page 72)

17. **e** Linus Pauling and Robert Corey elucidated and named the alpha helix and beta pleated sheet structures. (page 74)

18. **c** The primary level of protein structure is the amino acid sequence. The secondary level is a regular repetitive configuration resulting from hydrogen bonding between carbonyl and amino groups of the polypeptide backbone (e.g., the alpha helix and the beta pleated sheet). The tertiary structure is the three-dimensional structure resulting from interactions among R groups. The quaternary structure results from interactions between two or more polypeptide chains. (pages 74, 75, 76)

19. **a** Fibrous proteins typically have a regular, repeated sequence of amino acids that results in a regular, repetitious structure. Examples of fibrous proteins include collagen, silk, keratin, and elastin. (page 77)

20. **d** The molecular defect in sickle cell hemoglobin is the substitution of one valine for one glutamic acid at a critical point in the two beta chains. Since amino acids are distinguished by the R groups, this is essentially a difference of R groups between the normal and defective beta chains. (page 78)

21. **False** Ruminants, termites, and cockroaches cannot manufacture the enzymes needed to break the bonds between the beta-glucose units of cellulose. However, microorganisms living in their digestive systems can hydrolyze cellulose, thus allowing its use as an energy source by the host animal. (page 65)

22. **False** There are approximately 70 different fatty acids known. Fatty acids have an even number of carbons, typically between 14 and 22 in number. (page 67)

23. **True** The fact that women typically have more subdermal fat than men is thought to be an evolutionary adaptation that was advantageous during times of short food supply, because it acted as an energy reserve that enabled the mother to continue nourishing unborn and nursing infants. (page 68)

24. **False** In contrast to the carbon-oxygen double bonds of carbon dioxide, silicon forms single bonds with oxygen in forming silicon dioxide. This leaves two unpaired electrons on the silicon atom and one unpaired electron on each oxygen atom. These unpaired electrons are readily shared with electrons on adjacent silicon dioxide molecules, resulting in the formation of sand grains. (essay, page 58)

25. **True** All biochemical events in a cell depend upon enzymes. These enzymes "recognize" and interact only with molecules having specific structures. (essay, page 62)

26. **True** High-density lipoproteins (HDLs) carry excess cholesterol to the liver for degradation and excretion. Low-density lipoproteins (LDLs) transport dietary and newly synthesized cholesterol to various parts of the body. High levels of LDLs with low levels of HDLs would correspond to high levels of circulating cholesterol, a situation that would predispose a person to atherosclerosis. However, high levels of HDLs are associated with an efficient means of removing excess cholesterol from the bloodstream. (essay, page 71)

27. **False** The pyrimidines present in DNA are cytosine and thymine. Uracil is a pyrimidine that occurs in RNA instead of thymine. Adenine and guanine are purines present in both DNA and RNA. (page 80)

28. Biologists use the term strategy to refer to an adaptation (physical or behavioral) that an organism possesses due to the process of natural selection that makes the organism more successful in a particular environment. (Adaptations will be discussed in more detail in the chapters on ecology.) (page 59)

29. During fat synthesis, condensation reactions occur between the hydroxyl groups of glycerol and the carboxyl groups of the fatty acids. The resulting C—O—C bonds are nonpolar in character and are buried within a molecule that has no polar or charged functional groups. (page 67)

30. In drawing the straight chain and ring forms of sugars, the hydroxyl groups that appear to the left of the straight-chain representation of the carbon backbone are placed above the plane of the ring. Conversely, those appearing to the right of the carbon backbone are placed below the plane of the ring. (essay, page 63)

31. Humans can synthesize many of the amino acids needed by the body. However, there are some amino acids that cannot be synthesized and must be obtained in the diet. Meats, eggs, and dairy products contain all these essential amino acids in appropriate proportions. Vegetables and grains are typically rich in some amino acids but deficient in others. Vegetarians who consume no animal products must pay attention to the amino acid content of the plant products they consume so that they obtain adequate amounts of all essential amino acids. (essay, page 76)

REVIEW TEST 1

Introduction and Section 1: Chapters 1–3

This review test is *not* designed to cover all of the important information in these chapters. However, it does touch on the major topic categories in each chapter. It will also give you valuable practice in taking this type of test. When you are finished, check your answers with those provided in the following PERFORMANCE ANALYSIS section.

1. An atom is chemically unreactive when its:
   a. innermost energy level is completely devoid of electrons.
   b. innermost energy level is completely filled with electrons.
   c. outer energy level is completely devoid of electrons.
   d. outer energy level is completely filled with electrons.
   e. No atom is chemically unreactive under ordinary conditions.

2. You have just stepped out of a pool on a hot summer day. The wind is blowing and, even though it is 90 degrees, you feel chilled. This is due to water's high:
   a. specific heat.
   b. heat of vaporization.
   c. heat of fusion.
   d. ionization potential.
   e. surface tension.

3. The view that arms and legs move as levers, that the stomach acts as a mortar and pestle, and that the lungs function in a manner similar to a bellows was put forth by the:
   a. vitalists.
   b. catastrophists.
   c. mechanists.
   d. creationists.
   e. reductionists.

4. Which pair does NOT have the relationship of monomer/polymer?
   a. monosaccharide/trehalose
   b. glucose/cellulose
   c. fatty acid/steroid
   d. monosaccharide/polysaccharide
   e. amino acid/protein

5. Which factor does NOT influence the formation of polar covalent bonds?
   a. number of protons in the nucleus
   b. number of neutrons in the nucleus
   c. number of electrons between the nucleus and the outer electrons
   d. sharing of electrons between atoms that need to gain an electron
   e. closeness of the outer electrons to the nucleus

6. The work of which scientist did NOT contribute to the modern cell theory?
   a. Matthias Schleiden
   b. Robert Hooke
   c. Rudolf Virchow
   d. Theodor Schwann
   e. Louis Pasteur

7. Which molecule is composed of a carboxyl group, an amino group, a hydrogen atom, and an atom or group of atoms designated as "R," all bonded to a central carbon atom?
   a. amino acid
   b. monosaccharide
   c. fatty acid
   d. wax
   e. nucleotide

8. Which phenomenon is NOT a consequence of hydrogen bonding?
   a. adhesive attraction
   b. surface tension
   c. cohesive attraction
   d. capillary action
   e. hydrophobic interaction

9. Biological molecules that are soluble in nonpolar solvents and insoluble in polar solvents are classified as:
   a. monosaccharides.
   b. polysaccharides.
   c. lipids.
   d. proteins.
   e. nucleotides.

10. From the following groups of elements, select the group for which each member exists primarily as an ion in living systems.
    a. phosphorus, sodium, carbon
    b. phosphorus, sulfur, carbon
    c. potassium, sulfur, calcium
    d. potassium, sodium, calcium
    e. phosphorus, sulfur, calcium

11. Which of these scientists preceded Charles Darwin in proposing that species might undergo changes in the course of time?
    a. Anaximander
    b. Lucretius
    c. Georges-Louis Leclerc de Buffon
    d. Erasmus Darwin
    e. All of these men proposed some type of evolutionary theory prior to that postulated by Charles Darwin.

12. You can float a razor blade on the surface of a beaker of water because:
   a. of the surface tension of water.
   b. the razor blade is less dense than water.
   c. the razor blade and water repel one another.
   d. water is nonpolar.
   e. water is polar and the metal in the razor blade is polar.

13. Which ion is the major positively charged ion in most organisms?
   a. potassium
   b. sodium
   c. calcium
   d. magnesium
   e. chloride

14. The formation of which molecule does NOT involve condensation reactions?
   a. amino acid   d. disaccharide
   b. polysaccharide   e. polypeptide
   c. neutral fat

15. The tritium isotope of hydrogen has:
   a. one electron, one proton, and one neutron.
   b. one electron, one proton, and two neutrons.
   c. no electrons, one proton, and one neutron.
   d. one electron, no protons, and one neutron.
   e. one electron, one proton, and no neutrons.

16. Prokaryotes belong to kingdom:
   a. Fungi.   d. Monera.
   b. Plantae.   e. Animalia.
   c. Protista.

17. Which functional group confers the properties of a weak base on the molecule of which it is a part?
   a. methyl   d. aldehyde
   b. carboxyl   e. ketone
   c. amino

18. T or F   Fact: A bacterium consists of 9.94 percent hydrogen atoms and 73.68 percent oxygen atoms, by weight. Based on these data, it is therefore safe to say that bacteria consist mostly of water.

19. T or F   The stratification that occurs in deep lakes during the summer is beneficial to bottom-dwelling organisms in that it provides a constant oxygen source in the hypolimnion.

20. T or F   A variation that gives an organism a slight advantage makes that organism more likely to leave surviving offspring.

21. Match each scientist with his composition or concept. Use each answer only once. A blank will have only one answer.
   1. Carolus Linnaeus
   2. Aristotle
   3. Charles Darwin
   4. Thomas Malthus
   5. Charles Lyell

   a. _____ *Species Plantarum*
   b. _____ *The Origin of Species*
   c. _____ *Principles of Geology*
   d. _____ *Scala Naturae*
   e. _____ a sociological essay published in 1798

22. Explain why science, art, philosophy, and religion often come up with different answers to the same problem.

23. List in order of increasing complexity the levels of biological organization. Start with subatomic particles.

24. Discuss the significance of buffers to biological systems.

25. Distinguish between molecular formulas and structural formulas and give an example of each.

PERFORMANCE ANALYSIS

1. **d**   The most stable electron configuration is one in which the outer energy level is completely filled with electrons. Elements whose atoms have this configuration are called noble (implying disdainful) gases and are chemically unreactive. In the formation of molecules and ions, atoms generally interact (share, gain, or lose electrons) until the outer energy level is filled. (page 30)

2. **b**   The phenomenon of evaporative cooling is a consequence of water's high heat of vaporization. When water evaporates from a surface, a great deal of heat is removed. Perspiration is an effective means of cooling in humans. Dogs only perspire through their footpads. When a dog pants, it releases moisture and heat from its body via the tongue. Wind or air currents increase the rate of evaporation. (pages 43, 44)

3. **c**   Mechanists used analogies to various types of machines to describe the different parts of living organisms. Vitalists believed that living systems obeyed different physical and chemical laws from nonliving systems. Reductionists proposed that all living processes could be reduced to simpler, readily understandable processes. Catastrophists believed that the different species of organisms currently in existence appeared after a series of global catastrophes destroyed the previous sets of species. Creationists believed in the special creation of each individual species in its contemporary form. (pages 1, 3, 10)

4. **c**   Fatty acids are a component of fats (triglycerides); one fat molecule is composed of one glycerol molecule and three molecules of fatty acid. Steroids are not polymers but rather are based on a carbon skeleton composed of four connected rings. (pages 67, 70)

5. **b**   Since neutrons are neutral (have no electrical charge), they do not influence the attraction of the nucleus for electrons and therefore do not affect the formation of polar covalent bonds. (page 32)

6. **e**   Robert Hooke named the individual units of cork "cells." Matthias Schleiden concluded that all plants are composed of cells. Theodor Schwann proposed that all

life forms are composed of cells. Rudolf Virchow stated that all cells come from preexisting cells. (page 10)

7. **a** The fundamental structure of an amino acid is that of a central carbon atom bonded to (1) an amino group ($-NH_2$), (2) a carboxyl group ($-COOH$), (3) a hydrogen atom, and (4) an "R" group which determines the identity of the amino acid. (page 72)

8. **e** Hydrogen bonding between water molecules is responsible for the cohesive and adhesive properties of water. These properties (either separately or in combination) result in the following characteristics of water: surface tension, capillary action, imbibition, high specific heat, high heat of vaporization, high heat of fusion, and the lower density of ice relative to liquid water. (pages 42–45)

9. **c** By definition, lipids are organic molecules that are soluble in nonpolar solvents but insoluble in polar solvents such as water. (page 67)

10. **d** Potassium, sodium, and calcium ions are three of the most important ions in living systems. Potassium and sodium ions are essential to the initiation and propagation of nerve impulses and to a system that transports many types of molecules into and out of cells. Calcium ions are essential to the process of muscle contraction. The other elements listed are most often found in living systems in organic molecules. (pages 31, 55)

11. **e** All four of these scientists proposed that species might undergo changes over time. Anaximander and Lucretius were part of a school of Greek philosophy that devised an atomic theory in addition to a theory of evolution. Georges-Louis Leclerc de Buffon was among the first of the modern European scientists to suggest that species undergo change. Erasmus Darwin believed that species have historical connections amongst themselves, and that animals may change in response to the environment and pass these changes on to offspring. (page 2)

12. **a** A razor blade (which is more dense than water) can float on water because of the surface tension of water, one consequence of the hydrogen bonding between water molecules. (page 42)

13. **a** The most common positively charged ion in most living organisms is potassium (K^+), which is essential to many biological processes. (page 31)

14. **a** Condensation reactions are those in which two molecules join together and a water molecule is removed. They commonly occur when monomer units are linked together to make a polymer, as in the formation of polysaccharides, disaccharides, and polypeptides. The bonding of fatty acids to glycerol in the formation of neutral fats also involves condensation. Amino acids are the monomers of which polypeptides are made. (pages 64, 67, 72)

15. **b** The most common isotope of hydrogen has one proton and one electron with no neutrons. Deuterium, atomic weight of 2, has one proton, one neutron, and one electron. Tritium, with an atomic weight of 3, has one proton, two neutrons, and one electron. (page 25)

16. **d** All prokaryotic organisms belong to the kingdom Monera. The organisms in the other four kingdoms are composed of eukaryotic cells. (page 13)

17. **c** The amino group ($-NH_2$) can act as a weak base by removing H^+ from the solution and becoming $-NH_3^+$. Methyl groups are nonpolar. Carboxyl groups are hydrogen ion donors (weak acids). Ketone and aldehyde groups are polar but neither acidic nor basic. (page 48, Table 3–1)

18. **False** While it is true that bacteria consist mainly of water, that fact cannot be established by the data given; hydrogen and oxygen are found in many molecules other than water. (page 36, Table 1–4)

19. **False** During summer stratification, only the water in the epilimnion circulates. The cooler, denser thermocline lies beneath the warmer epilimnion and prevents circulation of water and oxygen to the hypolimnion. Stagnation is the result. (essay, page 46)

20. **True** A basic tenet of natural selection is that members of a species possessing traits that provide an advantage will leave more surviving offspring than those members that lack the advantageous characteristics. (page 7)

21. The answers are **a1, b3, c5, d2, e4**. (pages 1, 2, 5, 7)

22. Science answers questions by collecting data using objective techniques that can be duplicated by different workers to give repeatable results. These techniques include observation, experimentation, and logic. (The selection of subjects studied by scientists is made subjectively.) Art, philosophy, and religion all include subjectivity (e.g., beauty is in the eye of the beholder, right and wrong depend upon an individual's perspective) in arriving at answers to questions. (page 17)

23. The levels of biological organization from simplest to most complex are: subatomic particle, atom, molecule, cell, tissue, organ, organ system, organism, society, biosphere. (pages 36, 37)

24. The chemical reactions that occur in living systems are catalyzed by special protein molecules (enzymes) that only work within a narrow pH range. Buffers maintain this critical pH range. (page 52)

25. Molecular formulas show only the number and proportion of atoms in a molecule. The structural formula shows the bonding relationships of the different atoms. CH_4 is the molecular formula for methane; the structural formula is:

$$
\begin{array}{c}
\text{H} \\
| \\
\text{H} - \text{C} - \text{H} \\
| \\
\text{H}
\end{array}
$$

(essay, pages 62, 63)

CHAPTER 4

Cells: An Introduction

MAJOR CONCEPTS

According to current theory, the sun in our solar system originated from dust particles and gases that condensed together due to the force of gravity. During the condensation, hydrogen atoms collided with enough force to fuse their nuclei into helium atoms. This is a self-perpetuating reaction that is responsible for the tremendous amounts of energy radiated from the sun. The planets are presumed to have formed from the dust and gases remaining after the sun was created.

In 1922, A. I. Oparin hypothesized that, under the conditions present in the primitive environment, organic molecules were formed and accumulated in the earth's bodies of water. His hypothesis was rejected by his contemporaries because Louis Pasteur had put the concept of spontaneous generation to rest in 1864 by demonstrating that air is the source of contaminating particles responsible for the growth of microorganisms in broth. In the 1950s, Stanley Miller demonstrated the formation of amino acids and other organic molecules under conditions generally accepted to have been present in the primitive environment. In more recent experiments conducted by Sidney Fox, membrane-bound structures capable of conducting a few chemical reactions have been produced during a series of reactions that began with amino acids.

Of all the planets in our solar system, the earth is the one most suited to support life. The distance from the sun permits a temperature range compatible with the biochemical reactions essential to living systems. The size and mass of the earth (which determine its gravitational pull) maintain an atmosphere that is dense enough to protect living organisms from harmful radiation but thin enough to allow the passage of visible light, upon which photosynthesis depends.

Heterotrophs are organisms that require an outside source of organic molecules for energy and as raw materials for growth and repair. Autotrophs can manufacture their own organic molecules from simple inorganic compounds provided adequate energy is available. Autotrophs may be photosynthetic or chemosynthetic, depending on their energy source.

Prokaryotic cells have genetic material free in the cytoplasm, have few organelles, and exist only as single-celled organisms. Eukaryotic cells have genetic material in the form of chromosomes enclosed in a membrane (a true nucleus), have numerous complex membrane-bound organelles, may exist as single-celled organisms, and are the constituents of all multicellular organisms. The cells of multicellular organisms are typically specialized to perform one or a few specific functions.

The three major types of microscopes used to visualize cells are light, transmission electron, and scanning electron. Since electron microscopes use an electron beam to illuminate the subject, they have a greater resolving power. There are several special types of light microscopes including phase-contrast, differential-interference, and dark-field, which may be used to view living cells.

Specimens for microscopy are typically prepared for viewing by fixing, staining, and sectioning. Fixing stabilizes the cellular structures so that they remain intact during processing. Staining involves treating specimens with dyes or materials that differentially adhere to or react with cellular structures. For electron microscopy, "staining" is accomplished by treating specimens with compounds of heavy metals. Sectioning provides tissue sections thin enough for the unstained regions of the specimen to be transparent to the illuminating beam.

HOW TO STUDY THE CHAPTER

Read the entire chapter through quickly, focusing on the major concepts.

Use the GUIDED STUDY OF THE CHAPTER to help you identify the important details as you **reread** the chapter. Writing out the answers to these questions will help fix them in your mind as well as provide you with a valuable study aid.

Answer the questions in TESTING YOUR UNDERSTANDING without the aid of your text. Check your answers against those in PERFORMANCE ANALYSIS. Analyzing your answers will give you valuable feedback on your level of understanding and preparedness for classroom testing.

GUIDED STUDY OF THE CHAPTER

I. Introduction *(page 84)*

> *Focus:* **Living cells appear to have spontaneously self-assembled from molecules present in primitive bodies of water.**

II. The Formation of the Earth *(pages 84–85)*

1. Describe how the sun is believed to have originated.

2. What thermonuclear reaction is occurring in the sun that is the source of its radiant energy?

3. Describe the process by which the planets of our solar system are thought to have formed.

4. The oldest known rocks in the earth's crust are _____ old.

5. *Vocabulary:* Define the term biosphere.

III. The Beginning of Life *(pages 85–89)*

A. **General Remarks** *(pages 85–88)*

6. For a long time, the attitude of the scientific community regarding the origin of life was that it would remain an unsolvable mystery. Describe two developments that encouraged scientists to pursue investigations of the origin of life.

> *Focus:* **Scientists generally agree that in the primitive environment in which life originated (1) there was little or no free oxygen, (2) hydrogen, oxygen, carbon, and nitrogen were available in some form, and (3) multiple energy sources were present.**

7. a. Summarize Oparin's hypothesis concerning the origin of life.

   b. Explain why Oparin's hypothesis was not given serious consideration by his contemporaries.

8. a. Describe how Miller experimentally tested Oparin's hypothesis.

   b. Name two types of organic compounds that formed during Miller's experiment.

> *Focus:* **A form of natural selection was operating during chemical evolution in that the most stable compounds and aggregates of compounds persisted.**

B. *Essay:* **The Question of Spontaneous Generation** *(page 86)*

9. Describe Redi's experiment disproving the spontaneous generation of maggots.

10. a. Describe the experiments in which Louis Pasteur finally disproved the theory of spontaneous generation.

   b. What limitations apply to the conclusions drawn by Pasteur? (At the time of Pasteur's work, the scientific community did not recognize these limitations.)

C. **The First Cells** *(pages 88–89)*

11. What characteristics distinguish living cells from other chemical systems?

12. a. Describe Sidney Fox's experiments with proteinoid microspheres.

   b. What characteristics of proteinoid microspheres support the contention that they may represent an evolutionary link between Stanley Miller's spontaneously produced amino acids and the first cells?

13. The oldest known fossils of living cells are _____ years old.

D. **Why on Earth?** *(page 89)*

14. Name three factors that were critical to the development of life on the planet earth and discuss the importance of each factor.

IV. **Heterotrophs and Autotrophs** *(pages 89–90)*

15. *Vocabulary:* Distinguish between heterotrophs and autotrophs and identify examples of each type of organism.

16. *Vocabulary:* Distinguish between photosynthetic and chemosynthetic autotrophs.

17. a. Which type of cell, heterotrophic or autotrophic, was originally believed to have evolved first?

   b. Describe two recent pieces of evidence that autotrophs may have been the first living cells.

18. *Vocabulary:* What are methanogens? (Figure 4–7)

> *Focus:* **The capacity to perform photosynthesis characterizes the most successful autotrophs. It is through these organisms that energy from the sun is channeled to nearly all other life forms.**

V. **Prokaryotes and Eukaryotes** *(pages 90–94)*

A. **General Remarks** *(pages 90–92)*

19. List the four basic components of the cell theory.

Focus: There is apparently an unbroken continuity between all cells living on earth today and the first primitive cells.

20. Identify the two features common to all living cells.

21. *Vocabulary:* Distinguish between prokaryotes and eukaryotes.

22. a. *Vocabulary:* Describe a chromosome.

    b. What are histones?

23. *Vocabulary:* Distinguish among nuclear envelope, nucleus, and nucleoid.

24. a. *Vocabulary:* Define cytoplasm.

    b. What are organelles?

25. Name the function of ribosomes.

26. a. What is a cell wall?

    b. What types of cells may have cell walls?

27. How does the size of eukaryotic cells compare with that of prokaryotic cells?

28. Name two groups of organisms classified as prokaryotes.

29. *Vocabulary:* What are algae? (see footnote, page 91)

30. Which type of cell, prokaryote or eukaryote, evolved first?

31. a. Describe the general structure of *Chlamydomonas.*

    b. How is this organism classified?

32. Fill in the following table with the appropriate information on prokaryotic and eukaryotic cells.

Table for Question 32

	PROKARYOTIC CELLS	EUKARYOTIC CELLS
Presence of a nucleus (yes or no)		
Form of genetic material		
Relative cell size		
Number of organelles (few or many)		
Complexity of organelles		
Presence of a cell wall		
Present in multicellular forms? (yes or no)		

B. **The Origins of Multicellularity** *(pages 92–94)*

33. Compare and contrast the cells of modern multicellular organisms with single-celled eukaryotes.

34. How do cilia of tracheal cells (a) resemble and (b) differ from flagella of single-celled eukaryotic organisms? (Figure 4–13)

35. The prokaryotes all belong to the kingdom _____.

> *Focus:* Members of all kingdoms except Monera are made up of eukaryotic cells.

36. Fill in the table at the bottom of the page indicating which types of cells/organisms are present in each of the kingdoms.

VI. Viewing the Cellular World *(pages 94–100)*

A. General Remarks *(page 94)*

> *Focus:* Advances in understanding cell structure have generally been preceded by improvements in techniques for studying cells.

B. Types of Microscopes *(pages 94–96)*

37. *Vocabulary:* Define and distinguish between resolving power and magnification. (You must read through page 96 to complete this question.)

38. a. What is the resolving power of the human eye?

b. How does the size range of most eukaryotic cells compare?

39. What factor limits the resolving power of a light microscope?

40. a. Identify the illumination source of the transmission electron microscope.

b. What is its resolving power?

> *Focus:* The greater resolving power of the transmission electron microscope compared to the light microscope is a consequence of the much shorter wavelength of the electron beam compared to the wavelength of visible light.

41. Account for the value of the scanning electron microscope in spite of its limited resolution.

C. Preparation of Specimens *(pages 96–99)*

> *Focus:* Living cells are nearly completely transparent to visible light and to electrons; therefore, they must be stained before microscopic examination.

Table for Question 36

	PROKARYOTIC	EUKARYOTIC	HETEROTROPHIC	AUTOTROPHIC	MULTICELLULAR OR UNICELLULAR?
Monera					
Protista					
Fungi					
Plantae					
Animalia					

42. How do (a) dyes for light microscopy and (b) heavy metals for electron microscopy increase the "visibility" of the cells being stained?

43. Identify two important functions of fixation during processing tissues for microscopic examination.

44. Describe how (a) aldehydes and (b) osmium tetroxide function as fixation agents.

45. a. Describe how fixed and stained specimens are sectioned.

    b. Why is this step necessary?

46. *Vocabulary:* What is shadowing in reference to preparation of specimens for microscopic examination?

47. Explain why specimens for electron microscopy must be dehydrated and the beam chamber must be evacuated.

> *Focus:* An issue of major concern to cell biologists is this: Is the image produced by the microscope an accurate reflection of cellular structure or is it an artifact of processing?

48. Complete the following table comparing light, transmission electron, and scanning electron microscopy.

Table for Question 48

	RESOLVING POWER	CELLULAR STRUCTURES VISIBLE	TYPE OF STAIN USED IN PREPARATION OF SPECIMENS
Light			
Transmission electron			
Scanning electron			

49. a. List (in order) all the steps required to prepare a cell or tissue for examination with a transmission electron microscope.

    b. Compare this procedure with that used to prepare a specimen for examination with a light microscope.

D. **Observation of Living Cells** *(pages 99–100)*

> *Focus:* **Phase-contrast, differential-interference, and dark-field microscopy are variations of light microscopy that enable the study of living cells and provide information not obtained with conventional staining techniques.**

50. Describe how video cameras are being used to observe living cells and the potential of this technique to further our understanding of cells.

VII. Summary *(page 100):* Read the summary. If you are familiar with the essential features of the material presented there, you are ready to complete the section TESTING YOUR UNDERSTANDING.

TESTING YOUR UNDERSTANDING

After you have completed this examination, compare your answers with those in the section that follows.

1. Most biologists begin the story of life on earth with the origin of the earth about _____ years ago.
   a. 4.6 trillion
   b. 4.1 trillion
   c. 41 billion
   d. 4.6 billion
   e. 600 million

2. The age of the oldest fossilized cells known to humans is _____ billion years.
   a. 5.2
   b. 4.6
   c. 3.5
   d. 2.3
   e. 1.2

3. The first testable hypothesis about the events preceding and surrounding the origin of life was offered by:
   a. Stanley Miller.
   b. A. I. Oparin.
   c. Louis Pasteur.
   d. Robert Hooke.
   e. Sidney Fox.

4. Regardless of whether the first cells were heterotrophic or autotrophic, the flow of energy that sustains life on earth came to depend heavily on _____ cells.
   a. photosynthetic, autotrophic
   b. heterotrophic, prokaryotic
   c. heterotrophic, eukaryotic
   d. chemosynthetic, autotrophic, prokaryotic
   e. chemosynthetic, autotrophic, eukaryotic

5. Bacteria that obtain their energy from inorganic chemical reactions are MOST ACCURATELY classified as:
   a. heterotrophic.
   b. autotrophic.
   c. chemosynthetic.
   d. chemosynthetic, heterotrophic.
   e. chemosynthetic, autotrophic.

6. Prokaryotes differ from eukaryotes in that prokaryotes _____, whereas eukaryotes _____.
   a. are large cells; are small cells
   b. lack a nuclear membrane; have a nuclear membrane
   c. lack a cell membrane; have a cell membrane
   d. have cell walls; do not have cell walls
   e. have membrane-bound organelles; do not have membrane-bound organelles.

7. *Chlamydomonas* is a:
   a. multicellular alga.
   b. single-celled eukaryote.
   c. single-celled prokaryote.
   d. cyanobacterium.
   e. plant.

8. The first multicellular organisms are thought to have appeared _____ years ago.
   a. 4.6 billion
   b. 4.1 billion
   c. 1.1 billion
   d. 750 million
   e. 600 million

9. Which type of organism is multicellular, photosynthetic, autotrophic, and eukaryotic?
   a. a bacterium
   b. a plant
   c. a protist
   d. a fungus
   e. an animal

10. The transmission electron microscope has a resolution _____ times greater than that of the human eye.
    a. 5,000
    b. 50,000
    c. 100,000
    d. 150,000
    e. 500,000

11. Which step is NOT involved in preparing a biological specimen for electron microscopy?
    a. fixation
    b. staining
    c. dehydration
    d. embedding
    e. filtering

12. Most cells appear to be almost transparent because by weight they are:
    a. 95 percent water.
    b. 80 percent oxygen.
    c. 70 percent water.
    d. 50 percent protein.
    e. 35 percent lipid.

13. Fixing biological specimens prior to viewing with a microscope is done so that the:
    a. tissues become more transparent to the light or electrons.
    b. tissues adhere to the slide or grid.
    c. remaining water may be removed from the specimen.
    d. cell structures may be bound in place.
    e. heavy metal stains may bind to the cell surface.

14. Shadowing is a process that:
    a. involves coating a specimen with metal.
    b. prepares a specimen for light microscopy.
    c. increases the contrast between various subcellular organelles.

d. protects recessed regions of a cell from the electron beam.

e. increases the permeability of the specimen to electrons.

15. The resolution of a phase-contrast microscope is _____ that of a conventional light microscope.
   a. less than
   b. greater than
   c. about the same as
   d. not determined in the same way as
   e. The concept of resolution does not apply to phase-contrast microscopy.

16. The optical system of a differential-interference microscope produces contrast in the subject image based on:
   a. selective enhancement of some specimen dyes versus other dyes.
   b. the differential uptake of stains by various components of a cell.
   c. the differential interference of an electron beam versus that of visible light.
   d. the interference of electrons in the specimen with the transmission of visible light through the specimen.
   e. small differences in interference produced when light is bent by different components of the specimen.

17. The resolving power of a scanning electron microscope is about:
   a. 1,000 nanometers.
   b. 100 nanometers.
   c. 10 nanometers.
   d. 1 nanometer.
   e. 0.1 nanometer.

18. T or F   The experiments of Stanley Miller (and other similar studies) have proven that organic molecules actually did originate on earth through a process of chemical evolution.

19. T or F   Pasteur's disproof of spontaneous generation is applicable to all situations and experimental conditions.

20. T or F   Within a multicellular, autotrophic organism, such as a maple tree, some cells may be autotrophic while others are heterotrophic.

21. T or F   The human eye has a resolving power of 1/10 millimeter, which is 3 to 10 times that which is needed to resolve individual cells.

22. T or F   Although cells in multicellular eukaryotic organisms are remarkably self-sustaining, they differ from those of single-celled eukaryotes in performing a limited number of functions.

23. What evidence do we have that the interior of the earth is still hot?

24. How does a scientist interpret the light and dark areas of an image produced by (a) light, (b) transmission electron, and (c) scanning electron microscopy?

25. Why did the scientific community ignore Oparin's ideas at first?

PERFORMANCE ANALYSIS

1. **d**   According to current theory, the earth is believed to have formed from cosmic dust and gases swirling around the newly formed sun approximately 4.6 billion years ago. (page 85)

2. **c**   The oldest fossilized cells known to humans are 3.5 billion years old and were found in black chert deposits in South Africa. (page 89, Figure 4–6)

3. **b**   A. I. Oparin proposed in 1922 that a series of chemical reactions in the primitive environment preceded the formation of the first cells. Experiments performed by Stanley Miller in the 1950s demonstrated that organic molecules (including amino acids, nitrogenous bases, nucleotides, and ribose) could have formed under the conditions existing in the primitive environment from the inorganic compounds that were present. Sidney Fox carried the experiments further and achieved the formation of proteinoid microspheres in the laboratory. These microspheres can perform some chemical reactions, may increase in size by taking up proteinoid material from the environment, and pinch off buds. Louis Pasteur disproved the theory of spontaneous generation in a series of simple yet brilliant experiments (see essay, page 86). Robert Hooke was the scientist who coined the term "cells" (see Figure I-12, page 10). (pages 87, 88)

4. **a**   Most of the living organisms on earth today depend directly or indirectly on photosynthetic autotrophs that use the sun's energy to convert inorganic molecules into organic building blocks. (page 90)

5. **e**   Autotrophs obtain their energy either from photosynthesis or from chemical reactions and use inorganic compounds to make organic compounds. By definition, chemosynthetic autotrophs are those that depend on chemical reactions for their energy. (page 89)

6. **b**   Prokaryotes have their genetic material free in the cytoplasm (lacking a nuclear membrane), have few organelles (not enclosed in membranes), and have cell walls. Eukaryotes have genetic material in the form of chromosomes in a membrane-bound nucleus, have many complex membrane-bound organelles, may or may not have cell walls, and are generally larger than prokaryotic cells. (page 91)

7. **b**   *Chlamydomonas* is a single-celled photosynthetic alga that inhabits freshwater ponds and aquariums. It is classified as a eukaryote due to its complex cellular structure. (page 93, Figure 4–10)

8. **d**   According to current information, the first multicellular organisms evolved fairly recently, only 750 million

years ago. (This is recent considering the evidence that single-celled organisms have been on earth for 3.5 billion years.) (page 92)

9. **b** By definition, members of the plant kingdom are multicellular, photosynthetic autotrophs. All multicellular organisms are composed of eukaryotic cells. (page 94)

10. **e** The resolution of the transmission electron microscope is 0.2 nanometer or 0.2×10^{-9} meter. The resolution of the human eye is 100 micrometers or 100×10^{-6} meter.

$$\frac{100 \times 10^{-6}\,\text{m}}{0.2 \times 10^{-9}\,\text{m}} = 500 \times 10^3 = 500,000$$

This calculation shows that the resolution of the transmission electron microscope is 500,000 times that of the human eye. (page 96)

11. **e** In preparing a tissue for electron microscopic examination it must be fixed, dehydrated, embedded, and stained. Samples for transmission electron microscopy may be sectioned. (pages 96, 98)

12. **c** The high water content of cells (approximately 70 percent for most cells) causes them to be almost completely transparent to visible light and relatively transparent to electrons. Stains are used to increase the opacity of various components of the cell, thereby creating contrast in the microscopic image. (page 96)

13. **d** Cells and their components are fairly fragile structures and the manipulations that occur during processing for microscopic examination would destroy them if they were not fixed in some way to maintain internal structure. Aldehyde solutions link adjacent proteins together, thereby fixing them in position in cells. Osmium tetroxide binds lipid molecules together. Fixation also makes cells more permeable to stains and to rinsing solutions. (page 98)

14. **a** During shadowing, the surface of a sample is coated with heavy metal atoms. The coating is thickest on raised areas of the specimen. When scanned with a scanning electron microscope, the electron beam is reflected off this metal coating and produces an image of the specimen's surface. (page 98, Figure 4–17)

15. **c** The resolving power of a microscope is determined by the wavelength of the illumination beam. Since the illumination source for a phase-contrast microscope is visible light, the resolving power is similar to that of a conventional light microscope. (page 99)

16. **e** When light passes from one medium into another (such as from air into water), the beam is bent or diffracted. The degree to which the beam is bent depends on the properties of the medium. Each component inside a cell has slightly different diffractive properties. As light passes from one component to another, light is diffracted in a different manner. A differential-interference microscope has optical systems that enhance the interference resulting from this diffraction and thereby increase contrast in the image produced. (page 99)

17. **c** The resolving power of a scanning electron microscope is about 10 nanometers, which is considerably less than the 0.2 nanometer resolving power of the transmission electron microscope. (page 96)

18. **False** Miller's experiments only demonstrated that organic molecules *could have* originated spontaneously under the conditions present in the primitive environment. (There is no way to prove conclusively how the first organic molecules originated.) (page 87)

19. **False** Pasteur's disproof of spontaneous generation only applies to those circumstances defined in his experiments (i.e., broth sterilized by boiling in a swan-necked flask left open to air). (essay, page 86)

20. **True** The photosynthetic cells of maple tree leaves are autotrophic but the cells in the roots of the tree are heterotrophic and depend upon the leaf cells to supply them with organic molecules and energy. (page 94)

21. **False** Most eukaryotic cells are 10 to 30 micrometers in diameter. The resolving power of the human eye is 0.1 millimeter (or 100 micrometers). This is 3 to 10 times *below* that needed to identify individual cells. (pages 94, 95)

22. **True** One consequence of multicellularity is that individual cells become specialized to perform limited functions. These specialized cells do not carry out as many processes as do free-living single-celled organisms. (page 93)

23. Evidence that the center of the earth is still hot is provided by volcanic eruptions that spew molten lava across the land and by geysers that release steam and boiling water into the environment. (page 85)

24. (a) In light microscopy, the light areas of the image are transparent to visible light while the dark areas restrict the passage of visible light. (b) For transmission electron microscopy, a parallel situation exists in which the light areas are transparent to electrons and the dark areas of the image represent structures that scatter electrons. (c) In scanning electron microscopy, depressed areas of the surface appear dark while elevated areas are light. (page 96)

25. Oparin's contemporaries did not give credence to his ideas because the scientific community was still under the influence of Pasteur's disproof of spontaneous generation. At that time, the scientists did not take into account that Pasteur's results were valid for only a certain set of conditions and that Oparin's proposals centered around a completely different set of conditions. (page 87)

CHAPTER 5

How Cells Are Organized

MAJOR CONCEPTS

One important factor limiting cell size is the relationship between surface area and volume. As cell size increases, the ratio of surface area to volume decreases (i.e., there is less surface area per unit of volume). This means that for each unit of volume that needs to exchange materials with the external environment, there is less surface area available for the exchange.

Another factor that limits cell size is the capacity of the nucleus to direct cellular activities. Large single-celled organisms have multiple identical nuclei scattered throughout the cell.

All cells have a cell membrane, which acts as the primary barrier between a cell's internal and external environments. The eukaryotic cell membrane is a phospholipid bilayer; the phospholipids are arranged with their hydrophobic tails pointing toward the interior and their hydrophilic heads toward the periphery of the membrane. Numerous cholesterol and protein molecules are embedded in the cell membrane. Most of the lipid and some of the protein molecules can move about laterally in the membrane. The internal and external surfaces differ considerably in chemical composition. The cell membrane limits the movements of materials into and out of cells.

The cells of plants and many algae, fungi, and prokaryotes also have a cell wall which is constructed by the cell external to the cell membrane. Depending upon the organism, cell walls may contain cellulose, chitin, polysaccharides, or peptidoglycans.

The nucleus of a eukaryotic cell is bound by a membrane with pores, contains DNA in the form of chromosomes, and typically has two nucleoli. The nucleoli are the sites at which ribosomal subunits are constructed. Studies performed by Oscar Hertwig, Walther Flemming, and Joachim Hämmerling confirmed that the nucleus of a cell contains the genetic information and that it exerts an ongoing influence over the activities of a cell.

The cytoplasm of a eukaryotic cell, which was previously viewed as a nonstructured solution of molecules, ions, etc. in which organelles floated, has a complex and intricate structure that is maintained by the cytoskeleton. The cytoskeleton consists of a network of microtubules, actin filaments, and intermediate filaments.

Many membrane-bound organelles are present in the cytoplasm of a eukaryotic cell. Vacuoles lend structural support to plant cells and store materials until they are used by the cell. Vesicles store and transport materials within cells. Lysosomes and peroxisomes contain lytic enzymes that can be used to break down contents of vacuoles. Mitochondria are the sites at which the energy contained in organic molecules is repackaged into a form usable by individual cells. There are three main types of plastids: chloroplasts are the sites of photosynthesis; leucoplasts store starch, proteins, or oil; chromoplasts contain pigments. Two membranous networks (the endoplasmic reticulum and Golgi complexes) function in the manufacture, transport, processing, and packaging of materials (proteins, lipids, and derived substances) in the cell. Ribosomes (not membrane-bound) are the sites at which proteins are synthesized.

There are two known mechanisms for cellular movement. One involves assemblies of actin filaments within the cell that are associated with cytoplasmic streaming and amoeboid movement. The other mechanism depends on cilia and flagella, structures that arise from basal bodies within the cell and protrude through the cell membrane.

HOW TO STUDY THE CHAPTER

Read the entire chapter through quickly, focusing on the major concepts.

Use the GUIDED STUDY OF THE CHAPTER to help you identify the important details as you **reread** the chapter. Writing out the answers to these questions will help fix them in your mind as well as provide you with a valuable study aid.

Answer the questions in TESTING YOUR UNDERSTANDING without the aid of your text. Check your answers against those in PERFORMANCE ANALYSIS. Analyzing your answers will give you valuable feedback on your level of understanding and preparedness for classroom testing.

GUIDED STUDY OF THE CHAPTER

I. **Introduction** (*pages 102–103*)

> *Focus:* **Two remarkable and apparently contrasting features of the cells of living organisms are their great structural and functional diversity and their numerous similarities with one another.**

II. **Cell Size and Shape** (*pages 102–103*)

    1. Most plant and animal cells fall into what size range?

2. *Vocabulary:* Define metabolism.

3. a. Identify two major factors limiting cell size.

    b. Discuss how the relationship between surface area and volume influences cell size.

4. How do some of the larger one-celled organisms compensate for the limitations of the cell's nucleus with respect to cell size?

5. Describe and account for the relationship between cell size and level of metabolic activity. Include examples in your discussion.

6. Name three factors that cause variations in cell shape.

7. *Vocabulary:* What are axopods? (Figure 5–2)

III. **Subcellular Organization** (*page 104*)

    8. What organisms did van Leeuwenhoek refer to as "the most marvelous of all"?

9. In what ways are cell organelles analogous to the organs of multicellular organisms?

Focus: **A eukaryotic cell is highly compartmentalized with numerous specialized internal structures.**

Focus: **Most cellular activities are occurring simultaneously and affecting each other to varying degrees.**

IV. Cell Boundaries *(pages 104–107)*

A. The Cell Membrane *(pages 104–106)*

10. Name three characteristics common to all living cells.

11. What is the general function of the cell membrane?

12. The following figures represent individual components of a cell membrane. Use these drawings (you may wish to trace them) to construct a drawing of a cell membrane. Use as many copies of each component as you wish. When you finish, you may wish to color your drawing with colored pencils to make the relationships among the various components more apparent.

Phospholipid
(polar head, nonpolar tail)

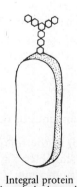
Integral protein
(with carbohydrate chain)

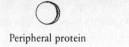
Peripheral protein   Cholesterol

13. How are integral membrane proteins arranged in a cell membrane? (Include the terms hydrophobic and hydrophilic in your answer.)

14. Describe the differences between the two surfaces of a cell membrane. Include the structural features that account for these differences.

15. Name two potential functions of the carbohydrate coat that exists on the surfaces of cells.

Focus: **The fluid-mosaic model of membrane structure recognizes the liquid nature of the phospholipid bilayer and the potential for movement among membrane components.**

---

16. Discuss the arrangement of the hydrophobic and hydrophilic portions of a membrane protein with alpha helix configuration.

17. Describe how a membrane protein with a globular configuration may form a pore through the membrane. (Include the terms hydrophobic and hydrophilic in your answer.)

18. How does the prokaryotic cell membrane differ from that of eukaryotic cells?

Focus: **All the membranes of a eukaryotic cell have the same basic structure.**

19. a. List three ways in which the various membranes of a eukaryotic cell differ.

b. Of what significance to the cell are these differences in membrane structure?

20. What is the typical proportion of lipid to protein found in cell membranes?

21. Name three functions of membrane proteins.

B. **The Cell Wall** *(pages 106–107)*

22. *Vocabulary:* What is the middle lamella?

23. Describe the formation and structure of a plant cell's primary cell wall.

24. What factor limits plant growth? (Figure 5–7)

25. What is the role of lignin in a secondary cell wall?

26. Identify the main ingredient of the cell walls of (a) plants and algae, (b) fungi, and (c) prokaryotes.

V. **The Nucleus** *(pages 108–110)*

A. **General Remarks** *(page 108)*

27. Describe the structure of a nuclear pore. (See also Figure 5–8.)

28. *Vocabulary:* Distinguish between chromatin and nucleolus.

29. What occurs at the nucleolus?

B. **The Functions of the Nucleus** *(pages 109–110)*

30. Fill in the following table comparing the studies of Oscar Hertwig, Walther Flemming, and Joachim Hämmerling. (See also Figures 5–10, 5–11, and 5–12.)

Table for Question 30

	HERTWIG	FLEMMING	HÄMMERLING
Organism studied			
Summary of observations/ experiments			
Conclusions			

Focus: **The nucleus of a cell (1) carries the hereditary information and (2) exerts an influence over the activities of the cell.**

VI. The Cytoplasm *(pages 110–120)*

A. General Remarks *(page 110)*

Focus: **The cytoplasm of a cell is highly organized and filled with organelles.**

31. Label the following drawings of a typical (a) animal and (b) plant cell. You may wish to color the various components with colored pencils to emphasize their structural relationships.

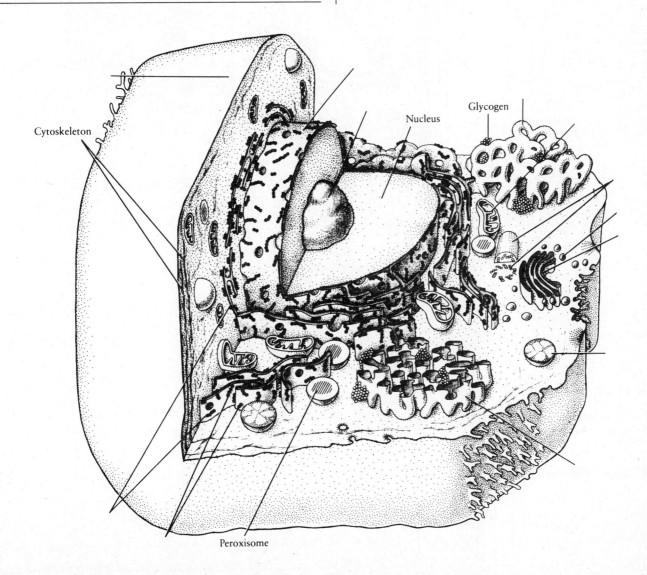

Cytoskeleton

Nucleus

Glycogen

Peroxisome

(a)

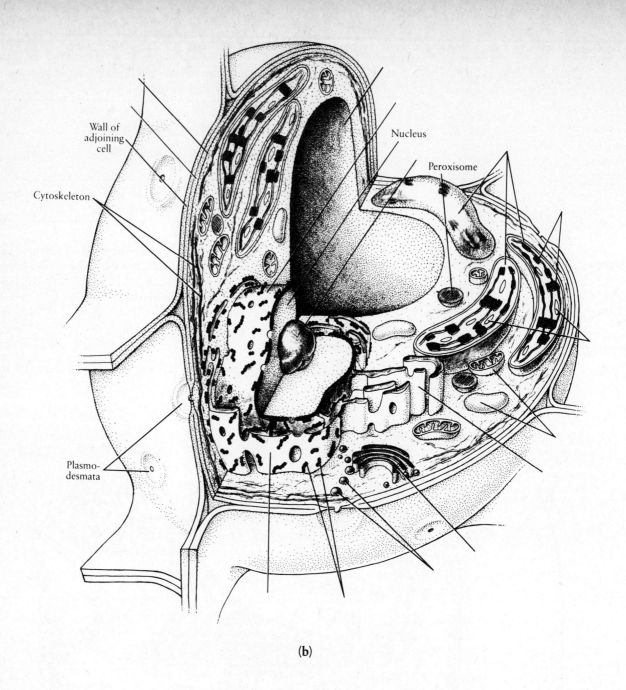

Wall of
adjoining
cell

Cytoskeleton

Nucleus

Peroxisome

Plasmo-
desmata

(b)

B. The Cytoskeleton *(pages 110–115)*

32. What technical development revealed the complexity of the cytoskeleton?

33. Name four functions of the cytoskeleton.

34. Identify three types of filaments that are major components of the cytoskeleton.

35. Fill in the following table comparing the three major filament types of the cytoskeleton.

Table for Question 35

	MICRO-TUBULES	ACTIN FILAMENTS	INTERMEDIATE FILAMENTS
Size			
Structure and composition			
Ease of disassembly			
Major functions			

 Spectrin dimer

  Actin

 Cell membrane

Ankyrin

Integral protein

 Actin-spectrin link protein

36. In what types of cells are intermediate filaments found in high density?

> *Focus:* Cells have an internal framework, the cytoskeleton, that continuously changes with the activities of the cell.

C. *Essay:* Spectrin and the Red Blood Cell *(page 114)*

37. Describe how the red blood cell is specialized for its function.

38. Describe the structure of spectrin.

39. Assemble the following cellular components to demonstrate their structural relationships. Use as many copies of each component as you wish.

D. **Vacuoles and Vesicles** *(page 115)*

40. a. *Vocabulary:* What is a vacuole?

b. What is its relationship to the tonoplast?

41. a. How does the large, central vacuole of plant cells develop?

b. Identify two roles of this vacuole.

42. In what three ways does a vesicle differ from a vacuole?

E. Ribosomes *(page 115)*

> *Focus:* Ribosomes are the most numerous of a cell's organelles; they are not bound by a membrane.

43. Compare the size and number of ribosomes per cell in prokaryotic and eukaryotic cells.

44. Name the essential cell function that occurs at ribosomes.

45. What is the relationship between number of ribosomes and amount of protein made by a cell?

> *Focus:* The location of ribosomes in a cell depends upon the ultimate use of the proteins being manufactured.

46. Where are ribosomes located in cells that mainly produce proteins (a) for their own use and (b) for use outside the cell?

F. Endoplasmic Reticulum *(pages 115–117)*

47. *Vocabulary:* What is endoplasmic reticulum?

48. Fill in the following table concerning rough and smooth endoplasmic reticulum.

Table for Question 48

	ROUGH	SMOOTH
Ribosomes present (yes or no)		
Cell type(s) in which each is abundant		

49. Describe the structural relationship between nuclear envelope, rough endoplasmic reticulum, and cisternae.

50. *Vocabulary:* Define and describe the function of the signal sequence during protein synthesis.

51. Describe the route taken by an export protein, from the ribosomes to the Golgi complex.

52. What types of processing might a newly synthesized protein molecule undergo before it reaches its ultimate destination?

53. a. Identify one role played by the smooth endoplasmic reticulum in liver cells.

    b. Name one function of a specialized transitional endoplasmic reticulum present in liver cells.

> *Focus:* There are apparently many different types of smooth endoplasmic reticulum, each of which has a separate, specialized function.

G. Golgi Complexes (pages 117–118)

54. Describe the structure of a Golgi complex.

55. Identify three functions of the Golgi complex.

56. a. Molecules of what type may be added to proteins and lipids in the Golgi complex?

    b. In addition to those mentioned in (a), molecules of what type may be added to proteins in the Golgi complex? (Figure 5–21)

57. What additional role do Golgi complexes play in plant cells?

Focus: **The Golgi complexes serve as centers for processing, packaging, and distributing proteins and lipids for use in the cell at the cell surface, or for export from the cell.**

H. Lysosomes (page 118)

I. Peroxisomes (pages 118–119)

58. Fill in the following table comparing lysosomes and peroxisomes.

Table for Question 58

	LYSOSOMES	PEROXISOMES
Definition		
Contents		
Function(s) (include one example)		

59. What event is thought to be related to some of the symptoms of rheumatoid arthritis and gout?

60. *Review Question:* Construct a flow diagram illustrating the pathway by which proteins are manufactured and exported from the cell. Use these compounds and cellular components as the basis for your diagram: amino acids, proteins, lipids, carbohydrates, glycolipids, glycoproteins, lipoproteins, transport vesicle, rough endoplasmic reticulum, transitional endoplasmic reticulum, ribosomes, Golgi complex, secretion vesicle, lysosome. Indicate the location of the following processes: protein synthesis, transport of proteins, chemical processing and packaging.

J. Mitochondria (page 119)

61. What cellular function occurs in mitochondria?

62. Describe the structure of a mitochondrion.

63. a. *Vocabulary:* Describe cristae.

    b. What essential functions are carried out on their surface?

64. What is the relationship between (a) energy requirements of a cell and number of mitochondria and (b) activity of a mitochondrion and number of cristae?

K. Plastids *(page 120)*

65. Plastids are found in the cells of what two types of organisms?

66. Fill in the following table concerning plastids.

Table for Question 66

	LEUCO-PLAST	CHROMO-PLAST	CHLORO-PLAST
Contents			
Function(s)			

VII. How Cells Move *(pages 121–123)*

A. General Remarks *(page 121)*

67. *Vocabulary:* What is cytoplasmic streaming?

68. Cite three examples of cellular movement.

69. List the two different molecular mechanisms of cellular movement that have been identified.

B. Actin and Associated Proteins *(pages 121–122)*

70. Name three cellular functions of actin.

71. *Vocabulary:* What is myosin?

> *Focus:* Actin, myosin, and associated proteins are active during cytoplasmic streaming, cell movement, and the muscle contractions of vertebrates.

72. Cite five examples of cellular movement in which actin plays a role.

C. Cilia and Flagella *(pages 122–123)*

73. *Vocabulary:* Distinguish between cilia and flagella.

74. a. How does the function of cilia differ in unicellular protists and multicellular organisms?

   b. Cite three examples in which cilia or flagella are associated with movement of the organism or cell.

   c. Cite two situations in which cilia function to move substances across a cell surface.

75. Name four types of organisms that have neither cilia nor flagella on any of their cells.

76. a. Diagram the structure of microtubules and summarize the mechanism by which they are thought to move.

   b. Identify the functions of the following microtubule components: the "arms," the "spokes," and the "links."

Basal Bodies and Centrioles (page 123)

77. Basal bodies give rise to _____ and _____.

78. Diagram cross-sectional views showing the arrangement of microtubules present in the following structures.

   Cilia   Flagella

   Basal Bodies   Centrioles

79. Describe the arrangement of centrioles in a cell.

80. *Review Question:* Match each cellular structure with its function. Answers may be used more than once. A blank may have more than one number, as indicated in parentheses.

   1. actin filaments
   2. cell membrane
   3. cell wall
   4. chloroplast
   5. chromoplast
   6. Golgi complex
   7. intermediate filament
   8. leucoplast
   9. lysosome
   10. microtubules
   11. mitochondria
   12. nucleolus
   13. nucleus
   14. peroxisomes
   15. ribosomes
   16. rough endoplasmic reticulum
   17. smooth endoplasmic reticulum
   18. vacuole

   a. _____ protein synthesis (2 answers)
   b. _____ energy production for use by the cell
   c. _____ storage of lytic enzymes (2 answers)
   d. _____ structural support (3 answers)
   e. _____ storage of hereditary information
   f. _____ chemical processing of proteins for export
   g. _____ controls flow of materials into and out of the cell
   h. _____ lipid synthesis and metabolism
   i. _____ storage of fats and oils
   j. _____ site of synthesis of export proteins
   k. _____ site of photosynthesis
   l. _____ ribosomal subunit construction
   m. _____ contains water and solutes and lends support to plant cells
   n. _____ responsible for the orange and yellow color of leaves in autumn
   o. _____ structural support of plant cells; contains cellulose

81. Organisms that have centrioles also have _____, _____, and _____.

82. Centrioles are thought to perform what function(s) for the cell?

VIII. Summary *(page 124):* Read the summary. If you are familiar with the essential features of the material presented there, you are ready to complete the section TESTING YOUR UNDERSTANDING.

TESTING YOUR UNDERSTANDING

After you have completed this examination, compare your answers with those in the section that follows.

1. Who first identified organelles using the light microscope?
   a. Robert Hooke
   b. Theodor Schwann
   c. Antony van Leeuwenhoek
   d. Matthias Schleiden
   e. Rudolf Virchow

2. One of the factors that limits cell size appears to be the:
   a. efficiency of its mitochondria.
   b. capacity of its nucleus to control its activities.
   c. size to which the nuclear membrane may expand.
   d. proportion of ribosomes attached to the endoplasmic reticulum.
   e. productivity of its endoplasmic reticulum.

3. Hämmerling contributed to our understanding of the function of the nucleus by studying the:
   a. fertilization of sea urchin eggs.
   b. "dance of the chromosomes."
   c. development of oak trees.
   d. effects of removing the nucleus from an amoeba.
   e. marine alga *Acetabularia*.

4. Which organelle, viewed using the electron microscope, appears to be a collection of fine granules and tiny fibers?
   a. mitochondrion
   b. endoplasmic reticulum
   c. chloroplast
   d. nucleolus
   e. Golgi complex

5. The middle lamella of plant cells is composed primarily of:
   a. pectins.
   b. chitin.
   c. cellulose.
   d. lignin.
   e. peptidoglycans.

6. In general, hydrophilic segments of integral proteins are found:
   a. on the outer surface of the cell.
   b. on the inner surface of the cell.
   c. lining the pores that pass through the cell membrane.
   d. in the helical zig-zag section of the protein.
   e. on both surfaces and lining the pores of the membrane.

7. Glucose is to cellulose as _____ is (are) to peptidoglycan.
   a. pectin
   b. chitin
   c. sugars and amino acids
   d. proteins and lipids
   e. lignin

8. A pore through the cell membrane may be created when:
   a. two or more peripheral proteins associate to form a channel through the membrane.
   b. two or more integral proteins associate to form a channel through the membrane.
   c. a lysosome releases a small, controlled amount of lytic enzyme that digests a hole in the membrane.
   d. globular proteins with alternating regular and irregular sequences fold like an accordion.
   e. linked cholesterol molecules penetrate both sides of the membrane.

9. Cellulose is a principal constituent of the cell walls of:
   a. bacteria only.
   b. fungi only.
   c. algae only.
   d. plants and fungi.
   e. plants and algae.

10. Which cytoskeletal elements radiate outward from an organizing center near the nucleus?
    a. microtubules
    b. wispy fibers
    c. fine protein threads
    d. intermediate filaments
    e. actin microfilaments

11. The signal sequences of proteins intended for export are removed from the proteins in:
    a. the nucleus.
    b. mitochondria.
    c. the cytoplasm by specialized ribosomes.
    d. the cisternae of the endoplasmic reticulum.
    e. transit from the endoplasmic reticulum to the Golgi complexes.

12. The site of lipid synthesis is the:
    a. lysosomes.
    b. mitochondria.
    c. cytoplasm.
    d. smooth endoplasmic reticulum.
    e. rough endoplasmic reticulum.

13. Select the correct sequence of organelles through which proteins manufactured for export move in a cell.
    1. transitional endoplasmic reticulum
    2. transport vesicles
    3. rough endoplasmic reticulum
    4. Golgi complexes
    5. ribosomes
    a. 5, 2, 3, 1, 4
    b. 5, 3, 1, 2, 4
    c. 2, 3, 5, 1, 4
    d. 1, 5, 3, 4, 2
    e. 5, 3, 1, 4, 2

14. Before they become part of the cell's membrane, lipids and carbohydrates are repackaged and modified (into glycolipids) in/on the:
    a. smooth endoplasmic reticulum.
    b. Golgi complex.
    c. mitochondria.
    d. ribosomes.
    e. tonoplast.

15. A cell would begin to hydrolyze itself if which structures broke apart?
    a. basal bodies
    b. mitochondria
    c. food vacuoles
    d. lysosomes
    e. leucoplasts

16. The enzyme that breaks down hydrogen peroxide (H_2O_2) into water and oxygen is found in the _____ of plant cells.
    a. chloroplasts
    b. mitochondria
    c. peroxisomes
    d. lysosomes
    e. central vacuoles

17. Cells that have a high energy demand (such as heart muscles) have numerous:
    a. nuclei.
    b. Golgi complexes.
    c. lysosomes.
    d. leucoplasts.
    e. mitochondria.

18. Chromoplasts are organelles found in _____ that contain _____ pigments.
    a. algae; green
    b. plants; green
    c. fungi; brown
    d. algae and plants; green
    e. algae and plants; yellow and orange

19. Which of these is an activity in which actin does NOT participate?
    a. whiplike action of flagella in the tails of sperm
    b. maintenance of cytoplasmic organization
    c. movement of chromosomes during cell division
    d. migration of cells during embryonic development
    e. streaming movements in slime molds

20. Which group of organisms has members that do NOT have cilia or flagella on any cells?
    a. ferns
    b. flatworms
    c. roundworms
    d. protists
    e. vertebrates

21. Which term does NOT belong with the others?
    a. actin filaments
    b. arms
    c. spokes
    d. links
    e. microtubules

22. T or F The interconnection of all structures within a cell by the network of wisplike fibers causes the cell to have a rigid, ordered, three-dimensional structure.

23. T or F Vacuoles and vesicles differ only with respect to the types of phospholipids in their membranes.

24. T or F A spindle, the group of microtubules that facilitates chromosome movement during cell division, can be constructed in a cell that does not contain centrioles.

25. T or F If a large cell and a small cell have the same rate of metabolism per unit volume, the large cell is better able to handle the transport of raw materials and products across the cell membrane.

26. T or F Peripheral membrane proteins and the carbohydrate portions of glycolipids constitute the major portion of the outer surface of a cell membrane.

27. T or F The nuclear envelope consists of two concentric lipid bilayer membranes that are studded with nuclear pores.

28. T or F Maximum cell size may be limited by the ability of the nucleus to produce enough copies of information to properly regulate the cell's metabolism.

29. List five activities in which a cell (for example a *Chlamydomonas*) may be engaged simultaneously.

30. Cite at least five examples of movement of cells or cellular components.

PERFORMANCE ANALYSIS

1. **c** Antony van Leeuwenhoek is credited with being the first scientist to observe subcellular organelles. The other scientists mentioned made contributions that have since been developed into the modern cell theory. (page 104)

2. **b** The two major factors that limit cell size are the surface area to volume relationship and the capacity of the nucleus to control cellular activities. (page 102)

3. **e** Joachim Hämmerling studied the influence of the nucleus on cap formation in the marine alga *Acetabularia*. He concluded that the nucleus directs the formation of substances involved in cap formation and that these substances persist in the cytoplasm. (page 109)

4. **d** The nucleolus has a fine granular/fibrous appearance under the electron microscope, which is thought to result from ribosomal subunits and chromatin threads. (page 108)

5. **a** The middle lamella that cements two adjacent plant cells consists of pectins and other polysaccharides. (page 106)

6. **e** The inner portion of a cell membrane is hydrophobic. Hydrophilic regions exist on the inner and outer membrane surfaces. The linings of membrane pores are also hydrophilic. (pages 105, 106)

7. **c** Peptidoglycans are polymers formed from amino acids and sugars. You will recall from Chapter 3 that cellulose is a polymer composed of glucose units. (page 107)

8. **d** Membrane pores may be created when globular proteins that possess regular helical sequences alternating with irregular sequences become folded back and forth through the cell membrane. The inner surface of these pores is hydrophilic. (page 106)

9. **e** The cell walls of plants and algae contain cellulose. The cell walls of fungi contain chitin. Prokaryotic cell walls contain polysaccharides and peptidoglycans. (pages 106, 107)

10. **a** Microtubules radiate out from an organizing center near the nucleus of many cells. Intermediate fibers may be closely associated with microtubules in some cells and radiate out from the nuclear membrane. (page 110)

11. **e** The signal sequence by which a cell keeps track of a protein during its synthesis is removed somewhere between the endoplasmic reticulum where it was constructed and the Golgi complex where it is further processed. (page 116)

12. **d** The presence of large quantities of smooth endoplasmic reticulum in cells involved in lipid synthesis and metabolism is taken as evidence that smooth endoplasmic reticulum has a major role in these processes. (page 117, Figure 5–21)

13. **b** Proteins manufactured for export are synthesized on ribosomes on rough endoplasmic reticulum. From the ribosomes, they travel into the lumen of the rough endoplasmic reticulum and move to a transitional endoplasmic reticulum where they are packaged into transport vesicles. These transport vesicles then proceed to the Golgi complexes. (pages 116, 117)

14. **b** In the Golgi complexes, cell products are processed and packaged for transport to other parts of the cell or for export out of the cell. Part of the processing includes the addition of carbohydrates to proteins and lipids to form glycoproteins and glycolipids. (page 117)

15. **d** Lysosomes contain hydrolytic enzymes that are capable of digesting the cell if the lysosomes are damaged and their contents are released into the cell. Lysosomes perform the important function of hyrolyzing the contents of vacuoles containing phagocytized particles. (page 118)

16. **c** In addition to lysosomes, peroxisomes also contain lytic enzymes, including those that break down purines and one that splits hydrogen peroxide (a harmful product of purine hydrolysis and other reactions) into oxygen and water. (page 119)

17. **e** Since mitochondria are the organelles in which the energy stored in organic molecules is converted into energy for use by the cell, they are numerous in cells with high energy demands. (page 119)

18. **e** Chromoplasts are plastids containing yellow and orange pigments found in the cells of algae and plants. Chloroplasts contain the main pigment, chlorophyll (which is green), found in these organisms. (page 120)

19. **a** Flagella are composed of microtubules constructed from globular protein units. The other functions mentioned are performed by actin. (page 121)

20. **c** Nearly all eukaryotic organisms have cells with either cilia or flagella at some point in their life cycle. Exceptions include red algae, fungi, flowering plants, and roundworms. (page 122)

21. **a** Arms, spokes, and links refer to structural components of flagella, which are composed of microtubules. (page 122)

22. **False** The cytoskeleton does contribute to the organization in a cell. However, its components are flexible and constantly changing. (page 115)

23. **False** Vesicles are distinguished from vacuoles primarily by size, function, and composition. (page 115)

24. **True** There has long been evidence that centrioles facilitate spindle formation during cell division. However, plant cells that contain no centrioles also form functional spindles. (page 123)

25. **False** The small cell has a greater surface area per unit volume than the larger cell and is therefore more efficient at moving metabolic products and waste materials across its cell membrane. (page 102)

26. **False** Peripheral membrane proteins are on the cytoplasmic side of the membrane and carbohydrates are attached to proteins on the outside of the cell membrane. These carbohydrates are thought to be important in the processes by which messenger molecules and cells recognize each other. (page 105)

27. **True** The nuclear envelope consists of two concentric membranes. Each of these membranes is a lipid bilayer similar in structure to that of the cell membrane. (page 108)

28. **True** See the answer to question 2.

29. Activities in which free-living cells such as *Chlamydomonas* may be engaged include: swimming, photosynthesis, absorbing nutrients, building cell walls, making proteins, converting sugars to starch, oxidizing food molecules, looking for a mate, and orienting in the sunlight. (page 104)

30. Examples of movement by cells or cell components include cytoplasmic streaming, beating of cilia on tracheal cells or cells of the oviduct, migration of embryonic cells, formation of axons on nerve cells, amoeboid movement, muscle cell contraction, and swimming of sperm cells by means of flagella. (pages 121–123)

CHAPTER 6

How Things Get into and out of Cells

CHAPTER ORGANIZATION

MAJOR CONCEPTS

The cell membrane plays an important homeostatic role by controlling the movement of substances into and out of cells. Internal cell membranes allow functional compartmentalization by regulating the flow of materials among various regions of the cell.

Water potential refers to the potential energy of water. Water tends to flow from regions of high water potential to regions of low water potential. Factors affecting water potential include gravity, pressure, and the concentration of solutes.

Water moves by the processes of bulk flow and diffusion. In living systems water also moves via osmosis, which is diffusion through a selectively permeable membrane in response to a solute concentration gradient. The rigidity of a mature plant cell results from the osmosis of water into the cell, which is hypertonic relative to plant fluids. The inflexible cell wall prevents an equilibrium state; the resulting internal pressure (turgor) keeps the plant body crisp. Diffusion is efficient only over short distances; however, the countercurrent arrangement of blood vessels maintains concentration gradients that promote the diffusion of substances into and out of the blood stream, where they are transported by bulk flow.

Many single-celled organisms are capable of detecting and responding to solute concentration gradients in their external environment.

The transport of substances within and between cells and between cells and their environment is accomplished by passive techniques that are driven by concentration gradients and by active techniques that require cellular energy. The passive processes are osmosis, simple diffusion, and facilitated diffusion. The active processes are active transport and vesicle-mediated transport. Active transport involves transport proteins in the cell membrane. Vesicle-mediated transport includes phagocytosis, pinocytosis, receptor-mediated endocytosis, and exocytosis.

Cells communicate with other cells by means of cell-cell connections, nerve impulses, and chemical signals. The cytoplasm of adjacent plant cells is connected by plasmodesmata. In animals, gap junctions allow the passage of materials between adjacent cells.

HOW TO STUDY THE CHAPTER

Read the entire chapter through quickly, focusing on the major concepts.

Use the GUIDED STUDY OF THE CHAPTER to help you identify the important details as you **reread** the chapter. Writing out the answers to these questions will help fix them in your mind as well as provide you with a valuable study aid.

Answer the questions in TESTING YOUR UNDERSTANDING without the aid of your text. Check your answers against those in PERFORMANCE ANALYSIS. Analyzing your answers will give you valuable feedback on your level of understanding and preparedness for classroom testing.

I. **Introduction** *(pages 127–128)*

> *Focus:* The exchange of materials between living and nonliving systems occurs at the level of the cell.

1. a. The cellular structure that controls the exchange of material between internal and external environments is the _____.

    b. What structures allow for functional compartmentalization within cells?

2. Name three reasons why the regulation of exchange of substances between cells is important.

3. a. Name two important but apparently contradictory functions of cell membranes.

    b. On what two factors do these functions depend?

II. **The Movement of Water and Solutes** *(pages 128–134)*

A. **General Remarks** *(page 128)*

4. *Vocabulary:* What is water potential?

5. Identify three factors that influence water potential.

6. Describe the relationship between the concentration of solute molecules and the water potential of a solution.

> *Focus:* No matter what factor causes water potential, water always moves from regions of greater to regions of lesser water potential.

7. *Vocabulary:* Define hydrostatic pressure.

8. What unit is usually used to measure hydrostatic pressure?

9. Distinguish between the functions of bulk flow and diffusion in living systems.

B. **Bulk Flow** *(page 129)*

10. *Vocabulary:* Define bulk flow.

11. List three examples of the movement of water or fluids by bulk flow.

C. **Diffusion** *(pages 129–130)*

12. *Vocabulary:* Define diffusion.

13. a. What situation is responsible for the diffusion of molecules?

    b. Describe the mechanism by which diffusion results in a uniform distribution of solute molecules in a solvent.

14. Distinguish between movement of molecules "down a gradient" and movement of molecules "against a gradient."

Focus: In the diffusion process, the net movement of individual molecules is always from a region of greater to a region of lesser potential energy.

---

15. How does the term dynamic equilibrium apply to the diffusion process?

16. Identify the two essential characteristics of diffusion.

Cells and Diffusion (page 130)

17. Name two characteristics of solutes that influence whether or not a solute molecule can penetrate a cell membrane.

18. How does the process of diffusion limit cell size?

19. List three examples of ways in which *cells* maintain steep concentration gradients.

Countercurrent Exchange (page 130)

20. Describe the role of concentration gradients in countercurrent exchange systems.

D. *Essay:* Sensory Responses in Bacteria: A Model Experiment *(page 131)*

Focus: Many single-celled organisms can move purposefully through their environments by detecting and responding to concentration gradients.

---

21. What feature of single-celled organisms determines whether or not they will respond to a concentration gradient?

22. Describe the two types of movement exhibited by flagellated bacteria and indicate which type is more common in the presence of a concentration gradient.

Focus: Bacteria perceive the presence of a concentration gradient by detecting changes in concentration over *time*, not distance.

---

23. What two components are essential to a process that answers questions scientifically?

24. How is Koshland's study of *Salmonella* a "model experiment"?

25. What frequently follows confirmation of a hypothesis?

E. Osmosis: A Special Case of Diffusion *(pages 132–134)*

26. *Vocabulary:* What is meant by a selectively permeable membrane?

27. How does osmosis differ from diffusion?

28. *Vocabulary:* Distinguish between the terms isotonic, hypotonic, and hypertonic as they refer to solutions. (Include the term "water potential" in your answer.)

Focus: The terms isotonic, hypotonic, and hypertonic can only be used when comparing the *relative* concentrations of two or more solutions.

Focus: During osmosis, water moves so that the more concentrated solution becomes diluted until the concentrations of both solutions are equal or until some external force prevents continued osmosis.

Osmosis and Living Organisms (pages 132–133)

29. Cite three examples of organisms/cells that are isotonic with their environment.

30. a. What is a consequence of osmosis to which fresh-water, single-celled organisms must adapt?

b. Describe how members of the genus *Paramecium* have adapted to the continuous osmosis that occurs in their fresh-water environment.

Osmotic Potential (pages 133–134)

31. *Vocabulary:* Define osmotic pressure and osmotic potential.

32. How can osmotic potential be measured?

33. Discuss the relationship between solute concentration, water potential, and osmotic potential.

34. The following diagram depicts a container divided into two parts by a membrane that is permeable to water but not to glucose. If compartment A contains a 5 percent glucose/water solution and compartment B contains a 10 percent glucose/water solution, in which direction would water flow? Explain your answer.

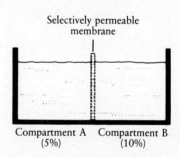

Selectively permeable membrane

Compartment A   Compartment B
(5%)   (10%)

Turgor (page 134)

35. Describe the role of osmosis in the elongation of plant cells.

36. *Vocabulary:* Define and distinguish between wall pressure and turgor.

37. Describe the mechanisms by which turgor is maintained in a plant.

III. Carrier-Assisted Transport *(pages 134–137)*

A. General Remarks *(pages 134–135)*

38. What component of the cell membrane is responsible for the fact that most ions and hydrophilic molecules cannot cross the membrane but hydrophobic molecules readily diffuse across the membrane?

39. Name two factors that prevent small ions from freely diffusing across cell membranes.

Focus: Molecules and ions that cannot diffuse freely across cell membranes are transported by proteins embedded in the cell membrane.

---

40. What feature of a transport protein determines which molecule(s) it will transport?

41. *Vocabulary:* What is a permease?

42. *Vocabulary:* Distinguish between facilitated diffusion and active transport.

B. Facilitated Diffusion *(page 136)*

43. a. Name two factors that determine the rate of diffusion for a molecule transported by facilitated diffusion.

    b. Which factor ultimately limits the rate of diffusion?

C. Active Transport *(page 136)*

Focus: In active transport, molecules or ions are transported against a concentration gradient.

The Sodium-Potassium Pump (page 136)

44. Describe the concentration gradients of sodium and potassium associated with the sodium-potassium pump.

45. The molecule providing the energy to power the sodium-potassium pump is ——————.

Focus: The transport of sodium and potassium by the sodium-potassium pump is thought to involve a single transport protein that changes configuration to accommodate intracellular Na^+ and extracellular K^+.

D. Types of Transport Molecules *(page 137)*

Focus: Several models have been proposed to describe the function of transport proteins. One current model suggests that these proteins have hydrophilic cores, and alteration of the protein configuration propels the transported molecule through the core.

---

46. a. Discuss two general mechanisms by which the configuration of a transport protein may be altered in the course of transporting a molecule or ion.

b. Which of these mechanisms operates in sodium ion transport across nerve cell membranes?

47. a. *Vocabulary:* Describe and distinguish between uniport, symport, and antiport transport mechanisms.

b. In the following diagram of a cell membrane, draw in examples of transported and cotransported solute molecules (including the direction of transport) to illustrate uniport, symport, and antiport systems.

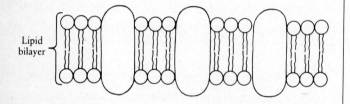

Lipid
bilayer

48. *Review Question:* Fill in the following table.

IV. Vesicle-Mediated Transport *(pages 138–140)*

49. *Vocabulary:* Distinguish between exocytosis and endocytosis.

50. *Vocabulary:* Distinguish between phagocytosis and pinocytosis and cite one example of each process.

51. How does receptor-mediated endocytosis differ from other forms of endocytosis?

52. Outline the process by which cholesterol enters animal cells.

53. *Vocabulary:* What is clathrin?

54. What is the relationship between clathrin and coated pits?

Table for Question 48

	SIMPLE DIFFUSION	FACILITATED DIFFUSION	ACTIVE TRANSPORT
Direction of movement along a concentration gradient			
Passive process? (yes or no) (What factor drives the process?)			
Active process? (yes or no)			
Carrier protein required? (yes or no)			

55. Describe the formation of a coated vesicle.

56. Describe the fate of the portion of the cell membrane that is internalized during endocytosis.

V. Cell-Cell Junctions *(pages 140–142)*

A. General Remarks *(pages 140–142)*

57. *Vocabulary:* Distinguish between tissues and organs.

58. Name the four main types of tissues found in animals.

> *Focus:* Cells in a multicellular organism may communicate with distant cells via nerve impulses and chemical signals.

> *Focus:* Cells may communicate with immediately neighboring cells through several types of cell junctions.

59. *Vocabulary:* Distinguish between plasmodesmata and gap junctions.

60. From what cell organelle do desmotubules arise?

61. *Vocabulary:* What are connexons? (see Figure 6–19)

62. There is one important factor common to the transport of materials via transport proteins, endocytosis and exocytosis, and cell-cell connections. What is this factor?

B. *Essay:* Communication in the Cellular Slime Mold *(page 141)*

> *Focus:* The compound cyclic AMP is an important mediator of cell-cell communication in plants and animals.

63. Outline the life cycle of a cellular slime mold.

VI. Summary *(pages 142–143):* Read the summary. If you are familiar with the essential features of the material presented there, you are ready to complete the section TESTING YOUR UNDERSTANDING.

TESTING YOUR UNDERSTANDING

After you have completed this examination, compare your answers with those in the section that follows.

1. Which statement about membranes is NOT correct?
   a. Cell membranes regulate the entry and exit of substances.
   b. Specific reactions are compartmentalized by internal cell membranes in prokaryotes.
   c. Membranes are necessary for osmosis to occur.
   d. Specific reactions are segregated by internal membranes in eukaryotes.
   e. Nonpolar molecules readily diffuse through the lipid bilayer of membranes.

2. The five solutions listed below are sitting on a laboratory table in beakers. Which one has the highest water potential?
   a. distilled water
   b. a 2 percent NaCl solution
   c. a 4 percent NaCl solution
   d. a 1 percent glucose solution
   e. a 6 percent glucose solution

3. You just knocked over your drink and the fluid is falling to the floor by:
   a. osmosis.
   b. diffusion.
   c. bulk flow.
   d. turgor pressure.
   e. active transport.

4. Oxygen and carbon dioxide move into and out of cells primarily by:
   a. active transport.
   b. bulk flow.
   c. diffusion.
   d. exocytosis.
   e. facilitated diffusion.

5. Diffusion occurs most efficiently when:
   a. the concentration gradient is steep.
   b. large molecules are involved.
   c. active transport systems are involved.
   d. transport occurs over long distances.
   e. molecules are moving against a concentration gradient.

6. When you place a dialysis bag containing a solution of 3 moles of NaCl per liter in a beaker of distilled water, a point will be reached when there is no more net movement of water. This point is referred to as _____ and the solution inside the bag is said to be _____ with respect to the solution outside the bag.
   a. turgor; hypotonic
   b. equilibrium; isotonic
   c. stagnation; hypertonic
   d. equilibrium; hypotonic
   e. turgor; hypertonic

7. An amoeba living in a freshwater pond is _____ relative to its environment and, therefore, water will move _____ the amoeba.
   a. hypotonic; into
   b. hypotonic; out of
   c. isotonic; neither into nor out of
   d. hypertonic; into
   e. hypertonic; out of

8. Osmosis will only occur if there is a(n) _____ between the two solutions. Which phrase will NOT correctly complete this sentence?
   a. selectively permeable membrane
   b. active transport mechanism
   c. difference in solute concentration
   d. difference in solvent concentration
   e. difference in water potential

9. You have forgotten to water your African violet for over a week and it has wilted due to a loss of:
   a. solutes.
   b. osmotic potential.
   c. selective permeability.
   d. turgor pressure.
   e. plasticity of the cell walls.

10. The primary barrier to the passage of ions and polar molecules through the cell membrane is the:
    a. boundary layer of carbohydrates.
    b. hydrophobic nature of the integral proteins.
    c. hydrophobic nature of the peripheral proteins.
    d. size of the mosaic protein aggregations.
    e. hydrophobic nature of the lipid bilayer.

11. Facilitated diffusion differs from simple diffusion in that facilitated diffusion:
    a. requires energy.
    b. moves materials against a concentration gradient.
    c. involves vesicles as facilitators.
    d. requires membrane transport proteins.
    e. requires helper cells.

12. When a molecule such as glucose is being moved into a cell by facilitated diffusion, the rate of glucose uptake is dependent upon:
    a. the glucose concentration gradient only.
    b. the amount of available cellular energy only.
    c. the number of glucose transport molecules only.
    d. both the glucose concentration gradient and the number of glucose transport molecules.
    e. both the amount of available cellular energy and the number of glucose transport molecules.

13. Which statement about the sodium-potassium pump is NOT correct?
    a. It maintains a double-ion concentration gradient.
    b. Approximately one-third of the ATP used by a resting animal is used to power the sodium-potassium pump.
    c. Nerve cells use the double-ion concentration gradient to generate nerve impulses.
    d. ATP causes a shape change in the protein of the pump mechanism.
    e. Sodium is pumped into the cell and potassium is pumped out of the cell.

14. In some nerve cells, vesicles containing acetylcholine fuse with the cell membrane and release their contents outside the cell in a process called:
    a. general endocytosis.
    b. pinocytosis.
    c. exocytosis.
    d. phagocytosis.
    e. receptor-mediated endocytosis.

15. Cholesterol enters a cell:
    a. as free cholesterol by simple diffusion.
    b. as free cholesterol by facilitated diffusion.
    c. as free cholesterol by active transport.
    d. in the form of LDL particles, which are transported across cell membranes via active transport.
    e. in the form of LDL particles, which cause vesicle formation when they attach to receptors.

16. Which statement is NOT true of coated pits?
    a. They represent areas where receptors are concentrated.
    b. They are lined on the cytoplasmic side by clathrin.
    c. The vesicle that forms from a coated pit is coated externally by clathrin.
    d. They are a site of phagocytic vesicle formation.
    e. They are depressions in the cell membrane where gap junctions occur.

17. Which statement is NOT true of plasmodesmata?
    a. They are composed of a ring of globular integral membrane proteins.
    b. They are 30 to 60 nanometers in diameter.
    c. They are lined by the cell membrane.
    d. They pass through the cell wall.
    e. They are found in plant cells.

18. T or F   Equilibrium of water concentration on the two sides of a plant cell wall is only reached when the cell matures and the cell wall stops stretching.

19. T or F   The membrane-bound transport proteins that carry out facilitated diffusion are the same molecules that carry out active transport.

20. T or F   Endocytic vesicles, which are derived from the cell membrane, are recycled after releasing their products into the cytoplasm, by fusing back into the cell membrane.

21. T or F   Communication between cells of a multicellular organism is crucial to the organism's survival. In higher vertebrates, such as mammals, communication between many cells is accomplished via hormones and cyclic AMP.

22. T or F   The manner in which a bacterial cell determines the particular direction it will move is by detecting differences in concentration of attractants (such as food) between its front and rear ends.

23. Cite two examples in which cells maintain concentration gradients of molecules that drive diffusion of these molecules across their membranes.

24. List the four principal types of tissues found in animals.

PERFORMANCE ANALYSIS

1. b   Prokaryotes do not have the complex internal membrane systems and membrane-bound organelles that characterize eukaryotes. The cell membrane is the feature of a cell that enables it to have an internal environment very different from its external environment. Cellular membranes have a lipid center through which nonpolar molecules readily diffuse. (page 127)

2. a   In a solution, water potential is determined by the total solute concentration. The lower the solute concentration (the higher the water concentration), the higher the water potential. Distilled water, which has no solutes, has the highest water potential. (page 128)

3. c   Bulk flow is the overall movement of a fluid. Diffusion is the net movement of particles from a region of relatively high concentration to a region of relatively low concentration. Osmosis is the diffusion of water through a selectively permeable membrane. Turgor refers to the internal pressure of plant cells which results from the osmosis of water into the cell. Active transport is a process whereby materials are transported into or out of cells and ATP is required as an energy source for the transfer. (page 129)

4. c   Since carbon dioxide and oxygen are small, nonpolar molecules, they can diffuse through the lipid bilayer of cell membranes. (page 130)

5. a   Diffusion is most efficient when the concentration gradient is steep and the distance is small. (page 129)

6. b   Dynamic equilibrium is reached when there is no longer a *net* movement of diffusing molecules. (The individual molecules continue to move, but they are evenly distributed on both sides of the membrane.) Two solutions are said to be isotonic when their osmotic potentials are equal and there is no net movement of water molecules from one solution to another. A solution is hypertonic with respect to another solution when it has a higher osmotic potential (higher solute concentration). The solution with the lower osmotic potential (lower solute concentration) is referred to as hypotonic. (page 132)

7. d   Freshwater organisms are hypertonic with respect to their external environments and, as a consequence, water moves into their cells. These organisms have special mechanisms for countering the influx of water. (page 132)

8. b   Osmosis is defined as the diffusion of water through a selectively permeable membrane from an area of higher water potential (lower solute concentration) into an area of lower water potential (higher solute concentration). (page 132)

9. d   See the answer to question 3. The osmotic potential, which depends upon the concentration of solutes in the plant, has *increased* due to water loss from the cells. (The concentration of solute increases as the concentration of solvent decreases.) (page 134)

10. e   The hydrophobic lipid bilayer of cell membranes is an effective barrier to the movement of ions and polar molecules. Hydrophilic compounds typically cross membranes via special membrane transport mechanisms. (page 134)

11. d   Facilitated diffusion is similar to simple diffusion in that molecules move down a concentration gradient but differs in that special membrane transport proteins are required. (page 136)

12. d   When facilitated diffusion is the transport mechanism, the rate of transport is influenced by the number of transport molecules present in the membrane and the concentration gradient of the substance being transported. (page 136)

13. e The sodium-potassium pump uses approximately one-third of a resting animal's ATP to pump sodium out of and potassium into cells. (page 136)

14. c Exocytosis is the process whereby vesicles fuse with the cell membrane and release their contents, acetylcholine in this case, outside the cell. (page 138)

15. e Cholesterol travels through the blood as LDL particles, which bind to receptors on target molecules, inducing the formation of phagocytic vesicles. (page 138)

16. e Coated pits are depressed regions of a cell membrane in which a number of receptors are aggregated. The cytoplasmic surface of a coated pit is covered with the peripheral membrane protein clathrin. When a phagocytic vesicle is formed, the clathrin coats the vesicle's outer surface. (pages 138, 139)

17. a Plasmodesmata are cell-cell connections in plants that range from 30 to 60 nanometers in diameter. They pass through the cell wall and are lined by cell membrane. (page 140)

18. False Plants cells are generally hypertonic with respect to their environment and water tends to diffuse into the cells; equilibrium is not established. Net water movement into cells is limited by the hydrostatic pressure that develops because the rigid cell wall cannot expand with the influx of water. The internal pressure resulting from these forces is called turgor; it is turgor that keeps the cell, and hence the plant body, stiff. (page 134)

19. False Facilitated diffusion and active transport processes use different membrane transport molecules. (page 136)

20. True The membranes of endocytic vesicles fuse to the cell membrane after their contents have been released into the cell, thereby recycling the membrane components from which they were formed. (page 140)

21. True Hormones and cyclic AMP are two important links in the information chain by which cells of higher vertebrates communicate with one another. (essay, page 141)

22. False Bacterial cells identify the presence of a concentration gradient by detecting differences in the concentration of a substance over *time* as they move through a medium. The distance from front to rear of a bacterial cell is not great enough for a detectable concentration gradient to exist. (essay, page 131)

23. Two concentration gradients maintained by cells are those of carbon dioxide and oxygen. The cell constantly uses oxygen, creating a gradient favoring the diffusion of oxygen into the cell. Conversely, the cell is constantly producing carbon dioxide, creating a gradient favoring its diffusion out of the cell. (page 130)

24. The four major tissue types present in animals are nerve, muscle, connective, and epithelial. (page 140)

CHAPTER 7

How Cells Divide

MAJOR CONCEPTS

The cell is the basic structural unit of all living things. Its capacity to produce two daughter cells from one parent cell permits one-celled organisms to increase in number and many-celled organisms to grow and replace injured or worn-out tissues.

During the process of cell division, each daughter cell receives a set of hereditary information that is identical with that of the parent cell. In prokaryotes, replicated chromosomes are attached to different sites on the interior of the cell membrane and separate into different daughter cells as the parent cell grows and divides. In eukaryotes, entire sets of replicated chromosomes are distributed to daughter cells by an elaborate system of fibers called the spindle.

The entire life history of a eukaryotic cell is known as the cell cycle and is composed of two major parts, interphase and cell division. During the three segments of interphase, the cell carries out its type-specific functions, synthesizes DNA in conjunction with the replication of chromosomes, and prepares for cell division.

Cell division is composed of two phases, mitosis (nuclear division) and cytokinesis (cytoplasmic division). Mitosis consists of four phases, during which the replicated chromosomes condense and sister chromatids migrate toward opposite poles of the cell. In animal cells, cytokinesis is achieved by the constriction of the cell membrane between daughter nuclei. In plant cells, a cell plate is laid down in the cytoplasm between daughter nuclei.

In general, cells must reach a minimum size and certain metabolic state before cell division occurs. Several factors inhibit cell division, including depletion of nutrients, changes in temperature or pH, and, in multicellular organisms, contact with adjacent cells.

We know a great deal about the general features of cell division, but many of the specific mechanisms still have to be worked out. For example, (1) Exactly how do sister chromatids migrate toward opposite poles during anaphase? (2) What makes a cancer cell disobey the normal mechanisms that control cell division? and (3) What are the metabolic clues that control cell division?

HOW TO STUDY THE CHAPTER

Read the entire chapter through quickly, focusing on the major concepts.

Use the GUIDED STUDY OF THE CHAPTER to help you identify the important details as you **reread** the chapter. Writing out the answers to these questions will help fix them in your mind as well as provide you with a valuable study aid.

Answer the questions in TESTING YOUR UNDERSTANDING without the aid of your text. Check your answers against those in PERFORMANCE ANALYSIS. Analyzing your answers will give you valuable feedback on your level of understanding and preparedness for classroom testing.

GUIDED STUDY OF THE CHAPTER

I. Introduction *(page 144)*

    1. What ends are achieved by one-celled and many-celled organisms in the process of cell division?

> *Focus:* **A cell divides when it reaches a certain critical size and metabolic state; each daughter cell receives about half of the parent cell's cytoplasm and organelles as well as an exact replica of the parent cell's hereditary information.**

    2. How often may a bacterium and a *Paramecium* divide?

II. Cell Division in Prokaryotes *(page 145)*

    3. Outline the sequence of events by which prokaryotes separate their replicated chromosomes into daughter cells.

III. Cell Division in Eukaryotes *(page 145)*

    4. In what two ways is eukaryotic cell division more complex than prokaryotic cell division?

    5. Indicate which statements refer to a prokaryotic cell and which to a eukaryotic cell.

        _____ Has a single, circular chromosome.

        _____ Contains much more DNA than the other cell type.

        _____ May divide as often as every six minutes.

        _____ Has linear DNA and a number of distinct chromosomes.

    6. *Vocabulary:* Distinguish between mitosis and cytokinesis.

IV. The Cell Cycle *(pages 146–148)*

A. General Remarks *(page 146)*

    7. *Vocabulary:* Define the term cell cycle.

    8. Name two external factors that influence the length of time required by a cell to complete its cell cycle.

    9. What four preparatory processes must a cell complete during interphase before it can actually engage in mitosis?

    10. Give a detailed account of the synthetic events that occur during the G_1, S, and G_2 phases of interphase.

        G_1

        S

        G_2

    11. Which cellular organelles replicate in G_1?

    Which are synthesized de novo?

    Which possess their own chromosome?

12. What two facts led many biologists to believe that mitochondria and chloroplasts originated as separate organisms?

13. What new cell structures begin to appear at the end of the G$_2$ phase?

B. **Regulation of the Cell Cycle** *(pages 146–148)*

> *Focus:* **The length of the cell cycle and the relative lengths of its phases differ among species and cell types.**

14. Identify three classifications of cells based on their ability to divide and cite one example of each type.

15. Under what circumstances might a cell, which does not usually divide once it is mature, undergo cell division?

16. a. Why is it important that cells divide no faster than the rate sufficient for organismal growth or tissue replacement?

    b. What other factor determines when cells divide?

> *Focus:* **The organization and function of normal tissues are disrupted when they are overwhelmed by rapidly dividing cancer cells.**

17. a. What environmental factors can cause cells to stop dividing?

    b. Name and describe the phenomenon in multicellular organisms, which is not exhibited by cancer cells, that prevents normal cells from dividing.

18. What is the significance of the G$_1$ restriction (R) point to the regulation of normal and cancerous cell division?

19. Summarize two different hypotheses related to R-point regulation of the cell cycle.

V. **Mitosis** *(pages 148–153)*

A. **General Remarks** *(page 148)*

20. What is the function of mitosis?

21. *Vocabulary:* What is the spindle?

B. **The Condensed Chromosomes** *(pages 148–149)*

> *Focus:* **In order for microtubules of the spindle to maneuver replicated chromosomes, the thread-like chromosomes must first undergo extensive condensation.**

22. *Vocabulary:* What is a chromatid?

23. *Vocabulary:* Distinguish between the terms centromere and kinetochore.

24. a. This diagram shows a replicated, condensed chromosome. Label both chromatids, the centromere, kinetochores, and the microtubules.

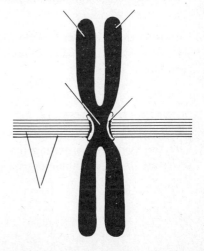

b. Which of the three types of spindle fibers appears in this illustration?

C. The Spindle *(page 149)*

25. a. Label this diagram of a spindle from an animal cell with the terms polar spindle fibers, astral spindle fibers, and centriole.

b. Which feature would be absent from a dividing plant cell?

26. What are the roles of the polar fibers, kinetochore fibers, and the aster fibers?

27. a. What is the source of tubulin dimers for spindle formation?

b. What happens to the dimers after mitosis?

Centrioles and the Microtubule Organizing Center (page 149)

28. Cite the evidence from *Chlamydomonas* suggesting that centrioles and basal bodies are identical structures.

29. Cite the evidence suggesting that centrioles themselves are not microtubule organizing centers.

D. The Phases of Mitosis *(pages 150–153)*

30. Summarize the cellular events that characterize each of the four mitotic phases. You may wish to include diagrams of chromosome movement in your answer. Be sure to include these terms: chromosomes, chromatids, centrioles, spindle, kinetochores, nuclear envelope, and cell membrane.

Prophase

Metaphase

Anaphase

Telophase

31. *Review Question:* Identify the phase of the cell cycle in which each event occurs.

———— DNA is synthesized.

———— Ribosomes, mitochondria, and chloroplasts are synthesized.

———— Individual chromosomes are visible with a light microscope.

———— Density-dependent inhibition arrests cell division.

———— Cell plate or cleavage furrow is formed.

———— Chromosomes are replicated.

The Mechanism of Chromosome Movement (pages 152–153)

> *Focus:* **The movement of chromosomes is the result of interactions among the kinetochore fibers and polar fibers of the spindle.**

32. Summarize two possible mechanisms for the chromosome movements of mitosis and indicate the evidence supporting each mechanism.

VI. Cytokinesis (pages 153–154)

33. Cite the experimental evidence supporting the proposed role of the spindle in positioning the structures responsible for cytokinesis.

34. Summarize the details of cytokinesis in plant and animal cells, emphasizing the differences between them.

35. *Vocabulary:* What is the cell plate?

36. List four cellular activities, other than division, that occur during the cell cycle.

37. Fill in the following table, using a plus (+) to indicate the presence and a minus (−) to indicate the absence of these structures in animals and higher plants.

Table for Question 37

STRUCTURE	ANIMALS	HIGHER PLANTS
Centrioles		
Preprophase band		
Aster		
Cell plate		
Cleavage furrow		

VII. Summary (page 154):
Read the summary. If you are familiar with the essential features of the material presented there, you are ready to complete the section TESTING YOUR UNDERSTANDING.

TESTING YOUR UNDERSTANDING

After you have completed this examination, compare your answers with those in the section that follows.

1. Chromosomes are separated during cell division in prokaryotes by the:
   a. elongation of the cell membrane.
   b. movement of centrioles to the poles.
   c. formation of the cell plate.
   d. polar spindle fibers attached to their kinetochores.
   e. action of actin filaments.

2. During which phase of the cell cycle is DNA replicated? (Give the most precise answer.)
   a. interphase
   b. anaphase
   c. G_1
   d. S
   e. G_2

3. Which organelles derive their membranes from the endoplasmic reticulum?
   a. Golgi complexes
   b. mitochondria
   c. ribosomes
   d. centrioles
   e. chloroplasts

4. Chromosomes become visible under the light microscope as a result of the coiling and condensation of:
   a. tubulin.
   b. actin filaments.
   c. chromatin.
   d. polar microtubules.
   e. microtubules of the cytoskeleton.

5. Nucleoli have disappeared from the nuclear area by the time _____ is complete.
   a. interphase
   b. prophase
   c. metaphase
   d. anaphase
   e. telophase

6. Asters are only found in cells that also have:
   a. spindles.
   b. nuclear membranes.
   c. vacuoles.
   d. kinetochores.
   e. centrioles.

7. Which condition is most characteristic of telophase?
   a. the spindle disperses into tubulin dimers
   b. chromosomes migrate toward the poles of the cell
   c. the endoplasmic reticulum reforms
   d. sister chromatids separate into daughter chromosomes
   e. mitochondria begin to replicate in preparation for cytokinesis

8. During which mitotic phase do identical sets of chromosomes migrate toward opposite poles of the cell?
   a. prophase   d. telophase
   b. metaphase   e. interphase
   c. anaphase

9. The fact that kinetochore fibers can lengthen and shorten without changing in diameter indicates that:
   a. chromosomal movement takes place by contraction of these fibers.
   b. fiber length is changed by the addition or removal of material.
   c. the spindle can contain no proteins other than those associated with microtubules.
   d. the polar fibers of the spindle are really not involved in chromosomal movement.
   e. the fibers are constructed in sections that telescope into one another, like a car radio antenna.

10. Our current understanding of the mechanism of chromosome movement is represented by which of these statements?
    a. Microtubules of the spindle are pulled past each other by the tractor-like walking of little arms.
    b. Microtubular shortening drags the chromosomes to the poles.
    c. Proteins other than tubulin bind to the chromatids, pulling them toward the poles.
    d. Actin filaments stretching between kinetochores of homologous chromosomes actively push the chromosomes apart.
    e. Although we have several good hypotheses, the exact mechanism is currently unknown.

11. A cell plate is formed in a(n) _____ cell during _____.
    a. plant; cytokinesis
    b. plant; the S phase of the cell cycle
    c. animal; cytokinesis
    d. animal; metaphase
    e. fungal; the S phase of the cell cycle

12. Rank these six events in the order of their occurrence.
    a. The nuclear envelope breaks down.
    b. The spindle is completely formed.
    c. Chromosomes become visible by light microscopy.
    d. Kinetochore microtubules shorten.
    e. Chromosomes become diffuse.
    f. Chromatid pairs become aligned.

13. T or F The typical eukaryotic cell contains about 10 times more DNA than a typical prokaryotic cell.

14. T or F It is important that all types of cells divide only after they have reached a critical size and metabolic state.

15. T or F After DNA replication in a prokaryotic cell, each daughter chromosome is attached to a different spot on the cell wall.

16. T or F The nucleolus disappears during prophase of mitosis and reappears during telophase.

17. Distinguish between mitosis and cytokinesis.

18. What is the significance of the G_1 restriction (R) point to the regulation of normal cell division? To the potential control of cell division in cancer cells?

19. Explain the role of the microtubule organizing center during mitosis and during interphase.

20. Explain why cells that lack cell walls tend to take on a rounded shape during mitosis.

PERFORMANCE ANALYSIS

1. **a** The key word in this question is "prokaryotes." Prokaryotic chromosomes are attached to the cell membrane and separate as the membrane elongates. Prokaryotes do not have a spindle or a cell plate. (page 145)

2. **d** The key phrase in this question is "most precise." DNA synthesis does not occur during one of the mitotic phases, such as anaphase, but during interphase, specifically during S phase. (page 146)

3. **a** During G_1 of interphase, a cell engages in growth and replication of molecules and organelles needed for the daughter cells of the next cell division. Some cellular structures, such as microtubules, are made from scratch (de novo); others, such as Golgi complexes, are constructed from membranes that are already part of the endoplasmic reticulum. (page 146)

4. **c** Chromosomes first become visible under the light microscope during prophase of mitosis. During the S stage of the previous interphase, DNA replication produced two identical chromatids per chromosome. In interphase, chromosomes have not yet condensed and so the nucleus appears to be filled with an amorphous substance called chromatin (uncondensed DNA combined with chromosomal proteins.) (page 148)

5. **b** Prophase is the first of the mitotic phases. During this time, the nuclear membrane and nucleoli are dispersed. (page 150)

6. **e** All eukaryotic cells possess nuclear membranes and form spindles during mitosis. Spindle fibers attach to a region of the chromosome called the kinetochore. Asters are composed of short microtubular arrays that radiate in all directions from a pair of centrioles. Since plant cells lack centrioles, they also lack asters. (page 149)

7. **a** Telophase is the last phase of mitosis. Sister chromatids have already separated and migrated to the poles of the cell. The endoplasmic reticulum provides material for the re-formation of the nuclear membrane. The spindle, which has performed its function of separating chromosomes, is no longer needed and disperses into tubulin dimers that rejoin the structure of the cytoskeleton. (pages 149, 152)

8. **c** During prophase the chromosomes condense; during metaphase they line up on the equatorial plane of the cell; during anaphase the identical sets of chromosomes migrate toward opposite poles of the cell; and during telophase the chromosomes redisperse into filamentous chromatin fibers. (page 151)

9. **b** The key concept in this question is that the fibers can change length without changing diameter. If *a* were true, the contracting fibers would increase in diameter. There is no way to evaluate choice *c* and therefore it cannot be the answer. Choice *d* is somewhat appealing but does not account for the fact that fibers do not change diameter. There is no evidence for fibers with telescoping sections (choice *e*). Only choice *b* satisfies the constant diameter provision. Adding or removing material from the ends of the fibers would not change their diameter. (page 152)

10. **e** Choices *a* and *b* are hypotheses mentioned in the text that are consistent with our current knowledge. Although it is known that proteins other than tubulin are associated with the spindle *(c)*, there is no current evidence that they pull the chromosomes. Choice *e* clearly represents our current understanding of this interesting problem. (pages 152, 153)

11. **a** Cell plates are found only in plant cells during cytokinesis. During metaphase of mitosis, cells line up along the equatorial plane, but this is an area of the cell, not a structure like the cell plate. (page 154)

12. The correct sequence of events is **c, a, b, f, d,** and **e.** (pages 150–152)

13. **False** A typical eukaryotic cell contains *1,000* times more DNA than a typical prokaryotic cell. (page 145)

14. **True** If cells divided before they reached roughly twice their normal size, each division would produce successively smaller daughter cells. If they divided before they were metabolically prepared to do so, the division attempt would fail. (page 144)

15. **False** The words "cell wall" should read cell membrane. As the membrane elongates, the chromosomes are separated. (page 145)

16. **True** The nucleolus is not present from late prophase through early telophase. (pages 150, 152)

17. Mitosis is that segment of the cell cycle in which sister chromatids become daughter chromosomes and migrate to the opposite poles of the cell. In other words, the hereditary material of the nucleus has been equally divided between the two new daughter nuclear regions. In cytokinesis, the entire cell divides in such a way that the two daughter nuclei and their surrounding cytoplasm are separated into distinct daughter cells. (page 145)

18. In order for a cell to enter mitosis, it must succeed in passing through G_1, S, and G_2. There is a point (the restriction point) late in G_1 that acts as a monitor of the cell's ability to carry out mitosis. If the cell has achieved a certain size and metabolic state, an unknown mechanism senses this and allows it to pass into the S phase. Once the cell passes beyond the restriction point, it is committed to undergoing mitosis. If the cell does not meet the requirement of the restriction point through lack of nutrients, insufficient size, or perhaps through density-dependent inhibition, it cannot pass out of G_1. The main feature of cancer cells is that their division is uncontrolled. They are not subject to density-dependent inhibition and therefore are capable of invading other tissues. If the mechanism of G_1 restriction could be discovered, perhaps the information could be used to develop a treatment that would prevent cancer cells from dividing. (page 148)

19. During mitosis, spindle fibers appear to originate in the microtubule organizing center. During interphase, this same center produces microtubules for the cytoskeleton. (page 149)

20. In a cell without a cell wall, the microtubules of the cytoskeleton help maintain the shape of the cell. During mitosis, these microtubules are disassembled to provide tubulin subunits for spindle production, and the cell takes on a rounded shape. (page 149)

REVIEW TEST 2

Section 1: Chapters 4–7

This review test is *not* designed to cover all of the important information in these chapters. However, it does touch on the major topic categories in each chapter. It will also give you valuable practice in taking this type of test. When you are finished, check your answers with those provided in the following PERFORMANCE ANALYSIS section.

1. Proteinoid microspheres are:
   a. precursors of chemosynthetic autotrophs.
   b. precursors of photosynthetic autotrophs.
   c. nonliving entities that share some properties with cells.
   d. organelles that bind to the DNA of eukaryotes.
   e. subunits of the eukaryotic cytoskeleton.

2. In the process of _____, vesicles that are formed by the _____ move to and fuse with the cell membrane, expelling their contents outside the cell.
   a. exocytosis; Golgi complexes
   b. endocytosis; endoplasmic reticulum
   c. pinocytosis; ribosomes
   d. pinocytosis; Golgi complexes
   e. exocytosis; endoplasmic reticulum

3. During which phase of the cell cycle or mitosis does a cell increase its amount of cytoplasm and its number of organelles?
   a. cytokinesis   d. G_2
   b. S phase   e. M phase
   c. G_1

4. Which statement is NOT true of all cells?
   a. They have numerous diverse and complex organelles.
   b. They are self-contained units.
   c. They are surrounded by a membrane.
   d. They have the ability to control movement of materials between internal and external environments.
   e. They have an information and control center containing genetic material.

5. Which of the following pairs does NOT have the relationship prokaryotic/eukaryotic?
   a. bacteria/animals
   b. no nuclear membrane/nuclear membrane

   c. one molecule of DNA/many molecules of DNA
   d. RNA as genetic material/DNA as genetic material
   e. DNA associated with little protein/DNA associated with much protein

6. The diameter of a typical eukaryotic cell is:
   a. 1–3 millimeters.
   b. 100–300 micrometers.
   c. 10–30 micrometers.
   d. 1–3 micrometers.
   e. 100–300 nanometers.

7. Membrane-bound proteins that aid in the movement of molecules across the membranes are called _____ proteins and are used in _____.
   a. receptor; active transport only
   b. transport; active transport only
   c. transport; facilitated diffusion only
   d. receptor; both facilitated diffusion and active transport
   e. transport; both facilitated diffusion and active transport

8. The fluid-mosaic model of the cell membrane refers to the fact that:
   a. the lipid molecules and integral proteins can move laterally within the membrane.
   b. membranes contain a high percentage of water.
   c. under the electron microscope, the appearance of the membrane surface resembles a tile floor.
   d. integral proteins must be anchored tightly in position by cytoplasmic filaments.
   e. peripheral proteins "float" on the surface of the membrane.

9. Which statement is NOT true of spindle fibers?
   a. They attach to the kinetochores of chromosomes.
   b. They arise de novo from tubulin subunits.
   c. Centrioles are not essential for their formation.
   d. They are elaborated from basal bodies.
   e. When the spindle is no longer required, the fibers disassemble.

10. The first test of Oparin's hypothesis was performed by:
   a. Stanley Miller.
   b. Sidney Fox.
   c. Louis Pasteur.
   d. Robert Hooke.
   e. Francesco Redi.

11. Because they allow the passage of sodium ions between cells, _____ contribute to the synchronization of muscle cell contraction in the heart.
   a. desmosomes
   b. gap junctions
   c. tight junctions
   d. coated pits
   e. plasmodesmata

12. Because the chromosome in bacteria is attached to the cell membrane:
   a. it can only replicate when the membrane is in the process of pinching off two daughter cells.
   b. the membrane is capable of exerting a low level of genetic control over the chromosome.
   c. it can provide genetic information directly to the membrane without first entering the cytoplasm.
   d. there is no need for a nuclear membrane.
   e. daughter chromosomes are separated into daughter cells as the membrane elongates.

13. When looking at a prokaryote, you would NOT expect to find:
   a. a cell wall.
   b. cytoplasm.
   c. ribosomes.
   d. genetic material.
   e. mitochondria.

14. The age of the oldest known rocks on the surface of the earth is:
   a. 5.2 billion years.
   b. 4.6 billion years.
   c. 4.1 billion years.
   d. 460 million years.
   e. 41 million years.

15. The nucleoli disappear during _____ of mitosis.
   a. interphase
   b. prophase
   c. metaphase
   d. anaphase
   e. telophase

16. The membrane that surrounds a vacuole is called a:
   a. blastocyst.
   b. tonoplast.
   c. plastid.
   d. vesicle.
   e. plasmodesma.

17. In order to create sufficient contrast in specimens prepared for light microscopy, they are often treated with:
   a. compounds containing heavy metals.
   b. an increasingly concentrated series of alcohols.
   c. solutions containing atoms of low molecular weight (such as CHNOPS).
   d. dyes that differentially adhere to specific subcellular components.
   e. Natural specimen contrast is sufficient. No special treatment is needed.

18. In 1864, this biologist claimed a prize for disproving the hypothesis of spontaneous generation.
   a. Francesco Redi
   b. Louis Pasteur
   c. Rudolf Virchow
   d. Matthias Schleiden
   e. Theodor Schwann

19. Actin filaments play a role in cytokinesis by:
   a. creating the cell plate in animal cells.
   b. creating the cell plate in plant cells.
   c. causing elongation of the primary plant cell wall.
   d. acting as a "purse string" in plant cells.
   e. acting as a "purse string" in animal cells.

20. T or F   Cilia differ from flagella in that cilia are more numerous and longer than flagella.

21. T or F   If the osmotic potential of a solution is high, then its water potential is also high.

22. T or F   Fact: A protein has properties not possessed by its constituent amino acids. This is an example of the observation that new properties appear at each level of organization in living systems.

23. T or F   Cells that are making proteins that will be used outside the cell have a large amount of rough endoplasmic reticulum.

24. T or F   Living matter differs from nonliving matter in the kinds and amounts of chemical substances of which it is composed.

25. Summarize the current theory of formation of our sun and the planets, including earth.

26. What two generalizations can be made about the role of nuclei in cells based on the work of Hertwig, Flemming, and Hämmerling?

27. Diffusion is the process whereby molecules move from an area of high concentration to an area of low concentration. Yet, any given molecule has the same probability of moving in one direction as in another. Explain this apparent contradiction.

PERFORMANCE ANALYSIS

1. c   Sidney Fox and coworkers created proteinoid microspheres from amino acids under conditions simulating those believed to be present on earth during its first one billion years. Proteinoid microspheres are membrane-bound structures that can perform a few chemical reac-

tions, may grow slowly by adding proteinoid material from the environment, and may bud off smaller microspheres. (page 88)

2. **a**   There are two major categories of vesicle-mediated transport: endocytosis and exocytosis. Exocytosis is a process in which vesicles formed by the Golgi complexes fuse with the cell membrane and expel their contents outside the cell. Endocytosis is the process in which the cell membrane invaginates around an extracellular particle (or fluid) and forms a vesicle, thereby carrying the substance into the cell. (page 138)

3. **c**   The G_1 phase follows cytokinesis and is characterized by an increase in cellular size and an increase in the number of cellular organelles and structures. (page 146)

4. **a**   Only eukaryotic cells have numerous diverse and complex organelles. The organelles of prokaryotic cells are fewer in number and simpler in design. (pages 90, 91 and Chapter 5)

5. **d**   Prokaryotic cells have cytoplasmic DNA with little associated protein, have no nuclear membrane, and are exemplified by bacteria and cyanobacteria. Eukaryotic cells have DNA associated with proteins in a membrane-bound nucleus and are typical of all kingdoms except Monera. (pages 90, 91)

6. **c**   The diameter of most eukaryotic cells is between 10 and 30 micrometers. Prokaryotic cells are typically smaller. (page 102)

7. **e**   Transport proteins are membrane-bound proteins that facilitate the movement of specific molecules across cell membranes during active transport and facilitated diffusion. (pages 134–136)

8. **a**   The fluid-mosaic model of cell membrane structure derived its name from the fluid nature of the lipid bilayer (which allows components to move in the membrane) and the mosaic pattern formed by the proteins embedded in the membrane. (page 106)

9. **d**   Spindle fibers are constructed of tubulin subunits and are disassembled when the spindle is no longer needed. They attach to the kinetochores of chromosomes and aid in chromosome movements during mitosis. Although they were once thought to form from centrioles, the presence of centrioles is not required for spindle fiber assembly. Basal bodies are not involved in spindle fiber formation. (pages 149, 152)

10. **a**   Stanley Miller simulated conditions believed to be present on the primitive earth and demonstrated that organic molecules could be produced under those conditions. The organic molecules in Miller's "soup" included nucleotides and nearly all the common amino acids. (page 87)

11. **b**   Gap junctions enable the passage of materials between animal cells. They appear to be clusters of small channels surrounded by proteins. Plasmodesmata serve a similar function in plant cells. (pages 140, 141)

12. **e**   The attachment of the two daughter chromosomes of prokaryotic cells to different spots on the cell membrane ensures their separation into the two forming daughter cells. Chromosome replication occurs before cell division. (page 145)

13. **e**   Prokaryotes have cytoplasm, ribosomes, and genetic material, and may have cell walls. They do not have membrane-bound organelles, of which mitochondria are an example. (page 91)

14. **c**   According to modern methods of dating rock layers, the oldest known rocks in the earth's crust are 4.1 billion years old. (page 85)

15. **b**   The nucleoli disappear during prophase, the phase during which the chromosomes condense and the spindle is formed. (page 150)

16. **b**   The tonoplast is the single membrane surrounding a vacuole in plant cells. (page 115)

17. **d**   Since 70 percent of a cell is water and water is transparent to visible light rays, cells need to be stained in order to be studied under the light microscope. The various stains bind preferentially to different components of a cell. (page 96)

18. **b**   Louis Pasteur disproved the hypothesis of spontaneous generation in a series of simple but well-designed experiments. However, his conclusions were only valid for the conditions under which his experiments were conducted. (essay, page 86)

19. **e**   Actin filaments are thought to act as a "purse string" during animal cell cytokinesis, constricting the cytoplasm of the dividing cell about its equator in the formation of the cleavage furrow. (page 153)

20. **False**   Cilia are typically more numerous and shorter in length than are flagella. The internal structures of cilia and flagella are identical. (page 122)

21. **False**   Osmotic potential refers to the tendency of water to move across a membrane into a solution. The higher the solute concentration, the greater is its osmotic potential. In solutions, water potential is related to solute concentration. The higher the solute concentration, the *lower* the water potential. Conversely, the lower the solute concentration (or the higher the water concentration) the higher the water potential. (pages 128, 133)

22. **True**   At each level of biological organization, the members of that level have properties not found on the lower levels. A protein (a polymer of amino acids) has properties not characteristic of individual amino acids. Similarly, a multicellular organism has abilities not found in the individual constituent cells. (page 84)

23. **True**   Proteins that are being manufactured for export are synthesized on rough endoplasmic reticulum. Cells

that are specialized to make exported proteins contain large amounts of rough endoplasmic reticulum. (pages 115, 116)

24. **True**   As you learned in Chapter 1, ninety-nine percent of all living matter is composed of various combinations of six elements. These combinations of elements are not found in nonliving matter. A living organism is able to maintain an internal environment that is very different from its external environment because the cell membrane regulates the flow of materials into and out of cells. (page 127)

25. Our sun and solar system are believed to have originated from cosmic dust and gases that were present in the universe following the "big bang." It is thought that the sun was born when matter and gases condensed in the presence of enough energy for hydrogen nuclei to fuse to form helium nuclei. This reaction releases enough energy to be self-perpetuating and is the main source of the sun's radiant energy. The planets formed from condensations of the remaining dust and gases that were orbiting the newly formed sun. As matter condensed to form planets, the gravitational pull of the growing planets attracted and held cosmic dust particles. Any planet that reached a certain minimum mass (and therefore gravitational pull) could attract and maintain a gaseous atmosphere. (pages 84, 85)

26. Two conclusions about the nucleus that can be drawn from these studies are that (1) the nucleus contains hereditary information and (2) the nucleus exerts a continuing influence over cellular activities. (pages 109, 110)

27. Molecules are constantly in motion and this motion is random in direction. The direction of a molecule's motion changes when it collides with another particle. When two types of molecules mix in a gaseous or liquid phase, they will eventually become evenly distributed throughout the container because of the random motion of the molecules. Diffusion, the movement from a relatively high to a relatively low concentration, refers to the *net* movement of molecules over time, not the movement of individual molecules. (page 129)

S E C T I O N 2

Energetics

CHAPTER 8

The Flow of Energy

CHAPTER ORGANIZATION

MAJOR CONCEPTS

The sun's rays striking the earth provide the energy upon which life depends. This energy is transferred among living organisms when animals eat plants or other animals. Less than 10 percent of the energy stored at one level is stored at the next level.

The first law of thermodynamics states that energy can be neither created nor destroyed, it can only be changed from one form to another. This law applies to living organisms as well as to nonliving systems. In all cases of energy conversion, some energy is converted to an unusable form and is dissipated through the environment. The second law of thermodynamics states that all natural processes occur in a manner that results in increasing randomness or disorder (entropy) in the system. The order characteristic of living systems is maintained through continual input of energy originating from the sun.

Energy transfer in living systems accompanies the transfer of electrons (and in some cases the transfer of hydrogen atoms) in oxidation-reduction reactions. The loss of electrons is oxidation whereas the gain of electrons is reduction.

Metabolism consists of all the biochemical processes performed by an organism. Catabolic processes are those in which substances are broken down. Anabolic processes are synthetic. Biochemical processes in living organisms are typically conducted via pathways that consist of multiple reactions occurring sequentially (i.e., a product of one reaction becomes a substrate of the next reaction). Complex proteins known as enzymes act as catalysts for each step in the pathway. Enzymes are typically synthesized only when they are needed.

In metabolic pathways, enzymes function by bringing the reactants together, thereby lowering the energy of activation of a reaction so that the reaction can proceed under the conditions present within living organisms. Many enzymes have associated ions or vitamins that are essential to their function. Factors that influence enzyme activity include temperature, pH, enzyme concentration, and substrate concentration. Enzymatic reactions are often regulated by feedback inhibition, in which a product inhibits a step in its synthetic pathway. The mechanisms by which enzymatic reactions are regulated include allosteric interactions, competitive inhibition, noncompetitive inhibition, and irreversible inhibition.

Membrane transport proteins are similar to enzymes in several respects. Both are large globular protein molecules whose tertiary or quaternary structures produce a precisely configured region—the active site—into which a specific molecule or ion fits. The binding of the particle to be transported (or the substrate) to the active site causes the configuration of the protein molecule to change; the result is transport of the substrate or catalysis of a chemical reaction. The rate at which membrane transport proteins (and enzymes) function is determined by the concentration of particles to be transported (or substrate concentration), the temperature, and the pH. The mechanisms that regulate enzyme function also regulate membrane transport proteins. In addition to membrane transport proteins, membranes also contain protein receptors at which specific molecules or ions bind, triggering chemical reactions in the cell.

The energy needed to maintain order in living systems is directly supplied by adenosine triphosphate (ATP). The hydrolysis of each phosphate-phosphate bond of ATP releases 7 kilocalories per mole of ATP. This energy is made available to the cell when the hydrolysis reaction is coupled with a cellular reaction requiring an input of energy.

HOW TO STUDY THE CHAPTER

Read the entire chapter through quickly, focusing on the major concepts.

Use the GUIDED STUDY OF THE CHAPTER to help you identify the important details as you **reread** the chapter. Writing out the answers to these questions will help fix them in your mind as well as provide you with a valuable study aid.

Answer the questions in TESTING YOUR UNDERSTANDING without the aid of your text. Check your answers against those in PERFORMANCE ANALYSIS. Analyzing your answers will give you valuable feedback on your level of understanding and preparedness for classroom testing.

GUIDED STUDY OF THE CHAPTER

I. Introduction *(page 161)*

1. a. Describe what happens to the solar energy that approaches the earth.

   b. What percentage of energy reaching the earth is used to drive the processes of living organisms?

Focus: **Each time energy flows from one stage in a biological system to the next, less than 10 percent of the chemical energy stored at the first level is converted to energy and stored at the second level. (Figures 8–1 and 8–2)**

---

2. How can (a) evolution and (b) a cell be described in terms of energy flow?

3. What factor determines the structure of an ecosystem or the biosphere?

II. The Laws of Thermodynamics *(pages 162–166)*

A. General Remarks *(page 162)*

4. *Vocabulary:* Define thermodynamics.

5. Name five forms of energy.

B. The First Law *(pages 162–164)*

6. State in your own words the first law of thermodynamics.

7. Describe what happens at the subatomic level when electrical energy is converted to light energy. (Figure 8–3)

8. Name five types of energy into which chemical energy can be changed in living organisms.

Focus: **In any energy conversion, some energy is inevitably converted to heat energy and dissipates in the environment.**

---

9. Explain why only 25 percent of the energy present in gasoline produces work in an automobile engine.

10. Identify three examples that illustrate the conversion of potential energy into other forms of energy.

11. Restate the first law of thermodynamics as applied to chemical reactions.

C. Essay: $E = mc^2$ (page 163)

12. State in your own words the law of conservation of mass.

13. a. Under what condition(s) can mass be "lost"?

    b. What actually happens to this "lost" mass?

Focus: The energy emitted by the sun results from reactions in which four hydrogen nuclei fuse to form one helium nucleus. Enough energy is released to maintain the extremely high temperature needed to perpetuate this reaction.

D. The Second Law (pages 164–165)

14. State in your own words the second law of thermodynamics.

Note: You may wish to review the discussion of potential energy on pages 28 and 29 before continuing this section.

15. *Vocabulary:* Distinguish between exergonic and endergonic reactions.

16. *Vocabulary:* Explain the term "spontaneous reaction."

Focus: If a reaction is to proceed without a net input of energy, the potential energy of the products must (with a few exceptions) be less than that of the reactants. Energy input is almost always required if the potential energy of the products exceeds that of the reactants.

---

17. *Vocabulary:* Define entropy.

18. *Vocabulary:* What are exothermic reactions?

Focus: All natural processes tend to proceed in a manner that increases randomness (or disorder) in the system.

---

19. *Review Question:* Name two factors that contribute to the overall change in energy during a reaction.

20. a. Describe two endothermic reactions that occur spontaneously.

    b. Explain this apparent contradiction.

21. *Vocabulary:* Define free energy change (ΔG) and indicate its relationship to ΔH and ΔS.

22. Restate the second law of thermodynamics in terms of entropy.

23. Match the following vocabulary words with their definitions. Use each answer only once. Each blank may have only one answer.

1. exergonic   3. endergonic
2. entropy   4. exothermic

a. _____ measurement of randomness or disorder in a system

b. _____ describes a reaction that gives off *heat*

c. _____ describes a reaction that requires energy to proceed

d. _____ describes a reaction that releases energy

E. **Living Systems and the Second Law** *(page 166)*

24. What is meant by a closed system?

25. a. What is meant by the authors' statement that "the universe is running down"?

b. What immediate implications does this concept have for living systems?

III. **Oxidation-Reduction** *(pages 166–167)*

Helpful Hint: **LEO** and **GER** can be useful in remembering oxidation and reduction reactions. **LEO**: Loss of Electrons is Oxidation. **GER**: Gain of Electrons is Reduction.

26. a. *Vocabulary:* Distinguish between oxidation and reduction with respect to gain and loss of electrons.

b. Define oxidation and reduction reactions in terms of transfer of hydrogen atoms.

> *Focus:* **An oxidation reaction cannot occur unless a reduction reaction occurs simultaneously.**

27. Describe glucose oxidation and photosynthesis in terms of hydrogen atom transfer.

28. Reactions that involve energy transfer in living systems are _____ reactions.

> *Focus:* **Living organisms have a number of mechanisms for releasing the energy contained in molecules in small quantities that can be used by cells; these mechanisms dissipate minimal amounts of heat.**

IV. **Metabolism** *(pages 167–168)*

29. *Vocabulary:* Define metabolism.

30. a. What is a metabolic pathway?

b. Of what significance are metabolic pathways to living organisms?

> *Focus:* **Many metabolic pathways are common to very diverse forms of life.**

31. Cite two examples of pathways unique to certain cell types and two examples of pathways occurring in virtually all cell types.

32. *Vocabulary:* Distinguish between anabolism and catabolism.

33. Identify two functions of catabolism.

Focus: **The structural compartmentalization of individual cells allows for functional compartmentalization of different cell processes.**

---

34. In order for a chemical reaction to occur between two molecules, what events must occur?

V. Enzymes *(pages 169–180)*

A. General Remarks *(pages 169–170)*

35. *Vocabulary:* Define energy of activation.

36. Name two reasons why heat is not an acceptable mechanism for providing the energy of activation for chemical reactions in cells.

37. Describe the role of a catalyst in a biochemical reaction.

38. What influence does a catalyst have on the overall energy change of a reaction? (Figure 8–8)

39. Label the following diagram showing the overall energy change of catalyzed and uncatalyzed reactions.

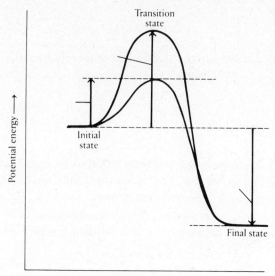

Focus: **A catalyst is not permanently changed during a reaction it catalyzes.**

Focus: **Enzymes are extremely efficient catalysts; catalyzed reactions occur at rates many thousands of times faster than the corresponding uncatalyzed reactions.**

---

40. Describe the catalytic activity of carbonic anhydrase.

Focus: **Cells are specialized with respect to the types of enzymes they can manufacture.**

---

41. *Vocabulary:* What is a substrate?

42. If you see a biological term ending in "-ase" it most likely refers to a(n) _____.

B. Enzyme Structure and Function *(page 170)*

43. How can the number of amino acids in a protein be estimated?

44. *Vocabulary:* Define and describe the general structure of an active site.

45. How does the active site of an enzyme relate to the various levels of protein structure? (If necessary, review primary, secondary, tertiary, and quaternary structures in Chapter 3.)

46. In addition to shape, what other characteristics of the active site complement the substrate?

Focus: **A consequence of the complementary structures of active site and substrate is that the substrate is oriented in a particular direction.**

The Induced-Fit Hypothesis (page 170)

47. Identify the essential components of the induced-fit hypothesis and the potential consequences of induced fit for the catalyzed reaction.

C. **Cofactors in Enzyme Action** *(pages 170–173)*

48. *Vocabulary:* What are cofactors?

Ions as Cofactors (page 171)

49. Identify two specific roles of ions as enzyme cofactors.

Coenzymes and Vitamins (pages 171–173)

50. *Vocabulary:* Distinguish coenzymes from cofactors.

51. One role of coenzymes in oxidation-reduction reactions is that of:

52. *Vocabulary:* What is a nucleotide?

A dinucleotide?

53. a. Name three molecules that contain adenine.

b. What is the other nitrogenous base in NAD?

54. Describe the relationship between the vitamin niacin and the coenzyme nicotinamide adenine dinucleotide (NAD).

55. Arrange the following components in a diagram showing their relationships in an NAD molecule: ribose, ribose, phosphate, phosphate, nicotinamide, and adenine. Label the nicotinamide nucleotide and adenine nucleotide portions. (You may use geometric shapes to represent the components rather than their structural formulas if you wish.)

56. What is the role of the nicotinamide ring of NAD in biochemical reactions?

57. Write out a simple equation showing the reversible oxidation-reduction of NAD. Include electron(s), proton(s), and electrical charge(s).

D. Enzymatic Pathways *(pages 173–174)*

> *Focus:* When equilibrium is reached in a chemical reaction, the *rate* of the forward reaction equals the *rate* of the reverse reaction. The relative concentrations of the products and reactants do not influence equilibrium.

58. What does the concentration of products relative to reactants indicate about the free energy change of the reaction?

59. The factor that determines how strongly a reaction will proceed in a given direction is:

60. What characteristics of enzymatic pathways prevent reactions from ever attaining equilibrium?

61. What is meant by an open system?

62. What is the role of exergonic reactions in enzymatic pathways?

63. Name three advantages to cells of enzyme reactions working in series, i.e., enzymatic pathways.

E. *Essay*: Auxotrophs *(page 173)*

64. *Vocabulary:* What are auxotrophs?

65. Of what value are auxotrophs to researchers studying enzymatic pathways?

66. How are auxotrophs produced, identified, and isolated by researchers?

F. Regulation of Enzyme Activity *(pages 175–178)*

67. Identify the three major factors limiting enzyme action.

68. List two major mechanisms by which cells regulate enzyme activity.

69. Show graphically the effects of (a) increasing substrate concentration and (b) increasing enzyme concentration on the rate of an enzymatic reaction. For each graph, indicate what factor is limiting the rate of the enzymatic reaction. (Figure 8–16)

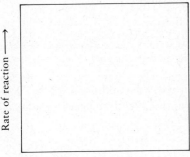

(a)

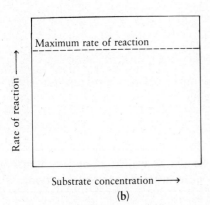

(b)

Effects of Temperature and pH (pages 175–176)

70. Describe and account for the influence of temperature on enzymatic reactions.

71. a. *Vocabulary:* What is a denatured enzyme?

b. Distinguish between reversible and irreversible denaturation.

Focus: **The temperature range through which a particular enzyme is active varies with the type of enzyme and the organism involved.**

Allosteric Interactions (pages 176–177)

72. Define allosteric interaction and allosteric effector.

73. Discuss the roles of allosteric interactions in the regulation of enzyme activity.

74. a. *Vocabulary:* Define feedback inhibition.

b. How does feedback inhibition regulate enzymatic activity? Use a simple diagram in your answer.

Competitive Inhibition (page 177)

75. What is competitive inhibition?

Focus: **The relative concentrations of substrate and competitor determine whether a reaction will occur or be inhibited.**

76. Describe one medically important example of competitive inhibition.

Noncompetitive Inhibition (page 177)

77. Distinguish noncompetitive from competitive inhibition.

78. a. Cite one example of noncompetitive inhibition.

b. How can this noncompetitive inhibition be reversed?

c. How does this mechanism differ from the reversal of competitive inhibition?

Irreversible Inhibition (page 177)

79. Name two mechanisms by which enzymes may be irreversibly inhibited.

80. Cite two examples of irreversible inhibition.

81. *Review Question:* Match the following mechanisms regulating enzyme function with the correct description. Each answer may be used only once. A blank may have only one letter.

1. allosteric interaction
2. competitive inhibition
3. noncompetitive inhibition
4. irreversible inhibition

a. _____ A compound binds irreversibly to the enzyme, rendering it nonfunctional.

b. _____ The end product of a pathway binds reversibly to the active site of an enzyme at some point earlier in the pathway. The relative concentrations of substrate and product are important.

c. _____ The binding of a compound to an enzyme at a site other than the active site results in a change in the configuration of the active site, either activating or inactivating the enzyme.

d. _____ A substance binds reversibly to a site on the enzyme other than the active site and the enzyme is rendered nonfunctional as long as the substance is present.

G. *Essay:* **Some Like It Cold** *(page 179)*

82. *Vocabulary:* Define thermolability.

83. a. Describe three examples of thermolabile enzymes and their effects.

b. Explain what the authors mean by "evolution is opportunistic," with respect to thermolability.

84. a. Discuss the effect of pH on enzymatic reactions.

b. How might pH play a role in restricting enzyme activity?

> **Focus:** The pH at which an enzyme functions optimally varies with the particular enzyme.

H. Membrane Transport Proteins and Receptors Revisited *(pages 178–180)*

85. How does the activity of membrane transport proteins (permeases) parallel that of enzymes? (Name four areas of similarity.)

86. Name four factors affecting the rate of membrane transport activity that also affect the rate of enzymatic activity.

87. a. What four regulatory mechanisms are common to permeases and enzymes?

b. Use a specific transport protein to illustrate one of these regulatory mechanisms.

> *Focus:* Binding of an external effector to some cell membrane receptors triggers activation of an intracellular enzymatic reaction.

VI. The Cell's Energy Currency: ATP *(pages 180–183)*

A. General Remarks *(pages 180–181)*

> *Focus:* Adenosine triphosphate (ATP) is the major immediate source of the energy needed by cells for metabolic processes.

88. Compare the roles of glucose and ATP as energy supply molecules.

89. a. Identify the subunits of which ATP is composed.

b. Assemble the following components to show their correct relationship in an ATP molecule: adenine, ribose, phosphate, phosphate, phosphate. You may use geometric figures to represent these components if you wish. (Figure 8–23)

90. *Review Question:* Distinguish between factors affecting the *rate* and the *direction* of a biochemical reaction.

91. a. *Vocabulary:* What are coupled reactions?

b. Hydrolysis of which portion of the ATP molecule supplies energy for coupled reactions?

92. How much energy per mole of ATP is released when the second or third phosphate-phosphate bond is broken?

93. When a phosphate group is removed from ATP or ADP, what happens to the electrons in the remaining molecule?

94. Distinguish between high-energy bonds (such as C—H) and energy-rich bonds (such as the $P \sim P$ bonds in ATP).

B. ATP in Action *(pages 181–183)*

95. a. *Vocabulary:* What are ATPases?

b. Cite two examples of ATPases.

96. a. *Vocabulary:* Define phosphorylation.

b. *Vocabulary:* What are kinases?

97. Describe how phosphorylation and ATP hydrolysis are coupled to supply energy to drive a particular reaction. Use simple equations.

98. Use the formation of sucrose from its component sugars to demonstrate the coupling of endergonic and exergonic reactions.

99. In general terms, what is the source of a cell's ATP?

VII. **Summary** (*pages 183–184*): Read the summary. If you are familiar with the essential features of the material presented there, you are ready to complete the section TESTING YOUR UNDERSTANDING.

TESTING YOUR UNDERSTANDING

After you have completed this examination, compare your answers with those in the section that follows.

1. The amount of solar energy that strikes the earth in one _____ is 1.5 billion times greater than the amount of electricity generated in the United States in _____.
   a. hour; one week
   b. day; one year
   c. week; one month
   d. month; one year
   e. year; the same period of time

2. The harnessing of running water to turn turbines, to generate electricity, or to grind flour is an illustration of:
   a. entropy.
   b. enthalpy.
   c. oxidation-reduction.
   d. the first law of thermodynamics.
   e. the second law of thermodynamics.

3. Regardless of the specific type of energy transformation, there is always some energy that is:
   a. lost to the surroundings in the form of heat.

   b. trapped and never again available.
   c. given off in the form of light.
   d. converted to electrical energy.
   e. destroyed.

4. If no energy enters or leaves a system during an energy conversion, then the _____ of the final state will always be _____ that of the initial state. This is a statement of the _____ law of thermodynamics.
   a. kinetic energy; less than; third
   b. potential energy; greater than; second
   c. kinetic energy; less than; first
   d. kinetic energy; equal to; first
   e. potential energy; less than; second

5. Which of these situations represents an increase in potential energy?
   a. a campfire burning dried logs
   b. leaves falling off trees in the autumn
   c. an exploding firecracker
   d. children sliding down a snowy hillside on cardboard
   e. a crate of books being moved up to a fifth floor dorm room

6. In which of these situations is the entropy of the final state likely to be greater than the entropy of the initial state?
   a. the enzyme salivary amylase converts starch to glucose in the mouth
   b. plants use the sun's energy to form glucose from carbon dioxide
   c. polypeptides are synthesized from amino acids
   d. DNA synthesis takes place during the S phase of the cell cycle
   e. cells in the liver and muscles convert glucose into glycogen

7. The free energy change (ΔG) of a system is defined by the relationship among absolute temperature (T) and change in:
   a. kinetic energy (K), and potential energy (P).
   b. entropy (S), and endergonic energy (En).
   c. heat energy (H), and entropy (S).
   d. endergonic energy (En), and exergonic energy (Ex).
   e. kinetic energy (K), and entropy (S)

8. The carbon atom of which of these molecules is most highly reduced?
   a. CO_2
   b. CH_4
   c. CH_2O
   d. CH_3OH
   e. $HCOOH$

9. The reactions in an organism that result in the breakdown of molecules to release energy and the building blocks for other chemical reactions are known collectively as:
   a. metabolism.
   b. anabolism.
   c. catabolism.
   d. allosteric interactions.
   e. endergonic reactions.

10. An enzyme lowers the activation energy of a chemical reaction by:
    a. bringing the substrates and products close to each other.
    b. supplying ATP energy to the reacting molecules.
    c. exciting the substrate's electrons into a higher energy level.
    d. weakening the chemical bonds of the substrate.
    e. absorbing the extra energy into its own molecular structure.

11. The role of the magnesium ion in enzyme-catalyzed reactions that involve the transfer of phosphate groups is to:
    a. change the tertiary structure of the phosphate group so that it will fit the active site.
    b. alter the tertiary structure of the enzyme so that the active site can be exposed to the phosphate group.
    c. bind the two negative charges on the phosphate group and hold it in position within the active site.
    d. bind to the active site of the enzyme, keeping it open until the phosphate group has an opportunity to displace it.
    e. neutralize the positive charges on the phosphate group so that it can bind to the active site.

12. Why do humans need to obtain the vitamin niacin in their diets?
    a. Niacin is a component of nicotinamide.
    b. We cannot synthesize niacin ourselves.
    c. Nicotinamide is needed by our cells to make NAD.
    d. NAD is a coenzyme involved in essential oxidation-reduction reactions.
    e. Taken together, all four responses constitute a complete answer to the question.

13. The binding of an allosteric effector molecule to an enzyme:
    a. either activates or inactivates the enzyme.
    b. occurs at the active site.
    c. is an irreversible process.
    d. is a necessary event in the activity of all enzymes.
    e. denatures the enzyme and prevents further activation.

14. Lead poisoning results when lead binds to enzymes at positions other than the active site and deactivates them. This is an example of:
    a. competitive inhibition.
    b. denaturation.
    c. irreversible inhibition.
    d. noncompetitive inhibition.
    e. product-mediated inhibition.

15. The energy available from the hydrolysis of ATP is released on splitting of the bond between the:
    a. adenine and the ribose.
    b. ribose and the first phosphate.
    c. second and third phosphates.
    d. adenine and the first phosphate.
    e. second phosphate and the adenine.

16. Which structure does NOT have ATPase activity?
    a. the protein arms in cilia
    b. the nicotinamide portion of the vitamin, niacin
    c. membrane transport proteins that perform active transport
    d. the protein arms of flagella
    e. the proteins arms on spindle microtubules

17. T or F   All exergonic (spontaneous) chemical reactions are also exothermic (give off heat).

18. T or F   In some instances in which molecules lose electrons, the electrons leave in the form of hydrogen atoms.

19. T or F   One example of the great diversity of living organisms is that the cells of different organisms have more differences than similarities in their metabolic pathways.

20. T or F   Even though every cell contains the genetic information for all enzymes of that organism, any given cell expresses only those enzymes it needs for its own particular function.

21. T or F   All enzymes within the same organism function optimally within the same narrow pH range.

22. T or F   Some membrane transport proteins require the input of energy before they can transport their substrate into a cell because the potential energy of the substrate molecules outside the cell is lower than the potential energy of the substrate molecules inside the cell.

23. T or F   When one enzyme in a metabolic pathway becomes defective due to a mutation, the substrate of that enzyme may accumulate and can lead to disease or death.

24. When a mole of glucose is completely oxidized (burned), it yields 686 kilocalories of free energy. In a living system, the same amount of energy is released on oxidation of glucose, but in small increments through a long series of chemical reactions. Explain why living systems take this longer approach to oxidation.

25. You are studying the following enzymatic pathway in a mutant bacterium. After you culture these cells, you find only orange colonies, but no red or yellow ones. Explain these findings in terms of your knowledge of enzyme pathways and the regulation of enzymes.

$$\text{Reactants} \xrightarrow{E_1} \underset{\text{Yellow}}{\text{Product A}} \xrightarrow{E_2} \underset{\text{Orange}}{\text{Product B}} \xrightarrow{E_3} \underset{\text{Red}}{\text{End product}}$$

26. If the reaction A + B ⟶ C + D is endergonic, how might a chemist and a cell use different techniques to force the reaction to proceed?

27. Discuss how the thermolability of the enzyme tyrosinase provides protective coloration for baby seals and Arctic foxes.

PERFORMANCE ANALYSIS

1. b The amount of solar energy striking the earth in one year is approximately 13×10^{23} calories. The amount delivered every day is equivalent to 1.5 billion times the amount of electricity generated by the United States in one calendar year. (page 161)

2. d The first law of thermodynamics governs energy conversions and states that energy can be neither created nor destroyed but only changed from one form to another. (page 162)

3. a Energy conversions are never 100 percent efficient and some energy is always lost to the environment as heat. (page 162)

4. e According to the second law of thermodynamics, if the total energy content of a system does not change, the potential energy of the final state will always be less than the potential energy of the initial state following an energy transfer or conversion. (page 164)

5. e When the crate of books is carried up to the fifth floor, it gains potential energy that would be released if the crate were to slip and fall back down to the ground level. All the other situations represent a decrease in potential energy in the system described. (page 164)

6. a Entropy is the amount of randomness or disorder present in a system. Entropy is greater in a system that has more particles than in one with fewer components. In the conversion of starch to glucose, the starch (a polymer of glucose) is broken down into its glucose subunits. The other reactions described are synthetic; energy is used to manufacture complex molecules from simpler components, thereby decreasing entropy. (pages 164, 165)

7. c The free energy change of a system can be described with the following equation, where ΔG is the free energy change, ΔH is the change in heat content, T is the absolute temperature, and ΔS is the change in entropy. (page 165)

$$\Delta G = \Delta H - T\Delta S$$

8. b Reduction can be defined as a gain of electrons. In some oxidation-reduction reactions, these electrons are gained (or lost) in the form of hydrogen atoms. Among the choices listed, the most highly reduced carbon atom is the one with the most hydrogen atoms bonded to it. The most highly oxidized carbon atom in the group is that in carbon dioxide, which has no hydrogen atoms. (page 166)

9. c Catabolic reactions are those in which molecules are broken down by cells to release energy and/or component atoms and molecules. Anabolic reactions are those in which substances are synthesized. The total anabolic and catabolic reactions in an organism are referred to as metabolism. (page 168)

10. d An enzyme lowers the activation energy of a system by bringing the reactants into close proximity and by weakening their chemical bonds. (page 169)

11. c The magnesium ion acts as an enzyme cofactor in the transfer of phosphate groups. The magnesium ion binds the negative charges of the phosphate group, holding it in position on the active site of the enzyme. (page 171)

12. e The vitamin niacin is a source of the nicotinamide portion of the NAD molecule. NAD is an important hydrogen (and electron) acceptor in oxidation-reduction reactions in living cells (to be studied in Chapter 9). Humans are unable to synthesize niacin and must obtain it in the diet. (pages 171, 172)

13. a An allosteric effector is a molecule that binds to an enzyme at a site other than the active site and either activates or deactivates the enzyme. (page 176)

14. d Noncompetitive inhibition occurs when a substance deactivates an enzyme by binding to a site other than the active site. Competitive inhibition occurs when a compound inhibits enzyme function by binding to the active site. An enzyme is denatured when its tertiary structure is irreversibly altered. Irreversible inhibition may occur from denaturation or from the irreversible binding of an inhibitory compound. (page 177)

15. c When ATP is hydrolyzed, the bond between the second and third phosphate groups is broken, releasing approximately 7 kilocalories per mole. (page 181)

16. b ATPases are enzymes that catalyze the hydrolysis of ATP. ATPases are present in some membrane transport proteins and as the protein "arms" of cilia, flagella, and spindle microtubules. (page 181)

17. False *Most* exergonic reactions are also exothermic. However, one notable exception is the spontaneous decomposition of dinitrogen pentoxide, in which heat (26.18 kilocalories per mole) is absorbed from the environment. This reaction proceeds spontaneously because of the large increase in entropy that results. (page 164)

18. True In some oxidation-reduction reactions, the transferred electron travels with a proton as a hydrogen atom. (page 166)

19. False The similarities among biochemical pathways in different living organisms far outweigh the differences. This is one piece of evidence supporting the relatedness of all living organisms. (page 168)

20. True Every cell of a multicellular organism (except for the reproductive cells) contains all the genetic information of that organism. At any one time, only those enzymes needed by the cell will be produced. (page 170)

21. False The optimal pH range for enzymes of living organisms varies with the individual enzyme and its function. Digestive enzymes active in the human stomach function at a much lower pH than most of the other enzymes in the body. (page 176)

22. **True**   Membrane transport proteins that actively transport substances against a concentration gradient to a state of higher potential energy require input of energy. (page 180)

23. **True**   If an enzyme in a metabolic pathway is defective owing to genetic mutation, the substrate of the enzyme accumulates and its presence may have serious to lethal consequences. Failure in production of the end product of the pathway may also have lethal consequences. (essay, page 173)

24. Living systems are unable to capture energy when it is released in large quantities, as occurs when organic molecules are burned in a calorimeter. By using a series of energy transfer (oxidation-reduction) reactions, the energy stored in organic molecules can be released in small increments that can be transferred to ATP and other molecules, which then make the energy directly available for cellular functions. (page 167)

25. The presence of only orange colonies suggests that a mutation has occurred rendering enzyme 3 (E_3) nonfunctional. Product B accumulates in the cells, producing their orange color. (page 173)

26. Endergonic reactions are those that need an energy input to proceed. A chemist in a laboratory might heat reactants A and B together in a container. A living cell might couple the reaction to the hydrolysis of ATP. A phosphorylated intermediate is formed when ATP reacts with one of the reactants (A). This intermediate then interacts with the other reactant (B), completing the reaction. (pages 169, 182)

27. The enzyme tyrosinase produces dark coat pigment in northern seals and Arctic foxes. However, it only functions in cold temperatures and the result of this thermolability is coat coloration that provides camouflage when the animal's need is greatest. Newborn seals, which cannot swim and must remain on ice floes, are born white, a consequence of developing inside the mother. The coat that develops by the time they are able to swim is brown and makes them less conspicuous in the water. Arctic foxes grow dark fur during the winter when cold temperatures permit tyrosinase activity. These coats are revealed when their white winter coat sheds in the spring. During the summer, the white coat is produced and becomes apparent when the dark summer coat sheds in the fall. (essay, page 179)

How Cells Make ATP: Glycolysis and Respiration

CHAPTER ORGANIZATION

MAJOR CONCEPTS

Living cells make ATP via the catabolic processes of glycolysis (occurring in the cytoplasm) and cellular respiration (occurring in mitochondria). Glycolysis is the breakdown of glucose into two molecules of pyruvic acid. Cellular respiration consists of two pathways: the Krebs cycle and the electron transport chain. During cellular respiration, the products of carbohydrate, fat, and protein breakdown are completely oxidized to carbon dioxide and water.

Glycolysis and the Krebs cycle consist of many steps, each step being catalyzed by a separate enzyme. The electron transport system involves a series of enzymes and iron-containing molecules called cytochromes that are present on mitochondrial membranes.

The production of pyruvic acid during glycolysis is anaerobic but the reactions of cellular respiration require oxygen as the final electron acceptor. In the absence of oxygen, living organisms can produce ATP by anaerobic pathways, but the energy yield is much less than that of aerobic respiration.

ATP is produced in a process called oxidative phosphorylation, which is catalyzed by ATP synthetase. A proton gradient across the inner mitochondrial membrane powers the phosphorylation of ADP to ATP. The proton gradient is formed as electrons move along the electron transport chain. The mechanism by which a proton gradient is used to power oxidative phosphorylation is called chemiosmotic coupling.

The energy yield of glycolysis is in the form of 2 NADH, 2 ATP, and 2 pyruvic acid molecules per molecule of glucose. Each pyruvic acid formed during glycolysis is oxidized to form an acetyl group which is then bound to coenzyme A, forming acetyl CoA. This reaction produces 2 NADH per glucose. Acetyl CoA enters the Krebs cycle and is oxidized to carbon dioxide. The energy yield of the Krebs cycle per glucose molecule is 6 NADH, 2 ATP, and 2 $FADH_2$. The hydrogens (and accompanying electrons) of NADH and $FADH_2$ enter the electron transport chain of mitochondria, and the maximal energy yield is 3 ATP per NADH and 2 ATP per $FADH_2$.

Many of the intermediate molecules in glycolysis and the Krebs cycle serve as precursors for biosynthesis. Although the products of biosynthetic (anabolic) pathways are often molecules catabolized in the reactions of glycolysis and respiration, certain key steps in synthetic pathways differ from those in catabolic pathways. Both anabolic and catabolic processes require a steady supply of molecules that can be broken down to supply energy and molecules that act as organic building blocks.

HOW TO STUDY THE CHAPTER

Read the entire chapter through quickly, focusing on the major concepts.

Use the GUIDED STUDY OF THE CHAPTER to help you identify the important details as you **reread** the chapter. Writing out the answers to these questions will help fix them in your mind as well as provide you with a valuable study aid.

Answer the questions in TESTING YOUR UNDERSTANDING without the aid of your text. Check your answers against those in PERFORMANCE ANALYSIS. Analyzing your answers will give you valuable feedback on your level of understanding and preparedness for classroom testing.

GUIDED STUDY OF THE CHAPTER

I. Introduction *(page 186)*

1. List at least five activities of cells or organisms in which ATP participates.

II. An Overview of Glucose Oxidation *(pages 186–187)*

> *Focus:* **During the oxidation of glucose in the presence of oxygen, carbon dioxide, water, and energy are released.**

2. How does the efficiency of energy conversion in a cell compare with that of an automobile?

3. *Vocabulary:* Distinguish between glycolysis, respiration, the Krebs cycle, and electron transport.

4. Fill in the following table comparing glycolysis and cellular respiration.

Table for Question 4

	GLYCOLYSIS	RESPIRATION
Starting compound(s)		
End product(s)		
Energy change (ΔG)		
Intracellular location of reactions		

5. What is the ultimate fate of the electrons and protons removed from the carbon atoms of the original glucose molecule?

6. What coenzyme plays a role in transfer of the electrons and protons?

7. Approximately how many ATP molecules are produced from the breakdown of one molecule of glucose during glycolysis and respiration?

III. Glycolysis *(page 187)*

A. The Steps of Glycolysis *(pages 187–189)*

8. *Vocabulary:* Define glycolysis.

> *Focus:* **Glycolysis is accomplished by a series of reactions, each of which requires a separate enzyme.**

9. Identify the two molecules that represent the cell's net energy harvest from the glycolytic pathway.

10. *Step 1:* Describe how ATP supplies the energy input needed to initiate glycolysis.

11. *Step 2:* What two factors favor the formation of fructose 6-phosphate over that of glucose 6-phosphate?

Focus: After completion of *step 3,* two molecules of ATP have been used to supply energy to the system but no useful energy has been produced.

---

12. Describe the mechanism by which phosphofructokinase activity is inhibited by ATP.

13. *Step 4:* What factor favors the formation of glyceraldehyde phosphate over dihydroxyacetone phosphate?

Focus: **Steps 1** through **4** are the preparatory reactions: each glucose molecule is converted to two molecules of glyceraldehyde phosphate, which requires energy.

---

14. *Step 5:* In what form(s) and in what quantity does the cell harvest energy at this step?

15. *Step 6:* In what form(s) and in what quantity does the cell harvest energy at this step?

16. "[This reaction] pulls all the preceding reactions forward." Explain what the authors mean by this statement.

17. *Steps 7 and 8:* What happens in these two steps and what are the consequences of these events?

18. *Step 9:* In what two ways does this reaction resemble step 6?

Focus: **Steps 5** and **6** are the energy-yielding reactions of glycolysis.

B. **Summary of Glycolysis** *(pages 189–190)*

19. Supply the following facts concerning glycolysis.

    Starting molecule

    End product(s)

    Total energy input required per glucose molecule:
    Number of ATP: _____
    Energy yield per glucose molecule:
    Number of ATP: _____
    Number of NADH: _____

20. By what process can carbohydrates other than glucose undergo glycolysis to release energy for the cell? (Figure 9–4)

IV. **Anaerobic Pathways** *(pages 190–191)*

21. Name two molecules commonly produced by the anaerobic metabolism of pyruvic acid.

22. Describe how grape juice is converted to wine.

23. In the making of wine, the fermentation process will cease when:

24. *Vocabulary:* Define fermentation.

25. a. Describe how NAD$^+$ is regenerated during ethanol fermentation. (Figure 9–5)

   b. Of what significance is this process to the cell? (Figure 9–5)

26. Describe the conditions under which muscle cells convert pyruvic acid into lactic acid.

27. a. *Vocabulary:* Define an oxygen debt, describing the conditions under which one develops.

b. How is an oxygen debt "repaid"?

28. Lactic acid metabolism is important to an organism for what reason?

Focus: In the evolution of biological processes, glycolysis is thought to have evolved fairly early, before the accumulation of free oxygen in the atmosphere.

29. Compare the energy yield of anaerobic glycolysis with that of aerobic pathways.

V. Respiration *(pages 191–200)*

A. General Remarks *(pages 191–193)*

30. *Vocabulary:* Distinguish between the two meanings of the term respiration as used in biology.

Focus: Cellular respiration begins where glycolysis ends, with pyruvic acid, and completes the oxidation of glucose to carbon dioxide and water.

31. In eukaryotic cells, the reactions of respiration occur in what organelle?

32. List the molecular components and properties of the following mitochondrial structures as they relate to cellular respiration. (See also Figure 9–7.)

Cristae/inner membrane

Matrix

Outer membrane

33. How does the composition of the inner mitochondrial membrane differ from that of a "typical" cell membrane mentioned in Chapter 6?

34. Mitochondrial processes supply _____ percent of the ATP produced within heterotrophic cells.

B. *Essay*: **Dissecting the Cell** *(pages 192–193)*

35. What feature of cells enables them to conduct such a great variety and number of chemical reactions within their boundaries?

36. a. Describe in general terms the procedure by which cell biologists can obtain pure samples of cellular organelles.

    b. What is the usefulness of isolating organelles for biochemical study?

37. *Vocabulary:* What is a supernatant?

38. *Vocabulary:* How does zonal centrifugation differ from differential ultracentrifugation?

39. Outline the method for separation of mitochondria from lysosomes.

C. **A Preliminary Step: The Oxidation of Pyruvic Acid** *(pages 193–194)*

> *Focus:* In order to enter the Krebs cycle, pyruvic acid must enter the mitochondria and be oxidized to acetyl groups, which are then bound to coenzyme A.

40. The total yield (including glycolysis) from one molecule of glucose at this stage is (see also page 190):

    _____ ATP

    _____ NADH

    _____ acetyl groups

    _____ CO_2

41. Coenzyme A consists of what two components?

D. **The Krebs Cycle** *(pages 194–195)*

42. Why is this series of reactions referred to as a cycle?

43. The energy yield per "turn" of the Krebs cycle in terms of these molecules is:

    _____ ATP

    _____ NADH

    _____ $FADH_2$

44. What vitamin is a component of FAD? (Figure 9–10)

45. What is the fate of the carbon atoms of (a) the acetyl groups and (b) oxaloacetic acid in one turn of the cycle?

E. Electron Transport (*pages 195–197*)

> *Focus:* During electron transport, the high-energy electrons of NADH and FADH$_2$ are passed through a series of electron carriers to oxygen. The energy of these electrons is used to generate ATP from ADP.

46. *Vocabulary:* What is an electron transport chain?

47. Fill in the following table summarizing the yield of energy-rich molecules to this point in the oxidation of one molecule of glucose. (Remember there are two pyruvic acid molecules oxidized and two "turns" of the Krebs cycle per molecule of glucose.) (See also pages 190, 193–194.)

Table for Question 47

	GLYCOLYSIS	OXIDATION OF PYRUVIC ACID	KREBS CYCLE
ATP			
NADH			
FADH$_2$			

48. *Vocabulary:* Define cytochromes and describe their structure.

49. What does oxidative phosphorylation involve?

50. The final electron acceptor at the end of the electron transport chain is:

51. How many molecules of ATP are produced for each molecule of (a) NADH and (b) FADH$_2$ that passes electrons to the electron transport chain?

F. The Mechanism of Oxidative Phosphorylation: Chemiosmotic Coupling (*pages 197–200*)

> *Focus:* Oxidative phosphorylation is powered by a proton concentration gradient established across the inner mitochondrial membrane.

52. a. *Vocabulary:* Explain the term chemiosmotic.

b. Define chemiosmotic coupling.

53. Name two events occurring during the chemiosmotic coupling of oxidative phosphorylation.

54. a. At which three points in the electron transport chain does a significant drop in potential energy of the electrons occur?

b. What is the consequence of these steps?

55. Describe two separate hypotheses proposed to explain the establishment of the proton gradient that drives chemiosmotic coupling.

56. a. Identify two factors contributing to the difference in potential energy across the inner mitochondrial membrane.

b. Explain the role played by the permeability properties of the membrane.

57. Discuss the structure of ATP synthetase and its role in oxidative phosphorylation. Include its relationship with the electron transport chain.

58. Name three other ways in which chemiosmotic power is used in living systems.

Control of Oxidative Phosphorylation (page 200)

59. How is the production of ATP via oxidative phosphorylation regulated?

VI. **Overall Energy Harvest** *(pages 200–201)*

A. **General Remarks** *(pages 200–201)*

60. Fill in the table below showing the maximum yield of ATPs when one molecule of glucose is completely oxidized in the presence of oxygen.

61. Cite two reasons why the amount of ATP produced from the oxidation of a glucose molecule is not constant. (Check pages 200 and 201.)

62. Nearly all of the ATPs produced are generated in reactions occurring in what organelle?

Table for Question 60

	GLYCOLYSIS	OXIDATION OF PYRUVIC ACID	KREBS CYCLE	TOTAL ATP
ATP				
ATPs from NADH (during electron transport)				
ATPs from FADH$_2$ (during electron transport)				
Totals				

Table for Question 67

	MODIFICATIONS	ENTRY-LEVEL MOLECULE	LEVEL OF ENTRY
Polysaccharide			
Fat			
Protein			

63. a. What percentage of the energy stored in a glucose molecule is captured and available for use in ATP molecules?

b. How does this compare with the energy yield of anaerobic glycolysis?

Focus: As each ATP molecule is exported from the mitochondrion, one ADP molecule is shuttled into the mitochondrion.

B. *Essay:* Ethanol, NADH, and the Liver *(page 202)*

64. Identify the breakdown product of ethanol that is responsible for (a) the intoxicating effects and (b) liver damage.

65. a. Describe how the function of the electron transport chain and the Krebs cycle is altered when ethanol is consumed.

b. What effect does this altered function have on the catabolism of sugars, proteins, and fats?

66. Summarize the changes that occur in the liver cells of heavy drinkers and the eventual consequences of these changes if heavy drinking continues.

VII. Other Catabolic Pathways *(page 203)*

67. For each type of compound in the table above, indicate (1) how it is modified so that it can be oxidized completely, (2) the molecule that enters the system, and (3) the level at which the molecule enters the system (glycolysis, Krebs cycle).

68. What is the fate of the amino acid groups of proteins?

69. *Vocabulary:* Define catabolism.

VIII. Biosynthesis *(page 203)*

70. Discuss in general terms the relationship between anabolic and catabolic processes in a living organism.

> *Focus:* Although anabolic processes can use many of the intermediates and products of catabolism, a steady input of organic molecules is required to supply energy and building materials for biosynthetic processes.

71. Outline the relationship between metabolic activities of heterotrophic and autotrophic cells.

IX. **Summary** *(pages 204–205):* Read the summary. If you are familiar with the essential features of the material presented there, you are ready to complete the section TESTING YOUR UNDERSTANDING.

TESTING YOUR UNDERSTANDING

After you have completed this examination, compare your answers with those in the section that follows.

1. What is the disposition of the hydrogen atoms removed from glucose during glycolysis?
   a. They are all picked up by NAD^+.
   b. They are all picked up by ATP.
   c. Two are picked up by NAD^+ and two are picked up by ATP.
   d. Two are picked up by NAD^+ and two remain in solution as hydrogen ions.
   e. Two are picked up by ATP and two remain in solution as hydrogen ions.

2. If a cell has enough ATP molecules, the process of glucose oxidation will be stopped by the allosteric interaction of ATP with an enzyme in:
   a. glycolysis.
   b. the conversion of pyruvic acid to acetyl CoA.
   c. the Krebs cycle.
   d. any catabolic pathway.
   e. the electron transport chain.

3. The first glycolytic reaction from which the cell harvests energy is:
   a. step 1, in which glucose is phosphorylated to yield glucose 6-phosphate.
   b. step 2, in which glucose 6-phosphate is converted into fructose 6-phosphate.
   c. step 3, in which fructose 6-phosphate is phosphorylated to yield fructose 1,6-diphosphate.
   d. step 4, in which fructose 1,6-diphosphate is split into dihydroxyacetone phosphate and glyceraldehyde phosphate.
   e. step 5, in which NAD^+ is reduced to NADH as glyceraldehyde phosphate is oxidized.

4. In glycolysis, glucose is enzymatically converted into two molecules of the three-carbon compound _____, which are then further oxidized to release energy in the Krebs cycle.
   a. diphosphoglycerate
   b. pyruvic acid
   c. fructose 1-phosphate
   d. dihydroxyacetone phosphate
   e. glyceraldehyde phosphate

5. Anaerobic pathways of pyruvic acid metabolism produce _____ so that the cell can continue to oxidize glucose.
   a. oxygen
   b. carbon dioxide
   c. inorganic phosphate
   d. NAD^+
   e. pyruvic acid

6. When yeast cells convert pyruvic acid to ethanol, they require (choose the most complete answer):
   a. the presence of oxygen.
   b. anaerobic conditions.
   c. a source of NADH.
   d. anaerobic conditions and a source of NADH.
   e. an investment of 2 ATP molecules per pyruvic acid metabolized.

7. Muscle fatigue may be caused by:
   a. alcohol formed during anaerobic glycolysis in muscles.
   b. an increase in muscle pH as a result of glycolysis.
   c. the accumulation of lactic acid in the absence of oxygen.
   d. excess NADH produced during glycolysis.
   e. an inadequate supply of NADPH.

8. Protruding into the _____ of mitochondria are folds of inner membrane called _____.
   a. stroma; thylakoids
   b. matrix; cristae
   c. stroma; cristae
   d. matrix; stroma
   e. thylakoid space; thylakoids

9. Which of these statements is NOT true of the conversion of pyruvic acid to acetyl CoA?
   a. Carbon dioxide is produced.
   b. Coenzyme A receives the acetyl group.
   c. This reaction takes place in the cytoplasm.
   d. A three-carbon molecule is converted into a two-carbon molecule.
   e. Some of the energy released by the reaction is trapped in a molecule of NADH.

10. In the oxidation of one molecule of glucose, _____ molecules of carbon dioxide are released during the oxidation of pyruvic acid and _____ molecules of carbon dioxide are released in the Krebs cycle.
    a. 0; 6
    b. 6; 0
    c. 2; 4
    d. 4; 2
    e. 3; 3

11. The energy output of the Krebs cycle is _____ per original glucose molecule oxidized.
    a. 3 ATP and 3 NADH
    b. 1 ATP and 3 NADH
    c. 3 NADH and 2 $FADH_2$
    d. 3 ATP, 3 NADH, and 3 $FADH_2$
    e. 2 ATP, 6 NADH, and 2 $FADH_2$

12. The maximum output of the electron transport chain is _____ ATP per NADH and _____ ATP per $FADH_2$.
    a. 2; 2
    b. 3; 2
    c. 2; 3
    d. 3; 1
    e. 4; 2

13. The final electron acceptor in the mitochondrial electron transport system is:
    a. oxygen.
    b. hydrogen.
    c. cytochrome *a*.
    d. cytochrome *b*.
    e. porphyrin.

14. Which statement is NOT true of the chemiosmotic coupling mechanism?
    a. A proton gradient that drives oxidative phosphorylation is established across the inner membrane of the mitochondrion.
    b. The potential energy released from the mitochondrial proton gradient is captured in ATP.
    c. The proton gradient provides energy for the rotation of bacterial flagella.
    d. Integral proteins of the inner mitochondrial membrane play an important role in oxidative phosphorylation.
    e. It depends on the free permeability of the inner mitochondrial membrane to most ions and molecules.

15. Most of the ATP molecules formed in the oxidation of glucose are produced during:
    a. electron transport.
    b. glycolysis.
    c. the Krebs cycle.
    d. the production of lactic acid.
    e. the conversion of pyruvic acid to acetyl CoA.

16. The maximum energy yield from the oxidation of one molecule of glucose is _____ ATP molecules, for an efficiency of _____ percent.
    a. 24; 40
    b. 24; 60
    c. 32; 80
    d. 38; 40
    e. 38; 60

17. T or F ATP is the principal energy carrier in living systems. Its stored energy is released on hydrolysis of its terminal phosphate bonds.

18. T or F Whether the net ATP yield of glycolysis is 6 or 8 molecules per molecule of glucose depends on the energetic cost of transporting the electrons of NADH across the inner mitochondrial membrane.

19. T or F The enzymes involved in the catabolism of carbohydrates, lipids, and proteins are the same ones involved in their synthesis.

20. Draw and label the various parts of a mitochondrion. Where do the reactions of the Krebs cycle and electron transport take place?

21. Identify the two major events of the chemiosmotic coupling mechanism of oxidative phosphorylation.

22. The energy contained in proteins and fats may be released via the same cellular respiration pathways that oxidize glucose. Describe the modifications needed and the site of entry into the catabolic scheme for (a) proteins and (b) fats.

PERFORMANCE ANALYSIS

1. **d** Of the four hydrogen atoms removed during glycolysis, two are picked up by NAD^+ (two protons plus four electrons) and two remain in solution as hydrogen ions. (page 190)

2. **a** Glucose oxidation is inhibited by the allosteric interaction of ATP with phosphofructokinase, the enzyme that catalyzes the phosphorylation of fructose 6-phosphate to fructose 1,6-diphosphate during glycolysis. (page 188)

3. **e** The first reaction of glycolysis that produces energy is the fifth step, in which glyceraldehyde phosphate is oxidized to 1,3-diphosphoglycerate and NADH is produced. Prior to this reaction, two ATP molecules have been used per molecule of glucose to power endergonic reactions. (page 189)

4. **b** The end product of glycolysis is pyruvic acid. Before entering the Krebs cycle, the pyruvic acid is oxidized to an acetyl group, which is bound to coenzyme A. (page 190)

5. **d** The anaerobic pathways that produce lactic acid and ethanol from pyruvic acid regenerate NAD^+, which is essential for the continuation of glycolysis with production of ATP. (page 190)

6. **d** Under anaerobic conditions, yeast cells convert pyruvic acid to ethanol. In the process, NAD^+ is regenerated from NADH. (Figure 9–5)

7. **c** The sensation of muscle fatigue results from the accumulation of lactic acid in muscles that work strenuously enough for oxygen demand to outstrip oxygen supply. Under such conditions, the cells anaerobically regenerate

the NAD$^+$ needed in glycolysis by converting pyruvic acid to lactic acid. (page 190)

8. **b** Cristae are folds of the inner mitochondrial membrane that contain many of the enzymes functioning in cellular respiration. The matrix of a mitochondrion is the dense solution surrounding the cristae in the inner compartment. (page 191)

9. **c** During the conversion of pyruvic acid to acetyl CoA, carbon dioxide and NADH are produced, and coenzyme A is bound to the two-carbon acetyl group formed from the three-carbon pyruvic acid. These reactions occur in the mitochondria. (page 193)

10. **c** For each pyruvic acid molecule that is converted to acetyl CoA, one molecule of carbon dioxide and one molecule of NADH are released. The products of the oxidation of one acetyl CoA in the Krebs cycle are: two molecules of carbon dioxide, three NADH, one FADH$_2$, and one ATP. Since two acetyl CoA molecules are produced from each glucose molecule broken down in glycolysis, double these figures to calculate numbers per glucose molecule. (pages 193, 194)

11. **e** See the answer to question 10. (pages 194, 195)

12. **b** The maximum output of the electron transport chain is three ATP molecules per NADH and two ATP molecules per FADH$_2$. In some cells, the energy cost of moving the electrons of NADH across the inner mitochondrial membrane is greater and each NADH yields only two ATP molecules. (pages 197, 201)

13. **a** Oxygen is the final electron acceptor of the electron transport chain. If oxygen is absent, cellular respiration cannot occur. (page 197)

14. **e** Like other cellular membranes, the inner mitochondrial membrane is not freely permeable to most ions and molecules. The electrons of NADH must be transported across the inner mitochondrial membrane before they can enter the electron transport chain. (pages 197–200)

15. **a** All but 4 of the 38 ATP molecules produced during the oxidation of one molecule of glucose are produced during electron transport. (page 201)

16. **d** The maximum energy yield from the oxidation of one molecule of glucose is 38 ATP and the efficiency of energy conversion is approximately 40 percent. (page 201)

17. **True** The bond between the second and third phosphate groups of ATP can be hydrolyzed fairly easily to release 7 kilocalories of energy, a quantity sufficient to power many reactions in living organisms. (page 186)

18. **True** See the answer to question 12. (pages 197, 201)

19. **False** Although *some* enzymes are common to both catabolic and anabolic pathways, the pathways differ in certain enzymes and in critical control steps. (page 203)

20. Compare your sketch to Figure 9–7. The reactions of the Krebs cycle occur in the matrix and on the cristae. The components of the electron transport chains are built into the cristae. (pages 191–193)

21. (1) The electron transport chain enables the establishment of a proton gradient across the inner mitochondrial membrane. (2) The potential energy stored in this proton gradient drives the phosphorylation of ADP to ATP. (page 197)

22. (a) The polypeptides of proteins are first hydrolyzed, releasing individual amino acids. The amino groups are removed and the carbon skeletons are (1) converted to acetyl groups that enter the Krebs cycle or (2) converted into molecules that are intermediates in glycolysis or the Krebs cycle. (b) Fats (triglycerides) are hydrolyzed to glycerol and fatty acids. Carbons are removed from the fatty acids two at a time as acetyl groups, which enter the Krebs cycle. (page 203)

C H A P T E R 10

Photosynthesis, Light, and Life

MAJOR CONCEPTS

In photosynthesis, light energy is converted to electrical energy and then to chemical energy. Visible light is just a small portion of the much broader electromagnetic spectrum that includes gamma rays, x-rays, microwaves, and radio waves. The various forms of radiation are classified according to their wavelengths, and the wavelength of an electromagnetic beam determines its energy level. The wavelengths in the visible spectrum contain enough energy to move electrons to higher energy levels but not so much energy that atomic structure is disrupted. Thus, visible light is well-suited to supply the constant input of energy needed to maintain order in living systems.

During the process of photosynthesis, light energy initiates a series of oxidation-reduction reactions when it strikes certain pigment molecules in the chloroplasts of photosynthetic organisms. The resulting transfer of electrons produces the energy-containing molecules ATP and NADPH. The energy contained in ATP and NADPH is then used to fix carbon from carbon dioxide into the organic molecule glyceraldehyde phosphate, from which all other organic molecules can be constructed. The production of ATP during photosynthesis is termed photophosphorylation and requires the establishment of a proton gradient across a membrane that is analogous to the gradient required during mitochondrial phosphorylation.

The light-requiring reactions of photosynthesis take place in photosystems located on the thylakoids of chloroplasts. The main pigments that capture light energy in photosynthetic organisms are chlorophyll *a*, chlorophyll *b*, and carotenoids. Halobacteria utilize a retinal-protein complex known as bacteriorhodopsin as the photosensitive pigment.

There are two major mechanisms by which carbon from carbon dioxide is fixed into organic molecules by plants. The three-carbon pathway is simpler and more direct but the four-carbon pathway has definite advantages for plants in arid climates. In C_3 plants, the carbon-fixing reactions occur in leaf mesophyll cells. In the mesophyll cells of C_4 plants, carbon is incorporated into a four-carbon compound (oxaloacetic acid), which is transported to the bundle sheath cells deeper in the leaf—the site of the Calvin cycle reactions.

Carbon is continuously cycling between inorganic carbon dioxide and organic molecules through the processes of photosynthesis and respiration.

HOW TO STUDY THE CHAPTER

Read the entire chapter through quickly, focusing on the major concepts.

Use the GUIDED STUDY OF THE CHAPTER to help you identify the important details as you **reread** the chapter. Writing out the answers to these questions will help fix them in your mind as well as provide you with a valuable study aid.

Answer the questions in TESTING YOUR UNDERSTANDING without the aid of your text. Check your answers against those in PERFORMANCE ANALYSIS. Analyzing your answers will give you valuable feedback on your level of understanding and preparedness for classroom testing.

GUIDED STUDY OF THE CHAPTER

I. Introduction *(page 206)*

> *Focus:* Since the evolution of photosynthesis, living organisms have played significant roles in shaping the face of the earth.

1. a. Before oxygen accumulated in the atmosphere, what two processes most likely provided the energy needed by living organisms?

   b. What effect did these processes have on the atmosphere?

2. Identify three potential sources of organic molecules that were probably available to organisms before significant accumulation of oxygen in the atmosphere.

3. The carbon source for most modern photosynthetic organisms is:

4. The electron transport chain may have originally served what function for a cell?

> *Focus:* The evolution of more efficient aerobic processes for releasing the energy stored in organic molecules allowed the evolution of more active and more complex life forms.

5. Summarize the importance of photosynthesis to life as we know it on earth.

6. Write out the equation for the overall process of photosynthesis.

II. The Nature of Light *(pages 207–211)*

A. General Remarks *(pages 207–209)*

> *Focus:* Sir Isaac Newton demonstrated that visible light is composed of light of a number of different colors.

7. What characteristic makes it possible to separate visible light into its component colors?

8. What conclusion did Newton reach about the *nature* of light?

9. Summarize the contribution of James Clerk Maxwell to the understanding of the nature of light.

10. *Vocabulary:* What is a wavelength?

> *Focus:* In a vacuum, electromagnetic radiation of various wavelengths travels at the same speed (300,000 kilometers per second).

11. a. *Vocabulary:* Define photoelectric effect.

    b. How did the observation of the photoelectric effect modify opinions on the nature of light?

12. What is unique about the photoelectric effect for individual metals?

13. Describe the effect of (a) the brightness or intensity of light and (b) the wavelength of light on the ejection of electrons from metal.

14. Who reintroduced the idea of the particle nature of light?

15. *Vocabulary:* What is a photon?

16. Describe the relationship between energy content and the wavelength of light.

Focus: **Both the wave model and the photon (particle) model are necessary to explain the behavior of light.**

B. **The Fitness of Light** *(pages 209–211)*

17. a. Name two reasons offered by George Wald to explain the suitability of the visible light spectrum to living organisms.

    b. What is meant by "the fitness of the environment"?

III. *Essay:* **No Vegetable Grows in Vain** *(pages 208–209)*

18. a. Describe Jan Baptista van Helmont's experiment to identify the material from which a plant is produced.

b. What conclusion did he draw from his results and what was faulty about this conclusion?

19. a. What is meant by "injured" air?

    b. How did Joseph Priestley formulate the opinion that vegetation "restores" air?

20. How did Jan Ingenhousz modify the conclusion drawn by Priestley?

21. Summarize the experiments conducted by Antoine Lavoisier leading to his statement that respiration is slow combustion.

Focus: **The quantitative studies by Nicholas Theodore de Saussure confirmed that the substance of a plant is derived from carbon dioxide *and* water in the presence of light.**

IV. **Chlorophyll and Other Pigments** *(pages 211–213)*

22. *Vocabulary:* What is a pigment?

Focus: **An object appears to be a certain color because it reflects light waves of that wavelength. Black objects absorb all wavelengths of visible light whereas white objects reflect all wavelengths.**

---

23. Why do leaves appear green?

24. *Vocabulary:* Define absorption spectrum.

25. How is the absorption spectrum of a substance determined? (Figure 10–5)

26. Name the pigment in plants that is directly involved in transforming light energy into chemical energy.

Focus: In addition to chlorophyll *a*, other plant pigments are present that absorb light at different wavelengths and pass the energy on to chlorophyll *a*.

27. What other group of pigments, besides chlorophylls, is found in photosynthetic organisms?

28. *Vocabulary:* Distinguish an action spectrum from an absorption spectrum.

29. Similarity between an action spectrum and an absorption spectrum is considered to be evidence that:

Focus: The action spectrum of photosynthesis parallels the absorption spectra for chlorophyll *a*, chlorophyll *b*, and carotenoids, providing evidence that these pigments are important in photosynthesis. (Figure 10–9)

30. Identify three possible fates of the energy that enters a system when a pigment absorbs light and electrons move to a higher energy level.

31. *Vocabulary:* Define fluorescence.

32. Chlorophyll molecules must exist in what state in order for the energy they absorb to be used by living systems?

V. **Photosynthetic Membranes: The Thylakoid** *(pages 213–214)*

A. **General Remarks** *(pages 211–213)*

33. *Vocabulary:* What is a thylakoid?

34. Describe how the thylakoid may exist structurally in (a) cyanobacteria, (b) other photosynthetic prokaryotes, and (c) eukaryotes.

B. **The Structure of the Chloroplast** *(page 214)*

35. *Vocabulary:* Define and distinguish between the stroma and the thylakoid space of a chloroplast.

36. Compare and contrast the membrane systems and spaces of a chloroplast and a mitochondrion. (Review Chapter 9 if necessary.)

37. *Vocabulary:* What are grana?

38. Of what advantage to a cell is the arrangement of thylakoids in parallel orientation?

39. Describe the location, structure, and function of palisade cells in a leaf. (Figure 10–11)

VI. The Stages of Photosynthesis *(pages 215–216)*

> *Focus:* F. F. Blackman concluded that photosynthesis consists of two sets of reactions: one set is light-dependent and the other is temperature-dependent.

40. a. Describe the effect on rate of photosynthesis of dim versus intense light.

    b. What is the significance of the plateau in photosynthetic rate at a certain light level? (Figure 10–12)

41. Describe the effect on rate of photosynthesis of increasing the temperature when the light is (a) dim and (b) fairly intense.

42. What evidence suggests that one set of reactions is enzymatic?

43. Why are the terms "light" and "dark" reactions misleading?

44. a. Briefly summarize the events of the (a) light-capturing and (b) carbon-fixing reactions.

    b. Where in the chloroplast does each set of reactions occur?

45. What are the two energy-rich molecules produced by photosynthesis?

VII. The Energy-Capturing Reactions *(pages 216–221)*

A. The Photosystems *(page 216)*

46. *Vocabulary:* Define and describe a photosystem.

47. *Vocabulary:* What is a reaction center in a photosystem?

48. Describe what happens when a photon of light strikes the reaction center.

49. Describe and distinguish between Photosystem I and Photosystem II.

50. Arrange the following components of Photosystems I and II in a diagram showing their correct functional relationships. (The I or II indicates to which photosystem each component belongs.) (Figure 10–14)

electron transport system I
electron transport system II
P_{700}
P_{680}
primary electron acceptor I
primary electron acceptor II
light energy I
light energy II
$H_2O \longrightarrow 2H^+ + 1/2\ O_2$
$ADP \longrightarrow ATP$
$NADP^+ + H^+ \longrightarrow NADPH$

B. *Essay:* **Van Niel's Hypothesis** *(page 217)*

51. State in your own words van Niel's hypothesis.

52. What type of bacterial photosynthesis was studied by van Niel and contributed to formulation of his hypothesis?

53. Describe how his hypothesis was eventually confirmed.

C. **The Light-Trapping Reactions** *(pages 218–219)*

54. Outline the flow of energy and electrons through Photosystems I and II of the light-trapping reactions.

55. The energy gain from Photosystems I and II is represented by what two molecules?

56. *Vocabulary:* What is photophosphorylation?

57. What is the source of electrons that replaces those boosted to a higher energy level from (a) P_{680} and (b) P_{700}?

Focus: **The molecules ATP and NADPH represent the energy yield of the light-trapping reactions and are the main sources of energy for the reduction of carbon.**

D. **Cyclic Electron Flow** *(page 219)*

58. Fill in the following table contrasting cyclic electron flow and the light-trapping reactions just described.

Table for Question 58

	CYCLIC ELECTRON FLOW	LIGHT-TRAPPING REACTIONS
Photosystems and electron acceptors		
Energy-storage molecules produced		
Oxygen release		
Carbon reduction		
Cell type(s) involved		

59. Under what conditions does cyclic electron flow occur in photosynthetic eukaryotic cells?

E. **Photosynthetic Phosphorylation** *(pages 219–221)*

60. List four characteristics common to photosynthetic phosphorylation and oxidative phosphorylation.

61. How does the movement of protons in a chloroplast differ from that in a mitochondrion?

62. a. What are the two sources of protons accumulated in the thylakoid space?

b. What is the ultimate energy source for formation of the proton gradient?

> *Focus:* The key factor in photophosphorylation is the establishment of a potential energy gradient (and flow of protons) between the thylakoid space and the stroma.

F. *Essay:* **Photosynthesis without Chlorophyll** *(page 221)*

63. Describe the environment required by halobacteria.

64. a. Name the photoactive pigment responsible for photosynthesis in halobacteria.

b. Where is this pigment located in vertebrates?

65. *Vocabulary:* Define and describe the function of bacteriorhodopsin.

> *Focus:* Halobacteria use a retinal-protein complex called bacteriorhodopsin to establish a proton gradient that drives photophosphorylation.

VIII. The Carbon-Fixing Reactions *(pages 222–228)*

A. General Remarks *(page 222)*

66. What two types of energy conversions occur in the energy-capturing reactions of photosynthesis?

> *Focus:* In the second phase of photosynthesis, the energy captured in the light-trapping reactions is used to reduce carbon.

67. *Vocabulary:* What are stomata?

B. The Calvin Cycle: The Three-Carbon Pathway *(pages 222–223)*

68. a. Name the compound that starts (and ends) the Calvin cycle.

b. What enzyme catalyzes the incorporation of carbon dioxide into this compound?

69. Identify the immediate product of the Calvin cycle.

70. The energy to drive the Calvin cycle is supplied by:

71. For each turn of the Calvin cycle, _____ carbon dioxide molecule(s) is (are) reduced.

72. a. How many revolutions of the Calvin cycle are required to produce one six-carbon molecule?

b. How much ATP and NADPH is required?

C. The Four-Carbon Pathway *(pages 223–228)*

73. a. Name the compound into which carbon dioxide is initially incorporated in C_4 plants.

b. What enzyme catalyzes this reaction?

74. Name the two cell types involved in carbon-fixation in C_4 plants and indicate which reactions occur in which cells.

75. How are the C_4 pathway and the Calvin cycle interrelated?

76. Describe the role of PEP carboxylase in maintaining a carbon dioxide concentration gradient that promotes carbon dioxide uptake in leaf cells of C_4 plants.

77. What factors make C_4 plants better adapted to dry climates than C_3 plants?

78. Fill in the following table comparing mitochondrial respiration and photorespiration. (See also pages 191, 225, 226.)

Table for Question 78

	MITOCHONDRIAL RESPIRATION	PHOTORESPIRATION
Location of reactions in the cell		
Starting compound		
Energy-rich products NADH		

79. How does photorespiration reduce the photosynthetic efficiency of C_3 plants?

80. Name two aspects of C_4 metabolism that limit photorespiration.

81. How do C_3 and C_4 plants differ with respect to optimum temperature for photosynthesis? Explain.

82. Name four types of C_4 grasses and four types of C_3 grasses.

83. *Review Question:* Fill in the following table comparing C_3 and C_4 plant metabolism.

Table for Question 83

	C_3	C_4
a. Initial compound formed on binding of carbon dioxide		
b. Enzyme catalyzing the reaction in (a)		
c. Location (cell type) of: Reaction (a) Components of the Calvin cycle		
d. Degree to which photorespiration occurs (high or low)		
e. Adaptation for arid conditions? (yes or no)		

D. *Essay:* **The Carbon Cycle** *(pages 226–227)*

> *Focus:* **Carbon cycles between the living and nonliving components of the biosphere (and between organic and inorganic molecules) through the processes of photosynthesis and respiration.**

84. Trace the flow of carbon from the atmosphere to living organisms and back to the atmosphere, identifying the participants and the processes at each step.

> *Focus:* **Atmospheric carbon dioxide absorbs heat from solar radiation; this could have serious global ramifications if carbon dioxide levels continue to increase.**

85. Name three human activities that are causing the levels of atmospheric carbon dioxide to increase.

86. Identify potential global consequences of increased atmospheric carbon dioxide levels.

IX. The Products of Photosynthesis *(page 228)*

> *Focus:* **Glyceraldehyde phosphate is the link between inorganic carbon dioxide and all organic molecules manufactured by photosynthetic organisms (and their predators).**

87. Name four possible fates of glyceraldehyde phosphate.

88. *Review Question:* Match each scientist with the correct hypothesis or conclusion. Use each number only once. Each blank should have only one answer.

1. Albert Einstein   3. Sir Isaac Newton
2. James Maxwell   4. George Wald

a. _____ White light is composed of light of a number of different colors.

b. _____ Visible light is only a small part of the electromagnetic spectrum.

c. _____ Light is composed of particles called photons.

d. _____ If life exists elsewhere in the universe, it probably depends on visible light for energy.

X. **Summary** *(pages 228–230):* Read the summary. If you are familiar with the essential features of the material presented there, you are ready to complete the section TESTING YOUR UNDERSTANDING.

TESTING YOUR UNDERSTANDING

After you have completed this examination, compare your answers with those in the section that follows.

1. It was a good thing that no _____ was available to early cells because it would have been poisonous to them.
   a. oxygen   d. carbon dioxide
   b. hydrogen   e. nitrogen
   c. methane

2. This English physicist found that white light is a combination of all colors of light.
   a. Albert Einstein
   b. James Maxwell
   c. Joseph Priestley
   d. Jan Ingenhousz
   e. Sir Isaac Newton

3. The photoelectric effect is dependent upon the:
   a. wavelength of the light.
   b. duration of exposure of the metal to the light.
   c. intensity of the beam of light.
   d. brightness of the beam of light.
   e. interaction of the photons of the light with the protons of the metal.

4. Wavelengths of light longer than those of the visible spectrum:
   a. have no effect on organisms.
   b. exhibit more energy than those of the visible spectrum.
   c. are absorbed by water, increasing molecular motion.
   d. have the capacity to break bonds in important molecules.
   e. are absorbed by oxygen and ozone in the atmosphere.

5. What wavelengths of visible light might be used to power photosynthetic reactions in a plant whose sole pigment is reddish-yellow?
   a. red and yellow
   b. red and violet
   c. yellow and violet
   d. green, blue, and violet
   e. red, green, and yellow

6. A particular pigment may be considered responsible for a specific plant process if:
   a. it is isolated from the plant in large amounts.
   b. its absorption spectrum matches the action spectrum of the process.
   c. the pigment occurs in the same cells as the process.
   d. both the pigment and the process reside within the same organelle.
   e. the process requires donated electrons from another molecule.

7. Matrix is to mitochondria as _____ is (are) to chloroplasts.
   a. algae
   b. cristae
   c. stroma
   d. thylakoids
   e. grana

8. The typical leaf cell has about _____ chloroplasts.
   a. 5
   b. 50
   c. 500
   d. 5,000
   e. 50,000

9. The electrons lost by chlorophyll a of Photosystem II when it absorbs light are replaced by electrons from:
   a. carbon dioxide.
   b. chlorophyll b.
   c. oxygen.
   d. carotene.
   e. water.

10. Chlorophyll a and its associated molecules are packaged in units called _____, which are located in the _____.
    a. photosystems; stroma
    b. cristae; intermembrane space
    c. stomata; cells of the leaf surface
    d. photosystems; thylakoid membranes
    e. chlorosacs; grana

11. Cyclic electron flow involves Photosystem I and:
    a. produces NADPH.
    b. releases oxygen.
    c. produces ATP.
    d. requires dark.
    e. uses carbon dioxide.

12. The production of one ATP and one NADPH during the energy-capturing reactions of photosynthesis requires the energy of _____ photon(s) of light.
    a. one
    b. two
    c. three
    d. four
    e. eight

13. The concentration gradient required for photosynthetic phosphorylation is generated as _____ are pumped from the _____.
    a. electrons; stroma to the cytoplasm
    b. electrons; stroma to the thylakoid space
    c. electrons; cytoplasm to the stroma
    d. protons; stroma to the thylakoid space
    e. protons; stroma to the cytoplasm

14. The carbon-fixing reactions of the Calvin cycle begin when a carbon dioxide molecule is joined to a molecule of:
    a. glucose.
    b. ribulose bisphosphate.
    c. phosphoglycerate.
    d. glyceraldehyde phosphate.
    e. oxaloacetic acid.

15. C_4 plants are so named because during carbon-fixation the carbon dioxide molecule is attached to _____, forming a four-carbon molecule.
    a. oxaloacetic acid
    b. ribulose bisphosphate
    c. glyceraldehyde phosphate
    d. malic acid
    e. phosphoenolpyruvate

16. In C_3 plants, when oxygen levels are high and carbon dioxide levels are low, _____ occurs, but this process produces no _____.
    a. photorespiration; carbon dioxide
    b. cyclic electron flow; NADH
    c. photorespiration; end products that are usable by the cell
    d. cyclic electron flow; ATP
    e. photorespiration; ATP and NADH

17. In photorespiration, oxygen combines with RuBP to produce _____, which is subsequently oxidized.
    a. glycolic acid
    b. oxaloacetic acid
    c. malic acid
    d. pyruvic acid
    e. phosphoglyceric acid

18. C_4 plants differ in several ways from C_3 plants. Which statement does NOT represent such a difference?
    a. They can live in drier areas.
    b. They have a higher rate of photorespiration.
    c. They can survive at higher temperatures.
    d. They do not need to open their stomata as often.
    e. They can fix carbon dioxide at lower carbon dioxide concentrations.

19. **T or F** The thylakoids within a chloroplast are randomly oriented in order to utilize light from any direction.

20. **T or F** The anaerobic breakdown of carbon-containing compounds yields more energy than aerobic breakdown because they are not completely oxidized in the aerobic process.

21. **T or F** James Maxwell was the first person to demonstrate that the light we see is only part of a much larger electromagnetic spectrum, which includes radio waves.

22. **T or F** Within the spectrum of visible light, violet light has a longer wavelength than red light.

23. **T or F** Because of the efficiencies of their enzymes, C_4 plants are able to obtain more carbon dioxide per "gasp" than C_3 plants.

24. Name two types of photosynthetic pigments that are found in plants. Discuss how the presence of several different pigments is beneficial to plants.

25. Discuss how Photosystems I and II interact under normal conditions to carry out the energy-capturing reactions.

26. Which steps in glycolysis cannot be reversed and must be bypassed when glyceraldehyde phosphate is being converted into glucose?

27. Briefly summarize the contribution of each of these scientists to our understanding of the overall process of photosynthesis: Jan Baptista van Helmont, Joseph Priestley, Jan Ingenhousz, Antoine Lavoisier, and Nicholas Theodore de Saussure.

PERFORMANCE ANALYSIS

1. **a** Since oxygen has such a high attraction for electrons, it would have disrupted chemical bonds in early cells, thereby interfering with cellular functions. (page 206)

2. **e** By passing light through a prism, Sir Isaac Newton discovered that visible light is a combination of light of different colors. (page 207)

3. **a** The photoelectric effect is a phenomenon in which metals acquire a positive electrical charge upon exposure to light, which forces electrons from the metal atoms. For each metal, a specific wavelength of light induces the photoelectric effect. (page 207)

4. **c** Wavelengths of light longer than those in the visible spectrum are absorbed by water and other molecules, increasing the motion (and heat) of these molecules, but not affecting electron configuration. (page 210)

5. **d** Objects appear to be certain colors because they contain pigments reflecting light of those wavelengths. A reddish-yellow object is reflecting red and yellow light, and is absorbing light in the green, blue, and violet wavelengths. (page 211)

6. **b** If the absorption spectrum of a pigment matches the action spectrum of a plant process, it is considered to be evidence that the pigment plays a role in the process. (page 213)

7. **c** The matrix is the fluid-filled space within mitochondria. The stroma is the fluid-filled space within chloroplasts. (page 214)

8. **b** A typical plant leaf cell has about 50 chloroplasts. (page 213)

9. **e** Water is the source of electrons that replace those boosted from the chlorophyll *a* molecule of Photosystem II. (page 218)

10. **d** Chlorophyll *a* molecules are packaged in photosystems, which are located in the thylakoid membranes. (page 216)

11. **c** Cyclic electron flow occurs when Photosystem I works independently. ATP is produced when the electrons boosted from P_{700} pass along the electron transport chain connecting Photosystems I and II and return to P_{700}. This differs from the complete reactions of photosynthesis in that oxygen and NADPH are not produced and carbon dioxide is not reduced. (page 219)

12. **d** In order to produce one molecule of NADPH during the light-trapping reactions, four photons of light must be absorbed. (page 219)

13. **d** ATP is generated in a chemiosmotic process known as photophosphorylation, which is similar to oxidative phosphorylation in mitochondria. A proton gradient is produced across chloroplast membranes as protons are pumped from the stroma to the thylakoid space. (page 220)

14. **b** The first step of the Calvin cycle is the binding of carbon dioxide to ribulose bisphosphate by the enzyme RuBP carboxylase. (page 222)

15. **e** The initial reaction of the C_4 (or Hatch-Slack) pathway is the binding of carbon dioxide to phosphoenolpyruvate by the enzyme PEP carboxylase. (page 223)

16. **e** When oxygen is abundant but carbon dioxide levels are low, C_3 plants conduct photorespiration, a process in which carbohydrates are oxidized in the presence of light and oxygen but ATP and NADH are not produced. This process greatly reduces the efficiency of photosynthesis in C_3 plants. (page 225)

17. **a** In photorespiration, RuBP carboxylase catalyzes the combination of RuBP and oxygen to form glycolic acid, which is then oxidized. (page 225)

18. **b** C_4 plants can fix carbon dioxide at much lower concentrations than can C_3 plants; the enzyme PEP carboxylase of C_4 plants has a much greater affinity for carbon dioxide than the RuBP carboxylase of C_3 plants. Therefore, C_4 plants do not need to open their stomata as often, can survive at higher temperatures, and thrive in arid climates. (pages 224–227)

19. **False**   The thylakoids in a chloroplast are arranged in parallel stacks aligned in the same direction. Within a cell, the chloroplasts may be oriented to obtain maximum light exposure. (page 214)

20. **False**   Organic molecules are completely oxidized to CO_2 and water during aerobic metabolism. Anaerobic metabolism results in only partial oxidation of organic compounds and yields far less energy. (page 206)

21. **True**   James Maxwell demonstrated that visible light is only a small part of the electromagnetic spectrum and that the radiations in this spectrum act as if they travel in waves. (page 207)

22. **False**   Within the visible spectrum, the shortest wavelength is violet and the longest is red. (page 207)

23. **True** C_4 plants contain enzymes that maintain a steep carbon dioxide concentration gradient between the inside and the outside of the leaf. Because of this, they can take in more carbon dioxide in the time stomata are open than a C_3 plant. (page 225)

24. Photosynthetic pigments found in plants include chlorophylls *a* and *b* and carotenoids. The carotenoids can absorb light of different wavelengths from those absorbed by the chlorophylls. This absorbed energy can then be passed to the chlorophylls of the photosystems, thereby capturing additional energy for photosynthesis. (pages 211, 213)

25. Electrons boosted from the reactive chlorophyll *a* molecule of Photosystem II are passed along an electron transport system to replace the electrons boosted from the reactive chlorophyll *a* molecule of Photosystem I. As electrons travel along the electron transport chain between the two photosystems, ATP is produced from ADP. (Figure 10–14)

26. The highly exergonic steps that pull glycolysis forward cannot be reversed when glyceraldehyde phosphate is being converted into glucose. These steps are bypassed by enzyme pathways that differ from those of glycolysis. (page 228)

27. **Jan Baptista van Helmont** documented that the increased weight of a growing plant is greater than the amount of soil lost during the growth of that plant. He concluded that all the substance of the plant came from the water applied to the soil and none from the soil. (He was only partially correct.) **Joseph Priestley** demonstrated that plants restore (i.e., add oxygen to) air "that has been injured by the burning of candles." **Jan Ingenhousz** confirmed Priestley's work but revealed that the purification only occurs in the presence of sunlight. **Antoine Lavoisier** measured the amount of oxygen consumed and carbon dioxide produced by animals and showed that respiration is the slow combustion of carbon compounds. **Nicholas Theodore de Saussure's** experiments documented that the increase in dry plant weight reported in van Helmont's experiments is due to two main sources: water, and the carbon of carbon dioxide. He also reported that equal volumes of carbon dioxide and oxygen are exchanged during photosynthesis. (essay, pages 208, 209)

REVIEW TEST 3

Section 2: Chapters 8–10

This review test is *not* designed to cover all of the important information in these chapters. However, it does touch on the major topic categories in each chapter. It will also give you valuable practice in taking this type of test. When you are finished, check your answers with those provided in the following PERFORMANCE ANALYSIS section.

1. In the first step of photosynthesis, light energy is changed to:
   a. chemical energy.
   b. the energy of motion.
   c. the energy of position.
   d. kinetic energy.
   e. electrical energy.

2. Glycolysis occurs in the _____ of eukaryotic cells.
   a. peroxisomes
   b. mitochondria
   c. lysosomes
   d. cytoplasm
   e. membrane

3. A graph that shows the wavelengths of light needed for flowering to occur in plants is an example of a(n) _____ spectrum.
   a. electromagnetic
   b. prizmatic
   c. action
   d. photon
   e. absorption

4. Which process is an example of a catabolic process?
   a. peptide bond formation between two amino acids
   b. photosynthesis
   c. condensation of monosaccharides to form cellulose
   d. cellular respiration
   e. triglyceride synthesis from fatty acids

5. How many $FADH_2$ molecules are produced per acetyl group oxidized by the Krebs cycle?
   a. one
   b. two
   c. three
   d. four
   e. five

6. In his studies of photosynthesis, Blackman observed the effects of _____ and concluded that photosynthesis occurs in two stages.
   a. light intensity and substrate concentration
   b. temperature and light intensity
   c. substrate concentration and pigment absorption spectrum
   d. NADP+
   e. light intensity on the development of chloroplasts

7. The main difference between permeases (membrane transport proteins) and enzymes is that:
   a. enzymes induce a chemical change in their substrates, whereas permeases do not.
   b. permeases require the input of energy in the form of ATP, whereas enzymes do not.
   c. enzymes have allosteric effector binding sites, whereas permeases do not.
   d. enzymes are affected by temperature and pH, whereas permeases are not.
   e. permeases abide by the laws of thermodynamics, whereas some enzymes do not.

8. Which event does NOT occur in the carbon-fixing reactions of the Calvin cycle?
   a. ATP is used as an energy source.
   b. Oxygen is released.
   c. Carbon dioxide is fixed.
   d. Glyceraldehyde phosphate is formed.
   e. NADPH formed during the energy-capturing reactions is used as an energy source.

9. **T or F** The end products of cellular respiration are the same molecules that are the reactants in photosynthesis.

10. **T or F** Oxidizing one mole of glucose consumes 686 kilocalories of energy.

11. **T or F** Living things use only a small part of the electromagnetic spectrum that we call visible light. This is partly because light of longer wavelengths would drive electrons out of chemical bonds and thus destroy organic molecules.

12. **T or F** In any reversible reaction, the point of equilibrium will lie in the direction for which ΔG is positive.

13. **T or F** During chemiosmotic coupling, protons are pumped out of the mitochondrial matrix and into the intermembrane space, creating both a pH and an electrical gradient.

14. **T or F** The 7 kilocalories of energy released when one mole of ATP is hydrolyzed result from the movement of electrons to lower energy levels and from the rearrangement of electron orbitals in the ATP molecule.

15. State the second law of thermodynamics. Discuss the relevance of this law to life, focusing particular attention on the role of entropy.

16. Discuss the conditions under which lactic acid is formed from pyruvic acid in muscle cells. In terms of energy production, why is this reaction beneficial? What happens to the lactic acid once oxygen is again available?

17. Write the overall equation for photosynthesis in plants. Which of the reactants is reduced?

18. Taking into account the environmental conditions under which C_3 and C_4 plants thrive, explain why it is significant that PEP carboxylase has a higher affinity for carbon dioxide than does RuBP carboxylase.

PERFORMANCE ANALYSIS

1. **e** When light strikes the reactive chlorophyll *a* molecule of a photosystem, electrons are boosted to a higher energy level and then are transferred along a series of electron carriers. Thus, light energy is transformed into electrical energy. (pages 162, 218)

2. **d** The biochemical pathways that constitute glycolysis are located in the cytoplasm of eukaryotic cells. Cellular respiration occurs in the mitochondria. (page 187)

3. **c** An action spectrum shows the wavelengths of light (or electromagnetic radiation) involved in a particular process. An absorption spectrum shows the wavelengths of light absorbed by a pigment. (page 213)

4. **d** Catabolic reactions are those in which compounds are broken down. Anabolic reactions are synthetic in nature. Cellular respiration is the process in which organic molecules are broken down, transferring the energy contained in their bonds to the bonds of ATP. (page 168)

5. **a** Each time an acetyl group is oxidized in the Krebs cycle the output is two carbon dioxide molecules, two ATP molecules, three NADH molecules, and one $FADH_2$ molecule. (page 195)

6. **b** The experiments of F. F. Blackman revealed the effects of temperature and light intensity on photosynthesis. From his results, Blackman concluded that there are two stages of photosynthesis: one depending on light and the other influenced by temperature. (page 215)

7. **a** Enzymes catalyze chemical reactions. Permeases, not true enzymes, facilitate the transport of molecules across cell membranes. (pages 169, 178)

8. **b** During the carbon-fixing reactions of photosynthesis, carbon dioxide is fixed into organic molecules. ATP and NADPH produced during the energy-capturing reactions provide the energy for the Calvin cycle. The immediate product of the Calvin cycle is glyceraldehyde phosphate. Oxygen is released during the light-capturing reactions. (page 222)

9. **True** Photosynthesis involves the production of glucose from carbon dioxide and water. In glycolysis and cellular respiration, glucose is catabolized into carbon dioxide and water. (pages 166, 167)

10. **False** The oxidation of one mole of glucose *releases* 686 kilocalories of energy. (page 186)

11. **False** The wavelengths of visible light contain just enough energy to raise electrons to higher energy levels without disrupting chemical bonds. Wavelengths of light shorter than the visible spectrum contain enough energy to break bonds between and within molecules, destroying their structure and function. Wavelengths longer than the visible spectrum increase the motion of molecules but do not excite electrons to higher energy levels. (page 210)

12. **False** The ΔG of a reaction is the net free energy change for the reaction. If ΔG is negative, the reaction is exergonic (energy is released during the reaction). If ΔG is positive, the reaction is endergonic (there is a net input of energy into the reaction system). In a reversible reaction, the point of equilibrium will lie in the direction of the exergonic reaction, the direction in which ΔG is negative. (page 174)

13. **True** During oxidative phosphorylation, a proton (pH) gradient and an electrical gradient are produced when protons are pumped out of the mitochondrial matrix into the intermembrane space. (page 197)

14. **True** When ATP is hydrolyzed, energy is released when the bonding electrons move to lower energy levels and when the electrons in other orbitals of the ATP molecule rearrange themselves. (page 181)

15. The second law of thermodynamics states that in energy transformations in which there is no change in the total energy content of the system, reactions will proceed so that the potential energy of the final state is less than the potential energy of the initial state. One consequence of this law is that, without a continual input of energy, the entropy (disorder or randomness) of a system will increase. Living systems require a constant input of energy to maintain the intricate organization needed to conduct the biochemical processes associated with life. (pages 164, 166)

16. Lactic acid is produced in muscle cells during strenuous exercise when oxygen demand by the cells exceeds oxy-

gen supply. Under these circumstances, pyruvic acid can be converted to lactic acid to regenerate NAD^+, thereby allowing energy production via glycolysis to continue. When oxygen becomes available again, the lactic acid is converted back to pyruvic acid, which enters the Krebs cycle. (page 190)

17. $$CO_2 + H_2O + \text{light energy} \longrightarrow \underset{\text{Carbohydrates}}{(CH_2O)} + O_2$$

In the overall equation of photosynthesis, the carbon of carbon dioxide is reduced in the formation of glucose. (pages 167, 206)

18. PEP carboxylase (of the C_4 pathway) and RuBP carboxylase (in the Calvin cycle) are the enzymes that bind carbon dioxide and catalyze its fixation into an organic molecule. The higher affinity of PEP carboxylase for carbon dioxide allows PEP carboxylase to bind carbon dioxide at much lower CO_2 concentrations than does RuBP carboxylase. This is a major adaptation of C_4 plants to arid environments, since these plants keep their stomata closed to conserve water and therefore take in less carbon dioxide than do C_3 plants. PEP carboxylase enables C_4 plants to manufacture organic molecules at reduced carbon dioxide levels relative to C_3 plants. (pages 224, 225)

SECTION **3**

Genetics

CHAPTER 11

From an Abbey Garden: The Beginning of Genetics

MAJOR CONCEPTS

Early ideas concerning heredity ranged from the accurate if simplistic knowledge of plant breeding possessed by the ancient Egyptians and Babylonians to numerous myths and legends in which two often dissimilar species produced offspring. In their attempts to explain the origin of some unusual animal species, even the early naturalists proposed that these species resulted from a cross of two other species.

Hippocrates and Aristotle were two of the earliest scientists to propose mechanisms of heredity. From their time until the observation of sperm and ova in the 1670s, no new ideas on this topic were espoused by the scientific community. Even though the reproductive cells had been identified, the roles of sperm and egg were unclear and were debated vigorously.

The concept of blending inheritance was proposed in the mid-1800s and stated that the characteristics of both parents were blended together in the offspring in a fashion analogous to the mixing of two different colors of ink. This flawed con-

cept did not account for characteristics skipping generations or for the variations that persist after generations of breeding.

Gregor Mendel was the first scientist to elucidate the fundamental mechanism of heredity by demonstrating that characteristics are inherited as discrete units, later termed genes. In his studies of garden peas, he showed that each gene for a trait exists in more than one form, now referred to as an allele, and that each individual carries a pair of alleles for each trait. If an organism possesses two identical alleles for a trait, it is said to be homozygous for that trait. If the two alleles for a trait are different, the organism is said to be heterozygous.

Mendel referred to a characteristic as dominant if it was expressed in the heterozygous state and recessive if it was not expressed in the heterozygote. Heterozygote carriers of a recessive allele may be identified by a testcross: an individual with a dominant trait is mated with an individual exhibiting the recessive trait.

The genotype of an organism is its genetic make-up, or the composition of its alleles if one is discussing specific traits. The phenotype is the outward appearance of an organism, or of the specific trait under consideration.

Mendel's principle of segregation of alleles states that the two alleles for each trait separate during gamete formation. Thus, there is a 50 percent chance that a particular gamete will have one of a pair of alleles.

Mendel's principle of independent assortment contends that the alleles for different genes separate into gametes independently of the alleles for other genes. (Later geneticists realized that this only holds true if the alleles for different genes are located on different chromosomes or are widely separated on the same chromosome.)

One of the reasons Mendel succeeded in deciphering the mechanism of heredity is that he quantified his results. This enabled him to identify patterns in types of offspring and to recognize that the laws of probability were operating.

Hugo de Vries first proposed the concept of mutations, sudden changes in the genetic make-up of an organism. Mutations are now recognized as the ultimate source of the variation upon which natural selection acts.

Mendel's principles of segregation and independent assortment combined with de Vries's concept of mutation filled an important gap in Darwin's theory of evolution.

HOW TO STUDY THE CHAPTER

Read the entire chapter through quickly, focusing on the major concepts.

Use the GUIDED STUDY OF THE CHAPTER to help you identify the important details as you **reread** the chapter. Writing out the answers to these questions will help fix them in your mind as well as provide you with a valuable study aid.

Answer the questions in TESTING YOUR UNDERSTANDING without the aid of your text. Check your answers against those in PERFORMANCE ANALYSIS. Analyzing your answers will give you valuable feedback on your level of understanding and preparedness for classroom testing.

GUIDED STUDY OF THE CHAPTER

I. Introduction (page 235)

1. *Vocabulary:* Define heredity.

II. Early Ideas About Heredity (page 236)

2. Summarize the understanding of heredity exhibited by (a) the ancient Egyptians and Babylonians and (b) the Greeks.

> *Focus:* Many myths and legends centered around offspring produced by mating two different species.

3. Cite one example in which early naturalists proposed the mating of two different species as an explanation for a third species.

4. a. Who was the first scientist known to have contemplated the mechanism of heredity?

   b. Summarize his hypothesis.

5. a. Who rejected the hypothesis mentioned in question 4 and what was the basis for the rejection?

   b. Outline the alternative hypothesis made by this individual.

> *Focus:* Prior to the convincing disproof of spontaneous generation offered by Pasteur in 1864, many scientists believed that lower forms of life arose spontaneously.

6. Describe the recipe concocted by Jan Baptista van Helmont for the production of mice.

III. The First Observations (page 237)

7. a. Who first visualized living sperm cells?

   b. *Vocabulary:* What was the homunculus thought to represent?

   c. What did the discoverers of the homunculus believe the female contributed to the offspring?

8. a. Régnier de Graaf made what contribution to reproductive knowledge?

b. Contrast the positions of ovists and spermists regarding the roles of egg and sperm.

9. a. How did ovists and spermists account for the production of successive generations of offspring?

b. According to this philosophy, what would be the eventual fate of the human race?

IV. Blending Inheritance (pages 237–238)

10. The facts that challenged the ideas of the spermists and ovists were derived from what studies?

11. a. Summarize the concept of blending inheritance.

b. Why was this concept unacceptable?

c. How was blending inheritance incompatible with Darwin's concept of natural selection?

V. The Contributions of Mendel (pages 238–245)

A. General Remarks (page 238)

12. What was Gregor Mendel's greatest contribution to the knowledge of heredity?

13. Vocabulary: Define genes.

B. Mendel's Experimental Method (page 238)

14. a. List five reasons why Mendel selected the common garden pea for his experiments in heredity.

b. Identify four aspects of Mendel's method that contributed to his successful formulation of the principles of heredity.

15. Vocabulary: Distinguish between self-pollination and cross-pollination. (Figure 11–5)

C. The Principle of Segregation (pages 239–241)

16. a. What characteristic did Mendel require of the traits whose inheritance he chose to study?

b. List the seven traits Mendel studied in garden peas.

17. a. In Mendel's studies, how was the F_1 generation produced?

b. Characterize the F_1 generation in terms of the traits of its parents.

18. *Vocabulary:* Define a dominant trait.

19. a. In Mendel's studies, how was the F_2 generation produced?

b. Characterize the F_2 generation in terms of the traits of its grandparents.

20. *Vocabulary:* What is a recessive trait?

21. How did Mendel explain the disappearance of recessive traits in the F_1 generation and their subsequent reappearance in the F_2 generation?

22. State in your own words the principle of segregation.

Consequences of Segregation (pages 240–241)

23. a. *Vocabulary:* What is an allele?

b. *Vocabulary:* Distinguish between homozygous and heterozygous.

c. Define dominant and recessive alleles in terms of the homozygous and heterozygous conditions.

24. a. *Vocabulary:* Distinguish between phenotype and genotype. (Include two definitions of phenotype.)

b. Describe the relationship between genotypic ratio and phenotypic ratio. (Figure 11–8)

25. Close your book and work this problem in the following Punnett square. List all possible genotypes that would result from crossing a heterozygous purple-flowering pea plant (*Ww*) with a white-flowering pea plant (*ww*). First list the *gametes* produced by:

The purple-flowering plant

The white-flowering plant

Now fill in the Punnett square.

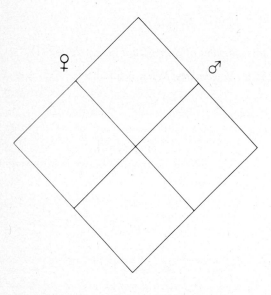

(Check your answer against Figure 11–9.)

26. a. *Vocabulary:* Define a testcross in terms of the parental genotypes.

b. What information is provided by a testcross?

Focus: If a testcross produces even one individual with the recessive phenotype, then the parent with the dominant phenotype is certainly heterozygous. However, concluding that the parent with the dominant phenotype is homozygous is not nearly so certain, especially if only a few offspring are produced. The likelihood that the parent is homozygous increases with the total number of offspring produced by testcrosses, all of which show the dominant phenotype.

D. The Principle of Independent Assortment *(pages 241–242)*

27. Summarize the principle of independent assortment.

28. Use this hypothetical example to trace the origin of the 9:3:3:1 phenotypic ratio characteristic of an F_2 generation. The parents that produced the F_1 generation were a tall plant with yellow seeds (*TTyy*) and a dwarf plant with green seeds (*ttYY*).

a. Indicate the genotype of the F_1 generation.

b. List all possible gametes produced by each F_1 parent.

c. Now fill in the Punnett square.

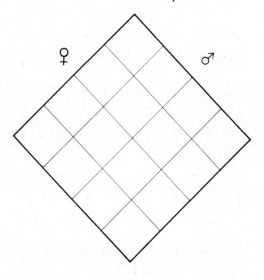

d. List all the genotypes that produce each phenotype.

Tall with green pods

Dwarf with green pods

Tall with yellow pods

Dwarf with green pods

A Testcross (pages 241–243)

29. What assumption is being made every time a Punnett square is constructed?

30. How does a testcross involving two different traits resemble and differ from one involving a single trait?

E. *Essay:* **Mendel and the Laws of Probability** *(pages 244–245)*

31. a. The product rule of probability estimates the probability that:

    b. How is this probability calculated?

32. What test results indicated to Mendel that he was dealing with the laws of probability?

33. a. The sum rule of probability estimates the probability when:

    b. How is this probability calculated?

34. List the three major assumptions made when Mendel designed his experiments.

VI. **Mutations** *(pages 244–246)*

A. **General Remarks** *(pages 244–245)*

35. a. Hugo de Vries observed the sudden appearance of variations in evening primroses. How did he explain this variation?

    b. *Vocabulary:* Distinguish between mutation and mutant.

    c. Even though de Vries made a significant contribution to thoughts on heredity, what flaw existed in the conclusions he drew from his observations on evening primroses?

B. **Mutations and Evolutionary Theory** *(page 245–246)*

36. How did the following concepts fill in the gaps in Darwin's theory of evolution?

    Segregation of alleles

    Independent assortment

    Mutations

Focus: Mutations are the ultimate source of the genetic variation upon which natural selection acts.

VII. **The Influence of Mendel** *(pages 246–247)*

Focus: In spite of Mendel's efforts to make his work known to his scientific contemporaries, it was ignored and its significance was not realized until after his death.

37. List three developments or discoveries important to unraveling the mechanism of heredity that were made during the 35 years that Mendel's work lay unappreciated in libraries.

VIII. Summary *(page 247):* Read the summary. If you are familiar with the essential features of the material presented there, you are ready to complete the section TESTING YOUR UNDERSTANDING.

TESTING YOUR UNDERSTANDING

After you have completed this examination, compare your answers with those in the section that follows.

1. The first man known to have pondered the mechanism of heredity was:
   a. Theophrastus.
   b. Homer.
   c. van Helmont.
   d. Hippocrates.
   e. Aristotle.

2. During the seventeenth century, the Dutchman _____ had a following of _____ who maintained that Eve contained all the yet unborn generations of humans.
   a. van Helmont; spermists
   b. van Helmont; ovists
   c. van Leeuwenhoek; spermists
   d. van Leeuwenhoek; ovists
   e. de Graaf; ovists

3. A logical conclusion of the concept of blending inheritance was that:
   a. all animals should be either black or white.
   b. the rate of evolution would greatly accelerate.
   c. natural selection would have no raw materials on which to act.
   d. hereditary material, once blended, could easily separate.
   e. hereditary variation would increase as new blends took place.

4. Mendel's success in formulating the basic principles of inheritance resulted from his:
   a. comparison of the numbers of offspring in the F_1 and F_2 generations.
   b. determination of the genotype of each gamete.
   c. analysis of the numbers and types of F_2 progeny.

   d. extensive training in biology at the University of Vienna.
   e. deep understanding of the mechanisms involved in meiosis.

5. Which of these is NOT a characteristic of pea plants that makes them ideal for Mendel's studies?
   a. Varieties could be obtained that bred true for certain characteristics.
   b. Garden peas were easy to cultivate.
   c. Garden peas grow and mature rapidly.
   d. The pea varieties Mendel used were generally unavailable to his competitors.
   e. If left unattended, garden peas self-pollinate.

6. Mendel called those traits that were not expressed in the F_1 generation:
   a. recessive.
   b. heterozygous.
   c. incompletely dominant.
   d. hybrids.
   e. null alleles.

7. Expressions of a trait, such as yellow versus green seeds, are referred to as:
   a. varieties.   d. factors.
   b. phenotypes.   e. alleles.
   c. genotypes.

8. Alternative forms of the same gene are known as:
   a. gametes.   d. homozygotes.
   b. heterozygotes.   e. genotypes.
   c. alleles.

9. Mendel tested the hypothesis that two alleles segregate during gamete formation by performing a:
   a. dissection and observation of the process of meiosis in pea plants.
   b. repeat of a famous experiment done years before by his mentor.
   c. new set of experiments that involved a different set of alleles.
   d. detailed analysis of the F_2 progeny of his crosses.
   e. testcross.

10. The information in the squares within a Punnett square is:
    a. progeny genotypes.
    b. parental phenotypes.
    c. gamete phenotypes.
    d. parental genotypes.
    e. gamete genotypes.

11. In a cross involving genes for two different traits, the _____ of the progeny should exhibit a _____ ratio.
    a. phenotypes; 9:3:3:1
    b. genotypes; 9:3:3:1
    c. phenotypes; 3:1
    d. genotypes; 3:1
    e. Insufficient information is given to answer the question.

12. You have a garden in which you would like to have only pea plants that have green seeds. (Green seed color is recessive to yellow.) You only have available yellow F_1 seeds that were obtained by crossing green and yellow varieties. If you allow plants that arise from these seeds to self-pollinate, what proportion of the offspring could be used to initiate your pure-breeding green pea garden?
    a. all
    b. three-fourths
    c. half
    d. one-fourth
    e. none

13. If both parents are heterozygous for genes governing two different traits, what proportion of the offspring will be homozygous recessive for both genes?
    a. 3/4
    b. 9/16
    c. 3/16
    d. 1/4
    e. 1/16

14. How many different kinds of gametes may be produced by an organism with the genotype *Rryy*?
    a. one
    b. two
    c. three
    d. four
    e. five

15. Although Hugo de Vries was correct in saying that sudden changes in the phenotype of an organism may be due to mutation, most of his examples were not due to mutation but to:
    a. uncontrolled cross-pollination.
    b. new combinations of existing alleles.
    c. sloppy experimental procedure.
    d. failure to start with pure-breeding varieties of plants.
    e. environmental influences.

16. Which of the following is an example of a testcross involving TWO genes?
    a. *AaBb × aabb*
    b. *AAbb × aabb*
    c. *AaBb × AaBb*
    d. *aabb × aabb*
    e. *AABb × aaBB*

17. A cross involving two parents that are doubly heterozygous (e.g., *AaBb × AaBb*) will produce a 9:3:3:1 phenotypic ratio among the offspring. This means that _____ of the offspring will exhibit _____.
    a. 3/8; both dominant traits
    b. 3/8; both recessive traits
    c. 6/16; one dominant and one recessive trait
    d. 3/16; both recessive traits
    e. 3/16; one dominant and one recessive trait

18. An organism of unknown genotype is crossed with a homozygous recessive individual having a genotype of *bb*. Fifty percent of the progeny are phenotypically similar to the unknown. The genotype of the unknown is therefore:
    a. *BB*.
    b. *Bb*.
    c. *Aa*.
    d. *AA*.
    e. still undetermined.

19. This matching problem applies to the following cross:

$$WwDd \times wwDd$$

    where *W* is a dominant allele producing smooth seeds
    w is a recessive allele producing wrinkled seeds
    D is a dominant allele producing tall stems
    d is a recessive allele producing short stems

    Match the phenotypic or genotypic classes with the frequency of their occurrence in the progeny. Answers may be used more than once.

    1. 3/4   5. 1/4
    2. 9/16   6. 3/16
    3. 1/2   7. 1/8
    4. 3/8

    a. _____ tall
    b. _____ wrinkled
    c. _____ smooth and tall
    d. _____ wrinkled and short
    e. _____ carry an unexpressed allele for short stems
    f. _____ heterozygous for both traits

20. **T or F**   Although general ideas about the biological roles of males and females in inheritance go as far back as the ancient Egyptians and Babylonians, it wasn't until the second half of the nineteenth century that inheritance was actually studied from a scientific perspective.

21. What was Mendel's single most important contribution to the fledgling science of genetics?

22. Explain how the product rule of probability applies to genetics.

PERFORMANCE ANALYSIS

1. **d**   Hippocrates was the first scientist known to have contemplated the mechanism of inheritance. He proposed that each portion of the body produced "seeds" that were transmitted to the offspring at conception. (page 236)

2. **e**   Régnier de Graaf was the first to describe the ovarian follicle. (He did not see the egg itself, however.) His followers contended that the egg contained the miniature human being and therefore that Eve must have contained all the unborn humans for future generations. (page 237)

3. **c**   If blending inheritance were true, over several generations all members of an interbreeding group of organisms would come to resemble one another. This concept left

no room for the appearance and maintenance of variation within a species, which was a key factor in Darwin's theory of evolution by natural selection. (page 238)

4. **c** One of the reasons Mendel was successful in defining the nature of heredity when others were not was that he analyzed his results mathematically. For each trait under study, he recorded the number of each type in the offspring and looked for mathematical trends in his data. (page 238)

5. **d** The garden peas were ideal for Mendel's studies for several reasons. Varieties were readily available that bred true for specific, easily definable characteristics. The plants were easy to cultivate and grew quickly. Garden peas self-pollinate naturally. (page 238)

6. **a** Mendel referred to those traits that were not expressed in the F_1 generation as recessive, and traits that did appear in the F_1 generation as dominant. (page 239)

7. **b** By definition, the expression of a trait is the phenotype. The genetic composition of an organism (or the alleles it possesses for a specific trait under consideration) is its genotype. (page 240)

8. **c** Alleles are alternative forms of the same gene. A single gene generally has at least two alleles; many genes have more than two alleles, as you will learn in more detail later. (page 240)

9. **e** A testcross evaluating one locus involves mating a homozygous recessive individual with an individual exhibiting the dominant phenotype. If any of the progeny exhibit the recessive phenotype, the parent with the dominant phenotype is known to carry the recessive allele. (pages 240, 241)

10. **a** The squares within a Punnett square contain genotypes of the progeny. The types of gametes from each parent are placed at the head of each column. (Figure 11–8)

11. **e** The genotypic and phenotypic ratios of the progeny cannot be determined unless the genotypes of the parents are known. It is incorrect to *assume* that the parents are heterozygous for the two traits under study. (pages 242, 243)

12. **d** In garden peas, yellow seed color (*Y*) is dominant to green color (*y*). The genotype of the F_1 seeds is *Yy*. The genotype that is desired is *yy*. Crossing two *Yy* individuals will result in three-fourths yellow seeds (1/4 *YY* and 1/2 *Yy*) and one-fourth green seeds (*yy*). (Figure 11–8)

13. **e** The situation may be illustrated by assigning both parents the genotype *AaBb*. Possible parental gametes are *AB*, *Ab*, *aB*, and *ab*. In working a Punnett square, it becomes evident that only 1/16 of the offspring will have the genotype *aabb*. (pages 242, 243)

14. **b** The possible gametes of this organism are *Ry* and *ry*. (pages 242, 243)

15. **b** In de Vries's studies of evening primroses, most of the new characteristics that appeared resulted from new combinations of existing alleles. (page 244)

16. **a** See the answer to question 9 for the definition of a testcross. Response **a** meets this definition. When two loci are involved, each locus is considered independently. The cross *Aabb* × *aaBb* is also a testcross; the homozygous recessive genotype for each locus is located in different parents. (page 243)

17. **c** In this cross, 3/16 of the progeny will have one dominant trait and one recessive trait. Another 3/16 will have the alternative dominant trait and the alternative recessive trait. A total of 6/16 (or 3/8) will possess one dominant and one recessive characteristic. Further, 9/16 will have both dominant traits and only 1/16 will have both recessive traits. (pages 242, 243)

18. **b** This situation describes a testcross in which the individual with the unknown genotype is heterozygous. The phenotype of the unknown is implied to be the dominant phenotype because if the "unknown" possessed the recessive phenotype, the genotype would be known. (pages 240, 241)

19. The answers are **a1, b3, c4, d7, e3, f5**. (pages 242, 243)

20. **True** Mendel was the first scientist to analyze the results of breeding experiments mathematically. This approach was the key to his success at unraveling the basic pattern of heredity. (pages 236, 238)

21. The single most important contribution that Mendel made to the science of genetics was his demonstration that characteristics are inherited in discrete units. (page 238)

22. The product rule of probability states that the probability of two independent events occurring simultaneously equals the product of the probabilities of the two events occurring alone. For an example, see question 19. The probability that a plant will be tall (*D___*) is 3/4. The probability that a plant will have smooth seeds (*W___*) is 1/2. The probability that a plant will be tall with smooth seeds (*D___W___*) is 3/4 × 1/2 or 3/8. (The blank after the dominant allele indicates that the second allele may be either dominant or recessive.) (essay, page 244)

C H A P T E R **12**

Meiosis and Sexual Reproduction

MAJOR CONCEPTS

Sexual reproduction always involves the processes of meiosis and fertilization. During meiosis, the chromosome number is halved to form haploid gametes (in animals) or spores (in plants). During fertilization, haploid gametes fuse to form the diploid zygote, which develops into a diploid adult (usually).

Every diploid cell in an organism contains one copy of the chromosomes received from the male parent and one copy of the chromosomes received from the female parent. These chromosomes exist in homologous pairs. Since each haploid cell produced during meiosis contains only one set of homologues, the genetic composition of the haploid cells varies and every gamete contains some genes from each parent.

Meiosis occurs at different stages of the life cycle for different species. In organisms whose predominant form is haploid, meiosis occurs immediately after the formation of a diploid cell. Plants exhibit alternation of generations, in which a haploid life stage alternates with a diploid stage. In animals, gametes are produced immediately prior to fertilization, and the organism is diploid for most of its life cycle.

Although meiosis is thought to have evolved from mitosis, there are several major differences. Meiosis occurs only in diploid cells; mitosis can occur in either haploid or diploid cells. In meiosis, the nucleus divides twice but the chromosomes replicate only once. Meiosis results in haploid cells of varying genetic composition; mitosis produces diploid cells identical to the parent and each other.

The two nuclear divisions of meiosis are referred to as meiosis I and meiosis II for convenience. Each stage has four phases (prophase, metaphase, anaphase, and telophase) which roughly parallel the same phases of mitosis.

Prophase I sets the stage for meiosis. At the end of this stage, each chromosome exists as a tetrad consisting of a replicated homologous chromosome pair. During prophase I, crossing over may occur between members of the homologous pair, resulting in new genetic combinations in which individual chromosomes contain genes from *both* parents.

Meiosis I results in the formation of haploid cells in which each chromosome exists as a pair of sister chromatids joined at the centromere. In meiosis II, the sister chromatids separate.

In vertebrates, meiosis occurs in the reproductive organs—the testes in males and the ovaries in females. The result of one meiotic division in the male is four sperm cells. In the female, one meiotic division produces one egg and three polar bodies that degenerate.

Walter S. Sutton was one of the first scientists to connect the chromosome movements of meiosis with Mendel's principles of heredity and to propose convincingly that chromosomes carry the units of heredity. The existence of chromosomes as homologous pairs, with one member of each pair derived from each parent, provides a mechanism for Mendel's principle of the separation of alleles. Mendel's law of independent assortment holds true for genes on different chromosomes.

Although sexual reproduction is energetically expensive and inefficient, it perpetuates genetic variability within a species.

HOW TO STUDY THE CHAPTER

Read the entire chapter through quickly, focusing on the major concepts.

Use the GUIDED STUDY OF THE CHAPTER to help you identify the important details as you **reread** the chapter. Writing out the answers to these questions will help fix them in your mind as well as provide you with a valuable study aid.

Answer the questions in TESTING YOUR UNDERSTANDING without the aid of your text. Check your answers against those in PERFORMANCE ANALYSIS. Analyzing your answers will give you valuable feedback on your level of understanding and preparedness for classroom testing.

GUIDED STUDY OF THE CHAPTER

I. Introduction *(page 249)*

1. Name and define the two events always involved in sexual reproduction.

II. Haploid and Diploid *(pages 249–250)*

2. a. *Vocabulary:* Distinguish between haploid, diploid, and polyploid.

   b. Write the *shorthand* designation for the haploid and diploid chromosome numbers in an organism that has 78 pairs of chromosomes.

3. Distinguish between somatic cells and gametes.

4. *Vocabulary:* What is a zygote?

5. a. *Vocabulary:* Define a homologue.

   b. What is the origin of each member of a homologous pair?

6. State the effect of (a) meiosis and (b) fertilization on chromosome number.

> *Focus:* **The effect of fertilization on chromosome number counterbalances that of meiosis.**

7. List two consequences of meiosis.

III. Meiosis and the Life Cycle *(pages 250–252)*

> *Focus:* **Depending upon the organism, meiosis occurs at different points in the life cycle.**

8. Use three examples to illustrate this Focus statement.

9. What is alternation of generations?

10. Distinguish between meiosis in plants and animals with respect to the type of cell produced and the fate of that cell.

IV. Meiosis vs. Mitosis *(pages 252–253)*

11. Summarize the events that occur during the phases of the cell cycle and *mitosis*. (Review Chapter 7 if necessary.)

    Interphase

    Prophase

    Metaphase

    Anaphase

    Telophase

12. List four ways in which meiosis and mitosis differ.

V. The Phases of Meiosis *(pages 253–258)*

A. General Remarks *(pages 253–257)*

Focus: Meiosis consists of two successive nuclear divisions.

13. Distinguish between meiosis I and meiosis II on the basis of chromosome movement.

14. At the beginning of meiosis I, the chromosomes exist in what form?

15. What occurs to the chromatin/chromosomes in early prophase I?

16. *Vocabulary:* What is a tetrad?

17. *Vocabulary:* Distinguish between synapsis and crossing over.

Focus: Crossing over is an important mechanism of genetic recombination.

18. *Vocabulary:* Define chiasmata.

19. By the end of prophase I, what meiotic events have occurred?

Focus: At the end of prophase I, homologous chromosomes exist as tetrads.

20. a. How does metaphase I of meiosis differ from mitotic metaphase?

    b. What meiotic events have occurred by the end of metaphase I?

21. Anaphase I differs from mitotic anaphase in what way?

22. a. Describe the arrangement of chromosomes in telophase I.

b. Name three events that may occur during telophase I, depending on the species.

Focus: A chromosome may be defined by the presence of a centromere. Thus, a single chromatid with its own centromere that is not attached to a sister chromatid is termed a chromosome. Paired sister chromatids attached at a single centromere may also be called a chromosome.

---

23. Summarize the events that occur in each stage of meiosis II and in cytokinesis.

Prophase II

Metaphase II

Anaphase II

Telophase II

Cytokinesis

B. Meiosis in the Human Species *(page 258)*

24. Name the sites of meiosis in vertebrates.

25. a. Name the cell in the (1) male and (2) female that undergoes meiosis. (Figures 12–12 and 12–13)

b. How does meiosis in the female differ from meiosis in the male with respect to end products?

26. When in the life of a human female does each of the following events occur? (Figure 12–13)

First meiotic division begins

First meiotic division is completed

Second meiotic division

27. What is the value of the unequal meiotic divisions in oocytes?

VI. *Essay:* **The Consequences of Sexual Reproduction** *(page 259)*

28. a. Cite one mechanism by which plants may reproduce asexually.

b. Name two means by which animals may reproduce asexually.

Focus: All offspring produced by asexual reproduction are genetically identical to the parent.

---

29. Identify two major factors contributing to the genetic variability made possible by sexual reproduction.

30. What are the major disadvantages associated with sexual reproduction compared with asexual reproduction?

VII. Cytology and Genetics Meet: Sutton's Hypothesis *(pages 260–261)*

31. Summarize the observations of Sutton that led him to the conclusion that chromosomes contain the units of heredity first described by Mendel.

32. a. How is the "dance of the chromosomes" consistent with Mendel's principle of the segregation of alleles?

b. What relationship between genes and chromosomes must exist in order for Mendel's principle of independent assortment to be demonstrated?

VIII. Summary *(page 261):* Read the summary. If you are familiar with the essential features of the material presented there, you are ready to complete the section TESTING YOUR UNDERSTANDING.

TESTING YOUR UNDERSTANDING

After you have completed this examination, compare your answers with those in the section that follows.

1. Sexual reproduction involves two processes, the effects of which offset one another. These processes are:
   a. meiosis and mitosis.
   b. mitosis and cytokinesis.
   c. meiosis and fertilization.
   d. cytokinesis and fertilization.
   e. mitosis and fertilization.

2. Which statement does NOT apply to homologous chromosomes?
   a. They are the same size and shape.
   b. They undergo synapsis during prophase of mitosis.
   c. They separate during anaphase I of meiosis.
   d. They form a tetrad during prophase I of meiosis.
   e. They contain the same kinds of hereditary information.

3. Meiosis occurs immediately after fertilization in organisms whose predominant form is a:
   a. haploid gametophyte.
   b. diploid gametophyte.
   c. haploid sporophyte.
   d. diploid sporophyte.
   e. Meiosis never occurs immediately after fertilization.

4. A sexually reproducing organism with 28 chromosomes in each somatic cell receives _____ chromosomes from each parent.
   a. 7   d. 42
   b. 14   e. 56
   c. 28

5. Chiasmata:
   a. is the mechanism by which DNA is replicated during interphase.

b. is the mechanism by which a tetrad of chromatids is held together in synapsis.
   c. are the two parts of a chromosome that are produced by DNA replication.
   d. are the points along two chromatids where crossing over has taken place.
   e. is the only mechanism by which genetic recombination takes place during meiosis.

6. During the tetrad stage of meiosis, interchange of genetic material takes place between:
   a. identical alleles.
   b. male and female gametes.
   c. male and female pronuclei.
   d. nonhomologous pairs of chromosomes.
   e. chromatids of homologous chromosomes.

7. Pairing along the entire length of the chromatids of a homologous pair occurs during _____ of meiosis.
   a. prophase I   d. metaphase II
   b. prophase II   e. anaphase II
   c. anaphase I

8. Chromosome replication does NOT precede nuclear division in:
   a. mitosis involving haploid cells.
   b. mitosis involving diploid cells.
   c. meiosis I.
   d. meiosis II.
   e. Chromosome replication always precedes nuclear division.

9. In counting the number of chromosomes, it is always safest to count the number of:
   a. chromatids.
   b. spindle fiber attachments.
   c. kinetochores.
   d. centrioles.
   e. centromeres.

10. If the normal diploid number of chromosomes in an organism is 8, how many *chromosomes* are present in each of the two daughter cells at the end of meiosis I?
    a. 2   d. 16
    b. 4   e. 32
    c. 8

11. If crossing over does not occur, a human male can produce how many genetically unique sperm cells as a result of meiosis?
    a. 46^2 c. 2^{23} e. 46
    b. 23^2 d. 2^{46}

12. In human females, how many eggs are produced for every primary oocyte that undergoes meiosis?
    a. one   d. six
    b. two   e. eight
    c. four

13. **T or F** Since meiosis apparently evolved from mitosis and uses much of the same cellular machinery, the differences between these two processes are insignificant.

14. **T or F** Even though there are differences between mitosis and meiosis, the four cells produced by meiosis, like the cells produced by mitosis, are identical with each other.

15. **T or F** Anaphase II of meiosis resembles anaphase of mitosis in that sister chromatids separate into different daughter cells.

16. **T or F** Walter Sutton's hypothesis provided a cytological explanation of Mendel's principle of segregation.

17. **T or F** For almost 10 years, Walter Sutton was the only scientist to recognize that the movements of chromosomes during meiosis account for segregation of alleles and independent assortment of genes.

18. Arrange the following components in a circular flow diagram illustrating alternation of generations in ferns: gamete, spore, zygote, gametophyte, and sporophyte. Indicate whether each component is haploid (n) or diploid ($2n$). Identify where mitosis, meiosis, and fertilization occur.

PERFORMANCE ANALYSIS

1. **c** During meiosis, chromosome number is reduced to the haploid state. The process of fertilization involves fusion of gametes, which restores the diploid state. (page 249)

2. **b** Homologous chromosomes undergo synapsis (linear pairing of the chromatids of a homologous pair) during prophase I of *meiosis*. Homologous chromosomes do not pair during mitosis. (pages 250, 253)

3. **a** In organisms characterized by a gametophyte as the predominant form, meiosis occurs immediately after the zygote is formed and the diploid stage of the life cycle is very brief. (page 250)

4. **b** In a diploid organism that undergoes sexual reproduction, half the chromosomes (one set of homologues) come from each parent. (page 249)

5. **d** Chiasmata are the sites on a homologous chromosome pair that indicate where each chromatid broke and was rejoined to its homologue during crossing over. (page 253)

6. **e** In crossing over, the chromatids of homologous chromosomes exchange portions. This occurs during prophase I while the chromosomes are tetrads. (page 253)

7. **a** In a process known as synapsis, the chromatids of homologous pairs line up and pair during prophase I. (page 253)

8. **d** The reduction from diploid to haploid state occurs when the second meiotic division occurs without being preceded by chromosome replication. (page 255)

9. **e** The most sure way to keep track of chromosome number while tracing the process of meiosis is to count the centromeres. Each centromere indicates one chromosome. (page 255)

10. **b** At the end of meiosis I, chromosome number has been reduced to half by the separation of replicated homologous chromosomes. In meiosis II, sister chromatids separate. (page 255)

11. **c** Without considering crossing over, the number of genetically unique gametes that a diploid organism may produce equals 2^n, where n is the haploid number of chromosomes. (essay, page 259)

12. **a** In human females, one egg and three polar bodies are produced for each primary oocyte that undergoes meiosis. (page 258)

13. **False** The differences between mitosis and meiosis are quite significant. Mitosis may occur in either haploid or diploid cells; meiosis occurs only in diploid (or polyploid) cells. Meiosis is characterized by two divisions with only one replication of chromosomes. In mitosis, the chromosome number of daughter cells equals that of the parent cell; chromosome number is halved during meiosis. The genetic composition of daughter nuclei from mitosis is identical to that of the parent nucleus. The haploid nuclei of meiotic cells contain new combinations of genes. (pages 252, 253)

14. **False** The four cells produced by meiosis differ genetically from the parent cell and from each other. Homologous chromosomes are separated into different cells, during which the parental chromosomes are mixed, and crossing over in prophase I results in new genetic combinations. (page 253)

15. **True** In anaphase of mitosis and in anaphase II of meiosis, the sister chromatids separate into different cells. Anaphase I of meiosis is characterized by the separation of paired homologous chromosomes into two cells. (page 256)

16. **True** Walter Sutton noticed the parallels between chromosome movements in meiosis and Mendel's segregation of alleles, and hypothesized that genes are located on chromosomes. Alleles segregate when homologous chromosomes separate during anaphase I. Genes located on one chromosome assort independently of the genes on other chromosomes. (page 260)

17. **False** Two other biologists noticed the parallels between meiosis and Mendel's principles at about the same time as Sutton. However, Sutton's paper was published first and his argument was the most convincing, so he gets the credit. (page 261)

18. Compare your diagram with Figure 12–4b and Figure 12–6.

Genes and Gene Interactions

CHAPTER ORGANIZATION

MAJOR CONCEPTS

In the early 1900s, Thomas Hunt Morgan established a colony of *Drosophila melanogaster* at Columbia University for use in genetics research. The studies conducted in this laboratory confirmed Mendel's laws of heredity, established that genes are located on chromosomes, and revealed that patterns of inheritance are not nearly as simple and direct as Mendel's work suggested.

The chromosomes of a diploid organism are paired. There are two types of sex chromosomes, X and Y, and these are paired differently in males and females. For all other chromosomes (autosomes), the members of each chromosome pair appear the same.

The sex with dissimilar sex chromosomes (genotype XY) is heterogametic, since he or she can produce two kinds of gametes. The sex with identical sex chromosomes (XX) is homogametic. (The genotype YY does not occur.) In most species, the male is heterogametic. However, the female is heterogametic in birds, moths, butterflies, and a few other species.

When the male is heterogametic, the gender of offspring is determined by which sex chromosome (X or Y) is contained in the sperm cell that fertilizes the ovum, which always carries an X: if an X chromosome, the offspring will be female (XX); if a Y chromosome, the offspring will be male (XY).

The expression of genes carried on sex chromosomes differs from that of autosomal genes. For sex-linked traits, the terms dominant and recessive apply only to the homogametic sex. If a recessive allele exists on an X chromosome, a male mammal (XY) will express that trait. In order for the trait to be expressed in the female (XX), she must have the allele on both X chromosomes, which means that her sire *exhibited* the recessive trait and her dam was at least a carrier of the recessive allele.

The phenotypic effect of a particular gene is influenced by which alleles of that gene are present, interactions with other genes, and environmental factors.

The alleles of a gene may exhibit Mendelian dominance and recessiveness, incomplete dominance, or codominance. Although any diploid organism possesses only two alleles for a particular gene, more than two alleles for that gene may exist in a population.

The interaction of two or more genes may result in the appearance of new phenotypes or the masking of one gene by another (epistasis). Further, the cumulative effects of several genes influencing one trait, known as polygenic inheritance, may result in a continuum of phenotypic variability.

The expression of a gene may be influenced by temperature, nutrition, and many other environmental factors. Also, the degree to which a particular genotype is expressed may vary from one organism to the next. A gene may show incomplete penetrance—the proportion of individuals in a population that exhibit the phenotype may be less than expected from the known genotypic frequency.

A single gene may affect multiple characteristics, a property known as pleiotropy.

Genes located on the same chromosome tend to be inherited together and are said to be linked. Mendel's law of independent assortment only applies if two genes are located on different chromosomes, or are far enough apart on the same chromosome to be separated during crossing over.

By studying the frequency with which linked gene pairs recombine, chromosome maps can be constructed indicating the relative locations of and distances between various genes.

Abnormal chromosome structure may result if a chromosome fragments and (1) a piece is lost, (2) the pieces rejoin in a different manner, (3) one piece joins to its homologue, or (4) a fragment attaches to a nonhomologous chromosome.

HOW TO STUDY THE CHAPTER

Read the entire chapter through quickly, focusing on the major concepts.

Use the GUIDED STUDY OF THE CHAPTER to help you identify the important details as you **reread** the chapter. Writing out the answers to these questions will help fix them in your mind as well as provide you with a valuable study aid.

Answer the questions in TESTING YOUR UNDERSTANDING without the aid of your text. Check your answers against those in PERFORMANCE ANALYSIS. Analyzing your answers will give you valuable feedback on your level of understanding and preparedness for classroom testing.

GUIDED STUDY OF THE CHAPTER

I. Introduction (page 263)

1. Name four scientists of the early 1900s who engaged in confirming and extending Mendel's findings.

2. a. Which organism did Thomas Hunt Morgan select for his genetic studies?

   b. Name four reasons why *Drosophila* is well suited for use in genetic experiments.

Focus: The selection of relatively "insignificant" plants and animals for genetic studies is based on the premise that principles of heredity are universal among living organisms.

II. The Reality of the Gene (pages 263–267)

A. General Remarks (page 263)

3. What was the most important principle established by Morgan and his colleagues?

Focus: When Morgan founded his laboratory, most scientists were not convinced that Sutton's hypothesis (that genes are located on chromosomes) was valid.

B. Sex Determination (page 264)

4. a. *Vocabulary:* Distinguish between autosomes and sex chromosomes.

   b. Differentiate between the X and Y chromosomes.

5. a. *Vocabulary:* Define heterogametic and homogametic.

   b. Which sex is heterogametic in humans?

   c. Name three types of animals in which the female is heterogametic.

6. Describe the chromosome complements of human males and females in terms of autosomes and sex chromosomes.

7. a. How is the sex of an embryo determined in humans? (The same mechanism is involved in all mammals.)

b. What process governs the roughly equal production of male and female offspring and by what mechanism does this occur?

C. Sex Linkage *(pages 264–266)*

8. In addition to the features listed in the answer to question 2b, identify one other characteristic that made *Drosophila* well suited for genetic studies.

9. a. What was Morgan looking for when he began his studies with *Drosophila*?

   b. What was the initial step, and how long did it take to accomplish this?

10. a. In Morgan's cross of F_1 flies that resulted from mating a white-eyed male with a red-eyed female, how did his results differ from what he expected?

    b. The results of a testcross between the white-eyed male and an F_1 female differed from Morgan's expectations in what manner?

    c. These results led Morgan and coworkers to formulate what hypothesis?

    d. How did they confirm this hypothesis?

    e. How did the vigor of white-eyed mutants compare with that of the wild type?

11. a. *Vocabulary:* Define a sex-linked trait.

    b. *Vocabulary:* Distinguish hemizygous from homozygous and heterozygous.

12. Using strict definitions, the terms dominant and recessive apply only in what cases?

13. Why do sex-linked "recessive" traits appear most often in the heterogametic sex?

> *Focus:* **The results of Morgan's experiments with *Drosophila* convinced many geneticists that Sutton's hypothesis was correct.**

D. *Essay:* Tortoiseshell Cats, Barr Bodies, and the Lyon Hypothesis *(page 267)*

14. What is a Barr body and what does it represent?

15. State the Lyon hypothesis.

> *Focus:* **Because of the inactivation of one *X* chromosome in each somatic cell, every mammalian female is a genetic "mosaic."**

16. a. What determines the expression of a sex-linked characteristic in a mammalian female?

    b. Describe two examples that demonstrate the Lyon hypothesis in mammalian females.

III. **Broadening the Concept of the Gene** *(pages 268–273)*

A. **General Remarks** *(page 268)*

> *Focus:* Studies in Morgan's laboratory and in other centers revealed that patterns of inheritance are not nearly as simple and direct as Mendel's work suggested.

17. Identify three factors that influence the phenotypic expression of a particular gene.

B. **Allele Interactions** *(page 268)*

Incomplete Dominance and Codominance (page 268)

18. *Vocabulary:* Use examples to distinguish between incomplete dominance and codominance.

Multiple Alleles (page 268)

> *Focus:* A diploid organism can possess only two alleles for a particular gene, but more than two alleles for that gene may exist in a population.

19. How do multiple alleles arise?

20. Cite one example in which the multiple alleles for a gene exhibit (a) a dominant/recessive pattern and (b) a codominant pattern of inheritance.

C. **Gene Interactions** *(pages 268–271)*

> *Focus:* Most characteristics of an organism are due to the interaction of many genes.

Appearance of Novel Phenotypes (page 269)

21. Describe one example illustrating the appearance of a novel phenotype due to the interaction of genes.

Epistasis (pages 270–271)

22. a. *Vocabulary:* Define epistasis.

b. If gene *C* is epistatic to gene *D*, what is the result of this interaction?

c. Describe one example of epistatic gene interaction.

D. **Genes and the Environment** *(page 271)*

> *Focus:* The phenotypic expression of a gene is always influenced by environmental factors.

23. Cite three examples of this principle.

E. **Expressivity and Penetrance** *(page 271)*

24. a. *Vocabulary:* Define expressivity and penetrance.

b. Use examples to distinguish between these concepts.

F. **Polygenic Inheritance** *(pages 271–272)*

25. a. *Vocabulary:* Define polygenic inheritance and discuss its relationship to continuous variation.

b. Identify two examples of traits exhibiting continuous variation.

> *Focus:* Most normal human traits are thought to be polygenic.

G. **Pleiotropy** (*page 273*)

   26. a. *Vocabulary:* What is pleiotropy?

   b. Describe four examples of pleiotropy.

   27. *Review Question:* Match each situation (a to h) with the correct terminology (1 to 9). A term may be used more than once. A blank may have more than one answer, as indicated in parentheses.

   1. codominance
   2. environmental influence
   3. epistasis
   4. expressivity
   5. incomplete dominance
   6. multiple alleles
   7. penetrance
   8. pleiotropy
   9. polygenic inheritance

   a. _____ high mortality of white-eyed fruit flies compared with wild type

   b. _____ one member of a family has 6 fingers on each hand and 6 toes on each foot while another member of the same family just has 6 fingers on one hand

   c. _____ dark fur on the extremities of Siamese cats

   d. _____ the human ABO blood type alleles (2 answers)

   e. _____ crossing a red snapdragon and a white snapdragon produces pink snapdragons

   f. _____ parents with normal hands and feet have a child with 6 fingers on one hand

   g. _____ height in humans (2 answers)

   h. _____ F_1 purple pea plants produce purple and white offspring in a 9:7 ratio

IV. Genes and Chromosomes (*pages 274–277*)

A. **Linkage** (*page 274*)

   28. Mendel's law of independent assortment only applies if what situation exists?

   29. *Vocabulary:* What are linked genes and linkage groups?

   30. a. *Vocabulary:* Define a mutagen.

   b. Name three mutagens.

   c. Who first discovered the mutagenic effects of x-rays?

   > *Focus:* **In all organisms that have been studied, the number of linkage groups equals the number of chromosome pairs.**

B. **Recombination** (*pages 274–275*)

   31. a. In Morgan's studies of the simultaneous inheritance of body color and wing length, how did his results differ from what he expected?

   b. What happened when Morgan performed a testcross?

   c. These results led Morgan to what conclusion?

   32. Review the process of crossing over in Chapter 12 if necessary.

C. **Mapping the Chromosome** (*pages 275–277*)

   33. *Vocabulary:* What are loci on chromosomes?

34. Identify the three parts of Sturtevant's hypothesis.

35. a. *Vocabulary:* Define a map unit.

    b. Describe how recombination percentages are used to construct chromosome maps. (Figure 13–18)

    c. In terms of map units, at what point do genes on the same chromosome assort independently?

    d. What is the relationship between map units and the physical distance separating genes on a chromosome?

V. Abnormalities in Chromosome Structure *(pages 277–278)*

36. a. How are the giant chromosomes of *Drosophila* created and of what are they composed?

    b. The banding patterns of giant chromosomes are useful to geneticists for what reason?

37. a. *Vocabulary:* Distinguish between deletion, duplication, translocation, and inversion as applied to chromosome structure.

    b. Diagram and describe the shape of a chromosome that has undergone inversion. (Figure 13–21)

VI. **Summary** *(pages 278–279):* Read the summary. If you are familiar with the essential features of the material presented there, you are ready to complete the section TESTING YOUR UNDERSTANDING.

TESTING YOUR UNDERSTANDING

After you have completed this examination, compare your answers with those in the section that follows.

1. The most important of the genetic principles established by Morgan and his colleagues was that:
   a. genes are located on chromosomes.
   b. a genetic cross can be depicted in a Punnett square.
   c. not every gene on sex chromosomes is related to the sexual phenotype.
   d. most of the examples used by de Vries to illustrate his theory of mutation were invalid.
   e. the gene concept has no basis in physical reality.

2. In a certain chromosome, the normal gene sequence is ABCDEFGH. A scientist discovers a new sequence of ABFEDCDEFGH. It is apparent that _____ has (have) occurred.
   a. an inversion
   b. a deletion
   c. a translocation
   d. a duplication
   e. an inversion and a duplication

3. In humans, *Drosophila*, and grasshoppers, the female is the _____ sex.
   a. homozygous
   b. homogametic
   c. heterozygous
   d. heterogametic
   e. autosomal

4. If a red-eyed *Drosophila* female, heterozygous for white eyes, is mated to a red-eyed male, what will be the phenotypic ratios for sex and eye color in their offspring?
   a. All flies, both male and female, have red eyes.
   b. All the females have red eyes and the males have white eyes.
   c. All the males have red eyes; all the females have white eyes.
   d. All the males have red eyes; half the females have red eyes and half have white eyes.
   e. All the females have red eyes; half the males have red eyes and half the males have white eyes.

5. When defining dominant and recessive traits with respect to sex linkage, these terms only apply to alleles in:
   a. males.
   b. females.
   c. the hemizygous sex.
   d. the homogametic sex.
   e. the heterogametic sex.

6. In rabbits, agouti coat color (C) is dominant to chinchilla (c^{ch}), which is dominant to Himalayan (c^h), which is dominant to albino (c). If a male that is heterozygous for chinchilla and Himalayan is crossed with a female that is heterozygous for Himalayan and albino, their progeny will be _____ in a ratio of _____.
   a. Himalayan, chinchilla, and albino; 1:2:1
   b. chinchilla and Himalayan; 1:1
   c. chinchilla, Himalayan, and albino; 1:2:1
   d. Himalayan and albino; 3:1
   e. all chinchilla.

7. All of the following EXCEPT ONE relate to phenotypic expression due to the interaction of two or more genes. Identify the exception.
   a. incomplete dominance
   b. expressivity
   c. epistasis
   d. penetrance
   e. appearance of novel phenotypes

8. When a trait, such as height in humans, is continuously variable over a wide phenotypic range, it is reasonable to suspect that it is exhibiting:
   a. epistasis.
   b. polygenic inheritance.
   c. the effects of environmental interaction.
   d. variable expressivity.
   e. variable penetrance.

9. Primroses raised at room temperature produce red flowers, those raised at 86°F (30°C) are white. This is an example of:
   a. epistasis.
   b. the interaction of homologous alleles.
   c. the interaction of two or more different genes.
   d. the appearance of a novel phenotype.
   e. the effect of the environment on the expression of a phenotype.

10. Genes that occur on the same chromosome are said to be:
    a. epistatic.
    b. autosomal.
    c. homologous.
    d. linked.
    e. alleles.

11. What phenomenon is strong evidence for the presence of alleles at the same position on homologous chromosomes?
    a. Epistasis of an allele of one gene over an allele of another gene.
    b. Condensation of chromosomes at prophase I of meiosis.
    c. Recombination by crossing over of chromatids.
    d. The pleiotropic effect of one gene on many traits.
    e. Mutation at a given gene locus that results in the formation of multiple alleles.

12. Dominant alleles *P* and *S* are closely linked and do not separate during crossing over. Two parents are heterozygous for both of these genes; one chromosome is *PS*, the other is *ps*. What fraction of the offspring would express the recessive phenotype for both loci?
    a. 1/16
    b. 3/16
    c. 4/16 or 1/4
    d. 8/16 or 1/2
    e. 9/16

13. According to the chromosome map below, the pair of genes exhibiting the highest recombination rate is _____, and the pair with the lowest recombination rate is _____.

A	B	C

    a. A — C; B — C
    b. A — B; B — C
    c. A — C; A — B
    d. A — B; A — C
    e. B — C; A — B

14. T or F In humans, *Drosophila*, and grasshoppers, it is the male that determines the sex of the offspring by the chromosomal composition of the sperm.

15. T or F Barr bodies can be found in approximately half of the somatic cells of a mammalian female because of chromosome inactivation.

16. T or F In chickens, the appearance of the new phenotype, walnut comb, resulting from the interaction of the pea and rose genes, is a good example of epistasis.

17. **T or F** Most of the characteristics that constitute the phenotype of an organism result from the interaction of many different genes.

18. Describe the process by which larval *Drosophila* acquire giant chromosomes in their salivary glands.

19. Distinguish between expressivity and penetrance of a gene, citing one example of each phenomenon.

PERFORMANCE ANALYSIS

1. **a** The most important principle established by Morgan and his coworkers was that genes are located on chromosomes. (page 263)

2. **e** From the FED sequence, it is apparent that an inversion has occurred. It is also apparent that a duplication has occurred since the sequence DEF appears in the same chromosome. (pages 277, 278)

3. **b** In these three species, the genotype of the female is *XX*. She can only produce gametes with *X* chromosomes and is therefore termed homogametic. (page 264)

4. **e** In sex-linked inheritance, a recessive allele may be masked by the dominant allele in the homogametic sex but is always expressed when present in the heterogametic sex. All female offspring from this cross will have red eyes since they receive one red allele from their red-eyed father. Half the males will have white eyes since half the maternal gametes contain the allele for white eyes and half contain the allele for red eyes. (pages 265, 266)

5. **d** A dominant allele is one whose expression in the heterozygous state masks that of the recessive allele. For sex-linked genes, these terms only have meaning in the homogametic sex, since this is the only sex in which the heterozygous state exists. In the heterogametic sex, a gene present on a sex chromosome is expressed whether it is dominant or recessive in the homogametic sex. (page 266)

6. **b** The parent genotypes are $c^{ch}c^h$ for the male and c^hc for the female. Paternal gametes are c^{ch} and c^h. Maternal gametes are c^h and c. Offspring phenotypes are 2 chinchilla ($c^{ch}c^h$ and $c^{ch}c$) and 2 Himalayan (c^hc^h and c^hc) for a phenotypic ratio of 1:1. (Figure 13–9)

7. **a** Mechanisms by which the phenotypic expression of a gene may be influenced by other genes include epistasis, the appearance of novel phenotypes due to different gene combinations, expressivity, and penetrance. Incomplete dominance involves interactions between the alleles of a single gene. (pages 268–271)

8. **b** Characteristics that exhibit continuous variation over a wide phenotypic range are typically inherited by polygenic mechanisms. Such traits include height and skin color in humans, and color in wheat kernels. In some cases, environmental influence may modify the expression of polygenic traits (e.g., malnutrition results in stunted growth; exposure to ultraviolet light darkens skin). Variations in expressivity and penetrance may mimic polygenic inheritance in some cases. (page 271)

9. **e** The effect of temperature on a characteristic is clearly a case of environmental influence on the expression of a gene. (page 271)

10. **d** By definition, genes located on the same chromosome are said to be linked, or in the same linkage group. (page 274)

11. **c** The recombination of genes during crossing over is strong evidence that genes are located in specific positions or loci on homologous chromosomes. (page 275)

12. **c** If two genes do not recombine during crossing over, they travel through meiosis in the same manner as a single gene. Therefore, one fourth of the offspring will be homozygous recessive for both traits. (page 276)

13. **a** Recombination frequency is related to the distance between two genes on a chromosome. The higher the recombination frequency, the greater the distance between the two genes. (pages 276, 277)

14. **True** In these three species, and most vertebrates except birds, males are heterogametic and determine the sex of the offspring. (page 264)

15. **False** A Barr body is present in *all* somatic cells of mammalian females and represents an inactivated sex chromosome. Each somatic cell therefore has a single active sex chromosome. (essay, page 267)

16. **False** The walnut comb phenotype in chickens is an example of the appearance of a novel phenotype from a combination of existing genes. In epistasis, the expression of one gene is masked by the presence of another gene. (page 269)

17. **True** Most of the characteristics exhibited by an organism result from interactions of multiple genes. (page 268)

18. Larval *Drosophila* acquire giant chromosomes in their salivary glands because the chromosomes replicate repeatedly but the cells do not divide. Also, homologous chromosomes are tightly paired, increasing the size. The banding patterns that characterize giant chromosomes facilitate the identification of altered chromosome structure. (page 277)

19. Expressivity of a gene refers to the degree to which a particular genotype is expressed in an *individual*. An example is polydactyly in humans, in which some people have extra digits on only one limb while other people may have extra digits on more than one limb. Penetrance refers to the degree to which a gene is phenotypically expressed in a *population*. If the gene for polydactyly is known to be present in 100 individuals in a population, but only 90 of these people have extra digits, the gene is showing incomplete penetrance. (page 271)

REVIEW TEST 4

Section 3: Chapters 11–13

This review test is *not* designed to cover all of the important information in these chapters. However, it does touch on the major topic categories in each chapter. It will also give you valuable practice in taking this type of test. When you are finished, check your answers with those provided in the following PERFORMANCE ANALYSIS section.

1. The Greek philosopher Theophrastus exhibited his understanding of inheritance when he:
   a. said "the males should be brought to the females."
   b. produced the first mule by breeding a donkey and a horse.
   c. created the myth about the origin of the Minotaur.
   d. claimed that a giraffe came from crossing a camel with a leopard.
   e. said "Hapsburg after Hapsburg, the lips are all the same."

2. In ferns, _____ sporophytes produce gametes by _____.
   a. haploid; mitosis
   b. haploid; meiosis
   c. diploid; mitosis
   d. diploid; meiosis
   e. In ferns, gametophytes produce the gametes and sporophytes produce the spores.

3. Which of these is NOT an example of pleiotropy?
   a. deafness in blue-eyed, white cats
   b. different shades of skin color in humans
   c. increased mortality of white-eyed male fruit flies
   d. a gene in rats affecting the ribs, the tracheal passages, and the snout
   e. physiological abnormalities in chickens carrying the frizzled gene

4. The insights of three men were needed to complete a workable mechanism for evolution. Who were they?
   a. de Vries, Mendel, and van Helmont
   b. van Helmont, Darwin, and Mendel
   c. Mendel, Darwin, and de Vries
   d. Darwin, Pasteur, and Mendel
   e. de Vries, Pasteur, and Darwin

5. As late as 1916, _____ and others were still not convinced by the work of _____ that genes were on chromosomes.
   a. Bateson; de Vries and Hertwig
   b. Punnett; Sutton and de Vries
   c. Hertwig; Bateson and Morgan
   d. Bateson; Morgan and Sutton
   e. Punnett; Hertwig and Sutton

6. In birds, butterflies, moths, and a few other species, the male is the _____ sex.
   a. autosomal
   b. autologous
   c. heterogametic
   d. homogametic
   e. homologous

7. In the grasses from which wheat was derived, the diploid number of chromosomes is 14. How many chromosomes are present in polyploid wheat?
   a. 7
   b. 14
   c. 21
   d. 28
   e. The answer cannot be determined from the information provided.

8. Mendel's discovery of the nature of heredity was so revolutionary that it was:
   a. not understood until after his death.
   b. recognized by only a few of the scientists of his day.
   c. not publishable in the scientific journals of his day.
   d. widely acclaimed as a breakthrough in the field of genetics.
   e. the work for which he received the Nobel Prize.

9. In an organism of genotype $RrYy$, Mendel's principle of independent assortment predicts that which alleles will segregate independently?
   a. R and r
   b. Y and y
   c. R (or r) and Y (or y)
   d. Rr and Yy
   e. RY and ry

10. **T** or **F**   The F_2 progeny from both of these crosses will have the same genotypic and phenotypic ratios:

    $$RRYY \times rryy \text{ and } rrYY \times RRyy$$

(These are the parental generations, remember to consider the F_1 generation when you answer.)

11. **T or F** Metaphase I of meiosis and metaphase of mitosis are essentially identical in that homologous chromosomes line up together at the equator of the cell.

12. **T or F** The degree to which two genes are linked can be determined by the degree to which the results of a testcross deviate from the 1:1:1:1 ratio expected from independently assorting genes.

13. Distinguish between incomplete dominance and codominance, citing one example of each mode of inheritance.

14. Outline a single experimental cross (or perhaps a set of crosses) that could clearly distinguish between Mendelian inheritance and blending inheritance.

15. Summarize the observations made by Walter Sutton in his studies of grasshopper meiosis and trace the line of reasoning that led to the formation of his hypothesis.

PERFORMANCE ANALYSIS

1. **a** Theophrastus was a Greek philosopher and naturalist whose understanding of male and female flowers was evident from his writings: "The males should be brought to the females for the male makes them ripen and persist." (page 236)

2. **e** In alternation of generations in ferns, the diploid sporophyte undergoes meiosis to produce haploid spores, which develop into haploid gametophytes. Gametes are produced in the gametophyte by mitosis. In the process of fertilization, gametes fuse to form diploid zygotes, which develop into sporophytes. (page 250)

3. **b** Pleiotropy is a phenomenon in which one gene affects many different characteristics. All the examples are valid except for skin color in humans, which is determined by polygenic inheritance. (page 273)

4. **c** Mendel's discovery of the discrete nature of heredity and de Vries's concept of mutations filled in the gaps in Darwin's theory of evolution. Their work provided a mechanism by which the variations upon which natural selection acts arise and are passed on from generation to generation. (pages 245, 246)

5. **d** Many scientists, including William Bateson, were slow to believe that particles of chromatin with their uniform appearance could carry hereditary information responsible for the great diversity of living organisms. (page 263)

6. **d** The homogametic sex produces only one type of gamete with respect to sex chromosomes. The heterogametic sex produces two types of gametes. In birds, butterflies, moths, and a few other species, the sex genotype of females is XY and of males is XX. In other vertebrates, including mammals, the reverse is true: males are XY and females are XX. (page 264)

7. **e** Polyploid cells (and organisms) have more than two sets of chromosomes, but the number will vary depending on the species. The number of chromosomes present in wheat cannot be determined without knowing the ploidy, or number of sets present. (page 249)

8. **a** Mendel reported his research at scientific meetings and his work was published in a scientific journal. However, his contemporaries did not comprehend the significance of his work and it slipped into obscurity until it was discovered 35 years later. (page 246)

9. **c** The principle of independent assortment predicts that the alleles for each characteristic separate independently of one another. Thus R (or r) separates into gametes independently of Y (or y). In reality, this is only true for genes located on different chromosomes or for genes that are more than 50 map units apart and separate during crossing over. (pages 242, 274, 276)

10. **True** The F_1 genotype for both crosses is $RrYy$. Therefore, the F_2 generations will have identical phenotypic and genotypic ratios. (page 243)

11. **False** In mitosis, *replicated* chromosomes line up along the equator but homologues are not paired, as they are in meiosis. (page 254)

12. **True** In a testcross involving two genes, if the genes are on different chromosomes the expected phenotypic ratio is 1:1:1:1. If the genes are on the same chromosome, the phenotypic ratio may be altered depending upon the distance between the genes. If they are very close together, the offspring may exhibit only two phenotypes with a ratio of 3:1, indicating that crossing over did not separate the genes. If crossing over separates the genes, four phenotypes will appear in the offspring but the ratio will not be 1:1:1:1 unless the recombination frequency is greater than 50 percent, in which case the genes will act as if they are not linked. (page 276)

13. For a trait inherited by incomplete dominance, the phenotype of the heterozygote is intermediate between the phenotypes of the homozygotes. An example is pink flower color in snapdragons. In codominance, both alleles are expressed in the heterozygote. Human ABO blood types are determined by multiple alleles with codominant inheritance. (page 268)

14. Self-pollination of an F_1 plant (or crossing two F_1 animals) should do the trick. Set up an example and see for yourself. Blending inheritance would not allow the recovery of the recessive phenotype in the F_2 generation. (pages 237, 239)

15. Walter Sutton noticed that chromosomes are paired at the beginning of meiosis and that the two chromosomes in a pair resemble one another physically. The parallel between the separation of homologous chromosomes in meiosis and Mendel's principle of segregation of alleles led him to speculate that genes (Mendel's *Elemente*) are located on chromosomes. (pages 260, 261)

The Chemical Basis of Heredity: The Double Helix

CHAPTER ORGANIZATION

MAJOR CONCEPTS

Scientists realized that the material containing the hereditary information must meet certain criteria. It must be able to contain a vast amount of information, and to transmit this information accurately to the next generation. The mechanism of replication must be very accurate since replication occurs at every mitotic division. The hereditary material must be chemically stable. It must be capable of mutation.

The first step toward elucidating the molecular basis of heredity was to determine the composition of chromosomes. Eukaryotic chromosomes contain approximately equal amounts of proteins and DNA. Because of the complexity of proteins compared with nucleic acids, early investigators suspected that proteins carried the hereditary information.

In his efforts to develop a vaccine, Frederick Griffith studied two strains of the bacterium *Streptococcus pneumoniae*, one of which was encapsulated and virulent and the other of which was unencapsulated and nonvirulent. He discovered that extracts from killed encapsulated bacteria could convert live nonencapsulated bacteria into virulent encapsulated bacteria. The "transforming factor" in these extracts was later identified by O. T. Avery as DNA.

Delbrück and Luria characterized replication of bacteriophages. Expanding on their work, Hershey and Chase conducted a series of experiments that provided conclusive evidence that DNA carries the hereditary information. They used two groups of virus particles, one having DNA labeled with radioactive phosphorus and the other having protein labeled with radioactive sulfur. They demonstrated that the protein coat stayed outside the infected bacterial cell while the DNA core entered the cell and apparently directed viral replication.

Several scientists contributed information that led to elucidation of the structure of DNA. Alfred Mirsky discovered that all somatic cells of a single species contain the same amount of DNA, and that gametes contain exactly half this amount. Levene concluded (incorrectly) that the four nitrogenous bases of DNA occur in equal proportions. Erwin Chargaff revealed that the relative quantities of nitrogenous bases are consistent within a species but vary between species. He also discovered that the amount of adenine equals that of thymine and the amount of guanine equals that of cytosine. Maurice Wilkins and Rosalind Franklin revealed the helical configuration of DNA by means of x-ray diffraction studies.

James Watson and Francis Crick were the first to correctly model the structure of DNA. Their model was constructed after they collected and analyzed all the known information on DNA.

DNA exists as a helix resembling a twisted ladder. The linear side pieces are composed of alternating sugar and phosphate units of the nucleotides. Each rung is a pair of nitrogenous bases; one purine and one pyrimidine constitute each pair. The pairing of nitrogenous bases is maintained by hydrogen bonds and is specific; adenine only pairs with thymine and guanine only pairs with cytosine. Thus, the two strands of a DNA helix are complementary.

The ends of a DNA strand are identified as 5′ or 3′ depending on whether the phosphate group is attached to the fifth or third carbon in the sugar ring. The complementary strands of a DNA helix are antiparallel, meaning that the 5′ end of one strand apposes the 3′ end of the other strand.

During DNA replication, the hydrogen bonds between the two strands break and the strands separate. Each strand serves as a template directing the synthesis of a new strand. Replication is semiconservative: a daughter DNA molecule contains one original (parent) strand and one newly synthesized strand.

In order for DNA replication to commence, an RNA primer must be hydrogen bonded at the origin of replication. The addition of nucleotides by DNA polymerases always occurs in the 5′ to 3′ direction (nucleotides are added at the 3′ end). One of the new strands (leading strand) is synthesized in a continuous chain. The antiparallel strand (lagging strand) is synthesized in segments known as Okazaki fragments, each of which is lengthened by nucleotide addition at the 3′ end. These segments are then joined by DNA ligase to form the complete strand. One other function of DNA polymerases is proofreading: new nucleotides are only added if the previous nucleotides correctly complement the template. Several other enzymes perform various functions associated with DNA replication.

Nucleotides are assembled as triphosphates. The loss of $P \sim P$ when the nucleotide is added to a growing DNA chain and its subsequent hydrolysis to two phosphates make the reaction strongly exergonic and less likely to be reversed.

HOW TO STUDY THE CHAPTER

Read the entire chapter through quickly, focusing on the major concepts.

Use the GUIDED STUDY OF THE CHAPTER to help you identify the important details as you **reread** the chapter. Writing out the answers to these questions will help fix them in your mind as well as provide you with a valuable study aid.

Answer the questions in TESTING YOUR UNDERSTANDING without the aid of your text. Check your answers against those in PERFORMANCE ANALYSIS. Analyzing your answers will give you valuable feedback on your level of understanding and preparedness for classroom testing.

GUIDED STUDY OF THE CHAPTER

I. Introduction *(page 281)*

> *Focus:* **A major turning point in genetics occurred when scientists focused their attention on discovering the molecular composition of the hereditary material.**

II. The Chemistry of Heredity *(page 281)*

A. General Remarks *(page 281)*

> *Focus:* **Those scientists who believed that the nature of heredity at the molecular level could be unraveled thought that understanding the chemical nature of chromosomes was an important first step.**

1. *Vocabulary:* What does the study of molecular genetics include?

B. The Language of Life *(page 281)*

2. a. What did early chemical analyses of chromosomes reveal about their composition?

   b. Of these two components, which was originally thought to be the most likely candidate for the carrier of genetic information and why?

> *Focus:* **Many early investigators believed that the protein component of DNA contained the genetic information and that chromosomes contained models of all the proteins needed by the cell.**

III. The DNA Trail *(pages 282–287)*

A. Sugar-Coated Microbes and the Transforming Factor *(page 282)*

3. a. Distinguish between the virulent and nonvirulent strains of *Streptococcus pneumoniae* that were studied by Frederick Griffith.

b. What was the purpose of Griffith's studies?

4. Summarize the experiment in which Griffith became aware that hereditary information could be transmitted between two organisms (other than by parent to offspring transmission).

5. a. *Vocabulary:* Define transformation.

   b. O. T. Avery discovered that ———————— is the transforming factor.

B. **The Nature of DNA** *(pages 282–284)*

6. a. Who first isolated DNA?

   b. What did he name this compound?

   c. The name was later changed (twice) for what reason?

7. How did Robert Feulgen contribute to studies of DNA?

8. a. What did P. A. Levene discover about the components of DNA?

   b. Paraphrase his two deductions about DNA structure and indicate which was correct and which was incorrect.

9. Cite two reasons why Avery's discovery of DNA as the transforming factor was not immediately embraced by scientists.

C. **The Bacteriophage Experiments** *(pages 284–286)*

10. a. Describe the organisms that Delbrück and Luria used for their genetic studies.

    b. Identify five characteristics of bacteriophages that make them well suited for genetic studies.

11. a. How does the multiplication of bacteriophages differ from that of bacterial cells?

    b. Name a sixth characteristic of bacteriophages that made them ideal for genetic studies.

12. a. How did Hershey and Chase label viral DNA and viral protein so that they could be distinguished?

    b. Describe the means by which Hershey and Chase established that only the phage DNA enters an *E. coli* cell. (Include the mechanism by which a T_4 phage infects a bacterial cell.)

    c. What conclusions did these scientists draw based on these observations?

    d. What technique has now been used to confirm their conclusions?

D. Further Evidence for DNA *(pages 286–287)*

13. a. What contribution to elucidation of the hereditary material was made by Alfred Mirsky?

    b. How was his finding consistent with the observed products of meiosis?

Chargaff's Results (pages 286–287)

14. How did Chargaff's findings conflict with Levene's conclusions?

15. What pattern is evident in the data presented in Table 14–1? (Look at the figures for one species at a time.)

E. The Hypothesis Is Confirmed *(page 287)*

16. List the four requirements that must be met by the genetic material.

17. Identify three physical characteristics of the DNA molecule that eventually led to its universal acceptance as the carrier of hereditary information.

IV. The Watson-Crick Model *(pages 288–291)*

A. General Remarks *(page 288)*

> *Focus:* James Watson and Francis Crick gathered together all the known information about DNA, formulated the first correct model of its structure, and suggested a mechanism by which it might be replicated.

B. The Known Data *(page 288)*

> *Focus:* DNA was known to be a large molecule composed of nucleotides containing the nitrogenous bases adenine, cytosine, guanine, and thymine.

18. Summarize the information on DNA that was provided by these scientists.

    Levene

    Pauling

    Wilkins and Franklin

    Chargaff

C. Building the Model *(pages 288–290)*

19. Name three requirements of DNA structure in order for it to fulfill its biological role.

20. a. Diagram the basic structure of the DNA molecule, showing the relative positions of these components: sugar, phosphate, purines, pyrimidines.

    b. How does the bonding of the nitrogenous bases to each other differ from their binding to the sugar-phosphate unit?

    c. Summarize the reason, in molecular terms, that a purine always binds to a pyrimidine and, specifically, that thymine always binds to adenine and guanine to cytosine.

21. *Vocabulary:* Distinguish between complementary and antiparallel.

D. *Essay:* **Who Might Have Discovered It?** *(page 291)*

22. When Watson and Crick disclosed their model for the structure of DNA, how close to elucidating the correct structure was each of the following scientists?

    Linus Pauling

    Rosalind Franklin

    Maurice Wilkins

23. What is Gunther Stent's opinion on what might have happened had Watson and Crick not determined the structure of DNA when they did?

24. What does Crick think is the most important aspect of the history of the discovery of the structure of DNA?

V. **DNA Replication** *(pages 292–298)*

A. **General Remarks** *(page 292)*

25. What is semiconservative replication?

B. **A Confirmation of Semiconservative Replication** *(pages 292–294)*

26. a. Name and distinguish between the three theoretically possible mechanisms of DNA replication. (Figure 14–13)

    b. Who performed the experiments that elucidated the correct mechanism?

27. a. Describe how ultracentrifuging in a cesium chloride solution can separate DNA of different densities.

    b. How did Meselson and Stahl create "heavy" DNA for their experiments?

    c. Outline the experiment performed by Meselson and Stahl that confirmed the semiconservative mechanism of DNA replication.

C. **The Mechanics of DNA Replication** *(pages 294–298)*

28. DNA synthesis occurs during what stage of the cell cycle?

29. How does the rate of DNA synthesis in mammals compare with that in prokaryotes?

30. a. *Vocabulary:* Define the origin of replication.

b. What two components are needed to initiate DNA replication?

31. a. By what mechanism is supercoiling of the DNA helix prevented?

b. Once the DNA strands are separated, what holds them apart and prevents kinking?

> *Focus:* DNA polymerases catalyze the addition of nucleotides in the synthesis of new strands.

32. *Vocabulary:* Distinguish between replication bubble and replication fork.

33. How does DNA replication in prokaryotes differ from that in eukaryotes?

RNA Primers and the Direction of Synthesis (pages 296–297)

> *Focus:* Before DNA polymerases can begin adding nucleotides, an RNA primer must be hydrogen bonded to the template strand.

34. Summarize nucleotide pairing between RNA and DNA strands.

35. What is the function of RNA primase?

> *Focus:* DNA polymerases add nucleotides only to the 3′ end of the growing chain so that the new DNA strand grows in the 5′ to 3′ direction.

36. a. *Vocabulary:* Distinguish between the leading strand and the lagging strand.

b. What are Okazaki fragments?

37. a. What is the role of the RNA primers?

b. Where are the primers located for the leading strand?

For the lagging strand?

38. What happens when the growing end of an Okazaki fragment encounters the RNA primer of the adjacent fragment? (Figure 14–18)

39. Identify the function of DNA ligase.

Proofreading (pages 297–298)

> *Focus:* DNA polymerase can only add nucleotides to the 3′ end of a growing strand if the previously added nucleotides are correct complements of the template.

40. What happens if a mistake is made during DNA replication?

41. What is the function of DNA repair enzymes?

D. The Energetics of DNA Replication *(page 298)*

42. a. The nucleotides used in DNA synthesis are assembled in what form?

b. What supplies the energy needed by DNA polymerase to catalyze its reactions?

43. What aspect of DNA replication prevents the reversal of the DNA polymerase reaction, with the loss of newly added nucleotides from the growing chain?

VI. DNA as a Carrier of Information *(page 298)*

> *Focus:* **The structure of the DNA molecule as revealed by Watson and Crick has ample capacity for storing the genetic information needed by living organisms.**

44. Compare the number of base pairs in viral DNA and in human DNA.

VII. Summary *(pages 299–300):* Read the summary. If you are familiar with the essential features of the material presented there, you are ready to complete the section TESTING YOUR UNDERSTANDING.

TESTING YOUR UNDERSTANDING

After you have completed this examination, compare your answers with those in the section that follows.

1. The staining of DNA with the red dye fuchsin was discovered by _____ in 1914.
a. O. T. Avery
b. Friedrich Miescher
c. Robert Feulgen
d. P. A. Levene
e. Alfred Mirsky

2. P. A. Levene incorrectly deduced that DNA contained:
a. four different nitrogenous bases.
b. four different nucleotides.
c. phosphate groups attached to sugars.
d. a five-carbon sugar.
e. a regularly repeating series of four nucleotides.

3. The process by which nonvirulent live *Streptococcus pneumoniae* can become virulent in the presence of heat-killed virulent *Streptococcus pneumoniae* is called:
a. conversion.
b. transduction.
c. transformation.
d. mutation.
e. transvection.

4. In 1952, whose landmark experiments showed that DNA and not protein carries the genetic information of T-even bacteriophages?
a. Alfred Hershey and Martha Chase
b. Francis Crick and James Watson
c. Max Delbrück and Salvador Luria
d. Friedrich Miescher and Robert Feulgen
e. Maurice Wilkins and Rosalind Franklin

5. Which scientist demonstrated that the somatic cells of any given species contain equal amounts of DNA, whereas the gametes contain half as much DNA?
a. Maurice Wilkins
b. Alfred Mirsky
c. Martha Chase
d. Max Delbrück
e. Erwin Chargaff

6. Chargaff's experimental findings were important in documenting DNA as the genetic material because they:
a. confirmed the tetranucleotide deduction of Levene.
b. implied that DNA could carry great quantities of information.
c. formed the basis for the x-ray diffraction studies of Franklin and Wilkins.
d. indicated that purines bonded with purines, and pyrimidines with pyrimidines.
e. showed that all organisms contain equal proportions of the four nitrogenous bases in their DNA.

7. Which base pairing is correct?
a. A to C with 3 hydrogen bonds
b. C to G with 2 hydrogen bonds
c. G to T with 2 hydrogen bonds
d. A to T with 2 hydrogen bonds
e. T to C with 3 hydrogen bonds

8. Which of the following is NOT a characteristic of DNA?
a. It is a double-stranded molecule.
b. There is directionality to the molecule.
c. The two strands are antiparallel in orientation.
d. The backbones of the strands are composed of alternating, covalently linked phosphate groups and sugars.
e. The bridges between the two strands are formed by covalent bonds between nitrogenous bases.

9. The nitrogenous bases _____ are composed of one nitrogenous ring and are known as:
a. adenine and guanine; purines
b. guanine and cytosine; pyrimidines
c. thymine and adenine; purines
d. adenine and cytosine; purines
e. cytosine and thymine; pyrimidines

10. Prokaryotes are able to replicate their DNA at a rate of _____ nucleotides per second.

a. 50   d. 500
b. 100   e. 1000
c. 250

11. Enzymes that nick the DNA molecule and allow it to swivel to relieve strain in the molecule during replication are called:
    a. helicases.   d. topoisomerases.
    b. DNA polymerases. e. exonucleases.
    c. RNA polymerases.

12. Okazaki fragments are formed during DNA replication because:
    a. DNA polymerase cannot work in a 3′ to 5′ direction.
    b. RNA primers are only needed on the leading strand.
    c. RNA primers are only needed on the lagging strand.
    d. The DNA does not unwind fast enough to permit continuous synthesis.
    e. RNA primase cannot work in a 3′ to 5′ direction.

13. Another name for the region of DNA synthesis or "eye" of a replicating DNA molecule is the:
    a. leading strand.   d. replication fork.
    b. lagging strand.   e. theta structure.
    c. replication bubble.

14. DNA polymerases:
    a. synthesize DNA in a 5′ to 3′ direction.
    b. unwind the DNA helix during replication.
    c. seal together the loose ends of DNA strands.
    d. join the new and old strands after replication.
    e. link together the Okazaki fragments of the lagging strand.

15. Proofreading of DNA synthesis is done by:
    a. proofase as it moves in a 5′ to 3′ direction some 10 nucleotides behind DNA polymerase.
    b. proofase as it rides on the back of DNA polymerase.
    c. DNA polymerase as it backtracks (3′ to 5′) to remove incorrectly paired bases.
    d. a second DNA polymerase that moves in the same direction, but lags 10 nucleotides behind the first polymerase.
    e. DNA ligase as it rides along a strand of DNA looking for breaks to repair.

16. **T or F** Some scientists believed proteins were more likely to carry genetic information than DNA because the 20 amino acids could be arranged in many different ways.

17. **T or F** Upon chemical analysis, T-even bacteriophages were shown to consists of proteins, DNA, and ribosomes.

18. **T or F** The x-ray diffraction photographs taken by Maurice Wilkins and Rosalind Franklin revealed the helical nature of the DNA molecule.

19. **T or F** Watson and Crick hypothesized that DNA would replicate itself in a semiconservative manner, such that each strand of the existing double helix would act as a template to direct the synthesis of the other strand of the daughter molecule.

20. The importance of O. T. Avery's demonstration that DNA is the genetic material in bacteria was not readily recognized by biologists. Cite two reasons for this phenomenon, one of which relates to the work of P. A. Levene.

21. Summarize the five basic facts about DNA that Watson and Crick used as the foundation for their model of DNA structure.

22. Describe the Meselson-Stahl experiments. Include in your discussion the logic that supported these experiments.

23. Label the following portions in the schematic of DNA replication: leading strand, lagging strand, Okazaki fragments, RNA primer, and sites of action of RNA primase, DNA polymerase, DNA ligase, helicase, topoisomerase. Indicate the overall direction of replication.

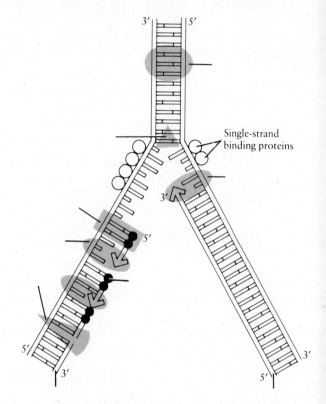

Single-strand binding proteins

PERFORMANCE ANALYSIS

1. **c** Robert Feulgen discovered that DNA is strongly stained when the dye fuchsin is applied to cells. (page 282)

2. **e** P. A. Levene correctly deduced the structural relationship of the nitrogenous base, sugar, and phosphate groups in nucleotides. However, he incorrectly concluded that DNA is composed of repeating sequences of the four nucleotides, units that he termed tetranucleotides. (page 284)

3. **c** The process by which living nonvirulent bacteria become virulent when incubated with nonliving virulent bacteria, as carried out in a living mouse in Griffith's

experiments, is termed transformation. O. T. Avery demonstrated that the transforming factor in these experiments was DNA. (page 282)

4. **a** The experiments of Hershey and Chase on T-even bacteriophages demonstrated that the DNA component of the viruses carries the hereditary information. (page 285)

5. **b** Alfred Mirsky revealed that, in any species, the somatic cells contain equal amounts of DNA and the gametes contain only half this amount. This fact is consistent with DNA as the carrier of hereditary information. (page 286)

6. **b** Chargaff's work, which refuted Levene's tetranucleotide proposal, was important because it showed that the DNA molecule is capable of great variation in the sequence of its nucleotides. This implied that DNA could carry the hereditary information. (pages 286, 287)

7. **d** Adenine base pairs with thymine with two hydrogen bonds. Guanine and cytosine base pair with three hydrogen bonds. (page 289)

8. **e** The two strands of the DNA molecule are held together by hydrogen bonds between purines and pyrimidines. In the ladder analogy, one purine and one pyrimidine constitute each rung of the ladder. (pages 288, 289)

9. **e** The pyrimidines, cytosine and thymine, contain one nitrogenous ring. The purines, adenine and guanine, have a double ring structure. (Figure 14–11)

10. **d** In prokaryotes, DNA replication can proceed at a rate of 500 nucleotides per second, a rate 10 times faster than that in mammals. (page 294)

11. **d** Topoisomerases break and reconnect one or both DNA strands during replication to relieve the stress of uncoiling when the strands are separated. (page 294)

12. **a** Okazaki fragments form during DNA replication because DNA polymerase cannot work in the 3' to 5' direction. The 3' to 5' (lagging) strand is synthesized in a series of 5' to 3' segments (Okazaki fragments) which are joined together by DNA ligase. (pages 296, 297)

13. **c** An active region of DNA synthesis in a replicating DNA molecule is called a replication bubble. A replication bubble arises at each origin of replication. A single eukaroytic chromosome may have many bubbles during replication but prokaryotes have only one origin of replication, and one replication bubble. (page 295)

14. **a** DNA polymerases have two major functions in DNA replication: addition of nucleotides to the new strand in the 5' to 3' direction, and proofreading with removal and replacement of incorrect nucleotides. (page 296)

15. **c** See the answer to question 14. (page 298)

16. **True** The number of amino acids (so similar to the number of letters in the English alphabet) suggested to some scientists that amino acids might be the "letters in the language of heredity." Further evidence that favored

proteins early in the search for the genetic material was that proteins exhibit great variety, whereas chromatin was observed to be a relatively homogeneous substance. (page 287)

17. **False** T-even bacteriophages consist of protein coats surrounding DNA cores. (page 285)

18. **True** Linus Pauling had speculated that DNA might have a helical structure, and the x-ray diffraction studies of Wilkins and Franklin confirmed this. (page 288)

19. **True** Watson and Crick hypothesized that DNA would replicate by a semiconservative mechanism, and the work of Meselson and Stahl confirmed this. (page 292)

20. Avery's work on the transforming factor was not readily accepted because prokaryotes were considered "lower" (and therefore different) forms of life and because the DNA molecule appeared too simple to contain the genetic information. Also, Levene was so respected as a biochemist that his tetranucleotide hypothesis (which supported the perception of DNA as a homogeneous substance) was considered true until it was decisively disproven. (page 284)

21. (1) DNA is a long thin molecule composed of nucleotides containing the four nitrogenous bases adenine, guanine, cytosine, and thymine. (2) Levene's hypothesis stated that the nucleotides were arranged in units of four repeating nucleotides. (3) Linus Pauling had suggested that DNA might have a helical structure. (4) The x-ray diffraction studies of Wilkins and Franklin confirmed the helical nature of DNA. (5) Chargaff's work revealed that the amount of guanine equals that of cytosine and the amount of adenine equals that of thymine. (page 288)

22. The experiments of Meselson and Stahl were designed to distinguish among three possible mechanisms for DNA replication: conservative, semiconservative, and dispersive. The researchers grew *E. coli* in a medium containing ^{15}N (heavy nitrogen) for several generations. When centrifuged in a cesium chloride density gradient, heavy DNA forms a band at a point that differs from the band formed by unlabeled, or light, DNA. (Light DNA contains the most common isotope of nitrogen, ^{14}N.) The bacteria with heavy DNA were then grown in a medium containing light nitrogen for one generation. The DNA isolated from these cells formed a band in the cesium chloride gradient at a density intermediate between that of heavy DNA and light DNA, suggesting that the daughter DNA contained one heavy and one light strand. After a second generation, two bands appeared, one that matched the band of the first generation (i.e., intermediate density) and one corresponding to light DNA. These data supported the semiconservative mechanism of replication. (pages 292, 293 and Figure 14–14)

23. Check your labels with Figure 14–19. (page 297)

The Genetic Code and Its Translation

CHAPTER ORGANIZATION

MAJOR CONCEPTS

Soon after their model of DNA structure was published, Watson and Crick correctly speculated that the sequence of nitrogenous bases constitutes the code carrying the genetic information.

Decades before the structure of DNA and its role as the hereditary material were elucidated, Sir Archibald Garrod postulated that certain diseases may be caused by hereditary enzyme deficiencies. In their studies of the red bread mold, Beadle and Tatum demonstrated that mutations may cause enzyme deficiencies that affect biochemical processes. They hypothesized that one gene codes for the production of one enzyme. As knowledge increased, this concept was modified to; one gene codes for the production of one polypeptide.

Linus Pauling postulated that a variation in the structure of hemoglobin might be the cause of sickle cell anemia. His electrophoretic studies confirmed this. Later studies by other investigators showed that the difference between normal and sickle cell hemoglobin is one amino acid substitution, which corresponds to one nitrogenous base substitution in the gene.

The central dogma of molecular genetics is that DNA specifies the production of RNA, which in turn specifies protein synthesis.

RNA differs from DNA in three major ways. (1) RNA contains ribose instead of deoxyribose. (2) RNA has uracil instead of thymine. (3) RNA is generally single-stranded.

Several clues pointed toward the role of RNA in protein synthesis. RNA is found mainly in the cytoplasm, the site of protein synthesis. Cells making large quantities of protein have high levels of RNA. When a DNA-containing bacteriophage infects a cell, RNA is synthesized before protein synthesis begins.

Protein synthesis is accomplished by the activities of three types of RNA and a number of enzymes. Messenger RNA (mRNA) provides the genetic code for protein synthesis. Ribosomes are the sites of protein synthesis and are composed of ribosomal RNA and protein. Transfer RNA (tRNA) carries amino acids to the ribosomes and positions them in correct orientation for peptide bond formation. A tRNA molecule has an amino acid binding site and an anticodon, which binds to the codon of mRNA.

The two major phases of protein synthesis are transcription and translation. During transcription, messenger RNA complementary to a single-strand DNA template is synthesized. Translation is the process during which proteins are synthesized according to the genetic code of messenger RNA.

Genetic information is carried in a triplet code: a sequence of three nucleotides designates one amino acid. Of the 64 possible codons, 61 actually specify amino acids and the other three are termination signals.

The triplet code was deciphered by synthesizing artificial mRNA with known nucleotide sequences and analyzing the amino acid sequences of the polypeptides produced from these mRNA templates.

There are three stages in translation: initiation, elongation, and termination. The first amino acid in a polypeptide chain is always N-formylmethionine (fMet). A polypeptide elongates by the addition of amino acids until a termination codon is reached, for which there are no complementary tRNA molecules. Several ribosomes may be translating the same mRNA molecule at the same time.

Molecular biologists define mutations in terms of alterations in the nucleotide sequence of DNA. A point mutation involves the substitution of one nucleotide, which may result in the substitution of one amino acid. Mutations may also result from the deletion or addition of a nucleotide, which results in frame-shifts and the production of entirely new (and most probably nonfunctional) polypeptides.

Except for the DNA of mitochondria, the genetic code is universal for virtually all organisms.

HOW TO STUDY THE CHAPTER

Read the entire chapter through quickly, focusing on the major concepts.

Use the GUIDED STUDY OF THE CHAPTER to help you identify the important details as you **reread** the chapter. Writing out the answers to these questions will help fix them in your mind as well as provide you with a valuable study aid.

Answer the questions in TESTING YOUR UNDERSTANDING without the aid of your text. Check your answers against those in PERFORMANCE ANALYSIS. Analyzing your answers will give you valuable feedback on your level of understanding and preparedness for classroom testing.

GUIDED STUDY OF THE CHAPTER

I. Introduction (page 301)

1. Watson and Crick speculated that the genetic information was carried in DNA in what manner?

II. Genes and Proteins (pages 301–304)

A. Inborn Errors of Metabolism (page 301)

2. a. Garrod suggested what situation caused inborn errors of metabolism?

b. Describe one example Garrod used to illustrate his hypothesis.

B. One Gene–One Enzyme (pages 301–304)

> *Focus:* **All the biochemical reactions in living cells depend on enzymes.**

3. a. State the hypothesis formulated by George Beadle while studying eye color mutations in *Drosophila*.

b. What strategy did Beadle and Tatum adopt to test this hypothesis?

4. a. Which organism did Beadle and Tatum use in their research?

b. List five advantages of using *Neurospora* for genetic research.

c. How did the nutritional requirements of *Neurospora* facilitate this research (forming the sixth advantage of using this organism)?

5. a. How were *Neurospora* spores treated to increase the mutation rate?

b. Outline the technique used by Beadle and Tatum to identify and isolate mutant fungi.

6. a. Cite two significant findings that resulted from the research of Beadle and Tatum.

b. What did Beadle and Tatum propose, based on these results?

c. What two revisions of detail (but not of basic principle) did this hypothesis undergo as more information was gained?

C. **The Structure of Hemoglobin** *(page 304)*

Focus: Based on the work of Beadle and Tatum, Linus Pauling suggested that human diseases involving hemoglobin might be due to changes in the normal protein structure.

7. Electrophoresis separates proteins based on what characteristic?

8. a. How did Pauling test his hypothesis? (See also Figure 15–4.)

b. What were the results?

9. a. Who identified the structural difference between normal and sickle cell hemoglobin?

b. What is this structural difference?

D. **The Virus Coat** *(page 304)*

Focus: Studies of bacteriophages provided additional evidence that DNA codes for the production of proteins.

III. **From DNA to Protein: The Role of RNA** *(pages 305–307)*

A. **General Remarks** *(page 305)*

Focus: The instructions carried in DNA are implemented through the production of proteins.

10. What type of molecule acts as an intermediary in the translation of DNA into protein?

11. Identify three ways in which RNA differs from DNA. (Figure 15–5)

12. Cite three clues that RNA plays a role in the translation of the genetic code into proteins.

B. **The Central Dogma** *(page 306)*

13. a. State the central dogma of molecular genetics as proclaimed by Crick.

b. How did the central dogma support Darwin's theory of evolution?

14. *Vocabulary:* What is reverse transcription? (footnote, page 306)

C. **RNA as Messenger** *(pages 306–307)*

15. a. Describe the process by which mRNA is formed. (See also Figure 15–8.)

b. What is this process called?

c. What enzyme catalyzes this process?

Focus: The mRNA strand is antiparallel to the DNA template that directs its synthesis.

Focus: Messenger RNA, which is the working copy of the genetic information, dictates the amino acid sequence in protein synthesis.

D. *Essay:* **The Elusive Messenger** *(page 308)*

16. a. Cite the observation that provided a major clue concerning the role of RNA in protein synthesis.

    b. What factor complicated the isolation of messenger RNA from cells?

17. How was mRNA identified as an intracellular carrier of genetic information? (Describe two sets of experiments.)

18. How has DNA-DNA and DNA-RNA hybridization facilitated research in molecular genetics?

IV. **The Genetic Code** *(pages 307–309)*

A. **General Remarks** *(page 307)*

19. Outline the logic leading to Gamow's suggestion that three nucleotides might code for one amino acid.

20. *Vocabulary:* What is a codon?

B. **Breaking the Code** *(pages 307–309)*

21. a. Describe the experiments of Nirenberg and Matthaei in which they identified the first codon.

    b. What was the first codon/amino acid pair to be identified?

22. Who developed the process for making synthetic RNA with a specified ribonucleotide sequence?

23. a. Of the 64 possible codons, how many code for amino acids?

    b. The others code for what event?

24. a. *Vocabulary:* Define degenerate in terms of molecular genetics. (footnote, page 309)

    b. How did this use of the term enter and persist in biological literature?

C. *Essay:* **AGA–GAG–AGA** *(page 310)*

25. a. What technique did Khorana use to crack the genetic code?

    b. Identify three aspects of the triplet code revealed by his work.

V. **Protein Synthesis** *(pages 309–314)*

A. **General Remarks** *(pages 309–312)*

> *Focus:* **Although there are differences in some details, the basic principles of protein synthesis are the same in prokaryotes and eukaryotes.**

26. *Vocabulary:* Distinguish between promoters and terminators.

> *Focus:* **RNA is transcribed in the 5′ to 3′ direction along one of the DNA strands.**

27. a. Name the three types of RNA needed for protein synthesis.

b. Which type is most abundant?

28. a. Describe the structure of a bacterial ribosome.

b. Identify the components that bind to the small and large ribosomal subunits.

Small subunit

Large subunit

29. a. What is the function of transfer RNA?

b. How does its length compare with that of rRNAs?

30. *Vocabulary:* What is an anticodon?

31. Transfer RNA has two attachment sites. What binds at each site?

32. Identify the sequence that is found at the 3′ end of every tRNA.

Focus: **Transfer RNA molecules contain unusual modified nucleotides.**

---

33. State the presumed function of the TφC loop of the tRNA molecule. (Figure 15–11)

34. How does the cloverleaf shape of the tRNA molecule suit its function?

35. a. By what mechanism is an amino acid attached to its corresponding tRNA?

b. Name the enzyme that catalyzes this reaction.

c. What molecule provides the energy required for this reaction?

d. Indicate how the amino acid is transferred from its tRNA to a growing polypeptide chain.

B. Translation *(pages 312–314)*

36. *Vocabulary:* Distinguish translation from transcription.

37. List the three stages of translation.

Note: Take time to study Figure 15–12, then read this entire section in the text before answering the questions.

38. a. Summarize the events of initiation. Include these components: small ribosomal subunit, large ribosomal subunit, mRNA, initiator codon, tRNA, fMet, initiation complex, P site, GTP.

b. Name the first amino acid in the new polypeptide.

c. What purpose is served by the presence of the formyl group on fMet? (Figure 15–13)

d. Identify the three components of the initiation complex.

e. What provides the energy for the attachment of the large and small ribosomal subunits?

39. a. Summarize the events of elongation. Include these components: mRNA, A site, tRNA, codon, anticodon, peptidyl transferase, and P site.

b. Distinguish between the A site and P site of the larger ribosomal subunit.

c. Name the enzyme that catalyzes peptide bond formation.

d. *Vocabulary:* What is a polysome?

40. a. List the three known termination codons.

b. By what mechanism is termination accomplished?

41. Study Figure 15–16 to review protein synthesis. Outline the events in your own words.

VI. Redefining Mutations *(pages 314–316)*

42. a. Indicate how the mRNA for sickle cell hemoglobin differs from that of normal hemoglobin.

b. What difference does this make in protein structure?

43. a. *Vocabulary:* Define a mutation in terms of molecular genetics.

b. Which mutations may be inherited by an organism's offspring?

c. *Vocabulary:* Define point mutations.

44. a. *Vocabulary:* What are frame-shifts?

b. Identify two mechanisms by which frame-shifts may occur.

VII. Universality of the Genetic Code *(page 316)*

Focus: **The genetic code is virtually universal among living organisms.**

45. Cite one exception to the universality of the genetic code. (footnote, page 316)

VIII. Summary *(pages 316–317):* Read the summary. If you are familiar with the essential features of the material presented there, you are ready to complete the section TESTING YOUR UNDERSTANDING.

TESTING YOUR UNDERSTANDING

After you have completed this examination, compare your answers with those in the section that follows.

1. What idea is implicit in Garrod's concept of inborn errors of metabolism?
   a. All metabolic disorders are inborn.
   b. Errors of metabolism are not due to genetic defects.
   c. Metabolic disorders can be corrected after birth.
   d. Metabolic pathways are independent of genetic control.
   e. Genes act by influencing the production of enzymes.

2. The specificity of an enzyme is determined by its:
   a. primary structure.
   b. nucleotide content.
   c. extent of mutation.
   d. biosynthetic pathway.
   e. helical structure.

3. The red bread mold, *Neurospora crassa*, was the organism used by Beadle and Tatum to study the relationship between mutations and enzymatic pathways. Which of these characteristics is NOT one of several advantages of using *N. crassa* in these studies?
   a. It has a very short life cycle.
   b. It is diploid throughout most of its life.
   c. It can be grown in large quantities in the laboratory.
   d. It can grow on a very simple medium.
   e. Many chromosome mapping studies had already been done.

4. Vernon Ingram found that sickle cell hemoglobin differs from normal hemoglobin by one _____ in the beta chains.
   a. nucleotide   d. phosphate group
   b. nitrogenous base   e. peptide bond
   c. amino acid

5. Evidence that DNA specifies the structure of proteins came from studies of protein coat synthesis in:
   a. prokaryotes.   d. protists.
   b. bacteriophages.   e. retroviruses.
   c. slime molds.

6. RNA molecules do NOT contain:
   a. thymine.   d. uracil.
   b. adenine.   e. cytosine.
   c. guanine.

7. Which of the following illustrates the "central dogma" of molecular genetics as proclaimed by Francis Crick?
   a. Protein $\longrightarrow$ RNA $\longrightarrow$ DNA
   b. DNA $\longrightarrow$ RNA $\longrightarrow$ Protein
   c. DNA $\longrightarrow$ Protein $\longrightarrow$ RNA
   d. RNA $\longrightarrow$ DNA $\longrightarrow$ Protein
   e. RNA $\longrightarrow$ Protein $\longrightarrow$ DNA

8. Ribosomal RNA could NOT be the link between the genetic message of DNA and proteins because it is:
   a. produced in the cytoplasm.
   b. double-stranded.
   c. too long.
   d. homogeneous.
   e. heterogeneous.

9. In the genetic code, a sequence of _____ nucleotides i[n] _____ constitutes a(n) _____, which specifies a par[-]ticular amino acid.
   a. two; mRNA; anticodon
   b. three; tRNA; codon
   c. three; DNA; codon
   d. three; mRNA; codon
   e. three; mRNA; anticodon

10. Which of these scientists developed a method of makin[g] synthetic RNA, thus making it possible to decipher th[e] genetic code?
    a. Ochoa   d. Crick
    b. Matthaei   e. Tatum
    c. Gamow

11. The triplet nucleotide sequences that do not code for [a] specific amino acid are known as:
    a. promoter signals.   d. elongation codons.
    b. initiation codons.   e. transcription signals.
    c. termination codons.

12. The type of mutation that results from the insertion of a nucleotide into a DNA sequence or the deletion of a nucleotide from DNA is known as a _____ mutation.
    a. transversion   d. conversion
    b. translation   e. frame-shift
    c. point

13. Which of the following was NOT a clue that RNA plays a role in the translation of genetic information from DNA into a polypeptide?
    a. RNA is found in the cytoplasm of eukaryotic cells.
    b. In a variety of organisms, cells of developing embryos contain high levels of RNA.
    c. Many different kinds of RNA exist in the nucleus of a cell.
    d. Ribosomes are two-thirds RNA and one-third protein.
    e. RNA is synthesized from viral DNA before viral protein synthesis begins.

14. Given the mRNA sequence, (5′)-AUGUACAAGGUCG-GAUGA-(3′), which of the following amino acid sequences would result from translation? (Refer to Figure 15–9.)
    a. tyr-met-val-lys-gly   d. met-val-lys-tyr-gly
    b. met-tyr-lys-val-gly   e. ser-arg-leu-glu-his-val
    c. gly-met-val-lys-tyr

15. Which type of RNA provides the crucial link between nucleic acids and proteins, the two languages of the living cell?
    a. mRNA   d. sRNA
    b. aRNA   e. rRNA
    c. tRNA

16. There are at least _____ different types of aminoacyl-tRNA synthetase enzymes present in a typical prokaryotic cell.
    a. 5   c. 15   e. 25
    b. 10   d. 20

7. Which component is NOT required for the attachment of an amino acid to a tRNA molecule?
   a. a ribosome
   b. a tRNA molecule
   c. ATP
   d. an amino acid
   e. an aminoacyl-tRNA synthetase

8. Which enzyme is part of the large ribosomal subunit and is involved in peptide bond formation?
   a. aminoacyl-tRNA transferase
   b. RNA polymerase
   c. polypeptide synthetase
   d. peptidyl transferase
   e. polypeptidase

9. During translation, what event must occur before the ribosome moves to the next codon along an mRNA molecule?
   a. The initiation complex is formed.
   b. A small ribosomal subunit attaches to the initiation codon.
   c. An aminoacyl-tRNA attaches to the A site.
   d. A peptide bond is forged.
   e. A tRNA-peptide chain attaches to the P site.

20. T or F   To test the hypothesis that genes govern the production of enzymes, Beadle and Tatum performed a detailed biochemical analysis of eye-color variations in *Drosophila*.

21. T or F   The correct sequence of events in translation is initiation, elongation, and termination.

22. "All enzymes are proteins and all proteins are enzymes." This statement is incorrect. Rephrase it in a correct form and discuss its significance or relevance to the one gene-one enzyme hypothesis of Beadle and Tatum.

23. Discuss why a frame-shift mutation is usually more detrimental to an organism than a point mutation. Then speculate about a situation in which a point mutation may be just as detrimental as a frame-shift mutation.

24. What three conclusions did Khorana draw from his work on the genetic code using artificial mRNAs?

25. Describe the protein and RNA composition of the large and small subunits of prokaryotic ribosomes.

PERFORMANCE ANALYSIS

1. **e**   Implicit in Garrod's concept of inborn errors of metabolism is the premise that genes act by influencing the production of enzymes. Studies performed by Beadle and Tatum confirmed this to be one mechanism by which genes influence phenotype. (page 301)

2. **a**   The primary structure (amino acid sequence) of an enzyme determines its secondary and tertiary structures, which in turn are responsible for the three-dimensional shape of the molecule and its specificity of action. (page 302)

3. **b**   The red bread mold, *Neurospora crassa*, was used by Beadle and Tatum for several reasons. It has a short life cycle and can be grown in large quantities in the laboratory. It is *haploid* through much of its life cycle, so that any recessive mutations are readily apparent. Many chromosome mapping studies had already been performed before Beadle and Tatum started their studies. The normal organism makes nearly all of the nutrients it needs and can grow on a very simple medium. Inability to grow on the simple medium indicates that a mutation has occurred, causing loss of ability to synthesize a required nutrient. (page 302)

4. **c**   Vernon Ingram was the scientist who discovered that sickle cell hemoglobin differs from normal hemoglobin by one amino acid substitution in the beta chain. It was later discovered that this corresponds to the substitution of a single nitrogenous base in the gene coding for the beta chain. (page 304)

5. **b**   Evidence that DNA specifies the structure of proteins was provided by the work of Beadle and Tatum on *Neurospora crassa* and by studies documenting the production of bacteriophage coat proteins when viral nucleic acid was introduced into a host cell. (page 304)

6. **a**   RNA molecules contain the nitrogenous bases adenine, guanine, uracil, and cytosine. DNA molecules contain thymine instead of uracil. (Figure 15–5)

7. **b**   The central dogma of molecular genetics, as formulated by Crick, stated that genetic information flows from DNA to RNA to protein. With the discovery of retroviruses, this was modified somewhat since these viruses can direct the synthesis of DNA from an RNA template. (page 306)

8. **d**   In their search for the carrier of the genetic message, scientists concluded that ribosomal RNA could not perform this role because it was too homogeneous, and the messenger must be heterogeneous. (essay, page 308)

9. **d**   In the genetic code, a sequence of three nucleotides in mRNA constitutes a codon and specifies a particular amino acid. The complementary sequence of nucleotides on tRNA is the anticodon. (page 307)

10. **a**   Severo Ochoa developed an enzymatic technique for making synthetic RNA. This technique was the basis for the studies of Nirenberg and Matthaei in which much of the genetic code was deciphered. (page 307)

11. **c**   Of the 64 mRNA nucleotide sequences, 61 code for amino acids and three are termination codons that signal the end of translation. (page 309)

12. **e**   A frame-shift mutation occurs when a nucleotide is either inserted into or deleted from a DNA sequence. A point mutation results when a nucleotide substitution occurs. (page 316)

13. **c** Several clues pointed toward the role of RNA in the translation of the genetic information from DNA into proteins. Unlike DNA, RNA is found mainly in the cytoplasm of eukaryotic cells, where protein synthesis occurs. The cells of developing embryos typically have high levels of RNA. Both prokaryotic and eukaryotic cells that are synthesizing large amounts of protein have numerous ribosomes. When a bacteriophage infects a cell, RNA is produced before viral protein synthesis occurs. (page 305)

14. **b** The leading end of mRNA (where protein synthesis begins) is the 5′ end. By consulting the genetic code table, it is apparent that the resulting amino acid sequence would be met-tyr-lys-val-gly. (page 309 and Figure 15–9)

15. **c** Transfer RNA (tRNA) transports amino acids to the ribosome. There is a different tRNA for each amino acid. A tRNA molecule can only bind to the ribosome if its anticodon complements the codon of the mRNA. Once the correct tRNA is bound to the ribosome, its amino acid is bonded to the growing polypeptide chain. Thus, tRNA acts as the link between information in nucleic acid and protein. (page 311)

16. **d** The joining of each amino acid to the correct tRNA molecule is catalyzed by a specific enzyme. There is one or more aminoacyl-tRNA synthetase molecule for each amino acid. Since there are 20 amino acids, there must be at least 20 aminoacyl-tRNA synthetase enzymes. (page 311)

17. **a** The attachment of an amino acid to its tRNA requires the amino acid, the correct tRNA molecule, ATP for energy, and the correct aminoacyl-tRNA synthetase. Ribosomes are not needed for this process. (page 312)

18. **d** Peptidyl transferase is the enzyme that catalyzes peptide bond formation between the growing peptide chain and incoming amino acids. It is part of the large ribosomal subunit. (page 312)

19. **d** A peptide bond must form before the ribosome moves to the next codon along an mRNA molecule. (page 312)

20. **False** Based on his work with *Drosophila*, Beadle formulated the hypothesis that each eye-color variation is due to changes in a single enzyme in a biosynthetic pathway. This hypothesis was tested by Beadle and his coworker Tatum in *Neurospora crassa*, by causing mutations and determining whether they affected biochemical pathways. (page 302)

21. **True** For convenience of study, translation has been divided into three phases: initiation, elongation, and termination. (pages 312–314)

22. "All enzymes are proteins, but not all proteins are enzymes." Although Beadle and Tatum were on the right track with their one gene-one enzyme hypothesis, it was too narrow. Genes code for all proteins, not just enzymes. This hypothesis has since been modified to the more accurate "one gene-one polypeptide." (page 304)

23. Because the reading frame is altered, frame-shift mutations result in the production of polypeptides that are completely different from those coded for by the unaltered gene. These polypeptides are therefore nonfunctional. Since point mutations involve a nucleotide substitution, the polypeptide for which the altered gene codes is only altered slightly (perhaps without affecting protein function) or may not be altered at all (owing to the degenerate nature of the genetic code). However, if a point mutation results in the creation of a new termination codon in the middle of a gene, the functional polypeptide will not be produced. Also, a point mutation leading to an amino acid change in the protein can have devastating consequences. (page 316)

24. (1) Messenger RNA is read sequentially. (2) The reading of mRNA depends upon the reading frame (i.e., upon the nucleotide at which translation begins). (3) The codon consists of an odd (as opposed to even) number of nucleotides. (essay, page 310)

25. Prokaryotic ribosomes are approximately two-thirds RNA and one-third protein. In *E. coli*, the large ribosomal subunit contains 5S rRNA, 23S rRNA, and 34 different proteins. The small ribosomal subunit contains 16S rRNA and one molecule of each of 21 different proteins. (page 310)

The Molecular Genetics of Prokaryotes and Viruses

MAJOR CONCEPTS

Recombinant DNA is a technology involving the modification and recombination of DNA from different sources. An understanding of the mechanisms by which genes are modified and exchanged in viruses and bacteria is essential to the study and use of this technology.

The chromosome of *E. coli* is one double-stranded loop of DNA, which replicates by theta replication from a single origin of replication.

A structural gene is a segment of DNA coding for a polypeptide. Structural genes that code for polypeptides with related functions are often transcribed onto one mRNA molecule.

There are two major mechanisms by which enzyme production is controlled in prokaryotes. Production of inducible enzymes occurs only in the presence of the substrate upon

which the enzyme acts. Repressible enzymes are produced continuously unless the product of the enzymatic pathway is present at a fairly high concentration, in which case enzyme production stops. Both mechanisms are mediated by allosteric interactions among molecules.

In the Jacob and Monod model of the regulation of transcription in prokaryotes, genes coding for functionally related proteins are organized into an operon, which consists of a promoter, structural genes, and the operator. A separate regulator gene codes for the production of a repressor protein. Binding of the repressor to the operator obstructs the promoter and prevents transcription. Allosteric effectors may either activate (corepressor) or inactivate (inducer) a repressor.

In some systems, catabolite activator protein (CAP) binds to cyclic AMP and the complex then binds to the operon promoter, enhancing transcription.

Virtually all types of bacteria contain plasmids, some of which may be incorporated into the bacterial chromosome as episomes. Plasmids may be replicated more or less frequently than the chromosome, or they may replicate only when the chromosome replicates.

The F plasmid of *E. coli* endows a cell with the ability to form pili, conjugate, and transfer the F plasmid to another cell. When this plasmid becomes an episome, the cell is known as an Hfr cell. During conjugation, an Hfr cell may transfer part or all of the replicating Hfr cell chromosome to the recipient cell. Recombination can occur between the recipient cell DNA and the donor DNA. In the laboratory, bacterial chromosomes have been mapped by using Hfr cells and interrupting conjugation at various time intervals.

Many prokaryotes carry R plasmids that contain genes for resistance to antibiotics and antimicrobial agents. These resistance genes can be transferred from one plasmid to other plasmids, to the bacterial chromosome, to viruses, and to bacteria of other species. R plasmids may be transferred at cell division, during conjugation, or by passing from cell to cell through the cell membranes. Drug resistance may result from the production of enzymes that (1) break down the drug or (2) establish a separate enzymatic pathway, circumventing the drug's effects.

The nucleic acid (DNA or RNA) of a virus codes for the coat proteins, all the enzymes needed to produce new virus particles, and the enzymes necessary to break out of the host cell. Viruses are obligate parasites.

Temperate bacteriophages are viruses that may establish a long-term relationship with a host cell, during which the cell appears uninfected. At some point, however, an infection cycle begins and the host cell is eventually lysed; this phenomenon is known as lysogeny. Temperate bacteriophages may become incorporated into the host cell chromosome as prophages. In such cells, lysis results when the prophage breaks out of the host chromosome.

During viral replication, host cell DNA may become incorporated into virus particles and be carried to new host cells, a process known as transduction. In general transduction, random fragments of host DNA are carried by the virus. Specialized, or restricted, transduction, which occurs with temperate bacteriophages, involves the transfer of host genes adjacent to the site where a prophage was inserted into the host chromosome.

Other movable genetic elements are transposons, DNA fragments that are integrated into the chromosome. Transposons contain a gene that codes for the enzyme responsible for their insertion into a new site. Simple transposons contain only those genes essential for the process of transposition. Complex transposons carry genes in addition to those essential for transposition.

DNA can be introduced into a bacterial cell by transformation, conjugation, viral infection, and transduction. Genetic recombination in prokaryotes may occur by exchange between homologous DNA segments and by the insertion of movable genetic elements.

HOW TO STUDY THE CHAPTER

Read the entire chapter through quickly, focusing on the major concepts.

Use the GUIDED STUDY OF THE CHAPTER to help you identify the important details as you **reread** the chapter. Writing out the answers to these questions will help fix them in your mind as well as provide you with a valuable study aid.

Answer the questions in TESTING YOUR UNDERSTANDING without the aid of your text. Check your answers against those in PERFORMANCE ANALYSIS. Analyzing your answers will give you valuable feedback on your level of understanding and preparedness for classroom testing.

GUIDED STUDY OF THE CHAPTER

I. Introduction (page 319)

> *Focus:* Studies with bacteria and viruses provided the foundation for work in modern molecular genetics.

> *Focus:* Recombinant DNA involves the modification and recombination of DNA from different sources.

> *Focus:* An understanding of the ways in which bacteria and viruses modify, recombine, and exchange genetic material is essential to the understanding of how geneticists have manipulated these processes in recombinant DNA technology.

II. The *E. coli* Chromosome (pages 319–320)

1. Describe the chromosome of *E. coli* (include shape and size).

2. Name the type of DNA replication in *E. coli* and summarize this process.

III. Transcription and Its Regulation (pages 320–326)

A. General Remarks (page 320)

3. Summarize transcription in prokaryotes.

4. a. *Vocabulary:* What are structural genes?

   b. How are structural genes often arranged on a bacterial chromosome?

   c. In what way are functionally related structural genes typically transcribed in prokaryotes?

5. a. Distinguish between leader and trailer sequences on mRNA.

   b. Characterize the coding region of an mRNA molecule.

B. The Need for Regulation *(pages 320–322)*

6. What is the doubling time for a culture of *E. coli* cells?

7. Cite two reasons for *E. coli*'s effectiveness in using available nutrients.

8. *Vocabulary:* Distinguish between inducible and repressible enzymes, describing one example of each type.

9. Contrast the reproductive capacity of mutant cells that cannot regulate enzyme production with that of normal cells.

10. a. In prokaryotes, the regulation of protein synthesis occurs mainly at what stage?

    b. What does this regulation involve?

    c. What features of protein synthesis maximize efficiency?

C. The Operon *(pages 322–326)*

11. a. Name the two scientists that developed the operon model.

    b. What phenomenon was studied by André Lwoff? (Figure 16-6)

12. a. *Vocabulary:* Define an operon.

    b. List three components of an operon.

13. a. *Vocabulary:* Distinguish among operator, regulator, and repressor.

    b. What happens when a repressor is bound to the operator?

    c. What evidence suggested the existence of the regulator gene?

14. a. Use examples to distinguish the effect of an inducer on an inducible operon from the effect of a corepressor on a repressible operon.

    b. By what mechanism do corepressors and inducers function?

Focus: **Both inducible and repressible operons function by negative control: the repressor turns transcription off in both systems.**

The CAP–Cyclic AMP System *(pages 325–326)*

15. a. *Vocabulary:* What is CAP?

    b. How does CAP work and what is its effect?

Focus: **The operon is under the negative control of the repressor and the positive control of the CAP-cAMP complex.**

---

c. How was the CAP system first discovered?

d. Describe the relationship between glucose supply, cAMP, and CAP.

e. In operons regulated by the CAP-cAMP system, what two distinct regions are included in the promoter? (See also Figure 16–10.)

16. a. Summarize all the conditions that must be met for *maximal* production of beta-galactosidase.

b. What three proteins are produced when these three conditions are met? (See also Figure 16–10.)

Focus: In order to induce bacteria to synthesize mammalian proteins in the laboratory, the mechanisms by which genes are regulated must be understood.

IV. Plasmids and Conjugation *(pages 326–329)*

A. General Remarks *(page 326)*

17. *Vocabulary:* Distinguish between plasmids and episomes.

18. Cite two ways in which plasmids resemble bacterial chromosomes.

Focus: Plasmid replication may occur more or less frequently than chromosome replication, or it may only occur when the chromosome replicates.

19. What differences in plasmid content in daughter cells would result from these three different modes of plasmid replication?

B. The F Plasmid *(pages 327–329)*

20. a. The F factor contains genes coding for what structures?

b. Distinguish between F+ and F− cells.

c. Receipt of the F factor endows a recipient cell with what two abilities?

21. *Vocabulary:* Define conjugation.

22. How does rolling-circle replication differ from chromosome replication? (See also Figure 16–13.)

23. a. *Vocabulary:* What is an Hfr cell?

b. How are Hfr cells created?

24. a. Summarize the transfer of genetic information from an Hfr cell to an F− cell. (See also Figure 16–14.)

b. In what way is bacterial conjugation a form of sexual recombination?

25. Carefully review Figure 16–14.

   a. What happens to the displaced recipient DNA and the noncomplementary portion of the donor DNA following recombination?

   b. What occurs immediately after incorporation of donor DNA into the recipient chromosome?

   c. Describe the character of the donor cell and the recipient cell with respect to F factor following recombination.

26. How can genetic recombinations of this type be detected in the laboratory?

Chromosome Mapping (pages 328–329)

27. Outline the procedure for mapping a bacterial chromosome.

28. Describe the arrangement of genes on a bacterial chromosome.

Focus: **The site and orientation of F plasmid insertion into a bacterial chromosome differs among different bacterial strains.**

---

29. The circularity of the *E. coli* chromosome has been confirmed by what technique?

C. R Plasmids *(page 329)*

Focus: **Genes that convey drug resistance are often carried on bacterial plasmids.**

---

30. a. Cite four locations to which the resistance genes of a plasmid may be transferred.

   b. Describe an example of transfer of drug resistance from an innocuous to a pathogenic bacterium.

   c. Name three means by which R plasmids may be transferred.

31. Identify two mechanisms of drug resistance resulting from the activity of genes carried on plasmids.

V. Viruses *(pages 330–335)*

A. General Remarks *(page 330)*

   32. *Vocabulary:* What is a capsid?

   33. Briefly outline the viral "life cycle."

   34. What property of a virus determines its attachment to a host cell membrane?

Focus: **Viruses are obligate parasites and cannot replicate outside their host cell.**

---

   35. Identify the forms in which viral chromosomes may occur.

36. Name three types of proteins coded for by viral nucleic acid.

37. When is a viral infection cycle complete?

B. Viruses as Vectors *(pages 331–334)*

> *Focus:* **Viruses can carry bacterial DNA from one cell to another.**

Temperance and Lysogeny (page 331)

38. *Vocabulary:* What are lysogenic cells and what causes lysogeny?

39. *Vocabulary:* Distinguish between temperate bacteriophages and prophages.

40. a. At what intervals do prophages spontaneously break free from the host chromosome?

    b. Name three factors (natural or experimental) that trigger lytic cycles.

41. List the similarities and differences between temperate phages and plasmids.

C. *Essay:* "Sir, I Am Entirely Lysed" *(page 332)*

42. The choice of ultraviolet light as a potential agent to trigger a lytic cycle was technically illogical. Why?

Transduction (pages 332–334)

43. *Vocabulary:* What is transduction?

44. a. Under what conditions does general transduction occur? (Figure 16–18)

    b. Summarize the mechanism by which general transduction takes place.

45. How does restricted transduction differ from general transduction? (See also Figure 16–18.)

46. How does transduction resemble and differ from conjugation?

D. Introducing Lambda *(pages 334–335)*

47. a. Describe the two forms in which lambda DNA can exist and the conditions under which it exists in each form.

    b. What is meant by "sticky" ends of a DNA molecule?

    c. Name the enzyme that catalyzes the union of the ends of lambda DNA. (Figure 16–19)

    d. *Vocabulary:* What is a COS region?

    e. Lambda DNA is replicated in what form?

48. a. Describe the mechanism by which lambda becomes integrated into a host cell chromosome. (Include the enzyme that catalyzes this reaction.)

b. What is the consequence of the "staggered cut" made by another enzyme?

VI. Transposons (pages 335–336)

49. How are transposons similar to and different from episomes and prophages?

50. Identify the function of transposase.

51. What are direct repeats and inverted repeats?

52. a. Describe the process by which a transposon becomes inserted into a host chromosome.

b. What is a target site?

53. Fill in the following table comparing simple and complex transposons.

Table for Question 53

	SIMPLE	COMPLEX
Alternative name (if given)		
Relative size		
Type(s) of information carried		
How they may be detected		

54. Explain what is meant by "selfish DNA."

55. *Vocabulary:* What are "jumping genes"?

56. Summarize one mechanism by which complex transposons might have arisen and cite the evidence supporting this proposal.

> *Focus:* Transposons may insert into bacterial chromosomes or plasmids. Drug-resistance genes are often part of transposons.

VII. Recombination Strategies (pages 336–337)

57. List and define four mechanisms by which new DNA may be introduced into a bacterial cell.

58. When and how does recombination occur in eukaryotic cells?

59. Outline the "single-strand switch" model of genetic recombination. (Figure 16–24)

60. Name three examples of movable genetic elements.

VIII. Summary *(pages 337–338)*: Read the summary. If you are familiar with the essential features of the material presented there, you are ready to complete the section TESTING YOUR UNDERSTANDING.

TESTING YOUR UNDERSTANDING

After you have completed this examination, compare your answers with those in the section that follows.

1. An *E. coli* chromosome is some 500 times longer than an *E. coli* cell. However, the chromosome fits within the cell because it is condensed into an irregularly shaped body known as the:
   a. nucleus.
   b. theta body.
   c. Okazaki fragment.
   d. nucleosome.
   e. nucleoid.

2. A segment of DNA that codes for a polypeptide product is known as a:
   a. regulatory gene.
   b. structural gene.
   c. polypeptide template.
   d. transcription codon.
   e. translation codon.

3. When an enzyme is produced in response to the presence of its substrate, that enzyme is said to be:
   a. repressible.
   b. evoked.
   c. activated.
   d. responsive.
   e. inducible.

4. Regulation of most prokaryotic genes occurs at the level of:
   a. transcription.
   b. translation.
   c. induction.
   d. replication.
   e. repression.

5. Both inducible and repressible systems are examples of _____, since they both involve _____ that turn off transcription.
   a. negative control; repressors
   b. positive control; activators
   c. negative control; activators
   d. positive control; repressors
   e. negative control; inactivators

6. Only when the CAP-cAMP complex is bound to the _____ is the operon transcribed at the maximal rate.
   a. operator
   b. repressor
   c. regulator
   d. promoter
   e. structural gene

7. When a plasmid becomes incorporated into the bacterial chromosome, it is known as a(n):
   a. episome.
   b. pilus.
   c. prophage.
   d. capsid.
   e. transposon.

8. In *E. coli* the genes that code for pili proteins and those that endow a bacterium with the ability to initiate conjugation are found on:
   a. recombinant plasmids.
   b. conjugation plasmids.
   c. R plasmids.
   d. F plasmids.
   e. prophages.

9. Which statement is NOT true of antibiotic resistance plasmids?
   a. They may be passed from cell to cell by conjugation.
   b. They may be passed from cell to cell through the cell membranes.
   c. They may integrate into the host chromosome and convert the cell into an Hfr cell.
   d. They may exist in multiple copies in a single cell.
   e. Resistance genes can be transferred from one plasmid to another.

10. When a prophage breaks loose from a host chromosome, it may take a fragment of the host chromosome along. When that fragment becomes incorporated into the chromosome of a new host, which process has occurred?
    a. general transduction
    b. general transformation
    c. specialized transduction

d. specialized transformation
e. specialized transposition

11. Enzymes that cleave apart the multiple copies of newly synthesized lambda phage DNA act at specific locations called _____ sites.
    a. DOA
    b. CUT
    c. CAP
    d. CNS
    e. COS

12. The enzyme that inserts lambda DNA into the host chromosome is called:
    a. lambdase.
    b. prophage ligase.
    c. prophage insertase.
    d. lambda integrase.
    e. lambda E. coli DNA ligase

13. Which process is NOT a means by which new DNA may be introduced into a bacterial cell?
    a. transduction.
    b. transvection.
    c. conjugation.
    d. transformation.
    e. transposon transfer.

14. T or F   The entire length of the mRNA sequence of a structural gene is translated into a sequence of amino acids that constitute a polypeptide.

15. T or F   In E. coli, the genes coding for all of the enzymes involved in the synthesis of some amino acids (e.g., tryptophan) are grouped together on the chromosome.

16. T or F   Like the bacterial chromosome, the F plasmid replicates by rolling circle replication.

17. T or F   Tobacco mosaic virus (TMV) and the T-even bacteriophages differ in that there are many different types of proteins in the coat of TMV, but only a single, repeating protein subunit in the T-even bacteriophage coat.

18. T or F   Because they code for at least one enzyme involved in nucleic acid replication, viruses can ensure the rapid replication of viral nucleic acid in preference to that of the host bacterium.

19. T or F   Transposase, which catalyzes the integration of a transposon into a new site on a chromosome, is actually encoded by the transposon itself.

20. E. coli can regulate the production of 1,700 different enzymes and proteins. Explain how this regulatory ability results in versatility and efficiency of nutrient utilization.

21. Outline the procedure in which a kitchen blender and a timer were used to map the E. coli chromosome.

22. Summarize the similarities and differences between a temperate bacteriophage and a plasmid.

23. Briefly summarize the two ways in which recombination can occur between DNA molecules.

PERFORMANCE ANALYSIS

1. e   In a nondividing cell, the circular E. coli chromosome is condensed into a body known as the nucleoid. (page 320)

2. b   Structural genes code for any polypeptide products, including enzymes, structural proteins, contractile proteins, transport proteins, etc. (page 320)

3. e   An inducible enzyme is one whose production is stimulated by the presence of its substrate. A repressible enzyme is one whose production is inhibited, usually by a product in the enzymatic sequence of which the repressible enzyme is a part. (page 321)

4. a   The regulation of most prokaryotic genes occurs at the level of transcription, thereby avoiding the production of superfluous mRNA. (page 322)

5. a   In prokaryotic operons, repressors turn off transcription. The activity of repressors is influenced by their interaction with molecules that act as allosteric effectors. In an inducible system, a substrate may act as an effector (inducer) that inactivates a repressor by binding to it and removing it from the operator. Protein synthesis then proceeds. In a repressible system, the enzymes are produced continuously unless adequate levels of a product of the pathway are present. The product (corepressor) binds to and activates the repressor, which then binds to the operator and blocks enzyme production. Both systems are examples of negative control. (page 325)

6. d   The binding of the CAP-cAMP complex to the promoter enables transcription to occur at a maximal rate. (page 325)

7. a   An episome is a plasmid that has become incorporated into a bacterial chromosome. A prophage is the DNA of a temperate bacteriophage that has become incorporated into a bacterial chromosome. (page 326)

8. d   The fertility genes of E. coli, which enable a bacterium to initiate conjugation and which code for the production of pili proteins, are located on F plasmids. (page 327)

9. c   Plasmids containing genes for antibiotic resistance (R plasmids) may be passed from cell to cell via conjugation or by passing through cell membranes. This transfer may involve cells of different species. The resistance genes can be transferred from one plasmid to another. The plasmids may exist in multiple copies in a single cell. A cell becomes an Hfr cell when the fertility (F) plasmid is incorporated into its chromosome. (page 329)

10. c   Specialized transduction occurs when a prophage breaks free from a host cell chromosome, carrying some of the host DNA with it to new host cells. The transferred DNA was located adjacent to the prophage insertion site. General transduction occurs when fragmented host DNA becomes randomly incorporated within the

protein coat of an infective virus and the host DNA is carried with the virus when it infects a new host. (page 334)

11. **e** When lambda phage DNA is replicated, multiple copies are synthesized in a long strand. The DNA is then cleaved into shorter lengths corresponding to individual viral chromosomes. Cleavage occurs at COS sites and leaves sticky ends on the DNA molecules. (page 334)

12. **d** Lambda integrase is the enzyme that inserts lambda DNA into a host chromosome. This enzyme is encoded by genes on the lambda DNA itself. (page 335)

13. **b** New DNA may be introduced into a bacterial cell by transduction, conjugation, transformation, and transposon transfer. (page 336)

14. **False** The mRNA for a structural gene includes leader and trailer sequences that do not code for polypeptide production. The 5′ leader sequence is thought to aid in binding the mRNA to the ribosome. The function of the 3′ trailer is not known. (page 320)

15. **True** In prokaryotes, the genes coding for enzymes involved in one particular pathway are typically grouped together on the chromosome and are transcribed onto the same mRNA molecule. (pages 321, 322)

16. **False** Plasmids replicate by rolling circle replication. The bacterial chromosome replicates by theta replication. (page 327)

17. **False** The protein coat of TMV is composed of multiple copies of a single protein subunit. The protein coat of T-even bacteriophages contains many different proteins. (page 330)

18. **True** The fact that viral nucleic acid carries genes for one or more enzymes involved in viral replication ensures that viral replication occurs in preference to host cell DNA replication. (page 330)

19. **True** The gene coding for transposase, an enzyme responsible for the integration of a transposon into its new site, is located on the transposon itself. (page 335)

20. The ability of *E. coli* to regulate gene expression maximizes efficient utilization of resources. Available raw materials are used to produce only those proteins that are needed by the cell. Mutant cells that are unable to regulate gene expression waste resources by producing unnecessary proteins. When in competition with more efficient cells, such cells are at a disadvantage. (pages 320–322)

21. Conjugating Hfr cells were used to map the *E. coli* chromosome. During conjugation, approximately 90 minutes (at 37°C) are required for a complete copy of the donor DNA to be transferred to the recipient cell. This transfer occurs at a constant rate. Conjugating cells were separated at different time intervals by whirling them in a blender. Analysis of the genes that were transferred to the recipient cell at each interval allowed mapping of the chromosome. (pages 328, 329)

22. Temperate bacteriophages are viruses whose DNA may become incorporated into the chromosome of the host cell. Such integrated DNA is called a prophage. Plasmids are small, extrachromosomal pieces of DNA that can also be inserted into the bacterial chromosomes. Integrated plasmid DNA is referred to as an episome. In both cases, the incorporated DNA may leave the chromosome. (page 331)

23. (1) Genetic recombination between chromosomes may result from crossing over, the exchange of sections of homologous chromosomes. (2) The insertion of movable genetic elements, including transposons, prophages, and plasmids, is another mechanism of genetic recombination. (page 337)

REVIEW TEST 5

Section 3: Chapters 14–16

This review test is *not* designed to cover all of the important information in these chapters. However, it does touch on the major topic categories in each chapter. It will also give you valuable practice in taking this type of test. When you are finished, check your answers with those provided in the following PERFORMANCE ANALYSIS section.

1. The white, sugary, slightly acidic, phosphorus-containing substance that Miescher isolated from cell nuclei was:
   a. an amino acid.
   b. a lipid.
   c. deoxyribonucleic acid.
   d. a carbohydrate.
   e. ribonucleic acid.

2. In the ladder analogy of DNA structure, each rung consists of:
   a. one purine (or pyrimidine) base and a phosphate group.
   b. one 5-carbon sugar and one phosphate group.
   c. two pyrimidine bases.
   d. two purine bases.
   e. one pyrimidine base and one purine base.

3. Which statement is NOT true of transcription?
   a. Ribonucleotides are added one at a time to the 3′ end of the growing RNA chain.
   b. The mRNA strand is complementary and parallel to the DNA template.
   c. RNA polymerase moves in a 3′ to 5′ direction along the template DNA strand.
   d. The strand of RNA nucleotides grows in a 5′ to 3′ direction.
   e. All these statements are true of transcription.

4. In *E. coli*, the first step in transcription is for RNA polymerase to bind to the:
   a. promoter.
   b. operator.
   c. structural gene.
   d. replication origin.
   e. Okazaki fragment.

5. After the DNA of *E. coli* was labeled with ^{15}N (heavy nitrogen), the cells were allowed to undergo two replications in medium containing ^{14}N (light nitrogen, the most common isotope). When the DNA was isolated and centrifuged in a density gradient, the experimenters found:
   a. a single band of DNA containing only heavy nitrogen.
   b. a single band of DNA containing only light nitrogen.
   c. a single band of DNA intermediate between the heavy and light bands.
   d. two bands of DNA: one intermediate and the other of light DNA.
   e. three bands of DNA: one of heavy DNA, one intermediate, and one of light DNA.

6. During translation, the energy required for the addition of the large ribosomal subunit to the initiation complex is supplied by:
   a. ATP.
   b. GTP.
   c. TTP.
   d. *N*-formylmethionine.
   e. reverse transcriptase.

7. The nucleic acids of viruses consist of several forms. Which of the following is NOT one of these?
   a. single-stranded DNA
   b. double-stranded RNA
   c. linear DNA
   d. circular RNA
   e. All of these forms of nucleic acids have been found in viruses.

8. T or F The foundation of modern molecular genetics has been laid primarily by studies involving *Drosophila*, mice, and humans.

9. T or F When portions of the Hfr chromosome recombine with the recipient chromosome, the donor genes totally replace the recipient genes in the recombining section of the chromosome.

10. T or F Jumping genes are found in a complex transposon.

11. T or F Lamarck contended that acquired characteristics could be inherited. If this was true, such traits would necessarily be associated with changes in the DNA of an individual's gametes.

12. List three clues that led scientists to conclude that RNA plays an important role in the translation of genetic information from DNA to proteins.

13. List the four major requirements that must be met by the genetic material in order to fulfill its biological role. Describe how DNA fulfills these requirements.

14. Compare and contrast the regulation of the lactose and tryptophan operons. Include the following terms in your discussion: operator, promoter, regulator, repressor, allosteric effector, inducible, and repressible.

PERFORMANCE ANALYSIS

1. **c** In 1869, Friedrich Miescher became the first scientist to isolate DNA from cell nuclei. He called the substance "nuclein." (page 282)

2. **e** In the ladder analogy of DNA structure, each rung consists of one pyrimidine base hydrogen-bonded with one purine base. Adenine always pairs with thymine and guanine always pairs with cytosine. The "rails" consist of alternating sugar and phosphate groups that are covalently bonded. (page 289)

3. **b** The mRNA strand is complementary and *anti*parallel to the DNA template directing its synthesis. (page 307)

4. **a** Transcription in prokaryotes is initiated by the binding of RNA polymerase to the promoter, which causes the double helix to open so that transcription may begin. (page 320)

5. **d** After the first DNA replication, one band of DNA would have appeared that was intermediate in density between heavy (labeled with ^{15}N) and light (unlabeled, containing ^{14}N) DNA. After the second replication, two bands appeared: one with an intermediate density (DNA with one ^{15}N-labeled strand and one unlabeled strand) and one band with the density of light DNA (both strands unlabeled). (page 293)

6. **b** The hydrolysis of guanosine triphosphate (GTP) provides the energy for the addition of the large ribosomal subunit to the initiation complex. (page 312)

7. **e** The nucleic acid of a virus may be either RNA or DNA, linear or circular, and single-stranded or double-stranded. (page 330)

8. **False** Most of the fundamentals of modern *molecular* genetics were derived from studies of bacteria and viruses. The general patterns of inheritance were worked out from studies of garden peas and fruit flies. (page 319)

9. **True** During genetic recombination involving an Hfr cell, the portion of the Hfr chromosome that enters the recipient cell may recombine with the recipient cell chromosome. In this case, the Hfr genes replace those of the recipient cell in the recombined portion of the chromosome. (pages 327, 328)

10. **True** The term "jumping genes" applies to those genes that travel as part of a complex transposon, which may leave a site on a chromosome and become inserted in another site on the same or on a different chromosome. (page 336)

11. **True** *All* heritable characteristics are coded for by the genetic information contained in DNA. An acquired characteristic could be inherited only if the acquisition of the trait was associated with a change in the DNA contained in the individual's gametes. (page 306)

12. Several clues pointed toward the role of RNA in translation of the genetic information of DNA into proteins. Unlike DNA, RNA is found mainly in the cytoplasm of eukaryotic cells, where protein synthesis occurs. The cells of developing embryos typically have high levels of RNA. Both prokaryotic and eukaryotic cells that are synthesizing large amounts of protein have numerous ribosomes. When a bacteriophage infects a cell, RNA is produced before viral protein synthesis occurs. (page 305)

13. (1) It must carry a great deal of genetic information from one cell generation to the next. (2) It must be able to produce a copy of itself with great precision. (3) It must be chemically stable. (4) It must be capable of mutation. The nitrogenous base sequences of DNA enable it to store vast amounts of information. Semiconservative replication provides an accurate mechanism for replication, but mutations occur at a low frequency when errors are made in the replication process. The chemical structure of DNA is quite stable. (page 287)

14. In prokaryotes, an operon consists of the promoter, the structural genes, and the operator. Transcription begins when RNA polymerase binds to the promoter sequence. An additional gene, the regulator, codes for a protein known as the repressor. Operon regulation depends on the activity of the repressor, which blocks protein synthesis by binding to the operator and obstructing the promoter. Allosteric effectors bind to repressors, either activating or inactivating the repressor by causing a change in its configuration.

The *lac* operon is an inducible operon that is activated when the substrate lactose is present. Allolactose, which is produced early in lactose metabolism, acts as an effector (an inducer) by binding to the repressor and removing it from the operator. Transcription of the gene coding for beta-galactosidase may then proceed.

Conversely, the *trp* operon is repressible. In the absence of tryptophan, the operon is in continuous operation, producing the enzymes needed for tryptophan production. When tryptophan is present, it acts as an effector (a corepressor) by binding to and activating the repressor. The activated repressor then binds to the operator and blocks the synthesis of enzymes that are unnecessary as long as tryptophan levels are adequate. (pages 323–325)

Recombinant DNA: The Tools of the Trade

MAJOR CONCEPTS

Recombinant DNA technology has facilitated research on the structure and function of genes. In addition to providing basic information, these techniques have practical applications including the production of large quantities of pure proteins (such as hormones for therapeutic use) and the genetic improvement of crop species through gene transfer.

The isolation of specific segments of DNA may be accomplished using restriction enzymes or reverse transcriptase. Restriction enzymes cleave DNA at specific nucleotide sequences to produce fragments (genomic DNA). Some restriction enzymes leave sticky ends on the cleaved DNA segments. Reverse transcriptase, produced by RNA viruses called retroviruses, catalyzes the synthesis of DNA (comple-mentary DNA) from an RNA template. It can be used in the laboratory to synthesize genes from mRNA. Specific segments of DNA (oligonucleotides) may also be produced synthetically by techniques that allow the researcher to designate any sequence of nucleotides.

To facilitate studies of specific genes, many copies (clones) of the gene of interest are needed. Genes can be cloned by splicing them into vectors that are then inserted into bacterial cells. Two common vectors are bacterial plasmids and viruses. Plasmids are easily taken up by bacterial cells, and small plasmids multiply rapidly. However, the larger a plasmid is, the longer it takes to replicate, so plasmids are only reliable vectors for DNA sequences up to 4,000 base pairs in length. By manipulating the lambda genome, DNA segments up to 40,000 base pairs long can be cloned in bacterial cells.

Nucleic acid hybridization is based on the hydrogen bonding between complementary nucleic acid strands. Radioactive probes made of labeled single-stranded RNA hybridize with denatured DNA bearing complementary sequences. Hybridization is used to identify genes of interest within a DNA molecule; the genes can then be isolated and cloned for further study.

The two most commonly used techniques for sequencing DNA involve cleavage of the DNA with restriction enzymes, separation of the different fragments, and cloning the fragments to obtain multiple copies. The nucleotide sequences of the fragments are then determined. The use of different restriction enzymes produces fragments whose nucleotide sequences overlap. The nucleotide sequence of the original DNA segment can be deciphered by matching the overlapping sequences.

The first mammalian protein to be produced in bacteria by biotechnological methods was somatostatin. Since then, bacteria have been induced to produce several other proteins. An important use of recombinant DNA technology is in vaccine production. Rather than using an entire virus to make a vaccine, the new biotechnology enables the production of vaccines using only synthetic viral protein coats.

One technical problem in gene transfer is determining whether or not the gene of interest has been taken up by the intended host. Recently researchers have transferred the gene for the enzyme luciferase, involved in firefly bioluminescence, to a tobacco leaf cell culture. They knew the gene transfer was successful when plants grown from the culture glowed in the presence of the enzyme's substrate, luciferin.

HOW TO STUDY THE CHAPTER

Read the entire chapter through quickly, focusing on the major concepts.

Use the GUIDED STUDY OF THE CHAPTER to help you identify the important details as you **reread** the chapter. Writing out the answers to these questions will help fix them in your mind as well as provide you with a valuable study aid.

Answer the questions in TESTING YOUR UNDERSTANDING without the aid of your text. Check your answers against those in PERFORMANCE ANALYSIS. Analyzing your answers will give you valuable feedback on your level of understanding and preparedness for classroom testing.

GUIDED STUDY OF THE CHAPTER

I. Introduction *(page 340)*

> *Focus:* **Recombinant DNA technology has greatly facilitated research into the structure and function of genes.**

1. How may pure samples of DNA fragments be obtained? (Figure 17–1)

2. List the four basic techniques used in recombinant DNA technology.

II. Isolation of Specific DNA Segments *(pages 340–343)*

A. General Remarks *(page 340)*

3. a. Name two types of enzymes used in the isolation of specific DNA segments.

   b. Identify the sources of these enzymes.

B. **Restriction Enzymes: gDNA** *(pages 340–341)*

4. a. What is the function of restriction enzymes?

   b. By what mechanism do restriction enzymes perform this function?

   c. *Vocabulary:* What are recognition sequences?

   d. How does a bacterium protect its own DNA from its restriction enzymes?

> *Focus:* **Some restriction enzymes cleave the two strands of DNA in a staggered fashion that leaves "sticky" ends.**

5. What is the significance to recombinant DNA technology of restriction enzymes that leave sticky ends? Include the role of DNA ligase.

6. *Vocabulary:* Define genomic DNA (gDNA).

> *Focus:* **Some viruses have restriction enzymes and methylating enzymes analogous to those of bacteria.**

C. **Reverse Transcriptase: cDNA** *(page 342)*

7. Outline the replication process for retroviruses. Include the function of reverse transcriptase in your answer. (See also Figure 17–3.)

8. What is the basic principle of the central dogma and why is it unaffected by the discovery of retroviruses? (Figure 17–4)

9. Cite three reasons why retroviruses are of great interest to scientists.

10. *Vocabulary:* Define complementary DNA (cDNA).

11. Describe the process by which cDNA may be spliced into other DNA molecules.

12. Cite one advantage and one disadvantage of using reverse transcriptase to make segments of DNA for research.

D. **Synthetic Oligonucleotides** *(page 343)*

13. a. Describe the process by which oligonucleotides with a specific nucleotide sequence may be synthesized.

    b. What step ensures that condensation reactions occur with the nucleotides in the desired orientation?

14. *Review Question:* Name three sources of uniform DNA or RNA segments in recombinant DNA technology.

III. **Clones and Vectors** *(pages 343–345)*

A. **General Remarks** *(page 343)*

B. **Plasmids as Vectors** *(page 344)*

15. *Vocabulary:* In recombinant DNA technology, what are clones?

16. Outline the use of pSC101 in the production of cloned DNA segments.

C. **Lambda and Cosmids** *(pages 344–345)*

17. What factor limits the use of plasmids as vectors in the production of DNA segment clones?

18. a. Summarize the use of lambda as a vector of DNA segments up to 20,000 base pairs long.

    b. How can lambda be used to clone even larger DNA segments?

    c. *Vocabulary:* What are cosmids?

IV. Nucleic Acid Hybridization *(page 345)*

A. General Remarks *(page 345)*

19. Hybridization is based on what property of nucleic acids?

20. a. Describe the procedure of nucleic acid hybridization in the laboratory.

    b. What information does nucleic acid hybridization provide?

    c. Cite two applications of nucleic acid hybridization.

B. Radioactive Probes *(pages 346–347)*

21. Describe the structure of a radioactive nucleic acid probe.

22. Answer these questions about the technique for isolating a particular gene using vectors, bacteria, and radioactive probes. (See also Figure 17–8.)

    a. How is the DNA containing the gene of interest fragmented?

    b. How are the DNA fragments cloned?

    c. How is the cloned DNA isolated from the bacteria?

    d. How is the radioactive probe used to identify the gene of interest?

Focus: Once the colonies containing the desired vector are identified, the vector can be transferred to other bacteria for cloning of the DNA segment.

23. Answer the following questions about the Southern blotting technique for isolating a particular gene. (See also Figure 17–9.)

    a. How are DNA fragments made?

    b. How are the DNA fragments separated?

    c. How is the radioactive probe used to identify the gene of interest?

Focus: Southern blotting is a very sensitive technique for identifying specific nucleotide sequences.

V. DNA Sequencing *(pages 348–350)*

A. General Remarks *(pages 348–349)*

24. In general terms, describe how the nucleotide sequence of a DNA molecule can be determined.

25. Answer these questions about the Maxam and Gilbert technique for DNA sequencing. (See also Figure 17–11.)

    a. What two specific techniques are used?

    b. Where are radioactive labels applied?

    c. How is the DNA fragmented? (Emphasize the crucial factor in this method.)

    d. How are the DNA fragments separated?

e. How is the nucleotide sequence determined from the resulting data?

26. How does Sanger's sequencing technique differ from that of Maxam and Gilbert?

B. *Essay:* Bacteriophage φX174 Breaks the Rules *(page 350)*

> *Focus:* The sequenced genome of bacteriophage φX174 reveals that the nucleotide sequences of some of its genes overlap.

VI. **Biotechnology** *(pages 350–352)*

> *Focus:* One major potential application of recombinant DNA technology is the mass production of desired proteins (such as hormones) by bacteria carrying the genes for those proteins.

27. Name the first mammalian protein to be synthesized in a bacterial cell and identify the scientist leading the team of researchers.

28. Answer these questions about Itakura's procedure.

    a. How was the nucleotide sequence of the somatostatin gene deduced, and how was the gene synthesized?

    b. What vector was used to carry the gene into *E. coli* cells?

    c. Even though no somatostatin was produced initially, how did the scientists know the gene had been incorporated into the bacterial cells?

    d. Where did they obtain a regulatory sequence to activate the gene?

e. How did the scientists circumvent the destruction of somatostatin by bacterial enzymes as it was produced?

f. How was somatostatin isolated from the hybrid protein?

29. a. Name four other proteins that have been successfully synthesized by genetically altered bacteria.

    b. Describe the medical importance of one of these proteins, and the economic importance of two of the proteins.

30. a. How is recombinant DNA technology used in the production of antiviral vaccines? (This is not explained in detail in the text, but you can deduce it from your reading thus far.)

    b. Why are these vaccines safer than those using killed or modified virus particles?

VII. **Gene Transfer: The Case of the Glowing Tobacco Plant** *(pages 352–353)*

31. Describe the one documented case of naturally occurring genetic recombination between prokaryotic and eukaryotic cells.

32. Identify two means by which the Ti plasmid exerts its effects on its hosts.

33. *Vocabulary:* What are opines?

34. Name three genes that are candidates for genetic recombination in plants.

35. Describe the components and products of the luciferase reaction.

36. Answer these questions about the experiment in which the gene for luciferase was transferred to tobacco leaf cells.

   a. How were clones of the gene produced?

   b. Where was a regulatory sequence for the gene obtained?

   c. What vector carried the gene into the infective bacterium?

   d. How did the researchers know that the transferred gene was operational?

   e. What implications does this experiment have for future DNA recombination experiments?

VIII. **Summary** *(pages 353–354):* Read the summary. If you are familiar with the essential features of the material presented there, you are ready to complete the section TESTING YOUR UNDERSTANDING.

TESTING YOUR UNDERSTANDING

After you have completed this examination, compare your answers with those in the section that follows.

1. In recombinant DNA research, some of the most commonly used tools, which are synthesized by bacteria, are:
   a. restriction enzymes.
   b. ribosomes.
   c. DNA polymerases.
   d. Okazaki fragments.
   e. sticky ends.

2. In order to prevent their restriction enzymes from cleaving their own DNA, bacteria add a(n) _____ group to the recognition sites of their DNA.
   a. chloride
   b. formyl
   c. carboxyl
   d. amino
   e. methyl

3. Which statement is NOT true of restriction enzymes?
   a. They are produced by many different bacteria.
   b. They recognize specific DNA sequences.
   c. They do not cleave the DNA present in their cell of origin.
   d. They methylate the DNA of their own recognition sites.
   e. They make either straight or staggered cuts.

4. Which type of organism can synthesize DNA from an RNA template?
   a. DNA viruses
   b. T-even bacteriophages
   c. retroviruses
   d. lambda bacteriophages
   e. all RNA viruses

5. The DNA molecules synthesized by reverse transcriptase are called:
   a. bDNA.   d. zDNA.
   b. gDNA.   e. cDNA.
   c. mDNA.

6. A crucial feature of synthetic oligonucleotide synthesis is the attachment of a(n) _____ to either the 5′ or 3′ end of a nucleotide.
   a. methyl group   d. base
   b. blocking group   e. RNA primer
   c. acid

7. When bacteriophage lambda is used as a vector, the DNA segment of interest can be packaged into a lambda protein coat if two _____ are separated by 35,000 to 40,000 base pairs.
   a. COS regions
   b. complementary regions
   c. recognition sites
   d. transposons
   e. CAP regions

8. The technique of sequencing DNA by chemical means was developed by:
   a. Frederick Sanger.
   b. Linus Pauling.
   c. Allan Maxam and Walter Gilbert.
   d. James Watson and Francis Crick.
   e. E. M. Southern.

9. What organism was used by Itakura as a "factory" to produce large quantities of the hormone somatostatin?
   a. bacteriophage lambda
   b. *E. coli*
   c. *Agrobacterium tumefaciens*
   d. bacteriophage φX174
   e. *Nicotiana tabacum*

10. T or F   Both bacteria and viruses may produce restriction and methylation enzymes.

11. T or F   Southern blotting is a sensitive technique for detecting specific DNA sequences using vector cloning in bacterial colonies.

12. T or F   In the DNA sequencing technique of Maxam and Gilbert, the Southern blotting technique is used to locate DNA fragments on the electrophoretic gel.

13. T or F   In crown gall disease of plants, opines facilitate spread of the disease by increasing bacterial conjugation.

14. Plasmids, viruses, and cosmids may all function as vectors. What characteristics common to all three vectors are important to this activity? How do they differ from one another with respect to cloning capabilities?

15. You are a graduate student studying the molecular biology of marine organisms. Your professor has just discovered a new species of fish. She has asked you to isolate its gene for hemoglobin and then check its homology (similarity) with human hemoglobin. (a) Outline the steps you would follow to isolate this gene. (b) Name the technique that would allow you to quickly and easily determine the sequence homology (at a qualitative level) of the fish and human genes.

16. Summarize the various steps that the San Diego group had to accomplish before their tobacco plants would glow "like fireflies."

PERFORMANCE ANALYSIS

1. **a**   Restriction enzymes, which are synthesized by bacteria and cleave DNA at specific recognition sites, are commonly used in recombinant DNA research. (page 340)

2. **e**   Many bacteria add a methyl group to their own recognition sites to prevent cleavage of their DNA by their restriction enzymes. (page 341)

3. **d**   Restriction enzymes are produced by many different bacteria and cleave DNA at specific nucleotide sequences. They cleave the two strands of DNA in either a straight or staggered fashion. They do not affect the DNA in the cell of origin because a bacterium has methylating enzymes that add methyl groups to its own recognition sites, protecting them from cleavage. (pages 340, 341)

4. **c**   Retroviruses are RNA viruses that contain the enzyme reverse transcriptase, which synthesizes DNA from the viral RNA template. (page 342)

5. **e**   Reverse transcriptase catalyzes the production of complementary DNA (cDNA). (page 342)

6. **b**   During synthetic oligonucleotide production, the attachment of a blocking group to either the 3' or the 5' end of the nucleotide is important to ensure the desired nucleotide sequence. (page 343, Figure 17–5)

7. **a**   The COS regions of bacteriophage lambda DNA produce "sticky ends" when the correct restriction enzyme cleaves the DNA at these sites to produce multiple copies of newly synthesized DNA. The single copies of viral DNA are then packaged into individual viruses. All that is required for the packaging of DNA into a lambda protein coat is the presence of two COS regions separated by 35,000 to 40,000 base pairs. The protein coat gains the viruses entry into the bacterial cell, where introduced DNA segments (like the normal lambda chromosome) become circular (thanks to the sticky ends) and replicate like plasmids. (page 345)

8. **c**   Maxam and Gilbert devised a technique for sequencing DNA by chemical means. Frederick Sanger's sequencing technique was enzymatic rather than chemical. (page 348)

9. **b**   When Itakura set out to produce somatostatin in large quantities by recombinant DNA techniques, he selected *Escherichia coli* as his "factory." (page 350)

10. **True**   Enzymes that function as restriction enzymes and some that function as methylating enzymes have been identified in both bacteria and viruses. (page 341)

11. **False**   Southern blotting is a sensitive technique for detecting specific DNA sequences. However, it does not require sequence cloning in bacterial colonies. DNA restriction fragments are separated on an electrophoretic gel and then blotted on a filter. The fragments on the filter are then exposed to a radioactive hybridization probe, which identifies the sequence(s) of interest. (page 347)

12. **False**   In the Maxam and Gilbert sequencing technique, the DNA fragments detected following separation by electrophoresis contain a radioactive phosphate on one end. The distance migrated indicates the length of the fragment. Since the DNA was cleaved at a specific nucleotide, scientists can determine the location of that nucleotide in the complete DNA strand. By repeating the electrophoretic procedure following cleavage at each of the four bases, the DNA sequence can be deduced. (page 349)

13. **True** Opines are unusual amino acids produced during crown gall disease and can be used by gall cells but not by normal cells. They increase bacterial conjugation, thereby promoting the spread of the Ti plasmid that causes the disease. (page 352)

14. Plasmids, viruses, and cosmids all are capable of entering bacterial cells and using the raw materials of the cell to produce copies of their nucleic acid. Plasmids can be used as vectors for DNA segments of up to 4,000 base pairs. Cloning by plasmid replication is slower than cloning using viruses or cosmids. Viruses can carry DNA segments of up to 20,000 base pairs. Cosmids carry DNA segments of 35,000 to 40,000 base pairs. (pages 344, 345)

15. (a) One approach in the isolation of the fish hemoglobin gene would be to either isolate or synthesize the mRNA that directs hemoglobin production. This RNA could then be labeled and hybridized with restriction fragments of the fish DNA until the hemoglobin gene is identified.

(b) Once the fish hemoglobin gene is isolated, it could be cloned and hybridized with the human hemoglobin gene to yield an overall impression of sequence homology. The fish hemoglobin gene could then be sequenced and compared with the human hemoglobin gene. (pages 340, 341; 345–349)

16. The gene for luciferase was isolated from fireflies and cloned in *E. coli*. The gene was then spliced into the chromosome of a plant virus to obtain a regulatory sequence. The modified viral chromosome was introduced into a Ti plasmid, which was transferred to *Agrobacterium* cells, which were then incubated with tobacco leaf cells in tissue culture. The callus that formed was treated to produce tobacco plants. When the plants were watered with a solution containing luciferin (the substrate for luciferase), they glowed. By the glowing of the plants, the scientists knew that the gene of interest had been incorporated into the plant and was producing functional enzyme. (pages 352, 353)

The Molecular Genetics of Eukaryotes

CHAPTER ORGANIZATION

MAJOR CONCEPTS

The molecular genetics of eukaryotes differs from that of prokaryotes in four important ways. (1) Eukaryotic cells have a much greater quantity of DNA. (2) There is a great deal of repetition within this DNA. (3) The DNA of eukaryotes is associated with proteins that play a major role in chromosome structure. (4) The organization of protein-coding sequences and the regulation of their expression is much more complex in eukaryotes.

Chromatin, which constitutes eukaryotic chromosomes, consists of DNA and proteins, the most abundant of which are basic proteins called histones. The fundamental packing unit of the chromosome, the nucleosome, is composed of a histone core around which the DNA filament is wrapped twice. The least condensed chromosome structure is a beads-on-a-string conformation in which the "beads" are nucleosomes and the "string" is the DNA filament between nucleosomes. As a chromosome condenses, it passes through these conformational stages: the 30-nanometer fiber, looped domains, and the compact chromosomes characteristic of mitosis and meiosis.

Chromosome replication in eukaryotes differs from that in prokaryotes in three ways: there are many replication origins in eukaryotes, replication is much slower, and eukaryotic DNA complexes with proteins as it is produced.

Several lines of evidence suggest a relationship between chromosome condensation and gene expression: highly condensed chromatin (heterochromatin) is apparently not transcribed, whereas dispersed chromatin (euchromatin) is expressed. One aspect of gene inactivation and activation is selective chromosome condensation. In addition to chromosome condensation, gene expression is regulated by cytosine methylation and by specific binding proteins.

It has been estimated that only a fraction (less than 10 percent and perhaps less than 1 percent) of the DNA in eukaryotic cells codes for proteins.

The genes of eukaryotic DNA consist of exons (regions that code for polypeptides) and introns (intervening sequences that do not code for protein). Introns are thought to persist because they promote genetic recombination during crossing over. In some cases, the different exons of a gene code for the various functional domains of a completed protein. During transcription, mRNA complementary to the entire DNA sequence is synthesized. Before the mRNA leaves the nucleus, the introns are removed and the exons are spliced together. This edited mRNA moves to the cytoplasm to direct protein synthesis.

Eukaryotic DNA can be classified into three categories based on the degree of repetition present. Simple-sequence DNA, which is present in great quantities, consists of fairly short, identical sequences arranged in tandem and found in specific locations on chromosomes. The sequences of intermediate-repeat DNA are longer than those of simple-sequence DNA, are similar but not identical to one another, and are scattered throughout the genome. Single-copy DNA consists of sequences that are not repeated or of which there are only a few copies. Nearly all the protein-coding sequences are single-copy DNA.

Gene families are protein-coding genes with similar but not identical nucleotide sequences that code for functionally related proteins. The most thoroughly studied example is the family of globin genes, which code for the various types of globins of the hemoglobin molecule that are active at different times in the life of a mammal.

The transcription of DNA into mRNA in eukaryotes is fundamentally similar to that in prokaryotes but there are two significant differences: eukaryotic genes are not grouped in operons, and the transcription of each structural gene is regulated individually. The enzymes involved in transcription differ (e.g., three different RNA polymerases catalyze RNA synthesis in eukaryotes).

Noteworthy moveable genetic elements in eukaryotes are antibody-coding genes, viruses, and transposons. In the differentiation of lymphocytes, segments that code for variable regions of antibody molecules move to new locations on the chromosome. The introduction of viral DNA into eukaryotic chromosomes may have no effect or may disrupt gene function. Eukaryotic transposons are basically analogous to those of prokaryotes. However, many eukaryotic transposons are copied to RNA then back to DNA before they are inserted in a new location in the host genome.

Cancer occurs when the mechanisms regulating normal cell growth go awry. Viruses may cause cancer by disrupting the function of a cell's regulatory genes, by encoding proteins needed for viral replication that interfere with regulatory genes, or by serving as vectors for oncogenes. Oncogenes are genes that closely resemble normal genes in the cells in which they are found. Of the oncogene products that have been studied, all are regulatory proteins involved in cell division or cell growth.

Perhaps the ultimate goal of recombinant DNA technology is the cure of human genetic diseases by replacing defective genes with functional genes. Many details must be worked out if this is to be accomplished. Thus far, mammalian genes of one species have been successfully transferred into tissue cell cultures and fertilized eggs of another species. Transposons have been used in successful genetic transfers in *Drosophila*. If the ultimate goal is to be reached, reliable vectors must be developed and more must be learned about gene regulation.

HOW TO STUDY THE CHAPTER

Read the entire chapter through quickly, focusing on the major concepts.

Use the GUIDED STUDY OF THE CHAPTER to help you identify the important details as you **reread** the chapter. Writing out the answers to these questions will help fix them in your mind as well as provide you with a valuable study aid.

Answer the questions in TESTING YOUR UNDERSTANDING without the aid of your text. Check your answers against those in PERFORMANCE ANALYSIS. Analyzing your answers will give you valuable feedback on your level of understanding and preparedness for classroom testing.

GUIDED STUDY OF THE CHAPTER

I. **Introduction** (*page 355*)

> *Focus:* The universality of the genetic code is compelling evidence that all living organisms share a common ancestor.

1. List four differences in the molecular genetics of prokaryotes and eukaryotes.

II. **The Eukaryotic Chromosome** (*pages 355–358*)

A. **General Remarks** (*pages 355–356*)

2. The total length of DNA present in each diploid human cell is estimated to be _____.

3. a. Name and distinguish between the three helical forms of DNA.

b. Which form is most common?

c. What function has been hypothesized for the helical variations?

B. Structure of the Chromosome *(pages 356–358)*

4. Identify the two components of chromatin.

5. a. *Vocabulary:* What are histones?

b. What is their function?

c. How many molecules of each type of histone are present in a diploid cell?

d. The amino acid sequences of which type of histone vary the most among different groups of organisms?

6. a. *Vocabulary:* Define and describe the structure of the nucleosome.

b. Describe the structural relationship between the DNA strand and the nucleosomes.

c. What is the structural relationship of H1 to the nucleosome?

7. Packaging of DNA in a nucleosome reduces its extended length by what factor?

8. What technique provided the first clue to the existence of nucleosomes?

> *Focus:* Forms through which a chromosome passes during condensation (from least to most condensed) are beads-on-a-string, 30-nanometer fibers, looped domains, and compact chromosomes.

9. In addition to histones, what other molecules are associated with the eukaryotic chromosome?

C. Replication of the Chromosome *(page 358)*

> *Focus:* The process of chromosome replication in eukaryotes is fundamentally the same as that in prokaryotes.

10. Outline the process of DNA replication.

11. Name three ways in which chromosome replication in eukaryotes differs from that in prokaryotes.

III. Regulation of Gene Expression in Eukaryotes *(pages 359–360)*

A. General Remarks *(page 359)*

12. Use an example to illustrate how the regulation of gene expression in multicellular eukaryotes is more complex than gene regulation in prokaryotes.

13. Cite two pieces of evidence (and the scientists involved) indicating that all the genetic information present in the zygote is also present in every diploid cell of the organism. (See also Figure 18–7.)

Focus: The differentiation of cells in a multicellular organism depends on selective gene activation and inactivation.

B. **Condensation of the Chromosome and Gene Expression** *(pages 359–360)*

14. *Vocabulary:* Distinguish between euchromatin and heterochromatin.

15. a. Indicate the relationship between chromosome condensation and transcription of RNA.

    b. Which type of chromatin may not be transcribed in some cells?

Focus: Some heterochromatin regions are constant from one cell to another and are never transcribed.

16. Name two examples illustrating this Focus statement.

17. a. How does the ratio of heterochromatin to euchromatin change as a cell differentiates?

    b. Account for this change in terms of cell function.

18. Summarize the evidence from studies of larval insects supporting the hypothesized relationship between chromosome condensation and gene expression. (Include "puffs," ecdysone, DNA, and RNA in your answer.)

C. **Methylation and Gene Expression** *(page 360)*

19. When does nucleotide methylation occur and which nucleotides undergo methylation?

20. What is methylation thought to accomplish?

21. Which DNA conformation has a high percentage of methylated cytosines?

D. **Regulation by Specific Binding Proteins** *(page 360)*

22. How does the regulation of gene expression in eukaryotes by specific binding proteins differ from similar regulation in prokaryotes?

IV. *Essay:* **The DNA of the Energy Organelles** *(page 361)*

23. a. What two generalizations have emerged at this point in your study of genetics?

    b. Exceptions to these two generalizations may be found in:

24. How does the DNA of mitochondria and chloroplasts differ from that of eukaryotic chromosomes?

25. Identify one hypothetical origin of mitochondria and chloroplasts and indicate what evidence led to this hypothesis.

26. How does transcription (and the genetic code) in mitochondria differ from eukaryotic transcription (and the eukaryotic genetic code)?

27. In animal species and in two-thirds of all plant species, what is the origin of the mitochondria (and chloroplasts) found in the zygote?

V. The Eukaryotic Genome (pages 361–368)

A. General Remarks (pages 361–362)

Focus: The amount of DNA per diploid cell differs greatly among the various species of organisms.

28. Is the amount of DNA per diploid cell directly related to the size of an organism? (Cite examples to illustrate your answer.)

29. a. What fraction of eukaryotic DNA is expressed (i.e., codes for protein)?

    b. How does this differ from DNA expression in prokaryotes and viruses?

Focus: Nearly half the DNA in a eukaryotic cell consists of nucleotide sequences that are repeated, i.e., multiple genes.

B. Introns (pages 362–363)

30. *Vocabulary:* Distinguish between introns and exons.

31. How were introns discovered?

32. *Review Question:* Name the four major "surprises" revealed by examination of eukaryotic DNA.

Focus: Most structural genes of multicellular eukaryotes contain introns, which are transcribed onto mRNA and later excised. The number of introns within different genes varies greatly.

33. Cite three examples of genes with introns, including the number of introns in each gene.

34. Describe the relationship between the complexity of an organism and the number of introns in its genes.

35. Which is thought to have evolved first, genes with introns or genes without introns? Explain your answer.

The Function of Introns (page 363)

36. What function might be served by the presence of introns?

37. What has been hypothesized regarding the various exons of a gene? (Illustrate your answer with an example.)

C. Classes of DNA: Repeats and Nonrepeats (pages 363–365)

38. What type of studies led to the discovery of DNA repeats in eukaryotic cells?

39. Read through this section (text pages 364–366) and fill in the following table on the classes of DNA.

Table for Question 39

	SIMPLE-SEQUENCE	INTERMEDIATE-REPEAT	SINGLE-COPY
Degree of repetition (length of sequence, number of repeats)			
Percentage of the genome			
Location(s)			

Simple-Sequence DNA (page 364)

40. What features of simple-sequence DNA facilitated its analysis?

41. Which types of repetitive sequences are likely to form Z-DNA?

42. What role has been proposed for the vast amount of repetitive sequences found in eukaryotic DNA?

43. What is the relationship between simple-sequence DNA and the centromere?

> *Focus:* The same simple, repeated sequence is found at the tips of chromosomes in a wide variety of eukaryotes.

Intermediate-Repeat DNA (page 365)

44. Identify four ways in which intermediate-repeat DNA differs from simple-sequence DNA.

45. Name two known functions of intermediate-repeat DNA.

46. Describe the arrangement of rRNA genes on the chromosome.

> *Focus:* All members of the *Alu* sequence family of intermediate repeats contain a recognition sequence for a specific restriction enzyme.

47. Where are *Alu* sequences frequently located?

D. *Essay:* The Nucleolus (page 366)

48. What forms the structure of the nucleolus?

49. Why does the nucleolus disappear at the beginning of mitosis or meiosis?

50. Identify the function of the nucleolus.

51. Where are the protein components of ribosomes synthesized?

52. Outline the assembly of ribosomal subunits from their protein and RNA components.

Single-Copy DNA (page 366)

53. What genes exist as single-copy DNA?

54. What fraction of single-copy DNA is apparently translated into protein?

55. a. *Vocabulary:* Define a transcription unit.

    b. What separates transcription units?

    c. Introns may constitute up to _____ percent of a transcription unit.

E. Gene Families *(pages 367–368)*

> *Focus:* The identical sequences of some gene families code for molecules needed in large quantities.

56. Name two molecules for which this Focus statement applies.

57. a. Gene families composed of similar but nonidentical sequences code for what types of molecules?

    b. Describe one example.

    c. In humans, what are the structural similarities among the five protein-coding genes for the beta branch of the globin gene family?

    d. Describe how fetal hemoglobin differs functionally from that of the mother and indicate the significance of this difference.

58. a. In the evolution of hemoglobin, the ancestral molecule is thought to have resembled what molecule?

    b. Summarize the evolutionary steps that are believed to have led to the present-day human globin gene family.

    c. What process was most likely responsible for the variations present in the present-day globin gene family?

59. a. *Vocabulary:* What are beta pseudogenes?

    b. What are they thought to represent?

60. Describe one other gene family that codes for related proteins with slightly different properties.

VI. **Transcription and Processing of mRNA in Eukaryotes** *(pages 368–370)*

A. **General Remarks** *(page 368)*

61. a. How is transcription in eukaryotes similar to that in prokaryotes?

    b. Summarize two ways in which transcription in eukaryotes differs from that in prokaryotes.

B. **mRNA Modification and Editing** *(pages 368–369)*

> *Focus:* In eukaryotes, newly transcribed mRNA molecules undergo extensive modification before they are transported to the cytoplasm—the site of protein synthesis.

62. a. Identify the function of the 5′ cap of mRNA.

    b. What is the composition of this cap?

63. How have molecular biologists made use of the poly-A tail of mRNA?

64. Describe the modifications made to an mRNA molecule before it leaves the nucleus.

65. a. What is the consequence of alternative splicing of identical mRNA transcripts?

b. Describe one example. (Figure 18–17)

66. a. In what form are mRNAs transported to the cytoplasm?

b. Name two potential functions of the associated proteins.

C. *Essay:* **RNA and the Origin of Life** *(page 370)*

67. Cite the arguments in favor of (a) DNA and (b) protein as the molecular precursor of life.

68. What evidence suggests that RNA was probably the molecular precursor of life?

69. Scientists currently believe that, in the origin of life, protein, RNA, and DNA appeared in what order? Justify your answer.

70. How can the intron be viewed, in evolutionary terms?

VII. **Genes on the Move** *(page 371–375)*

A. **General Remarks** *(page 371)*

Focus: Like the prokaryotic genome, eukaryotic chromosomes are subject to rearrangements, deletions, and additions.

71. Name three examples involving rearrangements of eukaryotic chromosomes.

B. **Antibody-Coding Genes** *(pages 371–372)*

72. a. *Vocabulary:* Distinguish between antibodies and antigens.

b. What types of molecules may act as antigens?

Focus: The interaction of antibody and antigen is closely analogous to the interaction between enzyme and substrate.

73. What puzzle did antibodies initially present to geneticists?

74. a. Diagram the basic structure of an antibody molecule, labeling heavy and light chains and constant and variable regions. (See also Figure 18–18.)

b. What is the function of the variable regions?

75. Describe the DNA that codes for the variable regions of heavy chains.

Focus: During lymphocyte differentiation, DNA segments coding for variable regions move to new locations on the chromosomes, producing functional genes.

76. Which scientist was given a Nobel Prize for his work on antibody-coding genes?

C. Viruses *(pages 373–374)*

77. List three features common to eukaryotic and prokaryotic viruses.

78. a. *Vocabulary:* What are proviruses?

    b. Identify the two types of proviruses in eukaryotes.

79. What mechanism is responsible for cancer development in newborn hamsters infected with SV40?

80. a. How do reverse transcriptase molecules enter the host cell?

    b. Identify two functions of reverse transcriptase in an infected cell.

81. *Vocabulary:* What are long terminal repeats (LTRs)?

82. Outline retroviral replication, including the factors provided by the virus and those provided by the host cell.

83. Identify two potential consequences of the insertion of DNA derived from a retrovirus into a host chromosome.

84. What is the consequence of retroviral infection of germ cells?

85. What fraction of the total DNA in a mouse may be of retroviral origin?

86. How much of a cell's mRNA may be retroviral?

D. Eukaryotic Transposons *(pages 374–375)*

> *Focus:* Transposons analogous to those of prokaryotes exist in eukaryotic cells.

87. Name three types of organisms in which transposons have been identified.

88. a. How do eukaryotic transposons resemble those of prokaryotes?

    b. Describe one significant difference between eukaryotic and prokaryotic transposons.

    c. Cite the evidence for the feature of eukaryotic transposons described in (b).

89. Study Figure 18–21.

    a. Identify and describe the four elements that may be carried on transposons.

b. Which of these is carried outside the transposon itself?

c. *Vocabulary:* Distinguish between inverted repeats and direct repeats, if you have not done so in (a) and (b).

> *Focus:* Scientists generally accept that the action of reverse transcriptase during retroviral infections has played an important role in the evolution of the eukaryotic genome.

E. *Essay: "It Was Fun . . ." (page 375)*

90. a. What did Barbara McClintock hypothesize based on her studies of corn kernels?

b. What observation prompted this deduction?

c. About how long did it take for McClintock's work to become widely appreciated?

VIII. **Genes, Viruses, and Cancer** *(pages 375–377)*

91. Identify the fundamental mechanism of cancer.

92. What determines the behavior of cancer cells and the prognosis of the resulting disease?

93. Cite three lines of evidence that cancer is associated with changes in the genetic material.

94. What three pieces of evidence slowed the development of a viral theory of cancer?

95. Identify two bits of evidence that confirmed a viral theory of cancer.

96. *Vocabulary:* In the context of eukaryotic cell growth, what are transformed cells?

97. a. *Vocabulary:* What are oncogenes?

b. Oncogenes are thought to cause cancer when either or both of two events occur. Identify these two events.

98. Summarize three mechanisms by which viruses can cause cancer.

99. How were oncogenes discovered?

100. How are the "viral theory" and the "mutation theory" of cancer compatible?

> *Focus:* The products of the oncogenes that have been studied all appear to be regulatory proteins involved in cell growth or cell division.

IX. **Transfers of Genes Between Eukaryotic Cells** *(pages 377–379)*

A. **General Remarks** *(page 377)*

> *Focus:* Perhaps the most ambitious goal of recombinant DNA technology is the replacement of "bad" genes with "good" genes.

101. What steps must be accomplished before the goal in the preceding Focus statement can be reached?

B. To Cells in Test Tubes *(pages 377–378)*

102. What were the transferred gene, the vector, and the host organisms in the first eukaryotic gene transfer experiment?

103. Name two advantages of using viruses as vectors of eukaryotic genes.

104. a. How may cells in tissue culture be induced to take up DNA not carried by a vector?

    b. Use a specific example to describe how cells that have taken up a specific gene may be identified.

Focus: **In a laboratory setting, eukaryotic cells have demonstrated the capacity to incorporate foreign DNA.**

C. To Fertilized Mouse Eggs *(page 378)*

105. a. Who were the first researchers to successfully insert a foreign DNA sequence into fertilized mouse eggs?

b. What gene was used?

c. How did they know the gene had been incorporated in the "right" place?

106. a. What gene was transferred into mouse eggs by Brinster and Palmiter?

    b. How did they know the transfer was successful?

107. *Vocabulary:* What is meant by "transgenic"?

Focus: **In both the experiments described above, the transferred gene was passed on to offspring of the recipient mice.**

---

108. a. In genetic transfers to the fertilized eggs of mice, what fraction of treated eggs survive?

    b. The foreign gene functions in what percentage of surviving recipients?

D. To *Drosophila* Embryos *(page 379)*

109. a. What vector did Spradling and Rubin use in their experiments?

    b. What gene did the vector carry into *Drosophila* embryos?

    c. How did the researchers know that the transfer was successful and that the new gene was incorporated in a stable fashion?

110. Name two areas in which further research is needed before attempts can be made to correct genetic defects.

X. **Summary** *(pages 379–381):* Read the summary. If you are familiar with the essential features of the material presented there, you are ready to complete the section TESTING YOUR UNDERSTANDING.

TESTING YOUR UNDERSTANDING

After you have completed this examination, compare your answers with those in the section that follows.

1. Loosely coiled, right-handed helical DNA is known as _____-DNA, whereas left-handed helical DNA is known as _____-DNA.
   a. A; B
   b. B; Z
   c. A; Z
   d. Z; C
   e. B; C

2. Histones are synthesized during:
   a. S phase of the cell cycle.
   b. telophase of mitosis and meiosis.
   c. M phase of the cell cycle.
   d. prophase of mitosis and meiosis.
   e. cytokinesis of the cell cycle.

3. When a segment of DNA is packaged into a nucleosome, its length is reduced to _____ of what it would be if fully extended.
   a. $1/2$
   b. $1/3$
   c. $1/4$
   d. $1/5$
   e. $1/6$

4. The DNA sequence coding for the hormone insulin is present in:
   a. pancreatic cells only.
   b. liver cells only.
   c. small intestinal cells only.
   d. gall bladder cells only.
   e. every diploid cell of the body.

5. Which of the following is NOT a characteristic of eukaryotic DNA?
   a. The amount of DNA per cell varies greatly among different species of eukaryotes.
   b. The amount of DNA per cell is the same for every normal diploid cell of any given species.
   c. There is a great excess of DNA, the function of which is not known.

   d. The protein-coding sequences of eukaryotic genes are continuous.
   e. Multiple copies of some genes are common.

6. One possible function of introns is to:
   a. reduce mutation by separating the coding regions of a gene.
   b. increase mutation by adding more nucleotides to a gene.
   c. promote recombination by crossing over.
   d. reduce recombination by crossing over.
   e. act as reserves of nucleotide sequences that can be expressed in the event of damage to the exons.

7. The genes for _____ are located in intermediate-repeat DNA.
   a. histones only
   b. globins only
   c. ribosomal RNA only
   d. ribosomal RNA and histones
   e. histones and globins

8. The two pseudogenes of the beta globin family are thought to be:
   a. a series of beta-family exons that are lacking the two beta-specific introns.
   b. a series of beta-family introns that lack the beta-specific exons.
   c. copies of the ancestral globin gene that were left behind as "genetic baggage."
   d. random stretches of DNA nucleotides that bear a superficial resemblance to beta globin genes.
   e. beta-family genes that have been inactivated by one or more mutations.

9. The nucleolus is made up of chromatin loops that code for the synthesis of:
   a. histones.
   b. ribosomal proteins.
   c. small ribosomal subunits.
   d. three of the four kinds of ribosomal RNA.
   e. all four types of ribosomal RNA.

10. DNA viruses that integrate into the chromosome of eukaryotes are called:
    a. viroids.
    b. proviruses.
    c. prophages.
    d. retroviruses.
    e. vectors.

11. Which of these scientists received a lifetime grant for work done on the genetics of "controlling elements" in corn?
    a. Bruce Alberts
    b. Barbara McClintock
    c. Frank Ruddle
    d. Allan Spradling
    e. Gerald Rubin

12. Early indications are that the vector(s) of choice in the manipulation of the human genome may be:
    a. transposons.
    b. cosmids.
    c. bacteriophage φX174.
    d. transgenic mice.
    e. DNA precipitated with calcium.

13. In experiments involving the uptake of DNA by eukaryotic cells, it is important to know which cells have actually taken up the DNA. Therefore, scientists have created a line of auxotrophic mouse cells that lack:
    a. reverse transcriptase.
    b. restriction enzymes.
    c. thymidine kinase.
    d. kinetochore-binding factor.
    e. DNA polymerase.

14. T or F   Biochemical analyses provided the first clue to the structure of the nucleosome.

15. T or F   The biosynthesis of different proteins by different types of cells is reflected by varying patterns of chromatin condensation among the different cell types.

16. T or F   The expression of eukaryotic DNA in which cytosines are not methylated cannot be regulated.

17. T or F   In a progression from simpler to more complex life forms (e.g., from prokaryotes to *Drosophila* to salamanders to humans), there is a direct relationship between the level of complexity and the amount of DNA possessed by a species.

18. T or F   Introns have been identified in prokaryotic, eukaryotic, and viral DNA.

19. T or F   In eukaryotes, there is a promoter for each structural gene, but in prokaryotes, one promoter functions for all the structural genes of an operon.

20. T or F   The genetic material of a mouse contains separate genes to code for 10 million different antibodies.

21. T or F   Once integrated into the chromosome, the DNA produced by the reverse transcriptase of retroviruses is transcribed by the host cell's own RNA polymerase.

22. T or F   It seems that many of the repeated sequences in the eukaryotic genome may have originated from transposons.

23. T or F   All known cancer-causing viruses are viruses that introduce genetic information into the host cell.

24. T or F   In Gordon's and Ruddle's experiments, the production of rabbit hemoglobin by mouse red blood cells indicates that the foreign rabbit globin gene was inserted into the correct place in the mouse genome and came under the control of the mouse gene regulatory system.

25. The differentiation of cells in multicellular organisms apparently depends on the inactivation of certain groups of genes and activation of others. What evidence led to this conclusion?

26. What evidence suggests that RNA evolved before either proteins or DNA?

27. What two techniques permitted Susumu Tonegawa to perform his Nobel Prize-winning work on the genetics of antibodies?

28. List three lines of evidence that link the development of cancer to changes in the genetic material.

29. How does the DNA of mitochondria and chloroplasts differ from the chromosomal DNA of eukaryotic cells?

PERFORMANCE ANALYSIS

1. **c**   The tightly coiled right-handed DNA helix first described by Watson and Crick is B-DNA. A-DNA is a right-handed helix that is less tightly coiled than the B form. Z-DNA is a left-handed helix that results in a zigzagging sugar-phosphate backbone. (page 356)

2. **a**   Histones are synthesized during the S phase of the cell cycle. (page 356)

3. **e**   When packaged into a nucleosome, a segment of DNA is reduced to one-sixth of its extended length. (page 357)

4. **e**   The gene for every protein that an organism can produce is present in every diploid cell in the body. However, every cell does not produce every protein. During the process of differentiation, some genes are inactivated and others are activated so that the fully specialized cell only produces a particular set of proteins. (page 359)

5. **d**   The protein-coding sequences (exons) of eukaryotic DNA are interrupted by noncoding sequences called introns. All the other statements are true of eukaryotic DNA. (page 361)

6. **c**   Introns are thought to have persisted because they promote genetic recombination through crossing over, which is more likely to occur in genes containing introns because of their greater length. (Remember that crossing over with recombination is more likely the greater the distance between genes.) (page 363)

7. **d**   The genes for rRNA and histones are located in intermediate-repeat DNA. All other protein-coding genes are single-copy DNA. (page 365)

8. **e**   The two pseudogenes in the beta globin family are thought to be beta-family genes in which mutations have occurred to such an extent that the genes are no longer functional. (page 368)

9. **d**   The nucleolus is the site of ribosome assembly. It is composed of chromatin loops that code for the synthesis of three of the four kinds of ribosomal RNA. (essay, page 366)

10. **b**   Proviruses are DNA viruses that integrate into the host chromosome. (page 373)

11. **b**   Barbara McClintock reported the existence of "jump-

ing genes" in corn decades before the scientific community accepted the presence of moveable genetic elements. She eventually gained recognition for her discovery and received a lifetime grant for her work. (essay, page 375)

12. **a** In the genetic transfer studies performed thus far, it appears that transposons may be the vectors of choice for transferring human genes. (page 379)

13. **c** As you will recall from Chapter 8, auxotrophic cells are those that can grow only on a supplemented medium. Scientists have created TK$^-$ mouse cells which lack the enzyme thymidine kinase and therefore can only grow on a special medium. In gene transfer studies, scientists insert the gene of interest *and* the gene for thymidine kinase into a vector. The cells that take up the gene of interest (and thus the thymidine kinase gene) can grow in a medium that would not support TK$^-$ cells. (page 378)

14. **False** Electron micrographs provided the earliest clues to nucleosome structure. The structure was further elucidated through analyses using biochemical techniques. (page 357)

15. **True** Gene activity is related to the degree of chromosome condensation. Highly condensed chromosome segments are inactive and therefore are not transcribed. Cells with different functions display different patterns of chromosome condensation. (page 359)

16. **False** Methylation of cytosines is one mechanism by which gene expression is thought to be inhibited. However, organisms in which cytosines are not methylated are quite capable of normal gene regulation. (page 360)

17. **False** There is no correlation between the total amount of DNA present in the cells of a eukaryotic species and its complexity. Salamanders have far more DNA per cell than do humans. (page 361)

18. **False** Introns have been identified in the DNA of eukaryotes and viruses, but not in bacterial DNA. (pages 362, 363)

19. **True** In prokaryotes, structural genes coding for proteins that have related functions may be grouped into operons in which the genes are under the influence of one promoter and are transcribed as a unit onto the same mRNA. In eukaryotes, each structural gene has its own promoter and its transcription is regulated separately from that of all other genes. (page 368)

20. **False** The genome of a mouse is capable of making 10 million different antibodies because of various combinations of V, D, and J genes. The various combinations are created during lymphocyte differentiation. (page 371)

21. **True** Once viral DNA is incorporated into a host chromosome, it is transcribed by the host cell's RNA polymerase as if it were host cell DNA. (page 373)

22. **True** There is evidence that many of the repeated sequences in the genomes of eukaryotes originated from transposons. (page 374)

23. **True** All known cancer-causing viruses introduce genetic information into host cell chromosomes. (page 376)

24. **True** A major goal in gene transfer is incorporating the gene of interest into the correct place in the genome so that regulation of gene expression is normal. The experiments in which mouse red blood cells produced rabbit hemoglobin accomplished this goal. (page 378)

25. (1) J. B. Gurdon removed the nuclei of tadpole intestinal cells and implanted these nuclei into egg cells in which the nuclei had been destroyed. Normal development occurred in some of the eggs. (2) F. C. Steward cultivated an entire carrot plant from a single *differentiated* cell of a carrot. Both of these experiments indicated that differentiated cells contain all the genetic information needed to form a complete, functioning organism. (Figure 18–7)

26. T. C. Cech and his coworkers discovered that the RNA of some cells has an intron that functions as an enzyme and catalyzes the excision and splicing of the RNA itself. Examples of RNA catalysts for other types of reactions have been found. These discoveries have led Bruce Alberts to speculate that an RNA capable of catalyzing its own replication was a crucial early event in the evolution of macromolecules and, subsequently, of life. (essay, page 370)

27. Susumu Tonegawa used restriction enzymes and refined hybridization techniques to test the hypothesis that constant and variable regions of antibody molecules might be encoded by separate genes. (page 372)

28. (1) Once a cell becomes cancerous, all of its daughter cells are also cancerous. (2) Gross chromosomal abnormalities are often observable in cancerous cells. (3) Most agents that are known to cause cancer are also known to cause mutations. (page 376)

29. Mitochondria and chloroplasts have their own DNA, which replicates when these organelles divide by binary fission. The triplet code of the DNA for these organelles differs from that of chromosomal DNA of both prokaryotes and eukaryotes. In animal species and in two-thirds of all plant species, the mitochondria and chloroplasts of the zygote are of maternal origin only. (essay, page 361)

CHAPTER 19

Human Genetics: Past, Present, and Future

MAJOR CONCEPTS

Historically, human genetics has been studied by tracing hereditary conditions through families. Recent advances in molecular genetics and recombinant DNA techniques have provided means for diagnosing human genetic disorders and offer hope for the development of cures and preventive measures.

Each human somatic cell has 44 autosomes and two sex chromosomes. A graphic representation of the chromosomes can be made by preparing a karyotype from cells in metaphase of mitosis. A karyotype enables a geneticist to determine the number, size, and shape of the chromosomes and to identify homologous pairs. It reveals abnormal chromosome structure and extra or missing chromosomes.

Nondisjunction occurs when homologous chromosomes or their chromatids do not separate. When nondisjunction occurs during meiosis, abnormal gametes result—some have one or more extra chromosomes and some are missing one or more chromosomes. Gametes with missing chromosomes do not produce viable offspring, but a gamete with one or more extra chromosomes *may* produce a viable embryo. Individuals with extra sex chromosomes are usually sterile.

An individual with Down's syndrome is characterized by mental retardation, a short, stocky body with a thick neck, a large tongue, and increased susceptibility to infections, and may have abnormalities of other organs. Down's syndrome results from an extra copy of chromosome 21 and may be caused by nondisjunction or by translocation in the chromosomes of a parent.

In addition to nondisjunction and translocation, small chromosome deletions can cause genetic disorders, such as Wilms' tumor.

The genetic abnormalities for many diseases can be detected prenatally by amniocentesis or by obtaining a biopsy sample of the chorion. Chorionic biopsy can be done as early as the eighth week of pregnancy, whereas amniocentesis is usually not performed until the sixteenth week.

Many genetic disorders are inherited as Mendelian recessives. In these cases, the disease occurs when the two abnormal recessive alleles are present in an individual. The abnormal allele typically codes for a defective or poorly functioning polypeptide, or the polypeptide is not produced at all. Heterozygotes are usually phenotypically normal because production of functional polypeptide is directed by the normal allele, but they are carriers of the genetic abnormality. Examples of such disorders are phenylketonuria, albinism, Tay-Sachs disease, and sickle cell anemia.

Two genetic disorders caused by the presence of a single dominant allele are achondroplastic dwarfism and Huntington's disease. An apparently high mutation rate contributes to the maintenance of achondroplastic dwarfism in humans. Huntington's disease persists largely because symptoms do not appear in affected individuals until they are in their thirties or forties, often after they have had children.

Most sex-linked disorders in humans are due to recessive alleles carried on the X chromosome. Such diseases are present in a much higher incidence in males than in females because males have only one X chromosome. If a man is affected by a sex-linked disorder, he received the allele from his mother and will pass the allele on to all of his daughters. Females can be affected by recessive sex-linked disorders if their father is affected and they also receive the defective allele from their carrier mother. Examples of sex-linked disorders in humans are color blindness, hemophilia, and Duchenne muscular dystrophy.

One technique for diagnosing genetic diseases is to identify genetic markers known as RFLPs, restriction-fragment-length polymorphisms, using radioactive probes. This technique has been used to identify RFLPs for sickle cell anemia and Huntington's disease. The ability to diagnose genetic disorders and, in some cases, carriers of genetic diseases raises ethical and moral dilemmas for individuals and societies.

Radioactive probes can also be used to identify the location of a gene coding for a particular protein. From the amino acid sequence of the protein, a radioactive RNA probe can be synthesized. Its hybridization with restriction fragments from the genome directs scientists to the correct chromosome. Similar probes can be used to identify the abnormal allele of sickle cell anemia. In Duchenne muscular dystrophy, the inability of a radioactive probe to pair with a restriction fragment indicates that the gene for the protein dystrophin is defective, and the disease exists.

Several groups of scientists are in the process of mapping the human genome. One approach taken by some is to identify the locations of genes that code for particular proteins. An alternative approach is to determine the DNA sequence of the entire genome, a process that has been facilitated by the development of automatic sequencing techniques.

HOW TO STUDY THE CHAPTER

Read the entire chapter through quickly, focusing on the major concepts.

Use the GUIDED STUDY OF THE CHAPTER to help you identify the important details as you **reread** the chapter. Writing out the answers to these questions will help fix them in your mind as well as provide you with a valuable study aid.

Answer the questions in TESTING YOUR UNDERSTANDING without the aid of your text. Check your answers against those in PERFORMANCE ANALYSIS. Analyzing your answers will give you valuable feedback on your level of understanding and preparedness for classroom testing.

GUIDED STUDY OF THE CHAPTER

I. Introduction (page 382)

1. Cite three ways in which studies of human genetics differ from genetic studies in other species.

2. What has historically been the source of most information on human genetics?

> *Focus:* Recent advancements in molecular genetics have revolutionized the study of human genetics.

II. The Human Karyotype (pages 383–388)

A. General Remarks (page 383)

3. How many chromosomes are present in a normal human diploid cell?

4. a. Why was the correct determination of human chromosome number so difficult?

   b. What technique finally enabled an accurate count of human chromosome number?

5. a. *Vocabulary:* Define a karyotype.

b. What information can be derived from a karyotype?

6. a. Chromosomes that have a similar size and shape can be distinguished by what technique? (See also Figure 19-3.)

b. Chromosomes in what mitotic phase are useful in determining banding patterns? (Figure 19-3)

B. **Essay: Preparation of a Karyotype** (page 384)

7. Chromosomes of a karyotype are in what mitotic phase?

8. a. What drug is used to interrupt mitosis at the correct phase for karyotype preparation?

b. How does this drug work?

> **Focus:** A karyotype is constructed from photographs of metaphase chromosomes that have been cut out and arranged according to size and shape.

9. How are chromosome pairs matched up?

C. **Chromosome Abnormalities** (pages 385–388)

10. a. *Vocabulary:* What is nondisjunction and how might it occur?

b. Identify the consequences of nondisjunction when it occurs during meiosis.

11. a. Under what conditions can a viable embryo result following nondisjunction?

b. What is usually the fate of this embryo?

12. What abnormalities are typically present in an individual with an extra autosomal chromosome?

13. *Vocabulary:* Distinguish between deletions and translocations.

Down's Syndrome (pages 385–386)

14. *Vocabulary:* Define a syndrome.

15. Identify five abnormalities characteristic of individuals with Down's syndrome.

16. Individuals with Down's syndrome who live into their thirties or forties are at risk for developing what condition?

17. a. What is the genetic basis for Down's syndrome?

    b. Identify two causes of this genetic abnormality and indicate which cause is most common.

18. a. In translocation Down's syndrome, what genetic abnormality is usually present in one parent?

    b. If a woman with this condition is impregnated by a man with a normal karyotype, what is the fate of the embryo? (Identify the six possible outcomes.) (See also Figure 19–5.)

    c. What is the usefulness of preparing a karyotype for parents who already have a child with Down's syndrome?

19. What is the relationship (known or postulated) between the incidence of Down's syndrome and the age of the parents? (See also Figure 19–6.)

Abnormalities in the Sex Chromosomes (page 386)

20. a. Identify three abnormal combinations of sex chromosomes that produce maleness.

    b. Characterize these males.

21. What abnormalities in the sex chromosomes produce females?

Focus: **Abnormalities in the sex chromosomes may be caused by nondisjunction. Such individuals are usually sterile.**

Chromosome Deletions (pages 386–387)

22. a. Cite one example of a chromosome deletion that causes heritable illness.

    b. What phenotypic feature alerts doctors to look for the chromosome deletion?

    c. Why is it important to determine whether a child with aniridia also has a chromosome 11 deletion?

Prenatal Detection (pages 387–388)

23. a. Outline the procedure for amniocentesis.

    b. What is the origin of the cells obtained by this technique?

    c. An alternative technique collects cells from the chorion. What is the advantage of this procedure?

24. If prenatal testing reveals a genetic abnormality in the fetus, what are the parent's options? (List them all, even if you personally object to one or more.)

III. PKU, Sickle Cell Anemia, and Other Recessives (*pages 388–390*)

A. General Remarks (*page 388*)

25. a. For disorders resulting from the presence of two recessive alleles, name two possible mechanisms by which the defective gene causes the disease.

    b. Why are heterozygotes usually phenotypically normal?

B. Phenylketonuria (*page 388*)

26. a. What defect characterizes PKU?

    b. Identify the consequences of this defect.

27. What is the incidence of PKU in the United States?

28. Describe the common course of events in a child with untreated PKU.

> **Focus:** Most states require routine testing of neonates to identify PKU homozygotes, which can be managed successfully by feeding a special diet.

29. What modification of the diet is necessary for the infant with PKU?

C. Albinism (*page 389*)

30. Identify two genetic mechanisms causing albinism in humans.

D. Tay-Sachs Disease (*page 389*)

31. Describe the usual course of events for an infant born with Tay-Sachs disease.

32. Describe the genetic and biochemical mechanisms responsible for Tay-Sachs disease.

33. a. The incidence of Tay-Sachs disease is higher in what ethnic group than in the general population?

    b. What fraction of this group is estimated to carry the Tay-Sachs allele?

> **Focus:** The incidence of Tay-Sachs disease in the American Jewish population has dropped dramatically since the development and use of a test to identify heterozygous carriers.

E. Sickle Cell Anemia (*page 390*)

34. Where did the allele that causes sickle cell anemia most likely originate?

35. What fraction of the American black population is (a) heterozygous and (b) homozygous for the allele?

36. Identify the physiological mechanism responsible for the anemia and the pain associated with the disease.

37. How may heterozygote carriers of the sickle cell allele be identified?

> *Focus:* Approximately 20 of the known 100 hereditary variations of the hemoglobin molecule cause disease.

38. *Review Question:* Of the four inherited disorders discussed in this section, which are treatable if detected at birth?

IV. Dwarfs and Other Dominants *(page 391)*

39. Why are serious medical problems caused by autosomal dominant alleles less common than those caused by autosomal recessives?

40. What factor contributes to the maintenance of achondroplastic dwarfism in the human population?

41. a. Characterize Huntington's disease.

b. What aspects of the disease contribute to its persistence in the human population?

V. Sex-Linked Traits *(pages 391–394)*

A. Color Blindness *(pages 391–392)*

42. What fraction of human males and females are red-green color-blind?

> *Focus:* The three genes responsible for color vision code for three visual pigments, each of which responds to a different portion of the visible spectrum.

43. a. Identify the two possible mechanisms of red-green color blindness.

b. Which type is more common?

44. a. Account for the much higher incidence of color blindness in men than in women.

b. Why can a woman be color-blind in only one eye?

c. Under what circumstances can a woman have complete red-green color blindness?

> *Focus:* If a man is color-blind, his mother carried the allele for color blindness and all of his daughters will also be carriers.

B. Hemophilia *(pages 392–393)*

45. What is hemophilia?

46. a. Specifically, what causes hemophilia A in humans?

    b. How is this condition inherited?

47. Identify two problems associated with treating hemophiliacs.

Focus: **Attempts are being made to use recombinant DNA techniques to produce human Factor VIII.**

---

48. What evidence indicates that Queen Victoria was the original carrier of the hemophilia allele that spread through several royal families of Europe?

C. Muscular Dystrophy *(pages 393–394)*

49. a. What health problems are associated with Duchenne muscular dystrophy?

    b. How is this condition inherited?

    c. When do the first symptoms appear and what is the life expectancy of an affected individual?

50. a. What is the apparent physiological abnormality in Duchenne muscular dystrophy?

    b. Describe the muscle changes that result from this abnormality.

51. What factor is complicating the cloning and sequencing of the gene coding for dystrophin?

Focus: **Duchenne muscular dystrophy is apparently caused by deletions in the gene coding for the muscle protein dystrophin.**

VI. Diagnosis of Genetic Diseases: RFLPs *(pages 394–397)*

A. Sickle Cell Anemia *(pages 394–395)*

52. Outline the use of a radioactive probe to detect the sickle cell allele.

53. RFLP is an acronym for:

Focus: **The use of RFLPs in tests detecting genetic disorders is based on the fact that mutations lead to restriction fragments of lengths that differ from those produced by the normal allele.**

---

54. a. *Vocabulary:* What is a genetic marker?

b. How has the concept of the genetic marker changed?

b. How was this probe used?

c. What is the nature of the Factor VIII gene?

B. Huntington's Disease *(pages 395–397)*

> *Focus:* Markers for a genetic disease can be identified even if the normal and abnormal alleles have not been identified.

55. With respect to RFLPs associated with Huntington's disease, how did the American family differ from the Venezuelan family?

56. What types of information were collected in the study of the Venezuelan family?

59. a. Name three genetic diseases for which radioactive probes are used to identify affected and/or carrier individuals.

b. How does the use of radioactive probes to diagnose Duchenne muscular dystrophy differ from their use in the other two diseases?

60. What has been the primary outcome of the development of these radioactive probes?

> *Focus:* Current studies of Huntington's disease are concentrating on identification of the normal and abnormal alleles with the hope that the mechanism of the disease will be elucidated.

57. RFLPs have also been found for which other genetic disease?

B. *Essay:* **Witness for the Prosecution** *(page 396)*

61. The use of "DNA fingerprinting" is based on what characteristic of DNA?

62. Outline the procedure for preparing a DNA fingerprint.

VII. Diagnosis of Genetic Diseases: Radioactive Probes *(pages 397–398)*

A. General Remarks *(pages 397–398)*

58. a. In the search for the gene coding for Factor VIII, how was a radioactive probe constructed?

63. a. *Vocabulary:* What is PCR?

b. Summarize the procedure for PCR.

64. What is unusual about the bacterium from which the DNA polymerase used in PCR originated?

65. Cite four situations in which gene amplification by PCR has been used.

C. *Essay:* **Some Ethical Dilemmas** *(page 398)*

Focus: **The ability to detect genetic disorders in humans creates many ethical dilemmas for which there are no simple answers.**

66. Identify four issues that arise from our ability to detect genetic diseases and present human society with difficult choices.

VIII. **The "Book of Man"** *(pages 398–399)*

67. Outline two different approaches to the mapping of the human genome.

68. What are the major reasons for the controversy over the plan to sequence the entire human genome?

69. What technological advancement is making the sequencing of the entire human genome more feasible?

Focus: **The sequencing of the human genome may provide clues to gene regulation, to the evolutionary relationships among genes, and to the evolutionary relationships among humans and other species.**

IX. **Summary** *(pages 399–400):* Read the summary. If you are familiar with the essential features of the material presented there, you are ready to complete the section TESTING YOUR UNDERSTANDING.

TESTING YOUR UNDERSTANDING

After you have completed this examination, compare your answers with those in the section that follows.

1. Each normal human diploid cell has two sex chromosomes and _____ autosomes.
   a. 40   d. 46
   b. 42   e. 48
   c. 44

2. By what mechanism does colchicine work?
   a. It blocks mitotic cells in metaphase by interfering with spindle microtubules.
   b. It prevents anaphase by altering centromere structure, thereby inhibiting the separation of sister chromatids.
   c. It holds chromosomes in early prophase by interfering with the formation of a spindle from the centrioles.
   d. It stops mitosis in early prophase by limiting chromosome condensation.
   e. It stops mitosis in metaphase by killing the cells.

3. During meiosis or mitosis, the phenomenon in which chromosomes or sister chromatids fail to separate is known as:
   a. translocation.   d. nondisjunction.
   b. crossing over.   e. Down's mutation.
   c. fusion.

4. Which two conditions are present in MOST individuals that have extra autosomes?
   a. heart defects and mental retardation
   b. mental retardation and sterility
   c. heart defects and sterility
   d. speech impairments and heart defects
   e. sterility and increased susceptibility to infections

5. Down's syndrome occurs when:
   a. an individual has any extra chromosome.
   b. an extra copy of chromosome 21 is present.
   c. an extra copy of chromosome 14 is joined to chromosome 21.
   d. a deletion error occurs in either chromosome 21 or chromosome 14.
   e. half of chromosome 21 is broken off and joined to chromosome 14.

6. One parent has a normal karyotype and the other parent is an asymptomatic carrier of the Down's syndrome translocation. Of the possible genetic combinations, _____ will be lethal and _____ will produce a normal child.
   a. 2 of 6; none
   b. 2 of 6; 1 of 6
   c. 3 of 6; none
   d. 3 of 6; 1 of 6
   e. 4 of 6; none

7. In cases of Down's syndrome due to nondisjunction, in what fraction does the extra chromosome come from the father?
   a. 10 percent
   b. 25 percent
   c. 30 percent
   d. 50 percent
   e. 70 percent

8. What phenotypic trait alerts doctors to look for the genetic anomaly associated with Wilms' tumor?
   a. absence of the iris
   b. extremely pale eye color
   c. squinted, slanted eyes
   d. increased urine production
   e. decreased urine production

9. What is a major advantage of chorionic sampling over amniocentesis for the prenatal detection of chromosome abnormalities?
   a. Chorion samples may be obtained as early as the eighth week of pregnancy.
   b. Since a sonogram is unnecessary, the procedure is much safer.
   c. Since fresh rather than sloughed cells are procured, the results of genetic analysis are more accurate.
   d. Chorionic sampling does not upset the osmotic balance between the amniotic fluid and the fetus.
   e. The premise is false, amniocentesis is far superior to chorionic sampling in every respect.

10. What is the inborn error of metabolism in phenylketonuria?
    a. The enzyme that synthesizes phenylalanine is missing or defective.
    b. Stored phenylalanine cannot be retrieved for use by the body and accumulates, with toxic effects.
    c. An abnormal enzyme is produced and it converts phenylalanine into ketones, which poison the body.
    d. Excess melanin is produced directly from phenylalanine, coloring the urine black.
    e. Phenylalanine cannot be converted to tyrosine.

11. The incidence of PKU in the United States is one affected (homozygous) infant per _____ born.
    a. 1,000
    b. 5,000
    c. 10,000
    d. 15,000
    e. 20,000

12. The symptoms of Tay-Sachs disease are caused by the:
    a. inability to make the pigment melanin.
    b. presence of an extra copy of chromosome 11.
    c. absence of the essential muscle protein, dystrophin.
    d. accumulation of phenylalanine and its abnormal breakdown products in the bloodstream.
    e. accumulation of GM_2 ganglioside in brain cell lysosomes.

13. In one form of albinism, the inborn error of metabolism is the:
    a. destruction of tyrosine by an abnormal enzyme, thereby preventing synthesis of melanin.
    b. premature breakdown of phenylalanine, which is then unavailable for tyrosine synthesis.
    c. inability of an enzyme required to produce melanin to enter the pigment cells.
    d. production of an abnormal enzyme that destroys melanin as it is being synthesized.
    e. production of an enzyme that adds a methyl group to melanin, rendering it a colorless molecule.

14. In sickle cell anemia, what causes the anemia characteristic of the disease?
    a. Affected individuals typically bleed into their joints, causing anemia and a painful arthritis.
    b. Persons with sickle cell anemia also have clotting abnormalities that result in blood loss anemia.
    c. An associated abnormality is increased fragility of capillaries, which are easily injured and result in an anemia of chronic blood loss.
    d. The production of red blood cells occurs at a much slower rate than normal and does not match the loss of aged red blood cells.
    e. Abnormal hemoglobin molecules form stiff fibers inside red blood cells that distort the cell, resulting in premature degradation.

15. How may heterozygous carriers of the sickle cell allele be identified?
    a. Their anemia will be mild compared to that of a homozygote.
    b. Heterozygotes have half the normal amount of N-acetyl-hexosaminidase in their blood.
    c. If a drop of fresh, untreated blood is smeared on a microscope slide, half the red blood cells of a heterozygote will be sickle-shaped.
    d. Some of the cells of a heterozygote will assume the sickle shape if all the oxygen is removed.
    e. At this time, there is no reliable technique for identifying sickle cell heterozygotes.

16. If the parents of a child are both heterozygous carriers of the sickle cell allele, there is a _____ percent chance that the child will also be a heterozygous carrier.
    a. 100
    b. 75
    c. 50
    d. 25
    e. 10

17. Which statement is NOT true of the relatives of a color-blind man whose parents both had normal color vision?
    a. All of his sons will be color-blind.
    b. All of his daughters will have at least one allele for color blindness.
    c. His mother had one allele for color blindness.
    d. There is a 50 percent chance that his sister has one allele for color blindness.
    e. All of these statements are true of the relatives of a color-blind man whose parents both had normal color vision.

18. What factor maintains Huntington's disease in the human population?
    a. Individuals with the Huntington's allele are resistant to several serious childhood diseases.
    b. By the time signs of the disease appear, an individual has often already had children.
    c. The condition is inherited by a recessive allele and there is not yet any way to detect heterozygous carriers.
    d. An increased conception rate and multiple births are associated with the allele for Huntington's disease.
    e. In the communities where the disease has been studied, marriages among relatives are common and this inbreeding concentrates the allele.

19. In most color-blind men, _____ cannot be distinguished from _____ because the gene for the _____ -responsive pigment is defective.
    a. green; blue; green
    b. green; red; red
    c. green; red; green
    d. red; blue; red
    e. blue; green; blue

20. Hemophilia A is inherited in what manner?
    a. a recessive allele on an autosome
    b. a recessive allele on the X chromosome
    c. a recessive allele on the Y chromosome
    d. a dominant allele on the X chromosome
    e. a dominant allele on the Y chromosome

21. The apparent abnormality in Duchenne muscular dystrophy is:
    a. absence of the muscle protein dystrophin.
    b. inability to synthesize Factor VIII muscle protein.
    c. a progressive degeneration of the nerves innervating skeletal muscle, resulting in muscle wasting.
    d. accumulation of GM_2 ganglioside in the nerves that control skeletal muscle contractions.
    e. long-term lactic acid accumulation due to chronic inadequate oxygen supply to muscles.

22. When the DNA from a person with sickle cell anemia is cleaved with restriction enzyme HpaI, the radioactive beta-globin probe hybridizes with a restriction fragment that is _____ nucleotides long.
    a. 4,000 or 4,600
    b. 5,600
    c. 7,000 or 7,600
    d. 10,000 or 11,000
    e. 13,000

23. In one American family in which Huntington's disease occurs, the disorder has been correlated with which restriction fragment pattern when the restriction enzyme HindIII is used?
    a. A
    b. B
    c. C
    d. D
    e. Both A and C restriction fragment patterns are associated with Huntington's disease in this case.

24. The most precise test to detect carriers of the sickle cell allele utilizes what technique?
    a. The blood level of Factor VIII is measured.
    b. The blood level of N-acetyl-hexosaminidase is measured.
    c. A genetic marker located near the sickle cell allele is detected by a radioactive probe.
    d. Synthetic radioactive probes detect the single base-pair difference between the alleles for normal and sickle cell beta globin.
    e. Radioactive probes detect both long and short restriction fragments.

25. What constitutes a positive result (i.e., the child will be affected) in the current prenatal test for Duchenne muscular dystrophy?
    a. Radioactive probes detect the characteristic restriction-fragment-length polymorphism.
    b. Assaying a sample of chorion reveals the absence of dystrophin from fetal tissues.

c. A synthetic radioactive probe hybridizes with the altered nucleotide sequence in the gene for dystrophin.

d. The abnormal, nonfunctional protein produced by the defective gene is identified in a chorion tissue sample.

e. A synthetic radioactive probe fails to hybridize with a segment of the gene coding for dystrophin.

26. The "DNA fingerprinting" technique is based on:

a. the arrangement of the intermediate-repeats in a person's DNA.

b. the nucleotide sequence of the gene that determines fingerprint patterns.

c. the number of repeated units in the simple-sequence DNA.

d. hybridization between restriction fragments of a suspect and restriction fragments of DNA found at the scene of a crime.

e. comparing the restriction-fragment-length pattern of a suspect with that of DNA found at the crime scene.

27. **T or F** Women with a sex chromosome genotype of *XXX* may be normal.

28. **T or F** Owing to the prevalence of the Tay-Sachs allele in Ashkenazic Jews, most Tay-Sachs infants born in the United States are born to Jewish parents.

29. **T or F** Attempts to clone and sequence the gene coding for dystrophin are complicated by the fact that it is the largest human gene known and it has numerous exons and huge introns.

30. Outline the slightly different approaches of Ray White and Helen Donis-Keller to the mapping of the human genome.

31. Some genetic disorders result from the presence of two recessive alleles for the same gene. In terms of the polypeptide products produced by such alleles, what two outcomes may result from the presence of the defective allele?

32. Cite two reasons why hemophilia is a life-threatening disorder.

PERFORMANCE ANALYSIS

1. **c** Each normal human diploid cell has 46 chromosomes: two sex chromosomes and 44 autosomes. (page 383)

2. **a** The drug colchicine is used to stop dividing cells in metaphase. It acts by interfering with spindle microtubules. (essay, page 384)

3. **d** Nondisjunction occurs during meiosis or mitosis when homologous chromosomes or sister chromatids do not separate. (page 385)

4. **b** The two conditions that are present in MOST individuals that have extra autosomes are mental retardation and sterility. Other conditions include heart defects, speech impairments, and increased susceptibility to infections. (page 385)

5. **b** Down's syndrome occurs when an extra copy of chromosome 21 is present. This may result from nondisjunction or from translocation involving chromosome 21 and chromosome 14. (page 385)

6. **d** There are six possible genetic combinations for the situation described, three of which will not produce a viable embryo. Of the other combinations, one will produce a normal individual, one will produce an asymptomatic carrier of the 14/21 translocation, and one will produce a child with Down's syndrome. (Figure 19–5)

7. **b** In cases of Down's syndrome due to nondisjunction, 25 percent are caused by an extra chromosome that came from the father. (page 386)

8. **a** A deletion in the short arm of chromosome 11 is associated with aniridia and Wilms' tumor. When a physician sees a child with aniridia, a test to look for the chromosome deletion should be performed. (page 387)

9. **a** Sampling the chorion can be done as early as the eighth week of pregnancy, providing time for the parents to obtain the test results and make an informed decision concerning maintaining or terminating the pregnancy. Amniocentesis cannot be performed with safety and good results until the 16th week of pregnancy. At this stage of the pregnancy, greater risks are associated with abortion. (page 388)

10. **e** In phenylketonuria (PKU), phenylalanine cannot be converted to tyrosine. Phenylalanine and abnormal metabolic products accumulate in the blood and urine. These substances are toxic to cells of the developing nervous system and cause mental retardation. If detected early enough, PKU can be successfully managed by dietary restrictions. (page 388)

11. **d** The incidence of PKU in the United States is one affected infant per 15,000 born. (page 389)

12. **e** Tay-Sachs disease is a progressive neurological disorder caused by inability to produce the enzyme N-acetyl-hexosaminidase, which breaks down the lipid GM_2 ganglioside. The lipid causes brain damage by accumulating in brain cell lysosomes. There is no effective therapy for this disorder. (page 389)

13. **c** There are two forms of albinism. In one form, an enzyme required to produce melanin is produced but cannot enter the pigment cells. In the other form, which is more common, one of the enzymes needed for the biosynthesis of melanin is absent. (page 389)

14. **e** The anemia of sickle cell anemia is caused by the premature destruction of red blood cells. Sickle cell hemoglobin molecules form stiff fibers that distort the shape of red blood cells, increase their fragility, and greatly reduce their flexibility. (page 390)

15. **d** There are at least three means by which carriers of the sickle cell allele may be identified. In the simplest test, some of the red blood cells of a heterozygote will assume the sickle shape if a blood sample is treated to remove the oxygen from the hemoglobin. A second test involves the detection of RFLPs. The third test uses a radioactive DNA hybridization probe to identify the sickle cell allele. (pages 390, 394, 397)

16. **c** Create a Punnett square: *S* is the normal allele; *s* is the sickle cell allele. The genotype of both parents is *Ss*. From your Punnett square, you will see that each child has a one in four chance of being normal, a one in four chance of having sickle cell anemia, and two chances in four (a 50 percent chance) of being a heterozygous carrier. (page 390)

17. **a** Since the genes for color blindness are carried on the X chromosome, they cannot be transmitted from father to son. They can be transmitted from father to daughter and from mother to either son or daughter. All the other statements are true in the scenario described. (page 392; Figure 19–14)

18. **b** The symptoms of Huntington's disease do not appear until an individual is in his or her thirties or forties. By this time, it is likely that the individual has already had children. (page 391)

19. **c** In the most common form of red-green color blindness, the gene for the green-responsive pigment is defective, and green appears to be red. In 25 percent of the cases, the gene for the red-responsive pigment is defective, and red appears to be green. (page 392)

20. **b** Hemophilia A is inherited by a recessive allele on the X chromosome. (page 392)

21. **a** Duchenne muscular dystrophy is an X-linked disorder associated with the absence of the muscle protein dystrophin. (page 394)

22. **e** One test for sickle cell anemia utilizes RFLPs. In 87 percent of persons with sickle cell anemia, the probe hybridizes with a restriction fragment of 13,000 base pairs. In normal individuals, the probe hybridizes with a fragment either 7,000 or 7,600 base pairs long. (page 394)

23. **a** In the study described, Huntington's disease was correlated with restriction fragment pattern A. In a study of a Venezuelan family, Huntington's disease was correlated with pattern C. (page 395)

24. **d** The most precise test to detect carriers of the sickle cell allele utilizes synthetic radioactive probes that detect the single base-pair difference between the normal and the sickle cell alleles. (page 397)

25. **e** Duchenne muscular dystrophy is apparently associated with deletions in the gene coding for dystrophin. In the current prenatal test for Duchenne muscular dystrophy, the synthetic radioactive probe is complementary to the fragment of the normal dystrophin gene that is missing in a child with the disease. If this probe fails to hybridize with the test DNA restriction fragments, it indicates that the gene deletion is present and the child will be affected by the disease. (page 397)

26. **c** The technique of "DNA fingerprinting" is based on the number of repeated units in the simple-sequence DNA. Since only identical twins have identical patterns, it is as accurate in identifying a person as the use of fingerprints. (essay, page 396)

27. **True** Although the presence of an extra sex chromosome is often associated with sterility, some *XXX* women may be normal. (page 386)

28. **False** In spite of the prevalence of the Tay-Sachs allele in Ashkenazic Jews, most Tay-Sachs births in the United States are to non-Jewish parents. This is due to wide use by American Jews of a blood test to detect heterozygote carriers. (page 389)

29. **True** Dystrophin is the largest human gene known, with 60 exons and large introns. These factors are complicating the cloning and sequencing of this gene. (page 394)

30. Ray White and colleagues are locating and tracing genetic markers through three generations of 60 Mormon families. Nearly 500 genes have been mapped by this group so far. Helen Donis-Keller has identified several hundred markers in 20 families not related to White's group. Her team has focused attention on locating the gene responsible for cystic fibrosis. They have identified 60 markers on chromosome 7, which carries the cystic fibrosis gene. (page 399)

31. A defective allele may produce no polypeptide product or a poorly functioning polypeptide. (page 388)

32. Hemophilia is life-threatening because of the obvious risk of bleeding to death. A more subtle risk is that of contracting a life-threatening disease, such as AIDS, from use of Factor VIII isolated form human blood. The risk of acquiring human immunodeficiency virus from human blood products has been greatly reduced with the development of screening tests to detect the presence of antibodies to the virus. (page 392)

REVIEW TEST 6

Section 3: Chapters 17–19

This review test is *not* designed to cover all of the important information in these chapters. However, it does touch on the major topic categories in each chapter. It will also give you valuable practice in taking this type of test. When you are finished, check your answers with those provided in the following PERFORMANCE ANALYSIS section.

1. Artificial sticky ends may be created on complementary DNA by the:
   a. action of DNA ligase.
   b. action of a restriction enzyme.
   c. enzymatic removal of a short piece of one strand at each end of the DNA molecule.
   d. addition of a synthetic AT nucleotide polymer.
   e. addition of a single strand of a single nucleotide, such as TTTTTT.

2. Studies of transformed cells have revealed a group of cancer-causing genes known as:
   a. mutagenes.   d. structural genes.
   b. oncogenes.   e. regulatory genes.
   c. transformogenes.

3. If nondisjunction occurs during meiosis, the result is gametes with:
   a. too few chromosomes, which can produce viable embryos.
   b. too many chromosomes, which cannot produce viable embryos.
   c. the correct total number of chromosomes, but not a complete set of homologues.
   d. the correct total number of chromosomes, some of which are fragmented.
   e. too few chromosomes and other gametes with too many chromosomes.

4. For disorders caused by the presence of two recessive alleles, the heterozygote usually:
   a. is asymptomatic because the normal allele produces enough normal polypeptide product to avert ill effects.
   b. is normal because, in the presence of the dominant normal allele, gene expression of the defective allele is blocked.

   c. shows only mild symptoms because the dominant normal allele is functioning properly.
   d. has a selective advantage over either homozygote, and this maintains the defective allele in the population.
   e. is less fertile than a normal individual, so there is less chance for the allele to be passed on to the next generation.

5. Plasmids are good vectors for the cloning of DNA fragments up to about _____ base pairs in length.
   a. 100   d. 4,000
   b. 500   e. 10,000
   c. 2,000

6. Transcription of DNA to RNA takes place during:
   a. telophase.   d. prophase.
   b. metaphase.   e. anaphase.
   c. interphase.

7. What step was essential to development of a diagnostic test for sickle cell anemia?
   a. preparation of a radioactive copy of the normal beta-globin nucleotide sequence to use as a probe
   b. sequencing of the gene coding for beta globin
   c. production of radioactive synthetic mRNA corresponding to the beta-globin gene to use as a probe
   d. isolation of a restriction enzyme that cleaves the beta-globin gene into fragments of uniform size
   e. production of a RFLP specific for the sickle cell allele

8. T or F   In karyotype preparation, chromosomes of the same size are initially paired according to banding patterns.

9. T or F   DNA-RNA hybridization is a technique that permits the synthesis of an mRNA from a cDNA molecule for use in recombinant DNA experiments.

10. T or F   It was the discovery of the universality of the genetic code that originally led scientists to believe that the molecular biology of prokaryotes and eukaryotes would be similar.

11. T or F   A given messenger RNA molecule can be edited

and spliced in only one way, so that the same polypeptide product is always produced.

12. **T or F** Frederick Sanger won his first Nobel Prize for sequencing a protein and his second Nobel Prize for sequencing a nucleic acid.

13. **T or F** One application of the new biotechnology is the production of vaccines using synthetic viral coat proteins rather than live or killed whole viruses.

14. **T or F** As much as 1 percent of the total DNA of mouse cells is estimated to be of retroviral origin.

15. Identify three of the four ways in which intermediate-repeat DNA differs from simple-sequence DNA.

16. A woman may be color-blind in one eye but have normal color vision in the other eye. Account for this phenomenon.

PERFORMANCE ANALYSIS

1. **e** The addition of a single strand of a single nucleotide, such as TTTTTT, is one means of producing an artificial "sticky end" on complementary DNA. (page 342)

2. **b** Oncogenes are those genes that have been found to cause cancer. All oncogenes studied to date appear to code for regulatory proteins involved in cell division or cell growth. (pages 376, 377)

3. **e** In nondisjunction, homologous chromosomes or sister chromatids do not separate during mitosis or meiosis. This results in one cell with too many chromosomes and one cell with too few chromosomes. In meiosis, the cells involved are gametes. (page 385)

4. **a** In genetic disorders caused by the presence of two recessive somatic alleles, the heterozygote is usually asymptomatic (not affected) because the normal allele produces enough of the normal polypeptide product and the effects of the abnormal allele are masked. (page 388)

5. **d** Plasmids are only efficient vectors of DNA fragments of up to 4,000 base pairs. Modified bacteriophage lambda chromosomes can carry segments of up to 20,000 base pairs. Cosmids are good vectors for DNA segments of 35,000 to 40,000 base pairs. (pages 344, 345)

6. **c** Transcription of DNA to RNA occurs during interphase of the cell cycle. (page 359)

7. **a** Radioactive copies of the normal beta-globin nucleotide sequence were prepared as the first step in the development of a diagnostic test for sickle cell anemia. These probes were allowed to hybridize with *Hpa*I restriction fragments of DNA from persons with sickle cell anemia. In normal DNA, the probe hybridizes with a restriction fragment either 7,000 or 7,600 base pairs in length. In 87 percent of individuals with sickle cell anemia, the restriction fragment identified by the probe is 13,000 base pairs in length. (page 394)

8. **False** In karyotype preparation, chromosomes of the same size are first paired according to centromere position. Once arranged in this fashion, banding patterns may be used to identify homologous pairs. (page 383 and Figure 19–3)

9. **False** DNA-RNA hybridization is a technique for comparing the homology (similarity) between a DNA strand and an RNA strand. (page 345)

10. **True** When the genetic code was first deciphered and its universality became apparent, scientists expected the molecular genetics of eukaryotes to be similar to that of prokaryotes. (page 355)

11. **False** A given messenger RNA molecule may be edited and spliced in several ways, producing a different protein product with each editing mode. (page 369)

12. **True** Frederick Sanger won his first Nobel Prize for determining the amino acid sequence of insulin. His second Nobel Prize was awarded for sequencing the genome of bacteriophage φX174. (page 349)

13. **True** The new biotechnology has enabled scientists to produce vaccines using only synthetic viral protein coats, the portion of the virus to which the host immune system responds. Since viral nucleic acid is not used in manufacturing these vaccines, they are much safer than vaccines made from either live or killed virus particles. (page 352)

14. **True** Studies of retroviral RNA have led scientists to estimate that 0.5 to 1.0 percent of the total DNA in mice is of retroviral origin. (page 374)

15. Simple-sequence DNA consists of short sequences (generally 5 to 10 base pairs) that are repeated and arranged in tandem on a chromosome. It may constitute 10 to 50 percent of the DNA of a multicellular organism. Such sequences are thought to be important in chromosome structure. They are located around the centromere and at the tips of chromosomes. Intermediate-repeat DNA sequences constitute 20 to 40 percent of the DNA of a multicellular organism. The sequences are 150 to 300 nucleotides long and are similar but not identical. Except for the genes for rRNA and histones, these sequences are scattered throughout the genome. Some of the sequences have known functions. (pages 364, 365)

16. The alleles for color blindness are carried on the X chromosome. A woman who is heterozygous for the color-blind allele has one normal allele and one allele for color blindness. In female mammals, one X chromosome is inactivated early in embryonic development. If the X chromosome bearing the normal allele is inactivated in the cell from which one eye is derived through differentiation, the woman will be color-blind in that eye. (page 392; Chapter 13, page 267)

PART **2**

Biology of Organisms

SECTION **4**

The Diversity of Life

CHAPTER 20

The Classification of Organisms

MAJOR CONCEPTS

Organisms are identified and classified in a systematic manner to facilitate their study and allow the exchange of information among biologists. The major categories of classification from least to most inclusive are species, genus, family, order, class, phylum or division, and kingdom. A genus is a collection of species; families are composed of several genera, etc. Ideally, each level of classification accurately represents evolutionary history. For example, the genus *Canis* contains several species believed to share a common ancestor.

The fundamental level of taxonomy (and the only level that has biological authenticity) is the species; all other levels of organization have been created by humans to facilitate study and communication. Species are groups of actually or potentially interbreeding organisms that are reproductively isolated from all other such groups. Species are named according to the binomial system developed by Carolus Linnaeus. The species name is always italicized and consists of a genus name and a specific epithet.

A monophyletic classification scheme is one in which all members of a single taxon are descended from one common ancestor. In a polyphyletic classification scheme, taxa contain organisms descended from more than one common ancestor. Although monophyletic classifications are ideal for accurately reflecting phylogeny, polyphyletic classifications exist because of inadequate information about evolutionary relationships and because convenience may be more important than phylogeny in some cases.

Organisms have traditionally been classified on the basis of structural similarities and differences. Homologous structures have a common evolutionary origin but not necessarily a common function. Analogous structures have a similar function and superficial appearance, but evolved independently.

Criteria traditionally used to classify organisms include superficial appearance, stages of the life cycle, and patterns of embryonic development. Two alternatives to traditional taxonomic methods are numerical phenetics and cladistics. Numerical phenetics classifies organisms based on superficial characteristics. Cladistics classifies organisms solely on the basis of phylogeny. Although both schemes have merit, neither is completely satisfactory. Newer techniques that aid in classification include several forms of molecular taxonomy: amino acid sequencing, nucleotide sequencing, and DNA-DNA hybridization.

In Linnaeus's time, living organisms were classified into kingdom Animalia or kingdom Plantae, primarily based on movement and growth patterns. Discoveries made in the twentieth century led to revision of this system. Although much is known (or suspected) about the evolutionary history of living organisms, most modern proposals regarding taxonomic kingdoms are based on cellular organization and mode of nutrition. The five kingdoms that are currently widely recognized are Monera, Protista, Fungi, Plantae, and Animalia.

HOW TO STUDY THE CHAPTER

Read the entire chapter through quickly, focusing on the major concepts.

Use the GUIDED STUDY OF THE CHAPTER to help you identify the important details as you **reread** the chapter. Writing out the answers to these questions will help fix them in your mind as well as provide you with a valuable study aid.

Answer the questions in TESTING YOUR UNDERSTANDING without the aid of your text. Check your answers against those in PERFORMANCE ANALYSIS. Analyzing your answers will give you valuable feedback on your level of understanding and preparedness for classroom testing.

GUIDED STUDY OF THE CHAPTER

I. The Need For Classification *(page 407)*

> *Focus:* A logical and orderly system of classifying organisms is necessary to facilitate their study and to allow communication between biologists.

II. What Is A Species? *(pages 407–410)*

A. General Remarks *(pages 407–408)*

1. a. *Vocabulary:* State in your own words the definition of a species presented by Ernst Mayr.

    b. What is the importance of the terms "groups" and "populations" in Mayr's definition of a species?

    c. Mayr's definition of a species is commonsensical because:

2. What characteristics of plants and bacteria complicate the definition of species in these groups?

3. How can species be defined in evolutionary terms?

> *Focus:* Speciation can occur when a group of organisms is separated from all other members of the species and becomes reproductively isolated.

B. The Naming of Species *(pages 408–410)*

4. *Vocabulary:* Define a genus.

5. a. What system of nomenclature was devised by Linnaeus?

    b. What two components constitute a scientific name?

    c. Why is the specific epithet alone inadequate identification for an organism?

    d. Under what conditions can the genus name be abbreviated?

6. Who is authorized to name a new species?

7. What criteria may be used in the naming of a new species?

8. Cite two reasons why common names are not adequate identification for a species.

III. Hierarchical Classification *(pages 410–412)*

    9. What is taxonomy?

   10. *Vocabulary:* Distinguish between taxon and category.

   11. Rank the following categories in order from *most to least* inclusive: class, family, order, kingdom, genus, phylum or division, and species.

   12. *Vocabulary:* Distinguish between phylum and division.

> *Focus:* **The names of genera and species are always italicized, the genus name with an initial capital and the specific epithet in lower case. Taxa at levels above the genus have initial capitals but are not italicized.**

   13. Describe the usefulness of a hierarchical classification system.

   14. "The only category in the hierarchical system that is a biological reality is the _____."
Complete and explain this statement. (Hint: consider the definition of this category.)

   15. On what philosophical grounds do "lumper" taxonomists and "splitter" taxonomists (a) agree and (b) disagree?

IV. Evolutionary Systematics *(pages 412–414)*

A. General Remarks *(page 412)*

   16. How did the goal of taxonomy change after 1859?

   17. *Vocabulary:* Distinguish between phylogeny and evolutionary systematics.

   18. How can taxonomic hypotheses be tested?

B. The Monophyletic Ideal *(pages 412–413)*

   19. *Vocabulary:* Distinguish between monophyletic and polyphyletic.

> *Focus:* **According to the monophyletic ideal, all organisms included in a specific taxon should be descendants of the same nearest common ancestor.**

   20. Identify two reasons why a monophyletic classification may not always work.

C. Homology and Phylogeny *(pages 413–414)*

   21. Explain why superficial phenotypic characteristics alone are not useful in taxonomic classifications.

22. Name two evolutionary possibilities for (a) similar characteristics in different species and (b) differences in characteristics between species.

23. a. *Vocabulary*: Distinguish between homologous and analogous characteristics.

   b. What features generally characterize homologous attributes?

V. Taxonomic Methods *(pages 414–418)*

A. General Remarks *(pages 414–415)*

24. Identify the first two stages involved in classifying an organism.

25. Name three types of information that are useful in identifying homologies. (Name an example for one of these.)

26. Traditional taxonomies contain what two types of information?

27. *Vocabulary*: What is a phylogenetic tree? (Figure 20–9)

B. Alternative Methodologies *(pages 415–418)*

28. What is the basic difference between the traditional taxonomic method and both alternative methods?

Focus: Proponents of numerical phenetics and cladistics realize that a single classification scheme cannot place equal emphasis on overall similarity and on genealogy.

Numerical Phenetics (page 416)

29. Explain the basis for numerical phenetics.

Focus: Numerical pheneticists disregard these distinctions: (1) homologous versus analogous structures and (2) characteristics subject to environmental influences versus those that are genetically determined.

30. How do numerical pheneticists justify the position presented in this Focus statement?

Cladistics (pages 416–418)

31. Explain the basis for cladistics.

32. Distinguish holophyletic taxa from monophyletic taxa.

33. Using the classification of the principal land-dwelling vertebrates as an example, describe how traditional classification and cladistic methods lead to radically different groupings.

34. a. According to traditional taxonomists, what do cladists ignore when classifying animals?

   b. How do cladists justify this position?

35. *Vocabulary:* Define a paraphyletic taxon.

Focus: When classifying organisms according to numerical phenetics or cladistics, some information about the organisms is ignored.

Focus: Neither phenetics nor cladistics is completely free of subjectivity.

---

36. If neither phenetics nor cladistics is the ideal approach to classification, what is the use of these methods?

C. *Essay:* How to Construct a Cladogram *(page 417)*

37. Describe how a cladogram fulfills the requirements of a testable hypothesis.

38. a. What are "nested sets"?

b. What type of absurdity do traditional taxonomists see in these constructions?

VI. Molecular Taxonomy *(pages 418–423)*

A. General Remarks *(page 418)*

39. Identify two advantages of molecular taxonomy over taxonomic methods based on anatomy.

B. Amino Acid Sequences *(pages 418–420)*

40. The use of amino acid sequences in taxonomy is based on what premise?

41. Cite two conflicting ideas about interpreting the significance of differences in amino acid sequences among organisms.

42. Cite two pieces of evidence that favor the perception of amino acid sequences as molecular clocks.

C. Nucleotide Sequences *(page 420)*

43. Name two advantages of nucleotide sequencing over amino acid sequencing in taxonomic studies.

44. How has nucleotide sequencing promoted the unraveling of microbial phylogeny?

45. Identify three factors influencing the rate of change in nucleotide sequences that must be considered by taxonomists using these sequences in classification schemes.

D. DNA-DNA Hybridization *(pages 420–423)*

46. a. Outline the procedure by which Sibley and Ahlquist determine the 50-percent-dissociation temperature of hybridized DNA.

b. The correlation of results from this method with evolutionary relationships is based upon what premise?

47. Cite two examples of questions in bird taxonomy that have been answered using DNA-DNA hybridization.

48. DNA-DNA hybridization methods for determining evolutionary relationships between groups of organisms are less useful the more widely separated the groups. Why?

Focus: DNA-DNA hybridization is the molecular technique of choice for studying the phylogeny of organisms that are not too distantly related.

E. *Essay:* **The Riddle of the Giant Panda** *(page 422)*

49. Describe how analogous characteristics between the giant panda and the lesser panda caused disagreement among taxonomists concerning classification of the giant panda.

50. Name five types of evidence supporting the classification of the giant panda as a bear.

VII. **A Question of Kingdoms** *(pages 423–426)*

51. What criteria originally distinguished members of kingdom Animalia and kingdom Plantae?

52. Name three technological developments that prompted the formation of additional kingdoms.

53. What problem arises when classification into kingdoms is based on evolutionary history?

54. The current establishment of kingdoms is based on what two characteristics of organisms?

55. What criteria distinguish members of the kingdom Monera from other living organisms?

56. Characterize the members of kingdom Protista.

57. Fill in the following table.

Table for Question 57

	FUNGI	ANIMALIA	PLANTAE
Mode of nutrition			
Ecological role(s)			

Focus: No single system of kingdoms can classify organisms according to strict phylogeny and retain maximum utility and convenience for the study of organisms.

Focus: Any difficulties or disagreements encountered during attempts at classifying groups of organisms into kingdoms do not affect most of the other hierarchical categories or the specific characteristics and evolutionary history of each group.

VIII. **Summary** *(pages 426–427)*: Read the summary. If you are familiar with the essential features of the material presented there, you are ready to complete the section TESTING YOUR UNDERSTANDING.

TESTING YOUR UNDERSTANDING

After you have completed this examination, compare your answers with those in the section that follows.

1. The process through which a group of organisms changes enough to become reproductively isolated from the original population is known as:
   a. divergence.
   b. differentiation.
   c. speciation.
   d. hybridization.
   e. colonization.

2. In the time of Linnaeus, these three categories were the only ones in common use.
   a. kingdom, genus, species
   b. kingdom, class, species
   c. phylum, genus, species
   d. order, class, species
   e. family, genus, species

3. Which of these lists is in descending order from the most to least inclusive category?
   a. class, order, division, genus
   b. kingdom, class, family, genus
   c. kingdom, order, class, species
   d. order, family, phylum, division
   e. phylum, family, class, order

4. The fundamental category in hierarchical classification is the:
   a. kingdom.
   b. phylum.
   c. family.
   d. genus.
   e. species.

5. A taxon that contains organisms descended from one common ancestor is:
   a. polyphyletic.
   b. homologous.
   c. primitive.
   d. holophyletic.
   e. monophyletic.

6. Which of these anatomical structures is (are) analogous to the tail of a blue whale?
   a. the tail of a monkey
   b. the hind legs of a race horse
   c. the tail of a great white shark
   d. the dorsal fin of a bass
   e. the tail feathers of a bird

7. A coding value of zero for a unit character in numerical phenetics means that:
   a. all representations of the unit character are equally abundant.
   b. the unit character is of no value in determining classification.
   c. data are not available for the unit character in question.
   d. the positive and negative aspects of the unit character exactly balance each other.
   e. the particular unit character does not exist in this organism.

8. In the DNA-DNA hybridization technique of Sibley and Ahlquist, a 1°C depression in the 50-percent-dissociation temperature represents an evolutionary divergence that occurred about _____ years ago.
   a. 4,500
   b. 45,000
   c. 450,000
   d. 4.5 million
   e. 45 million

9. Motility is exhibited at some stage in the life cycle by members of all the kingdoms EXCEPT:
   a. Fungi.
   b. Plantae.
   c. Animalia.
   d. Monera.
   e. Protista.

10. Unicellular organisms that are autotrophic or heterotrophic, and lack nuclear and most other internal membranes, characterize kingdom:
    a. Protista.
    b. Fungi.
    c. Animalia.
    d. Monera.
    e. Plantae.

11. Multicellular eukaryotic organisms that are strictly heterotrophic and contain a nervous system belong to kingdom:
    a. Protista.
    b. Fungi.
    c. Animalia.
    d. Monera.
    e. Plantae.

12. T or F   For many types of organisms, the term species is a category of convenience rather than one that conforms strictly to Ernst Mayr's definition.

13. T or F   The likelihood of two structures being homologous increases as the number of parts or components of those structures decreases.

14. T or F   Cladists maintain that the point at which two species diverge can be objectively determined because such a point is always represented by the appearance of an evolutionary novelty.

15. T or F   Different portions of the DNA within the same organism appear to be subject to different rates of mutation.

16. T or F   Chimpanzees and humans have only minor differences in amino acid sequences of their cytochrome c. This is evidence that humans and chimpanzees diverged fairly recently.

17. T or F   There is strong evidence that all eukaryotes arose from the same lineage of prokaryotic ancestors.

18. Explain why the specific epithet of a species name is insufficient for identifying the organism. Give an example.

19. Distinguish between monophyletic and holophyletic taxa.

20. Summarize two conflicting perspectives pertaining to the taxonomic interpretation of amino acid sequence differences.

21. In what way are numerical phenetics and cladistics "opposite" methods of taxonomic classification?

PERFORMANCE ANALYSIS

1. **c** Speciation, the formation of new species, occurs when a group of organisms becomes separated from all other members of its species and undergoes sufficient change that it becomes reproductively isolated. (page 408)

2. **a** Linnaeus developed the binomial system of naming organisms, which assigns them to a genus and species. In his time, all organisms were placed in either the animal or plant kingdom. (page 410)

3. **b** The order of taxonomic categories from most to least inclusive is kingdom, phylum or division, class, order, family, genus, species. (pages 410, 411)

4. **e** The fundamental category in hierarchical classification, and the only category that has biological authenticity, is the species. (page 412)

5. **e** In a monophyletic classification scheme, each taxon contains *only* those organisms descended from a single common ancestor. In polyphyletic classification, a taxon may contain organisms descended from different common ancestors. (pages 412, 413)

6. **c** Analogous structures are those that serve similar functions regardless of their evolutionary origin. Homologous structures are those that have similar evolutionary origin, but may or may not serve similar functions. (page 414)

7. **c** In numerical phenetics, a coding value of zero indicates that no data are available for that characteristic. (page 416)

8. **d** According to the work of Sibley and Ahlquist, a depression in the 50-percent-dissociation temperature of 1°C corresponds to a difference in nucleotide sequence of 1 percent. This indicates that the two organisms whose DNA is being compared diverged approximately 4.5 million years ago. (page 421)

9. **a** The only kingdom in which no members are motile at any stage in their life history is the kingdom Fungi. (Table 20–2)

10. **d** Kingdom Monera contains all the prokaryotes, unicellular organisms that may be autotrophic or heterotrophic, but which lack nuclear membranes and the complex internal membranes that characterize eukaryotic cells. (Table 20–2)

11. **c** The kingdom Animalia is characterized by multicellular, eukaryotic organisms that are heterotrophic; many have a nervous system. (Table 20–2)

12. **True** Many prokaryotes can exchange genetic information with unrelated organisms and therefore do not fit handily into Ernst Mayr's definition of a species. (page 408)

13. **False** Homologous characteristics are those that have a common evolutionary origin; they may or may not perform similar functions. The more complex two similar characteristics are (i.e., the more parts they contain), the more likely it is that they are homologous. (page 414)

14. **True** Cladists propose that the one evolutionary event that can be determined objectively is the branching of one line from another and that these branch points are characterized by the appearance of novel characteristics not observed in the common ancestor. (page 416)

15. **True** Mutations apparently occur at different rates within different regions of an organism's DNA. (page 420)

16. **False** The amino acid sequences of chimpanzee and human cytochrome *c* are identical. This is indeed evidence that these two species diverged fairly recently in evolutionary history. (page 420)

17. **False** From the evidence currently available, it is apparent that eukaryotes evolved from several different ancestral lines of prokaryotes. (page 424)

18. Different species may have identical specific epithets. Without the genus name, a specific epithet is meaningless. *Drosophila melanogaster* is a fruit fly, *Thamnophis melanogaster* is a semiaquatic garter snake. (page 409)

19. A monophyletic taxon contains organisms descended from a single common ancestor, not necessarily including all descendants or the ancestor. A homophyletic taxon must include all descendants plus the ancestor itself. (page 416)

20. (1) Some scientists argue that differences in amino acid sequence represent functional differences in proteins and therefore are subject to the pressures of natural selection. (2) Other scientists contend that differences in amino acid sequence arise from random genetic mutations and have no functional significance (are not subject to selection). These scientists promote the view that differences in amino acid sequence can be used as molecular clocks to estimate the time of evolutionary divergence of organisms. (page 419)

21. Numerical phenetics is based on degree of overall similarity of species' observable characteristics. Cladistics completely ignores overall similarity and uses only phylogeny. (page 416)

The Prokaryotes and the Viruses

MAJOR CONCEPTS

The prokaryotes represent the oldest group of living organisms and are currently classified in the kingdom Monera. This kingdom includes a wide diversity of unicellular organisms, all of which lack a nuclear membrane and have no membrane-bound organelles in the cytoplasm. They are ecologically important as decomposers and as key participants in nitrogen fixation.

Once classified mainly on the basis of staining characteristics and cell and colony appearance, prokaryotes are being reclassified based on their cell ultrastructure and biochemical activities. The current classification system is also under scrutiny following the discovery of another group of organisms that appear to represent a phylogenetic branch that differs from the prokaryotes previously known.

The structure of the prokaryotic cell membrane resembles that of eukaryotes except that the cell membrane of most prokaryotes does not contain cholesterol or other steroids. The cell walls of prokaryotes contain peptidoglycans, polymers of amino acids and sugars not found in eukaryotic cell walls. Prokaryotes have historically been classified as gram-positive and gram-negative based on staining characteristics that are determined by cell wall structure.

Prokaryotic flagella differ markedly from those of eukaryotes, with the only similarities being outward appearance and locomotor function. In addition to flagella, many bacteria contain projections known as pili, which are responsible for the attachment of bacteria to food sources or to other bacteria during conjugation.

There is considerable diversity of form among the prokaryotes. Most bacterial cells can be classified as bacilli (rod-shaped), cocci (spherical), spirilla (long, spiral rods), or vibrios (short, curved rods). The characteristic growth patterns of various bacterial species produce a diversity of cellular arrangements, including pairs, clusters, chains, and filaments.

Most prokaryotes reproduce by simple cell division. Many are capable of forming dormant, resting cells called spores that are resistant to unfavorable environmental conditions. Reduction of the food supply is a common stimulus for spore formation. Some of the more advanced prokaryotes live as colonies of closely associated individual cells that form fruiting bodies consisting of spores and slime. This activity parallels the reproductive functions of some multicellular fungi (to be studied in Chapter 23).

The majority of the prokaryotes are heterotrophs and obtain their nutrients from dead organic matter. However, many live in close association with other organisms; these associations may be beneficial or harmful to their hosts. The autotrophic bacteria include a variety of photosynthetic and chemosynthetic forms. Photosynthetic bacteria derive energy from the sun; chemosynthetic bacteria derive energy from chemical reactions occurring in the environment.

Viruses are composed of DNA or RNA surrounded by a protein coat. They exist independently but can replicate only inside living cells. When a virus infects a cell, the viral nucleic acid takes over the metabolic machinery of the cell and directs the production of new virus particles, using cellular components as building blocks and causing cellular damage in the process. The host cell is often destroyed when the new virus particles are released.

Viroids and prions are pathogenic particles even simpler in structure than viruses. Viroids are naked RNA molecules, and prions are protein particles that are not associated with a detectable nucleic acid. Viroids are thought to exert their effects by interrupting the control of cellular DNA expression. The mechanism of prion action is unknown.

Microorganisms may cause disease in their hosts by damaging host cells directly, by producing toxins that damage host cells, or by stimulating a response of the host that interferes with its normal function. The impact of microorganisms on human health has been reduced through improved sanitation practices, the development of antibiotics and other chemotherapeutic agents, and widespread immunization programs.

HOW TO STUDY THE CHAPTER

Read the entire chapter through quickly, focusing on the major concepts.

Use the GUIDED STUDY OF THE CHAPTER to help you identify the important details as you **reread** the chapter. Writing out the answers to these questions will help fix them in your mind as well as provide you with a valuable study aid.

Answer the questions in TESTING YOUR UNDERSTANDING without the aid of your text. Check your answers against those in PERFORMANCE ANALYSIS. Analyzing your answers will give you valuable feedback on your level of understanding and preparedness for classroom testing.

GUIDED STUDY OF THE CHAPTER

I. Introduction (pages 428–429)

1. Microbiology is the study of which groups of organisms?

Focus: The prokaryotes are the oldest and most abundant group of organisms on earth, and are the smallest unicellular organisms.

2. a. Name two factors that account for the success of the prokaryotes.

   b. How long does it take for an *E. coli* population to double in number?

3. Identify three "hostile" environments in which some prokaryotes thrive.

4. How do prokaryotes vary in their need for oxygen?

5. Describe the structure and function of prokaryotic spores.

6. What are two ecologically important roles of prokaryotes?

II. The Classification of Prokaryotes (pages 429–430)

7. What characteristics have, until recently, been used to classify prokaryotes, and in what way is this method of classification inadequate?

8. Name four recent developments that are enabling microbiologists to unravel evolutionary relationships among prokaryotes.

9. a. Identify four characteristics of methanogens, halobacteria, and thermoacidophiles that support their proposed placement in a kingdom separate from other prokaryotes.

b. What is the proposed name of this kingdom?

10. Characterize each of the following groups of prokaryotes.

Methanogens

Halobacteria

Thermoacidophiles

11. What has been suggested about the evolutionary origins of the three groups of prokaryotes listed in Question 10?

12. Why is it unlikely that the current prokaryote classification scheme will be abandoned?

III. The Prokaryotic Cell *(pages 431–435)*

A. General Remarks *(page 431)*

13. Contrast prokaryotes and eukaryotes with respect to the following features. (See also Chapter 4, pages 90–92.)

Table for Question 13

	PROKARYOTES	EUKARYOTES
Chromosome structure		
Cytoplasmic compartmentalization		
Complexity of organelles		

14. The only prokaryotes that have an extensive cytoplasmic membrane system are the _____, and these membranes contain _____.

B. The Cell Membrane *(page 431)*

15. How are the cell membranes of prokaryotes similar to and different from those of eukaryotes?

16. How do prokaryotes compensate for their lack of (a) mitochondria if they are aerobic and (b) chloroplasts if they are photosynthetic?

17. What is the role of the cell membrane in distributing genetic information to daughter cells during cell division?

C. The Cell Wall *(pages 432–433)*

18. What role does the cell wall play in the survival of a prokaryote?

19. a. Members of which group of prokaryotes do not have cell walls?

b. How do they compensate for their deficiency?

Chemical Structure of the Cell Wall (pages 432–433)

20. Peptidoglycans are responsible for what feature of prokaryotic cell walls?

21. Use a diagram to illustrate the difference between the cell walls of gram-positive and gram-negative prokaryotes. Show the relationships between the cell membranes and the cell walls and label your diagrams.

22. In what way is gram staining useful in the classification of prokaryotes?

23. Cite two examples in which the structure of the cell wall relates to the survival ability of the prokaryote.

24. What is the proposed function of the capsule that is secreted by some bacteria?

D. Flagella and Pili (pages 433–435)

25. Fill in the following table contrasting prokaryotic and eukaryotic flagella. (You may wish to refer to Chapter 5, pages 122, 123, to review eukaryotic flagella.)

Table for Question 25

	PROKARYOTIC FLAGELLA	EUKARYOTIC FLAGELLA
Basic subunit		
Arrangement of subunits		
Distribution on cell surface		
Enclosed by the cell membrane? (yes or no)		
Means of attachment to cell (distinguish between gram-positive and gram-negative bacteria)		
Means by which it increases in length		

26. a. How can the movement of bacterial flagella be observed?

b. What provides the energy for flagellar movement?

Focus: The direction of rotation of bacterial flagella determines whether the cell "runs" or "tumbles."

27. Discuss the relationship between direction of flagellar rotation and the running and tumbling of a bacterial cell.

28. What environmental factors influence the frequency with which bacterial cells run, tumble, and change from one activity to the other?

29. Distinguish bacterial pili from bacterial flagella with respect to subunit structure, size, quantity, and function.

E. *Essay:* Navigation by the Poles (page 435)

Focus: Several groups of bacteria have been discovered that move according to the earth's magnetic field.

30. Describe the structure within these bacteria that enables them to detect a magnetic field.

31. Why is this adaptation beneficial to bacteria? (Distinguish between bacteria in the Northern and Southern hemispheres and at the equator.)

IV. Diversity of Form (pages 436–437)

32. *Vocabulary:* Distinguish among bacilli, cocci, spirilla, and vibrios.

33. *Vocabulary:* Distinguish among diplococci, staphylococci, and streptococci.

34. Name three types of prokaryotes that form filamentous structures.

35. Describe the means by which gliding bacteria move from place to place.

36. a. *Vocabulary:* Define endoflagella and describe their structure and location.

    b. Endoflagella are present in what types of prokaryotes and perform what function?

37. Historically, what has been the effect of rickettsial organisms on public health? (Figure 21–13)

V. Reproduction and Resting Forms *(pages 437–438)*

38. *Vocabulary:* What is binary fission?

39. Name three processes by which prokaryotes may reproduce.

Focus: Two factors that account for the tremendous diversity and adaptability of prokaryotes are their short generation time and genetic mutations.

---

40. Name a third factor that contributes to adaptability of the prokaryotes.

41. *Vocabulary:* What are prokaryotic spores?

42. What circumstance stimulates sporulation in prokaryotes?

43. Diagram the stages involved in forming a spore.

44. Describe the composition of the spore coat.

45. List three events that occur during germination of a spore.

46. What is the value of spore formation?

47. How does reproduction in myxobacteria differ from that in *Bacillus* or *Clostridium*?

Note: Although the spores of *Clostridium botulinum* are not destroyed by hours of boiling, the *toxin* which causes botulism food poisoning is denatured by thorough cooking, which is why home-canned vegetables should be well boiled before they are eaten.

VI. Prokaryotic Nutrition *(pages 438–442)*

A. Heterotrophs *(pages 438–439)*

48. a. *Vocabulary:* What are saprobes?

b. What is their role in the transfer of energy and matter through living systems?

Focus: **Different types of prokaryotes can live in close association with each other because nutritional specialization minimizes competition. In some cases, one group of organisms makes nutrients available to another group.**

---

49. *Vocabulary:* Define pathogenic bacteria.

50. Cite three examples demonstrating how prokaryotes may benefit other organisms by living in close association with them.

51. Name one possible consequence of the alteration of the microbial population of the human intestinal tract.

B. **Chemosynthetic Autotrophs** *(page 439)*

52. What are the energy sources for chemosynthetic autotrophs?

53. Methanogens have only recently been studied extensively because:

54. What role do methanogens play in the decomposition of organic matter?

55. Summarize the role of chemosynthetic bacteria in converting ammonia or ammonium into a form usable by plants.

56. What two forms of sulfur can be used to supply energy to chemosynthetic bacteria, depending upon the type of bacterium?

C. **Photosynthetic Autotrophs** *(pages 440–442)*

Green and Purple Bacteria (page 440)

57. Name the three distinct phylogenetic lineages that are believed to represent the origin of the modern photosynthetic eubacteria.

58. Fill in the following table comparing the pigments of green and purple bacteria.

Table for Question 58

	GREEN BACTERIA	PURPLE BACTERIA
Reaction center pigment(s)		
Antennae pigment(s)		
Accessory pigments		

59. How closely do the photosynthetic pigments of bacteria resemble those of eukaryotes?

60. Name three types of compounds that may act as electron donors (instead of water) for photosynthesis in photosynthetic bacteria.

61. How does photosynthesis by green and purple bacteria (a) differ from and (b) resemble eukaryotic photosynthesis?

Cyanobacteria (pages 440–442)

62. a. Name two features that cyanobacteria have in common with photosynthetic eukaryotes.

    b. What prompted the change in classification from "blue-green algae" to cyanobacteria?

63. a. List at least four pigments that may be present in cyanobacteria.

    b. Where in the cell are the photosynthetic pigments located?

Focus: **Cyanobacteria have an outer polysaccharide sheath, which may be heavily pigmented or unpigmented.**

---

64. Discuss the importance of nitrogen-fixing cyanobacteria to other living organisms.

65. a. What features of cyanobacteria make them ideally suited as colonizers of barren rock and soil?

    b. What is the likely evolutionary significance of this ability to colonize?

D. *Essay:* **Two Unusual Photosynthetic Prokaryotes** (*pages 442–443*)

66. How does *Prochloron* (a) resemble and (b) differ from cyanobacteria?

67. What characteristics of *Prochloron* suggest that it might be descended from the prokaryotes that gave rise to chloroplasts?

68. In what two ways does *Heliobacterium chlorum* differ from all previously known photosynthetic bacteria?

69. Discuss the implications regarding the evolution of (a) *Heliobacterium chlorum* and (b) contemporary gram-positive bacteria that can be derived from the study of *H. chlorum*.

VII. **Viruses: Detached Bits of Genetic Information** (*pages 443–447*)

A. **General Remarks** (*pages 443–446*)

Focus: **Viruses cannot be classified into any of the kingdoms, but are usually studied along with prokaryotes.**

---

70. What evidence has prompted the suggestion that viruses are cellular fragments that have become partially independent?

71. Identify three characteristics of viruses that are used in their classification.

72. Describe the various forms of viral capsids.

73. a. The specificity of a viral particle—the cell type(s) it can infect—is determined by:

    b. The susceptibility of a cell to viral infection is determined by:

74. Construct a flow diagram to illustrate one means by which a DNA virus infects a cell and replicates. (Indicate the source of components involved in replication—viral or host cell.)

75. Describe two mechanisms of RNA virus replication. Emphasize how this differs from DNA virus replication.

76. Summarize three means by which viruses may acquire their outer layer.

B. **Viroids and Prions: The Ultimate in Simplicity** *(pages 446–447)*

77. *Vocabulary:* Distinguish between viroids and prions.

78. Characterize the nucleic acid of PSTV.

79. Where in infected cells are viroids located?

80. In what two ways does the RNA of viroids differ functionally from the genetic material of viruses?

81. a. Describe the mechanism by which viroids are currently believed to exert their effects.

    b. Summarize the evidence that supports this proposed mechanism.

82. What inferences can be drawn from the proposed mechanism of viroid activity?

83. Genes corresponding to the prion for scrapie, as well as their corresponding mRNAs, have been located in what two places?

Focus: **Viroids are thought to interfere with the regulation of gene expression, but the mechanism by which prions exert their effects is currently unknown.**

VIII. **Microorganisms and Human Ecology** *(pages 447–450)*

A. **Symbiosis** *(pages 447–448)*

84. *Vocabulary:* Distinguish among symbiosis, mutualism, commensalism, and parasitism.

85. Fill in the table to indicate which participant benefits (+), is harmed (−), or neither benefits nor is harmed (+/−). Show all the possible correct combinations.

Table for Question 85

	PARTICIPANT A	PARTICIPANT B
Mutualism		
Commensalism		
Parasitism		

86. Describe by citing an example how the same organism may enter into all three relationships with the same host depending upon the circumstances.

Focus: **It is advantageous for a parasite to cause minimal harm to its host, indicating that parasite and host are well-adapted to each other, a phenomenon that usually results from close association over a long evolutionary history. Disease results when there is a change in the parasite, the host, or the parasite-host relationship.**

B. How Microbes Cause Disease *(page 448)*

> *Focus:* Disease can be caused by microorganisms when (1) cells are destroyed directly by the organism, (2) toxins produced by the organism cause damage, or (3) the body's reaction to the invading organism interferes with normal functions.

87. Cite examples of each of the three situations listed in this Focus statement.

88. Describe examples of the following:

    a. A single organism causes more than one disease.

    b. A single disease is caused by more than one organism.

C. Prevention and Control of Infectious Disease *(pages 448–450)*

> *Focus:* The realization that microorganisms cause disease prompted the institution of sanitary practices in the areas of medicine and public health.

89. Name five public health measures designed to prevent the spread of infectious diseases.

90. *Vocabulary:* Define an antibiotic.

91. Identify two groups of organisms that have provided antibiotics for human use.

92. What property of antibiotics and other chemotherapeutic agents makes them so effective at combating disease?

> *Focus:* Two major factors underlying the current population boom are the effectiveness of antimicrobial agents in fighting disease and the prevention of disease by immunization (to be discussed in Chapter 38).

93. Why are hospitals again becoming sources of bacterial diseases?

94. Research to develop effective antiviral agents is currently focusing on what two potential mechanisms by which these agents might act?

IX. **Summary** *(pages 450–451):* Read the summary. If you are familiar with the essential features of the material presented there, you are ready to complete the section TESTING YOUR UNDERSTANDING.

TESTING YOUR UNDERSTANDING

After you have completed this examination, compare your answers with those in the section that follows.

1. A single gram of fertile soil can contain as many as _____ individual prokaryotes.
   a. 25,000
   b. 250,000
   c. 2.5 million
   d. 2.5 billion
   e. 2.5 trillion

2. With the discovery of the halobacteria, methanogens, and thermoacidophiles, certain taxonomists have proposed the formation of new taxa called:
   a. kingdom Archaebacteria and division Monera.
   b. kingdom Monera and subkingdom Archaebacteria.
   c. superkingdom Prokaryota and kingdom Archaebacteria.
   d. kingdom Prokaryota and subkingdom Monera.
   e. superkingdom Monera and subkingdom Prokaryota.

3. If a new classification scheme is worked out for the prokaryotes, it is likely to be used only by:
   a. evolutionary biologists.
   b. agricultural biologists.
   c. veterinarians.
   d. industrial microbiologists.
   e. medical researchers.

4. The most prominent feature of a prokaryotic cell is the:
   a. endoplasmic reticulum.
   b. nucleoid.
   c. Golgi complex.
   d. nucleus.
   e. ribosomes.

5. Which prokaryotes do NOT have some type of cell wall?
   a. true bacteria
   b. viruses
   c. prions
   d. cyanobacteria
   e. mycoplasmas

6. Gram-negative bacteria have a cell wall that is composed of _____, whereas the cell walls of gram-positive bacteria are composed of _____.
   a. a homogeneous peptidoglycan-polysaccharide layer; two separate peptidoglycan layers.
   b. two separate peptidoglycan layers; an inner peptidoglycan layer surrounded by a lipopolysaccharide layer.
   c. two separate peptidoglycan layers; a homogeneous peptidoglycan-polysaccharide layer.
   d. an inner peptidoglycan layer surrounded by a lipopolysaccharide layer; a homogeneous peptidoglycan-polysaccharide layer.
   e. an inner peptidoglycan layer surrounded by a lipopolysaccharide layer; two separate peptidoglycan layers.

7. Bacterial flagella are involved in _____, and pili are necessary for _____.
   a. motility; conjugation
   b. conjugation; drug resistance
   c. attachment; conjugation
   d. drug resistance; motility
   e. movement; drug resistance

8. The members of genus *Mycobacterium* form:
   a. grape-shaped clusters.
   b. funguslike filaments.
   c. long, curving rods.
   d. a mucuslike substance.
   e. filaments of cocci.

9. One bacterium that often causes pneumonia is _____, which consists of a chain of small, spherical cells.
   a. streptobacillus
   b. staphylobacillus
   c. diplospirilla
   d. staphylococcus
   e. streptococcus

10. Which process is NOT a means by which prokaryotic organisms accomplish genetic recombination?
   a. conjugation
   b. transformation
   c. plasmid transfer
   d. binary fission
   e. transduction

11. The smallest known cells belong to the:
   a. rickettsiae.
   b. viruses.
   c. cyanobacteria.
   d. mycoplasmas.
   e. mycobacteria.

12. The chemosynthetic prokaryotes known as methanogens were difficult to study initially because they are:
   a. facultative anaerobes.
   b. obligate aerobes.
   c. obligate anaerobes.
   d. facultative aerobes.
   e. unable to thrive at routine laboratory temperatures.

13. The reaction center of the green photosynthetic bacteria contains _____, which is similar to the photosynthetic reaction center pigment of eukaryotes.
   a. bacteriochlorophyll *b*
   b. chlorophyll *a*
   c. bacteriochlorophyll *d*
   d. bacteriochlorophyll *a*
   e. chlorophyll *c*

14. The photosynthetic pigments of the cyanobacteria are found in the:
   a. cell membrane.
   b. peripheral membrane system.
   c. nuclear membrane.
   d. thylakoid membranes.
   e. membranes of the grana.

15. Which information is NOT used to characterize viruses?
   a. their usual host
   b. their nucleic acid content
   c. their specific shape
   d. the presence or absence of additional proteins around their capsid
   e. the length of their life cycle

16. Of those host cell components listed below, which is (are) NOT used by an infecting virus during viral replication?
   a. ribosomes
   b. nucleotides
   c. mRNA
   d. amino acids
   e. enzymes

17. Viroids are believed by some scientists to represent:
   a. introns that have escaped from eukaryotic cells.
   b. naked DNA fragments that were once part of a prokaryotic chromosome.
   c. messenger RNA that was accidentally released from a prokaryotic cell.
   d. copies of exons from eukaryotic cells.
   e. fragments of eukaryotic proteins that interfere with gene regulation.

18. When tested with a DNA probe to search for the gene coding for scrapie, which location has NOT yielded positive test results?

a. the chromosomes of healthy animals.
b. the chromosomes of infected animals.
c. mRNA in the cells of healthy animals.
d. mRNA in the cells of infected animals.
e. purified preparations of the infectious agent itself.

19. A close, long-term association between organisms of two different species in which one benefits and the other is neither benefited nor harmed is SPECIFICALLY referred to as:
   a. mutualism.
   b. parasitism.
   c. commensalism.
   d. communism.
   e. symbiosis.

20. T or F   Prokaryotes are organized into a hierarchical classification scheme on the basis of the many evolutionarily relevant characteristics by which they may be distinguished.

21. T or F   The arrangement of flagella on the surface of a bacterium determines whether it runs in reverse or tumbles when the direction of flagellar rotation changes.

22. T or F   Certain chemosynthetic bacteria are capable of oxidizing elemental sulfur to sulfuric acid (H_2SO_4). This provides the sulfate ion, which is the major form in which plants take up sulfur.

23. T or F   Green and purple photosynthetic bacteria carry out photosynthesis that is nearly identical to that conducted by eukaryotic photosynthesizers.

24. T or F   On a worldwide scale, the nitrogen fixed by cyanobacteria is of equal importance agriculturally to that fixed by other prokaryotes.

25. T or F   Some bacteria are only able to cause illness if they harbor a certain prophage in their genetic material.

26. T or F   Strains of bacteria resistant to antibiotics are so widespread that, today, antibiotics are virtually useless.

27. T or F   Many scientists believe that the prokaryote *Prochloron* gave rise to the chloroplasts of the green algae and of the cyanobacteria.

28. Match each agent with the correct description. An answer may be used only once. Each blank may have only one answer.

   1. adenovirus and rhinovirus   4. HIV (AIDS virus)
   2. polio virus   5. influenza virus
   3. T-even virus   6. PSTV viroid

   a. _____ injects viral-coded proteins into host cell membrane
   b. _____ invades bacterial cells
   c. _____ invades respiratory, intestinal, and nervous tissues
   d. _____ can cause the common cold
   e. _____ is an RNA virus producing reverse transcriptase
   f. _____ causes disease in potato and tomato plants

29. Discuss the two primary ecological roles played by the prokaryotes.

30. Explain why so many bacteria can live in a small sample of soil.

31. Why is it advantageous for some bacteria to have an internal magnetic mechanism?

PERFORMANCE ANALYSIS

1. **d**   One gram (approximately 1/28 of an ounce) of soil can contain as many as 2.5 billion individual prokaryotes. So many can live in so little space because each type is nutritionally specialized to utilize a different set of raw materials. Also, the activities of one type of prokaryote often make nutrients available to other types. (pages 428, 439)

2. **c**   Some taxonomists have suggested the creation of superkingdom Prokaryota, which would include kingdom Monera and kingdom Archaebacteria. The other three kingdoms would be included in the superkingdom Eukaryota. (page 430)

3. **a**   The prokaryotic classification scheme currently in use is convenient for medical, industrial, and agricultural purposes. If the prokaryotes are reclassified according to molecular taxonomy, and the new classifications are less useful, only evolutionary biologists are likely to use them, at least at first. (page 430)

4. **b**   The most prominent feature of a prokaryotic cell is the nucleoid. (page 431)

5. **e**   The mycoplasmas are the only prokaryotes that lack a cell wall. Viruses and prions are not prokaryotes. (page 432)

6. **d**   Gram-negative and gram-positive bacteria are so classified because of their staining characteristics. Gram-positive bacteria will bind gentian violet and other similar stains; gram-negative bacteria will not bind these stains. The different staining behavior is due to differences in cell wall composition. The cell wall of a gram-positive organism is a single homogeneous layer of peptidoglycans and other polysaccharides. The cell wall of a gram-negative organism is composed of two layers: an inner layer of peptidoglycan and an outer layer of lipoproteins and lipopolysaccharides. (page 432)

7. **a**   Bacterial flagella function in locomotion; pili function in attachment to food sources and substrates, and in conjugation. (pages 434, 435)

8. **b**   The genus *Mycobacterium* was named for the fungus-like filaments characteristic of these organisms. (page 436)

9. **e**   Bacteria with a spherical cell shape are known as cocci. Chains of these bacteria are streptococci. A streptococcal bacterium causes one form of pneumonia. (page 436)

10. **d** Prokaryotes most commonly multiply by binary fission, an asexual process that does not involve genetic recombination. In prokaryotes, genetic recombination can be accomplished by conjugation, transformation, plasmid transfer, and transduction. (page 437)

11. **a** The rickettsiae are the smallest known living cells. (page 437)

12. **c** The methanogens are obligate anaerobes; they are poisoned by oxygen. They were difficult to study until successful culture techniques could be developed. (page 439)

13. **d** Green photosynthetic bacteria contain bacteriochlorophyll *a*, which closely resembles chlorophyll *a* of eukaryotic cells. (page 440)

14. **b** Prokaryotes lack chloroplasts; the photosynthetic pigments of cyanobacteria are located in the peripheral membrane system. (page 440)

15. **e** Viruses are typically classified on the basis of host organism(s), nucleic acid content, capsid shape, and presence or absence of extracapsular materials. Viruses have no predetermined life cycle length, and many can remain dormant outside their host for long periods of time. (page 444)

16. **c** The only component that is not used during viral replication is host cell mRNA. The virus directs the production of its own mRNA. (page 445)

17. **a** Evidence currently supports the proposal that viroids represent introns of mRNA that escaped from eukaryotic cells and established an independent existence. (page 447)

18. **e** When scientists used DNA probes to search for DNA that corresponded to the scrapie prion, such a gene was identified in the chromosomes of both healthy and affected animals. Also, mRNA transcribed from the gene was found in both types of animals. However, the gene has not been identified in any purified fractions of the infectious agent. (page 447)

19. **c** Symbiosis is the term applied to a close long-term association between two organisms. There are three types of symbiosis, classified according to the benefit or harm that results to each party in the relationship. Mutualistic relationships are those in which each party benefits. In commensalism, one party benefits and the other party neither benefits nor is harmed. Parasitism is defined as a relationship in which one party benefits and the other party is harmed. (Predation can be considered a form of parasitism.) (page 447)

20. **False** Prokaryotes are not currently classified according to phylogeny because the evolutionary relationships among the groups could not be accurately determined before the advent of molecular taxonomic methods. (page 429)

21. **True** Bacteria typically "run" when their flagella rotate in a counterclockwise direction. Species of bacteria in which the flagella originate all over the surface tumble when the flagella rotate in a clockwise direction. Those species in which flagella are arranged in tufts at the poles reverse their direction of movement by 180° when the flagella rotate in a clockwise direction. (page 434)

22. **True** Some chemosynthetic bacteria can oxidize elemental sulfur to sulfate, the form in which plants take up sulfur from the soil. (page 439)

23. **False** When green and purple photosynthetic bacteria conduct photosynthesis, hydrogen sulfide instead of water is the electron donor replacing the electrons lost from the bacterial photosystem. As a consequence, elemental sulfur is produced instead of oxygen. (page 440)

24. **False** Although the nitrogen fixed by cyanobacteria greatly increases the productivity of rice paddies, on a global scale the nitrogen fixed by other prokaryotes in association with crops (the nitrogen-fixing bacteria) is of greater significance. (page 441)

25. **True** Some bacteria only cause disease if they contain a specific prophage that codes for the production of a toxin, the effects of which result in disease. The causative agents of diphtheria and scarlet fever are examples of two such organisms. (page 448)

26. **False** The indiscriminate use of antibiotics has promoted the survival of strains of microorganisms that are resistant to these drugs, but antibiotics are still available to combat nearly every pathogenic organism. New drugs are being developed every year to combat those organisms that are resistant to previously used antibiotics. However, these new drugs are more expensive than the older agents and may have serious side-effects. (page 449)

27. **False** *Prochloron* is thought to represent the lineage of prokaryotes from which the chloroplasts of green algae and plants developed. (Cyanobacteria do not have chloroplasts.) (essay, page 442)

28. The answers are **a5, b3, c2, d1, e4, f6.** (pages 444–446)

29. (1) Prokaryotes decompose dead organic matter, releasing the compounds contained therein to be recycled through the biosphere. (2) Prokaryotes fix atmospheric nitrogen into a form that can be used by green plants. This is a critical link in the nitrogen cycle between the nonliving and the living components of the biosphere. (page 429)

30. See the answer to question 1. (page 439)

31. The presence of magnetic mechanisms in bacteria allows them to orient with respect to the earth's magnetic field, which has a vertical component in addition to the north-south component. Guided by their magnets, these bacteria swim down, toward sediments rich in decaying organic matter. (essay, page 435)

The Protists

CHAPTER ORGANIZATION

MAJOR CONCEPTS

The kingdom Protista includes a great diversity of organisms, all of which are composed of eukaryotic cells. The kingdom includes unicellular, coenocytic, and multicellular forms. Protists obtain nutrients in one of three ways: (1) by conducting photosynthesis (photosynthetic autotrophs), (2) from dead organic matter (saprobes), and (3) from living organisms (pred- ators, parasites, mutualists, and commensals). A few protists may utilize more than one nutritional mode.

According to the endosymbiotic hypothesis, eukaryotic cells evolved when specialized prokaryotic cells were engulfed by other prokaryotes and became established as intracellular residents. Several lines of evidence suggest that mitochondria and chloroplasts evolved in this manner. The structural and functional complexity of eukaryotic cells allowed the evolution of multicellularity.

Although the classification of protists is currently under debate, biologists generally agree that modern protists represent several evolutionary lines and that all higher organisms (fungi, plants, and animals) evolved from protists.

Protists can be conveniently grouped according to nutritional mode as photosynthetic autotrophs, unicellular heterotrophs, and multinucleate and multicellular heterotrophs. The photosynthetic autotrophs include euglenoids and unicellular and multicellular algae. Unicellular heterotrophs comprise mastigophores, sarcodines, ciliophores, opalinids, and sporozoans. The slime molds and water molds (not true fungi) are multinucleate or multicellular heterotrophs.

Protists exhibit predictable behavioral responses that enable them to move toward attractive stimuli and away from noxious stimuli.

HOW TO STUDY THE CHAPTER

Read the entire chapter through quickly, focusing on the major concepts.

Use the GUIDED STUDY OF THE CHAPTER to help you identify the important details as you **reread** the chapter. Writing out the answers to these questions will help fix them in your mind as well as provide you with a valuable study aid.

Answer the questions in TESTING YOUR UNDERSTANDING without the aid of your text. Check your answers against

GUIDED STUDY OF THE CHAPTER

I. Introduction *(page 452)*

> *Focus:* All protists consist of eukaryotic cells; most are unicellular. A protist may be heterotrophic, autotrophic, or capable of obtaining nutrients by either mechanism.

1. Contrast the complexity of a protist cell with the cells of a multicellular organism.

II. The Evolution of the Protists *(pages 452–455)*

A. General Remarks *(pages 452–454)*

2. Name five characteristics that distinguish eukaryotes from prokaryotes.

> *Focus:* The endosymbiotic hypothesis proposes that eukaryotic cells evolved as a consequence of symbiotic relationships in which prokaryotes lived inside other larger prokaryotes.

3. Summarize four lines of evidence supporting the hypothesis that mitochondria evolved from symbiotic prokaryotes.

4. a. What facts about mitochondria apparently contradict the endosymbiotic hypothesis?

   b. Mitochondria are thought to have developed from what type of prokaryote?

5. Name two as yet unexplained features common to all eukaryotic cells.

6. What role in the evolution of eukaryotic cells has been proposed for the hypothetical "urkaryotes"?

7. In addition to the evidence for the evolution of mitochondria, what other evidence supports the endosymbiotic hypothesis?

8. Identify two ways in which contemporary symbiotic cyanobacteria resemble chloroplasts.

9. List four advantages that eukaryotic cells have over prokaryotic cells.

B. Classification of Protists *(page 454)*

10. What two generalizations about protists and evolution are acceptable to most biologists?

11. Name four subjects relating to the classification of protists that are currently being debated.

> *Focus:* There is currently no protist classification scheme that is generally accepted by all biologists studying protists.

12. What is currently accepted to be the evolutionary relationship between modern protists and the lineages from which (a) plants, (b) animals, and (c) fungi arose?

13. Why have the terms "algae" and "protozoa" been abandoned as formal classification terms?

> *Focus:* **Study Table 22–1 carefully to gain an appreciation for the diversity within the kingdom Protista.**

III. Photosynthetic Autotrophs *(pages 455–465)*

A. General Remarks *(pages 455–456)*

14. a. Identify the three divisions of photosynthetic autotrophs that contain primarily unicellular forms.

    b. Which three divisions include many multicellular forms?

> *Focus:* **Nearly all members of the six divisions you just listed are photosynthetic and have fairly simple structures.**

15. Fill in the following table comparing unicellular and multicellular algae.

Table for Question 15

	UNICELLULAR ALGAE	MULTICELLULAR ALGAE
Most common location(s)		
Source(s) of nutrients		
Degree to which environmental conditions fluctuate (if known or your best guess)		
Degree of competition with other algae (if known or your best guess)		

16. *Vocabulary:* Distinguish between plankton and phytoplankton.

17. In what two ways do the phytoplankton contribute to other forms of life on earth?

18. How have multicellular algae adapted to the space limitations present along shorelines?

B. Characteristics of the Photosynthetic Protists *(pages 456–457)*

19. Name four ways in which the six divisions of algae differ.

20. a. In what way does the cell wall of algae resemble that of plants?

    b. What substance in the cell wall of algae produces its mucilaginous consistency?

21. For the pigments listed in the table, indicate their color and name at least one type of algae (or division of protists) in which they are present.

Table for Question 21

	PIGMENT COLOR	EXAMPLE ORGANISM(S)
Fucoxanthin		
Phycobilins		
Chlorophylls		

22. There is a wide variety of pigments in the chloroplasts of the various algae. Summarize two implications of this fact.

23. Identify the suggested evolutionary origin of the chloroplasts of:

Rhodophyta

Chlorophyta and Euglenophyta

Chrysophyta, Dinoflagellata, and Phaeophyta

24. a. *Vocabulary:* What is laminarin?

b. How is it similar to and different from starch?

25. How do the following organisms store food reserves?
Green algae

Dinoflagellates

Brown algae

Golden-brown algae

Diatoms

Red algae

26. Discuss the similarities and differences between cell division in algae and in other organisms.

C. **Division Euglenophyta: Euglenoids** *(pages 458–459)*

27. How are the chloroplasts of euglenoids similar to and different from those of green algae?

28. Cite two possible means by which euglenoids may have obtained chloroplasts.

29. How do the nonphotosynthetic euglenoids obtain their nutrients?

30. *Vocabulary:* What is paramylon?

31. a. Mention two mechanisms by which *Euglena* can move through the environment.

b. What structural features enable these types of movement?

32. What is the role of the contractile vacuole in *Euglena*?

33. Describe the means by which *Euglena* orients with respect to a light source.

34. a. Under what circumstances may nonphotosynthetic *Euglena* be produced?

b. What does this suggest about the possible origin of some modern heterotrophs?

D. **Division Chrysophyta: Diatoms and Golden-Brown Algae** *(pages 459–460)*

35. List four identifying characteristics of the chrysophytes.

36. What economic role has been suggested for chrysophytes?

37. *Vocabulary:* Define nanoplankton.

38. Of what importance are golden-brown algae to other aquatic organisms?

39. a. Describe the shells of diatoms.

    b. *Vocabulary:* What is diatomaceous earth?

    c. What is the economic importance of diatomaceous earth?

40. The amoeboid chrysophytes closely resemble species belonging to what protist phylum?

41. Summarize the sexual reproduction characteristic of some diatoms.

E. **Division Dinoflagellata: "Spinning" Flagellates** *(page 460)*

42. The sometimes bizarre appearance of dinoflagellates is a consequence of what structural characteristic?

43. Describe the mechanism of locomotion in dinoflagellates.

44. a. What causes "red tides"?

    b. What is the potential public health risk of the dinoflagellate, *Gessnerium catenellum*? (Include the conditions under which this organism becomes a potential health risk.)

F. **Division Chlorophyta: Green Algae** *(page 461)*

45. a. Name the three major classes of chlorophytes.

    b. Which class is thought to have originated from the same line as green plants?

    c. Identify four characteristics that at least some contemporary Charophyceae share with green plants.

46. Name six types of habitats in which chlorophytes live.

The Increase of Complexity (pages 461–462)

> *Focus:* The division Chlorophyta is the most diverse division of kingdom Protista and contains unicellular, multicellular, and intermediate forms.

47. How do colonies differ from true multicellular organisms?

48. Fill in the following table on the increasing complexity exemplified by the volvocine line.

Table for Question 48

	GONIUM	PANDORINA	VOLVOX
Colony size (number of cells)			
Colony shape			
Mode of movement			
Mode of reproduction			
Polarity of cells in the colony? (yes or no)			
Specialization of cells? (yes or no; if yes, tell how)			

49. Why is the volvocine line considered to be an evolutionary "dead end"?

50. *Vocabulary:* Define coenocytic.

51. Name two classes of green algae that contain several multicellular members.

52. How do the multicellular algae form, and what shapes can they take?

Life Cycles in the Green Algae (pages 462–464)

> *Focus:* For many organisms that reproduce sexually and asexually, sexual reproduction is triggered by a decreased food supply. Sexual reproduction in the face of a changing environment could result in organisms with new gene combinations that make them more suited to the altered environment.

53. *Vocabulary:* What is a zygospore?

54. Arrange the following components of the *Chlamydomonas* life cycle in a flow diagram to show the correct sequence of events. Indicate whether each component is haploid (n) or diploid ($2n$) and where mitosis and meiosis occur: zygospore, gametes, haploid cell type A, haploid cell type B, zygote. (Refer to Figure 12–5, page 251.)

55. a. *Vocabulary:* List four biological definitions of the term "spore." (footnote, page 462)

    b. Name six features that all spores have in common.

56. *Vocabulary:* Distinguish among isogamy, anisogamy, and oogamy.

57. In a life cycle characterized by oogamy, identify the functions for which the two different types of gametes are specialized.

58. Which genus of green algae demonstrates the entire range of gamete types?

Focus: The life cycles of some multicellular algae resemble those of plants by exhibiting alternation of generations.

---

59. Fill in the following table distinguishing gametophytes and sporophytes.

Table for Question 59

	GAMETOPHYTE	SPOROPHYTE
Haploid or diploid?		
Origin (cells from which it forms)		
Function in life cycle (include ploidy of cells produced)		

60. *Vocabulary:* Distinguish between spores and gametes.

61. Identify the two stages in alternation of generations where genetic recombination may occur.

62. *Vocabulary:* Distinguish between isomorphic and heteromorphic.

G. Division Phaeophyta: Brown Algae *(page 465)*

63. What features do the brown algae have in common with (a) chrysophytes, and (b) green algae and plants?

64. a. *Vocabulary:* Distinguish among thallus, holdfast, stipe, and blade.

b. In what way does the stipe functionally resemble stems in vascular plants?

65. a. Characterize the life cycle of brown algae.

b. Describe the resemblance between the life cycle of *Fucus* and that of higher animals. (Refer to Figure 12–4c, page 250.)

H. Division Rhodophyta: Red Algae *(page 465)*

66. How do red algae differ from brown algae with respect to preferred environment?

67. Red algae can grow at much greater water depths than other algae because of what characteristic?

68. Name three types of pigments found in red algae.

69. Discuss two ways in which the cell walls of red algae may differ from those of other algae.

70. How is the life cycle of red algae similar to and different from that of brown algae?

71. *Review Question*: Match each characteristic of photosynthetic protists with the division to which it applies. A division may be used more than once. A blank may have more than one answer, as indicated in parentheses.

1. Chlorophyta   4. Phaeophyta
2. Chrysophyta   5. Dinoflagellata
3. Euglenophyta   6. Rhodophyta

a. _____ paramylon as food reserve

b. _____ mostly marine forms (4 answers)

c. _____ cell wall absent

d. _____ laminarin and oils as food reserve (2 answers)

e. _____ phycobilin pigments present

f. _____ mostly freshwater forms (2 answers)

g. _____ no members have flagella

h. _____ starch as a food reserve (3 answers)

i. _____ cell walls may contain silicon

j. _____ some cell walls contain calcium carbonate

k. _____ no known sexual reproduction

IV. Multinucleate and Multicellular Heterotrophs *(pages 466–468)*

A. The Slime Molds *(pages 466–467)*

72. What are the two main groups of slime molds?

73. Where do slime molds live?

74. The pigments possessed by many plasmodial slime molds are thought to perform what function?

75. *Vocabulary:* What is a plasmodium? (See also Figure 22–20.)

76. Name four nutritional sources utilized by slime molds.

77. *Vocabulary:* What are sporangia?

78. What two situations can trigger the formation of sporangia?

79. Describe the process by which a new plasmodium develops.

80. How do the cellular slime molds differ from the plasmodial slime molds?

B. The Water Molds *(pages 467–468)*

81. These organisms are no longer classified as fungi for what two reasons?

Division Chytridiomycota (page 467)

82. *Vocabulary:* Define and name two functions of rhizoids.

83. a. Chytrids exhibit what types of reproduction?

b. What type of life cycle occurs in *Allomyces*?

84. *Vocabulary:* Define and identify the function of sirenin.

Focus: Chytrids exhibit two nutritional modes: parasitic and saprophytic.

Division Oomycota (pages 467–468)

85. *Vocabulary:* Distinguish between hyphae and gametangia.

86. How do the oomycetes resemble (a) fungi and (b) plants?

87. Fill in the following table comparing reproduction in chytrids and oomycetes.

Table for Question 87

	CHYTRIDS	OOMYCETES
Sexual Reproduction Isogamy, anisogamy, or oogamy		
Gametes motile or nonmotile		
Asexual Reproduction Spores motile or nonmotile		

88. Identify the two nutritional modes utilized by oomycetes.

89. Name two economically important oomycete parasites and identify the host of each pathogen.

V. Unicellular Heterotrophs *(pages 468–474)*

A. General Remarks *(page 468)*

90. Match each protozoan phylum with the correct description. An answer may be used more than once. A blank may have more than one answer, as indicated in parentheses. (See also Table 22–3.)

1. Ciliophora   4. Sarcodina
2. Mastigophora   5. Sporozoa
3. Opalinida

a. _____ movement by means of flagella

b. _____ ciliary movement (2 answers)

c. _____ only parasitic forms are included (2 answers)

d. _____ movement accomplished by pseudo-podia

e. _____ members are not highly motile

Focus: Protozoans reproduce mainly by binary fission. In addition, many forms are capable of recombining genetic information by means of sexual reproduction or conjugation.

B. Phylum Mastigophora *(pages 468–469)*

91. Describe two possible origins of the phylum Mastigophora.

92. Provide the requested information on members of phylum Mastigophora.

Means of locomotion (2)

Methods of reproduction (2)

Life-style types (2)

93. Name two genera of mastigophorans that include members that are parasitic to humans. What diseases do these organisms cause?

94. Identify one genus of mastigophorans whose members have a mutualistic relationship with their hosts and describe the relationship.

C. Phylum Sarcodina (page 470)

95. Provide the requested information on members of phylum Sarcodina.

Suspected origin

Means of locomotion (2)

Methods of reproduction (2)

Life-style types (2)

96. *Vocabulary:* What is the "test" of a sarcodine and what is its proposed function?

97. Name three types of materials of which a sarcodine test may be composed.

98. The white cliffs of Dover were formed by:

99. The shells of the Foraminifera are of what value to geologists?

D. Phylum Ciliophora (pages 470–471)

100. Provide the requested information on members of phylum Ciliophora.

Suspected origin

Means of locomotion (3)

Life-style types

101. *Vocabulary:* Distinguish among cilia, membranelles, and cirri.

102. *Vocabulary:* Distinguish between myonemes and trichocysts.

103. Fill in the following table comparing macronuclei and micronuclei.

Table for Question 103

	MACRONUCLEI	MICRONUCLEI
Relative size		
Relative content of DNA		
Undergo meiosis? (yes or no)		
Required for cell survival? (yes or no)		
Means by which daughter cells obtain them		

> *Focus:* The evolution of different and energetically expensive techniques for recombining genetic information indicates that genetic diversity is of great adaptive value to a species.

E. *Essay:* The Evolution of Mitosis (page 472)

104. Compare the patterns of mitosis in (a) prokaryotes, (b) dinoflagellates, and (c) the micronucleus of ciliated protozoa.

105. Summarize three stages thought to represent the evolution of mitosis in eukaryotes.

F. Phylum Opalinida (page 473)

106. Provide the requested information on members of phylum Opalinida.

    Suspected origin
    (3 proposed)

    Means of locomotion (2)

    Life-style types

107. What characteristic of opalinid reproduction leads some authorities to suggest that they are related to mastigophorans?

G. Phylum Sporozoa (pages 473–474)

> **Focus:** All members of phylum Sporozoa are parasitic, have complex life cycles, and possess neither cilia nor flagella.

108. Study Figure 22–29 closely then answer the following questions.

    a. Name the stage in the *Plasmodium* life cycle that is ingested by the mosquito. Is this stage haploid or diploid?

    b. What stage in the *Plasmodium* life cycle infects a human? Is this stage haploid or diploid?

    c. From what cells are merozoites produced and where do they multiply?

    d. What causes the recurring fevers associated with malaria?

    e. How are gametes formed?

109. a. Name two strategies used by humans to control malaria.

    b. Why is the effectiveness of these techniques decreasing?

110. How many humans are estimated to be infected with *Plasmodium*?

111. a. What percentage of malaria victims in Africa die as a direct result of the parasite?

    b. In young children with malaria, up to _____ percent die of the disease.

112. a. Describe how scientists hope to use recombinant DNA techniques to develop a vaccine against *Plasmodium*.

    b. What characteristic of *Plasmodium* complicates the development of an effective vaccine?

VI. Patterns of Behavior in Protists (pages 474–476)

A. General Remarks (page 474)

113. Name three types of stimuli to which some prokaryotes may respond.

114. Describe the response of amoebas to light.

115. *Vocabulary:* What is habituation and of what value is it to an organism?

116. Describe the chemotactic response of an amoeba to a food item.

B. Avoidance in *Paramecium* (pages 475–476)

117. Name two types of stimuli to which *Paramecium* is highly sensitive.

118. What structure of the *Paramecium* apparently functions in the detection of chemical stimuli?

119. a. The responses of *Paramecium* to noxious or attractive stimuli are "fixed." Explain this statement.

 b. Describe how the stereotyped movement pattern of *Paramecium* successfully moves it (a) away from noxious stimuli and (b) toward favorable stimuli.

120. *Paramecium* prefers a slightly acidic environment. What is the adaptive value of this preference?

121. *Review Question:* Match each division or phylum of heterotrophic protists with the correct description(s). Phyla may be used more than once. A blank may have more than one answer, as indicated in parentheses.

1. Acrasiomycota
2. Chytridiomycota
3. Ciliophora
4. Mastigophora
5. Myxomycota
6. Oomycota
7. Opalinida
8. Sarcodina
9. Sporozoa

a. _____ movement is by pseudopodia (4 answers)

b. _____ some members possess cilia (2 answers)

c. _____ all members exhibit oogamy

d. _____ exchange of genetic material by conjugation

e. _____ some members are coenocytic (3 answers)

f. _____ most or all members are parasitic (3 answers)

g. _____ cell walls contain chitin

h. _____ cell walls contain cellulose

VII. Summary (pages 476–477): Read the summary. If you are familiar with the essential features of the material presented there, you are ready to complete the section TESTING YOUR UNDERSTANDING.

TESTING YOUR UNDERSTANDING

After you have completed this examination, compare your answers with those in the section that follows.

1. Oxygen began to accumulate in the atmosphere about 2.5 billion years ago as a result of the photosynthetic activity of:
 a. green algae.
 b. purple algae.
 c. cyanobacteria.
 d. land plants.
 e. heterotrophic ciliates.

2. Mitochondria are thought to have arisen from _____ in which the capacity for photosynthesis had been lost.
 a. purple nonsulfur bacteria
 b. green nonsulfur bacteria
 c. cyanobacteria
 d. zygomycetes
 e. halobacteria

3. Cyanobacteria have been found living within several types of organisms. Which of the following are NOT such organisms?
 a. methanogens
 b. plants
 c. animals
 d. fungi
 e. other cyanobacteria

4. The following are divisions of photosynthetic protists. Which group includes only unicellular organisms?
 a. Dinoflagellata, Rhodophyta, Euglenophyta
 b. Chrysophyta, Rhodophyta, Chlorophyta
 c. Chlorophyta, Phaeophyta, Euglenophyta
 d. Euglenophyta, Dinoflagellata, Chrysophyta
 e. Rhodophyta, Dinoflagellata, Phaeophyta

5. The brown color of the Phaeophyta is due to the presence of:
 a. chlorophyll *e*.
 b. fucoxanthin.
 c. rhodoglobulin.
 d. phycobilin.
 e. xanthophyll.

6. The type of sexual reproduction in the euglenoids most closely resembles that of the:
 a. green algae.
 b. purple bacteria.
 c. slime molds.
 d. Rhodophyta.
 e. Euglenoids reproduce asexually by longitudinal division.

7. It is now thought that the major component of plankton is the _____ algae and that they may be the major food-producing organisms of the oceans.
    a. green
    b. golden-brown
    c. red
    d. brown
    e. flagellated

8. One of the most characteristic features of the dinoflagellates is their:
    a. chlorophyll *c*.
    b. paramylon food reserves.
    c. flexible pellicle.
    d. spinning motion.
    e. amoeboid movement.

9. Plants are thought to have arisen from the same ancestral group as class _____ of division _____.
    a. Chlorophyceae; Phaeophyta
    b. Chlorophyceae; Chlorophyta
    c. Charophyceae; Phaeophyta
    d. Charophyceae; Chlorophyta
    e. Ulvophyceae; Rhodophyta

10. Under what condition does *Chlamydomonas* engage in sexual reproduction?
    a. the presence of approximately equal numbers of + and − cells
    b. depletion of essential nutrients in the medium
    c. evaporation of the medium almost to dryness
    d. following each asexual reproductive event
    e. an excess of essential nutrients in the medium

11. What is unusual about the gametes of the red algae?
    a. The male gamete is larger than the female gamete.
    b. The female gamete is flagellated and the male gamete is not.
    c. Neither of the gametes is motile.
    d. The male gamete is fixed and the water brings the female gamete to it.
    e. Male gametes have centrioles, but female gametes do not.

12. Protozoans are classified into five divisions mainly on the basis of their:
    a. mode of nutrition.
    b. cellular biochemistry.
    c. means of reproduction.
    d. cellular size and shape.
    e. form of locomotion.

13. Which of these diseases is caused by a mastigophoran?
    a. malaria
    b. potato blight
    c. human amoebic dysentery
    d. downy mildew of grapes
    e. African sleeping sickness

14. The sarcodines are thought to have arisen from the mastigophorans because:
    a. they generally share the same mode of nutrition.
    b. their cellular biochemistry is almost identical.
    c. some sarcodines possess flagella during part of their life cycle.
    d. the parasitic forms cause almost identical diseases.
    e. neither has a cell wall outside its membrane.

15. Which phylum of protozoans consists exclusively of parasites?
    a. Sporozoa
    b. Sarcodina
    c. Ciliata
    d. Mastigophora
    e. Dinoflagellata

16. T or F Protists are relatively simple, unicellular, eukaryotic heterotrophs.

17. T or F In addition to single cells, the photosynthetic protista include members made up of filaments, plates, or solid masses of cells.

18. T or F Some of the multicellular green algae exhibit alternation of generations. The sporophyte generation produces haploid spores by meiosis; upon germination, the spores produce the gametophyte generation.

19. T or F Because the process of conjugation is so complex and is metabolically expensive, the variations produced thereby must be worth a great deal from an evolutionary point of view.

20. T or F *Paramecium* has two simple, preprogrammed behavioral responses, one that is triggered by harmful stimuli and one that is triggered by positive stimuli.

21. In an endosymbiotic relationship between unicellular organisms, what does each cell in the partnership gain?

22. Distinguish between the terms heteromorphic and isomorphic as they apply to alternating generations in some of the green algae.

23. Compare and contrast the Myxomycota and the Acrasiomycota with respect to common name, environment, nutrient source, and body organization.

24. In what ways is a chytrid similar to and different from a fungus?

PERFORMANCE ANALYSIS

1. c Oxygen began to accumulate in the primitive atmosphere as a result of the photosynthetic activity of cyanobacteria. (page 452)

2. a Mitochondria are thought to have descended from a line of purple nonsulfur bacteria that had lost the capacity for photosynthesis. (page 453)

3. **a**   Cyanobacteria have been found as endosymbionts in all of the listed types of organisms except the methanogens. (page 453)

4. **d**   The divisions Chrysophyta, Dinoflagellata, and Euglenophyta include only unicellular forms. The other three algal divisions (Chlorophyta, Phaeophyta, and Rhodophyta) include multicellular organisms. (Table 22–1)

5. **b**   Fucoxanthin is a xanthophyll pigment found in brown algae (Phaeophyta) and the golden-brown algae and diatoms (Chrysophyta). (page 465)

6. **e**   Euglenoids do not exhibit sexual reproduction. They reproduce asexually by longitudinal division. (page 459)

7. **b**   The golden-brown algae of division Chrysophyta are important components of the nanoplankton. Some authorities believe they may be the major food-producing organisms of the oceans. (page 459)

8. **d**   The spinning motion of dinoflagellates results from the beating of two flagella in grooves, the characteristic for which the division was named. (page 460)

9. **d**   The modern plants are believed to be descended from an ancestral group that also gave rise to members of class Charophyceae of division Chlorophyta. (page 461)

10. **b**   When nutrients become depleted in a growth medium, *Chlamydomonas* undergoes sexual reproduction. (page 462)

11. **c**   Unlike other types of algae, neither male nor female gametes of the red algae have flagella. (page 465)

12. **e**   Protozoans are classified into five divisions primarily based on mechanism of locomotion. Mastigophorans have flagella. Sarcodines move by pseudopodia. Ciliates possess cilia. Opalinids are parasites with flagella and sporozoans are basically nonmotile parasites. (page 468)

13. **e**   African sleeping sickness is caused by the mastigophorans *Trypanosoma gambiense* and *Trypanosoma rhodesiense*. The cause of potato blight is the oomycete *Phytophthora infestans*. Human amoebic dysentery is caused by sarcodines. Downy mildew of grapes is the result of *Plasmopara viticola*, an oomycete. Sporozoans of the genus *Plasmodium* cause malaria. (pages 467, 468, 469, 470, 473)

14. **c**   Some sarcodines possess flagella during part of their life cycle. For this reason, they are believed to have descended from the mastigophorans. Opalinids are also thought to be related to mastigophorans because they form gametes that fuse to form zygotes. (pages 470, 473)

15. **a**   All sporozoans and opalinids are parasitic. (page 473)

16. **False**   The kingdom Protista includes complex unicellular organisms, multicellular forms, and both heterotrophs and autotrophs. All members of the kingdom have eukaryotic cells. *Protozoans* are unicellular, heterotrophic eukaryotes. (page 452)

17. **True**   The photosynthetic protists include unicellular, colonial, and multicellular forms. (page 456)

18. **True**   In the alternation of generations exhibited by some multicellular green algae, the sporophyte generation produces haploid spores by meiosis, and the spores germinate to form the gametophyte. (page 463)

19. **True**   In the course of evolution, new characteristics are maintained in a species only if they contribute to its survival and reproduction. All the techniques that have evolved to accomplish genetic recombination are fairly expensive to the organism, but the advantage of genetic flexibility to the species outweighs the metabolic costs to the individual. (page 471)

20. **False**   *Paramecium* has only one programmed behavior pattern by which it responds to both negative and positive stimuli. However, the result is that *Paramecium* moves toward positive stimuli and away from negative stimuli. (page 475)

21. The smaller cell in an endosymbiotic relationship between unicellular organisms gains nutrients and protection. The larger cell gains a new energy source. (page 453)

22. In algal alternation of generations, a heteromorphic species is one in which the sporophyte and gametophyte generations are markedly different in physical appearance. Isomorphic species have sporophyte and gametophyte generations that look alike. (page 404)

23.

	MYXOMYCOTA	ACRASIOMYCOTA
Common name	Plasmodial slime mold	Cellular slime mold
Body organization	Amoeboid cells swarm to form coenocytic mass	Amoeboid cells swarm to form colonial/multi-cellular mass
Environment	Cool, moist places	Cool, moist places
Nutrient source	Dead organic matter	Dead organic matter

(pages 466, 467)

24. The chytrids superficially resemble fungi in their basic structure, which consists of a coenocytic sporangium and anuclear rhizoids. However, many chytrids have flagellated reproductive cells (fungi have no flagellated cells), and the biochemistry of chytrids differs substantially from that of the fungi. (page 467)

CHAPTER 23

The Fungi

MAJOR CONCEPTS

Kingdom Fungi includes unicellular, coenocytic, and multicellular forms. Coenocytic and multicellular fungi are composed of filaments known as hyphae. The hyphae (which collectively form the mycelium) may form a branching network that covers a large amount of space or they may be compressed to form structures specialized for spore production and dispersal. Fungi have rigid cell walls composed mainly of chitin. Most fungi reproduce by both sexual and asexual means.

All fungi are strict heterotrophs; individuals may be saprophytic, parasitic, or predatory. Fungi obtain nutrients by secreting digestive enzymes onto the substrate and then absorbing the constituent molecules. Some fungi have specialized hyphae, haustoria, that obtain nutrients directly from the host cells.

The fungi are believed to have evolved from unicellular eukaryotic organisms that apparently have no living counterparts. Three separate lines are thought to have given rise to the chytrids, oomycetes, and zygomycetes, all of which have a coenocytic organization. The ascomycetes and basidiomycetes are believed to have evolved from a common ancestor; this ancestor shared a common ancestor with the zygomycetes.

Fungi benefit human interests in several ways, including food production and the manufacture of therapeutic drugs. Conversely, fungi are detrimental to human interests in that they destroy crops and stored agricultural products, cause human and animal disease, and produce toxins that cause disease in humans and animals. Along with bacteria, fungi are the principal decomposers of organic matter in nature and play important roles in the cycling of nutrients through the biosphere.

Fungi are classified into three divisions (Ascomycota, Basidiomycota, and Zygomycota) on the basis of their structure and pattern of sexual reproduction. A fourth division (Deuteromycota) includes fungi in which sexual reproduction is unknown and a few closely related fungi that exhibit sexual reproduction.

Fungi exist in two important symbiotic relationships, lichens and mycorrhizae. A lichen consists of a fungus and either a cyanobacterium or a green alga. Lichens need only air, water, and a few minerals to exist and are often the first colonizers of barren terrain. Mycorrhizal associations occur between a fungus and the roots of a vascular plant. The fungus is thought to make soil nutrients available to the plant, and the plant is believed to provide organic molecules for the fungus.

HOW TO STUDY THE CHAPTER

Read the entire chapter through quickly, focusing on the major concepts.

Use the GUIDED STUDY OF THE CHAPTER to help you identify the important details as you **reread** the chapter. Writing out the answers to these questions will help fix them in your mind as well as provide you with a valuable study aid.

Answer the questions in TESTING YOUR UNDERSTANDING without the aid of your text. Check your answers against those in PERFORMANCE ANALYSIS. Analyzing your answers will give you valuable feedback on your level of understanding and preparedness for classroom testing.

GUIDED STUDY OF THE CHAPTER

I. Introduction (page 479)

1. Name four properties of eukaryotic cells that made possible the structural and functional diversification of protists.

2. a. Identify two factors that limit cell size.

   b. Name two means by which living organisms circumvent these limiting factors.

II. Characteristics of the Fungi (pages 480–481)

A. General Remarks (page 480)

3. Describe the three physical forms in which a fungus may exist.

4. *Vocabulary:* Distinguish between hypha and mycelium.

5. a. What is the major component of fungal cell walls?

   b. In what three ways does this substance resemble cellulose? (Figure 23–2)

   c. This substance is also a major component of:

6. a. The portions of a fungus that project above the surface on which they grow are specialized for what function?

   b. Where does the bulk of the mycelium reside?

7. Summarize the process by which mycelia develop and grow.

8. Fungi are nonmotile. How do they compensate for this limitation?

9. Name the two nutritional modes present in kingdom Fungi.

10. Describe the mechanism by which a fungus obtains nutrients.

11. *Vocabulary:* What are haustoria?

> *Focus:* Fungi and bacteria are responsible for the decomposition of dead organic matter, which releases the raw materials contained therein to cycle throughout the ecosystem.

12. Name six ways in which fungi are considered harmful to human interests.

13. Identify two major ways in which fungi benefit human interests.

B. Reproduction in the Fungi (page 481)

14. Asexual reproduction in fungi may be accomplished in what two ways?

15. What is the adaptive value of spore formation to a fungal species?

16. *Vocabulary:* Distinguish among sporangia, sporangiophores, and gametangia.

17. What characteristic of sporangia facilitates the dispersion of spores?

18. *Vocabulary:* What is a septum and what purpose does it serve?

19. List three means by which sexual reproduction may be accomplished in fungi.

20. *Vocabulary:* What is a dikaryon and how is it created?

III. **Classification of the Fungi** *(pages 481–482)*

21. Identify the two criteria used to classify fungi into the three main divisions.

22. What characteristic distinguishes the deuteromycetes from the other three fungal divisions?

23. The pattern of asexual reproduction in deuteromycetes closely resembles that present in:

Focus: The fungi are believed to have evolved from unicellular eukaryotic organisms for which there are no modern counterparts.

---

24. a. What is the proposed evolutionary relationship between chytrids, oomycetes, and modern fungi?

b. Why are chytrids and oomycetes not classified with the fungi?

25. a. *Vocabulary:* What are septate hyphae?

b. Which two groups of fungi have septate hyphae?

26. Describe the presumed evolutionary relationship between ascomycetes, basidiomycetes, and zygomycetes.

IV. **Division Zygomycota** *(pages 482–484)*

A. **General Remarks** *(pages 482–483)*

27. Provide the following information on zygomycetes.

Environment

Nutritional modes (2)

28. *Vocabulary:* Distinguish among rhizoids, stolons, and sporangiophores. (See also Figure 23–6.)

29. Identify three functions of rhizoids.

30. What fungal structure is responsible for the color of fungi?

31. Arrange the following terms in a flow diagram to illustrate sexual and asexual reproduction in zygomycetes: gametangium, sporangiophore, sporangium, spore, zygosporangium. Indicate whether each structure is haploid or diploid and identify where mitosis and meiosis occur. (Be sure to indicate the two possible origins of a sporangiophore.)

32. Describe what happens to the nuclei of + and − strains during sexual reproduction.

B. *Essay*: **Ready, Aim, Fire!** *(page 484)*

33. *Vocabulary:* Define phototropic.

34. Describe how the structural adaptations present in the zygomycete, *Pilobolus*, facilitate the dispersal of spores.

35. *Pilobolus* grows on manure. How does its spore dispersal technique increase the chances of spores coming in contact with manure?

V. **Division Ascomycota** *(pages 484–486)*

Focus: **The ascomycetes are the most numerous of the fungi and are characterized by the formation of ascospores within an ascus.**

36. Name several examples of ascomycetes.

37. Identify two types of trees that are susceptible to fungal diseases.

38. What is the importance of ascomycetes to modern medicine?

39. a. Define the following components of the ascomycete life cycle and indicate whether each component is haploid or diploid, or contains mixed nuclei. (See also Figure 23–8.)

    Ascocarp

    Ascospore

    Ascus

    Conidia

    Dikaryotic hyphae

    Gametangia

    Monokaryotic hyphae

    b. Arrange these components in a flow diagram that illustrates sexual and asexual reproduction in ascomycetes. Indicate where mitosis and meiosis occur and distinguish sexual from asexual stages.

40. *Vocabulary:* What are yeasts and to what type of environment are many yeasts adapted?

41. Describe how asexual and sexual reproduction in yeasts is similar to and different from the ascomycete life cycle diagramed in question 39b.

42. What characteristic of ergot infestations indicates that it is a parasite well-adapted to its host? (Review the discussion of symbiosis in Chapter 21 if necessary.)

Focus: **Ergotism results from the ingestion of a fungal toxin produced by *Claviceps purpurea*, a parasite of rye.**

---

43. a. Name three symptoms of ergotism in humans.

b. Identify two ways in which ergot affects specific organs in the body.

VI. Division Basidiomycota *(pages 486–488)*

44. Why do mushrooms often appear as "fairy rings"?

45. Explain why fairly large mushrooms can appear overnight.

46. a. Define each of the following components of the basidiomycete life cycle and indicate whether each component is haploid or diploid, or contains mixed nuclei. (See also Figure 23–10.)

Basidiocarp

Basidiospore

Basidium

Dikaryotic mycelium

Monokaryotic mycelium

b. Arrange these components in a flow diagram that illustrates reproduction in basidiomycetes. Indicate where hyphae fusion, mitosis, and meiosis occur.

Focus: **Reproduction in basidiomycetes closely parallels ascomycete reproduction.**

---

47. Where are the spores of gill fungi located?

48. a. Identify two species of toxic mushrooms.

b. Describe the mechanism by which the toxin of *Amanita bisporigera* poisons an animal that ingests this mushroom. (Figure 23–11)

49. Give the common names for two types of basidiomycetes that cause economically important crop losses.

50. White rot fungus is being considered as a potential instrument for disposing of toxic wastes because:

VII. Division Deuteromycota *(page 488)*

A. General Remarks *(page 488)*

51. What characteristic is the foundation for this fungal division?

52. Identify ways in which deuteromycetes are (a) beneficial and (b) harmful to human interests.

B. *Essay*: Predaceous Fungi *(page 489)*

53. Describe two mechanisms by which fungi trap prey items.

54. How does a fungus obtain nutrients from trapped prey?

55. Name four types of prey that may be captured by fungi.

VIII. Symbiotic Relationships of Fungi *(pages 488–491)*

A. General Remarks *(page 488)*

> *Focus:* Lichens and mycorrhizae are symbiotic combinations of fungi with some other type of organism. They are important ecologically as colonizers of barren terrain.

B. The Lichens *(pages 488–490)*

56. *Vocabulary:* What is a lichen?

> *Focus:* The conditions under which lichens can survive differ from the conditions under which the individual components can exist.

57. Identify six environments in which lichens may be found.

58. In what way are lichens "colonizers"?

59. a. What three substances do lichens need to survive?

    b. The requirements for survival of lichens differ from those of the individual components in what ways and for what reasons?

60. Name two sources of the minerals that lichens need.

61. What characteristic of lichens makes them very sensitive indicators of air pollution?

62. What is the basis for naming and classifying lichens?

63. Members of the division _____ are the most common fungi found in lichens.

64. *Vocabulary:* Distinguish among crustose, foliose, and fruticose lichens. (Figure 23–13)

65. Describe two means by which new lichens form.

C. Mycorrhizae *(pages 490–491)*

66. *Vocabulary:* What are mycorrhizae?

67. Describe the circumstances under which mycorrhizae were first discovered.

68. How common are mycorrhizae thought to be?

69. Fill in the following table distinguishing endomycorrhizae from ectomycorrhizae.

Table for Question 69

	ENDOMYCORRHIZAE	ECTOMYCORRHIZAE
Hyphae associated with plant roots (describe)		
Type(s) of fungus(i) involved		

70. Identify the presumed benefits derived from a mycorrhizal association by (a) the plant and (b) the fungus.

Focus: **The evolution of mycorrhizal associations has been suggested as a key factor in the initial colonization of the primitive landscape by plants.**

IX. Summary *(pages 491–492):* Read the summary. If you are familiar with the essential features of the material presented there, you are ready to complete the section TESTING YOUR UNDERSTANDING.

TESTING YOUR UNDERSTANDING

After you have completed this examination, compare your answers with those in the section that follows.

1. The fungal cell wall is composed primarily of:
   a. cellulose.
   b. pectin.
   c. lignin.
   d. chitin.
   e. several different types of polysaccharides.

2. Sporangiophores are specialized hyphae that:
   a. can fuse to produce a zygote.
   b. absorb nutrients directly from host cells.
   c. bear the sporangia, in which spores are produced.
   d. bear the structures that produce gametes.
   e. secrete digestive enzymes onto food sources.

3. The three groups of organisms that are thought to have arisen from the unicellular, eukaryotic ancestor of the fungi share which of these characteristics?
   a. septate hyphae with pores
   b. septate hyphae without pores
   c. life cycles with no apparent sexual stage
   d. multicellular organization
   e. coenocytic organization

4. Place these events in the order of their occurrence in sexual reproduction in the ascomycetes. (The process begins with modified hyphae of different mating strains.)
   1. meiosis   3. fusion of gametangia
   2. mitosis   4. fusion of nuclei

   a. 3, 4, 1, 2
   b. 4, 3, 2, 1
   c. 1, 4, 2, 3
   d. 2, 1, 3, 4
   e. 2, 1, 4, 1

5. Another name for ergotism is:
   a. LSD psychosis.
   b. St. Vitus' dance.
   c. delusions of grandeur.
   d. St. Anthony's fire.
   e. catcher in the rye.

6. Which types of fungi have hyphae subdivided by perforated septa?
   a. ascomycetes
   b. basidiomycetes
   c. zygomycetes
   d. zygomycetes and ascomycetes
   e. ascomycetes and basidiomycetes

7. Which of the following are NOT basidiomycetes?
   a. rusts
   b. yeasts
   c. puffballs
   d. stinkhorns
   e. white rot fungi

8. Which of the following is NOT true of deuteromycetes?
   a. Some cause ringworm in humans.
   b. They may infect plants and animals.
   c. They form mushrooms as reproductive bodies.
   d. They may produce antibiotics.
   e. They lack a sexual phase to their life cycle.

9. The first living things to colonize bare rock and initiate its breakdown to soil are:
   a. lichens.
   b. the smuts and rusts.
   c. yeasts.
   d. certain species of mushrooms.
   e. the zygomycetes because of their tough spores.

10. What do these three genera have in common: *Trebouxia, Trentepohlia,* and *Nostoc*?
   a. They all produce psychedelic drugs.
   b. They are the most common photosynthetic components of lichens.
   c. They are the primary genera involved in mycorrhizal associations.
   d. They all produce antibiotics of one form or another.
   e. They are all saprophytic algae or cyanobacteria.

11. Match the fungus with the division to which it belongs.
   1. Zygomycota
   2. Deuteromycota
   3. Basidiomycota
   4. Ascomycota

   a. _____ *Rhizopus stolonifer*, the black bread mold
   b. _____ *Neurospora*, the red bread mold
   c. _____ *Claviceps purpurea*, the cause of ergotism
   d. _____ *Agaricus campestris*, an edible mushroom
   e. _____ *Amanita bisporigera*, a poisonous mushroom
   f. _____ *Arthrobotrys dactyloides*, the nefarious noose fungus

12. T or F Fungi secrete digestive enzymes onto a food source and then absorb the digestive products, which are generally small molecules.

13. T or F The decomposers are just as important as the producers in the functioning of an ecosystem.

14. T or F The black bread mold, *Rhizopus stolonifer*, can infect only bread.

15. T or F When the gametangia of zygomycetes fuse, the + and − nuclei then fuse, forming a multinucleate cell containing a number of zygotes.

16. T or F In the ascomycetes the hyphae are divided by septa; a single nucleus is trapped between each pair of septa, since these septa have no pores.

17. T or F Basidiomycetes differ from other fungal types in that the dikaryotic mycelium may persist for years before the nuclei fuse.

18. T or F Most of the fungi that give rise to antibiotics belong to division Basidiomycota.

19. T or F It has recently been shown that mycorrhizae can actually transfer substances between plants that are growing adjacent to one another.

20. Distinguish between endomycorrhizae and ectomycorrhizae on the basis of structure and common participants.

21. Some fungi are able to trap live organisms of other species and use them for food. Describe the two adaptations used by fungi to trap nematodes (roundworms).

PERFORMANCE ANALYSIS

1. d The major component of the fungal cell wall is chitin, which is also a primary constituent of the exoskeletons of insects. (page 480)

2. c Sporangiophores are specialized hyphae that elevate the sporangia above the level of the mycelium, an adaptation that facilitates spore dispersal. (page 481)

3. e The fungi, chytrids, and oomycetes are all thought to have arisen from the same ancestral group, of which there are no living representatives. These three modern taxa all include coenocytic forms. (page 482)

4. a In ascomycete sexual reproduction, gametangia of different mating types fuse. Next, nuclei of different mating types fuse and undergo meiosis and then mitosis to form ascospores. (page 486)

5. d Ergotism is a disease caused by the ingestion of a toxin produced by the fungal parasite of rye, *Claviceps purpurea*. Another name for this disorder is St. Anthony's fire. (page 486)

6. e Both ascomycetes and basidiomycetes have hyphae divided by perforate septa. (page 482)

7. b Yeasts are classified as ascomycetes. The other organisms listed are basidiomycetes. (page 488)

8. c Basidiomycetes are the fungi that form mushrooms as spore-producing bodies. All the other statements characterize at least one or more deuteromycete. (page 488)

9. a Lichens, which require only air, water, and a few minerals, are often the first living organisms to colonize bare rock. (page 488)

10. b *Trebouxia* and *Trentepohlia* are genera of green algae. *Nostoc* is a cyanobacterium genus. Approximately 90 percent of all lichens contain a member of one of these four genera. (page 489)

11. The correct answers are **a1, b4, c4, d3, e3, f2**. (pages 482, 484, 486, 487, and essay, page 489)

12. **True** Fungi obtain nutrients by secreting digestive enzymes and then absorbing the products released by those enzymes. (page 480)

13. **True** The decomposers are responsible for releasing nutrients bound in organic matter, making them available for living organisms. (page 480)

14. **False** The black bread mold, *Rhizopus stolonifer*, can live on fruit and other organic matter in addition to bread. (page 482)

15. **True** Zygomycete sexual reproduction is characterized by the fusion of gametangia of different mating types. Next, the nuclei of the different types fuse, resulting in a multinucleate cell containing a number of zygotes. (page 483)

16. **False** The septa of ascomycetes are perforated and the pores are large enough for nuclei to pass through them. (page 485)

17. **True** The dikaryotic mycelium characteristic of basidiomycetes may persist for years before nuclear fusion and meiosis occur. (page 486)

18. **False** Antibiotics are produced by ascomycetes and deuteromycetes, but not by basidiomycetes. (pages 484, 488)

19. **True** The fungus of a mycorrhiza may in some instances act as a bridge between two plants, allowing the passage of materials from one plant to the other. (page 491)

20. Endomycorrhizae are characterized by fungi that penetrate the plant root cells. The most common fungus involved is a zygomycete. In ectomycorrhizae, the fungal hyphae form a sheath around the root. Basidiomycetes and ascomycetes are common participants in ectomycorrhizae. (page 491)

21. One deuteromycete, *Arthrobotrys dactyloides*, has specialized hyphae that form a noose. When a nematode wriggles into the loop, the cells constrict as a result of osmotic changes and the nematode is trapped. *Dactylella drechsleri* has small knobs on its hyphae that secrete an adhesive substance. When prey contact the adhesive, they are trapped. (essay, page 489)

C H A P T E R **24**

The Plants

MAJOR CONCEPTS

In terrestrial environments, it is through the plants that the sun's energy enters the biosphere, and inorganic substances are transformed into living matter.

Modern plants are believed to have evolved from the green algae. Both plants and green algae contain chlorophylls *a* and *b* and beta-carotene as photosynthetic pigments, and food reserves are accumulations of starch stored in plastids.

The 10 divisions of modern plants may be divided into two main groups: bryophytes, which have no internal transport systems and must live in moist environments; and vascular plants, which have complex vascular systems for transporting water and nutrients throughout the plant body. For the purposes of study, vascular plants can be further divided into seedless and seed-producing categories, and the seed plants form two groups: the gymnosperms and the angiosperms.

When plants adapted to a terrestrial existence, the main problems to be solved related to water balance: obtaining adequate water, conserving it within the plant body, and transporting it throughout the plant. In the vascular plants, root systems, waxy external coverings, and complex networks of vessels evolved as solutions to these problems.

Several major trends characterize the evolution of vascular plants: (1) evolution of increasingly efficient conducting systems, (2) reduction in the size of the gametophyte, (3) progression from homospory to heterospory, and (4) evolution of the seed. The evolution of the seed was one of the most significant adaptations contributing to the success of plants in terrestrial environments.

The life cycles of all plants are characterized by the alternation of a haploid gametophyte stage with a diploid sporophyte stage. The variations on this basic theme are numerous. In bryophytes, the gametophyte is independent of and often larger than the sporophyte; the sporophyte may be nutritionally dependent upon the gametophyte. In the vascular plants, the sporophyte is the dominant generation; there is an evolutionary trend toward reduction of the gametophyte. In primitive vascular plants such as ferns, the gametophyte may be independent. In more advanced vascular plants (gymnosperms and angiosperms), the gametophyte is microscopic and dependent upon the sporophyte.

The gymnosperms are naked-seed plants, the most familiar of which are the conifers. Conifer seeds are produced in ovules of the female cones; the eggs are fertilized by wind-borne pollen. The vascular systems of gymnosperms contain xylem and phloem, but the cells of which these tissues are made differ from those in angiosperms.

flowers and fruits, which are produced exclusively by angiosperms, are adaptations that make use of animals to facilitate plant reproduction. Plants that have flowers to attract animals as pollinators require fewer reproductive cells per living offspring than do plants that depend on wind pollination. When animals eat fruits (or when the fruits cling to an animal's fur) the seeds are distributed over a wider area than would otherwise be possible.

Plants that are susceptible to predation by animals produce a number of poisonous or noxious chemicals as defense mechanisms.

HOW TO STUDY THE CHAPTER

Read the entire chapter through quickly, focusing on the major concepts.

Use the GUIDED STUDY OF THE CHAPTER to help you identify the important details as you **reread** the chapter. Writing out the answers to these questions will help fix them in your mind as well as provide you with a valuable study aid.

Answer the questions in TESTING YOUR UNDERSTANDING without the aid of your text. Check your answers against those in PERFORMANCE ANALYSIS. Analyzing your answers will give you valuable feedback on your level of understanding and preparedness for classroom testing.

GUIDED STUDY OF THE CHAPTER

I. Introduction (page 493)

> *Focus:* Beginning approximately 500 million years ago, plants diversified from their aquatic environments and became adapted to terrestrial habitats.

1. Name three advantages to photosynthetic organisms of a terrestrial environment over an aquatic environment.

2. What major "problem" confronted plants when they moved to the land?

3. List five problems that accompany the increase in size of an organism made possible by specialization.

II. The Ancestral Alga (pages 493–495)

4. a. Identify three characteristics common to green algae and plants.

   b. Cite four pieces of evidence that the alga *Coleochaete* evolved from the same ancestor as green plants.

   c. What characteristic of plants is not found in *Coleochaete*?

> *Focus:* The presence of alternation of generations in various types of algae suggests that this life cycle evolved independently several times.

5. How does the production of (a) the zygote and (b) haploid cells differ in *Coleochaete* and *Chlamydomonas*?

6. The *Coleochaete* life cycle would exemplify alternation of generations if what modifications were present?

III. The Transition to Land (pages 495–497)

A. General Remarks (page 495)

7. Describe two adaptations of terrestrial plants that enable them to conserve water and yet exchange gases.

8. *Vocabulary:* Define cutin.

9. What two structural adaptations developed in the reproductive tissues of plants when they became land dwellers?

10. *Vocabulary:* Distinguish between archegonia and antheridia.

B. **Subsequent Diversification** *(page 496)*

11. Distinguish between bryophytes and vascular plants with respect to basic structure and time of appearance in the fossil record.

IV. **Classification of the Plants** *(pages 497–498)*

12. What uncertainty characterizes the classification of the three classes of bryophytes?

Focus: **The nine divisions of vascular plants are examples of monophyletic taxa.**

13. *Vocabulary:* What are tracheophytes?

14. *Vocabulary:* Distinguish between gymnosperms and angiosperms.

V. **Division Bryophyta: Liverworts, Hornworts, and Mosses** *(pages 498–500)*

A. **General Remarks** *(page 498)*

15. a. What structural limitations of bryophytes dictate the environment in which they live?

b. Describe the environment in which bryophytes may be found.

16. Name two features that bryophytes have in common with lichens.

17. Identify the bryophyte structure adapted for each of the following functions.

a. Attachment to a substrate

b. Photosynthesis

c. Support and food storage

B. **Bryophyte Reproduction** *(pages 499–500)*

Focus: **Bryophytes differ from vascular plants in that the gametophyte of bryophytes is larger than the sporophyte.**

18. Arrange the following components of the moss life cycle in a flow diagram illustrating alternation of generations. Indicate whether each component is haploid or diploid. Identify where fertilization, mitosis, and meiosis occur: archegonium, antheridium, calyptra, egg, gametophyte, protonema, sperm, sporangium, spore, sporophyte, zygote. (See also Figure 24–8.)

19. Sperm cells are attracted to the archegonium by:

20. Asexual reproduction in bryophytes may be accomplished by what two means?

The Vascular Plants: An Introduction *(pages 501–502)*

21. What feature of *Rhynia major* determines its classification as a vascular plant?

22. For what functions are (a) roots and (b) leaves specialized?

23. Fill in the following table. (See also Figure 24–11.)

Table for Question 23

	MICROPHYLLS	MEGAPHYLLS
Vascular system (describe)		
Evolutionary origin		

24. List the three divisions of modern plants characterized by microphylls.

25. Fill in the following table.

Table for Question 25

	XYLEM	PHLOEM
Substances transported (and direction of transport)		
Conducting elements		

26. Describe the two different arrangements of conducting tissues found in the stems of vascular plants.

> *Focus:* **The basic problems of obtaining adequate water and food supplies and distributing these items to all cells of the plant were solved by the evolution of roots, leaves, and vascular systems in terrestrial plants.**

27. Describe how the gametophyte of primitive vascular plants differs from that of more recently evolved vascular plants.

28. *Vocabulary:* Distinguish between heterospory and homospory.

29. a. Describe the structure of a seed.

    b. From what tissue is the seed coat derived?

> *Focus:* **The seed was probably the single most important innovation responsible for the success of vascular plants on land.**

VII. **The Seedless Vascular Plants** *(pages 502–503)*

A. **General Remarks** *(page 502)*

> *Focus:* **The four groups of seedless vascular plants of which there are living examples are whisk ferns, club mosses, horsetails, and ferns.**

30. What feature sets *Psilotum* apart from all other living vascular plants? (Figure 24–12)

31. The stems of horsetails contain what inorganic compound that is not commonly found in living organisms? (Figure 24–12)

B. **Division Pterophyta: The Ferns** *(pages 502–503)*

> *Focus:* **All ferns have flagellated sperm and need free water for fertilization.**

32. A "fiddle head" represents what stage of the fern life cycle? (Figure 24–13)

33. *Vocabulary:* Distinguish among rhizome, frond, and pinna.

34. What characteristic of fern leaves makes them well suited to growth on the forest floor?

35. *Vocabulary:* Distinguish between sporophylls and sori.

> *Focus:* **The gametophytes of ferns are small, nutritionally independent, and most commonly homosporous.**

36. Review the fern life cycle depicted in Figure 12–6 (page 251) and the pertinent Study Guide questions in Chapter 12.

VIII. The Seed Plants *(pages 503–515)*

A. General Remarks *(pages 503–505)*

37. What conditions resulted in the formation of coal deposits?

38. During the Permian period, selection pressures for what types of adaptations were imposed on plants as a result of the climate?

39. *Vocabulary:* Distinguish between gymnosperms and angiosperms.

B. *Essay:* Coal Age Plants *(page 504)*

40. *Vocabulary:* What is peat?

41. How may peat be converted into fossil fuels?

42. Describe the conditions present on earth during the Carboniferous period that facilitated the extensive deposition of materials that became fossil fuels.

43. a. What two types of plants dominated the Carboniferous forests?

    b. What two types of plants in the Carboniferous forests have surviving descendants?

C. Gymnosperms *(pages 505–508)*

> *Focus:* **The gymnosperms diversified during the Permian period, which was characterized by widespread glaciers and drought.**

44. What is the largest division of living gymnosperms?

Formation of the Seed (pages 507–508)

45. In what ways does a seed resemble a bacterial spore and an algal zygospore?

46. Name the three basic components of a seed.

47. How does the relationship between gametophyte and sporophyte in seed plants differ from that in (a) algae and (b) seedless vascular plants?

48. a. Define the following components of a pine tree life cycle.

Microspore

Microspore mother cell

Microsporangium

Megaspore

Megaspore mother cell

Megasporangium

Ovule

Gametophyte

Sporophyte

Pollen grain

Zygote

Embryo

b. Arrange these components in a flow diagram illustrating alternation of generations. Indicate where fertilization and meiosis occur, and the haploid and diploid stages.

49. a. Describe the course of events from the arrival of a pollen grain on a female pine cone to the release of a pine seed.

b. What time period has elapsed?

50. Why can conifers reproduce sexually under conditions that would not allow ferns or bryophytes to do so?

51. A pine seed can be thought of as "three generations under one roof." Explain this statement.

52. *Vocabulary:* What are cotyledons?

The Conifer Leaf (page 508)

53. Label the following cross section of a conifer leaf.

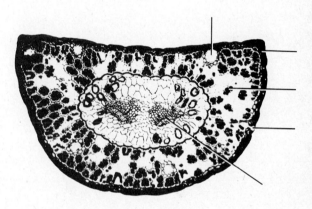

54. Name two characteristics of a conifer leaf that protect the leaf from water loss. (Figure 24–19)

55. *Vocabulary:* What is resin and what is its apparent purpose?

56. Identify three environmental conditions to which conifer leaves are adapted.

D. *Essay:* **The Ice Ages** *(page 510)*

57. a. *Vocabulary:* Define glaciation.

    b. What situation causes a glaciation?

58. Summarize the significant changes in living organisms associated with the following glaciations:

    End of the Paleozoic era (248 million years ago)

    End of the Mesozoic era (65 million years ago)

59. *Vocabulary:* What are interglacials?

60. Describe the events of the four Pleistocene glaciations.

Focus: **The dramatic changes in climate associated with glaciations placed severe selection pressures on plant and animal populations.**

---

61. Name seven proposed causes of the temperature fluctuations responsible for glaciations and interglacials.

62. What are the two possibilities for where we now stand with respect to the most recent Ice Age?

E. **Angiosperms: The Flowering Plants** *(pages 509–515)*

Focus: According to current information, angiosperms are thought to have evolved from a group of gymnosperms of which there are no living representatives.

---

63. What types of environments are the angiosperms believed to have inhabited initially?

64. Describe the climatic conditions present when angiosperms were increasing in number and diversifying during the mid-Cretaceous period.

65. a. Identify three characteristics that angiosperms share with gymnosperms.

    b. Name three angiosperm features that represent evolutionary advancements over gymnosperms.

66. a. What are tracheids? (Figure 24–20)

    b. Describe the probable evolutionary relationships between tracheids, wood fibers, and vessel members. (Figure 24–20)

Focus: **The angiosperms, or flowering plants, include all vegetable and fruit crops, all grains and grasses, and the hardwood trees.**

---

67. Use the following table to distinguish between monocots and dicots.

Table for Question 67

	MONOCOTS	DICOTS
Approximate number of species		
Number of cotyledons		
Leaf venation pattern		
Arrangement of vascular bundles in a young stem		
Presence or absence of woody growth		
Number of flower parts		
Structure of pollen grains		

The Flower (page 511)

68. a. Define the following flower components:

Carpel

Pistil

Ovary

Ovule

Stigma

Style

Stamen

Anther

Filament

b. Use these components to label this diagram of a simple flower. (You may wish to color this figure with colored pencils.)

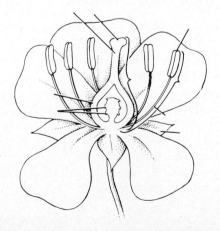

69. *Vocabulary:* Distinguish between imperfect and perfect flowers. (Figure 24–21)

70. Study Figure 24–22. Describe how angiosperm life history is similar to and different from conifer life history. (See also Figure 24–16.)

71. *Vocabulary:* What is endosperm and what is its ploidy? (Figure 24–22)

Evolution of the Flower (pages 512–515)

72. a. How may insect pollination have arisen?

b. What are the advantages of insect pollination to the plant?

To the insect?

73. *Vocabulary:* What are nectaries?

74. Name three evolutionary developments in the flower that favored insect pollination.

75. *Vocabulary:* Distinguish between simple and compound ovaries. (Figure 24–23)

76. List four trends in flower evolution.

77. *Vocabulary:* Distinguish between superior and inferior ovaries.

> *Focus:* In the evolution of flowers and their pollinators, the most advanced relationships are those in which only one or a few kinds of pollinators visit a particular kind of plant.

78. What are the advantages of the situation described in the focus statement (a) to the plant and (b) to the pollinator?

79. Name several adaptations of plants that discourage indiscriminate pollination.

Evolution of the Fruit (page 515)

80. *Vocabulary:* What is a fruit?

81. How did the evolution of fruits benefit both plants and animals?

> *Focus:* The seeds of some plants require passage through an animal digestive tract before they can germinate.

82. What purpose is served by seeds having a bitter taste or being toxic?

Biochemical Evolution (page 515)

> *Focus:* One adaptation of plants that discourages predation is the production and accumulation of poisonous or noxious chemicals.

83. a. Name several compounds that evolved as defenses against animal predation.

   b. Why is the term "secondary plant substances" no longer considered an accurate description for these compounds?

84. Identify three criteria by which angiosperms are determined to be the most successful of all plants.

IX. **The Role of Plants** *(pages 515–516)*

> *Focus:* All but a few simple life forms depend either directly or indirectly on plants.

85. Describe how plants make inorganic substances available to other organisms.

86. What event was necessary before animals could make the transition from water to land?

X. **Summary** *(pages 516–517)*: Read the summary. If you are familiar with the essential features of the material presented there, you are ready to complete the section TESTING YOUR UNDERSTANDING.

TESTING YOUR UNDERSTANDING

After you have completed this examination, compare your answers with those in the section that follows.

1. Which statement is NOT true of plants and green algae?
   a. They contain chlorophylls *a* and *b* as photosynthetic pigments.
   b. Beta-carotene and starch are present in the cytoplasm.
   c. Chlorophylls *a* and *b* and starch are contained within plastids.
   d. Beta-carotene is a photosynthetic pigment.
   e. They possess the same photosynthetic pigments and store food in the same form.

2. In the evolution of plants, one of the first adaptations to land was the formation of:
   a. a sporophyte generation.
   b. a protective cuticle.
   c. a water-vascular system.
   d. stabilizing roots.
   e. supportive stems.

3. Mosses and liverworts can engage in sexual reproduction by:
   a. fragmentation.
   b. binary fission.
   c. conjugation.
   d. spore formation.
   e. protonema formation.

4. What do the fossilized *Rhynia major* and the modern whisk fern *Psilotum* have in common?
   a. absence of leaves and roots
   b. absence of a vascular system
   c. absence of a sporophyte generation
   d. presence of primitive leaves
   e. presence of primitive roots

5. The earliest known seeds were fossilized about 360 million years ago in the _____ period.
   a. Cambrian
   b. Permian
   c. Carboniferous
   d. Silurian
   e. Devonian

6. The underground stems of ferns are:
   a. more complex than those of angiosperms.
   b. nonvascular structures.
   c. woody with a thin bark.
   d. called rhizomes.
   e. largely the products of secondary growth.

7. *Ginkgo* is the only representative of its division alive today, and many of these trees are found in city parks because:
   a. the rancid smell of their fruit drives off unwanted birds.
   b. they are resistant to air pollution.
   c. the kernel of the seed is a highly prized delicacy.
   d. they do not lose their leaves in the winter.
   e. the fishy taste of the seeds keeps insects away from other vegetation.

8. The two major groups of seed plants are the:
   a. gymnosperms and whisk ferns.
   b. conifers and ginkgos.
   c. gymnosperms and angiosperms.
   d. monocots and dicots.
   e. angiosperms and cycads.

9. Angiosperms appeared in the fossil record about 120 million years ago during the _____ period.
   a. Cretaceous
   b. Tertiary
   c. Jurassic
   d. Quaternary
   e. Triassic

10. Which of the following plants are monocots?
    a. tomatoes and cucumbers
    b. hickory trees and walnut trees
    c. apple trees and potato plants
    d. grasses and palm trees
    e. maple trees and rose bushes

11. A flower that possesses both stamens and carpels is known as:
    a. simple.
    b. complex.
    c. perfect.
    d. imperfect.
    e. hermaphroditic.

12. The female gametophyte of angiosperms is located in the _____ and typically consists of _____ cell(s).
    a. carpel; 5
    b. ovary; 1
    c. ovule; 7
    d. pistil; 3
    e. stigma; 9

13. Which of the following is NOT a trend in flower evolution?
    a. fusion of floral parts
    b. elevation of free floral parts above the ovary
    c. change from radial to bilateral symmetry
    d. reduction in the number of floral parts
    e. replacement of nectaries with complex, colored floral parts

14. Of the following elements that plants make available to living organisms, which is incorporated into plant tissues in the greatest abundance?
    a. magnesium
    b. phosphorus
    c. sulfur
    d. nitrogen
    e. carbon

15. By the beginning of the Permian period, _____ had not yet appeared.
    a. flowering plants
    b. ferns
    c. gymnosperms
    d. lycopods
    e. horsetails

16. T or F   As one would expect, bryophytes, being nonvascular plants, appear in the fossil record earlier than do the structurally more complex vascular plants.

17. T or F   Because bryophytes do not have a vascular system, they must absorb water through the surface of their bodies. It is therefore essential that they live in moist environments.

18. T or F   Ferns are generally homosporous, meaning that the same gametophyte produces both archegonia and antheridia.

19. **T or F** The design of the conifer leaf enables the plant to withstand cold and dry conditions.

20. **T or F** Glaciations occur when the summers are not hot enough or long enough to melt the ice that accumulates during the winters. A significant global temperature decrease is required to initiate a glaciation.

21. Summarize the problems plants had to circumvent in becoming adapted to terrestrial environments.

22. Distinguish between xylem and phloem with respect to (a) the substances they carry and (b) the type of conducting cells of which they are composed.

23. Why does it take two years or more from the time of megaspore and microspore formation to the time a new seed is ready to be dispersed in some pine species?

PERFORMANCE ANALYSIS

1. **b** Plants and the green algae have the following characteristics in common: they contain chlorophylls *a* and *b* and beta-carotene as photosynthetic pigments, and they accumulate food reserves in the form of starch stored in plastids. (page 493)

2. **b** The formation of a protective cuticle to prevent water loss from leaves and stems was one of the first adaptations made by plants to terrestrial living. The evolution of cuticle necessitated the development of pores (stomata) through which gas exchange occurs. (page 495)

3. **d** In bryophytes, gametes are formed in archegonia (eggs) and antheridia (sperm). Fertilization occurs in the archegonium and the zygote develops into the sporophyte. Haploid spores, which develop into the gametophyte, are formed by meiosis in the mature sporophyte. (page 500 and Figure 24–8)

4. **a** Both *Rhynia major* and *Psilotum* lack true leaves and roots. However, both groups have primitive vascular systems. (Figures 24–10, 24–12)

5. **e** The earliest known fossils of seeds were formed during the Devonian period, approximately 360 million years ago. (page 502)

6. **d** The underground stems of ferns (and other plants) are called rhizomes. (page 502)

7. **b** *Ginkgo* (the only living representative of a primitive line of gymnosperms) is a popular ornamental tree in urban areas because it is resistant to air pollution. (Figure 24–15)

8. **c** The two major groups of seed plants are gymnosperms and angiosperms. The two major groups of angiosperms are monocots and dicots. (pages 505, 511)

9. **a** Angiosperms appear abundantly in the fossil record approximately 120 million years ago during the Cretaceous period. Daniel Axelrod believes that angiosperms probably evolved earlier, during the Permian period, and

colonized upland habitats first (where conditions did not favor fossil formation), later moving into the lowlands (where conditions favored fossil formation). (page 509)

10. **d** Monocots (e.g., grasses, palm trees, lilies, orchids) are characterized by one cotyledon, scattered vascular bundles in the stem, no woody secondary growth, parallel arrangement of major leaf veins, and flower parts in multiples of three. Dicots are characterized by two cotyledons, vascular bundles arranged in a ring in the stem, woody secondary growth, branching arrangement of leaf veins, and flower parts in multiples of four or five. (page 511)

11. **c** A flower with both stamens and carpels is a perfect flower. A flower with *either* stamens *or* carpels is imperfect. (Figure 24–21)

12. **c** In angiosperms, the female gametophyte is located in the ovule and typically contains seven cells. (page 511)

13. **e** In the evolution of flowers, the following trends are apparent: (1) reduction in the number of floral parts, (2) fusion of floral parts, (3) elevation of the free floral parts above the ovary, and (4) change from radial to bilateral symmetry. Nectaries and complex, colored floral parts are innovations that evolved together to attract animal pollinators. (page 513)

14. **e** A major role of plants in the biosphere is the conversion of the carbon in carbon dioxide to organic molecules during photosynthesis. Except for a few chemosynthetic bacteria, all organic molecules present in living organisms are formed directly by photosynthesis or indirectly from photosynthetic products. (page 516)

15. **a** The angiosperms (flowering plants) had not yet appeared at the beginning of the Permian period, which was characterized by gymnosperm diversification. Flowering plants may have originated during this period, but there is no fossil evidence supporting this proposal. See also the answer to question 9. (essay, page 504)

16. **False** Bryophytes appeared in the fossil record 370 million years ago, but primitive vascular plants were alive 430 million years ago. (page 496)

17. **True** The absence of a vascular system and protective cuticle in bryophytes means that each cell of the organism must obtain water from the environment or from adjacent cells. Therefore, bryophytes live only in moist habitats. (page 498)

18. **True** Most fern gametophytes produce both archegonia and antheridia, a situation defined as homospory. (page 503)

19. **True** The conifer leaf is needlelike, has a hard epidermis, and releases resin to seal wounds. It is well adapted to withstand drought and cold conditions in which water is in the form of ice and unavailable to plants. (page 508)

0. False Glaciations do in fact occur when summers are not hot enough or long enough to melt the accumulated ice and snow from winter. However, in some regions, only a few degrees change in temperature can initiate or end a glaciation. (essay, page 510)

1. The major problems plants had to circumvent when adapting to terrestrial habitats were those of conserving water within the plant body and supplying all the cells with water. The most significant innovations that evolved to address these problems were the cuticle and the vascular transport systems. The evolution of specialized tissues permitted an increase in plant size, which created another series of problems. These were associated with supplying individual cells with nutrients, waste removal, physical support of a larger body, coordinating the functions of different plant tissues, and protection of immature plant stages until development is complete. (pages 493, 495)

22. The conducting elements of the xylem, the tracheids and vessel members, transport water and ions from the roots to the leaves. The sieve-tube members of the phloem carry dissolved sucrose and other products of photosynthesis from the leaves to the nonphotosynthetic cells of the plant. (page 501)

23. In some pine species, slightly over one year is required for the development of the megaspore into the female gametophyte with eggs. About three months pass between the pollen grain being drawn to the ovule and maturation of the male gametophyte. Passage of the sperm through the pollen tube and into the ovule takes almost another year. Following fertilization, the seed forms and is released from the mature cone, two years or more after formation of the megaspore and microspore. (page 508)

REVIEW TEST 7

Section 4: Chapters 20–24

This review test is *not* designed to cover all of the important information in these chapters. However, it does touch on the major topic categories in each chapter. It will also give you valuable practice in taking this type of test. When you are finished, check your answers with those provided in the following PERFORMANCE ANALYSIS section.

1. Which of the following is NOT a reason why taxonomy based on protein homology was largely abandoned when nucleotide sequencing became possible?
   a. There are fewer kinds of nucleotides than amino acids.
   b. Nucleic acids are easier to sequence than proteins.
   c. Changes in the nucleotide sequence may go undetected at the amino acid level.
   d. Not all changes in the amino acid sequence reflect changes in the nucleotide sequence.
   e. Nucleic acid data are more sensitive than amino acid data.

2. Bacteria that are rod-shaped, such as *E. coli*, are known as:
   a. cocci.
   b. spirilla.
   c. spirochetes.
   d. bacilli.
   e. virons.

3. The needs of algae are simple and consist of:
   a. light only.
   b. light and minerals only.
   c. light and carbon dioxide only.
   d. light, water, and carbon dioxide only.
   e. light, water, minerals, and carbon dioxide only.

4. Sporangia are raised in the air above the mycelial mat on:
   a. rhizoids.
   b. sporangiophores.
   c. sporangiocytes.
   d. zygophores.
   e. haustoria.

5. Most scientists today agree on a _____ kingdom system of classification.
   a. 2
   b. 3
   c. 5
   d. 6
   e. There is no generally accepted number of kingdoms.

6. It is necessary for bacteria to have a cell wall because the environment in which they live is _____ and the cell wall _____.
   a. hypertonic; prevents dehydration
   b. isotonic; maintains the cell's shape
   c. hypotonic; prevents the cell from rupturing
   d. isotonic; maintains the cell's size
   e. fluctuating between hypertonic and hypotonic; prevents both dehydration and rupturing

7. Which of these statements does NOT describe a feature of eukaryotic cells that contributed to the development of multicellularity?
   a. Because they are larger than prokaryotic cells, eukaryotic cells can carry much more genetic material— enough to specify a whole multicellular organism.
   b. Because they can continue to add internal membranes, there is theoretically no limit to the size of eukaryotic cells.
   c. The functional compartmentalization by internal membranes enables eukaryotic cells to be metabolically more efficient.
   d. Since they can accommodate increased membrane surface area, eukaryotic cells can engage in more photosynthesis.
   e. Because they are larger, eukaryotic cells are more adaptable to changes in the environment.

8. In which of the following settings have lichens NOT been found?
   a. on bare rock

b. in arid deserts
c. on tree trunks
d. in rock crevices below the frost in the antarctic
e. Lichens have been found in almost every conceivable ecological setting.

9. Which of the following is NOT considered to be a public health measure to control disease?
a. eradication of flies, fleas, and mosquitoes.
b. sanitary disposal of sewage.
c. pasteurization of milk.
d. garbage collection.
e. All of the above are public health measures designed to control disease.

10. The chloroplasts of Chlorophyta and Euglenophyta are thought to have been derived from organisms similar to:
a. *Prochloron.*   d. certain archaebacteria.
b. *Heliobacterium.*   e. the halobacteria.
c. the methanogens.

11. Some members of division Deuteromycota _____, whereas others _____.
a. produce antibiotics; cause disease in plants and animals
b. produce antibiotics; produce edible mushrooms
c. cause diseases of plants and animals; produce edible mushrooms
d. produce edible mushrooms; produce poisonous mushrooms
e. produce edible mushrooms; produce Roquefort and Camembert cheese

12. Several characteristics of eukaryotic cells have led to diversification of protista. Which of the following is NOT such a trait?
a. the capacity to carry a great deal of genetic information
b. the specialization of different cell compartments for different functions
c. the capacity to carry many nuclei in a common cytoplasm
d. the ability to acquire more food
e. the ability to adapt to environmental change

13. Mosses and liverworts can engage in asexual reproduction by:
a. conjugation.   d. spore formation.
b. binary fission.   e. protonema formation.
c. gemmae formation.

14. The conducting elements of xylem are:
a. sieve-tube members only.
b. tracheids only.
c. vessel members only.
d. vessel members and sieve-tube members.
e. tracheids and vessel members.

15. Selection pressures strongly favoring the evolution of seed plants first occurred in the _____ period.
a. Permian
b. Carboniferous
c. Devonian
d. Silurian
e. Cambrian

16. T or F Modern gymnosperms have a more highly evolved vascular system than modern angiosperms.

17. T or F All scientific names of organisms must conform to the binomial pattern in which both the genus name and the specific epithet describe a particular characteristic of the organism.

18. T or F The chemosynthetic autotrophic prokaryotes are capable of using inorganic compounds as energy sources.

19. T or F All protista that respond behaviorally to light are photosynthetic.

20. T or F Some taxonomists include the ascomycetes as a class within division Deuteromycota because of similarities in asexual reproduction.

21. T or F The cladistic classification scheme is based entirely upon the phylogeny of organisms, whereas numerical phenetics is based solely upon the similarities of the organisms.

22. T or F During the replication of all viruses, the capsids are manufactured first and then a nucleic acid is inserted into each one.

23. Match these common names with the appropriate division. Use each answer only once. Each blank may have only one answer.

1. Chlorophyta   4. Myxomycota
2. Chytridiomycota   5. Phaeophyta
3. Chrysophyta

a. _____ water molds
b. _____ diatoms
c. _____ green algae
d. _____ kelp
e. _____ slime mold

24. What are the two ways in which similarities and differences between organisms of different taxa may be interpreted from the perspective of evolutionary systematics?

25. Describe sporulation in bacteria and the conditions under which it takes place. Cite one example.

26. Describe the increase in complexity exhibited by the volvocine line as represented by *Chlamydomonas*, *Gonium*, *Pandorina*, and *Volvox*.

27. Describe the three ways in which sexual reproduction can take place in fungi.

28. Explain why many scientists believe that the genus *Coleochaete* is closely related to the ancestor of the plants.

PERFORMANCE ANALYSIS

1. **d** Amino acid sequencing was abandoned as a taxonomic technique because the smaller number of nucleotides present in nucleic acids (four nucleotides versus 20 amino acids) made nucleic acid sequencing technically easier. Also, nucleic acid sequencing is a more sensitive technique since differences in nucleic acid sequences may not be revealed as changes in amino acid sequence. Of course, changes in amino acid sequence *always* reflect changes in nucleic acid sequence. (page 420)

2. **d** Rod-shaped bacteria are bacilli. Spherical bacteria are cocci. Comma-shaped bacteria are vibrios. Bacteria that have a spiral appearance are spirilla. (page 436)

3. **e** All algae are photosynthetic and need only light, water, minerals, and carbon dioxide. (page 456)

4. **b** Sporangiophores are specialized hyphae that elevate the sporangia above the level of the mycelium, an adaptation that facilitates spore dispersal. (page 481)

5. **c** The five kingdoms currently agreed upon by most scientists are Monera, Protista, Fungi, Plantae, and Animalia. (page 424)

6. **c** Most bacteria are hypertonic with respect to the environment in which they live (the environment is *hypotonic* with respect to the cell). Under these conditions, water enters through the cell membrane by osmosis, and without the cell wall, the cell would swell and eventually rupture. (page 432)

7. **b** Maximum cell size is limited by the surface area-to-volume ratio and by the ability of the nucleus to control cytoplasmic activities. The ability to produce internal membranes has no significant influence on maximum cell size. (page 454)

8. **e** Lichens, which require only air, water, and a few minerals, have been found in nearly all locations. (page 438)

9. **e** All of the precautions listed are public health measures that have been instituted to control disease. (pages 448, 449)

10. **a** The recently discovered organism *Prochloron* is thought to represent the lineage of prokaryotes from which the chloroplasts of Chlorophyta and Euglenophyta are descended. (page 456)

11. **a** Division Deuteromycota includes members that manufacture antibiotics and other chemotherapeutic agents, function in cheese production, and cause diseases in plants and animals. Edible and poisonous mushrooms belong to division Basidiomycota. (page 488)

12. **c** Characteristics of eukaryotic cells that allowed the diversification of protists include the capacity to carry a considerable amount of genetic information and to transmit it to offspring, the compartmentalization of the cytoplasm by membranes and membrane-bound organelles, the development of adaptations for obtaining food, and the ability to adapt to environmental changes. The evolution of coenocytic organisms was an adaptation to circumvent the limitations imposed on cell size by the limited ability of the nucleus to influence cytoplasmic activities. (page 479)

13. **c** Bryophytes can accomplish asexual reproduction by fragmentation or by the formation of gemmae. (page 500 and Figure 24–9)

14. **e** The conducting elements of xylem are called tracheids and vessel members; the conducting elements of phloem are called sieve-tube members. (page 501)

15. **a** The Permian period (248 to 286 million years ago) was characterized by widespread glaciation and drought, two conditions strongly favoring the evolution of plants (and animals) with adaptations to resist water deprivation. One of the most important plant adaptations during this time was the seed. (page 503)

16. **False** The vascular systems of modern angiosperms are more highly evolved than those of modern gymnosperms. (page 509)

17. **False** A scientific name is not required to have any meaning with respect to the characteristics of the organism. The person who first describes a new species has the right to name it. The only limitation is that the organism may not be named after the discoverer. (page 409)

18. **True** Prokaryotes that are chemosynthetic autotrophs are capable of using inorganic compounds as energy sources. (page 439)

19. **False** Many heterotrophic protists, including amoebas, have photoreceptors and respond to light stimuli. (page 474)

20. **False** Some taxonomists believe that members of division Deuteromycota are a class of ascomycetes because of the similarities in patterns of asexual reproduction among the two groups. (page 482)

21. **True** Cladistics classifies organisms solely on the basis of phylogeny, whereas numerical phenetics classifies organisms strictly on the basis of observable characteristics. Neither system is ideal and both systems omit certain information about the organisms. (page 416)

22. **False** Some, but not all, viruses replicate in the manner described. Other viruses manufacture the capsid around the nucleic acid core. (page 445)

23. The answers are **a2, b3, c1, d5, e4.** (Table 22–1)

24. Similarities and differences between organisms of different taxa may be viewed as (1) modifications of a single ancestral structure or (2) adaptations that arose independently because of similar (or different) selection pressures. (page 413)

25. Bacteria most often form resistant spores as food supplies decrease. Spores are formed following chromosome replication when a new cell membrane is formed around one

chromosome. Outside this new cell membrane, two more layers are deposited, forming the spore coat. This mature spore is then released into the environment. (Some types of bacteria retain the mature spore within the parent cell.) (pages 437, 438)

26. *Chlamydomonas* is a unicellular alga. *Gonium* is a colonial organism composed of 4 to 32 *Chlamydomonas*-like cells arranged in a flat disk. The flagella of each cell beat separately and each cell can produce a new colony. The 16 or 32 cells in a *Pandorina* colony are packed into an ovoid or ellipsoid shape. The colony is polar; the eyespots at one pole of the colony are larger than those at the other pole. The flagella of individual cells all point outward. Reproduction is accomplished when the colony sinks to the bottom of its environment, each cell divides to form a new colony, and the mother colony breaks open. *Volvox* colonies are in the form of a hollow sphere composed of one layer of cells. The flagella beat in a coordinated fashion, causing the colony to spin on its axis. The colony is polar with only some cells in the lower hemisphere forming daughter colonies. Daughter colonies remain inside the hollow mother colony until the mother colony breaks apart. (page 461)

27. Sexual reproduction in fungi can be accomplished by (1) the fusion of gametes released from gametangia, (2) fusion of gametangia of different mating types, and (3) fusion of unspecialized hyphae of different mating types. (page 481)

28. The genus *Coleochaete* is thought to be closely related to the ancestors of green plants for several reasons. (1) The cells of plants and *Coleochaete* contain peroxisomes and have cellulose in their cell walls. (2) In both groups, cytokinesis is accomplished by the formation of a cell plate at the equator of the spindle. (3) The pattern of microtubules underlying flagella is similar in flagellated plant cells and *Coleochaete* cells. (page 494)

The Animal Kingdom I: Introducing the Invertebrates

MAJOR CONCEPTS

Members of the animal kingdom are heterotrophs that depend either directly or indirectly on photosynthetic autotrophs for their nutrition. They store food reserves as glycogen or fat. As a consequence of their nutritional mode, they have evolved integrated nervous and muscular systems that assist in the acquisition of food. Of the known animal species, 95 percent are invertebrates, animals without backbones.

Animals reproduce sexually; many species are also capable of asexual reproduction. Animal life cycles may be quite complex and may include immature forms (larvae) that differ dramatically from the adult. One advantage of producing very different larvae is that it reduces competition for food and other resources between immature and mature forms of a species.

Of the two distinct lines of animals that diverged by the Cambrian period, only one survived and gave rise to all contemporary animals. Modern animals are classified into phyla that are believed to be monophyletic. They are classified on the basis of (1) number of tissue layers, (2) basic body plan, (3) presence, absence, or type of body cavity, and (4) pattern of embryonic development.

The phylum Porifera contains the sponges, organisms whose body organization is intermediate between a colony and true multicellularity. Sponges are aquatic organisms that apparently traveled a different evolutionary route from the other animals. Sponges reproduce asexually by fragmentation and sexually by internal fertilization and the formation of larvae. The larvae are free-living but adults are sessile. Nutrients are obtained by filtering water.

Members of the phylum Mesozoa are parasites of marine invertebrates. Their very simple body plan consists of reproductive cells surrounded by a single layer of ciliated cells. Mesozoans reproduce sexually and asexually.

Two phyla, Cnidaria and Ctenophora, exhibit radial symmetry in which the body parts are arranged around a central axis. Members of these phyla have a body plan consisting of two tissue layers sandwiching a jelly-like mesoglea. A gastrovascular cavity with one opening to the outside functions in food digestion, gas exchange, and waste removal.

Phylum Cnidaria contains both polyp and medusa body forms. The distinguishing feature of this phylum is the presence of cnidocytes on tentacles. Cnidocytes contain nematocysts whose release can be triggered by certain stimuli. Nematocysts function to lasso, harpoon, and/or paralyze prey. Cnidarians have complex life cycles and reproduce sexually and asexually. Larvae are free-swimming planulas. The three main classes are Hydrozoa (including *Hydra*), Scyphozoa (jellyfishes), and Anthozoa (sea anemones and corals).

Phylum Ctenophora includes comb jellies and sea walnuts. They possess comblike plates of fused cilia in longitudinal bands on their bodies, which function in locomotion. All ctenophorans are hermaphrodites and reproduce sexually.

All other animal phyla exhibit bilateral symmetry. Two evolutionary consequences of bilateral symmetry are cephalization (the concentration of sensory organs and nerve cells toward the anterior end) and the development of more efficient means of locomotion.

These phyla have three basic tissue layers: mesoderm is present in addition to the endoderm and ectoderm found in radially symmetrical animals. During the development of an organism, ectoderm gives rise to nervous tissues and to the coverings and linings of organs and cavities, endoderm develops into digestive organs, and mesoderm gives rise to muscles and most other body parts. Triploblastic organisms have three basic body plans: acoelomate—no body cavity other than the digestive cavity; pseudocoelomate—a body cavity between the endoderm and the mesoderm that is not lined with epithelial cells; and coelomate—a body cavity (coelom) lined by epithelial cells within the mesoderm.

Members of phylum Platyhelminthes (flatworms) exhibit the organ level of biological organization. They are acoelomates, have a branched digestive cavity with one opening, and lack a circulatory system. The phylum includes organisms that are free-living (class Turbellaria, the planarians) and parasitic (class Cestoda, the tapeworms, and class Trematoda, the flukes).

Two other acoelomate phyla are Gnathostomulida, tiny marine worms that live along coastal shore lines and possess hard jaws, and Rhynchocoela, the ribbon worms. The ribbon worms have two developments not previously encountered: a one-way digestive tract and a circulatory system.

The pseudocoelomates include several phyla, the most familiar of which is phylum Nematoda, the roundworms. In these organisms, the pseudocoelom acts as a hydrostatic skeleton that lends rigidity to the animal body and facilitates purposeful locomotion. Pseudocoelomates have a one-way digestive tract but no circulatory system (the fluid circulating in the pseudocoelom compensates for this deficiency). Pseudocoelomates may be free-living or parasitic.

Other pseudocoelomate phyla are Nematomorpha (horse-hair worms) and Acanthocephala (spiny-headed worms); Loricifera, Kinorhyncha, Gastrotricha, and Rotifera, which contain organisms found on the shores of fresh or salt water; and Enteroprocta.

HOW TO STUDY THE CHAPTER

Read the entire chapter through quickly, focusing on the major concepts.

Use the GUIDED STUDY OF THE CHAPTER to help you identify the important details as you **reread** the chapter. Writing out the answers to these questions will help fix them in your mind as well as provide you with a valuable study aid.

Answer the questions in TESTING YOUR UNDERSTANDING without the aid of your text. Check your answers against those in PERFORMANCE ANALYSIS. Analyzing your answers will give you valuable feedback on your level of understanding and preparedness for classroom testing.

GUIDED STUDY OF THE CHAPTER

I. Introduction (*page 518*)

1. Provide the following information about animals.

   Principal mode of nutrition

   Type of organism upon which they depend for nutrition

   Food reserves in the form of

   Mechanism of movement

2. How do animal cells differ from the cells of most other eukaryotes?

Focus: **Animals have integrated systems for obtaining and responding to information about their environments.**

---

3. What percentage of known animals are invertebrates?

4. Cite two practical reasons for studying invertebrates.

II. The Diversity of Animals (pages 518–519)

5. What selection pressure favored the evolution of muscles and nerves?

> *Focus:* Within the animal kingdom, there is great diversity in locomotion and reproductive patterns, as a few basic mechanisms have been taken and modified over the course of evolution.

6. Diversification in methods of reproduction has provided mechanisms for what three processes?

7. The only haploid cells in an animal life cycle are the:

8. *Vocabulary:* Distinguish between external and internal fertilization.

9. What are larvae?

> *Focus:* Larvae often differ dramatically in form and lifestyle from the adults, a factor that decreases competition between larvae and adults for essential resources such as food and shelter.

> *Focus:* Animals have adapted to a wide range of environments, and a great diversity of animals can coexist in one habitat because of their different nutritional and space requirements.

III. The Origin and Classification of Animals (page 519)

10. Identify the two locations where the earliest evidences of animal life have been found. How old are these fossil deposits?

11. a. How did the body plan of Ediacaran animals differ from the fundamental body structure of nearly all living animals?

    b. What characteristic of Ediacaran animals increased the functional surface area of their bodies?

    c. Identify the relationship between the Burgess Shale fossils and modern animals.

> *Focus:* In the current classification of animals, each phylum is believed to be monophyletic.

12. Name four criteria used in classifying animals.

IV. Phylum Porifera: Sponges (pages 522–524)
A. General Remarks (pages 522–523)

> *Focus:* Sponges have apparently evolved separately from all other animal forms.

13. Why were sponges once classified as "zoophytes"?

14. *Vocabulary:* Define sessile.

15. Describe the basic body plan of a sponge.

16. Identify the function(s) of the following sponge cell types.

Choanocytes

Epithelial cells

Amoebocytes

17. *Vocabulary:* What is the osculum of a sponge? (Figure 25-8)

18. What are the possible evolutionary origins of sponges?

19. Characterize the level of biological organization represented by a sponge.

Focus: **The unique relationship between individual sponge cells provides a model for the study of cell recognition, adhesion, and differentiation.**

---

20. Describe the process by which sponges obtain nutrients.

21. How is body structure modified as sponge size increases?

22. a. Sponges are assigned to a class based on what characteristic?

b. For each class listed, describe the skeletal composition of its members.

Calcarea

Hexactinellida

Demospongiae

Sclerospongiae

23. Identify two mechanisms responsible for the flow of water through a sponge body. (Figures 25-8 and 25-10)

B. **Reproduction in Sponges** *(pages 523-524)*

24. a. Describe two mechanisms by which asexual reproduction may be accomplished in sponges.

b. In what type of environment are sponges most likely to give rise to resistant, asexually produced offspring?

25. a. What two sponge cell types may produce gametes?

b. How are sperm cells transported from one sponge to another?

c. Describe how fertilization is accomplished in sponges.

26. How does the larval sponge differ from the adult?

27. a. *Vocabulary:* What is a hermaphrodite?

b. Of what advantage is hermaphroditism to sponges?

V. Phylum Mesozoa: Mesozoans *(page 524)*

28. Where do mesozoans live?

29. Describe the basic body plan of a mesozoan.

30. What two evolutionary origins have been proposed for mesozoans?

31. How do mesozoans reproduce? (Figure 25–11)

VI. Radially Symmetrical Animals *(pages 524–531)*
A. General Remarks *(pages 524–525)*

32. Which two phyla consist of radially symmetrical, gelatinous animals?

33. a. *Vocabulary:* What is a gastrovascular cavity?

b. Describe how the gastrovascular cavity functions in obtaining nutrients and oxygen and in waste removal.

B. Phylum Cnidaria *(pages 525–529)*

34. In what two ways do the cells of cnidarians differ from those of sponges and mesozoans?

35. Describe the basic body plan of a cnidarian.

36. *Vocabulary:* Distinguish between polyp and medusa forms.

37. *Vocabulary:* Distinguish between epidermis, gastrodermis, and mesoglea. (Indicate the origin or composition of each layer.)

38. *Vocabulary:* Define diploblastic.

39. a. *Vocabulary:* Distinguish between cnidocyte and nematocyst.

b. What is the function of the cnidocyte?

c. What triggers the release of a nematocyst and in what three ways may a nematocyst function?

> *Focus:* Once a cnidocyte has released its nematocyst, it cannot function again; a new cnidocyte must be synthesized. (Figure 25–13)

40. How do some nudibranchs and flatworms acquire nematocysts?

41. Define the following components of the cnidarian life cycle. (See also Figure 25–14.)

Planula

Blastula

42. Arrange these components of the *Aurelia* life cycle in a flow diagram to illustrate sexual and asexual reproduction: blastula, planula, gametes, medusa, zygote, polyp. (Figure 25–14)

43. Although superficially resembling alternation of generations, the cnidarian life cycle does not qualify as such. Why?

44. Cite three adaptive advantages of the cnidarian life cycle and indicate which form (planula, polyp, or medusa) is responsible for each advantage.

45. Name the three major classes of cnidarians and indicate which body form is most common in each class.

Class Hydrozoa (pages 527–528)

46. Fill in the following table.

Table for Question 46

	EPITHELIOMUSCULAR CELLS	GASTRODERMIS
Function(s)		
Orientation of contractile fibers (myonemes)		

47. a. Name two functions of gland cells. (Figure 25–15)

b. What is the function of the nutritive cell of the gastrodermis? (Figure 25–15)

48. a. Name the three types of sensory cells present in cnidarians.

b. How do independent effectors differ from the other two cell types?

49. Sensory receptor cells are particularly sensitive to what types of stimuli?

50. How does the function of the nerve net differ from that of (a) independent effectors and (b) sensory receptor cells?

51. *Hydra's* sensory and motor abilities permit what three types of motility? (Figure 25–16)

Focus: **Most hydrozoans are colonial and live in marine waters.**

52. What two forms do colonial hydrozoans exhibit in their life cycle?

53. a. How does an *Obelia* colony form?

    b. Identify the two types of polyps present in this colony.

54. How does fertilization occur in *Obelia*?

55. How do colonies of the order Siphonophora differ from those of *Obelia*? (Figure 25–17)

56. a. *Vocabulary:* What is a pontoon? (Figure 25–17)

    b. Distinguish between the blue and purple strands present on a Portuguese man-of-war in terms of composition and function. (Figure 25–17)

Class Scyphozoa (pages 528–529)

Focus: In class Scyphozoa, the jellyfishes, the medusa form is dominant and colonies are not formed.
---

57. By what mechanism are nutrients transported through the mesoglea of some jellyfish?

58. How do the epitheliomuscular cells of scyphozoan differ from those of *Hydra*?

59. The nerve cells in the bell margin of scyphozoan innervate what three tissue types?

60. a. Name two multicellular sensory organs that ar present in scyphozoans.

    b. Describe how a statocyst enables a scyphozoan to orient with respect to gravity. Include the term "statolith" in your discussion. (See also Figure 25–19.)

Focus: Sensory organs that enable an organism to orient itself with respect to gravity are essential to motile animals and apparently have evolved independently many times.
---

61. What type of stimulus is detected by ocelli?

Class Anthozoa (page 529)

62. In what way do anthozoans resemble *Hydra*?

Focus: Polyps of the class Anthozoa reproduce either sexually or asexually.
---

63. Describe the portion of the anthozoan life history from zygote to adult animal.

64. How does the gastrovascular cavity of anthozoans differ from that of hydrozoan polyps?

65. a. *Vocabulary:* What are corals?

b. What forms the basic structure of a coral reef?

C. *Essay:* **The Coral Reef** *(page 530)*

> *Focus:* The coral reef represents the most diverse marine community.

66. What two types of organisms conduct photosynthesis in the coral reef?

67. How is the reef community supplied with carbon dioxide, oxygen, and minerals?

68. What environmental conditions are essential for the development of a coral reef?

D. **Phylum Ctenophora** *(page 531)*

69. Identify the common names for this group of animals.

70. Provide the following information about ctenophorans.

Distinguishing structural feature

Means of locomotion (2)

Mechanism for capturing food

Reproductive style

71. How is the basic body plan of ctenophorans similar to that of cnidarians?

72. *Review Question:* Name two types of organisms discussed thus far in the chapter that are components of plankton at some stage in their life cycle.

VII. **Bilaterally Symmetrical Animals: An Introduction**
(pages 531–532)

73. How does bilateral symmetry differ from radial symmetry?

> *Focus:* Two evolutionary consequences of bilateral symmetry were cephalization and the evolution of more efficient means of locomotion.

74. *Vocabulary:* Distinguish among dorsal, ventral, anterior, and posterior.

75. a. *Vocabulary:* Define cephalization.

b. Where are the most sensory cells aggregated in animals demonstrating cephalization and why is this of adaptive value?

c. What other two types of structures are usually located at the anterior end of an animal?

76. *Vocabulary:* Distinguish triploblastic from diploblastic animals.

77. Fill in the following table on the three germ layers.

Table for Question 77

	ENDODERM	ECTODERM	MESODERM
Location relative to the other two layers			
Types of organs derived from each layer			

78. How does the pattern of tissue derivation from the germ layers make "functional sense"?

79. *Vocabulary:* What is a coelom and where is it located with respect to the three germ layers?

80. a. Distinguish among acoelomates, pseudocoelomates, and coelomates.

b. Label each diagram with one of the following terms: acoelomate, coelomate, diploblastic, and

pseudocoelomate. For each diagram, correctly label the layers and structures with as many of these terms as apply: coelom, digestive cavity, ectoderm, endoderm, gastrovascular cavity, mesentery, mesoderm, mesoglea, pseudocoelom.

VIII. Phylum Platyhelminthes: Flatworms *(pages 533–537)*

A. General Remarks *(page 533)*

> *Focus:* The flatworms are the simplest triploblasts and demonstrate the organ level of biological organization.

81. What is one consequence of having a digestive cavity with only one opening?

82. Identify two features of flatworms that compensate for the absence of a circulatory system.

83. Discuss three possible evolutionary origins of flatworms.

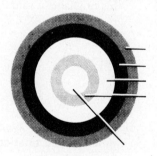

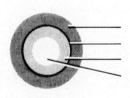

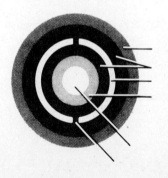

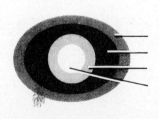

B. **Class Turbellaria** *(pages 533–535)*

84. *Vocabulary:* What are ganglia and what is their evolutionary importance? (Figure 25–24)

85. Describe how a planarian moves.

86. What two roles can cilia play in animals?

87. How does a planarian obtain food?

88. a. Describe the excretory system of a planarian. (Include the role of flame cells in your discussion.)

    b. What is excreted by the excretory system? (See also Figure 25–26.)

89. a. Discuss sexual reproduction in planarians.

    b. *Vocabulary:* Distinguish between oviduct and vas deferens. (See also Figure 25–27.)

The Planarian Nervous System (page 535)

90. How does the planarian nervous system differ from that of *Hydra*?

91. Identify three types of stimuli to which planarians may respond.

92. Describe the form and function of the planarian ocellus.

93. a. Which region of the planarian body has a particularly high concentration of chemoreceptors?

    b. How does a planarian locate and move toward food?

C. **Classes Trematoda and Cestoda** *(page 536)*

94. Name three characteristics that are common to members of both classes.

95. Contrast the nutritional modes of trematodes and cestodes.

96. Cite three means by which tapeworms may cause disease.

97. The parasitic trematodes and cestodes are believed to have evolved from what type of ancestor?

98. *Vocabulary:* Define proglottid and genital pore. (Figure 25–29)

99. a. How do tapeworms reproduce? (Figure 25–29)

b. Where do (a) eggs develop into larvae and (b) larvae develop into adults? (Figure 25–29)

D. *Essay:* The Politics of Schistosomiasis *(page 537)*

100. Arrange the following components of the *Schistosoma* life cycle in a flow diagram: humans, snails, larvae, eggs, adults. Indicate where (and in which host) each stage in the life cycle occurs and identify sexual and asexual phases.

101. What causes the symptoms of schistosomiasis?

102. In what regions of the world is schistosomiasis most common?

103. a. What factors have led to the spread of schistosomiasis?

b. What factors contribute to the lack of progress in finding effective treatment and control measures?

IX. Other Acoelomates *(page 538)*

A. Phylum Gnathostomulida *(page 538)*

104. Identify the distinguishing characteristic of animals in this phylum.

105. Where are gnathostomulids abundant?

106. Upon what do gnathostomulids feed?

107. What two evolutionary origins have been proposed for the gnathostomulids?

B. Phylum Rhynchocoela *(page 538)*

108. Members of this phylum are known by what common names?

109. What adaptation do ribbon worms possess for capturing prey?

110. Identify two innovations characteristic of this phylum that are not found in more primitive acoelomates.

111. Name two consequences of a one-way digestive system.

112. How is reproduction in ribbon worms accomplished?

X. Pseudocoelomates *(pages 539–541)*

A. General Remarks *(page 539)*

113. a. Describe how the pseudocoelom functions as a hydrostatic skeleton.

b. How does the pseudocoelom compensate for the absence of a circulatory system?

B. Phylum Nematoda (pages 539–540)

114. What lifestyle (free-living or parasitic) characterizes most nematodes?

115. Identify the function(s) of the nematode cuticle.

116. A unique feature of the muscular system of nematodes is:

117. What special feeding adaptations characterize nematodes?

118. a. List the five most common nematode parasites in North America.

b. What is the source of *Trichinella*?

c. Name the causative genus for each of the following diseases:

Filariasis

River blindness

Whipworm infection

Hookworm infection

Pinworm infection

Trichinosis

Intestinal roundworm infection

d. Parasitic nematodes possess unique enzymes and metabolic pathways that function in what three general areas?

C. Other Pseudocoelomate Phyla (pages 540–541)

119. Fill in the following table.

Table for Question 119

	NEMATOMORPHA	ACANTHOCEPHALA
Common name		
Habitat of adults		
Habitat of larvae		
Resembles what other group of organisms?		

120. Fill in the following table concerning members of the pseudocoelomate phyla located between sand and silt particles of shorelines. (If the text does not contain the requested information for a phylum, write "no info.")

Table for Question 120

	LORICIFERA	KINORHYNCHA	GASTROTRICHA	ROTIFERA
Larval life style and habitat				
Adult life style and habitat				
Separate sexes or hermaphrodites?				
Source of nutrients				
Other characteristics				

121. *Vocabulary:* Define parthenogenesis.

122. a. Members of phylum Entoprocta superficially resemble what two groups of organisms?

b. What is unusual about the digestive tract of entoprocts?

c. Identify three possible reproductive modes of entoprocts.

d. *Vocabulary:* What is sequential hermaphroditism?

XI. Summary *(pages 542–543):* Read the summary. If you are familiar with the essential features of the material presented there, you are ready to complete the section TESTING YOUR UNDERSTANDING.

TESTING YOUR UNDERSTANDING

After you have completed this examination, compare your answers with those in the section that follows.

1. The development of efficient systems to coordinate nerve and muscle activity in animals was required because of their mode of:
   a. reproduction.   d. nutrition.
   b. locomotion.   e. digestion.
   c. respiration.

2. Which of the following activities did NOT contribute to the great diversity present in the animal kingdom?
   a. reproduction
   b. locomotion
   c. prey capture
   d. self-defense
   e. All these activities contributed to the diversification of animals.

3. Water moves through the channels of sponges under the influence of:
   a. rhythmic pulsations of the sponge body.
   b. flagella of their choanocyte cells.
   c. a simple, tubular "heart."
   d. muscular contractions of the osculum.
   e. cilia that extend from channel epithelial cells.

4. Sponges of class _____ have either unfused silica spicules or _____.
   a. Calcarea; spicules of calcium carbonate
   b. Hexactinellida; keratin-like protein skeletons
   c. Demospongiae; a keratin-like protein called spongin
   d. Sclerospongiae; spicules of calcium carbonate
   e. Hexactinellida; fused spicules of calcium carbonate

5. Mesozoans are placed in their own subkingdom because they:
   a. are so different from all other animals.
   b. are single-celled eukaryotes.
   c. have only one layer of cells in their bodies.
   d. are the only organisms with a cilia-lined body cavity.
   e. have no ecological relationships with other organisms.

6. In which cells of *Hydra* do the contractile fibers run in a longitudinal direction?
   a. cnidocytes
   b. cells of the epidermis
   c. cells of the gastrodermis
   d. cnidocytes and epitheliomuscular cells
   e. cells of the gastrodermis and cells of the epidermis

7. Within a colony of *Obelia* there are two types of _____, a feeding form and a reproductive form from which _____ bud off. The forms that bud off produce gametes that fuse to form _____.
   a. polyps; other polyps; zygotes
   b. medusas; other medusas; polyps
   c. medusas; polyps; medusas
   d. polyps; medusas; polyps
   e. polyps; medusas; zygotes

8. Members of phylum Ctenophora move through the water by the:
   a. muscular contractions of their tentacles.
   b. coordinated beating of their cilia.
   c. current of water expelled from the osculum.
   d. contraction of amoeboid cells in their mesoglea.
   e. currents of water in which they are "swimming."

9. Both ocelli and statocysts are present in:
   a. *Obelia.*   d. *Hydra.*
   b. corals.   e. jellyfish.
   c. sea anemones.

10. Animals in which a cavity develops between the endoderm and mesoderm are known as:
    a. diploblastic.   d. coelomates.
    b. acoelomates.   e. gastromates.
    c. pseudocoelomates.

1. Which of the following are characterized by the presence of a digestive cavity with a single opening?
   a. flatworms and cnidarians
   b. nematodes and flatworms
   c. rhynchocoelates and cnidarians
   d. ctenophorans and rhynchocoelates
   e. gnathostomulids and nematodes

12. Why are planarians classified in the order Tricladida?
   a. Their digestive cavity has three main branches.
   b. Their flame cells have three main functions.
   c. They have three protective plates on their dorsal surfaces.
   d. They can use any of three forms of locomotion.
   e. They have three germ layers.

13. Which class consists primarily of parasites?
   a. Turbellaria
   b. Anthozoa
   c. Scyphozoa
   d. Trematoda
   e. Hydrozoa

14. Reproduction in the phylum Rhynchocoela is fairly simple and usually involves:
   a. the asexual production of gemmules.
   b. separate sexes that engage in external fertilization.
   c. hermaphrodites that alternately take on the role of male and female.
   d. separate sexes that engage in internal fertilization.
   e. the asexual production of buds that detach and grow into adults.

15. What feature is present in *Hydra* but not in nematodes?
   a. a continuous cuticle
   b. sexual reproduction
   c. a circulatory system
   d. circular muscle cells
   e. longitudinal muscle cells

16. Which phylum consists of spiny-headed worms whose larvae parasitize arthropods and whose adults parasitize vertebrates?
   a. Nematomorpha
   b. Gnathostomulida
   c. Acanthocephala
   d. Gastrotricha
   e. Rotifera

17. T or F   All of the most complex animals are vertebrates.

18. T or F   All 30 of our modern animal phyla are thought to be monophyletic.

19. T or F   A coral reef is mainly composed of the discarded skeletons of colonial anthozoans.

20. T or F   Compared with animals that are bilaterally symmetrical, radially symmetrical animals move more slowly or are sedentary.

21. T or F   Scientists generally agree that gnathostomulids are closely related to free-living flatworms.

22. T or F   One characteristic of pseudocoelomates is that all phyla have some members exhibiting sequential hermaphroditism.

23. Match the stage of the cnidarian life cycle with the corresponding function(s). An answer may be used more than once. A blank may have more than one answer, as indicated in parentheses.
   1. planula larva
   2. medusa
   3. polyp

   a. _____ asexual reproduction
   b. _____ sexual reproduction
   c. _____ habitat selection
   d. _____ dispersal (2 answers)
   e. _____ genetic recombination

24. Describe the means by which cnidarians capture their prey using nematocysts. How do certain nudibranchs and flatworms obtain nematocysts?

25. What are the advantages of the flattened body form to platyhelminthes?

26. Discuss the advantages to an organism of having a fluid-filled pseudocoelom.

PERFORMANCE ANALYSIS

1. d   Efficient systems to integrate muscle and nerve function developed as a consequence of the heterotrophic nutritional mode of animals, which requires them to actively seek prey. (page 518)

2. e   All of the activities listed contributed to the tremendous diversity exhibited by members of the kingdom Animalia. (pages 518, 519)

3. b   Water moves through the sponge body as a consequence of two forces: action of the choanocyte flagella and the suction effect created by local water currents. (page 522 and Figure 25–8)

4. c   Members of class Demospongiae have either spicules composed of unfused silica, or a tough protein called spongin, or a combination of these two components. In class Calcarea, the skeleton is constructed of individual calcium carbonate spicules. The glass sponges (class Hexactinellida) have spicules made of fused silica. (page 523)

5. a   Mesozoans are very different from all other animals. They are parasites of marine invertebrates and their bodies consist of 20 to 30 cells arranged in two layers. The body plan is a mass of reproductive cells surrounded by a layer of ciliated cells. (page 524)

6. b   The contractile fibers of *Hydra* epidermal cells run in a longitudinal direction. The contractile fibers of gastrodermal cells run in a circular direction. (page 527)

7. e   The two types of polyps in an *Obelia* colony are spe-

cialized for feeding and reproduction. Medusas, which bud off the reproductive polyps, form gametes that fuse to form zygotes. (page 528)

8. **b**   Ctenophorans move through water by the coordinated beating of their fused cilia, which are arranged in eight longitudinal rows on their bodies. (page 531)

9. **e**   Jellyfish are members of class Scyphozoa. They have light-sensitive ocelli and statocysts, which enable the organisms to orient with respect to gravity. (pages 528, 529)

10. **c**   Triploblastic animals have three germ layers: endoderm, ectoderm, and mesoderm. They may be divided into three categories based on the presence or absence of a body cavity and the type of body cavity. Acoelomates have no internal body cavity (other than the digestive cavity). Pseudocoelomates are characterized by a pseudocoelom, a cavity that forms between the endoderm and mesoderm. A true coelom is a cavity that forms within the mesoderm. (page 532)

11. **a**   Flatworms and cnidarians both have digestive cavities with a single opening. The other groups listed all have a one-way digestive tract with a mouth and an anus. (page 533)

12. **a**   The planarians are classified into order Tricladida because of the presence of a digestive cavity with three main branches. (page 534)

13. **d**   The platyhelminth classes Trematoda (tapeworms) and Cestoda (flukes) are composed mainly of parasitic organisms. (page 536)

14. **b**   The most common form of sexual reproduction in Rhynchocoela involves separate sexes and external fertilization. They may also reproduce asexually by fragmentation. (page 538)

15. **d**   Nematodes are unusual in that they contain no circular muscles. You will remember from question 6 that the contractile fibers of *Hydra* gastrodermal cells run in a circular fashion. (page 539)

16. **c**   The acanthocephalan life cycle is entirely parasitic, with larvae living in arthropods and adults parasitizing vertebrates. (page 540)

17. **False**   Some very complex animals, including octopuse and insects, are invertebrates. (page 518)

18. **True**   According to available evidence, all 30 contemporary animal phyla are monophyletic. (page 520)

19. **True**   The fundamental structure of a coral reef is composed of skeletons of colonial anthozoans known a corals. (page 529)

20. **True**   Consequences of bilateral symmetry include cephalization and the development of more efficien means of locomotion. (page 531)

21. **False**   There is some disagreement among scientists concerning the origin of gnathostomulids. Some believe the are closely related to free-living flatworms. However others contend that they are degenerate forms of pseudo coelomates. (page 538)

22. **False**   Among the pseudocoelomates, sequential her maphroditism is present only in phylum Entoprocta (page 541)

23. The answers are **a3; b2; c1; d1,2; e2.** (page 526)

24. When a cnidarian is touched or detects a certain chemica stimulus, nematocysts on the tentacles are discharged The nematocyst threads are poisonous or sticky and ca lasso, harpoon, or paralyze prey. Some nudibranchs and flatworms may acquire nematocysts because they can ea cnidarians without triggering the discharge of their nema tocysts. The nematocysts are transported to the body sur face of the predator and can be used in its defense. (page 526)

25. Flatworms are acoelomate animals with no circulatory systems. Therefore, gas exchange, waste removal, and nutrient acquisition and distribution must be accomplished by diffusion either through the body surface or through the walls of the digestive cavity. A flat body form minimizes the distance over which diffusion must occur. (page 533)

26. The fluid-filled pseudocoelom acts as a hydrostatic skeleton that facilitates purposeful movement and also provides a means by which materials may move from one body region to another. (page 539)

CHAPTER **26** _____

The Animal Kingdom II: The Protostome Coelomates

CHAPTER ORGANIZATION

MAJOR CONCEPTS

The evolution of the coelom was a significant development because it allowed room for organs to move about and fold on each other, increasing their functional surface area. Coelomates can be classified into two major groups based on pattern of embryonic development: in protostomes, the mouth forms at or near the blastopore; in deuterostomes, the anus forms at or near the blastopore. Protostomes undergo schizocoelous coelom formation in which the coelom forms from a slit that develops in the mesoderm. In enterocoelous development, which occurs in deuterostomes, the coelom forms from outpouchings of the embryonic gut.

Phylum Mollusca is characterized by organisms with soft bodies. Many members have external shells, but the shell has been lost or modified in several classes. The basic molluscan body plan has four regions: head-foot, visceral mass, mantle,

and mantle cavity. A radula is present in all classes except the bivalves. Molluscans have either a closed or open circulatory system with one or more hearts. Gas exchange occurs through gills and across the moist surface of the mantle. The ciliated digestive tract has several functional regions and produces mucus-coated solid feces. Nephridia remove nitrogenous wastes, influence the chemical composition of body fluids, and function in water balance. There are three major classes of mollusks: Bivalvia, Gastropoda, and Cephalopoda.

Members of class Bivalvia are aquatic organisms with shells divided into two parts. The body has become flattened between the halves of the shell. Both sessile and motile forms are included, and the bivalves are filter-feeders. The nervous system consists of three pairs of ganglia. Bivalves have receptors to detect gravity, touch, chemicals, and light. Reproductive patterns vary from separate sexes to hermaphroditism; fertilization may be internal or external.

The gastropods include terrestrial and aquatic forms. A gastropod either has a single shell or the shell has been lost during evolution. The body of a gastropod no longer exhibits the bilateral symmetry characteristic of the hypothetical molluscan ancestor; it has undergone a 180° torsion. Gastropods are motile and may be herbivores, omnivores, carnivores, scavengers, or parasites. The nervous system has up to six pairs of ganglia connected by nerve cords. The nerve cells are concentrated at the anterior end of the body. Some gastropod species have separate sexes, others exhibit simultaneous or sequential hermaphroditism. Fertilization is either internal or external.

In class Cephalopoda, the shell has either been lost during evolution or has become internalized. (The one exception is _Nautilus_ whose shell is only partially occupied and which functions as a flotation chamber.) The cephalopods are all marine animals who feed by predation. The loss of the shell enables them to move by jet propulsion. They can release a dark fluid into the water if threatened. Cephalopods have organized brains protected by cartilaginous cases. The sexes are always separate, and fertilization is internal.

The mollusks and the annelids apparently evolved from the same unsegmented coelomate ancestor. Evidence supporting this conclusion includes the fact that most marine mollusks (except the cephalopods) and marine annelids have a trochophore larval stage.

The phylum Annelida is characterized by organisms with a segmented body plan. Annelids have a segmented coelom, tubular gut, closed circulatory system, excretory nephridia in each segment, and a centralized nervous system with receptors that detect touch, chemicals, light, and moisture. Class Oligochaeta includes mostly terrestrial worms with some aquatic relatives. Class Polychaeta comprises mainly marine worms. Members of class Hirudinea are leeches.

The oligochaetes (earthworms) are characterized by the presence of four pairs of setae on each segment which anchor the worm during locomotion. The compartmentalized coelom acts as a hydrostatic skeleton and enables fine motor control of movements. Earthworms feed on organic matter in the soil, aerating and enriching the soil in the process. They are hermaphrodites, and some species are capable of parthenogenesis.

The distinguishing feature of the polychaetes is the presence of a variety of appendages, including tentacles, antennae, and specialized mouthparts. There are two parapodia on each segment, which function in locomotion and gas exchange. Tagmosis, the grouping of segments into different functional regions, characterizes the polychaetes. These organisms may be motile or sedentary as adults and are predators or filter feeders, depending upon the motility pattern. Fertilization is external.

Members of the class Hirudinea have no setae and either creep or swim with undulating movements. Leeches are hermaphrodites.

There are seven minor protostome phyla. Four of these include bottom-dwelling marine worms: Sipuncula, Echiura, Priapulida, and Pogonophora. Members of the three other phyla (Pentastomida, Tardigrada, and Onychophora) all possess a mixture of annelid and arthropod characteristics.

The lophophorates are a taxonomic puzzle. As in protostomes, the mouth develops at or near the blastopore. However, radial cleavage and enterocoelous coelom formation are characteristic of the deuterostomes. There are three phyla, all of which are aquatic and all characterized by the presence of a food-gathering organ known as a lophophore. These phyla are Brachiopoda, Phoronida, and Bryozoa.

HOW TO STUDY THE CHAPTER

Read the entire chapter through quickly, focusing on the major concepts.

Use the GUIDED STUDY OF THE CHAPTER to help you identify the important details as you **reread** the chapter. Writing out the answers to these questions will help fix them in your mind as well as provide you with a valuable study aid.

Answer the questions in TESTING YOUR UNDERSTANDING without the aid of your text. Check your answers against those in PERFORMANCE ANALYSIS. Analyzing your answers will give you valuable feedback on your level of understanding and preparedness for classroom testing.

GUIDED STUDY OF THE CHAPTER

I. Introduction *(pages 544–545)*

    1. Cite two reasons why the coelom is an evolutionarily important innovation.

    2. The phyla of coelomate animals are studied in groups based on what characteristic?

    3. Fill in the following table distinguishing protostomes and deuterostomes.

Table for Question 3

	PROTOSTOMES	DEUTEROSTOMES
Pattern of early embryonic cleavage		
Developmental fate of the blastopore		
Manner in which the coelom forms (name and describe)		
Representative phyla for each group		

    4. *Vocabulary:* Distinguish among blastula, blastopore, and blastocoel. (See also Figure 26–2.)

II. Phylum Mollusca: Mollusks *(pages 545–554)*

A. General Remarks *(page 545)*

    5. The mollusks were named based on what body characteristic?

6. Identify three fates of the molluscan shell over evolutionary history.

7. Name the three classes of mollusks and identify at least one example of each class.

8. Which are the most intelligent invertebrates?

B. **Characteristics of the Mollusks** *(pages 546–547)*

9. Summarize how the members of each molluscan class differ from the hypothetical ancestor. (See also Figure 26–3.)

10. Identify the function of each molluscan body region.

   Head-foot

   Visceral mass

   Mantle

   Mantle cavity

11. What substances are carried (a) into and (b) out of the mantle cavity in the flow of water?

12. a. *Vocabulary:* What is a radula and what functions does it perform?

   b. The radula is constructed of what type of material?

Supply Systems (page 547)

> **Focus:** In small and simple animals, the needs of individual cells for oxygen, food, and waste removal are met by diffusion, by amoeboid cells, and/or by the movement of fluids in a body cavity.

13. How are the needs of individuals cells for oxygen, food, and waste removal met in larger animals?

14. Identify the functions of the atria and the ventricle in the molluscan heart.

15. *Vocabulary:* What is meant by an open circulation?

16. *Vocabulary:* What is a hemocoel and what is its role in blood circulation in mollusks?

17. Describe the coelom of mollusks.

18. In what two ways does the circulatory system of cephalopods differ from that of other mollusks?

19. a. *Vocabulary:* What is a gill?

   b. Describe the factors that favor gas exchange (oxygen and carbon dioxide) across gill surfaces.

> **Focus:** In order for gas exchange to occur in animals, a moist membrane must be present.

20. Describe the process by which food particles are taken up by mollusks.

21. a. In what form are solid wastes expelled from the molluscan body?

b. What is the advantage of this type of waste disposal?

22. a. *Vocabulary:* What are nephridia?

b. A nephridium opens into what two cavities?

23. What is accomplished as coelomic fluid is passed through a nephridium?

> *Focus:* The excretory system of mollusks functions in water balance and in the regulation of internal fluid composition.

C. **Minor Classes of Mollusks** *(page 548)*

24. For each descriptive phrase, write the name of the molluscan class(es) to which that phrase applies. (See also Figure 26–5.)

Aplacophora (solenogasters)
Monoplacophora (includes *Neopilina*)
Polyplacophora (chitons)
Scaphopoda (tusk or tooth shells)

a. Five pairs of gills, six pairs of nephridia _____

b. Tubular shell, live a sedentary life _____

c. Resemble the hypothetical molluscan ancestor _____

d. No shell, foot greatly reduced _____

e. Single dorsal shell retracted with eight pairs of retractor muscles _____

f. Obtain food by means of tentacles _____

g. Dorsal shell formed from eight plates _____

D. **Class Bivalvia** *(pages 549–550)*

25. Bivalves are named for what anatomical feature?

26. What are two functions of the adductor muscles of a scallop?

27. Why are bivalves sometimes termed Pelecypoda?

> *Focus:* Class Bivalvia includes both sessile and motile species.

28. a. By what mechanism do some sessile bivalves remain anchored in place?

b. Of what interest is this mechanism to biomedical researchers?

29. Describe the process by which bivalves obtain nutrients.

30. Identify the names and locations of the three pairs of ganglia typical of bivalves.

31. Name four types of stimuli for which bivalves have sensory receptors. (Name the receptors if possible.)

32. How developed are the eyes of bivalves in terms of function?

33. a. What is the adaptive value of brood pouches?

   b. In which types of bivalves are brood pouches found?

E. **Class Gastropoda** *(pages 550–551)*

34. Name five different nutritional modes of gastropods.

35. a. *Vocabulary:* Define torsion as it applies to gastropods.

   b. What phenomenon is responsible for torsion?

36. Label the following diagrams of the hypothetical ancestral mollusk and a snail. Pay close attention to the effects of torsion on the arrangement of organs. (You may wish to color various structures to make this more apparent.) Labels: anus, foot, gill, gut, head-foot, mantle, mantle cavity, mouth, radula, shell, visceral mass.

37. What two structures have been lost in some species over evolutionary history owing to the crowding of internal organs that results from torsion?

38. Describe what happened in the process of detorsion in the evolution of slugs.

39. a. By what mechanism do land-dwelling snails accomplish gas exchange?

   b. Some aquatic snails lack gills. What is the presumed origin of these snails and how do they compensate for their deficiency?

40. Name the six pairs of ganglia and identify which structures are supplied by nerves from each ganglion. (Figure 26–8)

41. *Vocabulary:* Define and describe the function of the operculum. (Figure 26–8)

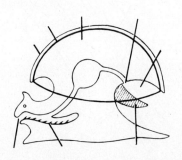

Hypothetical ancestor

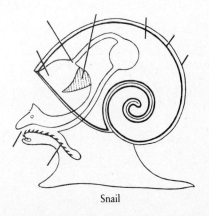

Snail

42. How do gastropods exemplify cephalization?

43. a. Name three types of sensory receptors in gastropods.

   b. How does the eye of a gastropod compare with that of a scallop in form and function?

44. List the three reproductive modes present in gastropods.

F. **Class Cephalopoda** *(pages 551–553)*

> *Focus:* **The cephalopods are the most evolutionarily advanced invertebrates.**

45. All cephalopods live in what type of environment?

46. Label the following diagrams of the hypothetical ancestral mollusk and a squid. Pay close attention to the structural modifications that have occurred. (You may wish to color various structures to make this more apparent.) Labels: anus, arms with tentacles, eye, gill, gut, head-foot, mantle, mantle cavity, mouth, radula, shell, siphon, visceral mass.

Hypothetical ancestor

47. How does the shell of (a) *Nautilus* and (b) the squid differ functionally from that of the hypothetical ancestral mollusk?

48. Describe how cephalopods move through the water.

> *Focus:* **The absence of a hard external shell allowed the evolution of the jet propulsion movement of cephalopods.**

49. What advantage is the dark fluid that can be released at will from special sacs in cephalopods?

> *Focus:* **The cephalopods have well-developed brains and nerve fibers.**

50. Why are squids useful in studies of nerve impulse conduction?

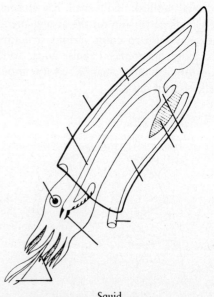

Squid

51. a. Describe how fertilization is accomplished in the octopus.

b. *Vocabulary:* What are spermatophores?

> *Focus:* The octopus demonstrates complex behavior patterns associated with reproduction, including courtship and protection of young.

G. *Essay:* Behavior in the Octopus *(page 552)*

52. Describe how an octopus obtains food.

53. How does the octopus compensate for its lack of a protective shell?

54. How does an octopus apparently compensate for its lack of stereoscopic vision?

55. What does an octopus do when confronted by a threat?

56. Describe the *mechanism* by which an octopus may change color when it is excited.

> *Focus:* By using negative and positive reinforcement techniques, an octopus can readily be taught simple exercises.

---

57. The arms of an octopus are sensitive to what two types of stimuli?

58. How does octopus anatomy make the processing and coordinating of sensory data difficult or impossible?

H. **Evolutionary Affinities of the Mollusks** *(page 554)*

59. a. What developmental stage is common to most marine mollusks and marine annelids?

b. What is the proposed significance of this fact in terms of the evolutionary history of these two groups?

60. Describe the process by which a trochophore larva develops into an annelid. (Figure 26–11)

III. **Phylum Annelida: Segmented Worms** *(pages 554–558)*

A. **General Remarks** *(page 554)*

61. The phylum was named for what characteristic common to all annelids?

62. *Vocabulary:* Define metamere.

63. What is the proposed evolutionary connection between annelids and arthropods?

64. Describe the following annelid structures or systems.

Coelom

Digestive system

Circulatory system

Excretory system

Nervous system and sensory receptors

> *Focus:* Class Oligochaeta includes terrestrial, freshwater, and marine worms. Class Polychaeta contains mostly marine worms. Class Hirudinea comprises the leeches.

---

65. What are the proposed evolutionary relationships between the three classes of annelids?

B. Class Oligochaeta: The Earthworms *(pages 555–557)*

66. a. Name five components present in each identical body segment of an earthworm.

    b. How do the anterior segments of an earthworm differ from the other body segments?

67. a. *Vocabulary:* What are setae?

    b. What is their function?

68. a. Describe locomotion in earthworms, including the muscles involved.

    b. What features of the coelom permit fine control of earthworm movement?

Digestion in Earthworms (page 555)

69. Trace the path of food through an earthworm digestive tract beginning with the mouth and naming each section. Indicate what happens to the food in each section.

70. What influence do earthworms have on the soil through which they travel?

Circulation in Earthworms (page 556)

71. Arrange the following components of the earthworm circulatory system in a flow diagram illustrating the flow of blood through the body: five hearts, dorsal longitudinal vessel, ventral longitudinal vessels, parietal vessels, capillaries (in each segment). Identify the two structures that exert a pumping action on the blood.

72. What prevents the backward flow of blood in the circulatory system?

73. The presence of a closed circulatory system allows:

Respiration in Earthworms (page 556)

74. How is gas exchange accomplished in earthworms?

75. The surface of an earthworm is kept moist by what two mechanisms?

Excretion in Earthworms (page 556)

76. Name two functions of earthworm nephridia (in general terms).

77. Describe the structure of a nephridium.

78. Summarize in specific terms the fate of coelomic fluid that enters a nephridium.

The Nervous System of Earthworms (pages 556–557)

79. List four types of stimuli that earthworms can detect and identify the location(s) of receptors for each type of stimulus.

80. *Vocabulary:* What are mechanoreceptors?

81. Nerves for what two functions are present in each segment of an earthworm?

82. What directs the movement of each body segment?

83. Describe how the cerebral ganglia influence movement.

84. What is the function of the ventral double nerve cord?

Reproduction in Earthworms (page 557)

> *Focus:* **Earthworms are hermaphrodites, and some species can undergo parthenogenesis.**

85. a. *Vocabulary:* What is a clitellum?

b. Describe its role in earthworm reproduction.

86. *Vocabulary:* Distinguish between gonopores and spermathecas.

C. Class Polychaeta *(page 558)*

87. Name three ways in which polychaetes differ structurally from oligochaetes.

88. a. *Vocabulary:* What are parapodia?

b. Name two functions of parapodia.

89. *Vocabulary:* Define tagmosis.

90. Describe three possible lifestyles that may be exhibited by polychaetes.

91. How does polychaete reproduction differ from that of earthworms?

D. Class Hirudinea *(page 558)*

92. a. How do the hirudineans differ from oligochaetes in terms of structure and locomotion?

b. In what way do hirudineans resemble oligochaetes?

93. Describe the means by which bloodsucking leeches attach to and obtain nutrients from their hosts.

94. Discuss the potential use in human medicine of two substances produced by leeches.

Table for Question 95

	SIPUNCULA	ECHIURA	PRIAPULIDA	POGONOPHORA
Common name or representative type(s)				
Distinctive features				
Segmentation (yes, no, or describe)				
Method of obtaining food				
Reproductive features				

IV. Minor Protostome Phyla *(pages 559–561)*

95. Fill in the table above. (If the text or figures do not contain the requested information, write "no info.") (See also Figures 26–18 and 26–19)

96. What two characteristics are common to members of phyla Pentastomida, Tardigrada, and Onychophora?

97. Fill in the following table. (See also Figures 26–20, 26–21, and 26–22.)

Table for Question 97

	PENTASTOMIDA	TARDIGRADA	ONYCHOPHORA
Common name or representative type(s)			
Environment			
Reproductive features			
Distinctive features of phylum			(See the next question)

98. List four characteristics that onychophorans have in common with (a) annelids and (b) arthropods.

99. What feature of onychophoran reproduction resembles mammalian reproduction?

100. What environment was inhabited by the earliest onychophorans?

V. The Lophophorates *(pages 562–563)*

101. What feature of lophophorates technically classifies them with protostomes?

102. Name three characteristics that lophophorates have in common with deuterostomes.

103. Name the three phyla that constitute the lophophorates.

104. a. *Vocabulary:* What is a lophophore?

    b. Describe the functional parts of a lophophore.

105. a. Identify two reasons why the brachiopods were reclassified with the lophophorates.

    b. How were brachiopods originally classified?

106. Fill in the following table.

Table for Question 106

	PHORONIDA	BRYOZOA (ECTOPROCTA)
Common names		
Environment		
Colonial (yes or no)		
Protective structure		
Modes of reproduction		

107. Describe the features of the bryozoan life cycle that enable new colonies to become established at distances from the parent colony even though the larvae are poor swimmers.

VI. Summary *(pages 563–564):* Read the summary. If you are familiar with the essential features of the material presented there, you are ready to complete the section TESTING YOUR UNDERSTANDING.

TESTING YOUR UNDERSTANDING

After you have completed this examination, compare your answers with those in the section that follows.

1. The acoelomates, pseudocoelomates, and some coelomates have what characteristic in common?
   a. book lungs
   b. mesenteries
   c. trochophore larvae
   d. radial cleavage of the early embryo
   e. spiral cleavage of the early embryo

2. The blastopore becomes the mouth in:
   a. echinoderms.
   b. platyhelminths.
   c. coelomates.
   d. protostomes.
   e. deuterostomes.

3. The hypothetical primitive mollusk displayed ———— and most resembled modern ————.
   a. radial cleavage; sea urchins
   b. spiral cleavage; clams
   c. bilateral symmetry; chitons
   d. radial symmetry; starfish
   e. spiral cleavage; squids

4. Except for the bivalves, all classes of mollusks have a:
   a. mantle.
   b. head-foot.
   c. radula.
   d. shell.
   e. visceral mass.

5. An open circulation is characterized by:
   a. a continuous system of vessels.
   b. a four-chambered heart.
   c. gas exchange through the body surface only.
   d. the absence of a pumping organ, the heart.
   e. a hemocoel.

6. Which of the following is (are) NOT released by mollusks into their mantle cavities to be dispersed by water currents?
   a. nitrogenous wastes
   b. mucus-coated fecal pellets
   c. carbon dioxide
   d. oxygen
   e. gametes

7. Water balance and the regulation of the chemical composition of molluscan body fluids take place in the:
   a. gills.
   b. nephridia.
   c. anterior intestine.
   d. contractile vacuole.
   e. flame cells.

8. The smallest class of mollusks is ————, with only eight living species.
   a. Aplacophora
   b. Polyplacophora
   c. Monoplacophora
   d. Bivalvia
   e. Scaphopoda

9. Scallops may move quickly to escape predators by:
   a. shooting water from their siphons.
   b. rapidly clapping their two shell halves.
   c. reeling in the protein strands by which they attach to a substrate.
   d. rapidly contracting their mantle cavity.
   e. digging themselves into the mud with their hatchet feet.

10. The protein strands by which sessile bivalves attach themselves to a substrate are being studied because:
    a. the adhesive may be useful in mending broken bones.
    b. the substance may be useful in making false teeth.
    c. cables made from this substance are stronger than steel.
    d. the adhesive may have economical industrial applications.
    e. the substance is stable in saltwater, but dissolves in freshwater.

11. Bivalves feed largely on:
    a. microscopic algae.
    b. decaying matter on the ocean floor.
    c. members of phylum Cnidaria.
    d. free-swimming larvae of other organisms.
    e. raw or processed sewage that humans release into the water.

12. Which of the following is NOT a secondary evolutionary development found among members of class Gastropoda?
    a. reinvention of the gill by land species returning to water
    b. loss of the shell
    c. use of the mantle cavity as a lung
    d. reduction in the number of internal organs
    e. detorsion, resulting in repositioning of the mantle cavity

13. The only cephalopod that has a shell is:
    a. *Eledone*.   d. *Neopilina*.
    b. *Nautilus*.   e. *Sepia*.
    c. *Octopus*.

14. The separate body segments of annelids are called _____ and are separated internally by _____.
    a. metameres; septa
    b. blastomeres; septa
    c. clitella; muscle bands
    d. nephridia; setae
    e. nephridia; muscle bands

15. In which part of the earthworm digestive system is food ground up with the help of soil particles?
    a. crop   d. esophagus
    b. gizzard   e. intestine
    c. pharynx

16. What do members of phylum Sipuncula and phylum Echiura have in common?
    a. setae
    b. a retractable proboscis
    c. segmentation
    d. separate sexes
    e. lack of a mouth and digestive tract

17. What do members of phyla Pentastomida, Tardigrada, and Onychophora have in common?
    a. a retractable proboscis
    b. a wormlike body form
    c. jointed appendages
    d. trochophore larvae
    e. unjointed appendages bearing claws

18. Which of these phyla contain members that exhibit a striking combination of annelid and arthropod characteristics as well as bearing live young?
    a. Tardigrada
    b. Pogonophora
    c. Pentastomida
    d. Onychophora
    e. Priapulida

19. T or F Gastropods have a more highly developed nervous system than bivalves, as evidenced by their greater number of ganglia and eyes that are capable of forming focused images.

20. T or F Exquisite control over earthworm movement is a result of the hydrostatic skeleton (formed from the pseudocoelom) being divided into right and left halves.

21. T or F Each segment of an earthworm contains two ganglia composed of the cell bodies of sensory neurons and neurons causing muscle contraction.

22. T or F Most polychaetes are marine, exhibit tagmosis, have separate sexes, and produce trochophore larvae.

23. Courtship and mating behavior in cephalopods is complex. Describe the events that lead to fertilization of the cephalopod egg mass. Describe the degree of parental care rendered by some cephalopods.

24. Describe the physical arrangement of the earthworm circulatory system.

25. List the similarities and differences between brachiopods and bivalves.

PERFORMANCE ANALYSIS

1. **e** Spiral cleavage in the early stages of embryonic development characterizes protostomes, which include acoelomates, pseudocoelomates, and some coelomates. Radial cleavage is characteristic of deuterostomes. (page 544)

2. **d** In protostomes, the mouth develops at or near the blastopore. In deuterostomes, the anus develops at or near the blastopore. (page 544)

3. **c** The hypothetical ancestral mollusk is thought to have displayed bilateral symmetry. Modern chitons are believed to resemble this ancestor most closely. (page 546)

4. **c** Members of all molluscan classes except Bivalvia possess a radula. (page 546)

5. **e** An open circulatory system has a hemocoel, which is a large blood-filled cavity in which blood from the tissues collects. (page 547)

6. **d** Mollusks release several items into their mantle cavities, including nitrogenous wastes, mucus-coated feces, gametes, and carbon dioxide. Oxygen is taken *into* the molluscan body across gill membranes and across the mantle surface. (page 546)

7. **b** The molluscan nephridia regulate the chemical composition of body fluids and water balance. (page 547)

8. **c** Class Monoplacophora has only eight living species. (page 548)

9. **b** Scallops can achieve quick bursts of movement by clapping their two valves together abruptly using their strong adductor muscles. A clam can dig into sand or mud rapidly using it "hatchet-foot." (page 549)

10. **a** The amino acid polymer by which mussels adhere to their substrate is being studied as a possible adhesive for orthopedic and dental procedures. Attempts are being made to use DNA recombinant techniques to synthesize quantities adequate for research. (page 549)

11. **a** Bivalves are primarily filter-feeders, and the major component of their diet is microscopic algae. (page 549)

12. **a** As members of class Gastropoda evolved, some of the modifications that occurred included loss of the shell, modification of the mantle cavity as a lung, reduction in the number of internal organs, and detorsion. Those snails that were once terrestrial and have returned to aquatic environments have not regained gills. (page 550)

13. **b** Members of the genus *Nautilus* are the only cephalopods that have shells. (page 551)

14. **a** The individual body segments of annelids are called metameres. Adjacent metameres are separated internally by septa. (page 554)

15. **b** In earthworms, food is ground up with the help of soil particles in the gizzard. (Incidentally, this function is analogous to that of the avian gizzard.) (page 555)

16. **d** Members of phylum Sipuncula and phylum Echiura are characterized by separate sexes, external fertilization, and trochophore larvae. (page 559)

17. **e** Members of phyla Pentastomida, Tardigrada, and Onychophora have combinations of arthropod and annelid characteristics, including an external cuticle that is molted (as in arthropods) and unjointed appendages (as in annelids). (page 560)

18. **d** Most members of phylum Onychophora give birth to live young. Some of these species have internal connections between mother and developing young that are analogous to the mammalian placenta. (page 561)

19. **False** The eyes of gastropods only detect changes in light intensity, they are not able to form focused images. (page 551)

20. **False** Oligochaetes are coelomates. Each segment of the body has a right and left coelomic cavity, which functions as a hydrostatic skeleton and contributes to the controlled movements of these creatures. (page 555)

21. **True** Each identical segment of an earthworm contains two nephridia, three pairs of nerves branching off the ventral nerve cord, part of the digestive system, left and right coelomic cavities, four pairs of setae, and two ganglia. These ganglia contain nerve cell bodies for sensory neurons and for nerves that stimulate muscle contraction. (page 557)

22. **True** Most polychaetes are marine animals exhibiting a marked degree of tagmosis compared with oligochaetes. They have separate sexes and produce trochophore larvae. (page 558)

23. Cephalopods undergo complex courtship and mating behavior that may be associated with males fighting for access to females. Sperm is transferred from the male to the female in sperm packets (spermatophores) that the male carries with one arm from his mantle cavity to the mantle cavity of the female. The female produces a gelatinous mass around the fertilized eggs. The embryos develop in this mass until they hatch out as miniature adults. In some genera of cephalopods, the female guards, cleans, and aerates the egg mass. (page 553)

24. The earthworm circulatory system is characterized by one dorsal and several ventral longitudinal vessels. The largest ventral vessel supplies blood to the digestive tract; smaller ventral vessels supply the nerve cord and the rest of the body through numerous branches. Each segment contains parietal vessels that transport blood from the subneural vessels to the dorsal vessel. The dorsal vessel collects nutrients from the intestines. Five pairs of hearts connect the dorsal and ventral vessels and pump blood throughout the body through the ventral vessels. Blood returns to the hearts through the dorsal vessel. (page 556)

25. Brachiopods superficially resemble bivalves in that they possess a shell with two parts that opens in a fashion similar to that of a bivalve shell. However, brachiopods differ from bivalves in that brachiopods have lophophores, and the two halves of their shells are dorsal and ventral, rather than left and right. Brachiopods are attached to their substrates by stalks. (page 562)

The Animal Kingdom III: The Arthropods

CHAPTER ORGANIZATION

MAJOR CONCEPTS

Phylum Arthropoda is the largest of the animal phyla. Its members are characterized by jointed appendages and a jointed exoskeleton. The body plan of arthropods is segmented, suggesting common ancestry with the annelids. The three major branches are chelicerates, terrestrial mandibulates, and aquatic mandibulates.

The exoskeleton or cuticle is composed of chitin and proteins and is secreted by the underlying epidermis. It provides protection for the soft body, aids in water conservation, and lends support for coordinated movement. There are some major disadvantages of an exoskeleton: it does not grow and must be molted periodically; molting is metabolically expensive and leaves the organism vulnerable until the new exoskeleton hardens.

The internal organ systems of arthropods are fairly well developed. Arthropods possess a tubular gut and an open circulatory system. Respiration is accomplished through tracheae, book lungs, or book gills. Malpighian tubules are the most common excretory organs. The arthropod nervous system is well developed to coordinate responses to sensory input.

The chelicerates have no antennae and no mandibles. The first pair of appendages, chelicerae, is adapted for biting prey. The second pair of appendages, pedipalps, may bear pincers, be modified as walking legs, or serve as sensory organs. Horseshoe crabs, sea spiders, and arachnids (spiders, ticks, scorpions, etc.) are all chelicerates.

The major aquatic mandibulates are the crustaceans (lobsters, crabs, and crayfish). Two pairs of antennae are the first appendages. They have appendages attached to the abdomen as well as the thorax. Their mandibles move laterally and crush food. Many members possess a carapace, a shieldlike portion of the exoskeleton that covers the head and thorax. There are a few terrestrial to amphibious forms that have modified gills.

The terrestrial mandibulates include the myriapods and the insects. They are characterized by one pair of antennae and mandibles of a different evolutionary origin from those of the aquatic mandibulates. Myriapods (centipedes and millipedes) are segmented, many-footed organisms with paired appendages originating from nearly every body segment.

The class Insecta is the largest class of arthropods and its members constitute 70 percent of all animals species. They are characterized by three body regions (head, thorax, and abdomen), three pairs of legs attached to the thorax, one pair of antennae, and complex mouthparts. Most insects have chitinous wings but a few orders either never evolved wings or lost them during evolution. The digestive systems of insects, including their digestive enzymes, are highly specialized for their particular diets. Excretion is accomplished through Malpighian tubules. The respiratory system is a network of cuticle-lined tubules, which extend to all parts of the body.

Insects have complex life histories in which larvae specialize in feeding, growth, and nutrient storage while adults accomplish dispersal to new environments. Sexes are always separate and some groups demonstrate parental care of young. Nearly 90 percent of all insect species undergo complete metamorphois, passing through egg, larval, pupal, and adult stages. Insect growth and metamorphosis are under hormonal control.

There are several reasons for the success of insects, including the nature of the exoskeleton, their small size and specificity for diets and environments, the diverse microenvironments created by vascular plants, the mobility afforded by flight, their complete metamorphosis, and their well-developed nervous systems.

Insects have sensory organs to detect visual stimuli, touch, sound, chemicals, and proprioceptive data. The compound eye of insects has less resolution than the vertebrate eye but is more sensitive in detecting motion. Insects communicate by visual signals, sound, and chemicals (pheromones).

The highly complex behavior patterns characteristic of insects are genetically determined, a necessity since their short life spans do not allow time for learning.

HOW TO STUDY THE CHAPTER

Read the entire chapter through quickly, focusing on the major concepts.

Use the GUIDED STUDY OF THE CHAPTER to help you identify the important details as you **reread** the chapter. Writing out the answers to these questions will help fix them in your mind as well as provide you with a valuable study aid.

Answer the questions in TESTING YOUR UNDERSTANDING without the aid of your text. Check your answers against those in PERFORMANCE ANALYSIS. Analyzing your answers will give you valuable feedback on your level of understanding and preparedness for classroom testing.

GUIDED STUDY OF THE CHAPTER

I. Introduction (page 565)

> *Focus:* **Phylum Arthropoda is the largest of the animal phyla and its members occupy virtually all environments.**

1. How many species of arthropods are (a) known to exist and (b) suspected to inhabit the earth?

II. Characteristics of the Arthropods (pages 565–568)
A. General Remarks (pages 565–566)

2. What characteristic is common to all arthropods?

3. In what two general ways have appendages been modified over the course of arthropod evolution?

> *Focus:* **The appendages of evolutionarily advanced arthropods are highly specialized to perform specific functions.**

4. The arthropods share what characteristic with the annelids?

5. List three ways in which the arthropod body has become modified during evolution.

6. Contrast tagmosis in arthropods with that in polychaetes (refer to Chapter 26 if necessary).

7. The basic segmented body plan of arthropods may be discerned in what two ways?

8. What are the three principal types of arthropods?

9. *Vocabulary:* Distinguish between mandibles and chelicerae.

10. Distinguish between the mandibulates and the chelicerates on the basis of type and location of appendages.

11. What evidence suggests that the terrestrial and aquatic mandibulates do *not* share a common ancestor?

12. What respiratory structures are found in chelicerates but not in mandibulates?

B. **The Exoskeleton** (*pages 566–567*)

13. a. How is the exoskeleton formed?

    b. Describe the composition of the three layers of an arthropod exoskeleton.

14. Name three locations of the exoskeleton.

15. Identify six functions of the cuticle or exoskeleton proper.

16. The exoskeletons of some crustaceans may contain:

17. Describe the mechanism by which arthropods are capable of precise movements.

18. a. Name a major disadvantage of possessing an exoskeleton. Describe the strategy that arthropods have developed for coping with this.

    b. What are the drawbacks of the strategy described in (a) and how do arthropods minimize these drawbacks?

19. When an arthropod molts its old exoskeleton, by what means is it able to rapidly expand its body so that the new exoskeleton is larger than the one it just shed? (Figure 27–3)

20. In terms of evolutionary history, what was significant about the evolution of a waterproof exoskeleton?

21. How old are the oldest known fossils of animals adapted to semiarid environments?

> *Focus:* **Included in phylum Arthropoda are the only invertebrates adapted to withstand desiccation as adults.**

C. **Internal Features** (*pages 567–568*)

22. Briefly describe each arthropod characteristic listed and indicate whether it most closely resembles the same feature of mollusks or of annelids.

    Circulatory system

    Coelom

    Gut

23. Study Figure 27–4.

    a. *Vocabulary:* Distinguish between tracheae and spiracles.

    b. What feature of tracheae keeps them open?

    c. This system is very efficient because:

24. How does the respiratory system of insects limit maximum body size?

25. Name one other device by which arthropods may accomplish gas exchange.

26. a. Describe the means by which terrestrial arthropods excrete wastes.

    b. Wastes may be excreted from arthropods in what two forms? (Figure 27–4)

D. **The Arthropod Nervous System** *(page 568)*

27. Sketch the nervous system in the following drawing of a bee.

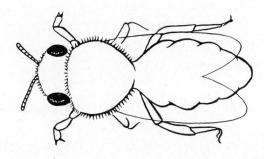

> *Focus:* **Although arthropods have a simple brain, many functions are controlled at the segmental level.**

28. What is the apparent functional mode of the arthropod brain? Cite the evidence for your answer.

III. **Subdivisions of the Phylum** *(pages 568–578)*

A. **The Chelicerates** *(pages 568–571)*

29. a. Describe the two forms of chelicerae.

    b. Chelicerae are used for what purpose?

30. a. *Vocabulary:* What are pedipalps?

    b. Name three possible functions for which pedipalps may be specialized.

31. Provide the following information about chelicerates.

    Names of the two tagmata present

    Nutritional mode

    Reproductive mode

32. Fill in the following table. (See also Figures 27–6 and 27–7.)

Table for Question 32

	MEROSTOMATA	PYCNOGONIDA
Common name of representative species		
Means of respiration		
Source of nutrition		
Common location		

33. *Vocabulary:* Distinguish between carapace and operculum. (Figure 27–6)

Class Arachnida (pages 570–571)

34. Cite common names for five arachnids.

35. How are the chelicerae of spiders modified for a specific purpose?

36. The pedipalps of (a) scorpions and (b) spiders are adapted for what function?

37. Describe how spiders obtain nutrients.

38. a. Name the mechanisms by which arachnids may respire.

b. Describe the structure and function of book lungs.

39. *Vocabulary:* Define spinnerets.

40. Identify nine uses that spiders have for silk.

Focus: **The construction of a spider web is genetically programmed and unique for each species.**

B. **The Aquatic Mandibulates: Class Crustacea** *(pages 571–572)*

41. Give the common names for eight crustaceans.

42. Crustaceans differ from terrestrial mandibulates in what two ways?

43. Cite two exceptions to the generalization that the sexes are usually separate in crustaceans.

44. What immature life cycle stage occurs in most marine crustaceans?

The Lobster (pages 571–572)

45. Describe the segmentation of the lobster's body.

46. Identify the function of each type of lobster appendage.

Antennae

Mandibles

Walking legs:
    First pair

    Second and third pairs

    Fifth pair

Swimmerets

47. a. How does the action of lobster mandibles differ from that of vertebrate jaws?

b. What is the presumed evolutionary origin of crustacean mandibles?

48. Describe two mechanisms for locomotion in the lobster.

49. *Vocabulary*: Define autotomize.

Focus: If a lobster has a damaged appendage or one that is trapped by a predator, the lobster can purposefully drop off that appendage from its body and a new one will grow over a series of molts.

---

50. Identify the lobster structures responsible for the following functions.

Food absorption

Respiration

Excretion

51. By what means are fertilized lobster eggs protected until they hatch?

52. Name the group of crustaceans that contains more individual members than any other group of animals. (Figure 27–11)

Terrestrial Crustaceans (page 572)

53. How is the respiratory mechanism of (a) amphibious crabs and (b) land crabs modified for terrestrial existence?

C. The Terrestrial Mandibulates: Myriapods *(page 572)*

54. Characterize the myriapods according to the following features.

Number of antennae

Type of mandible

Respiration

Excretion

Degree of tagmosis

55. In what way is the term "myriapod" descriptive of this group?

56. Distinguish among centipedes and millipedes in the following table.

Table for Question 56

	CHILOPODA	DIPLOPODA
Common name		
Preferred environment		
Diet		
Apparent number of appendages per segment		

57. What is deceiving about the segments of a millipede body?

58. How do pauropods and symphylans differ from other myriapods?

D. The Terrestrial Mandibulates: Class Insecta *(pages 573–578)*

Focus: The evolution of flight in insects permitted their diversification to the extent that they are now the dominant terrestrial animals with respect to number of species and number of individuals.

---

59. What percentage of the animal species on earth are insect species?

60. Name the four largest orders of insects and identify examples of each order.

61. a. Describe the wings of beetles.

  b. Of all the classified insect species, beetles account for what proportion?

Insect Characteristics (page 574)

62. List the four characteristics common to all insects.

63. Identify five functions for which insect mouthparts may be adapted.

64. *Vocabulary:* Distinguish among labium, labrum, maxillae, and palpi. (Figure 27–15)

65. a. Insect wings are made of what material?

  b. What feature of insect wings lends structural support?

66. Name two reasons for the absence of wings in some insect orders.

Digestive, Excretory, and Respiratory Systems (pages 574–575)

67. a. Where does digestion begin in an insect?

  b. Identify the major organ of nutrient absorption in an insect.

68. Name at least five nutrient sources for which insects may have specific digestive enzymes.

69. a. Excretion is accomplished by what organ?

  b. Nitrogenous wastes are excreted from many insects in the form of:

70. a. Describe the respiratory system of an insect.

  b. Movement of air through an insect body is facilitated by:

  c. Name two factors regulated by the opening and closing of spiracles.

Insect Life Histories (pages 575–578)

71. Which life stage is responsible for dispersal of the species in (a) marine invertebrates and (b) insects?

Focus: Insect larvae are adapted for feeding, growth, and storage of food reserves. Adults are adapted for sexual reproduction (including parental care in some species) and dispersal of the species.

---

72. Name four types of insects that live in complex societies.

73. a. *Vocabulary:* Define metamorphosis.

  b. Describe three types of metamorphosis that differ with respect to the change in form between immature and mature stages.

74. *Vocabulary:* Distinguish between nymphs, larvae, and instars.

75. Summarize the process of complete metamorphosis as an insect passes through each of the four stages: egg, larva, pupa, and adult.

76. *Vocabulary:* What is a hormone?

77. Fill in the following table. (See also Figure 27–18.)

Table for Question 77

	SITE OF PRODUCTION	FUNCTION
Brain hormone		
Molting hormone (ecdysone)		
Juvenile hormone		

78. *Vocabulary:* What is the imaginal molt? (Figure 27–18)

Tidbit: Insect hormones are currently being used in pest control products to interrupt pest life cycles by arresting development.

IV. Reasons for Arthropod Success *(pages 578–585)*

A. General Remarks *(pages 578–579)*

79. Discuss how each of these factors contributed to arthropod success.

Exoskeleton

Small size and highly specific requirements

Environmental diversity of vascular plants

Capacity for flight

Complete metamorphosis

Nervous system

B. Arthropod Senses and Behavior *(pages 579–585)*

Vision: The Compound Eye (pages 579–582)

80. a. *Vocabulary:* What is an ommatidium?

b. Label the following diagram of an ommatidium.

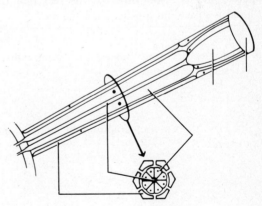

c. Identify the function of each part of the ommatidium. (See also Figure 27–20.)

Cornea

Lens

Retinular cells

Pigment cells

Rhabdom

81. Contrast the compound eye with the vertebrate eye with respect to size of photoreceptor, degree of resolution, ability to detect motion, and speed of response to stimuli.

82. Describe how the compound eye's ability to detect motion is determined in a laboratory.

Focus: **The highly refined perception of motion is critical to an insect's ability to fly safely at high speeds, and is responsible for the fact that the "hand is so seldom quicker than the fly."**

---

83. a. Some arthropods possess what other type of light-sensitive organ?

b. How does it compare functionally to a compound eye?

C. *Essay:* **Firefly Light: A Warning, an Advertisement, a Snare** *(pages 580–581)*

84. What type of insect is a firefly?

85. Briefly outline the life history of a firefly.

86. What is the adaptive value of larval luminescence to a firefly species?

87. Describe the means by which fireflies identify potential mates of the same species.

88. How does the nutritional mode of most adult fireflies differ from that of the larvae?

89. a. Summarize the behavioral change that occurs in females of the species *Photuris versicolor* about three days after mating and ensures a steady supply of nutrients.

b. What strategy has evolved in males of the species *Photinus macdermotti* that discourages competition while they establish the validity of the signals to which they are attracted?

Touch Receptors (page 582)

90. *Vocabulary:* Define and describe the structure of sensilla.

91. Identify two locations where sensilla are typically found.

92. a. Name three means by which sensilla may be stimulated.

b. Cite two specific examples of the role of stimulation of sensilla.

Proprioceptors (page 582)

93. a. *Vocabulary:* What are proprioceptors?

b. Describe the function and typical location of campaniform sensilla.

94. Summarize the mechanism that enables a praying mantis to strike moving prey accurately.

Communication by Sound (pages 583–584)

95. List the five distinct types of calls produced by insects and indicate (if possible) whether each type is long-range or short-range.

96. Identify three characteristics of sound that influence its recognition by an insect and the response of the insect to the stimulus.

97. How can the effectiveness of calling songs be enhanced?

Focus: **The production of and response to songs is an inherited trait that requires no learning on the part of the insect.**

---

98. a. How are sound waves generated?

b. The sound receptors of insects respond to what two types of changes?

99. Describe how modified sensilla function in sound detection. Cite one specific example.

100. To what specific stimulus do tympanic organs respond?

101. a. Describe how a tympanic organ responds to sound.

b. How are some insects able to identify the specific direction from which a sound originated? Describe one example.

Communication by Pheromones (page 585)

102. a. *Vocabulary:* What are pheromones?

b. Name one well-studied example of an insect pheromone.

103. What behavioral characteristic of a male gypsy moth increases its likelihood of encountering a female by detecting pheromones?

> *Focus:* **Male gypsy moths fly upwind until they detect female pheromone, at which point they continue to fly upwind. Only when the male is fairly close to the female can he locate her by detecting a concentration gradient and flying "up the gradient."**

Programmed Behavior (page 585)

> *Focus:* **The complex behavior patterns of insects are genetically programmed and not subject to learning. This is advantageous for organisms with very short life spans.**

V. Summary *(pages 585–586):* Read the summary. If you are familiar with the essential features of the material presented there, you are ready to complete the section TESTING YOUR UNDERSTANDING.

TESTING YOUR UNDERSTANDING

After you have completed this examination, compare your answers with those in the section that follows.

1. The most characteristic feature of arthropods is their:
   a. book lungs.
   b. compound eyes.
   c. jointed appendages.
   d. segmented abdomen.
   e. chitinous tracheae.

2. In arthropods, tagmosis has resulted in _____ body regions.
   a. two
   b. two or three
   c. three
   d. three or four
   e. four

3. Because arthropods are segmented, it is thought that they share a common ancestor with the:
   a. flatworms.
   b. kinorhynchs.
   c. nematodes.
   d. annelids.
   e. mollusks.

4. The mandibulates differ from the chelicerates in that the first pair of appendages in a mandibulate is modified as _____ whereas the first pair of appendages in a chelicerate is modified as _____.
   a. fangs or pincers; pedipalps
   b. jaws; claws
   c. antennae; sucking tubes
   d. claws; egg depositors
   e. antennae; fangs or pincers

5. Many arthropods minimize the metabolic expenditure of molting an exoskeleton by:
   a. reabsorbing all of its layers.
   b. eating the shed exoskeleton.
   c. shedding the exoskeleton in sections.
   d. reabsorbing parts of the exoskeleton and eating the rest.
   e. eating the sections of exoskeleton as they are shed at different times.

6. The arthropod brain consists of:
   a. three fused pairs of dorsal ganglia.
   b. two ventral nerve cords.
   c. two fused pairs of ventral ganglia.
   d. three dorsal nerve cords.
   e. segmental ganglia that encircle the esophagus.

7. Chelicerates generally have unsegmented abdomens. Which animals are exceptions to this rule?
   a. mites
   b. spiders
   c. scorpions
   d. sea spiders
   e. horseshoe crabs

8. The favorite prey of sea spiders (class Pycnogonida) is:
   a. cnidarians.
   b. nematodes.
   c. platyhelminths.
   d. annelids.
   e. mollusks.

9. Respiration in arachnids takes place through:
   a. tracheae only.
   b. book lungs only.
   c. the hemocoel.
   d. tracheae that lead to the hemocoel.
   e. either tracheae or book lungs or both.

10. Aquatic mandibulates differ from terrestrial mandibulates in that aquatic mandibulates have:
    a. no thorax.
    b. no antennae.
    c. no chelicerae.
    d. one pair of antennae.
    e. leglike appendages on the abdomen.

11. Some species of _____ are males when they are young and females when they are older and large enough to carry eggs.
    a. barnacles
    b. crabs
    c. sowbugs
    d. horseshoe crabs
    e. shrimp

12. In the lobster, the most posterior walking legs:
    a. are claws.
    b. function as antennae.
    c. are used to clean the abdominal appendages.
    d. have pincers for grasping food.
    e. cannot be regenerated following autotomization.

13. Waste products that are extracted from lobster blood accumulate in a bladder and are excreted from pores located at the base of the:
    a. gills.
    b. first pair of walking legs.
    c. swimmerets.
    d. second antennae.
    e. mandibles.

14. More than _____ percent of the animal species are insects. Of the approximately one million classified insects, some 300,000 are _____.
    a. 70; flies
    b. 70; beetles
    c. 80; moths
    d. 80; beetles
    e. 90; flies

15. Butterflies belong to insect order:
    a. Coleoptera.
    b. Hymenoptera.
    c. Lepidoptera.
    d. Diptera.
    e. Diplopoda.

16. Many insects excrete nitrogenous wastes as _____, which conserves water.
    a. uric acid
    b. urine
    c. bile
    d. urea
    e. ammonia

17. Air movement and water loss are regulated in insects by:
    a. Malpighian tubules.
    b. spiracles.
    c. tracheae.
    d. ocelli.
    e. ganglia.

18. How are the life histories of insects fundamentally different from those of marine invertebrates?
    a. Most insect larvae have the form of immature adults.
    b. The egg, larva, and pupa of all insects require parental care.
    c. Insect larval growth is largely a consequence of cell division.
    d. Most insect larvae are preoccupied with eating rather than dispersal.
    e. Insect larvae travel great distances from their site of origin in search of new habitats.

19. The function of pigment cells in the insect ommatidium is to:
    a. trap light of a given wavelength.
    b. prevent light from traveling from one ommatidium to another.
    c. provide eye-color recognition between members of the species.
    d. give the insect a measure of color vision.
    e. reflect light to the light-sensitive cells.

20. The fact that much insect behavior is genetically controlled and does not have to be learned correlates well with insects':
    a. short life spans.
    b. existence in larval and adult forms.
    c. ability to detect minute quantities of pheromones.
    d. relatively well-developed brain.
    e. skeletal dexterity.

21. T or F   Among all the invertebrate phyla, only arthropods contain significant numbers of species that can withstand the drying action of air.

22. T or F   Millipedes and centipedes prefer damp environments, are carnivores (they eat cockroaches), and belong to the same class of myriapods.

23. T or F   The digestive enzymes of an insect depend on the diet to which it is adapted.

24. T or F   All insects go through a pupal stage in which the larval cells of each functional system are modified to form the same system in the developing adult.

25. T or F   The big advantage of complete metamorphosis in insects is that adaptations for feeding and growth are separated from adaptations for reproduction and dispersal.

26. T or F   One reason that flies are so adept at escaping the swatter is that they can perceive minute changes in air currents made by the approaching instrument of doom.

27. T or F   All four life stages (egg, larva, pupa, and adult) of the firefly are luminescent.

28. Describe the role of campaniform sensilla in the capture of prey by a praying mantis.

PERFORMANCE ANALYSIS

1. c   Phylum Arthropoda derives its name from the jointed appendages that characterize all its members. (page 565)

2. b   Tagmosis is the fusion of body segments into specialized regions. Arthropods typically have three body regions: head, thorax, and abdomen. In some arthropods, the head and thorax are fused into a cephalothorax. (pages 565, 566)

3. d   The segmented body structure of arthropods suggests a common ancestor with the annelids. (page 565)

4. e   The first pair of appendages in mandibulates are antennae, whereas the first pair of appendages in chelicer-

ates are chelicerae, which may be in the form of pincers or fangs. (page 566)

5. **d** Molting an exoskeleton is a metabolically expensive adaptation that leaves an arthropod vulnerable to predators while the new exoskeleton is drying. Some arthropods reduce the energy loss of molting by reabsorbing the inner layers and by ingesting the shed exoskeleton. (page 567)

6. **a** The arthropod brain is composed of three pairs of dorsal ganglia that are fused at the anterior end of a double chain of ganglia. (page 568)

7. **c** Most chelicerates have unsegmented abdomens, but the abdomens of scorpions are distinctly segmented. (page 568)

8. **a** Members of class Pycnogonida (sea spiders) prey upon soft-bodied invertebrates, especially cnidarians. (page 569)

9. **e** Arachnids accomplish gas exchange through tracheae or book lungs or both. (page 570)

10. **e** Aquatic mandibulates have appendages on the abdomens as well as the thorax and possess two pairs of antennae. Terrestrial mandibulates have only one pair of antennae and have no abdominal appendages. Insects have legs only on the thorax. (page 571)

11. **e** Although the sexes are usually separate in crustaceans, some shrimp species exhibit sequential hermaphroditism —they are males when young and females when older and large enough to accommodate eggs. (page 571)

12. **c** In the lobster, the first of five pairs of walking legs are claws. The next two pairs have small pincers. The most posterior pair is used to clean the abdominal appendages, the swimmerets. (page 572)

13. **d** The lobster has pores located at the base of the second antennae through which wastes removed from the blood are excreted. (page 572)

14. **b** Approximately 1 million insect species (more than 70 percent of the classified animal species) have been identified. Of these, 300,000 are beetles of the order Coleoptera. (page 573)

15. **c** Butterflies and moths are classified in order Lepidoptera. (page 573)

16. **a** Many insects (and birds) excrete nitrogenous wastes as uric acid crystals, an adaptation that conserves water. (page 575)

17. **b** The opening and closing of insect spiracles, which are the openings into tracheae, regulate air intake air and water loss. (page 575)

18. **d** Most insect larvae are adapted for feeding and growth and the adults are specialized for dispersal. In marine invertebrates, the larvae typically seek new habitats and the adults, many of which are fairly stationary, feed and reproduce. (page 575)

19. **b** The pigment cells that line an ommatidium prevent the passage of light between adjacent ommatidia. (page 579)

20. **a** Animals with short life spans typically exhibit behavior patterns that are genetically transmitted. Learning occurs in species of animals that have longer life spans and extended periods of parental care. (page 585)

21. **True** Because of their waterproof exoskeletons, many species of arthropods can withstand the drying effects of air. (page 567)

22. **False** Millipedes are herbivores and belong to class Diplopoda. Centipedes are carnivores belonging to class Chilopoda. (page 573)

23. **True** One key to the success of insects is the great diversity of specialized diets. Each insect species has digestive enzymes suited to its particular diet. (page 574)

24. **False** In complete metamorphosis, which occurs in 90 percent of classified insects, the adult develops from cells that were set aside during embryonic development. The larval tissues are digested and function as a culture medium for the developing adult. (page 576)

25. **True** One major advantage of complete metamorphosis is that the adaptations for feeding and growth (in the larva) are separated from the adaptations for dispersal and reproduction (in the adult). A consequence of this separation is that competition between adults and larvae for scarce resources is eliminated or minimized. (page 579)

26. **True** Two adaptations of insects that account for their rapid response to a threat are their perception of small changes in air currents and their compound eyes, which are well adapted for detecting motion. (page 582)

27. **False** In the firefly, only the egg, larva, and adult stages are luminescent. (essay, page 580)

28. Campaniform sensilla are located on the head and thorax of the praying mantis. The eyes of a mantis do not move. To bring an object into binocular range, the mantis moves its head, stimulating the hairs of the campaniform sensilla. The impulses received from the campaniform sensilla are processed in the brain and enable the insect to strike its prey accurately. (page 582)

The Animal Kingdom IV: The Deuterostomes

CHAPTER ORGANIZATION

MAJOR CONCEPTS

The embryonic development of deuterostomes differs from that of protostomes in that deuterostomes exhibit radial cleavage, anus development at or near the blastopore (the mouth develops elsewhere), and enterocoelous coelom formation. Four phyla are characterized by deuterostome development: Echinodermata (starfish, sea urchins), Chaetognatha (arrow worms), Hemichordata (acorn worms), and Chordata (which includes the vertebrates).

Echinoderms are aquatic organisms characterized by an interior skeleton made of calcium-containing plates from which spines project. The coelom is modified into a water vascular system that functions in circulation of body fluids and in locomotion (in some groups). Adults exhibit either an imperfect radial symmetry or bilateral symmetry; larvae are bilaterally symmetrical. Reproduction is generally sexual, although some forms may reproduce asexually. Extensions of the water vascular system known as tube feet function primarily in locomotion (starfish, sea urchins, and sand dollars) or in gathering and handling food (brittle stars, sea cucumbers, and starfish).

Arrow worms are predators of marine plankton whose bodies are organized into head, trunk, and tail regions. They are hermaphroditic. There is no larval stage; the young resemble miniature adults.

The hemichordates are so named because they have characteristics of both echinoderms and chordates. Their echinoderm features include ciliated larvae resembling those of starfish, coelomic cavities that provide hydrostatic support, and a ventral nerve cord. Traits they have in common with chordates are a dorsal nerve cord (part of which is hollow), and the presence of a pharynx.

Four features, present at some point in the life cycle of all chordates, distinguish this phylum from all others: a notochord, a dorsal, hollow nerve cord, a pharynx with gill slits, and a tail. There are three subphyla of chordates: Cephalochordata (lancelets), Urochordata (tunicates), and Vertebrata (vertebrates). Lancelets and tunicates are aquatic organisms, vertebrates include aquatic and terrestrial forms.

Vertebrates are distinguished by the presence of a bony vertebral column, from which projections extend to protect the dorsal nerve cord. A major advantage of a bony skeleton is that it can grow with the animal. Two of the seven classes of living vertebrates (Agnatha and Chondrichthyes) have lost their bony skeletons during evolution and have skeletons made of cartilage.

The fishes are classified into three classes: Agnatha, the jawless fishes (lampreys and hagfish), Chondrichthyes (sharks and skates), and Osteichthyes (bony fishes). Fishes evidently evolved in fresh water and then moved to marine environments. Primitive bony fish had lungs or lunglike structures as accessory respiratory organs that enabled them to survive in waters with low oxygen content or during seasonal droughts. These lungs evolved into swim bladders in strictly aquatic fishes but retained their respiratory function in the modern lungfishes.

Amphibians evolved from lunged fishes. Modern amphibians exchange gases in lungs, through their moist skin, and/or through the mucous membranes of the throat. Many, but not all, amphibians have life cycles characterized by an aquatic larval stage and a terrestrial adult stage.

The innovation that freed animals from dependence upon water was the amniote egg, which first evolved in reptiles. The amniote egg is a self-contained aquatic environment in which an embryo develops. Reptiles have dry skin which is usually covered by protective scales. Most modern reptiles are strictly ectothermic. Dinosaurs have long been assumed to be ectothermic; however, recent evidence suggests that some groups of dinosaurs were endothermic. Reptiles were the ancestors of the modern birds and mammals.

The unique feature of birds is the presence of feathers, which function in flight and insulation. The bird's body is modified for flight, and many adaptations reflect selection for minimal body weight. All birds supply some degree of parental care for their young.

There are two opposing hypotheses concerning the evolution of flight. One hypothesis suggests that the ancestral reptile dwelt in trees, and wings evolved as a technique for gliding among the branches. An alternative hypothesis argues that feathers evolved as insulation first, and their function in flight was secondary; wings provided advantages in the pursuit of prey.

Mammals are distinguished by the presence of body hair, the provision of milk for young, and endothermy. All mammals exhibit some degree of parental care. The three groups of mammals are the monotremes (e.g., duckbilled platypus), the marsupials (e.g., opossum and kangaroo), and the placentals (most familiar mammals).

The primates are placental mammals characterized by all four kinds of teeth, opposable first digits, mammary glands in the chest, frontally directed eyes, and a well-developed brain. Humans are distinguished by their upright posture, long legs and short arms, sparse body hair, high forehead, small jaw, and an extremely specialized brain.

HOW TO STUDY THE CHAPTER

Read the entire chapter through quickly, focusing on the major concepts.

Use the GUIDED STUDY OF THE CHAPTER to help you identify the important details as you **reread** the chapter. Writing out the answers to these questions will help fix them in your mind as well as provide you with a valuable study aid.

Answer the questions in TESTING YOUR UNDERSTANDING without the aid of your text. Check your answers against those in PERFORMANCE ANALYSIS. Analyzing your answers will give you valuable feedback on your level of understanding and preparedness for classroom testing.

GUIDED STUDY OF THE CHAPTER

I. Introduction *(page 587)*

1. *Review Question:* Fill in the following table distinguishing protostomes from deuterostomes.

Table for Question 1

	PROTOSTOMES	DEUTEROSTOMES
Cleavage pattern		
Location where mouth develops		
Method of coelom development (name and describe)		

2. List the four deuterostome phyla and name at least one example of each phylum.

II. Phylum Echinodermata: The "Spiny-Skinned" Animals *(pages 587–590)*

A. General Remarks *(page 587)*

3. Name one example for each of the four classes of echinoderms.

4. What type of symmetry is exhibited by (a) adult and (b) larval echinoderms?

5. Outline the development of an echinoderm from larva to adult.

6. a. Trace the presumed evolution of echinoderms, describing the hypothetical ancestor and the two stages that have succeeded that ancestor.

b. What is a logical fourth evolutionary stage?

B. **Class Stelleroidea: Starfish and Brittle Stars** *(pages 588–589)*

7. Describe the basic body plan of a starfish.

Focus: **The phylum Echinodermata derives its name from the internal skeleton that typically bears spines projecting through the skin.**

---

8. Describe the structure of an echinoderm skeleton.

9. Identify the location and function of each component of the starfish nervous system.

Nerve ring

Nerve cords

Eyespots

Neurosensory cells

10. Name three structures present in each arm of a starfish.

11. Describe how starfish accomplish each of the following activities.

Reproduction (3 mechanisms)

Circulation of body fluids

Respiration

Waste removal

12. Outline the mechanism by which starfish move. Include these terms in your discussion: water vascular system, tube feet, sucker, ampulla.

13. Describe how a starfish obtains nutrients from bivalves.

14. The tube feet of brittle stars are used not for locomotion but for:

Focus: **If captured by a predator, starfish and brittle stars can autotomize an arm in order to escape.**

C. Other Echinoderms (pages 589–590)

15. Fill in this table on the other classes of echinoderms.

Table for Question 15

	CRINOIDEA	ECHINOIDEA	HOLOTHUROIDEA
Examples			
Lifestyle of adult/ mobility			
Nutrient source(s)			
Unique physical characteristics			

16. a. Which three echinoderm characteristics does *Xyloplax medusiformis* (the sea daisy of class Concentricycloidea) possess? (Figure 28–6)

    b. How are sea daisies thought to obtain nutrients? (Figure 28–6)

III. Phylum Chaetognatha: Arrow Worms (page 590)

17. Characterize the arrow worms according to the following features.

    Mode of nutrition

    Method of locomotion

    Basic body plan

    Mode of reproduction (including developmental stages of young)

IV. Phylum Hemichordata: Acorn Worms (page 591)

18. Describe the basic body plan of hemichordates.

19. Pterobranchs resemble which other group o organisms?

20. a. Identify three characteristics that at least som hemichordates share with echinoderms.

    b. Name two characteristics the hemichordates shar with chordates.

V. Phylum Chordata: The Cephalochordates and Urochor dates (pages 591–592)

21. List the three subphyla of phylum Chordata and name one example of each subphylum.

22. Is the term "chordate" synonymous with "vertebrate"? Explain your answer.

23. The best-known member of subphylum Cephalochordata is:

Focus: **The four features common to all chordates are a notochord, a dorsal, hollow nerve cord, a pharynx with gill slits, and a tail.**

---

24. a. What is a notochord?

    b. What is its function?

25. How does the location and structure of the chordate nerve cord differ from that of other animals?

26. a. Name two possible functions of the chordate pharynx.

   b. Describe the function of the pharynx in *Branchiostoma*.

27. *Vocabulary:* Define a tail in anatomical terms.

28. *Branchiostoma* has traditionally been considered a primitive member of phylum Chordata. What have some scientists suggested regarding its origin?

29. a. Which chordate characteristics are absent in *adult* tunicates? (See also Figure 28–11)

   b. Why are tunicates still considered to be chordates?

   c. The tunicates were given their common name based on what characteristic?

30. What has been proposed as the evolutionary origin of the vertebrates? (Figure 28–11)

VI. Phylum Chordata: The Vertebrates *(pages 592–604)*

A. General Remarks *(pages 592–593)*

31. Describe the developmental relationship between the vertebral column and the notochord.

32. What is the anatomical relationship between the vertebral column and the dorsal nerve cord?

33. How is the vertebrate brain protected?

34. Name two features that confer flexibility and movement upon the vertebral column.

35. a. Identify one major advantage of a bony endoskeleton.

   b. What feature enables the endoskeleton to grow with the animal?

Focus: **The evolution of the seven classes of modern vertebrates is clearly recorded in the fossil record.**

B. Classes Agnatha, Chondrichthyes, and Osteichthyes: Fishes *(pages 593–595)*

36. Cite at least one example of each class.

   Agnatha

   Chondrichthyes

   Osteichthyes

37. A common name for agnaths is:

38. Describe two means by which agnaths obtain nutrients.

39. a. Members of classes Agnatha and Chondrichthyes have what structural feature in common?

   b. How is the evolution of these two groups similar?

40. *Vocabulary:* What are denticles and where are they located?

41. How does the skeleton of adult Osteichthyes differ from that of agnaths and chondrichthyans?

42. a. From what structure did fish jaws evolve?

    b. Summarize the significance of the evolution of jaws in fish.

43. a. How is the evolution of Osteichthyes different from that of chondrichthyans?

    b. Identify the ancestral group that gave rise to both the osteichthyans and the chondrichthyans. (Figure 28–16)

44. Describe the journeys made by either salmon or eels in the course of their life histories.

The Transition to Land (pages 594–595)

45. What characteristics of freshwater environments favored the evolution of lunglike structures in fishes?

46. Describe the two evolutionary fates of fish lungs.

47. a. What are two functions of a swim bladder?

    b. By what mechanism may a fish regulate its depth in the water?

48. Name two types of environments inhabited by modern lungfish. (See also Figure 28–17.)

49. In addition to lungs, describe one behavioral and one anatomical adaptation that enable modern lungfish to survive periodic droughts. (Figure 28–17)

50. How did the evolution of skeletal supports facilitate the survival of some species of lungfish?

Focus: Some authorities believe that the evolution of lungs in fish was an adaptation enabling them to locate new water sources or to survive in water that would not support other fish life.

---

51. For animals, the transition to land may have begun in what manner?

C. **Class Amphibia** *(page 595)*

52. The amphibians evolved from what group of organisms?

53. What major characteristic distinguishes amphibians from reptiles?

54. Identify three organs that perform a respiratory function in amphibians.

55. Why are amphibians so susceptible to desiccation?

56. Arrange these components of the frog life cycle in a flow diagram illustrating the development of a frog from egg to adult: fertilization of eggs, hatching of eggs, laying of eggs, tadpole, adult. Indicate which stages are aquatic and which are terrestrial.

57. What change does the respiratory system undergo as the tadpole matures into an adult frog?

Focus: **The aquatic larval stage (tadpole) of frogs is *not* characteristic of all modern amphibians; some lay eggs in terrestrial locations and the eggs hatch into miniatures of the adults.**

58. In what type of amphibians does internal fertilization occur?

59. a. What is unusual about the life cycles of the mud puppy and the axolotl?

    b. Under what conditions can these species be stimulated to "mature"?

D. **Class Reptilia** *(pages 596–598)*

60. The vertebrates became truly terrestrial with the development of what evolutionary innovation?

61. Name the two structures of the amniote egg that retain its internal aquatic environment but allow gas exchange to occur. (Figure 28–19)

62. Identify the function of each of these components of the amniote egg.

    Yolk

    Albumen

    Amnion

Focus: **The embryos of all terrestrial animals develop in an aquatic environment during which they pass through a gill-like stage.**

63. In what way does the reptilian embryo resemble the mammalian embryo?

Focus: **The reptilian dry skin, usually covered with protective scales, is an adaptation that allows their terrestrial existence.**

Evolution of the Reptiles (pages 597–598)

64. Summarize the evolution of reptiles through each of the following periods in evolutionary history. (See also Table 24–1, page 496)

    Carboniferous

    Permian

    Triassic

    Jurassic

    Cretaceous

65. a. Which reptilian characteristic was modified most during the evolution of the Archosauria?

    b. Describe these modifications.

66. Name the three major groups of archosaurs.

67. *Vocabulary:* Distinguish between ectothermic and endothermic.

68. The only modern group of animals that descended from dinosaurs of the Mesozoic era is the:

> *Focus:* Extinction is a natural consequence of evolution. However, the *rate* at which animal species become extinct has increased dramatically since humans began modifying natural environments.

E. **Class Aves: Birds** *(pages 598–600)*

69. Name four anatomical features of birds that represent adaptations for flight.

70. Identify two functions of feathers.

71. What avian characteristic is a remnant of reptilian ancestry?

72. In what ways do the following fossil organisms resemble primitive reptiles and modern birds? (Figure 28–22)

Table for Question 72

	SIMILARITIES TO REPTILES	SIMILARITIES TO BIRDS
Archaeopteryx		
Protoavis		

> *Focus:* While some members of many animal taxa exhibit parental care, virtually *all* newly hatched birds require parental care.

Evolution of Flight (pages 598–600)

73. Briefly outline two hypotheses for the origin of flight.

74. a. According to John Ostrom, what was the first function of feathers?

b. What evidence supports Ostrom's hypothesis for the origin of flight?

c. What two functions might wings have served in the pursuit of prey?

75. What two methods are now being used to study the evolution and mechanics of flight?

76. Of what evolutionary significance was the evolution of flight?

F. **Class Mammalia** *(page 600)*

77. a. From what group are mammals descended?

b. What event permitted the diversification of the mammals?

78. Name three characteristics that distinguish mammals from all other animals.

79. *Vocabulary:* Distinguish between monotremes, marsupials, and placental mammals on the basis of reproductive strategies. Name one example of each type.

80. What is the advantage of placental over marsupial development?

81. a. Describe the presumed lifestyle of the earliest placental mammals.

   b. Which modern organisms are thought to resemble this group?

82. a. Name three ways in which the mammalian skull differs from that of other vertebrate groups.

   b. Mammals cannot move the upper jaw in relation to the brain case. How does this affect feeding practices?

83. a. Identify five characteristics that distinguish the primates from other mammals.

   b. List five traits that distinguish humans from other primates.

84. a. "Humans are among the least specialized of the mammals." Explain this statement using examples.

   b. What feature of humans is specialized to a greater extent than in any other animal?

VII. **Summary** *(page 604):* Read the summary. If you are familiar with the essential features of the material presented there, you are ready to complete the section TESTING YOUR UNDERSTANDING.

TESTING YOUR UNDERSTANDING

After you have completed this examination, compare your answers with those in the section that follows.

1. Which of the following is NOT a characteristic of deuterostomes.
   a. The early cleavage pattern is radial.
   b. The anus develops near the blastopore.
   c. The mouth forms secondary to the anus.
   d. Outpocketings of the embryonic gut form the coelom.
   e. The mouth develops near the blastopore.

2. In deep ocean waters, members of phylum _____ often make up the bulk of living tissue.
   a. Mollusca
   b. Echinodermata
   c. Annelida
   d. Arthropoda
   e. Chordata

3. In starfish, waste removal is accomplished by _____ cells.
   a. flame   d. renal
   b. amoeboid   e. radial canal
   c. gill

4. The food-gathering tentacles of sea cucumbers are modified:
   a. radial canals.
   b. skin gills.
   c. dermal spines.
   d. tube feet.
   e. arms.

5. The phylum Chaetognatha contains:
   a. acorn worms.
   b. sea cucumbers.
   c. flatworms.
   d. arrow worms.
   e. centipedes.

6. Which of the following is NOT true of hemichordates?
   a. The body is divided into a proboscis, a short collar, and a long trunk.
   b. These animals burrow into ocean sediment using their proboscises.
   c. Gill slits are found in the animal's pharynx.
   d. They have both a dorsal and a ventral nerve cord.
   e. Both of the nerve cords are hollow.

7. Chordates that are covered with a protective cellulose-containing layer belong to subphylum:
   a. Vertebrata.
   b. Hemichordata.
   c. Cephalochordata.
   d. Urochordata.
   e. Agnatha.

8. Subphylum Vertebrata contains seven classes. Three of these classes constitute the:
   a. reptiles.
   b. fishes.
   c. amphibians.
   d. mammals.
   e. birds.

9. The growing portions of the vertebrate endoskeleton consist largely of:
   a. bone.
   b. chitin.
   c. endothelial cells.
   d. cartilage.
   e. collagen.

10. One characteristic shared by members of classes Agnatha and Chondrichthyes, but not Osteichthyes, is:
    a. a cartilaginous skeleton.
    b. the lack of a jaw.
    c. a notochord.
    d. denticles.
    e. pharyngeal gills.

11. Denticles are characteristic of:
    a. sharks and skates.
    b. lampreys and hagfish.
    c. *Branchiostoma*.
    d. jawless fishes.
    e. bony fishes.

12. One of the major events in fish evolution is thought to be the development of jaws from:
    a. denticles.
    b. the posterior-most row of teeth.
    c. other bones in the skull.
    d. cartilaginous portions of the oral sucker.
    e. anterior gill arches.

13. A life cycle that contains an aquatic stage followed by metamorphosis and then a terrestrial stage is characteristic of members of class:
    a. Aves.
    b. Reptilia.
    c. Amphibia.
    d. Mammalia.
    e. Agnatha.

14. Which structure do reptiles have while still within the egg but not as adults?
    a. lungs
    b. coelom
    c. gill pouches
    d. dry skin
    e. scales

15. The environmental factor that may have been the selective force in the evolution of reptiles from amphibians was:
    a. a drier climate.
    b. coniferous forests.
    c. the ice age.
    d. cooling of the earth's atmosphere.
    e. tree ferns.

16. Around the beginning of the Triassic period, several groups of specialized reptiles arose, including:
    a. dinosaurs.
    b. mammal-like reptiles.
    c. turtles.
    d. birds.
    e. crocodiles.

17. Evidence that the original function of feathers may not have been flight is provided by the fact that:
    a. flight came before feathers.
    b. wing feathers of *Archaeopteryx* were not attached to the skeleton.
    c. early feathers were too heavy.
    d. the structure of early feathers was not aerodynamically sound.
    e. insects can fly and they do not have feathers.

18. Which feature is NOT characteristic of ALL primates?
    a. They retain all four kinds of teeth.
    b. They have two mammary glands in the chest.
    c. They exhibit upright posture and long legs relative to the arms.
    d. They have frontally directed eyes.
    e. They can oppose the thumb and fingers of the hand.

19. T or F The ancestral echinoderm is believed to have exhibited fivefold radial symmetry, motility, and a soft elongated body similar to that of sea cucumbers.

20. T or F The nerve cord in all phyla except Chordata is solid and ventrally located.

21. T or F Some species of amphibians undergo complete metamorphosis, while others skip the free-living larval stage altogether.

22. Describe the structural and functional differences and similarities between the tube feet of starfish and those of brittle stars.

23. Describe the various ways in which the body weight of birds has been reduced over the course of evolution.

PERFORMANCE ANALYSIS

1. **e** Deuterostomes are characterized by radial cleavage, enterocoelous coelom formation in which the coelom forms from outpocketings of the embryonic gut, and anus development at or near the blastopore, with the mouth developing elsewhere. In protostomes, the mouth develops at or near the blastopore. (page 587)

2. **b** Members of phylum Echinodermata often constitute the bulk of living tissue in deep ocean waters. Exemplary species include starfish, sea urchins, sea cucumbers, sea lilies, and sand dollars. (page 587)

3. **b** Amoeboid cells that migrate through coelomic fluid are responsible for waste removal in starfish. (page 588)

4. **d** Sea cucumbers are characterized by modified tube feet surrounding the mouth, which resemble tentacles and are used for gathering food. (page 590)

5. **d** The phylum Chaetognatha consists of arrow worms, which are among the most abundant predators in marine plankton. (page 590)

6. **e** Hemichordates have some features characteristic of echinoderms and some features characteristic of chordates. The chordate characteristics are a pharynx with gill slits and a dorsal nerve cord, part of which is hollow in some species. The echinoderm characteristics exhibited by some hemichordates are ciliated larvae similar to those of starfish, coelomic cavities that provide hydrostatic support, and a ventral nerve cord (in addition to the dorsal cord). They have bodies with three regions: proboscis, collar, and trunk. They use the proboscis to burrow into ocean sediments. (page 591)

7. **d** The Urochordates, or tunicates, which inhabit the plankton and ocean bottoms, are covered with a cellulose-containing tunic. (page 592)

8. **b** The three fish classes of subphylum Vertebrata are Agnatha, Chondrichthyes, and Osteichthyes. The other four vertebrate classes are Amphibia, Reptilia, Aves, and Mammalia. (page 593)

9. **d** The regions of the vertebrate skeleton that grow consist of cartilage. When the animal reaches its adult size, these growth regions become bony. (page 593)

10. **a** Members of classes Agnatha and Chondrichthyes are characterized by cartilaginous skeletons. Members of class Osteichthyes are the only fish that have bony skeletons. (page 593)

11. **a** Denticles are small, bony teeth covering the skin of chondrichthyans, which include sharks and skates. (page 593)

12. **e** The evolution of jaws from the anterior gill arches of filter-feeders greatly increased the variety of organisms on which fish could feed. (page 594)

13. **c** Many members of class Amphibia (especially frogs in cold climates) have a life cycle characterized by an aquatic stage followed by a terrestrial stage. Amphibians with this life cycle typically undergo metamorphosis from an aquatic larval form into an adult that is terrestrial and has a quite different form. (page 595)

14. **c** Reptiles have gill pouches during embryonic development in the amniote egg. (page 596)

15. **a** During the evolution of reptiles, a drier climate was thought to exert a major selective pressure favoring the amniote egg over amphibious reproductive strategies. The same climate favored the evolution of conifers and other seed plants. (page 597)

16. **c** Several groups of specialized reptiles evolved at the beginning of the Triassic period. These groups included turtles, lizards, and thecodonts. (page 597)

17. **b** The wing feathers of *Archaeopteryx* were not attached to the bones of the wing but were merely embedded in the skin. This is evidence that the original function of feathers was probably insulation rather than flight. (page 599)

18. **c** Upright posture and long legs with relatively short arms are two characteristics that distinguish humans from other primates. (page 601)

19. **False** The ancestral echinoderm is thought to have been bilaterally symmetrical and motile. It is hypothesized that this form settled down to a sessile life and then became radially symmetrical. In a third evolutionary stage, these radially symmetrical animals became motile, as exemplified today by starfish and brittle stars. (page 587)

20. **True** Except for phylum Chordata, which is characterized by a dorsal, hollow nerve cord, all animal phyla with developed nervous systems are characterized by major nerve cords that are ventrally located and solid. (page 592)

21. **True** Some amphibians, particularly frogs in cold climates, go through a free-living larval stage before undergoing metamorphosis into the adult. In other species (e.g., salamanders) the eggs hatch into miniature versions of the adult. (page 595)

22. The tube feet of starfish are part of the water vascular system. Each tube foot has an external sucker and an associated internal ampulla from which fluid is pumped into the tube foot. These tube feet function in locomotion and in opening bivalve prey. The tube feet of brittle stars have no suckers and function in gathering and handling food, but not in locomotion. (pages 588, 589)

23. Avian bones are hollow, and air sacs are present in the body. The female reproductive tract has only one ovary. In both males and females, the reproductive organs enlarge and become functional only during the courtship and mating seasons. (page 598)

REVIEW TEST 8

Section 4: Chapters 25–28

This review test is *not* designed to cover all of the important information in these chapters. However, it does touch on the major topic categories in each chapter. It will also give you valuable practice in taking this type of test. When you are finished, check your answers with those provided in the following PERFORMANCE ANALYSIS section.

1. How do the ventral epithelial cells of a planarian function in locomotion?
   a. They possess cilia on which the planarian "walks."
   b. They produce mucus on which the planarian "glides."
   c. They produce a muscular wave that drags the planarian along.
   d. The planarian's cilia gain traction in the trail of mucus they produce.
   e. Their muscular contractions propel the animal along the trail of mucus they produce.

2. Which phylum of worms contains members with a proboscis bearing spines that is used to capture soft-bodied prey?
   a. Annelida
   b. Sipuncula
   c. Echiura
   d. Priapulida
   e. Pogonophora

3. The lophophore of phyla Brachiopoda, Bryozoa, and Phoronida is used as _____ in addition to its primary function as a feeding device.
   a. a hole through which gametes may be released
   b. an organ of respiration
   c. an organ of locomotion
   d. an organ of excretion
   e. an organ of reproduction

4. A habitat in which arthropods are NOT very abundant is:
   a. tall grass.
   b. estuaries.
   c. forest canopies.
   d. underneath old houses.
   e. Arthropods are abundant in virtually all habitats.

5. Which of these echinoderms have (has) "arms"?
   a. sea lilies and feather stars
   b. sand dollars
   c. *Xyloplax*
   d. sea cucumbers
   e. sea urchins

6. The Ediacaran animals, which died out about 570 million years ago, were different from animals we know today in that all Ediacarans:
   a. were flat.
   b. were not heterotrophs.
   c. lacked tubular internal structures.
   d. had a low surface area-to-volume ratio.
   e. lacked any degree of diversification.

7. Chitons are somewhat flattened animals whose dorsal shell is formed from a series of eight plates. These grazers belong to class:
   a. Aplacophora.
   b. Polyplacophora.
   c. Gastropoda.
   d. Cephalopoda.
   e. Scaphopoda.

8. Which of the following is NOT descriptive of at least some members of class Gastropoda?
   a. omnivorous
   b. herbivorous
   c. carnivorous
   d. parasitic
   e. Each of these nutritional modes is exhibited by one or more members of class Gastropoda.

9. The first reptiles began to evolve in the _____ period.
   a. Carboniferous
   b. Permian
   c. Devonian
   d. Triassic
   e. Jurassic

10. T or F   Cnidarians exhibit alternation of generations just like plants. The medusa form is diploid and the polyp form is haploid.

1. T or F   It is clear from fossil evidence that oligochaetes gave rise to both polychaetes and leeches.

2. T or F   Every time a spider bites, poison flows automatically from its poison glands through ducts in the chelicerae.

3. T or F   The acorn worms are especially interesting to evolutionary biologists because they have some features characteristic of echinoderms and others that are characteristic of chordates.

4. T or F   Most species of plants and animals are parasitized by at least one species of nematode.

5. T or F   A lobster that has lost an appendage such as a claw or leg will regenerate that part completely by the next molt.

6. T or F   In insects that undergo metamorphosis, the adult body is formed from groups of larval cells that are set aside for that purpose early in development.

7. T or F   The most widely accepted hypothesis of the evolution of flight is that it developed from gliding in tree-dwelling reptiles.

8. Match each structure with the germ layer from which it develops. Answers may be used more than once. Each blank will have only one answer.

   1. mesoderm
   2. ectoderm
   3. endoderm

   a. _____ nerve cells
   b. _____ covering and lining tissues
   c. _____ digestive structures
   d. _____ muscles
   e. _____ reproductive structures
   f. _____ lining of the food-digesting cavity

9. Describe the several methods by which sponges can reproduce.

10. Distinguish between protostome and deuterostome coelomates based on embryonic cleavage pattern, blastopore development, and coelom formation.

11. List the five different types of insect calls and indicate (where possible) whether each one is a long-range or a short-range call.

12. Describe the four features of chordates that are so well illustrated in *Branchiostoma*.

PERFORMANCE ANALYSIS

1. d   The ventral epithelial cells of planarians produce mucus in which the cilia on the cell surfaces gain traction to propel the animal. (page 533)

2. d   Members of the protostome phylum Priapulida possess a spiny proboscis that is used to capture soft-bodied prey. (page 559)

3. b   The lophophore is not only a food-gathering organ, but also a respiratory organ: gases are exchanged across the membranes of the lophophore tentacles. (page 562)

4. e   Arthropods occupy nearly every imaginable environment on earth. (page 565)

5. a   Of the echinoderms listed, only sea lilies and feather stars of class Crinoidea possess "arms." Starfish and brittle stars of class Stelleroidea also have arms. (page 589)

6. c   The Ediacarans differed remarkably from nearly all other animals in that their basic body plan lacked tubular internal structures. Most were flat, although some forms had a quilted structure. (page 520)

7. b   Chitons are mollusks belonging to class Polyplacophora. (page 548)

8. e   Members of class Gastropoda have varied nutritional modes, including omnivorous, herbivorous, carnivorous, and parasitic. (page 550)

9. a   The first reptiles began to evolve from amphibians during the late Carboniferous period. Reptiles exploded in numbers and diversified during the following Permian period. (page 597)

10. False   The alternation of medusa and polyp forms only superficially resembles the alternation of generations in plants. Both medusa and polyp are diploid, and the only haploid cells are the gametes. (page 526)

11. False   Zoologists generally agree that the leeches evolved from oligochaetes, but the origin of the polychaetes is less clear. (page 554)

12. False   The release of poison from a spider's poison glands is under voluntary control. (page 570 and Figure 27–8)

13. True   The acorn worms of phylum Hemichordata possess some characteristics of echinoderms and other features that are characteristic of chordates. The chordate characteristics are a pharynx with gill slits and a dorsal nerve cord, part of which is hollow in some species. The echinoderm characteristics include ciliated larvae similar to starfish larvae, coelomic cavities that provide hydrostatic support, and a ventral nerve cord (in addition to the dorsal cord). (page 591)

14. True   Nematodes are ubiquitous organisms, most of which are free-living. However, most species of plants and animals have at least one nematode parasite. (page 539)

15. False   After a limb is autotomized, several molts are required before a fully functional limb is regenerated. (page 572)

16. True   In the pupal stage, the adult forms from cells that

were set aside during embryonic development. The larval tissues provide a culture medium for the growth of the adult. (page 576)

17. **True** The hypothesis of the evolution of flight that is currently the most widely accepted proposes that flight evolved from gliding activities of tree-dwelling reptiles. This hypothesis has been challenged by John Ostrom, who contends that feathers first evolved as insulation, rather than for flight. He argues that flight evolved as a consequence of prey pursuit. (page 598)

18. The answers are **a2, b2, c3, d1, e1, f2.** (page 532)

19. Sponges can reproduce asexually by fragmentation or by producing gemmules, clumps of amoebocytes within a protective outer layer. In sexual reproduction, gametes are produced by amoebocytes or choanocytes. Sperm are released into the water and the current carries them to another sponge. There the sperm are captured by choanocytes and transferred to amoebocytes, which then transfer them to eggs. Embryonic sponges develop into free-swimming larvae that attach to a substrate and develop into sessile adults when they locate an appropriate site. (pages 523, 524)

20. Protostomes undergo spiral cleavage and schizocoelous coelom formation, and the mouth develops at or near the blastopore. Deuterostomes undergo radial cleavage and enterocoelous coelom formation, and the anus develops at or near the blastopore, with the mouth developing elsewhere. (page 544)

21. Long-range calls include calling by males and calling by females. Short-range calls include courtship and aggressive sounds by males. The fifth type of call is alarm sounds, which may be made by males or females. (page 583)

22. *Branchiostoma*, which exhibits all four chordate characteristics, is a member of the subphylum Cephalochordata of phylum Chordata. The *notochord* is a rod that extends the length of the body and provides flexible support. The *dorsal hollow nerve cord* lies dorsal to the notochord. The *pharynx with gill slits* is used as a food collector in *Branchiostoma* but serves a respiratory function in many other chordates. The *tail* extends posterior to the anus and consists of muscle tissue. (pages 591, 592)

SECTION **5**

Biology of Plants

The Flowering Plants: An Introduction

CHAPTER ORGANIZATION

MAJOR CONCEPTS

Plants are the major force shaping natural terrestrial environments. The character of the plant life determines the nature of all other forms of life in an environment.

The angiosperms (flowering plants) are the most evolutionarily advanced plants. They are divided into monocots and dicots based on several characteristics including number of cotyledons, pattern of leaf venation, and arrangement of vascular tissues in the stem.

Flowers are exquisite adaptations for sexual reproduction and appear in a variety of forms. The basic male component is the stamen, which consists of a filament and an anther (in which pollen grains are produced.) The basic female component is the carpel, which consists of an ovary (in which the female gametophyte develops), style, and stigma (on which the pollen grain lands). Flowers differ greatly with respect to number and arrangement of these basic components.

The pollen grain is the male gametophyte and consists of two sperm cells within a tube cell. When a pollen grain lands on a stigma of the same species, it germinates and grows a pollen tube that extends through the style to the ovary. The two sperm nuclei are released into the ovary and one fertilizes the egg. The other sperm nucleus combines with the two polar nuclei of the female gametophyte to form a triploid ($3n$) cell that develops into endosperm, the nutritive material of the seed.

The fertilized egg develops into an embryo, in which three tissue layers form. Protoderm develops into the outer covering of the plant. The procambium gives rise to the vascular tissues. The ground meristem is the tissue from which the bulk of the young plant body is derived. As specialization progresses, new growth is restricted to the apical meristems of the plant shoots and roots.

A seed consists of the embryo, stored food in the form of endosperm, and an outer seed coat. Seeds are dormant stages of the plant life cycle and are designed to persist during adverse environmental conditions. Dormancy is maintained by the seed coat, which may serve as a mechanical barrier or may contain chemical inhibitors. Dormancy may be broken by the wearing down of the seed coat or by changes in the chemicals triggered by environmental conditions. In order for any seed to germinate, it must imbibe water.

Seeds of angiosperms are enclosed by fruits, which develop from the ovary wall. Fruits exist in a variety of forms and may be fleshy or dry at maturity. Fruits aid in the dispersal of seeds by various mechanisms.

Plants are classified as annuals, biennials, or perennials based on their patterns of growth, dormancy, and death.

HOW TO STUDY THE CHAPTER

Read the entire chapter through quickly, focusing on the major concepts.

Use the GUIDED STUDY OF THE CHAPTER to help you identify the important details as you **reread** the chapter. Writing out the answers to these questions will help fix them in your mind as well as provide you with a valuable study aid.

Answer the questions in TESTING YOUR UNDERSTANDING without the aid of your text. Check your answers against those in PERFORMANCE ANALYSIS. Analyzing your answers will give you valuable feedback on your level of understanding and preparedness for classroom testing.

GUIDED STUDY OF THE CHAPTER

I. Introduction *(page 613)*

1. According to the fossil record, when did plants first begin to utilize terrestrial environments?

> *Focus:* **Once plants invaded and altered terrestrial environments, other forms of life followed.**

2. Name five essential items that plants provide for terrestrial animals.

> *Focus:* **In the natural environments still present on the earth, plants determine the character of the environment and the type and amount of animal life it can support.**

3. Review the differences between monocots and dicots by completing the following table. (See Table 24–3, page 511.)

Table for Question 3

	MONOCOTYLEDONES	DICOTYLEDONES
Number of cotyledons		
Pattern of leaf venation		
Arrangement of vascular bundles in young stem		
Flower parts in multiples of:		
Secondary growth (yes or no)		
Pollen grain structure		

4. Of the 235,000 living species of angiosperms, more than two-thirds are:

5. a. Name three functions of flowers.

   b. Name two possible fates of flower parts after fertilization.

6. What event is necessary before fertilization can occur?

II. Sexual Reproduction: The Flower *(pages 613–618)*

A. General Remarks *(pages 613–614)*

7. Name the four floral parts.

8. Each floral part is believed to have evolved from what plant structure?

9. *Vocabulary:* Distinguish between complete flowers and perfect flowers. (Figure 29–2)

10. a. Distinguish between sepals and petals on the basis of color and function.

    b. Sepals are to _____ as petals are to _____ .

11. a. Name the two components of the stamen.

    b. In terms of the plant life cycle, what are pollen grains and where are they formed?

12. a. Name the three components of the carpel.

    b. What is the role of the stigma?

13. a. In terms of the plant life cycle, what is the embryo sac?

b. Where is the embryo sac located?

14. After fertilization, what is the fate of the following flower structures?

Ovule

Ovary

15. a. *Vocabulary:* Define an imperfect flower.

b. Distinguish between staminate and carpellate flowers.

c. *Vocabulary:* Distinguish between monoecious and dioecious flowers.

B. The Pollen Grain *(page 615)*

16. Name the three haploid cells contained in a pollen grain and identify their functions.

Focus: Over the course of plant evolution, the gametophyte has become reduced in size and in angiosperms is nutritionally dependent upon the sporophyte.

C. Fertilization *(pages 615–618)*

17. a. Identify one factor that is thought to be important to the germination of pollen grains on a stigma. (Figure 29–4)

b. What role has been proposed for sugary substances present on the stigma? (Figure 29–4)

18. a. Identify the role of the tube nucleus in fertilization.

b. What selection pressure is responsible for the evolution of rapid pollen tube growth?

19. Diagram the female gametophyte of an angiosperm. Label the following structures: antipodal cells, embryo sac, egg, polar nuclei, and synergids.

20. Outline the role of synergids in fertilization.

21. The union of what components forms the zygote?

22. a. The union of what components forms the endosperm?

b. What is unique about endosperm as a tissue?

c. Identify the function of endosperm.

Focus: In angiosperms, the fertilization of the egg and the formation of 3*n* endosperm are referred to collectively as double fertilization.

III. The Embryo *(page 619)*

23. Mitotic division of the 3*n* cell produces _____ and mitotic division of the zygote produces _____.

24. *Vocabulary:* What happens when the cells of an embryo differentiate?

25. a. Describe the structure of an early embryo.

b. Identify the function of the suspensor.

26. Name the three embryonic plant tissues and indicate what develops from each tissue. (Figure 29–7)

27. *Vocabulary:* Define morphogenesis.

28. a. What are apical meristems and what is their role in plant growth?

b. Where are apical meristems located?

IV. The Seed and the Fruit *(pages 619–621)*

A. General Remarks *(page 619)*

29. Name the three components of the seed.

30. The fruit develops from what specific portion of the flower?

31. *Vocabulary:* Distinguish between scutellum and coleoptile. (Figure 29–8)

B. Types of Fruits *(pages 620–621)*

32. a. Fruits are classified as simple, aggregate, or multiple based on what flower characteristic?

b. Describe the flower structure from which each type of fruit is derived.

Simple

Aggregate

Multiple

c. Name one example of each of the three types of fruits.

Focus: **Simple fruits may be fleshy or dry when mature.**

---

33. Name, describe, and cite at least one example of the three kinds of fleshy, simple fruits.

34. Identify the three layers of the drupe that develop from three distinct layers of the ovary wall.

35. What is the origin of a pome?

36. Name and distinguish between the two types of dry fruits.

37. Describe the two basic ways in which dehiscent fruits release their seeds.

38. a. For dehiscent fruits, what factor may determine the distance from the parent plant that seeds land? (Figure 29–11)

    b. Identify two means by which the seeds of indehiscent fruits may be transported some distance from the parent plant. (Figure 29–11)

39. *Vocabulary:* Distinguish between achenes and nuts.

C. *Essay:* The Staff of Life *(page 622)*

40. What has been the importance of grains in human agricultural history?

41. Wheat has low amounts of what essential amino acid?

42. a. What fraction of the wheat kernel is endosperm?

    b. What fraction of the total protein of the wheat kernel is in the endosperm?

43. Distinguish between the origins of white flour and wheat germ.

44. a. Identify the three components of wheat bran.

    b. Wheat bran consists mainly of what substance?

c. How does the presence of bran affect the character of flour?

d. What effect does bran have on the passage of food through the digestive tract? Account for this effect.

45. Where are most of the vitamins in the wheat kernel located?

V. Adaptations to Seasonal Change *(pages 621–624)*

A. General Remarks *(page 621)*

> *Focus:* A major adaptation to seasonally adverse environmental conditions was the evolution of dormant stages in the plant life cycle.

B. Dormancy and the Life Cycle *(pages 621–623)*

46. a. Describe the life cycle of an annual plant.

    b. What portion of an annual plant persists over the winter?

47. *Vocabulary:* What is a herbaceous plant?

48. a. Summarize the life cycle of a biennial plant. Emphasize the major accomplishment of the first year versus the second year.

b. The roots of biennials may be modified for what purpose?

49. a. *Vocabulary:* What are perennials?

b. Distinguish between herbaceous and woody perennials.

c. Identify the major advantage woody perennials have over other plants.

50. Name two adaptations of perennials to seasonally unfavorable environmental conditions.

51. a. *Vocabulary:* What is meant by the term deciduous when applied to plants?

b. What conditions are necessary for deciduous plants to persist? Explain your answer.

C. **Seed Dormancy** *(pages 623–624)*

52. Many seeds will not germinate until they have experienced a period of dormancy. Of what adaptive value is this trait?

53. What is the longest recorded time for seed dormancy?

54. a. The seed coat can maintain dormancy by one of two major mechanisms. Identify these two mechanisms.

b. Indicate the ways in which dormancy can be ended for each type of seed coat.

> *Focus:* **The mandatory dormancy of seeds apparently evolved fairly recently in the history of plants, as a response to the most recent Ice Age.**

55. In addition to the specific requirements a particular seed may need to break dormancy, what other condition is required if a seed is to germinate?

VI. **Summary** *(page 624):* Read the summary. If you are familiar with the essential features of the material presented there, you are ready to complete the section TESTING YOUR UNDERSTANDING.

TESTING YOUR UNDERSTANDING

After you have completed this examination, compare your answers with those in the section that follows.

1. According to the fossil record, plants invaded land approximately _____ years ago.
   a. 2 billion
   b. 1 billion
   c. 500 million
   d. 250 million
   e. 100 million

2. The most abundant plants on the earth today are the:
   a. gymnosperms.
   b. angiosperms.
   c. grasses.
   d. conifers.
   e. mosses.

3. Angiosperms are distinguished from all other plants by their:
   a. flowers.
   b. flattened leaves.
   c. alternation of generations.
   d. widespread distribution.
   e. mechanism of photosynthesis.

4. In angiosperms, pollen grains are produced within the:
   a. calyx.   d. stigma.
   b. carpel.   e. anther.
   c. filament.

5. The sticky surface to which pollen grains adhere is called the:
   a. filament.   d. ovule.
   b. stigma.   e. carpel.
   c. style.

6. Staminate flowers and carpellate flowers are _____. When these flowers occur on separate plants, the species is said to be _____.
   a. perfect; monoflorate
   b. imperfect; monoecious
   c. perfect; diflorate
   d. imperfect; dioecious
   e. imperfect; bisexual

7. What is the developmental fate of the ovary and ovules following fertilization of the egg in an angiosperm flower?
   a. The ovule develops into a seed and the ovary develops into a fruit.
   b. The ovule develops into a fruit and the ovary develops into a seed.
   c. Both structures disintegrate when no longer needed.
   d. The ovule develops into the embryo and the ovary becomes the seed coat.
   e. The ovary becomes the embryo of the next generation and the ovule disintegrates.

8. All of these fruits are commonly called nuts. Which one fits the botanical definition of a nut?
   a. almond   d. pine nut
   b. peanut   e. coconut
   c. acorn

9. Peaches and olives are examples of which type of fruit?
   a. aggregate fruit
   b. berry
   c. drupe
   d. multiple fruit
   e. pome

10. A pineapple is an example of a multiple fruit, which develops from:
    a. the fusion of carpels and stamens.
    b. a large flower with many ovaries.
    c. a large ovary with many ovules.
    d. a flower with an inferior ovary.
    e. an inflorescence.

11. As an angiosperm seed is maturing, the wall of the ovary develops into:
    a. a fruit.
    b. the seed coat.
    c. endosperm.
    d. the suspensor.
    e. an embryo sac.

12. Dormancy in plants permitted them to adapt to severe environmental conditions, including:
    a. high humidity.
    b. long summers.
    c. four seasonal changes per year.
    d. drought.
    e. flooding.

13. T or F The type and character of land plants determine the types and numbers of animals in a terrestrial environment.

14. T or F The female gametophyte of many angiosperms consists of seven cells with seven haploid nuclei, of which one is the egg.

15. T or F The annual replacement of leaves by deciduous plants is so expensive that this type of plant is found only in humid climates with lengthy growing seasons.

16. T or F Dormancy in seeds may be broken by changes in chemical inhibitors, triggered by environmental factors.

17. Describe the process of double fertilization. Include the fates of both resulting cell lines in your answer.

18. What are the distinguishing features of annual, biennial, and perennial plants?

PERFORMANCE ANALYSIS

1. **c** Plants are estimated to have invaded land approximately 500 million years ago. (page 613)

2. **b** The angiosperms are the dominant forms of terrestrial plant life on earth today. (page 613)

3. **a** Flowers distinguish the angiosperms from all other types of plants. (page 613)

4. **e** Pollen grains, the male gametophytes, are produced within the anthers of angiosperm flowers. (page 614)

5. **b** The sticky surface of the carpel upon which pollen grains land is the stigma. (page 614)

6. **d** Staminate flowers are those that have stamens but no carpels. Carpellate flowers have carpels but no stamens. Flowers that have either stamens or carpels, but not both, are imperfect. When the male and female flowers occur on different plants, the species is referred to as dioecious. When the flowers occur on the same plant, the species is monoecious. (page 614 and Figure 24–21, page 514)

7. **a** After fertilization, the ovule develops into a seed consisting of embryo, endosperm, and seed coat. The ovary develops into the fruit. (pages 614, 619)

8. **c** A nut is a single-seeded indehiscent fruit with a stony coat that is derived from a compound ovary. Examples are acorns and hazelnuts. Almonds and coconuts are drupes, peanuts are legumes, and pine nuts are conifer seeds, which are naked. (page 621)

9. **c** Peaches and olives are examples of drupes, which are

fruits with a stony inner wall that usually adheres tightly to the seed. (page 620)

10. **e** Multiple fruits develop from inflorescences, which are flower clusters. (page 620)

11. **a** The wall of an angiosperm ovary develops into the fruit. (page 619)

12. **d** The evolution of dormancy enabled plants to survive adverse environmental conditions, the most important of which was probably drought, including that caused by unavailability of water when frozen. (page 621)

13. **True** In a terrestrial environment, the plants are the single most important feature that determines the character of the environment and the kinds and numbers of animals that can live there. (page 613)

14. **False** The female gametophyte of angiosperms typically contains seven cells and eight haploid nuclei. (The large central cell contains two polar nuclei.) (page 617)

15. **False** Deciduous plants only exist in environments where the soil is fertile enough to provide the necessary nutrients and where growing seasons are long enough to make leaf-dropping energetically profitable. (page 623)

16. **True** Dormancy maintained by a seed coat acting as a mechanical barrier may be broken by the erosion of the seed coat by sand or soil, its partial digestion as it passes through an animal digestive tract, or its destruction by fire. Dormancy maintained by chemical inhibitors in a seed coat may be broken when those chemicals are washed away or when they undergo changes triggered by environmental factors, such as light or cold. (page 623)

17. During double fertilization, one sperm nucleus combines with the nucleus of the egg. The resulting zygote ($2n$) develops into the young sporophyte. The other sperm nucleus combines with the two polar nuclei to form a triploid ($3n$) cell. This cell undergoes mitosis to produce the nutritive endosperm. (page 617)

18. **Annual plants** are usually herbaceous (nonwoody) and the entire life cycle from seed to flower to seed occurs in one growing season. All vegetative parts of the plant die at the end of the growing season and only the seed persists. **Biennial plants** typically require two growing seasons to complete their life cycles. Food is stored during the first growing season. Flower, fruit, and seed production occur during the second growing season. The vegetative structures of **perennial plants** persist for years. Herbaceous perennials have modified underground structures that survive during unfavorable seasons. Woody perennials can increase in height and girth each growing season and have numerous adaptations for surviving during unfavorable seasons. (pages 621, 622)

CHAPTER **30**

The Plant Body and Its Development

MAJOR CONCEPTS

The three major regions of the vascular plant body are the roots, stems, and leaves. The roots anchor the plant and obtain water and minerals from the soil. The stems support the aboveground structures of the plant and contain tissues for transporting materials between the roots and the leaves. The leaves are the major sites of photosynthesis in most vascular plants.

The three major tissue systems in a plant are the dermal system, the vascular system, and the ground tissue. The dermal system includes cells that cover the outside of the plant body and produce a protective cuticle on the aboveground parts.

The vascular system is composed of xylem and phloem, the conducting tissues of the plant. Xylem transports water and dissolved minerals; phloem transports dissolved sugars and other organic molecules. The vascular tissues are embedded in the ground tissue, which is composed of parenchyma cells.

The structure of a leaf represents a compromise among three conflicting pressures: the need for maximum leaf exposure for photosynthesis, the need to conserve water, and the need to exchange gases. The photosynthetic cells of leaves are palisade parenchyma and spongy parenchyma, which together constitute the leaf mesophyll. The vascular bundles of leaves are called veins and are embedded in the mesophyll. The mesophyll is surrounded by one or more layers of epidermal cells, which secrete a waxy coating, the cuticle. Scattered through the epidermis are the stomata through which gas exchange occurs.

Leaves exhibit a great variety of adaptations and modifications reflecting the environmental conditions in which the plant grows.

Roots are adapted to obtain water and minerals from the soil. A central vascular bundle is surrounded by cortex in dicots and most monocots. Some monocot roots have vascular bundles arranged in a ring around a central pith. In order for water and materials to enter the vascular bundle, they must pass through the cell membranes of endodermal cells. Taproot systems are typical of dicots; fibrous root systems characterize monocots. Roots exhibit a variety of adaptations and modifications.

Stems are specialized to raise leaves to the light and to transport substances between the roots and the leaves. Although green stems are photosynthetic, the bulk of a stem is composed of vascular tissues embedded in ground tissue. The vascular bundles in dicot stems are arranged in a ring around a central pith. In monocot stems, the vascular bundles are scattered throughout the ground tissue. The ground tissue of stems may include collenchyma and sclerenchyma cells, which are specialized for support.

The vascular tissues of plants consist of conducting cells, supporting fibers, and parenchyma cells. In angiosperms, the conducting cells of xylem are tracheids and vessel members, both of which are dead at functional maturity. The conducting cells of phloem are sieve-tube members, which are alive at maturity.

Many plants have stems specialized for vegetative propagation, which results in the growth of a new plant that is genetically identical to the parent. Plants may be artificially propagated by means of stem cuttings or by grafting the stem of one plant onto the stem and root structure of another plant.

The growth of plants is analogous to the mobility of animals.

Primary growth of a plant involves differentiation of the three tissue systems, elongation of roots and shoots, and branching of roots and shoots. The growth region of a root tip has three zones: apical meristem, zone of elongation, and zone of differentiation. Branch roots arise from the pericycle. The growth regions of shoot tips have growth zones similar to those of the root tip. The apical meristems of shoots provide the meristematic tissue for all leaves, flowers, and branches.

Secondary growth occurs in woody dicots and involves an increase in the girth of stems and branches. Secondary growth is accomplished by the lateral meristems: vascular cambium and cork cambium, which exist in concentric rings within the stem. New (secondary) xylem is deposited to the interior and secondary phloem is formed to the exterior of the vascular cambium. Cork cambium produces cork, which is the outer protective layer and is dead at maturity. Secondary phloem and cork constitute bark.

The growth rings of a tree are formed of secondary xylem. Various environmental factors, including water availability and temperature, influence the width of the growth rings.

HOW TO STUDY THE CHAPTER

Read the entire chapter through quickly, focusing on the major concepts.

Use the GUIDED STUDY OF THE CHAPTER to help you identify the important details as you **reread** the chapter. Writing out the answers to these questions will help fix them in your mind as well as provide you with a valuable study aid.

Answer the questions in TESTING YOUR UNDERSTANDING without the aid of your text. Check your answers against those in PERFORMANCE ANALYSIS. Analyzing your answers will give you valuable feedback on your level of understanding and preparedness for classroom testing.

GUIDED STUDY OF THE CHAPTER

I. Introduction (page 625)

1. Name seven characteristics that modern plants have in common with the presumed ancestral green alga.

2. List the requirements of a plant's photosynthetic cells.

> *Focus:* **For most plants, the plant body provides the life-support system for the photosynthetic cells.**

3. a. Identify the three basic regions of the vascular plant body and indicate the function for which each region is specialized.

   b. How do the cells of a modern vascular plant differ from those of a multicellular green alga?

4. *Vocabulary:* Distinguish between nodes and internodes. (Figure 30–1)

5. What is a tuber? (Include the function for which it is specialized and the plant part from which it develops.) (Figure 30–1)

II. The Cells and Tissues of the Plant Body (pages 625–627)

6. Identify and describe the function of the tissue system that develops from each embryonic plant tissue.

   Protoderm

   Procambium

   Ground meristem

7. a. Describe the general structure of parenchymal cells.

b. Name three types of plastids that parenchymal cells may contain.

c. Identify three functions for which parenchymal cells may be specialized.

III. Leaves *(pages 627–630)*

A. Leaf Structure *(pages 627–629)*

8. Name three conflicting evolutionary pressures that have influenced leaf structure.

Focus: **Within the photosynthetic cells of a plant, the chloroplasts move about in the cytoplasm to achieve optimal exposure to the sun. (Figure 30–2)**

---

9. Fill in the following table concerning palisade and spongy parenchyma. (See also Figure 30–2.)

Table for Question 9

	PALISADE PARENCHYMA	SPONGY PARENCHYMA
Structure and packing of cells		
Location in leaf		
Function(s) of tissue type		

10. *Vocabulary:* Define the leaf mesophyll, including the cells of which it is composed.

11. *Vocabulary:* Distinguish between cutin and cuticle.

12. What feature of the epidermal cells and the cuticle enables light to reach the underlying photosynthetic cells?

13. Name the two structures through which substances move into and out of leaves.

14. a. *Vocabulary:* Define veins.

b. Use a sketch to distinguish between the pattern of veins in the leaves of monocots and dicots.

15. *Vocabulary:* What is a petiole?

16. a. Describe the structure of a stoma.

b. What substances pass through the stomata?

c. Of the water that is lost from the plant body, what fraction is lost through the stomata?

d. Where on a leaf are stomata most abundant?

17. The air spaces between the spongy parenchyma cells constitute what fraction of the total volume of the leaf? (Figure 30–4)

B. Leaf Adaptations and Modifications *(pages 629–630)*

18. Describe how the leaves of plants that grow on the floor of a tropical rain forest reflect adaptations to that environment. (Include two major environmental factors that influence the evolution of these plants.)

19. a. Describe three leaf adaptations for plants in dry environments.

b. *Vocabulary:* Distinguish between spines and thorns.

20. In addition to photosynthesis, identify three other functions for which plant leaves may be adapted, and cite examples for each function.

IV. Characteristics of Plant Growth *(pages 630–631)*

21. Name the three events that follow seed germination.

22. a. Name three processes involved in primary plant growth.

b. Where does all primary growth originate?

23. *Vocabulary:* What is the hypocotyl? (Figure 30–7)

24. How does the growth of plants differ from that of higher animals?

25. Explain the following statement, using examples: "Growth in plants is the counterpart . . . to motility in animals."

V. Roots *(pages 631–636)*

A. General Remarks *(page 631)*

26. Name two functions for which roots are specialized.

27. What structure is generally the first to break through the seed coat?

28. Identify two factors that affect root growth.

> *Focus:* **In a mature plant, the root system may constitute more than half the plant body, and the surface area of the root system may be many times greater than that of the leaves and stem.**

B. Root Structure *(pages 632–633)*

29. Name the three layers of a dicot root from outermost to innermost.

The Epidermis *(page 632)*

30. Cite two functions of the epidermis.

31. a. *Vocabulary:* What are root hairs and where do they form? (See also Figure 30–9)

b. Root hairs perform what function for the plant?

c. In many species, _____ apparently substitute for root hairs.

The Cortex (page 632)

32. a. The root parenchyma cells are specialized for what function?

b. The plastids of root parenchyma cells may contain what compounds?

33. How do root cells obtain the oxygen they need to conduct respiration?

34. a. Describe the structure and arrangement of endo-dermal cells.

b. What feature of endodermal cells prevents the direct movement of water into the vascular cylinder?

c. How do water and other materials enter the vas-cular cylinder?

Focus: **The movement of water and other sub-stances into the plant body is restricted by the Casparian strip of the endodermal cells. Materials enter the vascular cylinder only by passing through endodermal cell membranes, which therefore act as a regulator of substances entering the vascular tissues of the root.**

The Vascular Cylinder (page 633)

35. What is the pericycle and what structures arise from this tissue layer?

36. a. Diagram the arrangement of vascular tissues char-acteristic of dicot and most monocot roots.

b. Diagram the alternative arrangement of vascular tissues that occurs in the roots of some monocots.

c. What is the pith?

C. **Primary Growth of the Root** (*page 634*)

37. *Vocabulary:* Distinguish between radicle and root cap.

38. Identify three possible fates of the cells that descend from the root meristem.

39. What is the major cause of primary root growth?

40. Describe the way in which the various cells of the root differentiate.

41. a. From what tissue do branch roots arise? (Figure 30–13)

b. How are branch roots able to grow through the root cortex to break through the epidermis? (Fig-ure 30–13)

Focus: **The three growth regions in a plant root are the apical meristem, the zone of elongation, and the zone of differentiation. (Figure 30–14)**

D. **Patterns of Root Growth** (*pages 635–636*)

42. Fill in the following table.

Table for Question 42

	TAPROOT SYSTEM	FIBROUS ROOT SYSTEM
Origin		
Overall structure (a diagram may help)		
Monocot or dicot?		

43. a. *Vocabulary:* Distinguish between adventitious roots and aerial roots.

b. Name three functions of aerial roots. Identify an example for each type.

VI. Stems *(pages 636–640)*

A. General Remarks *(page 636)*

44. a. Identify the two main functions of stems.

    b. Name two additional functions for which stems may be adapted.

B. Stem Structure *(pages 636–640)*

45. In what four ways do green stems resemble leaves?

46. a. A young green stem is primarily composed of:

    b. What provides the chief support for young green stems?

47. a. *Vocabulary:* Distinguish between collenchyma and sclerenchyma.

    b. Where are collenchyma cells often located and what is their function?

48. a. *Vocabulary:* Distinguish between fibers and sclereids.

    b. Fibers are often associated with what plant tissues?

    c. Cite five examples of plant fibers.

    d. Where may sclereids be found?

    e. What feature of sclereids is responsible for their hardness?

49. Apart from the features already described, list three ways in which sclerenchyma cells differ from collenchyma cells.

Vascular Tissues (pages 638–639)

50. Name three general cell types of which the vascular tissues are composed.

51. What substances are transported in the phloem?

52. Identify the phloem conducting cells in gymnosperms and angiosperms.

53. *Vocabulary:* Distinguish between sieve tube, sieve-tube member, and sieve plate.

54. a. Name two substances that may be present in sieve-tube members.

    b. What function has been proposed for P-protein?

    c. What is callose and what is its function? (See also Figure 30–18.)

> *Focus:* **Sieve-tube members are alive at maturity but the nucleus and many of its organelles disintegrate during development.**

55. a. Name two functions of companion cells.

    b. Companion cells originate from what cells?

56. What substances are transported in the xylem?

57. Comment on the *direction* of transport in the phloem and xylem.

58. a. Tracheids and vessel members are similar in what two ways?

   b. Describe how tracheids and vessel members differ with respect to shape and the passage of water from one cell to the next.

59. What is a vessel in angiosperm vascular tissue?

60. Distinguish between the xylem conducting elements present in gymnosperms and those in angiosperms.

Stem Patterns (page 640)

61. Diagram the arrangement of vascular bundles in the stems of monocots and young dicots.

Focus: Note that the arrangement of vascular bundles in *young dicot stems* is similar to the arrangement in some *monocot roots.* (Compare Figures 30–20a and b with Figure 30–11b.)

---

62. What happens to the vascular bundles in the transition region of the plant axis?

VII. Primary Growth of the Shoot System *(pages 640–643)*

A. General Remarks *(pages 640–641)*

63. a. How does the pattern of shoot tip growth resemble the pattern of root tip growth?

   b. Identify two ways in which the development of shoots differs from that of roots.

Focus: The apical meristem of roots contributes only cells that lengthen the root. The apical meristem of shoots provides cells that will produce new leaves, branches, and flowers.

---

64. Epidermal hairs arise from what source? (Figure 30–21)

65. Describe the process by which a new leaf is formed. Use the terms primordia and apical meristem in your answer.

66. Describe two different patterns of leaf formation along the growing plant stem.

67. a. *Vocabulary:* What are axillary buds?

   b. Name two structures that may develop from axillary buds.

   c. What factors may influence the development of axillary buds?

68. A flower may form from what two types of buds?

69. What happens to the meristematic tissue from which a flower arises?

B. Modifications in the Pattern of Shoot Growth *(page 642)*

70. Primary shoot growth proceeds from what two regions?

71. a. Identify the two locations of shoot meristems in grass plants.

    b. This adaptation results from what selection pressure? (Figure 30–22)

> *Focus:* **The shoots of climbing plants are adapted for obtaining support from the structure upon which the plant is climbing.**

72. Tendrils are modified _____ or _____.

C. Vegetative Reproduction *(pages 642–643)*

73. a. *Vocabulary:* Distinguish between runners and rhizomes.

    b. In what two ways is reproduction by these two mechanisms similar?

    c. Cite examples of plants that reproduce by runners and rhizomes.

    d. Name three advantages of these reproductive mechanisms.

74. Describe the life history of creosote bushes from seed germination to clone formation. (Figure 30–25)

75. a. Describe two techniques by which plants may be propagated artificially.

    b. Identify six plants of economic value that can *only* be propagated vegetatively.

VIII. Secondary Growth *(pages 644–647)*

A. General Remarks *(pages 644–647)*

76. a. Which plants undergo secondary growth?

    b. What plant dimension increases during secondary growth?

77. a. Name the two meristematic tissues responsible for secondary growth.

    b. These tissues are collectively known as:

78. a. The vascular cambium is derived from what tissues in (1) stems and (2) roots?

    b. Describe the role of the vascular cambium in plant growth.

79. a. The cork cambium develops from what tissue?

    b. Name the tissue produced by cork cambium and identify its function.

    c. How does cork differ from the epidermis?

80. *Vocabulary:* What tissues constitute the bark?

Tidbit: If the bark is removed from a tree in a ring that completely circles the trunk, the tree will eventually die because of the interrupted transport of materials in the phloem.

81. Fill in the following table. (See also Figure 30–27.)

Table for Question 81

	HEARTWOOD	SAPWOOD
Composition		
Location		
Function		

82. What information can be derived from studying the width of growth rings?

83. Besides increasing its diameter, what other function is served by formation of new vascular tissue as a plant ages?

B. *Essay:* **The Record in the Rings** *(page 646)*

84. Account for the visibility of growth rings.

85. Name five factors that may influence the width of individual growth rings.

86. a. What do dendrochronologists study?

b. What type of trees have been particularly useful to these scientists?

Focus: Studies of the growth rings of living, dead, and fossilized woody plants are providing information that may enable scientists to identify cyclic patterns of climatic conditions.

87. What is the practical usefulness of the studies described in the Focus statement?

IX. **Summary** *(pages 647–648):* Read the summary. If you are familiar with the essential features of the material presented there, you are ready to complete the section TESTING YOUR UNDERSTANDING.

TESTING YOUR UNDERSTANDING

After you have completed this examination, compare your answers with those in the section that follows.

1. Of the three primary meristems, the SECOND one to differentiate in the embryonic plant is the:
   a. procambium.
   b. dermal tissue.
   c. protoderm.
   d. parenchyma.
   e. ground meristem.

2. The densely packed, columnar, photosynthetic cells that are specialized to trap light energy constitute the:
   a. palisade parenchyma.
   b. mesophyll.
   c. cuticle.
   d. sieve-tube members.
   e. spongy parenchyma.

3. The cuticle of a leaf is produced by the:
   a. mesophyll.
   b. epidermis.
   c. phloem.
   d. pallisade parenchyma.
   e. spongy parenchyma.

4. Mesophyll is the primary ground tissue of a leaf and consists of:
   a. parenchyma and epidermal cells.
   b. epidermis and cuticle.
   c. palisade parenchyma only.
   d. spongy parenchyma only.
   e. both palisade and spongy parenchyma.

5. Netted leaf venation is characteristic of:
   a. all angiosperms.
   b. all gymnosperms.
   c. dicots only.
   d. monocots only.
   e. both dicots and monocots.

6. Small openings in the leaf epidermis, the opening and closing of which are controlled by two specialized cells, are called _____ and are essential for _____.
   a. sclereids; photosynthesis
   b. stomata; gas exchange
   c. radicles; water absorption
   d. tracheids; water transport
   e. epidermal pores; water conservation

7. The bulb of an onion consists of _____ modified for _____.
   a. roots; water storage
   b. a stem; food storage
   c. roots; food storage
   d. leaves; food storage
   e. a stem; water storage

8. Which activity of plants is roughly equivalent to movement in animals?
   a. growth
   b. reproduction
   c. photosynthesis
   d. production of annual rings
   e. respiration

9. Root hairs function to:
   a. anchor the roots more firmly in the soil.
   b. increase the storage capacity of a root.
   c. increase the surface area of the root system.
   d. increase the effective length of a root.
   e. hold soil particles together, preventing erosion and loss of nutrients.

10. The Casparian strip is a(n) _____, which is a component of the _____ of the root.
   a. intracellular junction; epidermal cells
   b. continuous band of wax; vascular cylinder
   c. intercellular junction; cortical cells
   d. continuous band of wax; endodermal cells
   e. intercellular junction; epidermal cells

11. Branch roots arise from the _____, which is a layer of cells surrounding the _____.
   a. pericycle; vascular cylinder
   b. cortex; pith
   c. pericycle; endodermis
   d. cuticle; epidermis
   e. epidermis; cortex

12. In roots, the first tissue to differentiate as cells elongate is the:
   a. xylem.   d. endodermis.
   b. cortex.   e. pericycle.
   c. phloem.

13. Which statement is NOT true of the dermal tissue of young green stems?
   a. It is composed of epidermal cells.
   b. A waxy cuticle coats the surface.
   c. Stomata are scattered over the surface.
   d. It is protected by a thin, transparent, layer of cork.
   e. It is photosynthetic.

14. Which type of cells have a supportive function and are often associated with vascular tissues of stems?
   a. cambium cells   d. tracheids
   b. parenchyma cells   e. fibers
   c. collenchyma cells

15. Callose is a _____ found in _____ cells.
   a. protein; xylem
   b. protein; phloem
   c. carbohydrate; xylem
   d. carbohydrate; phloem
   e. lipid; phloem

16. Water passes from one tracheid to another through:
   a. stomata.
   b. pits.
   c. sieve plates.
   d. the secondary cell wall.
   e. openings in the primary cell walls.

17. The transition zone is the region in the:
   a. root in which cells differentiate into distinct tissue types.
   b. stem in which the stem pattern of vascular tissue changes to the root pattern.
   c. root in which cells undergo elongation.
   d. stem in which the xylem and phloem change directions.
   e. root in which cell division takes place.

18. Which root structure does NOT have an analogous structure in the stem?
   a. ground tissue   d. cuticle
   b. xylem   e. root cap
   c. phloem

19. Primary growth is the result of cell division in the _____, whereas secondary growth results from the activity of _____.
   a. cork cambium; vascular cambium
   b. vascular cambium; apical meristem
   c. apical meristem; cork and vascular cambiums
   d. vascular cambium and apical meristem; cork cambium
   e. apical meristem and cork cambium; vascular cambium

20. The tendrils of grape vines, English ivy, and Virginia creeper are all modified:
   a. thorns.
   b. spines.
   c. leaves.
   d. branches.
   e. adventitious roots.

21. Plants may reproduce vegetatively by structures that run along the surface of the ground, which are known as _____, or by structures that grow either along or just below the surface of the soil, which are known as _____.
   a. rhizomes; tubers
   b. tubers; rhizoids
   c. runners; rhizomes
   d. rhizomes; runners
   e. runners; rhizoids

22. The girth of a plant increases primarily because of the activities of what tissue(s)?
    a. cork cambium as it produces cork
    b. vascular cambium as it produces secondary xylem
    c. remnants of the apical meristem that were left behind as the plant grew
    d. both cork cambium and vascular cambium
    e. lateral meristems as they produce more parenchyma cells in the cortex

23. Which plant cell type or tissue is NOT dead at maturity?
    a. phloem companion cell
    b. xylem vessel member
    c. cork
    d. heartwood
    e. tracheid

24. Dendrochronologists have been able to construct a record of climatic patterns from the rings of fossilized, dead, and living trees that reaches back some _____ years.
    a. 2500   d. 6800
    b. 3600   e. 8200
    c. 4900

25. T or F   The algal ancestor of the plants depended upon sunlight for energy and obtained oxygen, carbon dioxide, and minerals from the water in which it lived.

26. T or F   In older plants, the root system may constitute more than half of the plant body.

27. T or F   Xylem transports water and minerals in the direction opposite to the pull of gravity, and phloem transports photosynthetic products toward the pull of gravity.

28. T or F   In all angiosperms, environmental conditions (e.g., day length) determine whether an axillary bud becomes a lateral branch or a flower.

29. Identify the primary function(s) of each major plant body part.

30. Describe the similarities and differences between the primary growth of stems and of roots.

31. How does the production of secondary xylem and phloem contribute to the overall lifespan of a woody tree?

PERFORMANCE ANALYSIS

1. **a**   Listed in the order in which they differentiate, the three primary meristems are protoderm, procambium, and ground meristem. Protoderm gives rise to the dermal tissue system. Procambium develops into the xylem and phloem of the vascular system. Ground meristem produces the ground tissues of the plant. (page 625)

2. **a**   The leaf mesophyll has two components: palisade parenchyma consists of densely packed, columnar, photosynthetic cells; spongy parenchyma is composed of loosely packed, irregular cells that have large air spaces between them. (page 627)

3. **b**   The cuticle of a leaf is produced by epidermal cells. (page 627)

4. **e**   See the answer to question 2. (page 627)

5. **c**   Dicots are characterized by netted leaf venation. Monocots typically have parallel leaf venation. (page 628)

6. **b**   Stomata are the openings in the leaf epidermis that allow gas exchange between the air spaces in the spongy mesophyll and the external atmosphere. The opening and closing of each stoma are regulated by its two guard cells. (page 628)

7. **d**   The bulb of an onion plant is composed of modified leaves that are specialized to store food. (page 630)

8. **a**   Plant growth is roughly analogous to movement in animals. Plant shoots grow to extend leaves toward a light source and facilitate seed dispersal. Roots grow to obtain nutrients and water. (page 630)

9. **c**   Root hairs are extensions of root epidermal cells. They greatly increase the surface area of the root, maximizing absorptive capacity. (page 632)

10. **d**   The Casparian strip is a continuous band of wax located within the cell wall of root endodermal cells. It prevents water and solutes from entering the root vascular system by passing between cells, forcing all materials to first pass through the endodermal cell membranes. (page 632)

11. **a**   Branch roots arise from the pericycle, which surrounds the vascular cylinder. The pericycle is adjacent to and internal to the endodermis. (page 633 and Figure 30–13)

12. **c**   As root cells elongate, the first cells to differentiate are the conducting cells of the phloem. (page 634)

13. **d**   The dermal tissue of young green stems is composed of epidermal cells and is covered by a waxy cuticle. Stomata are scattered over the surface and the cells are photosynthetic. Cork develops as a consequence of secondary growth and is not found on young green stems. (page 636)

14. **e**   There are two types of sclerenchyma cells, both of which have cell walls impregnated with lignin. Fibers are elongated, somewhat elastic cells often associated with the vascular tissues of stems. Sclereids are also common in stems, and form the hard outer coverings of nuts, seeds, and fruit stones. Sclerenchyma cells differ from collenchyma cells in that sclerenchyma cells have secondary cell walls, are often dead at maturity, and usually occur in regions of the plant body that have completed primary growth. (page 637)

15. **d**   Callose is a polysaccharide found in phloem cells. It is thought to function with P-protein in sealing sieve-plate pores when a plant is injured. (page 638)

16. **b**   Water passes from one tracheid to another through pits, which are thin areas of the cell that have no secondary cell wall. These pits are not perforations, and mate-

rials must pass through two cell walls and the middle lamella to pass from one tracheid to the next. (page 639)

17. **b** The transition region is the area of the stem in which the arrangement of the vascular system changes from the pattern in the stem to the pattern in the root. (page 640)

18. **e** Both stems and roots have ground tissue, xylem, and phloem. All stems and many roots have cuticle, although the cuticle of roots is much thinner than that of stems. The stem has no structure analogous to the root cap, which protects the apical meristem as the root pushes through the soil. (page 640)

19. **c** Primary growth is a consequence of apical meristematic activity. Secondary growth results from the activities of the lateral meristems: cork cambium and vascular cambium. (page 641, 644)

20. **d** The tendrils of grape vines, English ivy, and Virginia creeper are all modified branches. Remember that the tendrils of some plants may be modified leaves or leaflets. (page 642)

21. **c** Runners are vegetative structures that grow along the surface of the soil. Rhizomes may grow along or beneath the surface of the soil. Both structures originate from shoot axillary buds and are mechanisms of vegetative reproduction. (page 642)

22. **b** A plant increases in girth primarily by the deposition of secondary xylem by the vascular cambium. (page 644)

23. **a** Phloem companion cells are alive at maturity. However, xylem vessel members and xylem tracheids are dead at maturity. Cork, a component of bark which is produced by cork cambium, is also dead at maturity. Heartwood is the nonconducting xylem at the center of the trunk and major tree roots. (pages 639, 645)

24. **e** By studying growth rings of fossilized, dead, and living trees, dendrochronologists have constructed a record of climatic conditions that dates back 8200 years. (essay, page 646)

25. **True** The algal ancestor of plants was photosynthetic and obtained the oxygen, carbon dioxide, and minerals it needed from water. (page 625)

26. **True** The root system of an older plant may constitute more than half the mass of the plant body. Also, in mature trees, the root system typically extends beyond the reach of the crown of the tree. (page 631)

27. **False** Xylem transports water and minerals from the roots to all parts of the plant body. Similarly, phloem transports photosynthetic products from the photosynthetic cells to all parts of the plant body. Neither transport process is controlled by the pull of gravity. (page 639)

28. **False** The development of an axillary bud into a flower or a lateral branch may be influenced by environmental conditions and/or hormonal events, depending upon the species. (page 641)

29. Roots are adapted for anchoring the plant in the soil and for obtaining water and minerals. Leaves are adapted for photosynthesis (in most plants). Stems raise the leaves toward the light and contain vascular tissues that transport water, minerals, and photosynthetic products throughout the plant. In some plants, including cacti, stems may perform most of the photosynthesis. (page 625)

30. The primary growth of both shoots and roots involves the processes of cell division, cell elongation, and cell differentiation. The growth differs in two main ways. The growth zones of stems are not as distinct as those in roots. Stems have no structure analogous to the root cap, which protects the growing root as it pushes through the soil. (page 640)

31. The production of secondary xylem and secondary phloem by the vascular cambium of woody dicots not only increases their girth, it also renews the vascular system that may have been damaged by parasites and herbivores. (page 647)

Transport Processes in Plants

MAJOR CONCEPTS

Plants have two major transport systems: water and ions are transported in the xylem; and sugars are transported in the phloem.

Transpiration is the loss of water vapor from the plant body. More than 90 percent of the water entering a plant's roots is lost to the air, mostly through the stomata.

Nearly all of the water entering a plant is taken up by the roots by osmosis. The pressure created by the osmosis of water into roots is called root pressure and is sufficient to move water a short distance up the stem.

The main forces pulling water up plant stems are explained in the cohesion-tension theory. Water molecules are held together (cohesion) in continuous columns within the xylem by hydrogen bonds. The evaporation of water from stomata creates negative pressure (tension) within the xylem that pulls the columns of water up the stem.

Several factors affect the rate of transpiration, including temperature, humidity, air currents, and stomatal movements. The most important is stomatal movements.

Each stoma is surrounded by two guard cells that open and close the stoma in response to turgor changes within the cells. A stoma opens when potassium ions are transported into guard cells, increasing the osmotic potential within the cells. This draws water inside and causes the guard cells to bow open as turgor increases. Transport of potassium ions out of guard cells results in closure of the stoma. It has been proposed that energy for the active transport of potassium ions into and out of guard cells may come either from ATP or from the electrochemical gradient established when hydrogen ions move in the opposite direction from potassium ions.

The availability of water is the major factor influencing stomatal movements. Other factors include the hormone abscisic acid, carbon dioxide concentration, temperature, and light.

One adaptation to hot, dry climates is Crassulacean acid metabolism (CAM). CAM plants open their stomata at night and the carbon dioxide collected is converted into malic and isocitric acids. In a process analogous to the C_4 pathway, the carbon dioxide is released from these acids during the day for use in photosynthesis. CAM plants lose far less water per gram of carbon dioxide fixed than either C_3 or C_4 plants.

Minerals in solution are taken up by the roots and move through the plant in the xylem. Many minerals enter plants by active transport into root cells. Minerals are important in osmotic balance, as functional components of essential molecules, and in creating the ionic environment necessary for certain reactions to occur. Minerals are classified as macronutrients or micronutrients based on the amount present in the plant.

In translocation, sugars are transported in solution in the

phloem in a source-to-sink fashion explained by the pressure-flow hypothesis. At the source (photosynthetic or storage cells), sugars are actively transported into sieve tubes. Water moves into the phloem by osmosis, increasing the pressure. The solution moves by bulk flow down a sugar concentration gradient. At the sink, sugars move out of the phloem and water follows. Most of the water enters the xylem and is recirculated. The rate of sugar transport depends on the concentration gradient between source and sink.

Two major factors influencing plant nutrition are soil composition and symbiotic relationships.

The mineral content of a soil depends on the parent rock from which the soil was formed, biological factors relating to the interaction of plants and animals with the soil, the size of the soil particles, and soil pH.

Mycorrhizae are symbiotic relationships between fungi and the roots of plants. The fungi make soil nutrients available to the plant and the plants provide organic molecules to the fungi. In some cases, the fungi may also protect plants from toxic chemicals.

Atmospheric nitrogen is made available to living organisms during nitrogen fixation, which is performed mainly by prokaryotes. Symbiotic bacteria are the most important nitrogen fixers. Members of the genus *Rhizobium*, which are symbiotic with leguminous plants, are the most common nitrogen-fixing bacteria.

Researchers are currently attempting to endow economically important plants with the capacity to fix nitrogen. One approach is to use recombinant DNA techniques to incorporate the genes responsible for nitrogen fixation into these plants. An alternative approach is to provide the plants with the genes involved in forming a symbiotic relationship with nitrogen-fixing bacteria. Although some progress is being made, neither technique has yet produced the desired results.

Nitrogen may be lost from the soil by removal of plants (the major route), loss of topsoil (erosion), leaching by water percolation, and the breakdown of nitrates by soil bacteria in the absence of oxygen.

HOW TO STUDY THE CHAPTER

Read the entire chapter through quickly, focusing on the major concepts.

Use the GUIDED STUDY OF THE CHAPTER to help you identify the important details as you **reread** the chapter. Writing out the answers to these questions will help fix them in your mind as well as provide you with a valuable study aid.

Answer the questions in TESTING YOUR UNDERSTANDING without the aid of your text. Check your answers against those in PERFORMANCE ANALYSIS. Analyzing your answers will give you valuable feedback on your level of understanding and preparedness for classroom testing.

GUIDED STUDY OF THE CHAPTER

I. Introduction *(page 650)*

1. Identify the two problems whose solutions determined which plants would become the dominant land forms.

Review from Chapter 2 (page 42)

> *Focus:* **The hydrogen bonds that form between water molecules are responsible for several biologically important properties of water.**

2. Distinguish between cohesion and adhesion.

3. Define capillary action and imbibition.

Review from Chapter 6 (pages 128–134)

4. a. What is water potential?

   b. Name three factors that determine water potential.

5. Define bulk flow.

6. a. Osmosis is:

   b. Describe the relationship between water potential and osmosis.

   c. Distinguish osmotic potential from water potential.

7. Define turgor.

II. The Movement of Water and Minerals (pages 650–657)

A. Transpiration (page 650)

8. What fraction of the water taken up by a plant's roots is released into the atmosphere?

9. a. *Vocabulary:* Define transpiration.

   b. In addition to water loss, identify two accomplishments of transpiration. (Figure 31–1)

10. Loss of water to the atmosphere is an unavoidable consequence of what process?

11. For every gram of carbon dioxide that is fixed in organic compounds, how much water is lost from a C_3 plant? _____ a C_4 plant? _____

12. Explain the statement that a terrestrial plant is a "wick connecting the water reservoir of the soil with the atmosphere."

B. The Uptake of Water (pages 650–651)

13. Through what plant part does the majority of the water enter the body of most plants?

14. a. Name two mechanisms by which water moves through the soil to a region of water depletion around plant roots.

    b. In what other way can a plant obtain water beyond the region of water depletion?

15. a. *Vocabulary:* What is root pressure?

    b. How is root pressure created?

c. *Vocabulary:* Define guttation.

d. Guttation is a consequence of:

16. Cite two pieces of evidence that point toward transpiration as the driving force behind water movement up a plant body. (See also Figure 31–3.)

17. Why can suction force alone not be responsible for movement of water to the top of a 20-meter high tree?

C. The Cohesion-Tension Theory (page 651)

18. a. What is the tensil strength of a thin column of water?

    b. Name the force that creates this tensil strength.

19. a. Describe the mechanism by which water moves through xylem, according to the cohesion-tension theory.

    b. Water can be transported through (i) small vessels and (ii) large vessels at what rate?

    c. Cite two factors that prevent gas bubbles from forming and interrupting the water column in the vessels.

    d. What provides the power for the movement of water by this process?

D. **Factors Influencing Transpiration** *(pages 652–655)*

20. a. Name four factors that affect the rate of transpiration.

    b. Which factor is the most important?

    c. How do leaf hairs influence water loss from the plant?

The Mechanism of Stomatal Movements (page 653)

21. Diagram the structure of a stoma.

Focus: **Changes in the turgor of guard cells cause the stomata to open and close.**

---

22. a. Outline the mechanism by which the turgor in guard cells changes, causing the stomata to open and close. (See also Figure 31–6.)

    b. The movement of which ion is critical to the osmotic changes in the guard cells?

    c. Name two other ions that may be important in stomatal movements in some plants. (Indicate how they move relative to potassium ions.)

    d. What structural feature of guard cells makes them bow open when turgor increases? (Figure 31–6)

23. Identify two possible sources of the energy needed to transport the ions involved in stomatal movements.

Factors Influencing Stomatal Movements (pages 654–655)

24. Name the major factor influencing stomatal movements.

25. a. Identify one hormone that affects stomatal movements.

    b. Cite three lines of evidence confirming the influence of this hormone on stomatal movements.

    c. Describe the mechanism by which this hormone functions.

26. a. Summarize the influence of carbon dioxide concentration on stomatal movements.

    b. Where is the site that detects carbon dioxide concentration in the intercellular spaces?

27. a. By what mechanism are temperature changes thought to affect stomatal movements?

    b. Name two factors that cause some plants to routinely close their stomata at midday.

28. a. Describe the effect of blue light on stomata.

    b. This is thought to occur by what mechanism?

    c. What type of pigment is most likely involved in this process?

Crassulacean Acid Metabolism (page 655)

29. Species of plants that open their stomata only at night are adapted to what environmental conditions?

30. How is crassulacean acid metabolism similar to and different from the C_4 pathway? (Review Chapter 10, page 224, if necessary.)

31. For every gram of carbon dioxide that is fixed by a CAM plant, how much water is lost? (Compare this with your answer to question 11.)

E. **The Uptake of Minerals** *(pages 655–656)*

32. *Vocabulary:* Define a mineral.

33. Name the four elements that are most abundant in rocks and soil.

34. Which cells regulate the substances entering the vascular system of the root?

35. Cite two pieces of evidence that mineral ions enter plant cells by active transport.

36. Describe the function of carrier proteins in the transport of potassium ions across root cell membranes.

37. Name two possible functions of electric potentials in plants.

F. **Mineral Requirements of Plants** *(pages 656–657)*

38. a. Identify five functions of minerals in plants. (Distinguish between specific and nonspecific mineral requirements.)

    b. Cite two examples of minerals that are components of essential molecules.

39. *Vocabulary:* Distinguish between macronutrients and micronutrients.

40. a. How is the mineral content of a plant determined?

 b. What factors influence the composition of ash in a plant?

41. a. Summarize how growth tests are designed to determine mineral requirements for plants.

 b. These tests are sometimes complicated by what factor?

42. a. Name one mineral that is required by animals but is apparently not required by plants.

 b. Distinguish the principal osmoregulator in animals from that in plants.

43. Of the four most abundant minerals in the soil (see question 33), which is (are) (a) not required by any plants, and (b) required by only a few species?

G. *Essay:* Halophytes: A Future Resource? *(pages 658–659)*

 44. Cite three problems facing plants that grow in soils of high salinity.

45. a. Describe three major mechanisms by which halophytes adapt to living in saline soils.

 b. What must be present in the soil in order for the sodium-potassium pump of root cells to function?

 c. In halophytes that sequester salt in vacuoles, how does this influence water balance in the plant?

46. Describe how heavily irrigating land increases its salinity over time.

47. a. Why are halophytes currently being studied intensely?

 b. What are the most promising findings to date?

III. **The Movement of Sugars: Translocation** *(pages 657–662)*

A. **General Remarks** *(page 657)*

 48. *Vocabulary:* Define translocation.

B. **Evidence for the Phloem** *(page 660)*

 49. Early evidence that the phloem transports sugars was based on what observations?

50. Outline the technique of using radioactive tracers to study the role of phloem in sugar transport.

51. Describe the contribution of aphids to our understanding of the role of phloem and its contents. (Figure 31–11)

52. a. Describe the composition of sieve-tube sap.

b. How fast can sap move through the phloem?

C. *Essay:* **Radioactive Isotopes in Plant Research** *(page 661)*

53. a. Describe the means by which radioactive isotopes are incorporated into a plant.

b. Cite examples illustrating how the desired length of exposure time to radioactive material is determined.

54. a. Distinguish between whole-plant autoradiography and tissue radiography, focusing on technique and information gained.

b. Which technique reveals the exact location of radioactive materials in a plant?

55. In the study illustrated in this essay, radioactive carbon in carbon dioxide was incorporated into the plant and found in what location?

D. **The Pressure-Flow Hypothesis** *(page 662)*

56. a. Distinguish between sources and sinks as these terms apply to material transport in plants.

b. Explain how the same tissue may act as either a source or a sink under differing circumstances.

57. a. According to the pressure-flow hypothesis, what force is responsible for the bulk flow of solutions from source to sink in a plant body?

b. Cite two pieces of evidence that support this hypothesis.

c. By what mechanism do sugars move from photosynthetic cells into the sieve tubes?

> *Focus:* **As a consequence of the active transport of sugars into the sieve tubes, osmotic potential in the sieve tubes increases. Thus, water is drawn into the sieve tubes and the pressure within the phloem increases, forcing movement of the solution away from the source.**

58. a. What happens when the sugar solution arrives at a sink?

b. After leaving the phloem, what happens to (i) the sugars and (ii) the water?

c. What factor determines the rate of sugar transport from source to sink?

59. Name two functions of companion cells in sugar transport. (See also Figure 31–13.)

> *Focus:* Sugar enters the phloem by active transport, but the movement of the sugar solution in the sieve tubes is by bulk flow, which does not require energy.

IV. Factors Influencing Plant Nutrition *(pages 662–669)*

A. General Remarks *(page 662)*

60. Identify two factors that influence the availability of minerals to plants.

B. Soil Composition *(pages 663–664)*

61. a. *Vocabulary:* Define soil and describe its composition in general terms.

b. Name and distinguish among the three layers of soil.

> *Focus:* The depth and composition of the three soil horizons vary dramatically in different environments.

62. Fill in the following table. (Figure 31–14)

Table for Question 62

	NORTHERN CONIFEROUS FOREST SOIL	TEMPERATE DECIDIOUS FOREST SOIL	GRASSLAND SOIL
Relative amount of organic matter			
Acidity (if known)			
Nutrient content			

63. Discuss the role of the following four factors in determining the mineral content of a soil

a. Parent rock

b. Biological factors (Cite examples.)

c. Size of soil particles (Distinguish between sand, silt, and clay.)

Name two characteristics of clay particles that contribute to plant growth.

Why is a pure clay soil poorly suited for plant growth?

Vocabulary: What are loams?

What is their significance?

d. Soil pH

Cite examples of plants that grow best in (i) alkaline and (ii) acidic soils.

64. Identify two ways in which plants contribute to the soil.

> *Focus:* Under natural conditions, the composition and character of the soil, and the size, number, and kinds of plants growing in the soil, are constantly changing.

C. *Essay:* Carnivorous Plants *(pages 666–667)*

65. a. Carnivorous plants obtain what types of nutrients from their prey?

b. In what type of environment are carnivorous plants typically found?

66. Describe the means by which the following plants attract, trap, and digest their prey.

Sundew

Pitcher plant

Venus flytrap

D. The Role of Symbioses *(pages 664–669)*

Mycorrhizae (page 664)

67. In mycorrhizae, name two ways in which the fungus contributes to plant growth.

Rhizobia and Nitrogen Fixation (pages 664–665)

68. a. Plants have the greatest requirement for what nutrient?

b. Plants obtain nitrogen from the soil in what two forms?

69. Briefly outline how nitrogen "cycles" between soil nitrates, plant substance, and soil nitrates.

70. Identify five routes by which nitrogen may be lost from the soil. (Indicate the major route of loss.)

Focus: Atmospheric nitrogen is made available to plants by the process of nitrogen fixation.

---

71. a. What type of organism performs most of the nitrogen fixation and which specific group is the most important?

b. Name the most common genus of nitrogen-fixing bacteria.

72. How are legumes used to enrich the soil in agriculture?

The Symbiotic Relationship (pages 666–668)

Focus: A *Rhizobium* species is typically specific for a particular legume host.

---

73. By what mechanism are *Rhizobium* bacteria thought to recognize the correct host?

74. Describe the process by which Rhizobia enter plant roots and then root cells. (See also Figure 31–18.)

75. What does (a) the host plant and (b) the bacterium gain from the symbiotic relationship? (Figure 31–18)

76. a. Summarize the role of nitrogenase in nitrogen fixation.

b. What provides the energy to break the triple bond of the nitrogen molecule?

c. State the fate of the ammonia formed from the action of nitrogenase.

77. a. Sweet fern forms a symbiotic relationship with what nitrogen-fixing organism?

b. Because of this relationship, sweet fern can occupy what type of habitat?

78. a. Where do the nitrogen-fixing bacteria reside that are symbiotic with some grasses, such as sorghum?

b. Name one type of bacterium involved in this type of relationship.

Recombinant DNA and Nitrogen Fixation (page 669)

79. a. Summarize the process by which the capacity to fix nitrogen was transferred to *Escherichia coli* cells.

b. What is the status of similar studies in eukaryotes?

80. a. The most direct way to produce nitrogen-fixing plants would be to insert the genes for proteins involved in nitrogen fixation into plants. A less direct approach would be to:

b. Cite two reasons why many biologists believe this alternative approach is many years from achievement.

81. Of the total energy expended in producing an agricultural crop in the United States, what fraction is expended on chemically fertilizing the soil?

Focus: **At present, all dominant forms of life on earth depend either directly or indirectly on biological nitrogen fixation for the nitrogen needed in organic molecules.**

---

82. *Review Question* (Table 31–1): Match each macronutrient with the correct function(s). Use each answer only once. Each blank will have only one answer.

1. calcium
2. magnesium
3. nitrogen
4. phosphorus
5. potassium
6. sulfur

a. _____ component of ATP, phospholipids, nucleic acids

b. _____ component of cell walls; enzyme cofactor; necessary for cell membrane permeability and transport

c. _____ part of the chlorophyll molecule

d. _____ found in some amino acids, proteins, and coenzyme A

e. _____ functions in osmosis and ionic balance and in stomatal movements

f. _____ component of amino acids, nucleotides, chlorophyll, and coenzymes

83. *Review Question* (Table 31–1): Match each micronutrient with its function(s). Answers may be used more than once. A blank may have more than one answer, as indicated in parentheses.

1. boron
2. chlorine
3. cobalt
4. copper
5. iron
6. manganese
7. molybdenum
8. sodium
9. zinc

a. _____ required by many nitrogen-fixing microorganisms (3 answers)

b. _____ activator or component of enzymes (3 answers)

c. _____ influences calcium ion utilization, nucleic acid synthesis, and membrane integrity

d. _____ functions in osmosis and ionic balance; essential to oxygen-producing reactions of photosynthesis

e. _____ necessary for chloroplast development; component of cytochromes and nitrogenase

f. _____ required by some desert and salt-marsh species

g. _____ essential for nitrogen metabolism

h. _____ required for chloroplast membrane integrity and oxygen release

V. **Summary** *(pages 669–670)*: Read the summary. If you are familiar with the essential features of the material presented there, you are ready to complete the section TESTING YOUR UNDERSTANDING.

TESTING YOUR UNDERSTANDING

After you have completed this examination, compare your answers with those in the section that follows.

1. British ecologist John L. Harper was justified in describing a terrestrial plant as a "wick connecting the water reservoir of the soil with the atmosphere" because a plant such as corn requires _____ liters of water during its life from seed to harvest.
   a. 10 to 40
   b. 40 to 80
   c. 80 to 160
   d. 160 to 200
   e. 200 to 240

2. The driving force of root pressure is the:
   a. suction created by transpiration.
   b. negative pressure caused by guttation.
   c. active transport of water in root epidermal cells.
   d. osmosis of water from the soil into the roots.
   e. diffusion of water from phloem into xylem.

3. Simple suction, coupled with atmospheric pressure, is insufficient to move water through xylem to the tops of tall trees; even under a vacuum, atmospheric pressure can only push water up:
   a. 2 meters at sea level.
   b. 10 meters at sea level.
   c. 10 meters at an altitude of 2 kilometers (e.g., in Denver, Colorado).
   d. 20 meters at an altitude of 2 kilometers (e.g., in Denver, Colorado).
   e. Since atmospheric pressure is exerted on the entire tree, it cannot move water in the tree at all.

4. The technical term for the negative pressure that develops in xylem vessels as water evaporates from a plant is:
   a. cohesion.   d. tension.
   b. transpiration.   e. water stress.
   c. root pressure.

5. In a C_3 plant, for every gram of carbon dioxide incorporated into organic molecules, how much water is lost by transpiration?
   a. 4 to 5 grams   d. 4 to 5 liters
   b. 40 to 50 grams   e. 40 to 50 liters
   c. 400 to 500 grams

6. Which factor would have the LEAST effect on the opening and closing of stomata in the leaf of a plant at 25°C?
   a. a temperature increase of 5°C
   b. sunrise
   c. nightfall
   d. a brisk, steady breeze
   e. an increase in the carbon dioxide concentration within the leaf

7. When isolated guard cells are incubated in a potassium-containing medium, the cells swell if they are exposed to blue light. Which factor is NOT involved in this phenomenon?
   a. uptake of water
   b. stimulation of a blue-absorbing pigment
   c. transport of hydrogen ions out of the cells
   d. uptake of potassium ions
   e. activation of a nuclear gene by a blue-absorbing pigment

8. In CAM plants, the carbon dioxide is temporarily stored as:
   a. malic and citric acids.
   b. isocitric and malic acids.
   c. crassulacic acid.
   d. glucose and glyceraldehyde phosphate.
   e. malic acid and glucose.

9. As a consequence of the low permeability of root cell membranes to potassium ions and of the continuous action of the potassium pump, all of these conditions occur EXCEPT one. Identify the exception.
   a. A sodium ion concentration gradient is created across the membrane.
   b. An osmotic potential is established across the membrane.
   c. An electrical potential is created across the membrane.
   d. A potassium ion concentration gradient develops across the membrane.
   e. Carrier proteins transport potassium ions across the cell membrane.

10. In order to regulate water balance, plants have a specific requirement for:
    a. potassium.
    b. sodium.
    c. calcium.
    d. chloride.
    e. This ion requirement is nonspecific since several ions may contribute to water balance in plants.

11. Removing the bark from a tree in a band that completely circles the tree will:
    a. cut off the supply of water to the leaves.
    b. prevent the transport of sugar to the roots.
    c. leave the tree vulnerable to woodpeckers.
    d. release the tree from apical dominance.
    e. not harm the tree in any way if the strip of removed bark is fairly narrow.

12. The rate of sugar movement through the phloem has been recorded to be as fast as _____ centimeters per hour.
    a. 5
    b. 10
    c. 25
    d. 50
    e. 100

13. In the leaf, sugar moves from mesophyll cells into sieve tubes by:
    a. cotransport with hydrogen ions.
    b. cotransport with potassium ions.
    c. simple diffusion.
    d. facilitated diffusion.
    e. a process in which sugar follows water passively.

14. What is the fate of agricultural land from which crops are repeatedly harvested?
    a. It can bear crops indefinitely because the roots of each crop provide nutrients for the next crop.
    b. It will need fertilization if it is to remain productive.
    c. All the topsoil will eventually be lost by erosion.
    d. Even with fertilizers and the best management techniques, it will eventually become worthless.
    e. The humus will eventually be depleted and the soil will become infertile clay.

15. Clay contributes to the fertility of a soil because of its capacity to hold:
    a. water only.
    b. potassium ions only.
    c. magnesium ions only.
    d. water, and magnesium and calcium ions.
    e. water, and magnesium, calcium, and potassium ions.

16. As the pH of a soil increases (the soil becomes more basic), an element that becomes more available to plants is _____, and an element that becomes less available is _____.
    a. zinc; calcium
    b. potassium; zinc
    c. calcium; iron
    d. iron; potassium
    e. potassium; magnesium

17. The organism that inhabits the nodules of leguminous plants and converts atmospheric nitrogen to a form usable to the plant is a species of:
    a. mycorrhizal fungus.
    b. *Nitrobacter.*
    c. *Rhizobium.*
    d. *Nitrosomonas.*
    e. *Azospirillum.*

18. The nitrogen in the ammonia produced by the nitrogenase reaction is combined with organic compounds to produce:
    a. nitrates.   d. uric acid.
    b. nitrites.   e. urea.
    c. amino acids.

19. An angiosperm that may effectively colonize barren land is _____, whose roots are infected by the moldlike _____.
    a. sweet fern; actinomycetes
    b. sweet fern; azospirilla
    c. sweet clover; actinomycetes
    d. sorghum; azospirilla
    e. alfalfa; rhizobia

20. T or F   The movement of water through a plant body is largely a consequence of the unique properties of water.

21. T or F   When water supply is inadequate, stomata close because the leaf begins to wilt.

22. T or F   Because mineral elements are required by plants in such small quantities, the effects of mineral deficiencies are highly specific, and affect only one or a small number of functions.

23. T or F   Once sugar enters phloem sieve tubes, no metabolic energy is required for its transport to the roots.

24. T or F   The gene cluster controlling nitrogen fixation has been successfully transferred to *Escherichia coli* cells, which were then capable of synthesizing nitrogenase and fixing nitrogen.

25. Name the hormone that enables a plant to regulate stomatal movements and cite three lines of evidence supporting this role. Identify the mechanism of action of this hormone.

26. Cite two pieces of experimental evidence indicating that mineral uptake is a process supported by active transport.

27. Describe at least three ways in which halophytes cope with excess sodium.

28. Briefly outline the experimental technique that elucidated the role of phloem in plants.

PERFORMANCE ANALYSIS

1. **d**   A single corn plant requires 160 to 200 liters of water during its life from seed to harvest. This high water requirement results from evaporation (transpiration) of water from the leaves, which is the main force driving water movement from roots to leaves. (page 650)

2. **d**   The driving force of root pressure is the osmosis of water from the soil into roots. Root pressure is only strong enough to move water a short distance up plant stems. (page 651)

3. **b**   Suction will remove air from a system, allowing water to be forced up by atmospheric pressure. Under a vacuum at sea level, atmospheric pressure is only great enough to raise water 10 meters, and many trees are taller than 10 meters. (page 651)

4. **d**   Tension is the technical term for the negative pressure that develops in xylem subsequent to the evaporation of water from leaves. (page 651)

5. **c**   For each gram of carbon dioxide incorporated into organic molecules, a C_3 plant loses 400–500 grams of water, a C_4 plant loses 250–300 grams of water, but a CAM plant loses only 50–100 grams of water. (pages 650, 655)

6. **a**   Under moderate temperature conditions, temperature has little or no effect on stomatal movements. Above 35°C, temperature begins to influence stomatal movements. (page 654)

7. **e**   When blue light strikes a blue-absorbing pigment in the guard cells, hydrogen ions are transported out of the cells. Subsequently, potassium ions are transported into guard cells, creating the osmotic potential that draws water into the cells. The blue-absorbing pigment must be yellow in color and is thought to be a flavin. (page 654)

8. **b**   In CAM plants, carbon dioxide is taken into the plant at night when the stomata are open and is temporarily stored as isocitric and malic acids. When light is available, the carbon dioxide is released from these molecules and made available for photosynthesis. (page 655)

9. **a**   Root cell membranes are impermeable to most substances, including potassium. Root cells accumulate

potassium through the activities of transport proteins that function as potassium pumps. As a consequence of this active transport, a potassium ion concentration gradient, an electrical potential, and an osmotic potential are established across the cell membrane. (page 655)

10. **e** Although potassium is the major osmoregulator of plants, several ions may function in the regulation of water balance. Therefore, this ion requirement is nonspecific. (page 656)

11. **b** If a strip of bark that completely circles a tree is removed, the transport of photosynthetic products to the roots ceases and the tree will eventually die. (page 660)

12. **e** The rate of sugar movement through the phloem is a consequence of the sugar concentration gradient between source and sink. The greater the concentration difference, the faster the transport. Rates up to 100 centimeters per hour have been recorded. (page 660)

13. **a** Sugar moves against a concentration gradient from leaf mesophyll cells into sieve tubes by cotransport with hydrogen ions. This is performed by specific membrane transport proteins in the sieve-tube cell membrane. (page 662)

14. **b** The repeated harvesting of agricultural products from the same plot of land necessitates the use of fertilizers to replace soil nutrients. In an uncultivated, natural environment, nutrients contained in the plant matter are returned to the soil. (page 663)

15. **e** Clay contributes to soil fertility by holding water and positively charged ions such as magnesium, calcium, and potassium. However, a soil that is all clay becomes packed so tightly that oxygen cannot reach plant roots. (page 664)

16. **c** The pH of a soil affects the availability of ions for plant assimilation. Only soluble ions are available to plant roots. As the pH increases, calcium ions become more available to plants (i.e., more soluble) but iron becomes less available (i.e., less soluble). (page 664)

17. **c** Legumes are plants characterized by a symbiotic relationship between the plant roots and bacteria of the genus *Rhizobium*. The bacteria make atmospheric nitrogen available to the plant and in turn obtain their energy from carbohydrates released from plant roots. (page 665)

18. **c** The nitrogen from the ammonia formed during nitrogen fixation eventually combines with organic molecules in a plant to produce amino acids. (page 668)

19. **a** Sweet ferns are angiosperms that can colonize barren land. Their roots are infected with actinomycetes, funguslike bacteria that fix atmospheric nitrogen. (page 668)

20. **True** The movement of water through a plant body is largely a consequence of the hydrogen bonding of water molecules. This hydrogen bonding results in cohesion among water molecules and adhesion of water molecules to solid surfaces—two forces involved in capillary action. (page 650)

21. **False** When the water available to a leaf drops below a certain critical level, stomata close *before* a leaf begins to wilt. (page 654)

22. **False** Since minerals are involved in many fundamental plant processes, mineral deficiencies in plants typically have wide-ranging effects on plant growth and development. (page 656)

23. **True** Sugars enter sieve tube members by active transport but travel in the phloem by bulk flow, which requires no metabolic energy. (page 662)

24. **True** The gene cluster controlling nitrogen fixation from a free-living nitrogen-fixing bacterium has been successfully incorporated into *E. coli* cells, which subsequently synthesized nitrogenase and fixed nitrogen. To date, similar transfers into eukaryotic cells have been unsuccessful. (page 669)

25. (1) Abscisic acid levels increase dramatically during the initial stages of water stress. (2) Direct application of abscisic acid causes stomatal closure within minutes. (3) Individuals of the "wilty mutant" strain of tomato plants produce little abscisic acid and are unable to close stomata when exposed to water stress. Abscisic acid influences stomatal movements by binding to receptors in guard cell membranes, triggering a series of reactions that increase the permeability of the membrane to potassium ions. (page 654)

26. (1) Ion uptake is greatly reduced in roots in which cellular respiration has been blocked. (2) Plants that are deprived of light stop absorbing ions and eventually release them back to the soil. (page 655)

27. (1) In many halophytes, a sodium-potassium pump maintains low intracellular sodium concentrations while ensuring adequate levels of potassium. (2) Sodium may be sequestered in the plant away from the cytoplasm of the plant body. Sequestration may occur in intracellular vacuoles or in intercellular spaces. Salt in intercellular spaces may be secreted from the plant. (3) Palmer's grass plants have specialized cells that exude salt onto the surfaces of leaves. (4) In saltbush, salt is concentrated by special glands and pumped into bladders, which eventually burst. (essay, pages 658, 659)

28. Plants were grown in an atmosphere containing radioactive carbon dioxide (tagged with ^{14}C). Radioactive photosynthetic products were subsequently identified in the phloem using tissue autoradiography. In tissue autoradiography, plant tissue sections are mounted on microscope slides and exposed to photographic film. The processed film is then compared with the microscope slide to determine the exact location of the radioactivity in the plant. (essay, page 661)

Plant Responses and the Regulation of Growth

MAJOR CONCEPTS

A tropism is the growth of a plant part in response to a stimulus. If the plant grows toward the stimulus, it is a positive tropism. If it grows away from the stimulus, the tropism is negative. Plants exhibit tropisms in response to light and gravity. These responses are of great adaptive value.

Plant growth and activities are mediated by the interactions of hormones with each other and with other chemicals. The major classes of plant hormones are auxins, cytokinins, ethylene, abscisic acid, and gibberellins. Oligosaccharins are short carbohydrate chains that also function as plant hormones. Calcium ions and calmodulin are two substances that interact with hormones.

Auxins are produced primarily in apical meristems of shoots. Their effects include the stimulation of cell elongation, ethylene synthesis, fruit development, and adventitious root development, and inhibition of leaf abscission prior to abscission layer formation. Auxins play important roles in phototropism, gravitropism, apical dominance, and vascular differentiation.

Cytokinins stimulate cell division and are found in actively dividing tissues. They reverse apical dominance, delay leaf senescence, and are involved in shoot growth and fruit development. Cytokinins and auxins interact in the differentiation of plant tissues.

Ethylene is an unusual growth regulator in that it is a gas. Auxins stimulate ethylene production, and many auxin effects are now thought to be a consequence of ethylene release. Ethylene stimulates fruit ripening and leaf and flower senescence and abscission. It may also function in apical dominance.

Abscisic acid is considered the stress hormone since it prepares the plant for adverse environmental conditions. It stimulates stomatal closure and maintains seed and bud dormancy.

Gibberellins stimulate shoot elongation and can function with auxins to induce cellular differentiation. They also stimulate production of hydrolytic enzymes in germinating seeds.

Oligosaccharins are oligosaccharide molecules released from plant cell walls in response to different stimuli, including plant hormones. Current evidence indicates that oligosaccharins may be the actual regulators of physiological processes.

Gravitropism is the curving of shoots away from gravity and the curving of roots toward gravity. In roots and in monocot shoots, this growth is mediated by auxin and calcium ion concentration gradients established in a horizontal root or shoot.

Also, the inner cells of the root cap have amyloplasts that function as statoliths and enable the root to detect gravity.

Photoperiodism refers to the measuring of relative light and dark periods by plants. It is a mechanism by which plants anticipate changing seasons. The production of flowers by many plants is also influenced by photoperiods. Short-day plants require a light period shorter than a critical length whereas long-day plants require a light period longer than a critical interval. In actuality, the plant measures the length of the *dark* period rather than the light.

Circumstantial evidence supports the existence of a chemical produced in leaves that acts on apical meristems and stimulates flower production. However, this substance has not yet been isolated.

The chemical mediator of photoperiodism is a phytochrome that exists in two forms. P_r is converted to P_{fr} in the presence of red light. P_{fr} is converted back to P_r by slow degradation in the dark or by the presence of far-red light. P_{fr} is the biologically active form that stimulates flowering in long-day plants but inhibits flowering in short-day plants. Phytochrome is also important in seed germination and the early development of seedlings.

Circadian rhythms are activities that occur on a regular cycle independent of external influences. These rhythms are regulated by an internal biological clock, the exact mechanism of which is unknown. Circadian rhythms may be modulated by external factors, but if the external environment differs too greatly from the biological clock, the internal rhythm prevails. The primary function of biological clocks is to enable an organism to recognize seasonal changes by comparing the constant internal rhythms with changing external patterns.

Plants exhibit a variety of responses to touch. In general, the regular application of tactile stimuli to stems results in the growth of shorter, stockier plants. More specialized touch responses include the twining and coiling responses of climbing plants, the brisk leaf folding of the sensitive plant, and the rapid movements of carnivorous plants in trapping prey. These responses involve electrical signals and/or changes in cell membrane permeability.

Many plant species produce toxic or noxious chemicals when they are damaged by predators. Current evidence suggests that damaged plants release airborne chemicals that stimulate neighboring undamaged plants to produce protective chemicals.

HOW TO STUDY THE CHAPTER

Read the entire chapter through quickly, focusing on the major concepts.

Use the GUIDED STUDY OF THE CHAPTER to help you identify the important details as you **reread** the chapter. Writ-

ing out the answers to these questions will help fix them in your mind as well as provide you with a valuable study aid.

Answer the questions in TESTING YOUR UNDERSTANDING without the aid of your text. Check your answers against those in PERFORMANCE ANALYSIS. Analyzing your answers will give you valuable feedback on your level of understanding and preparedness for classroom testing.

GUIDED STUDY OF THE CHAPTER

I. Introduction *(page 671)*

> *Focus:* **In plants, growth accomplishes many of the same purposes served by motility in animals.**

1. a. Name five environmental factors that affect the rate of plant growth and differentiation.

   b. Growth rates affect what two plant characteristics?

> *Focus:* **Plants have mechanisms by which they anticipate and prepare for environmental changes.**

2. a. Regulation of physiological events in plants is influenced by what two types of factors?

   b. What are the major internal factors?

II. Phototropism and the Discovery of Plant Hormones *(pages 671–673)*

3. a. *Vocabulary:* Define tropism.

   b. Distinguish between positive and negative tropisms.

   c. "Plants exhibit positive phototropism." Explain this statement.

4. a. Summarize the phototropism experiments of Charles and Francis Darwin.

b. What did they conclude from their results?

5. *Vocabulary:* Distinguish between the botanical definitions of bending and curvature. (See footnote, page 672.)

6. a. Outline the steps by which Frits Went isolated the substance that caused plants to bend toward the light. Summarize his results.

b. How did he rule out the possibility that his results were due to the presence of an agar block rather than something in the agar block?

c. What did Went conclude from his experiments?

d. *Vocabulary:* What is auxin?

7. a. Describe the mechanism by which plants respond to and grow toward light.

b. What wavelength(s) of light will stimulate this event?

c. The pigment involved in phototropism is a:

III. Hormones and the Regulation of Plant Growth *(pages 673–682)*

A. General Remarks *(page 673)*

8. *Vocabulary:* Define a hormone.

> *Focus:* **Hormones are typically active in small quantities.**

9. Identify four factors that influence a plant's response to a hormone.

10. a. Name the five major types of plant hormones.

b. What other type of compound may also have hormonal action in plants?

> *Focus:* **The ultimate response of a plant to any stimulus results from the interaction of its hormones rather than the influence of just one chemical regulator.**

B. Auxins *(pages 673–675)*

11. a. The auxin isolated by Went is the chemical:

b. Where is IAA synthesized in a plant?

c. Name the organic molecule from which IAA is manufactured.

12. Summarize the effects of IAA on shoots and roots. Include in your discussion the effects of increasing concentrations of IAA.

13. a. What characteristic of synthetic auxins makes them useful commercially?

    b. Name two commercial uses of synthetic auxins.

Mechanism of Action of Auxin (pages 674–675)

14. a. The stimulatory effects of auxin result from what phenomenon?

    b. Outline the events that occur during acid growth. (Include the role of auxin in your answer.)

15. a. List three events stimulated by auxin in long-term growth.

    b. The "new" proteins synthesized in response to auxin probably have what function?

16. a. Cite two pieces of evidence that calcium ions are involved in auxin function.

    b. Summarize the proposed link between auxin, calcium ions, and calmodulin.

    c. Name three known effects of calmodulin in eukaryotes.

Apical Dominance and Other Auxin Effects (page 675)

17. a. *Vocabulary:* Define apical dominance.

    b. Describe one way in which you could demonstrate the role of auxin in apical dominance.

    c. Discuss the mechanism by which auxin maintains apical dominance.

18. What is the role of auxin in the seasonal growth of woody plants?

19. a. Other than the apical meristems, identify four sites of auxin production.

    b. Name two effects of auxin in fruit formation.

C. Cytokinins (pages 676–679)

20. a. What is one effect of cytokinins on plant cells?

b. Where in the plant are cytokinins found in the highest concentrations?

c. Cytokinins resemble what organic compound?

d. Name the most active naturally occurring cytokinin.

21. a. When in the cell cycle does cytokinin exert its effects?

b. How and by what mechanism do cytokinins influence protein synthesis?

Responses to Cytokinin and Auxin Combinations (pages 676–677)

22. a. Name the two possible fates of an undifferentiated plant cell.

b. Indicate the outcome of each developmental alternative.

23. a. How do the following hormones influence cell division and differentiation when added to tissue cultures?

Auxin alone

Kinetin alone

Auxin plus kinetin

b. What relative concentrations of auxin and kinetin can produce each structure?

Callus

Roots

Buds

24. a. Describe the effect of calcium ions on tissue cultures treated with auxin and cytokinin.

b. What mechanism is thought to cause this effect?

Focus: **The effects of hormones are modified by other hormones and by nonhormonal substances such as calcium ions.**

D. *Essay:* Plants in Test Tubes (page 678)

25. How can meristematic tissue be manipulated to produce a perfect plant?

26. a. *Vocabulary:* What are protoplasts?

b. Name two uses of protoplasts.

c. Describe how protoplasts may be used to create new plants from unrelated parents.

d. Describe one example of protoplast fusion that is of commercial interest.

e. Identify three features that may be incorporated into plants by creating hybrids from protoplasts.

Other Cytokinin Effects (pages 677–679)

27. a. Describe how cytokinins influence auxin-induced apical dominance in a growing plant.

    b. Where are cytokinins produced?

28. a. Describe the role of cytokinins in leaf senescence.

    b. Summarize the proposed mechanism by which cytokinins are thought to exert this effect. Include experimental evidence in your answer.

E. **Ethylene** *(page 679)*

29. What is unusual about the nature of ethylene compared with that of other plant hormones?

30. Describe how the effects of ethylene on fruit ripening were discovered.

31. Where in the plant is ethylene synthesized?

32. a. In addition to fruit maturation, name one other effect of ethylene.

    b. What is the relationship between auxin and ethylene in fruits, flowers, and axillary buds?

Ethylene and Leaf Abscission (page 679)

33. a. *Vocabulary:* Define abscission.

    b. Describe the structure of the leaf abscission zone in woody dicots.

34. a. Summarize the mechanism by which ethylene mediates leaf drop.

    b. What is the role of auxin in leaf abscission?

F. **Abscisic Acid** *(page 680)*

35. a. Why is this hormone misnamed?

    b. Identify three functions of abscisic acid.

    c. What other name for abscisic acid is more appropriate to its functions?

Focus: Abscisic acid is considered the stress hormone of plants since it stimulates the plant to prepare for adverse environmental conditions.

G. **Gibberellins** *(pages 680–682)*

36. a. Gibberellins were discovered while studying what plant disease?

    b. What effect did the fungal chemical have on infected plants?

37. a. Describe the effect of gibberellin on genetic dwarf plants.

    b. What does this suggest about the basis of dwarfism in these plants?

38. a. Identify the role of gibberellin in bolting.

    b. How do abscisic acid and cytokinins influence bolting?

39. Discuss the role of gibberellins and auxins in cellular differentiation.

Gibberellins and Seed Germination (pages 681–682)

40. What plant structure contains the highest concentration of gibberellins?

41. a. Summarize the means by which gibberellins affect the supply of nutrients to the embryo at germination.

    b. It has been hypothesized that this function results from what effect on certain genes?

H. Oligosaccharins *(page 682)*

42. a. *Vocabulary:* What are phytoalexins?

    b. What is the relationship between phytoalexins and oligosaccharins?

43. a. Oligosaccharins are released from what portion of a plant cell?

    b. Oligosaccharins stimulate what event in a cell?

    c. In cultures of plant cells, name four structures that have been produced by adding different oligosaccharins to the culture medium. (Figure 32–17)

44. What is the proposed relationship between oligosaccharins and other classes of plant hormones?

IV. Gravitropism *(pages 682–684)*

45. *Vocabulary:* Define gravitropism.

46. a. Describe the roles of calcium ions and auxin in gravitropism in monocot shoots.

    b. What are the findings of studies on the role of calcium ions and auxin in gravitropism of dicot shoots?

    c. How do auxin and calcium ions mediate root gravitropism?

47. a. *Vocabulary:* What are amyloplasts?

b. Summarize the mechanism by which the plant root "knows" it is horizontal or vertical.

c. What evidence indicates that amyloplasts are not the only mechanism by which a root can orient with respect to gravity?

48. The detection of gravity and the subsequent creation of chemical gradients most likely involves what two factors?

V. Photoperiodism *(pages 684–688)*

A. General Remarks *(page 684)*

49. a. *Vocabulary:* What is photoperiodism?

b. Identify five events that a plant can anticipate based on photoperiods.

B. Photoperiodism and Flowering *(pages 684–686)*

50. *Vocabulary:* Distinguish among day-neutral, short-day, and long-day plants.

51. For both short-day and long-day plants, initiation of flowering occurs only when the plant passes from its juvenile stage to:

52. Use examples to illustrate how photoperiodism may limit the geographic distribution of plants.

Focus: For short-day and long-day plants, the important factor in stimulation of flowering is not the absolute length of the photoperiod itself but whether or not the photoperiod is less than or greater than the critical photoperiod for that plant.

---

53. Which group first conducted field studies of photoperiodism in plants?

Measuring the Dark (pages 685–686)

54. What feature of cocklebur flowering makes it a useful research plant?

55. a. What did Hamner and Bonner discover about the importance of the dark period to flowering in short-day plants?

b. What are the findings of similar studies in long-day plants?

C. Photoperiodism and Phytochrome *(page 686)*

56. a. *Vocabulary:* What is phytochrome?

b. Distinguish between the two forms in which phytochrome exists. Include the wavelength of light to which each form responds, and the relationship between the two forms.

57. Account for each of the following phenomena in terms of the characteristics of P_r and P_{fr}.

a. A long-day plant requires 16 hours of light to flower; it can be induced to flower if it is exposed to 8 hours of light but its dark interval is interrupted by a brief exposure to red light.

b. Under the same conditions described in (a), a short-day plant will not bloom.

D. Other Phytochrome Responses *(page 687)*

58. a. Discuss the role of phytochrome in seed germination.

b. What is the adaptive value of this phenomenon?

59. a. Describe an etiolated seedling.

b. Under what conditions does a seedling exhibit etiolated growth?

c. Of what adaptive value is etiolated growth to the plant? (Figure 32-24)

d. What is the effect of phytochrome on the etiolated seedling?

e. Outline the mechanism by which phytochrome is suspected to influence seedling growth.

60. An increase in concentration of what substances is observed immediately following phytochrome activation?

E. *Essay:* Is There a Flowering Hormone? *(page 688)*

61. a. Chemicals involved in flowering are apparently produced in what plant tissue?

b. Cite three pieces of evidence supporting the existence of one or more chemical mediators that stimulate flowering.

c. Identify three plant chemicals for which evidence suggests a role in initiation of flowering.

VI. Circadian Rhythms *(pages 689–691)*

A. General Remarks *(page 689)*

62. Who first noted that daily rhythms of plants continue even if the lighting conditions are kept constant?

63. *Vocabulary:* What is a circadian rhythm?

Focus: Circadian rhythms have been identified in all four kingdoms of eukaryotes.

B. Biological Clocks *(pages 689–691)*

64. a. To what locations have organisms been transported in an attempt to identify an *external* factor controlling daily rhythms?

b. What were the findings of these investigations?

c. What characteristic of circadian rhythms strongly suggests that they are mediated by an *internal* biological clock?

65. *Vocabulary:* Define endogenous.

66. Identify two functions of biological clocks in living organisms.

Resetting the Clock (page 690)

67. a. Describe how circadian rhythms can be modified by external conditions.

b. What name is given to this phenomenon?

c. What happens if the external conditions deviate dramatically from a plant's internal rhythm?

The Nature of the Clockwork (pages 690–691)

> *Focus:* **The chemical nature of the biological clock has yet to be defined.**

68. What is the relationship between phytochrome and biological clocks?

69. Outline the most likely mechanism of the biological clock, according to current knowledge.

VII. Touch Responses *(pages 691–693)*

A. Twining and Coiling *(pages 691–692)*

70. *Vocabulary:* What are tendrils?

71. a. *Vocabulary:* Define circumnutation.

b. Identify the energy source for circumnutation.

c. Where are the touch sensors located?

d. Name two plant hormones that are probably involved in coiling responses.

72. How did M. J. Jaffe demonstrate that tendrils "remember" tactile stimulation?

73. Summarize the means by which a dodder plant parasitizes its host.

B. Rapid Movements in the Sensitive Plant *(page 692)*

74. a. Outline the events occurring when the leaves of the sensitive plant fold in response to touch. Include in your answer the method by which the stimulus spreads to other leaves.

b. The motor cells are involved in what other activity?

75. What is the presumed role of plant tannins?

> *Focus:* **In the course of evolution, the release of tannins may have been of more survival value to the plant than the accompanying response to touch. (Figure 32–29)**

C. Rapid Movements in Carnivorous Plants *(pages 692–693)*

76. a. Identify the touch receptors in the Venus flytrap.

    b. Describe in general terms what happens when an insect brushes against these receptors.

    c. Outline the mechanism by which the leaves of the flytrap close around prey.

    d. How does the trap open again?

Focus: **The increase in size of a Venus flytrap leaf that occurs when the leaf closes by acid growth is permanent, and the leaf is slightly larger after each catch.**

---

77. By what mechanism does the sundew trap its insect prey?

D. Generalized Effects of Touch on Plant Growth *(page 693)*

78. Explain why plants grown in greenhouses are taller and more spindly than plants grown outside.

Focus: **The response of plants to touch in general involves both electrical signals and changes in cell membrane permeability.**

---

79. Changes in cell permeability are thought to have what effect?

VIII. Chemical Communication among Plants *(pages 693–694)*

Focus: **Many angiosperms produce toxic or noxious chemicals in response to predation.**

---

80. a. What evidence is there that a damaged plant may "warn" nearby plants of potential danger?

    b. Scientists believe this communication takes what form?

IX. Summary *(pages 694–696)):* Read the summary. If you are familiar with the essential features of the material presented there, you are ready to complete the section TESTING YOUR UNDERSTANDING.

TESTING YOUR UNDERSTANDING

After you have completed this examination, compare your answers with those in the section that follows.

1. Charles Darwin and his son Francis performed some of the first experiments on:
   a. photoperiodism.
   b. geotropism.
   c. movement of sap through phloem.
   d. phototropism.
   e. the mechanism of phytochrome responses.

2. The chemical substance that causes bending in oat seedlings was isolated by Frits Went and transferred between experimental plants using:
   a. syringes with fine needles.
   b. radioactive tracers.
   c. stem grafts.
   d. agar blocks.
   e. lanolin paste.

3. Originally, scientists thought that all hormones _____;
   more recently, they have found that some hormones _____.
   a. must be proteins; are nucleic acids
   b. must be small molecules; are extremely large
   c. had stimulatory effects; are inhibitory
   d. had inhibitory effects; are inhibitory and some are stimulatory
   e. had many target organs; do not have any specific target organ

4. Auxin is produced mainly by the:
   a. apical meristems of roots.
   b. apical meristems of shoots.
   c. elongating tissues in shoots.
   d. apical meristems and elongating tissues of shoots.
   e. apical meristems of both shoots and roots.

5. According to current theory, indoleacetic acid (IAA) activates an ion pump in the cell membrane that:
   a. alters the distribution of sodium and potassium ions.
   b. acidifies the cell wall and activates a pH-dependent enzyme.
   c. changes the level of potassium ions in the cell wall.
   d. transports water molecules into the cell.
   e. decreases the internal solute concentration, causing water to leave the cell.

6. Which statement is NOT consistent with the current knowledge about calmodulin?
   a. It is activated by an auxin-induced increase in calcium ion concentration.
   b. It affects cell membrane permeability.
   c. It influences the activity of several enzymes.
   d. It affects hydrogen ion transport.
   e. It is an important regulator of intracellular calcium ion levels.

7. A tissue culture contains Jerusalem artichoke cells treated with auxin and cytokinin. When a high concentration of calcium ions is added to this culture, the growth pattern shifts from cell enlargement to cell division. Identify the presumed reason for this phenomenon.
   a. Calcium ions are antagonistic to the effects of auxin.
   b. Calcium ions in high concentrations prevent cell wall expansion.
   c. Calcium ions inhibit the flow of hydrogen ions into the cell wall.
   d. Calcium metabolism saps the energy reserves needed for cell enlargement.
   e. Calcium ions inhibit the sodium-potassium pump required for cell enlargement.

8. When the concentration of auxin and cytokinin are roughly equal in tissue culture media, undifferentiated cells form:
   a. buds.
   b. roots.
   c. a callus.
   d. both roots and shoots.
   e. The cells do not respond at all to this experimental treatment.

9. What activity does NOT appear to be a plant response to the hormone cytokinin?
   a. increased cell division
   b. reversal of apical dominance
   c. prevention of leaf senescence
   d. increased protein synthesis
   e. cell elongation

10. The ethylene-synthesizing system is most likely located in:
    a. cell membranes.
    b. cell walls.
    c. smooth endoplasmic reticulum.
    d. ribosomes.
    e. Golgi bodies.

11. Ethylene in a concentration as low as _____ part(s) per million can speed the ripening of fruit.
    a. 1   d. 125
    b. 25   e. 175
    c. 50

12. The plant hormone responsible for changing vegetative buds to winter buds and bringing about dormancy is:
    a. abscisic acid.   d. ethylene.
    b. auxin.   e. oligosaccharin.
    c. cytokinin.

13. The interaction of _____ promotes cellular differentiation in _____.
    a. cytokinin and auxin; trees
    b. gibberellin and ethylene; roots
    c. ethylene and cytokinin; secondary xylem and phloem
    d. auxin and gibberellin; secondary xylem and phloem
    e. auxin and ethylene; roots

14. When infected with certain bacteria, fungi, or protists, some plants are capable of producing antibiotics known as:
    a. phytokinins.   d. oligosaccharins.
    b. phytoalexins.   e. phytophthorans.
    c. phytoplanktons.

15. Which structures in plant root-cap cells are thought to play the same role as the statoliths in jellyfish?
    a. protoplasts
    b. chloroplasts
    c. leucoplasts
    d. chromoplasts
    e. amyloplasts

16. Plants such as spinach, potatoes, clover, and lettuce, are examples of _____ plants and will flower when exposed to a light period that is _____.
    a. day-neutral; variable in length
    b. long-day; greater than twelve hours
    c. short-day; less than twelve hours
    d. long-day; greater than a critical length
    e. short-day; less than a critical length

17. Why do small seeds, such as lettuce, need to be planted in shallow, loose soil? (Select the MOST COMPLETE explanation.)
    a. Red light can reach them and stimulate germination.
    b. Far-red light can reach them and stimulate germination.
    c. The seedling can reach the surface before stored food supplies are exhausted.

d. There is an adequate supply of oxygen for metabolism.

e. Red light can stimulate germination and the seedling can reach the surface before its stored food supplies are exhausted.

18. The synchronization of a plant's internal clock with external environmental circumstances is called:
   a. entrainment.
   b. derailment.
   c. adaptation.
   d. acclimation.
   e. circadian escape.

19. The hinged leaf of a Venus flytrap springs closed as a result of:
   a. the rapid active transport of potassium ions out of hinge cells, causing them to lose turgor and collapse.
   b. the influx of potassium ions into hinge cells, which then swell.
   c. acid growth triggered by electrical stimulation.
   d. leakage of calcium ions from cells in the hinge region.
   e. changes in turgor pressure caused by the influx of hydrogen ions into hinge cells.

20. T or F   The primary means by which a plant responds to environmental stimuli is to alter its pattern of growth.

21. T or F   In horizontal stems of both monocots and dicots, the increase in calcium ions in the upper part of the stem and the increase in auxin concentration in the lower part cause the gravitropic response.

22. T or F   Hamner and Bonner discovered that, under experimental conditions, short-day plants such as the cocklebur require days that are uninterrupted by even a short period of darkness.

23. T or F   An increase in the concentrations of all growth-promoting hormones can be detected almost immediately after phytochrome activation.

24. T or F   Biological clocks are the mechanisms by which circadian rhythms in plants are controlled.

25. T or F   Regular rubbing and bending of plant stems enhances their elongation and results in tall, spindly plants.

26. List the six different classes of plant hormones and name at least one function of each class.

27. Explain how cytokinin interacts with auxin in the lower axillary buds of plants. What phenomenon does this interaction overcome?

28. Describe the interconversion of P_r and P_{fr} and discuss how the biologically active form functions in long-day and short-day plants.

PERFORMANCE ANALYSIS

1. **d**   Charles and Francis Darwin performed some of the first experiments on phototropism. By placing light-proof cylinders around grass seedlings, they determined that a chemical produced in the coleoptile tip migrates down the stem and produces bending toward the light. (page 672)

2. **d**   Frits Went used agar blocks to isolate the chemical agent responsible for phototropism, which he named auxin. (page 672)

3. **c**   The term hormone originated from the Greek word meaning "to excite." However, it is now apparent that some hormones have stimulatory effects and others have inhibitory effects. (page 673)

4. **b**   Auxin is produced primarily in the apical meristems of shoots. It is also produced by leaves, flowers, embryos, and fruits. Auxin is not produced by roots. (page 674)

5. **b**   Current evidence indicates that indoleacetic acid initiates acid growth by activating an ion pump in the cell membrane that pumps hydrogen ions from the cytoplasm into the cell wall. This acidifies the cell wall and activates a pH-dependent enzyme that breaks cross-links between cellulose molecules. The cell wall can then expand under the turgor pressure of the cell, after which the cross-links reform. (pages 674, 675)

6. **e**   Calmodulin is thought to be activated when the calcium ion concentration increases under the influence of auxin. Calmodulin is a regulatory substance that affects cell membrane permeability, enzyme activity, and hydrogen ion transport. (page 675)

7. **b**   In experiments with Jerusalem artichoke cells in tissue culture, auxin and low concentrations of kinetin favored cell elongation rather than cell division. However, when calcium ions were added to the culture, cell division was preferred. Apparently, high calcium ion concentrations inhibit cell wall expansion. (page 677)

8. **c**   When equal concentrations of auxin and cytokinin are present in a plant tissue culture, the undifferentiated cells tend to form a callus, which is a mass of undifferentiated tissue. (page 677)

9. **e**   Cytokinins stimulate cell division in the presence of auxin, reverse apical dominance, prevent leaf senescence, and increase protein synthesis in plant cells. They do not appear to cause cell elongation. (page 676, 677)

10. **a**   The system that synthesizes ethylene is apparently located in plant cell membranes. (page 679)

11. **a**   An ethylene concentration as low as one part per million can speed the ripening of fruit. (page 679)

12. **a**   Abscisic acid, the "stress hormone," is responsible for initiating and maintaining dormancy in winter buds. (page 680)

13. **d**   The differentiation of secondary xylem and phloem is brought about by the interaction of auxin and gibberellin. (page 681)

14. **b**   Phytoalexins are antibiotics produced by some plants in response to infections by certain bacteria, fungi, and protists. The stimulus for the production of phytoalexins

is the release of oligosaccharins from the cell wall. (Oligosaccharins directly stimulate mRNA synthesis.) (page 682)

15. **e** Amyloplasts are plastids filled with starch that are present in root cells and are thought to enable a root to orient with respect to gravity. (page 683)

16. **d** Long-day plants (e.g., spinach, potatoes, clover, and lettuce) flower when exposed to a light period that is longer than a critical period. Short-day plants (e.g., poinsettias, strawberries, primroses, and ragweed) flower when exposed to a light period shorter than a critical period. The flowering of day-neutral plants is not affected by day length. (page 684)

17. **e** Small seeds should be planted in shallow, loose soil so that red light can stimulate germination and the seedlings can reach the surface before their food reserves are exhausted. (page 687)

18. **a** Entrainment is the synchronization of a plant's circadian rhythms with external environmental conditions. It can only occur when the environmental conditions do not differ too greatly from the plant's internal clock. (page 690)

19. **c** The closing of a Venus flytrap leaf is due to acid growth of cells on the outside of the hinge. An electrical stimulus initiates the acid growth. (page 693)

20. **True** Plants compensate for their lack of motility by their growth, which brings them closer to sources of light, water, and nutrients. (page 671)

21. **False** The role of calcium ion and auxin concentration gradients in gravitropism has only been documented in monocots. Similar studies in dicots have thus far yielded conflicting information. (pages 682, 683)

22. **False** In photoperiodism associated with flowering, plants measure the length of the dark period rather than the length of the light period. Short-day plants must have a period of uninterrupted darkness in order to flower. Long-day plants will flower in light periods shorter than their critical period if the dark period is interrupted by a brief exposure to light. (page 686)

23. **True** The mechanism by which the phytochrome system operates has not yet been elucidated. However, studies have demonstrated that the concentrations of all growth-promoting hormones increase following phytochrome activation. (page 687)

24. **True** Plant circadian rhythms are regulated by internal biological clocks. (page 689)

25. **False** Regular tactile stimulation of plants results in more compact plants with shorter, stouter stems. (page 693)

26. The major classes of plant hormones are auxins, cytokinins, ethylene, abscisic acid, and gibberellins. Oligosaccharins are oligosaccharide molecules released from the plant cell wall in response to a variety of stimuli. They stimulate the transcription of mRNA and are thought to be the actual effectors of many hormonally induced phenomena. See Table 32–1 for functions of the other classes of plant hormones. (page 682 and Table 32–1)

27. Auxin is produced in the apical meristem of the shoot and migrates downward. Auxin induces the synthesis of ethylene in axillary buds, which prevents their development. This phenomenon is known as *apical dominance*. Cytokinin is produced in the roots and migrates upward through the plant. In the lower axillary buds, the concentration of auxin is relatively low but the concentration of cytokinin is relatively high. Cytokinin stimulates development of these lower axillary buds, reversing apical dominance. (pages 675, 677)

28. Phytochrome is synthesized in the P_r form, which absorbs red light (660 nanometers) and is converted to P_{fr}. The P_{fr} form is the biologically active pigment and is converted back to P_r on absorption of far-red light (730 nanometers). In the dark, P_{fr} is relatively stable and is either slowly converted back to P_r or is degraded and replaced by new P_r. P_{fr} promotes flowering in long-day plants but inhibits flowering in short-day plants. (page 686)

REVIEW TEST 9

Section 5: Chapters 29–32

This review test is *not* designed to cover all of the important information in these chapters. However, it does touch on the major topic categories in each chapter. It will also give you valuable practice in taking this type of test. When you are finished, check your answers with those provided in the following PERFORMANCE ANALYSIS section.

1. What is the major role of flower petals in plant reproduction?
   a. attracting animal pollinators
   b. protecting the developing embryo
   c. providing raw materials for developing fruits
   d. providing nutrients to animal pollinators
   e. funneling wind-borne pollen grains to internal flower structures.

2. Which type of cell generally contains plastids?
   a. collenchyma   d. tracheid
   b. parenchyma   e. cork
   c. epidermal cell

3. As plants decay, they add _____ to the soil, which improves the soil's texture and its capacity to hold _____.
   a. carbon; humus
   b. water; nitrogen compounds
   c. nitrogen compounds; humus
   d. humus; minerals and water
   e. minerals; water and carbon dioxide

4. Plant tendrils often grow by _____, a pattern of growth that may be described as _____.
   a. circumnutation; spiral
   b. etiolation; spiral
   c. circumnutation; concentric
   d. etiolation; circular
   e. circumnutation; erratic

5. Which cells are structurally modified to provide support for the growing regions of young stems and branches?
   a. cortical cells   d. sclerenchyma
   b. collenchyma   e. pith cells
   c. parenchyma

6. Fruit trees and roses are most frequently propagated vegetatively by:
   a. harvesting runners to set out as individual plants.
   b. growing pollen grains in culture to yield many small plants.
   c. grafting cuttings onto stems of plants with established root systems.
   d. stimulating cuttings to develop adventitious roots in water.
   e. allowing cuttings to develop adventitious roots in moist sand.

7. Which statement concerning the phototropic response is NOT correct?
   a. Flavin, when stimulated by light, alters a cell membrane's permeability to auxin.
   b. Auxin migrates to the dark side of a stem.
   c. Cells with higher amounts of auxin elongate more rapidly.
   d. Both red and blue light stimulate flavin.
   e. Elongation on the dark side of the stem causes it to bend toward the light.

8. Which factor is NOT involved in apical dominance?
   a. the presence of the primary apical meristem
   b. ethylene
   c. abscisic acid
   d. auxin
   e. cytokinin

9. Cells of which structure are actively involved in delivering nutrition to the young angiosperm embryo?
   a. apical meristem
   b. integument
   c. suspensor
   d. carpel
   e. seed coat

10. The embryonic plant tissues differentiate into which of the following tissue systems?
    a. dermal, vascular, and ground
    b. vascular, meristematic, and fibrous
    c. fibrous, ground, and dermal

389

d. meristematic, dermal, and vascular

e. ground, connective, and dermal

11. As _____ guard cells, stomata _____.

a. hydrogen ions diffuse into; open

b. potassium ions diffuse into; close

c. chloride ions diffuse into; close

d. potassium ions are transported into; open

e. hydrogen ions are transported into; open

12. Which type of plants actually require the element silicon for growth?

a. grasses

b. hemlock trees

c. plants of the stonecrop family

d. cacti

e. creosote bushes

13. A surprising fact about circadian rhythms in plants is that they:

a. are exactly 24 hours in length, regardless of species.

b. persist even when environmental conditions are held constant.

c. cannot be reset once they are interrupted.

d. average 24 hours in length even though the length of an individual period is difficult to predict.

e. are not evident at all unless certain requirements for light and darkness are met.

14. The small leaves of conifers evolved as a means to survive:

a. the winter rains.

b. windy habitats.

c. the blazing summer sun.

d. cold, humid climates.

e. the drought of winter.

15. The selective transport of solutes into the vascular cylinder from the cortex of the root is regulated by the:

a. Casparian strip.

b. parenchyma cells.

c. cork cambium.

d. cortical cell membranes.

e. endodermal cell membranes.

16. All of these elements are essential for chlorophyll or chloroplast development or play an important role in photosynthesis EXCEPT:

a. magnesium.
d. manganese.

b. iron.
e. cobalt.

c. chlorine.

17. T or F Biennial plants may complete their life cycles in a single growing season if conditions are favorable, but they may require three years under adverse conditions.

18. T or F The driving force for the movement of sugars within plants is provided by differences in water potential caused by sugar concentration gradients.

19. T or F Low concentrations of auxin stimulate elongation in roots whereas high concentrations inhibit elongation.

20. T or F Wheat protein is considered a complete protein source because it contains all the essential amino acids needed by humans.

21. T or F Of the 250 million metric tons of nitrogen added to the soil each year, most comes from commercially produced nitrogen fertilizers.

22. T or F Auxin and gibberellin cause cellular elongation in plants by the same mechanism.

23. Describe three experiments with cockleburs that provided circumstantial evidence for the presence of a flowering hormone (or at least one or more chemical substances working together to stimulate flowering).

24. Define and describe the development of an aerial root. Cite two examples of aerial roots.

25. In terms of the anatomy of a grass plant, explain why a lawn continues to grow after mowing.

26. Identify and indicate the effects of four factors that influence the rate of water loss by transpiration. Which factor is most important?

PERFORMANCE ANALYSIS

1. **a** Flower petals attract animals that are likely to be carrying pollen to the flower. The more highly evolved flowers attract only one or a few types of pollinators. The flower is pollinated and the animal is rewarded with food. (page 614)

2. **b** Parenchyma cells, which are found in all three tissue systems but which predominate in the ground tissues, generally contain plastids. The plastids may be chloroplasts, leucoplasts, or chromoplasts, depending on the location of the parenchyma cells. (page 626)

3. **d** Decaying plants add organic matter (humus) to the soil, which improves the capacity of the soil to retain water and minerals. (page 663)

4. **a** Circumnutation is the spiraling movement characteristic of tendrils as they grow. This growth pattern enhances their chances of finding a support. (page 691)

5. **b** Collenchyma cells are specialized cells in the ground tissue that have primary walls thickened in an irregular fashion. They provide support for the growing regions of young stems and branches. (page 637)

6. **c** Commercially produced fruit trees and roses are often propagated by grafting cuttings onto the stems or trunks of woody plants with established root systems. (page 643)

7. **d** When *blue* light strikes cells on the side of a stem, a flavin pigment is stimulated. Consequently, the permeability of the cell membranes to auxin increases, and auxin migrates away from the light to the dark side of the stem. Under the influence of auxin, the cells on the dark side of the stem elongate more rapidly than cells on the light side, and the stem bends toward the light. (page 673)

8. **c** The production of auxin by the primary apical meristem is the key to apical dominance. As auxin migrates down the stem, it stimulates the production of ethylene in axillary buds, which inhibits bud development. Meanwhile, cytokinin is migrating up the stem from the roots, where it reaches the lowest axillary buds first and in the greatest concentration. Axillary buds that are distant from the apical meristem are stimulated to develop as the influence of cytokinin exceeds that of auxin. (pages 675, 677)

9. **c** The suspensor is the stalk on which the globular mass of cells that constitutes the early embryo is suspended. The cells of the suspensor function in nutrient supply to the embryo. (page 619)

10. **a** The embryonic plant tissues differentiate into the primary meristems, each of which gives rise to a major tissue system. Protoderm gives rise to the dermal system. Procambium differentiates into the vascular system. The ground tissues are derived from ground meristem. (pages 625, 626)

11. **d** Stomata open when the osmosis of water into guard cells causes these cells to swell. The osmotic potential resulting in water movement is caused by the active transport of potassium ions into the guard cells. Guard cells bow because there are cellulose microfibrils arranged in hoops around each cell that prevent radial expansion. As the cells lengthen, they bow away from each other, opening the stoma. (page 653, Figure 31–6)

12. **a** Grasses are among the few types of plants that require silicon for growth. (page 657)

13. **b** Circadian rhythms are endogenous and persist even if environmental conditions are kept constant. A circadian rhythm can be entrained to environmental conditions that are slightly different from the internal rhythm. However, if the external rhythm differs significantly from the internal one, the plant "escapes" and the internal rhythm prevails. (page 689)

14. **e** Conifer leaves are adapted to withstand drought, including that occurring in the winter when water is locked in ice and snow and is unavailable to plants. (page 629)

15. **e** All substances that enter the vascular cylinder of a root must pass through the cell membranes of endodermal cells. (page 632)

16. **e** *Magnesium* is part of the chlorophyll molecule. *Iron* is essential for chloroplast development. *Manganese* is required for the integrity of the chloroplast membrane and for oxygen release during photosynthesis. *Chlorine* is essential to the oxygen-producing reactions of photosynthesis. Cobalt is required by nitrogen-fixing microorganisms. (Table 31–1)

17. **True** Biennial plants typically store food reserves in the roots in their first year and flower and produce seeds in their second year. However, under exceptionally good conditions, they may complete their life cycle in one growing season. If conditions are extremely poor, the life cycle may require three growing seasons. (page 621)

18. **True** In the phloem, a sugar concentration gradient exists between the source and the sink, and this provides the driving force for translocation. (page 662)

19. **True** In roots, low auxin concentrations stimulate cell elongation but high concentrations are inhibitory. (page 674)

20. **False** Wheat protein is lacking in some amino acids that are essential to humans, notably lysine. (essay, page 622)

21. **False** Although a great deal of money is spent every year on commercial nitrogenous fertilizers and their application to croplands, most of the nitrogen added to the soil is biological in origin. (page 669)

22. **False** The mechanism by which gibberellins stimulate cell elongation is unknown but it does not involve transport of hydrogen ions and acid growth, which is characteristic of cell elongation under the influence of auxin. (page 680)

23. (1) Under light conditions appropriate for flowering, a cocklebur plant with even a fraction of one leaf flowers, whereas a plant with no leaves does not. (2) Exposure of a leaf on one branch of a plant to the correct light cycle stimulates flower development on other branches of the plant. (3) When two plants are grafted together, exposing one plant to the correct light cycle induces flowering in both plants. (essay, page 688)

24. Adventitious roots are those that develop from the shoot system rather than from the root system. Aerial roots are adventitious roots produced from aboveground structures. Aerial roots are produced by English ivy, corn, red mangrove trees, and the bald cypress. (page 636)

25. The leaf of a grass consists of a sheath, which wraps around the stem, and a blade, the broad portion that extends from the sheath. Grasses have meristematic tissue at the base of the leaf. When the blade of the leaf is cut, the cells in the sheath are activated to produce more blade. (page 642)

26. The most important factor influencing the rate of water loss from a plant by transpiration is **stomatal movements**. Water is lost whenever stomata are open. The rate of water loss from a leaf doubles for every increase in **temperature** of 10°C. **Humidity** affects water loss in that evaporation is much slower in air that is already laden with water. **Air currents** influence water loss because wind blows moist air near the leaf surface away from the leaf, causing a steeper concentration gradient of water vapor between the surface of the leaf and the surrounding air. Many plants growing in windy environments have leaf hairs that trap moist air near the leaf surface. (page 652)

SECTION **6**

Biology of Animals

The Vertebrate Animal: An Introduction

MAJOR CONCEPTS

As a vertebrate, a human has a bony endoskeleton that grows as the body grows. As a mammal, a human has a coelom divided into thoracic and abdominal cavities, mammary glands for nursing young, hair covering the body, highly developed sensory structures, and a complex nervous system. Humans are endothermic, give birth to live young, and have an extended period of parental care.

The fundamental structural unit of a multicellular organism is the cell. Cells that have similar structures and functions are organized into tissues. Different tissues may be united to form an organ, which has one or more specific functions. A group of organs performing a set of related functions is an organ system. An organism consists of organ systems whose functions are integrated.

There are four basic tissue types in mammals: epithelial, connective, muscle, and nerve. Epithelial tissues cover and line organs and the body surface; glands are composed of modified epithelial tissues. Two types of cell-cell junctions that are particularly important in epithelial tissues are desmosomes and tight junctions. Connective tissues are composed of cells and fibers in an acellular matrix; examples include blood, bone, and tendons. The two basic types of muscle tissue are smooth and striated, which includes skeletal and cardiac muscle. Nerve tissue consists of supporting glial cells and neurons, which transport nerve impulses. A neuron is composed of a cell body, one or more dendrites, which carry impulses toward the cell body, and one or more axons, which carry impulses away from the cell body. A nerve consists of axons from several different neurons.

Biological "problems" get "solved" over the course of evolution as the pressures of natural selection determine which adaptations persist in a species. For example, when the dry climate of the Permian period occurred, those plants and animals possessing adaptations that promoted survival in a dry climate persisted and left relatively more offspring than organisms lacking such adaptations.

There are four major problems that every animal faces. The most important problem is that of obtaining energy for body functions and supplying each cell in the body with energy, nutrients, and waste removal functions. A second problem is homeostasis, the maintenance of a relatively constant internal environment. Since animals are multicellular organisms composed of many organ systems, the activities of these organ systems must be integrated and controlled. Integration and control is accomplished by the nervous and endocrine systems acting in concert. Finally, every animal is faced with the challenge of reproduction.

HOW TO STUDY THE CHAPTER

Read the entire chapter through quickly, focusing on the major concepts.

Use the GUIDED STUDY OF THE CHAPTER to help you identify the important details as you **reread** the chapter. Writing out the answers to these questions will help fix them in your mind as well as provide you with a valuable study aid.

Answer the questions in TESTING YOUR UNDERSTANDING without the aid of your text. Check your answers against those in PERFORMANCE ANALYSIS. Analyzing your answers will give you valuable feedback on your level of understanding and preparedness for classroom testing.

GUIDED STUDY OF THE CHAPTER

I. Introduction *(page 701)*

> *Focus:* **Insights into important physiological principles and adaptations can be obtained by studying humans and other animals.**

II. Characteristics of *Homo sapiens* *(pages 701–702)*

> *Review Focus:* **The four characteristics of phylum Chordata are a dorsal hollow nerve cord, a notochord, pharyngeal gill slits, and a postanal tail.**

1. Identify and describe the characteristic that distinguishes vertebrates from other members of phylum Chordata?

2. a. Name the two largest compartments of the mammalian coelom.

   b. Describe the contents of the two compartments mentioned in (a).

3. Name two consequences of endothermy.

4. Identify four characteristics of mammals not previously mentioned.

5. In general, what is the relationship between parental care and learning?

III. Cells and Tissues *(pages 702–709)*

A. General Remarks *(page 702)*

6. How do the cells of a multicellular organism resemble and differ from those of a unicellular organism?

7. *Vocabulary:* Distinguish between tissues and organs.

8. a. The human body contains _____ distinct cell types.

   b. These cell types are classified into what four tissue types?

B. Epithelial Tissues *(pages 702–703)*

9. Identify four functions of epithelial tissues.

10. a. Epithelial tissues are classified on what basis?

    b. Name and distinguish between the three types of epithelial cells on the basis of shape and general function. (See also Figure 33–3.)

11. a. *Vocabulary:* What is the basement membrane?

    b. Identify the origin and content of the basement membrane.

12. What types of cells produce mucus and what is its function?

13. a. *Vocabulary:* What are glands?

    b. Glands are composed of what type of epithelial cells?

Desmosomes and Tight Junctions (page 703)

14. Fill in the table comparing desmosomes and tight junctions.

Table for Question 14

	DESMOSOMES	TIGHT JUNCTIONS
Structure		
Function(s)		
Common location(s)		

15. Tight junctions occur only in _____.

16. *Vocabulary:* What are septate junctions?

C. **Connective Tissues** *(page 704)*

17. Name three functions of connective tissue.

18. a. What is the matrix and what is its function?

    b. Identify the two most common components of the matrix.

    c. Name three different types of fibers and cite one example of each type.

19. a. The three major connective tissues are:

    b. Name the matrix of blood and lymph.

20. a. Describe the matrix of bone.

    b. Identify the cells that produce bone matrix. (Figure 33–6)

21. Study Figure 33–7.
    a. Distinguish between spongy and compact bone with respect to location and structure.

    b. What is the periosteum?

    c. The nutrient canals of bone serve what function?

D. **Muscle Tissue** *(pages 706–708)*

22. Muscle cells are specialized for what function?

23. a. *Vocabulary:* Distinguish between striated, smooth, and cardiac muscle.

    b. *Vocabulary:* Distinguish between voluntary and involuntary muscles.

24. *Vocabulary:* Define a muscle fiber. (Figure 33–8)

25. Name the two proteins upon which muscle contraction depends.

Striated Muscle (pages 706–708)

26. Striated muscle constitutes _____ percent of the adult human male body and _____ percent of the adult human female body.

27. *Vocabulary:* Distinguish between tendons and ligaments.

28. a. Use an example to illustrate how skeletal muscles function in antagonistic groups.

    b. By what mechanism does a skeletal muscle lengthen? (Figure 33–9)

29. a. Describe the structure of a skeletal muscle.

    b. A muscle fiber forms by what process?

30. How is cardiac muscle similar to and different from skeletal muscle?

E. **Essay: The Injury-Prone Knee** *(page 707)*

31. a. Which two bones are connected at the knee joint?

    b. What is the function of the patella?

32. State the function of these components of the knee joint.

    Collateral ligaments

    Cruciate ligaments

    Menisci

33. a. Why is the knee so vulnerable to injury?

    b. Name two common knee injuries.

Smooth Muscle (page 708)

34. Describe the cells of smooth muscle.

35. Contractions of smooth muscles differ from those of skeletal muscles in what two ways?

36. a. Describe the arrangement of smooth muscle fibers in most hollow organs.

    b. In general, what function is accomplished by muscles so arranged?

    c. Describe one example of smooth-muscle function.

F. **Nerve Tissue** *(pages 708–709)*

37. Name the functional unit of nerve tissue.

38. a. *Vocabulary:* Distinguish between neuroglia and Schwann cells.

    b. What are two functions of these glial cells?

39. a. Sketch the general structure of a neuron and label these parts: cell body, dendrites, axon.

b. Identify the functions of the three components labeled in your drawing.

c. *Vocabulary:* What are nerve fibers?

d. Study Figure 33–10 to appreciate the variety of shapes characteristic of neurons.

40. a. Identify three functions for which neurons are specialized.

b. Distinguish between the functions of sensory neurons, interneurons, relay neurons, and motor neurons.

41. a. Trace the route of a nerve impulse through a reflex arc beginning at the sensory receptor and ending at the effector organ. (Figure 33–11)

b. What is the role of relay neurons in a reflex arc? (Figure 33–11)

Focus: The axon of a single neuron may reach a great length, depending upon its function.

---

42. *Vocabulary:* Define a nerve.

IV. Levels of Organization *(pages 709–710)*

43. a. The basic structural unit of a living organism is the
_____.

b. *Vocabulary:* Distinguish among the following levels of organization: tissues, organs, organ systems.

Focus: Animals are composed of organ systems whose functions are integrated.

---

44. Study the structure of human skin in Figure 33–12.

a. Describe the two layers of the human epidermis.

b. Where are skin pigment cells located?

c. Name four components found in the dermis.

d. Identify the function of sebaceous glands.

V. Functions of the Organism *(pages 710–713)*

A. General Remarks *(page 710)*

45. Explain this statement: "Over the course of evolution, biological problems get solved."

B. Energy and Metabolism *(pages 710–711)*

46. What is the major problem facing living organisms?

47. Large, multicellular animals must solve what two problems that do not confront unicellular organisms and small multicellular organisms?

C. Homeostasis *(page 711)*

48. a. *Vocabulary:* Define homeostasis.

b. Homeostasis involves what three general functions?

49. Name three ways in which invading organisms may damage living cells.

50. Why is it more difficult for small organisms to achieve homeostasis than it is for large organisms?

51. Name four organ systems that are particularly important in homeostasis.

D. Integration and Control (pages 711–713)

> **Focus:** In an animal, integration and control involve coordinating internal activities and interacting with the external environment.

52. a. Name the two systems whose primary function is integration and control.

    b. Cite two examples that illustrate the functional relationship between the brain and the endocrine system.

    c. Identify one structure of the nervous system that functions as an endocrine gland.

53. *Vocabulary:* Distinguish between sensory and motor neurons. (Figure 33–14)

54. a. Distinguish between somatic and autonomic nervous systems.

    b. Summarize the functional differences between the sympathetic and parasympathetic divisions of the autonomic nervous system.

> **Focus:** Every organ system is under the influence of the endocrine system and the sympathetic and parasympathetic nervous systems.

Feedback Control (pages 712–713)

55. Using the production of antidiuretic hormone (ADH) as an example, construct a flow diagram illustrating a negative feedback loop. Include the factors that influence ADH production and whether they increase or decrease hormone production.

56. Using an example, explain what is meant by a "relay loop."

> **Focus:** In negative feedback control, either the hormone or the response by the target tissue inhibits further production of the hormone.

E. Continuity of Life (page 713)

57. In general terms, characterize reproduction in mammals.

VI. Summary (pages 713–714): Read the summary. If you are familiar with the essential features of the material presented there, you are ready to complete the section TESTING YOUR UNDERSTANDING.

TESTING YOUR UNDERSTANDING

After you have completed this examination, compare your answers with those in the section that follows.

1. The esophagus, lungs, and heart are contained in the:
   a. thoracic cavity.
   b. abdominal cavity.
   c. dorsal coelom.
   d. ventral coelom.
   e. central coelom.

2. An endotherm is to an ectotherm as a _____ is to a _____.
   a. robin; bat
   b. trout; frog
   c. rabbit; lizard
   d. cobra; chicken
   e. dog; cat

3. Which trait is NOT a general characteristic of mammals?
   a. hair or fur
   b. endothermy
   c. nursing and nurturing young
   d. highly developed sensory and response systems
   e. low metabolic rate

4. The approximately 200 different human cell types can be classified into which of the following four tissue types?
   a. bone, epithelial, connective, and nerve
   b. blood, epithelial, nerve, and muscle
   c. skin, connective, muscle, and nerve
   d. connective, epithelial, muscle, and nerve
   e. cartilage, muscle, connective, and nerve

5. Within glands, the cells that produce secretions such as saliva, milk, insulin, HCl, and thyroxine are nearly always _____ tissue cells.
   a. connective   d. nerve
   b. epithelial   e. hormonal
   c. muscular

6. The linings of body cavities and passageways frequently contain modified epithelial cells that secrete:
   a. mucus.
   b. hormones.
   c. digestive enzymes.
   d. plasma.
   e. saliva.

7. Septate junctions in invertebrates are functionally analogous to _____ of vertebrates.
   a. septate junctions
   b. desmosomes
   c. basement membranes
   d. elastic junctions
   e. tight junctions

8. Collagen, one of the connecting and supporting fibers, constitutes a major portion of:
   a. blood.   d. solid organs.
   b. ligaments.   e. epithelial tissue.
   c. vessel walls.

9. In connective tissue, the extracellular material that is composed of ground substance and fibers is known as the:
   a. plasma.
   b. middle lamella.
   c. basement membrane.
   d. matrix.
   e. stroma.

10. The walls of the blood vessels, digestive tract, uterus, and urinary bladder contain:
   a. striated muscle.
   b. cardiac muscle.
   c. smooth muscle.
   d. both striated and smooth muscle.
   e. no muscle tissue.

11. In most joints, the bones are held together by:
   a. tendons.   d. muscles.
   b. cartilage.   e. menisci.
   c. ligaments.

12. In nerve tissue, the cells that support and insulate the neurons are called _____ within the central nervous system and _____ in the peripheral nervous system.
   a. axons; dendrites
   b. Schwann cells; neuroglia
   c. dendrites; Schwann cells
   d. neuroglia; Schwann cells
   e. Schwann cells; axons

13. Which division of the nervous system is most involved in supporting everyday activities such as digestion and excretion? (Select the MOST SPECIFIC answer.)
   a. parasympathetic   d. autonomic
   b. sympathetic   e. central
   c. somatic

14. Which physiological parameter is NOT maintained by negative feedback control?
   a. blood volume
   b. blood thyroxine level
   c. body temperature
   d. blood pressure
   e. All of these parameters are maintained by negative feedback control.

15. T or F   Humans can stand upright because of the ability of antagonistic muscle groups to contract simultaneously.

16. T or F   The relationship between an organism and its environment creates certain "problems" that are "solved" during the evolution of species. The structures of an organism make sense only when viewed as adaptations that solve these problems.

17. T or F   The success of an organism in solving biological problems is measured by its longevity relative to other members of its species.

18. Identify the function(s) of each component of the knee joint: patella, menisci, collateral ligaments, and cruciate ligaments. Discuss why this joint is so prone to injury and indicate which structures are most often injured.

19. List the four functional classes of neurons. Describe the role of each type within the nervous system.

20. List three conditions that must be met in order for a cell to carry out its chemical reactions and maintain its constituent structures.

21. Distinguish between the sympathetic and parasympathetic nervous systems.

PERFORMANCE ANALYSIS

1. a The thoracic cavity contains the lungs, heart, and esophagus and is separated from the abdominal cavity by the diaphragm. The abdominal cavity contains the stomach, intestines, liver, and other organs. (page 701)

2. c An endothermic animal (birds and mammals) maintains its body temperature through metabolic activities. An ectotherm (e.g., reptiles, amphibians, and fish) has a body temperature similar to that of its external environment or uses behavioral means such as basking in the sun to regulate its body temperature. (pages 701, 702)

3. e All mammals have hair or fur, nurse their young, and have highly developed sensory and response systems. They are endothermic and consequently have a high metabolic rate. All mammals except the monotremes give birth to live young. (page 702)

4. d The four basic tissue types are connective, muscle, nerve, and epithelial. (page 702)

5. b Within glands, the cells that produce the secretory product are nearly always epithelial cells. The major exception is the glandular adrenal medulla, which is a collection of nerve cell bodies. (pages 703, 712)

6. a The linings of body cavities and passageways (e.g., the intestines) frequently contain modified epithelial cells that secrete mucus. (page 702)

7. e The septate junctions of invertebrates prevent the leakage of materials between the epithelial cells lining a hollow organ. This function is accomplished by tight junctions in vertebrates. (page 703)

8. b Collagen, which is a connecting and supporting fiber, is a major component of ligaments, tendons, bone, and skin. (page 704)

9. d The matrix is the acellular component of connective tissue and is composed of an amorphous ground substance and fibers. (page 704)

10. c Smooth muscles are present in the walls of many hollow organs, including the digestive tract, uterus, and urinary bladder. (page 706)

11. c Bones are held together by ligaments. Tendons attach muscles to bones. (page 706)

12. d The glial cells, which support and insulate neurons, are called neuroglia in the central nervous system and Schwann cells in the peripheral nervous system. (page 708)

13. a The parasympathetic division of the autonomic nervous system promotes restorative functions such as diges-

tion and excretion. The sympathetic division prepares the body to handle dangerous or stressful situations. (page 712)

14. e Most physiological parameters, including those listed, are controlled by negative feedback mechanisms. (pages 712, 713)

15. True Skeletal muscles typically function in antagonistic groups around joints, with one group flexing the joint and the other group extending the joint. The simultaneous contraction of antagonistic groups stabilizes a joint, as occurs when humans stand in one place. (page 706)

16. True Living organisms are not perfect—their characteristic structures are only as good as they need to be for the organism to survive in its environment and reproduce. (page 710)

17. False The success of an organism in solving biological problems is measured by its capacity to survive *and reproduce*, passing its particular "solutions" on to the next generation. (page 710)

18. The **patella** acts as a stop so that the knee (which is a hinge joint) cannot open beyond 180°. (It also protects the tendon in which it lies from excessive wear as it rubs across the joint.) The **menisci** are the cartilage pads between the bones that cushion the joint. The **collateral ligaments** are on the lateral (away from the midline) and medial (toward the midline) surfaces of the joint and prevent the joint from bending in a side-to-side plane. The **cruciate ligaments** stabilize the joint internally. The knee is vulnerable to injury because it represents a compromise between the need for flexibility and the need for stability. It has been compared to fastening two sticks together end-to-end with rubber bands, an inherently unstable structure. Torn ligaments and crushed menisci are common knee injuries. (essay, page 707)

19. Sensory neurons receive sensory information and relay it to the central nervous system. **Interneurons** transmit signals within localized regions of the central nervous system. **Relay neurons** transmit signals between different regions of the central nervous system. **Motor neurons** transmit signals from the central nervous system to effectors, such as muscles, organs, and glands. (pages 708, 709)

20. (1) The chemical environment of the cell must be strictly controlled. (2) The temperature must remain within a fairly narrow range. (3) The cell must be protected from invading organisms. (page 711)

21. The sympathetic division of the autonomic nervous system prepares the body to deal with dangerous or stressful situations. The parasympathetic division stimulates restorative activities such as digestion and excretion. (page 712)

Energy and Metabolism I: Digestion

CHAPTER ORGANIZATION

MAJOR CONCEPTS

Digestion involves the breakdown of ingested food and the absorption of the resulting molecules into the body. These molecules may provide energy, essential chemical elements, or raw materials for synthetic processes.

The digestive system of mammals consists of mouth, pharynx, esophagus, stomach, small intestine, and large intestine. Accessory digestive organs are the salivary glands, liver, pancreas, and gallbladder.

The wall of the digestive system has the same basic structure throughout. An inner mucosa lines the tract and contains secretory cells. The submucosa contains blood and lymph vessels and nerve fibers. The muscularis externa consists of two muscle layers, one arranged longitudinally and one arranged in concentric rings. The serosa is the outer coating of connective tissue.

In the mouth, food is mixed with saliva and torn or chewed into pieces small enough to swallow. The saliva of many mammals contains amylase, an enzyme that breaks down starch. The tongue bears taste buds and manipulates the food.

The dentition of humans is relatively unspecialized. Adult humans have 32 teeth; children have 20. Incisors are specialized for cutting; canines are used by carnivores for stabbing and tearing; premolars and molars are used for grinding.

The pharynx is the common passageway for the digestive and respiratory systems. During swallowing, the opening to the respiratory system is closed as the food passes into the esophagus. In humans, swallowing is initially voluntary and then involuntary after the food reaches a certain point. In animals that gulp their food, swallowing is totally voluntary.

In humans, food travels through the esophagus to the stomach by peristalsis. The esophagus penetrates the diaphragm just before entering the stomach. Sphincters prevent food from reentering the esophagus and regulate the flow of food into the small intestine.

The peritoneum, which is a layer of epithelial tissue, lines the abdomen and covers all abdominal organs. The organs of the digestive tract are attached to the posterior abdominal wall by mesenteries, which are folds of the peritoneum.

In the stomach, food is mixed with gastric juice, which contains hydrochloric acid and digestive enzymes secreted by glands in the stomach mucosa. The combination of gastric juices and the churning action of the stomach produces a semiliquid mass that passes into the small intestine. The pH of the stomach contents is normally between 1.5 and 2.5 in humans.

Most digestion occurs in the duodenum, the first portion of the small intestine. Alkaline secretions from the pancreas and gallbladder enter the duodenum and raise the pH to between 7 and 8, the range at which digestive enzymes work best. Digestive enzymes in the duodenum originate in the pancreas and the intestinal mucosa. The rest of the small intestine is concerned primarily with nutrient absorption.

Digestive enzymes are specialized to act on a single type of nutrient. Amylases break down starch and lipases break down lipids. Several different types of enzymes, including pepsin, trypsin, chymotripsin, and peptidases, break down proteins and polypeptides. For each disaccharide, there is a specific enzyme that cleaves between the monosaccharide units.

The small intestine has three features that greatly increase its absorptive surface area. The mucosa itself is folded and has numerous fingerlike projections known as villi. Mucosal epithelial cells have microvilli, which are cytoplasmic extensions through which nutrients are absorbed. Carbohydrates and proteins are absorbed by active transport processes into the blood. Small fatty acids enter intestinal blood vessels by diffusion. Large fatty acids, glycerol, and cholesterol diffuse into mucosal cells where they are processed before entering the bloodstream. Fats are resynthesized from fatty acids and glycerol and packaged into chylomicrons. Cholesterol is packaged into low-density lipoprotein (LDL) complexes. The chylomicrons and the LDLs are secreted into lymph vessels, which converge and eventually empty into the bloodstream.

Most of the water in the intestine is reabsorbed from the large intestine. Also, the large intestine harbors symbiotic bacteria that continue digesting food matter and incidentally produce vitamins needed by the host.

The entire digestive tract is coated with a layer of mucus, which protects the walls of the digestive organs from enzymes and other caustic substances.

The pancreas is both an exocrine and an endocrine gland. Digestive enzymes are secreted through ducts into the duodenum. Three hormones that function in the regulation of blood glucose (insulin, glucagon, and somatostatin) are released from the pancreas into the bloodstream.

The liver has many important functions including regulation of blood glucose, detoxification of toxic chemicals, production of plasma proteins, storage of fat-soluble vitamins, and production of bile.

Like many physiological processes, digestion is under the control of the autonomic nervous system and the endocrine system. Sympathetic stimulation inhibits digestion, whereas parasympathetic stimulation promotes digestive activities.

The influence of three hormones on digestion is well-documented. Gastrin is produced in the stomach in response to protein-containing foods and stimulates the release of gastric juices. It also initiates stomach and intestinal contractions. Secretin is produced in the duodenum when HCl is present and stimulates the release of alkaline fluids into the duodenum. When fats and amino acids enter the duodenum, cholecystokinin is released, stimulating the release of digestive enzymes.

Blood glucose is regulated in a negative feedback system in which the principal participants are pancreatic hormones and the liver. When blood glucose levels are high, glucose is converted to glycogen for storage. When blood glucose levels fall, glycogen is broken down to release glucose.

Animals can obtain the energy needed for life processes from either fats, carbohydrates, or proteins. On a dry matter basis, the energy content of fats is approximately twice that of carbohydrates or proteins.

In addition to energy, animals must obtain certain amino acids, fatty acids, vitamins, and minerals in their diets. Vitamins are often coenzymes or cofactors for enzymatic reactions. Minerals are functional components of molecules and are important in ionic balance.

The major nutritional problem facing North Americans is obesity, which has been associated with several serious health problems.

HOW TO STUDY THE CHAPTER

Read the entire chapter through quickly, focusing on the major concepts.

Use the GUIDED STUDY OF THE CHAPTER to help you identify the important details as you **reread** the chapter. Writing out the answers to these questions will help fix them in your mind as well as provide you with a valuable study aid.

Answer the questions in TESTING YOUR UNDERSTANDING without the aid of your text. Check your answers against those in PERFORMANCE ANALYSIS. Analyzing your answers will give you valuable feedback on your level of understanding and preparedness for classroom testing.

GUIDED STUDY OF THE CHAPTER

I. **Introduction** (page 715)

1. *Vocabulary:* Define digestion.

2. The molecular products of digestion serve what three functions?

3. Describe two evolutionary modifications of the digestive tract as animals became larger and more complex.

II. Digestive Tract in Vertebrates (pages 715–725)

A. General Remarks (pages 715–716)

4. Name four accessory organs of the digestive system.

> *Focus:* The lining of the digestive system is continuous with the skin. Consequently, the digestive cavity is external to the body tissues and gut contents are sequestered.

5. a. Identify one consequence of this sequestration of gut contents from the body tissues.

   b. What two processes are involved in digestion?

6. Arrange the following components of the digestive tract in the order in which food items pass through them: anus, esophagus, large intestine, mouth, pharynx, small intestine, stomach.

7. Name the layers of the digestive tract in order from the innermost to the outermost.

8. a. Identify the four components of the mucosa.

   b. In some parts of the digestive system, the epithelium contains cells that secrete what two substances?

9. a. The submucosa is composed of what tissue type?

   b. What other structures may be found in the submucosa?

10. a. Describe the orientation of the two layers of the muscularis externa.

    b. *Vocabulary:* What is peristalsis?

   c. What part of the gut wall is responsible for peristalsis?

   d. *Vocabulary:* Sphincters are modifications of what structure and serve what function?

11. a. Cite another name for the serosa. (Figure 34–2)

    b. *Vocabulary:* What is a mesentery? (Figure 34–2)

B. The Oral Cavity: Initial Processing (pages 716–718)

12. In vertebrates, the mouth typically performs what two functions related to nutrition?

13. a. The teeth of vertebrates are adapted for what functions?

    b. What structure in birds compensates for their lack of teeth?

    c. Describe the following components of the vertebrate tooth.

    Crown

    Root

    Dentine

    Pulp cavity

14. a. Identify the function for which each type of tooth is specialized.

    Incisors

    Canines

    Premolars

    Molars

    b. Describe modifications of dentition in at least three vertebrates. (Figure 34–4)

15. a. Name four functions of the tongue in vertebrates.

    b. Taste buds may be located in what three places? (Figure 34–5)

16. a. *Vocabulary:* What is saliva?

    b. Where is saliva produced? (See also Figure 34–6.)

    c. Identify the function of each component of saliva.

    Mucus

    Amylase

    d. Account for the alkaline nature of saliva.

    e. How does the saliva of carnivores differ from that of humans?

17. a. Name four factors that can stimulate saliva production.

    b. Under what conditions is salivation inhibited?

C. **The Pharynx and Esophagus: Swallowing** *(page 718)*

18. *Vocabulary:* Define swallowing.

19. a. *Vocabulary:* What is the pharynx? (Figure 34–7)

    b. What mechanism prevents food from entering the trachea? (Figure 34–7)

20. a. How does the structure of the human esophagus differ from that in animals that gulp their food?

    b. Use the terms voluntary and involuntary to distinguish between the act of swallowing in humans and in animals that gulp their food.

    c. What propels food along the esophagus in humans?

    > *Focus:* The esophagus passes through the thoracic cavity between the lung fields and then penetrates the diaphragm to enter the abdominal cavity, where it empties into the stomach.

21. a. Describe the composition of the peritoneum.

    b. Describe the structure of mesenteries.

D. *Essay:* **The Heimlich Maneuver** *(page 719)*

22. a. Describe what causes a person to choke on food.

    b. How can a choking victim be distinguished from a heart attack victim?

23. Describe the correct position of the hands to perform the Heimlich maneuver when the victim is (a) sitting or standing, and (b) lying down.

24. Why should a choking victim who has had a successful Heimlich maneuver performed see a physician immediately following the procedure?

E. The Stomach: Storage and Liquefaction *(page 720)*

> *Focus:* Among mammals, stomachs vary widely in their capacity depending on the dietary habits of the organism.

25. a. What are gastric pits?

    b. Name the two types of secretory cells in the gastric glands and identify the product of each type.

    c. What constitutes "gastric juice"?

26. a. The pH of gastric juice is normally _____ .

    b. Fluids this acidic would damage the stomach mucosa except for the presence of _____ , which is secreted by _____ .

27. Name four functions of HCl in the stomach.

28. a. In addition to HCl, what converts pepsinogen to pepsin?

    b. What is the digestive action of pepsin?

    c. What conditions are necessary if pepsin is to work?

29. How may stomach function be influenced by the nervous system?

30. a. Where is gastrin produced?

    b. What stimulates gastrin release?

    c. What is the effect of gastrin?

> *Focus:* Food leaving the stomach has been reduced to a semiliquid mass.

31. a. Food leaves the stomach through what structure?

    b. How long does it take after a meal for the human stomach to empty?

F. The Small Intestine: Digestion and Absorption *(pages 720–723)*

> *Focus:* In the small intestine, the breakdown of food particles is completed and the products are absorbed into the body.

32. a. Name three structural characteristics that increase the surface area of the small intestine.

    b. *Vocabulary:* Distinguish between villi and microvilli.

    c. The total surface area of the adult human small intestine is approximately _____ .

33. a. Which region of the small intestine is most active in digestion?

    b. The rest of the small intestine functions mainly in:

34. Fill in the following table concerning digestive components active in the small intestine.

Table for Question 34

COMPONENT	SOURCE(S)	FUNCTION(S)
Mucus		
Digestive enzymes		
Alkaline fluid		
Bile		

35. Why is it essential to neutralize the acidity of the food bolus as it enters the small intestine?

36. What is the major component of gallstones and under what conditions do they form? (Figure 34–11)

37. Identify the substrate and products for each group of enzymes.

   Pancreatic amylases

   Lipases

38. Describe the action of the three different types of digestive enzymes that act upon proteins.

39. For each hormone, state (a) the stimulus that increases its production and (b) its effect.

   Secretin

   Cholecystokinin

40. Summarize the effect on the intestinal tract of parasympathetic and sympathetic stimulation.

41. Fill in the table with the source and substrate of each enzyme. (Table 34–1)

Table for Question 41

ENZYME	SOURCE	SUBSTRATE
Chymotrypsin		
Dipetidase		
Lipase		
Pancreatic amylase		
Pepsin		
Sucrase		

Absorption of Nutrients (page 723)

42. How are (a) disaccharides, (b) amino acids and dipeptides, and (c) small fatty acids absorbed into the bloodstream?

43. Construct two flow diagrams.
   a. Trace the route by which large fatty acids and glycerol are absorbed and enter the bloodstream.

   b. Trace the route by which cholesterol is absorbed and enters the bloodstream.

44. Identify the fates of the chylomicrons, fats, and LDL particles circulating in the blood.

G. **The Large Intestine: Further Absorption and Elimination** *(page 725)*

45. a. Name three sources of the water in the large intestine.

   b. What happens if water absorption in the large intestine malfunctions, and how seriously may this affect a person's health?

46. a. How do symbiotic bacteria of the large intestine obtain nutrients?

    b. What do these bacteria provide to their host?

47. a. What is the origin of the human appendix?

    b. What is (i) the cause and (ii) the possible consequence of a ruptured appendix?

    c. The appendix is the site of what activity?

48. Identify three components of feces.

Focus: **Feces, which are lubricated by mucus, are stored in the rectum before elimination through the anus.**

H. *Essay:* **Aids to Digestion** *(page 724)*

49. a. What adaptation enables terrestrial herbivores to obtain nutritional value from cellulose and other structural polysaccharides of plants?

    b. The symbionts benefit from the association in what ways?

    c. Name two types of nutrients that the symbionts provide for their host.

50. Identify the structure(s) in which fermentation occurs in each animal.

    Horse

    Rabbit

    Cow (a ruminant)

51. How do rabbits recycle their symbiotic bacteria?

52. Explain what is happening when a cow is "chewing the cud."

53. Name two functions of the symbionts in the rumen.

Focus: **Ruminants obtain nutritional value from cellulose indirectly. The rumen microorganisms digest the cellulose, and the symbionts themselves are a source of energy and protein for the host.**

54. The rumen-stomach complex constitutes _____ percent of the body weight of a cow.

Tidbit: In ruminants, bloat is a life-threatening condition in which the rumen is distended with gas. Cattle are known to bloat quickly if the opening from the rumen to the esophagus is blocked.

55. Account for the phenomenon described in the Tidbit.

Focus: **Adult cattle secrete approximately 60 liters of saliva per day. This is a major source of bicarbonate and is important in acid-base homeostasis.**

III. **Major Accessory Glands** *(pages 725–726)*

A. **The Pancreas** *(page 725)*

56. The pancreas develops from what structure in the embryo?

57. a. Pancreatic tissue resembles what other tissue?

    b. See Figure 34–12. Name two characteristics, visible in the micrograph, that are typical of cells producing proteins for export.

58. a. Identify six digestive enzymes and three endocrine products produced by the pancreas. (See also Table 34–1.)

b. Name the cells that produce the pancreatic hormones.

B. The Liver *(pages 725–726)*

> *Focus:* The liver is the largest internal organ of the body.

59. List nine functions of the liver.

> *Focus:* The liver plays a role in the breakdown, synthesis, and storage of a vast variety of substances.

IV. Regulation of Blood Glucose *(page 726)*

60. a. State the major purpose of digestion.

b. What molecule is the major cellular energy source and the fundamental building block for organic molecules in the body?

61. By what route do substances absorbed in the intestine reach the liver?

62. a. Identify the fate of monosaccharides and amino acids in the liver.

b. What happens to the nitrogen and glucose that result from amino acid processing?

63. What factor determines whether the liver releases glucose to or removes it from the blood?

64. List three hormones produced by the pancreas that influence blood glucose levels, and indicate the action of each hormone.

V. Some Nutritional Requirements *(pages 726–730)*

A. General Remarks *(pages 726–728)*

> *Focus:* The energy requirements of the body can be met by either carbohydrates, proteins, or fats.

65. a. On a dry weight basis, compare the energy content of carbohydrates, proteins, and fats.

b. Which types of food molecules are always associated with water?

66. a. *Vocabulary:* What are essential amino acids?

b. Of the 20 amino acids used in protein synthesis, how many *cannot* be synthesized by humans?

c. Why does obtaining a diet balanced in amino acids require extra thought and planning if a person does not consume animal products?

67. a. Identify the importance of obtaining essential fatty acids.

b. How might these be obtained?

68. Many vitamins function as _____.

69. Name one consequence of severe vitamin A deficiency.

> *Focus:* The fat-soluble vitamins A, D, and K are stored in the body and are toxic if consumed in excess. Vitamin E is also a fat-soluble vitamin that is stored in the body.

70. a. Name three organs involved in the production and activation of vitamin D. (Figure 34–14)

b. From what chemical is vitamin D produced? (Figure 34–14)

c. Summarize the current hypothesis explaining the evolution of human skin color variations and the role of vitamin D in this process.

71. List six minerals essential to the body and indicate one function of each mineral.

72. Identify one good source for each of the following vitamins. (Table 34–3)

Vitamin A   Vitamin C

B vitamins   Vitamin D$_3$

Biotin   Vitamin E

Folic acid   Vitamin K

73. Match each vitamin with the system disorder(s) or type of syndrome resulting from its deficiency. Answers may be used more than once. A blank may have more than one answer, as indicated in parentheses. (Table 34–3)

1. vitamin A
2. vitamin C
3. vitamin D
4. vitamin E
5. vitamin K
6. vitamin B$_6$
7. vitamin B$_{12}$
8. biotin
9. folic acid
10. niacin
11. pantothenic acid
12. riboflavin
13. thiamine

a. _____ dermatitis, skin lesions (5 answers)

b. _____ nervous system disorders (3 answers)

c. _____ visual abnormalities (2 answers)

d. _____ digestive system disorders (2 answers)

e. _____ abnormalities of red blood cells (3 answers)

f. _____ defective bone formation

g. _____ inadequate blood coagulation

h. _____ connective tissue malformation

i. _____ heart or cardiovascular disorders (2 answers)

B. *Essay:* **Mother's Milk—It's the Real Thing** *(page 727)*

> *Focus:* Infant mammals, including humans, depend upon milk as their sole nutrient source until their digestive systems are fully developed.

74. Compare and contrast the milk of the following animals with respect to water, carbohydrate, fat, and protein content: human, cow, and harp seal.

75. a. In addition to nutrients, what other substances does a mammalian infant obtain in milk and what is their significance?

   b. What other features of mother's milk are important?

Focus: **The amount of time that an infant requires milk varies among the different mammalian species.**

---

76. a. What is the cause of "lactose intolerance"?

   b. What are the consequences of lactose ingestion by individuals with this intolerance?

77. Summarize the selective forces thought to be responsible for the presence of lactase in some human adults and its absence in others.

78. Identify the problems associated with bottle-feeding infants in Third World nations.

C. **The Price of Affluence** (pages 728–730)

   79. a. *Vocabulary:* Define obesity.

   b. Obesity has been associated with what health problems?

80. Summarize the health risks associated with the following features common to North American diets:

Excess salt

Excess animal fat

Low fiber

Fad dieting

VI. **Summary** (page 730): Read the summary. If you are familiar with the essential features of the material presented there, you are ready to complete the section TESTING YOUR UNDERSTANDING.

TESTING YOUR UNDERSTANDING

After you have completed this examination, compare your answers with those in the section that follows.

1. Which statement does NOT reflect a trend in the evolution of digestive systems?
   a. The tract became more convoluted within the coelom.
   b. Working surface area increased.
   c. Different regions became specialized for different digestive functions.
   d. All activities associated with digestion became localized in organs of the digestive tract itself.
   e. A one-way digestive tract developed.

2. Which structures act as valves to control the passage of food from one area of the digestive tract to another?
   a. villi   d. sphincters
   b. microvilli   e. islets of Langerhans
   c. mesenteries

3. A mesentery is a:
   a. thin film of connective tissue covering the external surface of all abdominal organs.
   b. thin film of connective tissue lining the abdominal cavity.
   c. thickened band of circular muscle tissue that regulates the passage of food from one portion of the digestive tract to another.
   d. layer of connective tissue within the mucosa of the digestive tract.
   e. fold of connective tissue connecting the digestive tract to the posterior abdominal wall.

4. The tearing and grinding of food items, performed by the teeth in mammals, are accomplished in modern birds by the:
   a. beak.   d. talons.
   b. gizzard.   e. teeth, which are actually modifications of the beak.
   c. crop.

5. The bulk of a mammalian tooth is composed of:
   a. bone.
   b. enamel.
   c. dentine.
   d. calcium phosphate.
   e. the pulp cavity.

6. Salivary secretion is NOT stimulated by which situation?
   a. the presence of a noxious substance in your stomach
   b. thinking about the first meal you will have after a seven-day backpacking trip
   c. letting a piece of peppermint candy melt in your mouth
   d. the odor of your neighbor's grilling steaks
   e. taking an exam for which you are poorly prepared

7. How is it possible to distinguish with certainty a choking victim from a person experiencing a heart attack?
   a. The choking victim cannot speak.
   b. The heart attack victim will clutch his or her chest.
   c. The choking victim will lose consciousness rapidly.
   d. The heart attack victim will be in obvious distress.
   e. The choking victim can only breathe with his or her mouth open.

8. The distended stomach of an adult human holds _____ liters of food.
   a. 0.5 to 1.5
   b. 1.0 to 2.5
   c. 2.0 to 4.0
   d. 3.0 to 4.5
   e. 4.0 to 5.0

9. When protein-containing food reaches the stomach, _____ is secreted by gastric cells.
   a. gastrin
   b. pepsin
   c. secretin
   d. carboxypeptidase
   e. cholecystokinin

10. In a human adult, the extended length of the small intestine is _____ meters.
    a. 10
    b. 8
    c. 6
    d. 4
    e. 2

11. In the presence of acidic food, the duodenum releases _____, which stimulates the pancreas and liver to secrete alkaline fluids.
    a. cholecystokinin
    b. secretin
    c. gastrin
    d. insulin
    e. glucagon

12. Cholesterol obtained in food is transported through the blood as:
    a. chylomicrons.
    b. protein-coated droplets.
    c. lipid-bile salt complexes.
    d. low-density lipoprotein complexes.
    e. high-density lipoprotein complexes.

13. What happens to chylomicrons in the bloodstream?
    a. They are gradually broken apart, yielding fatty acids.
    b. They are actively transported into muscle cells, where they are broken apart and the fatty acids are used for energy.
    c. They diffuse into fat cells, where they are stored.
    d. They are actively transported into liver cells, where the proteins are removed and the fatty acids are released back into the bloodstream.
    e. They are actively transported into fat cells, where the proteins are removed and the fatty acids are stored.

14. The symbiotic bacteria inhabiting the large intestine:
    a. constitute a major portion of fecal matter.
    b. are the primary source of vitamin K in humans.
    c. break down substances for which humans lack the appropriate digestive enzymes.
    d. provide some amino acids to their host.
    e. All of these statements are true of the bacteria inhabiting the large intestine.

15. The hormones insulin, glucagon, and somatostatin are produced by the:
    a. parietal cells.
    b. islets of Langerhans.
    c. chief cells.
    d. liver.
    e. mucosal glands of the duodenum.

16. Which of the following is NOT a function of the liver?
    a. regulation of blood cholesterol
    b. synthesis of digestive hormones
    c. storage of fat-soluble vitamins
    d. degradation of hemoglobin to bilirubin
    e. inactivation of hormones

17. Which hormone stimulates the release of glucose from liver cells, thereby raising blood glucose?
    a. insulin
    b. secretin
    c. glucagon
    d. somatostatin
    e. cholecystokinin

18. Severe deficiencies of vitamin A have recently been associated with:
    a. anemia.
    b. scurvy.
    c. failure of blood coagulation.
    d. cardiovascular disorders.
    e. increased mortality from certain childhood diseases.

19. Which vitamin is NOT among those listed in your text in association with blood or blood cell abnormalities?
    a. riboflavin, B_2
    b. cyanocobalamin, B_{12}
    c. folic acid
    d. tocopherol, E
    e. naphthoquinone, K

20. T or F Digestive enzymes function by catalyzing con-

densation reactions, in which the bonds between the units of a polymer are broken by the removal of water molecules.

21. **T or F**   In some animals, swallowing is an entirely voluntary process. In these animals, the entire length of the esophagus contains striated muscle.

22. **T or F**   Pepsinogen, which breaks down proteins into peptides, is active only at the high pH of the normal stomach.

23. **T or F**   Infant deaths due to diarrheal diseases are largely a consequence of dehydration.

24. **T or F**   On a dry weight basis, proteins and carbohydrates supply approximately the same amount of energy per gram, but fats supply twice as much energy per gram.

25. **T or F**   The only nutrients that mammals cannot synthesize and must obtain in their diet are the essential amino acids.

26. Match each type of mammalian tooth with the function for which it is specialized. Use each answer only once. A blank may have more than one answer, as indicated in parentheses.

    1. canines
    2. incisors
    3. premolars
    4. molars

    a. _____ cutting
    b. _____ grinding (2 answers)
    c. _____ stabbing and tearing

27. Match each mineral with the correct function or substance of which it is a component. An answer may be used more than once. A blank may have more than one answer, as indicated in parentheses.

    1. calcium
    2. chloride
    3. iodine
    4. iron
    5. phosphorus
    6. sodium

    a. _____ osmotic balance (2 answers)
    b. _____ hemoglobin
    c. _____ thyroid hormone
    d. _____ cytochromes
    e. _____ bone formation (2 answers)

28. Describe the activity of the three groups of enzymes that digest proteins in the small intestine.

29. Summarize the influence on intestinal activity of the parasympathetic and sympathetic divisions of the autonomic nervous system.

30. Name three major problems associated with the use of infant formulas in Third World nations.

PERFORMANCE ANALYSIS

1. **d**   In the evolution of digestive systems, a major step was development of the one-way digestive tract. The digestive tract lengthened and became more convoluted, resulting in an increased working surface area. In addition, the different regions became functionally specialized, and accessory digestive structures, which produce enzymes and digestive fluids, developed outside the tract itself. (page 715)

2. **d**   Sphincters are areas in which the circular muscle in the wall of a hollow organ is thickened. They regulate the flow of food from one region of the digestive tract to another. (page 716)

3. **e**   A mesentery is a fold of the peritoneum that connects the digestive tract to the posterior abdominal wall. Blood and lymph vessels and nerves are located between the two layers of peritoneum that form a mesentery. (Figure 34–2)

4. **b**   Modern birds have no teeth. The avian gizzard is a muscular structure that contains gravel and grinds food items so that they may be digested chemically. (page 716)

5. **c**   The bulk of a mammalian tooth consists of dentine, a bone-like material that forms throughout the life of the tooth. (page 716)

6. **e**   Salivary secretion may be stimulated by the smell or anticipation of food, by the presence of food in the mouth, or by the presence of a noxious substance in the stomach. Fear or great stress inhibits salivary secretion. (page 718)

7. **a**   A person who is choking cannot talk because the trachea (through which air must pass in order for the person to speak) is occluded by the food item. A person experiencing a heart attack can speak. (essay, page 719)

8. **c**   An adult human stomach, fully distended, may hold 2 to 4 liters of food. (page 720)

9. **a**   Gastrin is secreted by gastric cells when protein-containing food reaches the stomach. Secretin is secreted by the duodenum when HCl is present in the duodenum. Cholecystokinin is secreted by the duodenum when fats and amino acids are present in the duodenum. (page 720)

10. **c**   The extended length of the adult human small intestine is approximately 6 meters. (page 720)

11. **b**   Secretin, which is secreted when the duodenum contains HCl, stimulates the secretion of alkaline fluids by the liver and pancreas. (page 722)

12. **d**   Cholesterol that is absorbed by intestinal cells is packaged as low-density lipoproteins, which then travel in the lymphatic system to the blood. (page 723)

13. **a**   In the bloodstream, chylomicrons gradually break apart, yielding fatty acids that may be taken up by the body cells. (page 723)

14. **e** The symbiotic bacteria of the small intestine are the major source of vitamin K in humans and constitute a major portion of the fecal material. They break down food substances that the host is unable to digest and in the process provide some amino acids and vitamins to the host. (page 725)

15. **b** The islets of Langerhans are the sites in the pancreas in which the hormones insulin, glucagon, and somatostatin are produced. Pancreatic amylase is produced in a separate type of cell. (page 725)

16. **b** Although the liver plays many roles in digestion and nutrient processing, it is not known to produce any digestive hormones. (pages 725, 726)

17. **c** Glucagon, which is produced by the pancreas, stimulates the breakdown of glycogen to glucose and release of glucose from liver cells, thereby raising blood glucose. Insulin stimulates the uptake of glucose by cells, lowering blood glucose. (page 726)

18. **e** In the severe famines that struck Ethiopia in the mid-1980s, severe vitamin A deficiency increased deaths due to several infectious childhood diseases. (page 728)

19. **a** Vitamin B_{12} deficiency is characterized by anemia and malformed red blood cells. Folic acid deficiency results in the failure of red blood cells to mature, leading to anemia. Vitamin E deficiency results in increased fragility of red blood cells. Vitamin K deficiency is characterized by failure of blood clotting. Vitamin B_2 deficiency results in photophobia and fissuring of the skin. (Table 34–3)

20. **False** In the enzymatic digestion of a polymer such as starch or a polypeptide, the bonds are broken by hydrolysis, which is the splitting of a bond by addition of a water molecule. (page 718)

21. **True** The entire length of the esophagus of dogs contains striated muscle and swallowing is entirely voluntary. In humans, the upper portion of the esophagus contains striated muscle but the rest contains smooth muscle. Therefore, swallowing begins as a voluntary process but is involuntary for the most part. (page 718)

22. **False** Pepsinogen, which is the precursor for the active enzyme, is converted to pepsin in the presence of HCl. Pepsin is active only at the *low* pH of the stomach. (page 720)

23. **True** In a diarrheal disease, the major threat to life is severe dehydration resulting from loss of body fluids. Dehydration is a major cause of infant deaths due to diarrheal disease. (page 725)

24. **True** On a dry weight basis, fats supply twice as much energy per gram as do carbohydrates and proteins, which supply approximately equal amounts of energy per gram. (page 726)

25. **False** Mammals must obtain certain amino acids, fatty acids, minerals, and vitamins in the diet. (page 728)

26. The answers are **a2, b3, b4, c1**. (page 716)

27. The answers are **a2, a6, b4, c3, d4, e1, e5**. (page 728)

28. One type of protein-digesting enzyme breaks apart the long chains, with each enzyme cleaving the protein at specific amino acids. A second type removes dipeptides from the ends of polypeptides produced by the first group. The third type of enzyme cleaves the dipeptides into amino acids. (page 722)

29. The parasympathetic nervous system promotes intestinal contractions and digestive activities. The sympathetic nervous system inhibits these activities. (page 723)

30. (1) Poor sanitation is often the rule—neither clean water nor facilities for cleaning bottles are available. (2) The formula may be overdiluted in attempts to save money. (3) Children fed with infant formula may be more susceptible to infectious diseases since they do not obtain maternal antibodies. (essay, page 727)

Energy and Metabolism II: Respiration

MAJOR CONCEPTS

There are two meanings for the term respiration in biology. Cellular respiration involves the aerobic reactions by which ATP is made in mitochondria (Chapter 9). At the level of the organism, respiration is the exchange of oxygen and carbon dioxide between cells and the environment.

Oxygen consumption is directly related to energy expenditure, which in turn is directly related to metabolic rate.

Respiration involves the processes of bulk flow and diffusion. Gases move within the bloodstream, into lungs, and across gills by bulk flow. They move into and out of tissues and the blood by diffusion.

The concentration of a gas is defined in terms of the pressure it exerts. In a mixture of gases (such as the air) the total pressure equals the sum of the pressures of individual gases, which are known as partial pressures. Gases diffuse from regions of high partial pressure to regions of low partial pressure.

Atmospheric pressure is measured in terms of millimeters of mercury. At sea level, this value is 760 mm Hg. Air is a mixture of gases that consists mainly of nitrogen (77 percent) and oxygen (21 percent).

Diffusion alone is adequate to meet the gas exchange needs of unicellular organisms and very small animals, especially those with flattened bodies. Larger animals evolved respiratory systems for exchanging gases with the environment and circulatory systems for transporting them through the body. Respiratory systems in animals include body surfaces, insect tracheae, gills, and lungs.

Vertebrate gills are thought to have evolved from structures designed for filter-feeding. In the modern vertebrate gill, the flow of water across the gill membranes is in a direction opposite to the flow of blood in capillaries within the membranes. This countercurrent arrangement greatly increases the efficiency of gas exchange.

Respiration by lungs has two major advantages over gills: air is a much better source of oxygen than water; and oxygen diffuses far more rapidly through air than through water. Respiration by lungs is advantageous over respiration through the body surface because lungs can be kept moist without losing large quantities of water through evaporation.

In lungfish, lungs evolved from the pharynx. The evolution of respiratory systems in amphibians, reptiles, and other air-breathing vertebrates has resulted in a trachea guarded by a valve (the epiglottis) and in nostrils, which enable the animal to breathe with its mouth closed. In addition to lungs, amphibians accomplish a significant amount of gas exchange through the skin.

In humans, air is inspired through the nasal cavities (where it is filtered, humidified, and warmed) and travels through the pharynx, trachea, bronchi, and bronchioles to the alveoli, in which gas exchange occurs.

The trachea, bronchi, and bronchioles are lined with ciliated cells coated with mucus. Foreign particles are trapped in the mucus and carried out of the lungs by the sweeping action of the cilia.

Inspiration occurs when the diaphragm and intercostal muscles of the rib cage contract, increasing the volume of the thoracic cavity. This creates negative pressure in the lungs, and air rushes in. When the muscles relax, thoracic volume decreases and air is expelled.

In all animals that depend on a circulatory system for transporting oxygen to the tissues, the blood contains respiratory pigments to which molecular oxygen binds. All respiratory pigments have a metal-containing unit (the oxygen-binding portion) and a protein chain. The respiratory pigments of vertebrates and echinoderms are located in red blood cells. Those of most invertebrates are dissolved in blood plasma.

The major respiratory pigment of humans is hemoglobin, each molecule of which consists of four subunits. One hemoglobin molecule can combine with up to four oxygen molecules. The binding of the first molecule increases the affinity of the hemoglobin molecule for oxygen, thereby facilitating the binding of subsequent oxygen molecules.

The partial pressure of oxygen in the surrounding plasma determines whether hemoglobin will bind oxygen or release it. At partial pressures greater than 60 mm Hg, oxygen remains bound to hemoglobin. The partial pressure of oxygen decreases as blood flows through tissue capillaries, and hemoglobin releases its bound oxygen. In alveolar capillaries, the partial pressure of oxygen is high and oxygen binds to hemoglobin.

The affinity of hemoglobin for oxygen is also influenced by blood pH—at the lower pH present in tissue capillaries, the affinity for oxygen is reduced, encouraging the release of oxygen.

Myoglobin is a respiratory pigment present in skeletal muscle. It has a higher affinity for oxygen than does hemoglobin. The oxygen bound to myoglobin acts as a reserve for use in times of great oxygen demand, as occurs during exercise.

Carbon dioxide exists in the blood in three forms: dissolved in the plasma, bound to amino groups of hemoglobin molecules, and as bicarbonate ion (the largest fraction). In a reaction catalyzed by carbonic anhydrase, dissolved carbon dioxide combines with water to form carbonic acid, which dissociates to release bicarbonate ions and hydrogen ions. The reactions are reversible and the direction depends upon the partial pressure of carbon dioxide in the blood.

Respiratory rate and depth are controlled by respiratory neurons in the brainstem. These neurons respond to information received from receptors that detect blood levels of hydrogen ions, carbon dioxide, and oxygen, and the degree to which lungs and chest are stretched.

HOW TO STUDY THE CHAPTER

Read the entire chapter through quickly, focusing on the major concepts.

Use the GUIDED STUDY OF THE CHAPTER to help you identify the important details as you **reread** the chapter. Writing out the answers to these questions will help fix them in your mind as well as provide you with a valuable study aid.

Answer the questions in TESTING YOUR UNDERSTANDING without the aid of your text. Check your answers against those in PERFORMANCE ANALYSIS. Analyzing your answers will give you valuable feedback on your level of understanding and preparedness for classroom testing.

GUIDED STUDY OF THE CHAPTER

I. Introduction (page 732)

1. *Vocabulary:* State the two biological meanings of respiration.

2. a. *Vocabulary:* Define basal metabolism.

   b. What is the relationship between oxygen consumption, energy expenditure, and metabolic rate?

II. Diffusion and Air Pressure (page 733)

3. Gas exchange between living cells and the environment is accomplished by the process of:

4. *Review Vocabulary:* Define diffusion.

5. The concentration of a gas is described in what terms?

Focus: In a mixture of gases, the total pressure of the mixture equals the sum of the pressures of the individual gases, each of which is referred to as a partial pressure.

6. a. What is the value of atmospheric pressure at sea level? (See also Figure 35–2.)

   b. Why is atmospheric pressure usually measured using a column of mercury rather than of colored water?

7. a. When a value is assigned to the partial pressure of a gas in a liquid, what does that value actually represent?

   b. The partial pressure of oxygen (Po_2) in the lungs is 100 mm Hg and the Po_2 in the blood is 40 mm Hg. Knowing these facts, summarize the mechanism by which oxygen exchange occurs in the lungs.

8. a. At increasing altitudes, does atmospheric pressure increase or decrease?

   b. Does Po_2 increase or decrease?

9. a. Under what conditions do the "bends" develop?

   b. What causes the pain and other symptoms associated with the bends?

III. Evolution of Respiratory Systems (pages 734–737)

A. General Remarks (page 734)

> **Focus:** Oxygen enters cells and carbon dioxide leaves cells by diffusion, which is efficient only over very short distances.

10. a. What types of organisms obtain oxygen by direct simple diffusion from the external environment?

    b. In large animals, oxygen enters and is transported through the body by what mechanism in what two body systems?

11. a. Describe the anatomical adaptations by which gas exchange is accomplished in the earthworm.

    b. How do some worms vary their surface area depending on oxygen availability?

12. a. Outline the mechanism(s) by which gas exchange is accomplished in insects.

    b. Where (specifically) does gas exchange occur in insects? (Figure 35–3)

    c. What is one of the major limitations on the size of insects?

B. Evolution of Gills (pages 734–735)

13. Distinguish between gills and lungs in terms of development.

14. a. Describe the "working surface" of a gill.

b. What was probably the original purpose of the vertebrate gill?

15. a. Primitive vertebrates obtained oxygen primarily through:

b. The major selection pressure favoring the evolution of gills was:

c. List three specific changes (either unrelated or indirectly related to respiration) that occurred in animals as a consequence of this selection pressure.

16. The gill became the respiratory site as animals developed a greater need for:

17. a. Describe the flow of water across fish gills.

b. Name two factors that influence the rate of diffusion. (Figure 35–5)

c. How does the countercurrent arrangement of blood vessels with respect to water flow maximize oxygen diffusion?

18. By what means may fish regulate the rate of water flow across gills?

Focus: Some fish are so dependent upon swimming to provide adequate water flow across the gills that they will suffocate if their movement is restricted.

C. Evolution of Lungs (page 736)

19. a. What disadvantage do lungs have compared with gills?

b. What is their major advantage?

20. a. Contrast the amount of energy spent on respiration by aquatic animals versus air-breathers.

b. Identify three characteristics of air versus water that account for this discrepancy.

21. What advantage do lungs have over other mechanisms for breathing air?

22. a. Summarize the adaptive value of lungs to lungfishes.

b. The lungs of lungfishes developed from what anatomical structure?

23. a. How did the evolution of the respiratory system in amphibians and reptiles differ from that in lungfishes?

b. What structures enable reptiles and amphibians to breathe with closed mouths?

Focus: Although the respiratory structures are similar in amphibians and reptiles, amphibians rely largely on the skin for gas exchange, whereas reptiles respire nearly entirely through their lungs.

24. a. Vocabulary: Define ventilation lungs.

b. How is air movement into and out of lungs accomplished in frogs?

In reptiles, birds, and mammals?

25. a. What is the role of avian air sacs? (Figure 35–7)

b. How does respiration in birds differ from that in mammals as a consequence of this adaptation? (Figure 35–7)

D. **Respiration in Large Animals: Some Principles** *(page 737)*

26. Summarize the major differences between the following types of respiratory systems: body surface, external gills, internal gills, tracheae, and lungs. (Figure 35–8)

27. Identify the four stages involved in gas exchange in animals, focusing on the roles of bulk flow and diffusion.

IV. **The Human Respiratory System** *(pages 738–740)*

A. **General Remarks** *(pages 738–740)*

28. *Vocabulary:* Distinguish between inspiration and expiration.

> *Focus:* Inspiration and expiration normally occur through the nostrils in humans.

29. Name the function of each feature of the nasal cavities.

Hairs and cilia

Mucus

Rich blood supply

Tidbit: The epiglottis folds over and occludes the opening of the trachea during swallowing of food and fluids.

30. a. Describe the structure and location of the vocal cords.

b. Vocal cord development is influenced by what hormones?

c. What is laryngitis?

31. a. The trachea is lined with cells of what type?

b. What function is served by the rings of cartilage in the tracheal wall?

32. a. *Vocabulary:* Distinguish between bronchi and bronchioles.

b. What feature of their wall structure alters resistance to air flow?

33. By what mechanism are foreign particles removed from lung air passages?

34. a. Where does the actual exchange of gases occur in lungs?

b. List in order from the cavity of the alveolus to the lumen of the capillary the three layers that gases must cross.

c. Gas exchange between the air and the blood occurs by _____.

d. How does the total respiratory surface of the alveoli compare with total body surface area?

35. a. *Vocabulary:* What is the pleura and what is its function?

b. What is pleurisy?

B. *Essay:* Cancer of the Lung *(page 739)*

Focus: From 1950 to 1985, the increase in the death rate from all types of cancer was due entirely to an increase in the number of deaths from lung cancer.

36. a. What percentage of lung cancer patients will die within five years of diagnosis?

b. Most lung cancer patients have a history of:

37. How does cigarette smoke affect the mechanisms that normally clear the lung of foreign particles?

V. **Mechanics of Respiration** *(page 741)*

Focus: Air flows into and out of lungs because of differences between alveolar air pressure and atmospheric pressure.

38. a. By what mechanism is alveolar air pressure altered in humans?

b. Distinguish between inspiration and expiration in terms of activity of the diaphragm and intercostal muscles.

39. a. What fraction of lung volume is exchanged during an average breath?

b. What fraction of lung volume may be exchanged during intentional deep breathing?

40. Why do large aquatic mammals suffocate when they are beached?

VI. **Transport and Exchange of Gases** *(pages 741-745)*

A. **Hemoglobin and Its Function** *(pages 741-743)*

41. a. Why are respiratory pigments important to the survival of most animals?

b. Respiratory pigments are found in virtually all active animals EXCEPT:

c. All respiratory pigments have what two structural features in common?

42. Fill in the following table comparing hemoglobin and hemocyanin.

Table for Question 42

	HEMOGLOBIN	HEMOCYANIN
Example phyla		
Metal involved		
Color when combined with oxygen		

43. Where are most respiratory pigments located in (a) invertebrates and (b) vertebrates and echinoderms?

44. a. Describe the structure of the hemoglobin molecule.

b. What is a heme unit?

c. Oxygen binds to what *atom* of a hemoglobin molecule?

45. a. How many iron atoms are present in one hemoglobin molecule?

b. How many oxygen molecules can be transported by one hemoglobin molecule?

> *Focus:* The combination of an oxygen molecule with hemoglobin increases the affinity of the hemoglobin molecule for another oxygen molecule.

46. Interpret the sigmoid shape of the oxygen-hemoglobin dissociation curve in terms of hemoglobin subunits and oxygen affinity.

47. The presence of hemoglobin increases the oxygen-carrying capacity of the blood by what degree relative to plasma?

48. a. What factor determines whether oxygen combines with or is released from hemoglobin?

b. Summarize the conditions present in the alveoli and in the tissues that cause the oxygenation of hemoglobin in the lungs and the release of oxygen in the tissues.

49. a. At what Po_2 does oxygen leave hemoglobin?

b. What is the adaptive value of this feature?

> *Focus:* Blood that has passed through tissue capillaries is still 70 percent saturated with oxygen.

50. The extra oxygen mentioned in the Focus statement serves what function?

51. Describe how the blood of the following organisms is specially adapted to meet their oxygen demands. (Figure 35–15)

Small animal

Llama

Human fetus

52. List three forms in which carbon dioxide exists in the blood and indicate which form predominates.

53. a. Write the two-stage reaction by which carbon dioxide is converted to bicarbonate ion. Name the intermediate compound.

b. Indicate the step at which carbonic anhydrase functions.

c. Where is carbonic anhydrase located?

d. What factor determines which direction will be favored in the reaction?

e. Describe the situation in the (i) lungs and (ii) tissues that influences the reaction, and indicate which direction is favored in each location.

54. By what mechanism does the partial pressure of carbon dioxide in blood influence the affinity of hemoglobin for oxygen?

B. *Essay:* **Diving Mammals** *(page 744)*

55. Other than the diving reflex, name four physiological adaptations of diving mammals that contribute to their diving ability.

56. List five physiological events that are part of the diving reflex.

57. What is the survival value of the diving reflex to humans?

58. Summarize the observations of Martin Nemiroff that led to his conclusion that the diving reflex may be activated in humans. (Include the conditions under which the reflex is activated.)

Focus: Even if a person exhibits no signs of life following drowning, if he or she has been immersed in cold water, resuscitation procedures should be started immediately and continued indefinitely.

C. **Myoglobin and Its Function** *(page 745)*

59. a. Where is myoglobin found?

b. Describe the structure of a myoglobin molecule.

60. a. How does the affinity of myoglobin for oxygen compare with that of hemoglobin?

b. What are the consequences of this situation?

61. a. Myoglobin releases its oxygen when tissue P_{O_2} falls to what level?

b. Under what conditions might this occur?

c. How would you describe the role of myoglobin?

VII. Control of Respiration (pages 745–747)

A. General Remarks (pages 745–747)

> *Focus:* Respiratory neurons in the brainstem control the rate and depth of respiration, which is an involuntary activity over which a degree of voluntary control can be exerted.

62. Redefine inspiration and expiration in terms of the neural events controlling these activities.

63. Respiratory neurons receive signals from receptors that detect what four types of information?

64. Where are the chemoreceptors that detect changes in pH and in carbon dioxide and oxygen concentrations in the blood?

> *Focus:* The respiratory control centers are extremely sensitive to changes in the hydrogen ion concentration (pH), which reflects the carbon dioxide concentration.

65. *Vocabulary:* What is hyperventilation and what is its effect on the acid/base status of the blood?

66. Why is it impossible to commit suicide by holding one's breath?

67. If the hydrogen ion detectors malfunction, what system assumes control of respiration?

68. What are the effects of a massive barbiturate overdose on respiratory function?

B. *Essay:* High on Mt. Everest (page 746)

69. a. What factor presents the major threat to life in extremely high altitudes?

    b. Survival at high altitudes depends largely upon what activity?

70. How do extreme altitude conditions affect the P_{O_2} of blood and the work capacity?

71. Identify five changes in metabolism or brain function that occurred in the climbing scientists.

VIII. Summary (pages 747–748): Read the summary. If you are familiar with the essential features of the material presented there, you are ready to complete the section TESTING YOUR UNDERSTANDING.

TESTING YOUR UNDERSTANDING

After you have completed this examination, compare your answers with those in the section that follows.

1. Respiration at the cellular level refers to oxygen-requiring chemical reactions that take place in the:
   a. mitochondria.
   b. endoplasmic reticulum.
   c. chloroplasts.
   d. ribosomes.
   e. Golgi bodies.

2. Air pressure at sea level is 760 millimeters of mercury (mm Hg). If dry air at sea level is 77 percent nitrogen, the partial pressure of nitrogen is _____ mm Hg.
   a. 7.6
   b. 159
   c. 585
   d. 770
   e. The answer cannot be determined from the information provided.

3. Oxygen moves into cells by _____, but is transported in the blood by _____.
   a. simple diffusion; bulk flow
   b. bulk flow; simple diffusion
   c. active transport; bulk flow
   d. facilitated diffusion; simple diffusion
   e. bulk flow; active transport

4. At 15°C, air contains about _____ times more oxygen than does water. Further, oxygen diffuses _____ times faster through air than through water.
   a. 21; 300
   b. 42; 300,000
   c. 100; 300
   d. 300; 100
   e. 300,000; 21

5. Oxygen needs are met by simple diffusion without bulk flow in:
   a. ants.
   b. fish.
   c. earthworms.
   d. cockroaches.
   e. jellyfish.

6. In insects, air is piped directly into the tissues by a series of chitin-lined tubules called:
   a. tracheae.
   b. bronchioles.
   c. alveoli.
   d. bronchi.
   e. spiracles.

7. Lungs are more advantageous than gills because:
   a. air contains 21 percent oxygen whereas water contains only 0.5 percent oxygen.
   b. more muscular work is required to aerate gills than lungs.
   c. oxygen diffuses more rapidly in air than in water.
   d. water has a much higher viscosity than air.
   e. All of these statements represent advantages of lungs over gills.

8. Although lungs are largely a vertebrate "invention," some invertebrates, such as _____, also have lungs.
   a. shrimp
   b. grasshoppers
   c. snails
   d. beetles
   e. slugs

9. Which of the following is the correct (although incomplete) order of structures through which an inhaled oxygen molecule would pass on its way from the nose or mouth to the capillaries of the lungs?
   a. larynx; pharynx; trachea; alveoli
   b. pharynx; trachea; bronchioles; bronchi
   c. larynx; pharynx; trachea; bronchi
   d. larynx; pharynx; bronchioles; alveoli
   e. pharynx; larynx; trachea; bronchi

10. In mammals, the vocal cords are located in the:
    a. bronchi.
    b. trachea.
    c. esophagus.
    d. pharynx.
    e. larynx.

11. What structures within the human respiratory system are the actual sites of gas exchange between the lungs and the circulatory system?
    a. bronchi
    b. alveoli
    c. bronchioles
    d. trachea/pharynx
    e. nose/mouth

12. The presence of respiratory pigments in blood can increase its oxygen-carrying capacity by a factor of:
    a. 25.   d. 70.
    b. 40.   e. 85.
    c. 55.

13. The maximum number of oxygen molecules that can be carried by a vertebrate hemoglobin molecule is:
    a. two.
    b. four.
    c. six.
    d. eight.
    e. ten.

14. Most invertebrates:
    a. do not contain respiratory pigments in their blood.
    b. have respiratory pigments in red blood cells.
    c. have respiratory pigments dissolved in their plasma.
    d. have several different respiratory pigments in their blood.
    e. do not even have blood.

15. Which of the following is NOT a respiratory pigment?
    a. hemoglobin
    b. hemocyanin
    c. myoglobin
    d. myocyanin
    e. All of these molecules are respiratory pigments.

16. In the tissues of a person at rest, the P_{O_2} is about _____ mm Hg, and oxygen _____.
    a. 100; binds to hemoglobin
    b. 75; is released from myoglobin
    c. 65; leaves hemoglobin and binds to myoglobin
    d. 40; is released from hemoglobin
    e. 20; is released from myoglobin

17. Carbon dioxide is _____ than oxygen, and some may be carried in the blood simply dissolved in the plasma.
    a. larger
    b. smaller
    c. more soluble
    d. less soluble
    e. The premise is false; less carbon dioxide than oxygen is carried in blood in the dissolved state.

18. In the lungs, the partial pressure of carbon dioxide is _____, and _____.
    a. high; bicarbonate and hydrogen ions are formed in the blood
    b. high; carbon dioxide diffuses into the blood
    c. low; carbonic acid in the blood dissociates into bicarbonate and hydrogen ions
    d. low; carbon dioxide diffuses out of the blood into alveoli
    e. low; the affinity of hemoglobin for oxygen decreases

19. Myoglobin does not begin to release a significant amount of its oxygen to the tissues until the Po_2 in the tissues falls below _____ mm Hg.
    a. 20   d. 50
    b. 30   e. 60
    c. 40

20. When hyperventilating (breathing deeply and rapidly), a person may feel faint and dizzy because of increased:
    a. acidity of the blood and brain.
    b. alkalinity of the blood and brain.
    c. carbon dioxide concentration of the blood and brain.
    d. oxygen concentration of the blood and brain.
    e. stimulation of the respiratory center.

21. The respiratory center that controls normal breathing by causing the diaphragm and intercostal muscles to contract is located in the:
    a. cerebral cortex.   d. pituitary gland.
    b. pineal gland.   e. hypothalamus.
    c. brainstem.

22. **T or F** Exhalation occurs when the diaphragm and intercostal muscles relax.

23. In terms of the partial pressures of carbon dioxide in air and liquid, explain what happens when a carbonated drink "goes flat." Relate this to the phenomenon that causes the bends in divers.

24. Identify the major selection pressure to which aquatic vertebrates were exposed that favored the evolution of more efficient respiratory mechanisms. List three consequences of this selection pressure that favored the "capture" of the gills as a respiratory device.

25. Describe the roles of the diaphragm and intercostal muscles in inhalation and exhalation.

26. Why do large aquatic mammals suffocate when they are beached, even though they are air breathers?

27. Identify two means by which opiate drugs (e.g., morphine) and barbiturates influence the control of respiration.

PERFORMANCE ANALYSIS

1. **a** Cellular respiration, which involves the reactions of the Krebs cycle and the electron transport chain, takes place in mitochondria. (page 732)

2. **c** The total pressure of a mixture of gases equals the sum of the partial pressures of the individual gases. The partial pressure of each individual gas is analogous to its concentration. At sea level, atmospheric pressure is 760 mm Hg, and 77 percent of the mixture of gases is nitrogen. Therefore, the partial pressure of nitrogen is 760 mm Hg $\times$ 0.77 or 585 mm Hg. (page 733)

3. **a** Oxygen crosses the respiratory surface (e.g., lung epithelium, gill epithelium, moist integument) to enter the blood by diffusion. Once in the blood, oxygen is transported by bulk flow to all regions of the body. In the tissues, oxygen diffuses from the blood into individual cells. (page 734)

4. **b** Air is a much better source of oxygen than is water. At 15°C, air has 42 times as much oxygen as does water, and oxygen diffuses through air 300,000 times faster than through water. (page 736)

5. **e** Jellyfish have no circulatory system and their oxygen needs are met by simple diffusion through the body surface. (page 734)

6. **a** The tracheae of insects are chitin-lined tubes through which air is pumped when the body moves. Gas exchange takes place at the internal ends of the tubes, which are not lined with chitin. (page 734)

7. **e** Even though lungs do not allow a constant flow of air across the respiratory surface, they are more efficient respiratory structures than gills for all of the listed reasons. (page 736)

8. **c** Some terrestrial snails have lungs that evolved independently but resemble the lungs of amphibians in some respects. (page 736)

9. **e** Air flows through structures of the human respiratory system in the following order: nose/mouth, pharynx, larynx, trachea, bronchi, bronchioles, alveoli. (pages 738, 740)

10. **e** The vocal cords of mammals are located in the larynx. (page 738)

11. **b** In mammals, gas exchange between the external atmosphere and the blood occurs in the alveoli, the presence of which greatly increases the respiratory surface area of the lungs. (page 740)

12. **d** Molecular oxygen is not very soluble in plasma. The presence of respiratory pigments can increase the oxygen-carrying capacity of the blood by as much as seventyfold. (page 741)

13. **b** Each hemoglobin molecule of vertebrates carries a maximum of four oxygen molecules, one attached to each of four heme units. (page 742)

14. **c** Most invertebrates have respiratory pigments dissolved in the plasma. The notable exception is the echinoderms, whose respiratory pigments are carried in red blood cells, as are those of vertebrates. (page 742)

15. **d** Hemoglobin is found in vertebrates and a variety of invertebrates. Myoglobin is a respiratory pigment found in vertebrate muscles. Hemocyanin, which contains copper and is blue when oxygenated, is commonly found in mollusks and arthropods. (pages 741, 742, 745)

16. **d** In a human at rest, the P_{O_2} in the tissues is about 40 mm Hg. Oxygen is released from hemoglobin when the P_{O_2} falls below 60 mm Hg. (page 742)

17. **c** Carbon dioxide is more soluble in plasma than is oxygen. Some of the carbon dioxide present in the plasma is dissolved, and dissolved carbon dioxide is in equilibrium with carbonic acid, which dissociates into bicarbonate and hydrogen ions. Most of the carbon dioxide in blood is in the form of bicarbonate. Some carbon dioxide is bound to the protein of the hemoglobin molecule. (page 743)

18. **d** The partial pressure of carbon dioxide in the lungs is fairly low, and dissolved carbon dioxide diffuses out of the blood into the alveoli. (page 743)

19. **a** Myoglobin in the muscles acts as reserve of oxygen in case tissue demand exceeds oxygen supply. Myoglobin does not release its bound oxygen until the P_{O_2} in the tissues falls to 20 mm Hg. In a human at rest, P_{O_2} in the tissues is approximately 40 mm Hg. (page 745)

20. **b** When a person hyperventilates, carbon dioxide is removed from the blood at a faster rate than normal and the partial pressure of carbon dioxide in the blood falls. The equilibrium of carbon dioxide/carbonic acid/carbonate and hydrogen ions is shifted toward carbon dioxide, resulting in a decrease in hydrogen ion concentration (increase in pH and alkalinity). The dizzy, faint feeling associated with hyperventilation is due to increased alkalinity of the blood passing through the brain. (page 747)

21. **c** The respiratory center regulating breathing is located in the brainstem. (page 745)

22. **True** When the diaphragm and intercostal muscles contract, the thoracic cavity expands, creating negative pressure which draws air into the lungs. Exhalation occurs when the diaphragm and intercostal muscles relax, reducing the volume of the thoracic cavity and forcing air out of the lungs. (page 741)

23. In a carbonated drink, carbon dioxide has been forced into the liquid under pressure and the drink sealed under pressure. The partial pressure of carbon dioxide in a sealed drink is greater than the partial pressure of carbon dioxide in the atmosphere. Opening a carbonated beverage allows the pressure in the container to equalize with the atmosphere, and the carbon dioxide in the drink consequently bubbles out. An opened carbonated beverage "goes flat" as the carbon dioxide in the drink diffuses into the atmosphere. Diffusion continues until the partial pressure of carbon dioxide in the drink equals that in the air.

The tanks used by divers contain compressed air—air at a pressure greater than atmospheric pressure. The air regulator through which a diver breathes delivers the compressed air at a pressure equal to that of the surrounding water pressure. When breathing compressed air, nitrogen is dissolved in the blood to equal the partial pressure of the delivered nitrogen. The force of the water on the diver's body also maintains the nitrogen partial pressure in the blood. If the diver ascends too quickly, the rapid decrease in pressure allows the dissolved nitrogen to return to the gas phase, forming bubbles in the bloodstream. These circulating bubbles may lodge in capillaries of vital organs, interrupting blood supply and causing the tissue damage and pain characteristic of "the bends." (page 733)

24. The major selection pressure favoring the evolution of more efficient respiratory systems was predation. Three consequences of this pressure were a trend toward armored or scaled skin, selection for larger body size and swifter movement, and greater energy requirements accompanying larger, swifter bodies. (page 735)

25. See the answer to question 22. (page 741)

26. Large aquatic mammals suffocate when they are beached because the muscles responsible for thoracic expansion are not strong enough to adequately expand the rib cage against the force of gravity compressing their heavy bodies. In their aquatic environment, their body weight is supported by the water and the respiratory muscles can expand the thorax. (page 741)

27. Barbiturates and opiates such as morphine depress the brainstem receptors that are sensitive to hydrogen ion concentration. In this situation, signals from receptors sensitive to oxygen concentration provide the information needed by respiratory neurons for the regulation of respiration. At high doses, these drugs may depress the activity of the respiratory neurons themselves. (page 747)

CHAPTER 36

Energy and Metabolism III: Circulation

MAJOR CONCEPTS

The circulatory system is the route by which oxygen, nutrients, chemical messengers, and other substances are transported to cells and waste products are carried away from cells. The system is composed of a network of vessels through which blood is pumped by the heart. Blood travels away from the heart through arteries and then smaller arterioles to capillary beds in organs and tissues. The actual function of the circulatory system (exchange of materials between individual cells and the external environment) occurs in the capillaries. From the capillaries, blood travels through venules that converge to form veins as they carry blood back to the heart.

The blood has two major fractions: the cellular fraction contains red blood cells, white blood cells, and platelets; the fluid fraction (plasma) is 90 percent water in which proteins, chemical messengers, ions, and other substances being transported are suspended or dissolved. Human blood is approximately 60 percent plasma and 40 percent cells. The osmotic potential of plasma is maintained by plasma proteins which, when present in normal amounts, prevent the loss of fluid from the vessels to the tissues.

The major production site for all types of blood cells is the bone marrow. The function of red blood cells (also called erythrocytes) is to transport oxygen to all cells of the body. Oxygen is bound to the hemoglobin molecules within erythrocytes as blood passes through the lungs. Mature erythrocytes do not have a nucleus and have a life span of only 120 to 130 days. White blood cells (leukocytes) defend the body against invasion by pathogenic microorganisms or foreign bodies. There are several types of leukocytes, each of which has a specific role in defense. Platelets are fragments of megakaryocytes and are essential to the formation of blood clots.

Blood clot formation is initiated when a substance called tissue factor reacts with a specific plasma protein. This sets in motion a series of reactions resulting in the conversion of fibrinogen to fibrin, which forms an insoluble meshwork entrapping red blood cells and platelets to form a clot. There are several known hereditary bleeding disorders caused by the absence or inactivity of one or more factors in the clotting pathway.

The cardiovascular system includes the heart and the blood vessels through which blood is pumped. Vertebrates have closed circulatory systems and hearts with two to four chambers.

The simplest vertebrate heart is that of fishes. Consisting of two chambers, the fish heart pumps blood directly to the gills, where it is oxygenated before continuing on to the body.

Amphibians have three-chambered hearts; two atria to receive blood and one ventricle to pump it. Oxygen-poor blood from the body enters one atrium, is passed to the ventricle, and is pumped to the lungs and to vessels going to the skin, where it is oxygenated. At the same time, oxygenated blood enters the other atrium, passes to the ventricle, and is pumped to the body. Even though the ventricle receives oxygenated and deoxygenated blood at the same time, very little mixing occurs.

The most complex heart structure is exhibited by birds and mammals, whose hearts have four chambers. Functionally, they have two hearts; the left heart pumps oxygenated blood to the body (systemic circuit) and the right heart pumps deoxygenated blood to the lungs (pulmonary circuit). This arrangement provides maximal oxygen to the tissues, which is essential to maintain the high metabolic rate associated with endothermy in these animals.

The sounds associated with a heartbeat result when two sets of valves close. These valves are located between the atria and ventricles and between the ventricles and the arteries exiting the heart.

Cardiac output is defined as the volume of blood pumped by the heart per minute. Cardiac output depends on the heart rate and the volume pumped by each ventricular contraction.

The electrical impulse stimulating cardiac muscle contraction originates in the sinoatrial node located in the right atrium. This impulse spreads from cell to cell through the atria, which contract simultaneously. The impulse travels to the ventricles through the atrioventricular node and the bundle of His. The atrioventricular node contains slow-conducting fibers, the presence of which imposes a delay on impulse transmission so that the ventricles contract after atrial contraction is completed.

Blood pressure is the force with which blood pushes against vessel walls. Pressure is greatest in the aorta, the vessel carrying blood from the left ventricle to the body. As blood flows through the arteries, blood pressure decreases. Blood pressure is lowest in the right atrium (which receives blood from the body). Systolic pressure is blood pressure during ventricular contraction; diastolic pressure is blood pressure during ventricular relaxation.

Blood pressure and the flow of blood through specific regions of the body are controlled by the cardiovascular regulating system in the medulla of the brain. This center receives information on blood pressure from specialized stretch receptors in the walls of vessels. This information is integrated with other data from the body. The cardiovascular regulating system sends signals to the heart and blood vessels, influencing heart rate and shunting blood to some capillary beds while bypassing others.

Cardiovascular disease is the major cause of death in the United States. Factors associated with cardiovascular disease include atherosclerosis, high levels of circulating low-density lipoproteins, high blood pressure, cigarette smoking, inactivity, and heredity (in some cases).

The lymphatic system is an open network of vessels that begin in the tissues. These vessels collect surplus interstitial fluid, filter it through lymph nodes, and return it to the bloodstream. The lymph nodes are also the sites of proliferation of lymphocytes, specialized leukocytes that perform important immune functions.

HOW TO STUDY THE CHAPTER

Read the entire chapter through quickly, focusing on the major concepts.

Use the GUIDED STUDY OF THE CHAPTER to help you identify the important details as you **reread** the chapter. Writing out the answers to these questions will help fix them in your mind as well as provide you with a valuable study aid.

Answer the questions in TESTING YOUR UNDERSTANDING without the aid of your text. Check your answers against those in PERFORMANCE ANALYSIS. Analyzing your answers will give you valuable feedback on your level of understanding and preparedness for classroom testing.

GUIDED STUDY OF THE CHAPTER

I. **Introduction** *(page 749)*

    1. a. In general terms, what is the role of the circulating blood?

        b. Name seven substances transported through the body in the bloodstream.

II. **The Blood** *(pages 750–752)*

A. **General Remarks** *(page 750)*

    2. a. *Vocabulary:* What is plasma?

        b. Plasma constitutes what percentage of the blood volume?

        c. What fraction of plasma is water?

    3. Name three components that form the nonplasma portion of blood.

    4. Blood is classified as a type of _____ tissue.

B. **Plasma** *(page 750)*

5. Identify three types of substances that travel in plasma.

6. a. What are three functions of plasma proteins?

   b. Name the three types of plasma proteins and identify the function of each type.

C. **Red Blood Cells** *(page 750)*

7. What is another name for red blood cells?

8. The shape of a red blood cell facilitates what two activities? (Figure 36–2)

9. a. How does a mature erythrocyte differ structurally from less specialized eukaryotic cells?

   b. What occupies most of the volume of a mature erythrocyte?

10. What is the concentration of red blood cells in human blood?

11. a. What is the average life span of a human erythrocyte?

    b. Explain why red blood cells have such a short life span.

12. Where are replacement erythrocytes made?

D. **White Blood Cells** *(pages 750–751)*

13. a. *Vocabulary:* What are leukocytes?

    b. What is the ratio of leukocytes to erythrocytes?

14. Name five ways in which leukocytes differ from erythrocytes.

15. a. Identify the general function of white blood cells.

    b. How do white blood cells move through tissues?

16. *Vocabulary:* Define pus.

17. Cite two specific places where leukocytes are produced.

E. **Platelets** *(page 751)*

18. a. *Vocabulary:* What are platelets?

    b. Platelets originate from what cell type?

19. Platelets play an important role in what body function?

F. **Blood Clotting** *(pages 751–752)*

20. a. Identify two events that can trigger the initiation of the clotting cascade.

    b. Where is tissue factor located and how is its location important to the initiation of the clotting cascade?

21. Fill in the blanks in the flow diagram of the clotting cascade. Use these components: fibrin, fibrinogen, prothrombin, thromboplastin.

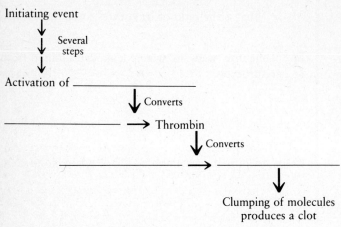

Initiating event

↓ Several steps

Activation of _____

↓ Converts

_____ → Thrombin

↓ Converts

_____ → _____

↓

Clumping of molecules produces a clot

22. The mature blood clot is a barrier preventing the passage of what two materials? (Figure 36–4)

23. What triggers the clotting of blood in a test tube?

Focus: Plasma is the noncellular component of blood; serum is plasma without clotting components.

Focus: In the body, the clotting reactions are controlled so that the clot is no larger than necessary to close the defect in the damaged tissue.

---

24. a. *Vocabulary:* What is hemophilia?

b. Some people argue that research for the sake of knowledge ("basic research") does not contribute to society. Cite one example in which techniques developed by molecular biologists are being applied to solve a real problem.

III. The Cardiovascular System *(pages 752–753)*

25. Compare and contrast the circulatory systems of earthworms, mollusks, and vertebrates. (Define the systems as open or closed, and describe the pumping mechanisms.)

26. *Vocabulary:* What is a cardiovascular system?

IV. The Blood Vessels *(pages 753–754)*

A. General Remarks *(page 753)*

27. Arrange these vessel types in order to trace the path of blood from the heart to the capillaries and back to the heart: capillaries, venules, veins, arterioles, arteries.

28. *Vocabulary:* Distinguish between aorta and vena cava.

Focus: The aorta is the artery through which blood exits from the heart and the vena cava is the vein through which blood returns to the heart. (Actually, there are two venae cavae, one from the anterior regions of the body and the other from the posterior regions.)

---

29. *Vocabulary:* What is endothelium?

30. Describe the composition of the three layers of an arterial wall.

31. Fill in the table comparing arteries and veins.

Table for Question 31

	ARTERIES	VEINS
Thickness of wall		
Number of wall layers		
Degree of elasticity		
Shape retention when empty		
Diameter of lumen relative to thickness of wall (study Figures 36–5 and 36–6)		

B. The Capillaries and Diffusion (pages 753–754)

32. How is the structure of the capillary wall well suited to its function?

33. Name two factors that contribute to the slow movement of blood through capillaries.

34. a. Identify two *routes* by which materials may leave or enter capillaries.

    b. Describe three *mechanisms* by which materials may travel along these routes.

35. What is the maximum distance of any one body cell from a capillary and what is the significance of this fact?

V. The Heart (pages 754–757)

A. Evolution of the Heart (pages 754–755)

36. *Vocabulary:* Distinguish between atrium and ventricle.

37. a. Construct diagrams to compare and contrast the hearts and circulatory systems of (i) fish, (ii) amphibians, and (iii) birds/mammals. Indicate the immediate destination(s) of blood leaving the heart and the immediate destination(s) of newly oxygenated blood.

    (i)

    (ii)

    (iii)

b. For (i), (ii), and (iii), above, describe the means by which oxygenated blood is prevented from mixing with deoxygenated blood.

Focus: The separation of oxygen-rich from oxygen-poor blood in birds and mammals is essential to supply adequate oxygen levels to cells functioning at a high metabolic rate.

B. The Human Heart *(pages 755–756)*

38. Trace the flow of blood through the heart and body beginning with the entry of blood into the right atrium.

39. a. What devices in the heart prevent the backflow of blood?

    b. Identify the events causing each of the two heart sounds.

    c. What causes a heart murmur?

40. *Vocabulary:* Define cardiac output in words and by an equation.

41. a. Name the peptide hormone produced and released by the atrial muscle cells and identify four organs on which receptors for this hormone have been identified.

    b. What is the apparent role of this hormone?

C. Regulation of the Heartbeat *(pages 756–757)*

42. a. What tissue initiates heart muscle contraction?

    b. How does this differ from the initiation of skeletal muscle contractions?

    c. Where is this "pacemaker" located?

43. a. Describe the spread of the impulse originating in the sinoatrial node and the mechanism by which coordinated contraction of the atria and ventricles is accomplished.

    b. What structure imposes a delay on the conduction of the impulse as it approaches the ventricles?

    c. Name the structure that is the only electrical bridge between the atria and the ventricles.

Focus: An electrocardiogram is a record of the *electrical* activity of the heart.

---

44. Explain how electrodes placed on a person's skin can detect electrical activity in the heart.

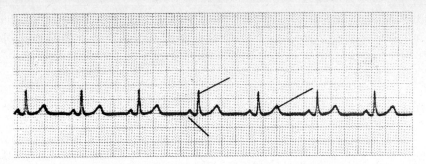

45. Label each peak on the electrocardiogram above to indicate which phase of heart function is represented. Use these labels: atrial excitation, ventricular excitation, ventricular relaxation. Indicate the peak in which the activity of atrial relaxation is hidden. (Figure 36–11)

46. a. Distinguish between the effect of parasympathetic and sympathetic nerve stimulation on the sinoatrial node.

    b. Which nerve carries parasympathetic fibers to the pacemaker?

47. What effect does adrenaline have on heart rate?

VI. The Vascular Circuitry (pages 758–759)

> *Focus:* **The circulatory system of air-breathing vertebrates consists of a systemic and a pulmonary circuit.**

48. a. Describe the flow of blood through the pulmonary circuit.

    b. What is unique about the (i) pulmonary arteries and (ii) pulmonary veins?

49. The first two branches off the aorta supply what organ?

50. a. An interruption of oxygenated blood supply to the brain results in unconsciousness after what time period?

    b. Irreversible brain damage may occur if the oxygenated blood supply is inadequate for what period of time?

51. a. *Vocabulary:* Define a portal system.

    b. The hepatic portal system enables:

    c. The liver receives blood through what two routes?

VII. Blood Pressure (pages 759–762)

A. General Remarks (pages 759–762)

52. a. *Vocabulary:* Define blood pressure.

    b. Distinguish between systolic and diastolic blood pressure.

    c. How is blood pressure usually expressed on a physician's chart?

53. Name four factors that influence blood pressure.

54. What is the relationship between blood pressure and rate of flow?

55. a. *Vocabulary:* Distinguish between vasoconstriction and vasodilation.

    b. Describe the means by which arterioles influence the flow of blood to capillaries.

    c. Identify five factors that influence the smooth muscles of arterioles.

Focus: The regulation of blood flow through arterioles enables the body to shunt blood toward areas that need it most and to restrict flow to less needy areas.

---

56. Describe three examples in which blood flow is *increased* in specific areas of the body.

57. What is the adaptive value of fainting (assuming an individual does not land on his or her head)?

58. List three examples in which emotional events influence the flow of blood through capillaries.

59. a. Name two factors responsible for the decrease of blood pressure as blood flows through the vessels

    b. The most dramatic decrease in blood pressure occurs as blood passes through the

    _____. (Figure 36–15)

60. a. Identify two characteristics of veins that promote return of blood to the heart.

    b. What two active processes move blood through veins toward the heart?

B. *Essay:* **Diseases of the Heart and Blood Vessels** *(pages 760–761)*

61. a. What is the most prevalent type of disease in the United States?

    b. What percentage of the U.S. population is thought to have some variety of this disease?

62. a. *Vocabulary:* Define a heart attack.

    b. What is ischemia?

    c. What are the three principal causes of a heart attack?

d. Distinguish between thrombus and embolus.

63. Identify three factors that influence recovery from a heart attack.

64. What fraction of all deaths in the United States are due to ischemic heart disease?

65. a. How does angina pectoris differ from a heart attack?

   b. What symptoms are associated with both angina pectoris and heart attacks?

66. a. *Vocabulary:* What is a stroke?

   b. Name three potential causes of a stroke.

   c. What two factors influence the effect of a stroke?

67. a. Describe the abnormalities in blood vessels affected by atherosclerosis.

   b. How does atherosclerosis contribute to heart attacks and strokes?

68. a. Name two conditions apparently associated with atherosclerosis.

   b. What four factors contribute to increased LDL levels and decreased HDL levels, an undesirable situation?

69. What evidence suggests that female hormones (a) protect against and (b) increase the risk of atherosclerosis?

70. a. Summarize the trend in mortality from ischemic heart disease since the mid-1960s.

   b. What environmental factors may be contributing to this trend?

71. a. *Vocabulary:* Define hypertension.

   b. How many people in the United States are affected by hypertension?

   c. Name two ways in which hypertension may contribute to heart disease.

   d. How is hypertension currently controlled?

e. What evidence supports a potential racial predisposition to cardiovascular disease?

C. **Cardiovascular Regulating Center** *(page 762)*

72. a. Where is the cardiovascular regulating center located?

b. What types of nerves are controlled by this center?

c. Identify four locations of receptors that send information about blood pressure to the cardiovascular regulating center.

73. a. Describe what happens when stretch receptors in the carotid artery indicate that blood pressure is falling.

b. This type of regulation is an example of:

VIII. **The Lymphatic System** *(pages 762–763)*

74. Outline two mechanisms by which the fluid component of the blood may leave the vessels.

75. How can plasma proteins be lost from the blood?

76. *Vocabulary:* Define edema.

77. How is the lymphatic system (a) similar to and (b) different from the blood venous system?

78. Where does the lymphatic system empty its contents into the bloodstream?

79. Identify three mechanisms by which lymph may be moved through lymphatic vessels in vertebrates.

80. a. What are lymph nodes?

b. What two functions do they perform?

IX. **Summary** *(pages 763–764)*: Read the summary. If you are familiar with the essential features of the material presented there, you are ready to complete the section TESTING YOUR UNDERSTANDING.

TESTING YOUR UNDERSTANDING

After you have completed this examination, compare your answers with those in the section that follows.

1. The straw-colored liquid that remains after all the cells and platelets have been removed from whole, uncoagulated blood by centrifugation is called:
   a. serum.
   b. complement.
   c. lymph.
   d. plasma.
   e. saline.

2. The major components preventing excessive loss of fluid from the bloodstream are:
   a. small organic molecules.
   b. plasma proteins.
   c. ions, such as sodium and chloride.
   d. white blood cells.
   e. platelets.

3. In an average 24-hour day, approximately _____ red blood cells die in your body and are replaced by new ones.
   a. 2 million   d. 170 billion
   b. 120 million   e. 1.7 trillion
   c. 7 billion

4. Which statement is NOT true of white blood cells?
   a. They can migrate through the tissues.
   b. There are only 1 or 2 white blood cells per thousand red blood cells.
   c. They can engulf bacteria, viruses, and other foreign particles.
   d. They stimulate the loss of interstitial fluid as pus.
   e. They move by means of pseudopodia in the tissues.

5. Initiation of the blood clotting reactions is the primary function of:
   a. platelets.   d. fibrinogen.
   b. immunoglobulins.   e. leukocytes.
   c. lymphocytes.

6. Which factor catalyzes the conversion of fibrinogen to fibrin in the final step of the blood clotting cascade?
   a. prothrombin
   b. thromboplastin
   c. thrombin
   d. tissue factor
   e. fibrinogen kinase

7. In mammals, oxygenated blood bound for the body leaves the heart via the:
   a. aorta.
   b. vena cava.
   c. carotid artery.
   d. jugular vein.
   e. coronary artery.

8. The actual function of the circulatory system is carried out by the:
   a. arteries.   d. lungs
   b. heart.   e. capillaries.
   c. veins.

9. The distinguishing feature of all portal systems is that they:
   a. carry unoxygenated blood.
   b. consist of two capillary beds in series.
   c. carry nutrients from the intestines to other organs.
   d. converge on the liver.
   e. are the only blood supply for the organ in which they are located.

10. Which animals possess a heart that functions as two distinct organs?
    a. fish and amphibians
    b. amphibians and birds
    c. birds and mammals
    d. mammals and reptiles
    e. reptiles and birds

11. When you listen to your heartbeat, the second sound ("dup") you hear is due to the _____ of the valves between the _____.
    a. closing; venae cavae and the right atrium
    b. opening; atria and ventricles
    c. closing; atria and ventricles
    d. opening; ventricles and the aorta and the pulmonary artery
    e. closing; ventricles and the aorta and the pulmonary artery

12. If a human heart beats 70 times a minute and expels 0.072 liters of blood with each beat, what is its cardiac output?
    a. 0.72 liters per minute
    b. 5 liters per minute
    c. 7.2 liters per minute
    d. 49 liters per minute
    e. 70 liters per minute

13. The right and left ventricles of the human heart contract almost simultaneously because they are connected to the atrioventricular node by:
    a. the bundle of His.   d. desmosomes.
    b. the sinoatrial node.   e. gap junctions.
    c. sympathetic nerves.

14. Which agent is NOT known to affect the heart rate?
    a. parasympathetic nervous system
    b. sympathetic nervous system
    c. adrenaline
    d. sinoatrial node
    e. cardiac peptide

15. Brain cells are irreversibly damaged if they are deprived of oxygen for:
    a. 30 seconds.
    b. 90 seconds.
    c. 2 to 3 minutes.
    d. 4 to 6 minutes.
    e. Brain cell damage is reversible if oxygen therapy is given within 10 minutes following deprivation.

16. Blood pressure is greatest in the:
    a. capillaries.   d. ventricles.
    b. atria.   e. venae cavae.
    c. aorta.

17. Special valves and skeletal muscle contractions facilitate blood flow in:
    a. arteries.   d. capillaries.
    b. venules.   e. veins.
    c. arterioles.

18. The cardiovascular regulating center is located in the:
   a. pons.
   b. medulla.
   c. sinoatrial node.
   d. atrioventricular node.
   e. cerebrum.

19. T or F For every 2 million molecules of thrombin produced in the blood clotting cascade, almost 40 million fibrin molecules are produced.

20. T or F The heart of the earthworm is more highly evolved than that of the mollusk.

21. T or F The atria of the human heart secrete a hormone that acts on the brain, kidneys, blood vessels, and adrenal glands.

22. T or F Even if it is removed from the body, a vertebrate heart will continue to beat if supplied with oxygen and a nutrient solution because cardiac muscle cells pulsate spontaneously.

23. T or F Blood flow back to the heart may be assisted by inspiratory movements that dilate the veins within the chest.

24. T or F Fluid leaves the blood at the venule end of capillaries by hydrostatic pressure and reenters the arteriole end by osmosis.

25. List three differences between arteries and veins and discuss the functional significance of these differences.

26. Distinguish between pulmonary arteries and pulmonary veins with respect to function and type of blood (oxygenated vs. deoxygenated) transported.

27. Name three factors that influence the smooth muscles controlling the diameter of arterioles.

28. What is the mechanism by which heart cells are damaged during a heart attack? Name three specific factors that may cause a heart attack.

29. Identify two functions of lymph nodes.

PERFORMANCE ANALYSIS

1. **d** When the cells are removed from uncoagulated whole blood by centrifugation, the liquid that remains behind is plasma; it contains all the blood clotting factors. Serum is the fluid portion of the blood that remains after a blood clot has contracted. (The clotting factors are removed during the formation of the blood clot.) (page 750)

2. **b** The plasma proteins, particularly albumin, maintain the osmotic potential of the blood, thereby preventing excessive fluid loss. (page 750)

3. **d** Approximately 2 million red blood cells die every *second*. A simple calculation reveals that 170 billion red blood cells die within a 24-hour period (2 million red blood cell deaths/sec × 60 sec/min × 60 min/hr × 24 hours = 170 billion). (page 750)

4. **d** White blood cells circulate in the blood and migrate through tissues in response to infection or inflammation. In the tissues, they move by pseudopodia and engulf bacteria, viruses, and foreign particles. They are the major component of pus. There are only 1 or 2 white blood cells per 1000 red blood cells. (pages 750, 751)

5. **a** Platelets, which are formed by fragmentation of megakaryocytes in the bone marrow, are involved in the initiation of blood clot formation. (page 751)

6. **c** Thrombin converts the soluble plasma protein fibrinogen into the insoluble protein fibrin, strands of which form a mesh that is the basis of a blood clot. (page 751)

7. **a** Oxygenated blood leaves the left ventricle through the aorta to travel through the systemic circulation. Deoxygenated blood leaves the right ventricle to travel through the pulmonary artery to the lungs. (page 755)

8. **e** The actual function of the circulatory system, which is to supply all cells of the body with nutrients, oxygen, and waste removal services, occurs in the capillaries as materials are exchanged between the bloodstream and individual cells. (page 753)

9. **b** A portal system consists of two capillary beds in series, i.e., blood flows from one capillary bed into another capillary bed before returning to the heart. In the hepatic portal system, blood from intestinal capillaries flows through the liver capillaries before returning to the heart. (page 759)

10. **c** Birds and mammals have four-chambered hearts in which oxygenated blood is physically separated from deoxygenated blood. Functionally, there is one circulatory system (and one "heart") for each type of blood. (page 755)

11. **e** The first sound of a heartbeat is made by the closing of the valves between the atria and ventricles; the second sound is made by the closing of the valves between the ventricles and the aorta and pulmonary artery. (page 755)

12. **b** Cardiac output is defined as heart rate (per minute) times stroke volume (volume of blood pumped per contraction). A simple calculation reveals a cardiac output of 5 liters per minute for the situation described. (page 756)

13. **a** The sinoatrial node of the right atrium is the heart's pacemaker. Its spontaneous activity initiates each heartbeat. Electrical impulses traveling through the heart from the sinoatrial node stimulate the atrioventricular node, which transmits the signal to the bundle of His. The bundle of His is the only electrical connection between the atria and the ventricles and carries the signal that stimulates the ventricles to contract nearly simultaneously. (page 757)

14. **e** The heart rate is affected by the sinoatrial node (the pacemaker), parasympathetic and sympathetic nerves, and adrenaline. At the present time, cardiac peptide is not known to affect heart rate. (pages 756, 757)

15. **d** If brain cells are deprived of oxygen for four to six minutes, they will suffer permanent damage. (page 759)

16. **c** In the cardiovascular system, blood pressure is greatest in the aorta and lowest in the right atrium. (page 760)

17. **e** The movement of blood through veins is aided by skeletal muscle contractions and by the presence of valves that prevent backflow of blood. (page 762)

18. **b** The cardiovascular regulating center, which coordinates information on blood pressure, blood flow, and heart rate, is located in the medulla of the brainstem. (page 762)

19. **False** The blood clotting cascade is an example of the amplification effects of an enzyme cascade. In the formation of a typical clot, 2 million thrombin molecules are produced at the beginning of the cascade and 160 *billion* fibrin molecules are produced in the last reaction. (page 752)

20. **False** In the earthworm, five "hearts," which are muscular portions in certain vessels, propel blood through a closed circulatory system. The mollusk heart is more complex, consisting of several chambers that pump blood through an open circulatory system. (page 752)

21. **True** Natriuretic factor (or cardiac peptide) is secreted by the atria and influences the regulation of blood volume and blood pressure. Receptors for this hormone have been identified in several organs, including the brain, kidneys, blood vessels, and adrenal glands. (page 756)

22. **True** Since the electrical impulses initiating the heartbeat begin in the heart itself, a vertebrate heart that has been removed from the body can continue to beat if it is supplied with oxygen and a nutrient solution. (page 756)

23. **True** Inspiratory movements of the thoracic cavity dilate vessels in the chest, lowering venous pressure and facilitating the flow of blood back to the heart. (However, taking a very deep breath and holding it will temporarily constrict the veins in the chest.) (page 762)

24. **False** At the arteriole end of a capillary bed, fluid loss from the blood is due to hydrostatic pressure. At the venule end of the capillary bed, most of the fluid returns to the blood by osmosis. The extra fluid is picked up by the lymphatic system. (page 762)

25. The walls of arteries are thicker, stronger, and more elastic than those of veins. Arteries expand when blood pulses through them, absorbing the shock of the pulse, and recoil when the pulse moves on. The lumen of an artery remains open even when empty. Veins are more pliable and less elastic than arteries and collapse when empty. They contain valves that prevent the backflow of blood. Blood moves through arteries because of blood pressure created by the pumping action of the heart. Blood moves through veins mainly as a result of muscular contractions. (page 753)

26. Pulmonary arteries carry deoxygenated blood from the right ventricle to the lungs. Pulmonary veins carry oxygenated blood from the lungs to the left ventricle. (page 758)

27. The flow of blood through a capillary bed is regulated by rings of smooth muscle in the wall of the arteriole feeding the capillary bed. The contraction and relaxation of these muscles is influenced by autonomic nerves (mainly sympathetic), adrenaline, noradrenaline, cardiac peptide, and other chemicals produced locally in the tissues. (page 759)

28. During a heart attack, cardiac muscle cells are damaged when their blood supply is inadequate (ischemia). If the oxygen supply is interrupted for a certain period, the cardiac cells may die. Occlusion of the coronary blood vessels may result from atherosclerosis, thrombus formation, or the lodging of an embolus in a vessel. (essay, page 760)

29. Lymph nodes filter foreign particles and cellular debris from the lymph before it enters the bloodstream. They are also the sites at which lymphocytes proliferate during an immune response. (page 763)

Section 6: Chapters 33–36

This review test is *not* designed to cover all of the important information in these chapters. However, it does touch on the major topic categories in each chapter. It will also give you valuable practice in taking this type of test. When you are finished, check your answers with those provided in the following PERFORMANCE ANALYSIS section.

1. There is a strong tendency among _____ for smaller litters and increased periods of parental care.
   a. vertebrates in general
   b. fish
   c. mammals
   d. reptiles
   e. amphibians

2. In multicellular organisms, cells that carry out the same function are organized to form:
   a. organs.
   b. tissues.
   c. organ systems.
   d. epithelia.
   e. cell aggregates.

3. The process of digestion involves two components:
   a. ingestion of food particles and elimination of wastes.
   b. breakdown of food molecules and absorption of nutrients into the body.
   c. enzymatic hydrolysis of polymers and bacterial fermentation.
   d. mechanical breakdown of food particles and chemical breakdown of food molecules.
   e. enzymatic processes and hormonal processes.

4. The lining of the abdominal cavity is known as the:
   a. pleura.
   b. mesentery.
   c. serosa.
   d. peritoneum.
   e. abdominal epithelium.

5. Which of these statements does NOT apply to respiration?
   a. Respiration is essential for the oxidation of energy-containing organic molecules.
   b. Respiration at the cellular level occurs in mitochondria.
   c. Respiration at the organismal level involves the intake of oxygen and the release of carbon dioxide.
   d. Respiration at the cellular level is independent of respiration at the level of the whole organism.
   e. Respiration that utilizes the electron transport chain always requires oxygen.

6. In the human respiratory system, the movement of debris-containing mucus is accomplished by:
   a. contractions of the alveoli.
   b. peristalsis of the tracheal rings.
   c. sweeping movements of cilia.
   d. spastic contractions of the larynx.
   e. rhythmic oscillations of flagella.

7. The life span of a red blood cell is only about _____ days because it _____.
   a. 20 to 30; has no nucleus
   b. 20 to 30; has no mitochondria
   c. 60 to 90; has no nucleus
   d. 60 to 90; has no mitochondria
   e. 120 to 130; cannot repair itself

8. Why is blood flow so sluggish through the body tissues of a fish?
   a. The propulsive force of the heart has been largely dissipated in the gills.
   b. Since the fish is a cold-blooded animal it is naturally sluggish.
   c. The fish heart is too simple to generate much power.
   d. The constant movement of the fish's body restricts blood flow.
   e. The fish has no lungs, the expansion of which would help pull blood back to the heart.

9. Neurons that transmit impulses within local regions of the central nervous system are called _____, whereas neurons that transmit signals from one region to another of the central nervous system are called _____.
   a. motor neurons; sensory neurons
   b. relay neurons; motor neurons
   c. interneurons; relay neurons
   d. relay neurons; interneurons
   e. sensory neurons; interneurons

10. Larger animals have a greater resistance to foreign invaders than do smaller animals because they have:
    a. more complex immune systems.
    b. larger surface-to-volume ratios.
    c. protective outer coverings.
    d. more complex immune systems and protective outer coverings.
    e. smaller surface-to-volume ratios and protective outer coverings.

11. What two substances are secreted into the small intestine to neutralize the gastric contents received from the stomach?
    a. alkaline fluid from the gallbladder and bile from the pancreas
    b. acidic fluid from the pancreas and alkaline fluid from the gallbladder
    c. acidic fluid from the gallbladder and bile from the pancreas
    d. alkaline fluid from the pancreas and bile from the gallbladder
    e. alkaline fluid from the pancreas and acidic fluid from the gallbladder

12. Monosaccharides are converted to glycogen and fat in the:
    a. liver.
    b. pancreas.
    c. small intestine.
    d. duodenum.
    e. pylorus.

13. Which animals have NO respiratory pigments in their blood?
    a. frogs
    b. insects
    c. earthworms
    d. lizards
    e. whales

14. Tissue fluid that travels through a separate vascular system before entering the bloodstream is called:
    a. plasma.
    b. edema.
    c. serum.
    d. interstitial fluid.
    e. lymph.

15. T or F Because it is constantly filled with blood, the heart is one of the few organs that does not possess capillaries.

16. T or F Integration and control within an organism involve two aspects: coordinating internal physiological processes; and receiving, processing, and correctly responding to information about the external environment.

17. T or F Sensors in the carotid arteries that measure carbon dioxide and hydrogen ion concentrations serve as backups to those that measure oxygen concentration.

18. Match each vitamin with the correct function. Use each answer only once. Each blank will have only one answer.
    1. vitamin A
    2. vitamin B$_{12}$, cyanocobalamin
    3. vitamin C
    4. vitamin B$_6$, pyridoxine
    5. vitamin K

    a. _____ essential for connective tissue formation
    b. _____ maturation of red blood cells
    c. _____ necessary for fatty acid metabolism
    d. _____ formation of visual pigments
    e. _____ synthesis of blood clotting factors

19. Compare and contrast the structure and function of skeletal and cardiac muscle.

20. By what behavioral means do fish regulate the delivery of oxygen-containing water to their gills?

21. Name four factors that influence the ratio of LDLs to HDLs, which is correlated with a person's tendency to develop atherosclerosis.

PERFORMANCE ANALYSIS

1. c Of all the vertebrates, mammals and birds tend to produce fewer young per reproductive effort and have extended periods of parental care. (page 702)

2. b Cells that conduct the same function are grouped together to form tissues. Different tissues may be organized into a functional unit known as an organ. (page 702)

3. b The two components of the digestive process are the breakdown of food molecules and the absorption of nutrients into the body. (page 715)

4. d The lining of the abdominal cavity is the peritoneum, which is composed of a sheet of epithelialized connective tissue. (page 718)

5. d Cellular respiration involves the reactions of the Krebs cycle and the electron transport chain, which occur in mitochondria. Since the final electron acceptor of the electron transport chain is molecular oxygen, cellular respiration depends upon respiration at the level of the organism. (page 732)

6. c The rhythmic, sweeping movements of cilia lining the airways of the respiratory system move the debris-trapping mucus layer up out of the airways to the pharynx. (page 740)

7. e In humans, a red blood cell lives about 120 to 130 days. Since mature red blood cells have neither nuclei nor mitochondria they cannot repair or sustain themselves. (page 750)

8. a Blood flows through a fish from the heart to the gill capillaries. By the time the blood leaves the gill capillaries, most of the propulsive force of the heart has been dissi-

pated because of the resistance of the capillaries. (page 754)

9. **c** Interneurons transmit signals within localized regions of the central nervous system, whereas relay neurons transmit signals between different regions of the central nervous system. (page 709)

10. **e** Two factors that contribute to larger animals' greater resistance to foreign invaders are their lower surface-to-volume ratios and their protective outer coverings. (page 711)

11. **d** The alkaline fluid from the pancreas and the sodium bicarbonate of bile from the gallbladder neutralize the acidic gastric contents when they enter the small intestine. (page 722)

12. **a** Monosaccharides are converted into glycogen and stored in that form in the liver. Fats produced from monosaccharides in the liver are stored in fat cells. (page 726)

13. **b** The blood of virtually all active animals except insects has respiratory pigments that transport oxygen. (page 741)

14. **e** Lymph is the interstitial fluid that is collected in the tissues and travels in lymphatic vessels before emptying into the bloodstream. (page 763)

15. **False** The muscles of the heart are supplied by capillaries fed by the coronary arteries, which are the first branches off the aorta. (page 754)

16. **True** The integration and control activities of a multicellular organism must include regulating internal activities and correctly receiving, processing, and responding to information about the external environment. (page 711)

17. **False** The sensors (chemoreceptors) that send signals to respiratory neurons in the brainstem are *most* sensitive to hydrogen ion concentration (which is a function of carbon dioxide concentration). Control of respiration by blood oxygen concentration only occurs if the hydrogen ion/carbon dioxide system is not functional. (pages 746, 747)

18. The answers are: **a3, b2, c4, d1,** and **e5.** (Table 34–3)

19. The assemblies of actin and myosin are similar in skeletal and cardiac muscle, and impart a striated appearance to both muscle types. Cardiac muscle cells are usually mononucleate whereas skeletal muscle cells are typically multinucleate. Skeletal muscle cells contract in response to signals transmitted by motor neurons. Certain cardiac muscle cells are capable of spontaneous electrical activity. The electrical impulses that stimulate heart muscle contractions are transmitted through cell-to-cell connections and by muscle cells specialized for impulse conduction. (page 708)

20. Fish regulate the flow of water across their gills by oscillating movements of the gill cover and by opening and closing their mouths as they swim. Some fish must swim constantly with their mouths open to obtain adequate oxygen. Such fish die if kept in a small tank. (page 735)

21. Factors that influence the ratio of LDLs to HDLs are cigarette smoking, level of physical activity, dietary intake of cholesterol and saturated fats, and perhaps hereditary factors. (essay, page 761)

Homeostasis I: Excretion and Water Balance

CHAPTER ORGANIZATION

MAJOR CONCEPTS

In animals, total body water constitutes approximately 70 percent of body weight; two-thirds of this water is located within cells (intracellular fluid) and one-third is extracellular. The extracellular fluid component of vertebrates consists of plasma and interstitial fluid. Water moves among these three compartments by osmosis, and the direction of movement is determined by relative solute concentrations.

An animal is in water balance when the acquisition of water from all sources equals the total lost. Animals gain water by drinking, by eating foods containing water, and by the release of water during metabolic reactions. Means by which animals lose water include excretion (urine and feces), respiration, and perspiration. The major route of water loss is through urine.

The earliest function of the vertebrate kidney was probably removal of excess water from the bodies of freshwater fishes while conserving salt and other essential solutes. When fish moved to saltwater environments, mechanisms evolved to prevent loss of body water to the environment.

In vertebrates, the kidney is the major organ influencing the chemical composition of body fluids. It excretes waste products—particularly nitrogenous wastes—while conserving essential nutrients and ions, functions that are critical to homeostasis. Birds, terrestrial reptiles, and insects excrete nitrogenous wastes in the form of uric acid or uric acid salts; mammals excrete urea, which is synthesized from ammonia by the liver.

The kidney has three main regions. The cortex is the outer layer where the glomeruli are located and plasma is filtered. The medulla is the inner zone where portions of the renal tubules pass through a concentration gradient, a process resulting in the formation of urine that is hypertonic with respect to plasma. Urine collects in the renal pelvis before it leaves the kidney through the ureter to travel to the urinary bladder. After storage in the urinary bladder, urine is voided from the body through the urethra.

The functional unit of the mammalian kidney is the nephron, consisting of two capillary beds in association with a renal tubule, which is composed of Bowman's capsule, proximal and distal convoluted tubules, and the loop of Henle. In filtration, blood is filtered from the first capillary bed (the glomerulus) into Bowman's capsule, the portion of the renal tubule that surrounds the glomerulus and collects the filtrate. As the filtrate moves through the different regions of the renal tubule, some additional substances are actively transported from the peritubular capillary network into the lumen of the tubule (secretion). Other substances are reabsorbed from the filtrate into the peritubular capillaries (reabsorption). The terminal portions of renal tubules empty into collecting ducts, which carry the urine to the renal pelvis (excretion).

The function of the kidney is regulated by several hormones. Antidiuretic hormone causes the collecting ducts to be permeable to water, thereby allowing water to leave the urine and be conserved in the body. Aldosterone stimulates conservation of sodium and secretion of potassium. Cardiac peptide inhibits the reabsorption of sodium.

HOW TO STUDY THE CHAPTER

Read the entire chapter through quickly, focusing on the major concepts.

Use the GUIDED STUDY OF THE CHAPTER to help you identify the important details as you **reread** the chapter. Writing out the answers to these questions will help fix them in your mind as well as provide you with a valuable study aid.

Answer the questions in TESTING YOUR UNDERSTANDING without the aid of your text. Check your answers against those in PERFORMANCE ANALYSIS. Analyzing your answers will give you valuable feedback on your level of understanding and preparedness for classroom testing.

GUIDED STUDY OF THE CHAPTER

I. Introduction *(page 765)*

1. Name six activities that contribute to body homeostasis in animals.

II. Regulation of the Chemical Environment *(pages 765–767)*

A. General Remarks *(page 765)*

2. a. What percentage of body water is intracellular (within cells)?

   b. What percentage of body water is extracellular (outside of cells)?

   c. How does the extracellular fluid resemble the external environment of unicellular organisms?

3. What percentage of total body fluids exists as plasma?

> *Focus:* The regulation of plasma composition is central to the regulation of water balance and the chemical environment of the vertebrate body.

4. How does the removal of wastes from the blood differ from waste elimination from the digestive tract?

5. Name three excretory organs in the animal kingdom that function to regulate the internal chemical environment, and indicate which groups of animals possess them.

> *Focus:* Major advances in vertebrate evolution, particularly adaptation to terrestrial environments, are correlated with increasing efficiency of kidney function.

B. Substances Regulated by the Kidneys *(pages 766–767)*

6. Identify three problems associated with regulating the internal chemical environment.

7. a. Name the two major metabolic waste products released into the blood and identify their origin.

   b. Why must ammonia be rapidly eliminated or converted into another compound soon after its removal from an amino acid?

8. Identify three groups of animals that convert nitrogenous wastes to uric acid or uric acid salts.

9. *Vocabulary:* What is the cloaca of a bird?

10. *Vocabulary:* Define guano.

11. a. In mammals, ammonia is converted to

_____ in the _____,

and in this form is carried by the

_____ to the _____

for excretion.

b. Urea is synthesized from what constituents? (Figure 37–3)

12. How does the amount of water required for urea excretion differ from that required for uric acid excretion?

13. Cite three examples that illustrate the selective nature of the excretory process.

14. The presence of glucose in the urine is indicative of what disease?

Focus: **The kidneys regulate the excretion and retention of many ions.**

---

15. Name five *functions* of the ions whose blood levels are regulated by the kidney.

16. Identify the two factors that influence the concentration of a particular substance in the body.

Focus: **Many techniques for dealing with the problem of water balance have evolved in animals. A major factor influencing this evolution was the availability of water to the animal.**

III. Water Balance *(pages 767–769)*

A. An Evolutionary Perspective *(page 767)*

17. a. How did the earliest organisms probably resemble their environment in terms of salt and mineral composition and relative total solute concentration?

b. What was the consequence of transition to freshwater environments?

18. Fish most probably evolved in what type of environment? Substantiate your answer.

19. What was most likely the first function of the vertebrate kidney?

20. Contrast freshwater fish, cartilaginous fish, and marine bony fish with respect to the body solute concentration relative to solute concentration in the environment.

21. a. Name two means by which freshwater fish lose solutes.

b. How do freshwater fish compensate for this loss?

22. Distinguish between the mechanisms by which hagfish and sharks maintain isotonicity with their environment.

23. a. How do marine bony fish compensate for continual water loss to the environment?

b. What adaptations have developed to remove excess ingested (i) salt and (ii) minerals?

24. Describe the mechanism by which some marine turtles excrete excess salt. (Figure 37–6)

B. Sources of Water Gain and Loss in Terrestrial Animals (*pages 767–768*)

25. *Vocabulary:* Define water balance. (Figure 37–7)

26. Identify three ways by which terrestrial animals gain water.

27. Indicate how much water is released during the catabolism of 1 gram of each compound.

Glucose

Protein

Fat

28. a. Explain how the diet of the kangaroo rat enables it to live without a source of drinking water.

b. What happens if kangaroo rats are placed on high-protein diets and not provided with supplemental water?

c. Name five adaptations that promote water conservation in the kangaroo rat.

29. Name four sources of water loss from the human body and indicate which one is the major route of loss.

30. a. What is the minimal urinary output required to maintain health for an adult human?

b. Why is this level of urine production essential?

C. Water Compartments (*pages 768–769*)

31. a. Identify the three main body water compartments and indicate the fraction of total body water represented by each compartment.

b. How can the rate of water transfer between these compartments be measured?

32. a. What is the major source of water gain?

b. By what mechanism and into what compartment does this water move into the body?

c. What maintains the osmotic potential that draws water from the digestive tract?

33. a. By what mechanism does water move out of the plasma into the interstitial fluid?

b. What happens to this interstitial fluid?

34. Name four factors that affect the movement of water between body compartments.

35. a. *Vocabulary:* What is dehydration?

b. Describe water movement between compartments in a dehydrated animal.

36. a. *Vocabulary:* Define edema.

b. Identify two situations that may result in edema.

37. What are two possible reasons for the excretion of salt by human sweat glands?

Focus: Water intoxication may occur if excess salt is lost from the body by perspiration and the water, but not the salt, is replaced.

---

38. a. Describe water movement during water intoxication.

b. What system may be seriously affected by water intoxication and what are the symptoms of this intoxication?

IV. The Kidney *(pages 769–775)*

A. General Remarks *(pages 769–770)*

39. Which organ in vertebrates is primarily involved in regulating the internal chemical environment?

40. *Vocabulary:* Distinguish between the cortex and medulla of the kidney in terms of location and function. (Figure 37–9)

41. a. *Vocabulary:* What is a nephron?

b. Label the following diagram of a nephron. Use these labels: afferent arteriole, Bowman's capsule, collecting duct, distal convoluted tubule, efferent arteriole, glomerulus, loop of Henle, proximal convoluted tubule.

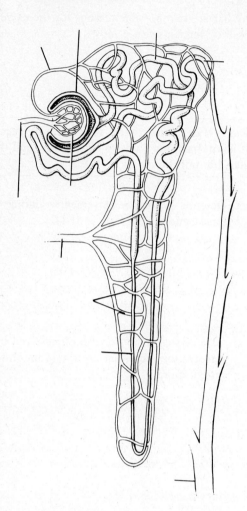

42. a. Trace the flow of urine from the glomerulus through the nephron to the urinary bladder. Identify the three structures that constitute the renal tubule. (See also Figure 37–9.)

b. Through what structure is urine excreted outside the body?

B. **Function of the Kidney** (*pages 770–774*)

43. a. What is unusual about the glomerulus compared with most other capillary beds?

b. Describe the anatomical relationship between the glomerulus and the peritubular capillaries.

c. Describe the means by which blood pressure in the glomerulus is regulated.

44. a. *Vocabulary:* Define filtration and filtrate.

b. What fraction of the plasma becomes filtrate as blood passes through the nephron?

c. How does the composition of the filtrate compare with that of plasma?

45. a. *Vocabulary:* Distinguish secretion from reabsorption.

b. Secretion and reabsorption occur in what part of the nephron?

c. Cite examples of substances that undergo secretion and reabsorption.

Focus: The secretion and reabsorption of many materials are accomplished by active transport mechanisms which are responsible for the high energy requirements of the kidney.

	PERMEABLE TO:	IMPERMEABLE TO:
Loop of Henle: Descending branch		
Ascending branch		
Collecting duct		

46. Identify two characteristics of proximal tubule cells that are essential for their functions of secretion and reabsorption. (Figure 37–14, page 774)

47. Name the fourth stage in the formation of urine by the nephron.

Water Conservation: The Loop of Henle (pages 772–774)

Focus: The glomeruli of the nephrons are located in the renal cortex. The concentration gradient through which the loops of Henle and the collecting ducts pass, and which is responsible for the formation of hypertonic urine, is located in the renal medulla.

48. The major means by which body water is regulated is:

49. Compare the type and volume of urine produced by an amphibian and a mammal.

50. Which section of the nephron is associated with a mammal's ability to excrete hypertonic urine?

51. Identify two structural features that are crucial to the formation of hypertonic urine.

52. Fill in the following table summarizing the permeability of the various regions of the nephron. (You will need to finish reading this entire section to complete this table.)

53. Summarize the permeability properties of the (a) proximal and (b) distal convoluted tubules.

54. Discuss two factors that maintain the concentration gradient in the medulla through which the loops of Henle and the collecting ducts pass.

55. Match the following statements with the correct region(s) of the nephron. Answers may be used more than once. A blank may have more than one answer as indicated in parentheses. (See also Figure 37–13.)

1. proximal convoluted tubule
2. descending loop of Henle
3. ascending loop of Henle
4. distal convoluted tubule
5. collecting duct

a. _____ Na$^+$ pumped out; Cl$^-$ follows passively

b. _____ walls freely permeable to water; water moves out by osmosis, independent of antidiuretic hormone (2 answers)

c. _____ 60–70 percent of the fluid is removed and returned to the blood

d. _____ pass through the zone of high solute concentration (3 answers)

e. _____ function to maintain the zone of high solute concentration (2 answers)

f. _____ permeability to water influenced by antidiuretic hormone

g. _____ Na$^+$ and Cl$^-$ BOTH pumped out

h. _____ filtrate hypotonic to plasma

56. Indicate the concentration of filtrate relative to plasma at each portion of the nephron. Use the terms isotonic, hypertonic, and hypotonic. (Figure 37–13)

Proximal convoluted tubule

Descending loop of Henle

Ascending loop of Henle

Distal convoluted tubule

Collecting duct, close to distal tubule

Collecting duct, passing through the medulla

57. Discuss the mechanism by which antidiuretic hormone (ADH) influences the concentration of excreted urine.

58. Why is the loop of Henle considered to be a countercurrent multiplier?

59. What is the primary factor limiting the concentration of excreted urine?

C. **Control of Kidney Function: The Role of Hormones**
(*pages 774–775*)

60. Fill in the following table on hormones involved in water and ion balance.

Table for Question 60

HORMONE	SITE OF PRODUCTION	SITE OF RELEASE (IF DIFFERENT FROM SITE OF PRODUCTION)	FUNCTION
Antidiuretic hormone (ADH)			
Aldosterone			
Cardiac peptide			

61. a. Name two factors that influence the release of ADH.

b. Where are receptors located that monitor plasma solute concentration?

c. Identify three locations for receptors monitoring blood pressure.

62. a. Describe how ADH release is altered by:

Increasing solute concentration

Decreasing solute concentration

Increasing blood pressure

Decreasing blood pressure

b. Cite one condition that can cause each of these situations.

63. Identify the effect of each event on ADH secretion and urine production.

Cold stress

Alcohol consumption

Pain

Emotional stress

64. Removal or malfunction of the adrenal glands has what effect on sodium balance?

65. a. What is Addison's disease and what type of therapy is available?

b. What are the consequences of untreated Addison's disease?

66. What two factors affect aldosterone production?

67. a. Name two mechanisms by which cardiac peptide exerts its effects.

b. There is evidence for what additional mechanism of action?

V. Summary *(pages 775–776):* Read the summary. If you are familiar with the essential features of the material presented there, you are ready to complete the section TESTING YOUR UNDERSTANDING.

TESTING YOUR UNDERSTANDING

After you have completed this examination, compare your answers with those in the section that follows.

1. In vertebrates, the transition to land was in large part facilitated by increasing efficiency of the:
   a. liver.   d. heart.
   b. kidney.   e. muscle.
   c. brain.

2. The excreta of seabirds, which is called _____, has been harvested commercially as a fertilizer.
   a. cliff manure   d. urea
   b. uric acid   e. guano
   c. night soil

3. In mammals, the highly toxic nitrogenous waste product, _____, must be converted to a less toxic form before it is removed from the blood.
   a. ammonia   d. guano
   b. urea   e. uric acid
   c. free amino acids

4. The body fluids of marine bony fish are _____ to (with) their environment.
   a. hypertonic   d. isosmotic
   b. hyperosmotic   e. hypotonic
   c. isotonic

5. The blood of sharks is isotonic to sea water because they accumulate:
   a. uric acid.   d. glycogen.
   b. urea.   e. albumin.
   c. glucose.

6. How do marine bony fish get rid of the excess magnesium and sulfate ions they take in when they drink sea water?
   a. The ions are secreted from glands in their nasal passages.
   b. Magnesium ions are removed by the gills and sulfate ions by the kidneys.
   c. The ions are removed from the blood by the kidneys.
   d. The ions leave the body in the feces as the salt, magnesium sulfate.
   e. Marine bony fish do not drink sea water.

7. If a kangaroo rat is fed only soybeans and is not given water, it will eventually die of dehydration because soybeans have a high _____ content.
   a. carbohydrate   d. protein
   b. fat   e. oil
   c. nucleic acid

8. In vertebrates, regulating the composition of body fluids is the chief responsibility of the:
   a. lungs.   d. stomach.
   b. liver.   e. kidneys.
   c. pancreas.

9. Which of these is NOT a symptom or potential consequence of water intoxication?
   a. death   d. coma
   b. convulsions   e. tissue edema
   c. disoriented behavior

10. Identify the correct order of structures through which urine passes in its excretion from the body.
   a. nephrons, renal pelvis, ureters, bladder, urethra
   b. renal pelvis, nephrons, ureters, bladder, urethra
   c. nephrons, ureters, renal pelvis, bladder, urethra
   d. ureters, nephrons, bladder, urethra, renal pelvis
   e. nephrons, renal pelvis, ureters, urethra, bladder

11. The _____ arteriole divides into _____, which surround the renal tubule and then merge to form a venule that eventually empties into the renal vein.
   a. afferent; peritubular capillaries
   b. efferent; collecting ducts
   c. efferent; peritubular capillaries
   d. afferent; portal venules
   e. efferent; peritubular venules

12. Glucose and amino acids are returned to the bloodstream during which phase of urine formation?
   a. secretion   d. excretion
   b. collection   e. filtration
   c. reabsorption

13. The ability of an animal to secrete urine that is hypertonic to its body fluids is associated with the:
   a. length of the loop of Henle.
   b. pressure within the glomerulus.
   c. extent of the peritubular capillary bed.
   d. length of the ureter.
   e. number of nephrons per kidney.

14. The wall of the proximal tubule is freely permeable to _____ and contains transport proteins that pump _____ out of the tubule by active transport.
    a. water; chloride ions
    b. both water and sodium; chloride ions
    c. water; both sodium ions and chloride ions
    d. water; sodium ions
    e. both sodium ions and chloride ions; water molecules

15. Fluid entering the proximal tubule is _____ relative to blood plasma, whereas the fluid entering the ascending portion of the loop of Henle is _____ relative to blood plasma.
    a. hypotonic; hypertonic
    b. isotonic; hypertonic
    c. isotonic; hypotonic
    d. hypertonic; hypotonic
    e. hypotonic; isotonic

16. The volume of glomerular filtrate is reduced by _____ percent in the proximal convoluted tubule of the nephron.
    a. 5 to 10   d. 60 to 70
    b. 15 to 25   e. 75 to 90
    c. 35 to 50

17. Which factor increases the synthesis and/or release of ADH?
    a. emotional stress
    b. cold stress
    c. ingestion of large amounts of water
    d. alcohol consumption
    e. increased blood pressure

18. Aldosterone stimulates the distal tubule and collecting duct to reabsorb _____ ions while secreting _____ ions.
    a. potassium; sodium   d. potassium; chloride
    b. chloride; potassium   e. calcium; sodium
    c. sodium; potassium

19. Receptors for cardiac peptide have been found in the:
    a. atria of the heart and the nephron.
    b. adrenal cortex and the adrenal medulla.
    c. adrenal medulla and the nephron.
    d. nephron and the adrenal cortex.
    e. atria of the heart and the adrenal cortex.

20. T or F Before multicellular organisms could evolve mechanisms to produce extracellular fluid, they first had to evolve mechanisms to regulate its composition.

21. T or F Regulation of the chemical composition of blood by the kidney involves not only nutrient molecules, but ions such as potassium and calcium.

22. T or F Water tends to follow dissolved food molecules passively from the intestines into the blood plasma.

23. T or F Humans and other mammals excrete hypertonic urine because the wall of the collecting duct is always permeable to water.

24. Cite at least two conditions that influence the movement of water between the various fluid compartments of the body.

25. Name two features of the blood that influence the release of ADH and indicate the effect of each factor.

26. "Although the kidneys have an excretory function, they are more accurately regarded as regulatory organs." Explain this statement.

PERFORMANCE ANALYSIS

1. **b** Major advances in vertebrate evolution, especially the transition to land, can be correlated with increasing efficiency of kidney function. (page 765)

2. **e** Guano, which is the excreta of seabirds, was once harvested commercially for use as a fertilizer because of its high nitrogen content. (page 766)

3. **a** The mammalian liver converts highly toxic ammonia to much less toxic urea, which is removed from the blood by the kidneys. (page 766)

4. **e** The body fluids of marine bony fish are hypotonic to their environment and these fish must drink sea water to prevent dehydration. They have adaptations that enable them to eliminate the excess salt ingested. (page 767)

5. **b** Sharks maintain body fluids isotonic with sea water by accumulating urea, for which they have evolved a high tolerance, in their blood. (page 767)

6. **c** The excess magnesium and sulfate ions taken in when marine bony fish drink sea water are removed from the blood by the kidneys. (page 767)

7. **d** The kangaroo rat living in its natural environment eats a diet of high-fat seeds, feeds at night when it is cool, and excretes an extremely concentrated urine. It does not need to consume extra water in its diet because the metabolism of the fat in the seeds provides adequate water. If a kangaroo rat is fed only soybeans, which are high in protein, it will eventually die of dehydration if supplemental water is not provided. (The oxidation of 1 gram of fat yields 1.1 grams of water, whereas the oxidation of 1 gram of protein yields only 0.3 gram of water.) (page 768)

8. **e** In vertebrates, the organ primarily responsible for regulation of the composition of body fluids is the kidney. (page 769)

9. **e** During profuse sweating, a great deal of salt may be lost from the body. If the lost water is replaced without replacing the salt, the water moves into body cells, diluting their contents. The effects of this water intoxication on the central nervous system are particularly serious, and disorientation, convulsions, coma, and even death may result. (page 769)

10. **a** Urine is formed in the nephrons and moves through collecting ducts to the renal pelvis. It travels from the

kidneys through ureters to the urinary bladder, where it is stored before being eliminated from the body through the urethra. (page 770)

11. **c** Blood enters the glomerulus of the nephron via the afferent arteriole and leaves the glomerulus through the efferent arteriole. The efferent arteriole then branches to form the peritubular capillaries, which surround the renal tubule and then merge into venules and veins that eventually lead to the renal vein. (page 770)

12. **c** Glucose and amino acids are returned to the blood by active transport during the reabsorption phase of urine formation. (page 771)

13. **a** The length of the loop of Henle is directly correlated with the maximum possible urine concentration that a species can produce; the longer the loop of Henle, the more concentrated the urine. When the concentration gradient extending from the proximal convoluted tubule to the bend in the loop of Henle is longer, the countercurrent multiplier effect produces a greater interstitial solute concentration in the medulla. As the collecting ducts pass through the medulla, water is removed from the urine until the urine concentration equals that of the medullary interstitial fluid. (page 772)

14. **d** The wall of the proximal convoluted tubule contains transport proteins that pump sodium ions out of the tubule. Water follows these ions passively since this portion of the renal tubule is freely permeable to water. (page 772)

15. **b** Filtrate that enters the proximal convoluted tubule is isotonic with blood plasma. Filtrate entering the ascending branch of the loop of Henle is hypertonic to blood plasma because water leaves by osmosis as the filtrate travels through the descending branch of the loop of Henle. (page 773)

16. **d** During passage through the proximal convoluted tubule, 60 to 70 percent of the solutes and water of the glomerular filtrate are removed and returned to the bloodstream. (page 773)

17. **a** Pain and emotional stress increase the synthesis and/or release of ADH, thereby reducing urine production. All the other factors listed decrease ADH release, thereby increasing urine production. (page 775)

18. **c** Aldosterone stimulates the reabsorption of sodium ions and the secretion of potassium ions by the distal convoluted tubule and collecting duct. (page 775)

19. **d** Receptors for cardiac peptide, which is produced by the atria of the heart and influences blood volume and blood pressure, have been located in the nephrons and the adrenal cortex. (page 775)

20. **False** The composition of the extracellular fluid of early multicellular organisms resembled that of sea water. Mechanisms for regulating the composition of extracellular fluid evolved as the animals evolved. (page 765)

21. **True** The chemicals whose blood concentrations are regulated by the kidney include nutrient molecules, ions, and metabolic wastes. (page 766)

22. **True** As dissolved food molecules are absorbed from the lumen of the intestines into the bloodstream, water follows passively by osmosis. (page 769)

23. **False** The permeability of the wall of the collecting duct to water is regulated by ADH; the wall is permeable to water when ADH is present and impermeable when ADH is absent. Mammals can secrete hypertonic urine because the interstitial solute concentration of the medulla is hypertonic to plasma. (page 773)

24. Dehydration, kidney disease, and protein malnutrition all influence the movement of water between various fluid compartments of the body. During dehydration, the solute concentration of the extracellular fluid increases and water moves out of cells. Interstitial fluids may accumulate when salts are retained because of kidney disease or when plasma proteins are inadequate to maintain the normal osmolality of plasma, as occurs during protein malnutrition. (page 769)

25. (1) Increasing blood pressure inhibits ADH production, thereby increasing urine production and lowering blood volume and blood pressure. Decreasing blood pressure has the reverse effect. (2) The total solute concentration of the blood also influences ADH production. If osmotic receptors detect a decreasing solute concentration, ADH production is inhibited. Conversely, increasing solute concentration stimulates ADH production. (pages 774, 775)

26. In addition to their function of waste elimination, the kidneys regulate ion concentrations (including potassium, sodium, bicarbonate, hydrogen, magnesium, and calcium ions), retain nutrient molecules (i.e., glucose and amino acids), and regulate the water content of body fluids. (page 766)

Homeostasis II: Temperature Regulation

MAJOR CONCEPTS

The biochemical processes essential to most animals occur within the fairly narrow temperature range in which enzymes are active. This temperature range is from just below freezing to between 45 and 50°C. With only a few exceptions, animals must either occupy environments within this temperature range or have mechanisms for maintaining body temperature within this range.

Effective temperature regulation requires that heat gained by the body equal heat lost. The two main sources of heat gain are the radiant energy of the sun and exothermic metabolic reactions. Heat is lost from animal bodies by the processes of conduction, convection, evaporation, and radiation.

The rate of heat loss depends on the surface-to-volume ratio of an animal or a body part. The greater the surface-to-volume ratio, the faster the rate of heat loss. If similar animal species have a wide geographic distribution, the larger animals are found in cooler environments; large body size facilitates heat conservation. Conversely, many desert animals are smaller than their counterparts in cooler climates and have large ears, an adaptation facilitating heat radiation from the body.

Poikilotherms are animals whose body temperatures fluctuate. Homeotherms are animals that maintain fairly constant body temperatures. Poikilotherms are generally ectothermic animals, while all homeotherms are endotherms. Homeotherms are characterized by the presence of feathers or fur as insulation; most mammals are also insulated with subcutaneous fat. (The terms poikilotherm and homeotherm have replaced the much less satisfactory terms "cold-blooded" and "warm-blooded.")

Body temperature in most poikilotherms is determined by environmental factors. Most aquatic animals (aquatic mammals are the main exception) have body temperatures identical to that of the water in which they live. Terrestrial poikilotherms have elaborate behavioral strategies by which they increase body temperature by absorbing solar radiation while they are active, and retreat to a safe den during periods of inactivity (or when the environment is too cold for them to move).

Homeotherms have elaborate physiological and sometimes behavioral adaptations for maintaining a fairly constant body temperature in the face of changing environmental temperatures. These are expensive innovations, and a homeotherm requires approximately ten times as much energy as does a poikilotherm of the same size. Some homeotherms reduce the total energy cost by experiencing daily or seasonal drops in metabolic rate and body temperature.

Of the homeotherms, the birds have the highest metabolic requirements. They maintain a higher body temperature than most other homeotherms, and spend much of the time exposed. Because of the need to minimize body weight for flight, they cannot store large amounts of energy. Consequently, birds must eat constantly to maintain their high-energy lifestyle.

The thermostat of homeotherms is located in the hypothalamus. The hypothalamus receives temperature input from skin receptors and from its own receptor cells that monitor blood temperature. The hypothalamus responds to the input by triggering the appropriate response to increase or decrease body temperature. Increasing body temperature triggers dilation of blood vessels near the skin and evaporative cooling mechanisms (i.e., panting or sweating). Decreasing body temperature triggers constriction of superficial blood vessels and acceleration of metabolic processes, which is a consequence of increased muscle activity and, in some cases, stimulation of the endocrine and nervous systems. The hormones thyroxine and adrenaline influence metabolic rate and therefore are important in regulating body temperature.

Animals living in extremely harsh climates (very cold or very hot and dry) have evolved specialized adaptations that enable them to tolerate the severe environmental conditions. Animals adapted to cold environments typically have abundant insulation in the form of fat, fur, and/or feathers. In the limbs of Arctic mammals, blood vessels exhibit a countercurrent arrangement, which facilitates body core heat retention. The camel is adapted to hot environments by its tolerance of significant body water loss and body temperature increases, and by its hump, which is a localized fat deposit that allows heat to leave the body core. Most small desert animals are nocturnal, avoiding the hottest environmental temperatures by staying sheltered and inactive.

HOW TO STUDY THE CHAPTER

Read the entire chapter through quickly, focusing on the major concepts.

Use the GUIDED STUDY OF THE CHAPTER to help you identify the important details as you **reread** the chapter. Writing out the answers to these questions will help fix them in your mind as well as provide you with a valuable study aid.

Answer the questions in TESTING YOUR UNDERSTANDING without the aid of your text. Check your answers against those in PERFORMANCE ANALYSIS. Analyzing your answers will give you valuable feedback on your level of understanding and preparedness for classroom testing.

GUIDED STUDY OF THE CHAPTER

I. Introduction (pages 777–778)

> *Focus:* The ability to regulate body temperature is critical to the survival of an animal. Animals have physiological, anatomical, and behavioral adaptations related to temperature regulation.

1. a. Summarize the relationship between temperature and the rate of an enzymatic reaction.

  b. What factors determine the temperature range at which enzymes are functional?

2. a. Identify the temperature range to which physiological processes are generally confined.

  b. How do most animals ensure such a temperature range?

  c. What special adaptations enable some animals to exist outside this temperature range?

3. *Vocabulary:* Define absolute zero. (Figure 38–2)

II. Principles of Heat Balance (pages 778–779)

A. General Remarks (pages 778–779)

4. a. Name the two principal sources of heat for animals.

  b. Identify four mechanisms of heat loss from animals, one of which may also be a route of heat gain.

5. *Vocabulary:* Distinguish between conduction, convection, and radiation.

6. a. Why does a metal knife feel colder than a cloth napkin when both are at room temperature?

   b. Why would you feel more uncomfortable in water at 20°C than in air at the same temperature?

7. What phenomenon is responsible for convection?

8. a. Air and fat are useful as insulators because they are:

   b. Name three insulators used by animals.

9. Why does evaporation of sweat from your skin make you feel cooler?

10. Cite two fates of light energy falling on an object.

B. **Body Size and the Transfer of Heat** *(page 779)*

11. a. What is the relationship between body surface-to-volume ratio and the ease with which a constant body temperature is maintained?

    b. Identify one technique utilized by small animals to circumvent this phenomenon.

III. **"Cold-Blooded" vs. "Warm-Blooded"** *(page 780)*

12. a. Why are the terms cold-blooded and warm-blooded not very useful from a scientific perspective?

    b. How do these terms correlate with endothermic and ectothermic?

13. a. *Vocabulary:* Distinguish between poikilotherms and homeotherms.

    b. Correlate these terms (generally) with cold-blooded/warm-blooded and ectotherm/endotherm.

c. Cite two examples illustrating how the terms poikilotherm and homeotherm cannot always be strictly correlated with cold-blooded and warm-blooded.

IV. Poikilotherms *(pages 780–781)*

14. a. What internal mechanisms contribute to the relatively constant body temperature of most fish?

b. How does the environment contribute to this constant body temperature?

c. What behavioral changes can fish make when their environment undergoes changes in temperature?

Focus: **In the transition from aquatic to terrestrial environments, animals had to adapt to environments that are subject to wide fluctuations in temperature.**

---

15. a. Describe the behavioral adaptations by which reptiles maintain stable body temperatures during their active periods.

b. What happens to reptiles when, because of environmental conditions, they are unable to maintain a body temperature that permits activity?

c. What is the adaptive value of pigment changes that some reptiles undergo while regulating body temperature? (Figure 38–6)

d. How rapidly can reptile bodies heat up under optimum conditions?

V. Homeotherms *(pages 781–786)*

A. General Remarks *(pages 781–782)*

16. a. Which two groups of animals are the only true homeotherms?

b. All homeotherms are also:

17. What is the major source of heat for homeotherms?

18. Name two advantages of homeothermy.

19. a. Identify the major disadvantage associated with homeothermy.

b. Which animal, a mouse or a moose, would pay the highest price for homeothermy? Explain your answer.

c. How do homeotherms obtain enough energy to support their high metabolic rates?

d. What features of the digestive tract make the process in (c) possible?

20. a. What problems are unique to birds in meeting their energy requirements?

b. How do birds solve these problems?

21. Describe how heat is transferred from the body core to the body surface of a homeotherm.

Focus: **Temperature regulation in homeotherms involves balancing production of heat with the loss of heat from the body surface.**

B. *Essay:* **Avian Mechanical Engineers** *(page 783)*

22. In general terms, summarize the mechanisms by which mammals and birds maintain developing young at the correct body temperature.

23. Describe the elaborate behavioral strategy by which mallee fowl maintain correct egg temperature during incubation in spite of wide fluctuations in environmental temperatures.

24. How dependent are the newly hatched chicks on their parents? (Compare this with newborn mammals.)

C. **The Thermostat** *(pages 784–786)*

25. Where is the physiological thermostat in homeotherms?

26. a. Name two locations of temperature receptors that provide information to the hypothalamus.

b. If signals from the two locations are in conflict, the hypothalamus "believes" the information from which receptors?

27. a. Name five events that may be triggered by the hypothalamus in the regulation of human body temperature. (Figure 38–8)

b. Outline the hormonal pathway that is of major importance in the temperature regulation of some animals. (Figure 38–8)

c. Indicate how thyroxine exerts its effects. (Figure 38–8)

28. a. What happens to the temperature regulation system during a fever?

b. Explain why a person with a fever may have chills.

c. What substance resets the thermostat?

29. In what two ways may the immune system be stimulated by a fever?

Regulating as Body Temperature Rises (page 785)

30. Name two mechanisms by which heat may be unloaded as body temperature rises.

31. What anatomical adaptation that functions in heat dissipation characterizes animals inhabiting hot climates? (see also Figure 38–9)

32. a. Discuss the different ways in which horses, dogs, and cats utilize evaporative cooling as a means of regulating body temperature.

    b. What is the cost of evaporative cooling to an animal?

    c. How does evaporation of water from the tongue differ from evaporation from the skin? (Figure 38–10)

Regulating as Body Temperature Falls (page 786)

33. a. Name two mechanisms by which body temperature is maintained as the external temperature falls.

    b. Metabolic processes increase in what two ways?

c. Identify two hormones that increase metabolism and state their effects.

d. How does autonomic nerve stimulation influence metabolism?

D. Cutting Energy Costs *(page 786)*

34. How does a hummingbird reduce its overall energy requirements?

35. a. *Vocabulary:* What is hibernation?

    b. Most hibernators are:

    c. Identify four physiological parameters that change when an animal hibernates.

    d. Name four conditions that may stimulate arousal of a hibernating animal.

    e. What is the source of the body heat gained when a hibernating animal is aroused?

VI. **Adaptations to Extreme Temperature** *(pages 787–789)*

A. **General Remarks** *(page 787)*

B. Adaptations to Extreme Cold *(pages 787–788)*

36. What is the major technique by which animals adapt to extreme cold?

37. In addition to insulation, what other adaptation is exhibited by marine animals living in cold waters?

38. Name two mechanisms by which large Arctic mammals unload excess heat when they are on land.

Countercurrent Exchange in Heat Conservation (pages 787–788)

Focus: **Many Arctic animals have countercurrent exchange systems that conserve energy by allowing the temperature of extremities to drop to a level dramatically lower than core body temperature.**

---

39. The foot of the Arctic fox exhibits what special adaptation to its environment?

40. a. Describe how a countercurrent exchange system enables limb temperature to be much lower than core body temperature.

    b. Explain why such a system is of adaptive value to an animal.

C. Adaptations to Extreme Heat *(pages 788–789)*

41. Name five adaptations of camels to the hot, arid conditions of the desert. Indicate the adaptive value of each characteristic. (See also Figure 38–14)

42. How do most small desert animals cope with the harsh environment?

VII. **Summary** *(pages 789–790):* Read the summary. If you are familiar with the essential features of the material presented there, you are ready to complete the section TESTING YOUR UNDERSTANDING.

TESTING YOUR UNDERSTANDING

After you have completed this examination, compare your answers with those in the section that follows.

1. Based on chemical principles, the rate of an enzymatic reaction approximately doubles for every _____ °C rise in temperature.
   a. 1   d. 15
   b. 5   e. 20
   c. 10

2. The temperature range within which life exists on earth roughly parallels the temperatures at which:
   a. enzymes function optimally.
   b. water exists as a liquid.
   c. oxidation of carbon compounds occurs.
   d. oxygen will dissolve in water.
   e. carbon dioxide will dissolve in water.

3. The transfer of heat between two bodies by radiation requires that they be:
   a. in motion.
   b. dark in color.
   c. in direct contact.
   d. at different temperatures.
   e. extremely close to one another.

4. Small animals lose heat more rapidly than do large ones because they:
   a. have higher metabolic rates.
   b. contain less total body water.
   c. have less volume in which to produce heat.
   d. have a higher surface-to-volume ratio.
   e. are generally ectotherms and therefore cannot regulate their internal temperatures effectively.

5. Terrestrial poikilotherms maintain relatively stable body temperatures during periods of activity by regulating:
   a. evaporation.
   b. shivering.
   c. blood flow to the skin.
   d. their metabolic rate.
   e. the amount of solar energy absorbed.

6. The correct term for an animal that has a constant body temperature is:
   a. homeotherm.   d. endotherm.
   b. poikilotherm.   e. warm-blooded.
   c. exotherm.

7. Relative to a reptile of similar body size and temperature, the metabolic rate of a mammal is about _____ times higher.
   a. 2   d. 10
   b. 5   e. 15
   c. 8

8. The integration of signals regarding body temperature and the initiation of appropriate responses to these signals is carried out by the _____ in the brain.
   a. cerebrum
   b. medulla
   c. pituitary gland
   d. cerebellum
   e. hypothalamus

9. You may feel cold and have chills even though you have a fever because:
   a. your body temperature is lower than the normal thermostat set point.
   b. your body temperature is lower than the new thermostat set point.
   c. the hypothalamus has temporarily malfunctioned.
   d. your body temperature is higher than the normal thermostat set point.
   e. your metabolic rate is lower and you are losing body heat during the fever.

10. When the air temperature is higher than body temperature, a major mechanism by which the internal temperature of humans is regulated is:
    a. profuse sweating.
    b. constriction of peripheral blood vessels.
    c. convectional heat loss.
    d. a decrease in the metabolic rate.
    e. an increase in the metabolic rate.

11. Which of the following is NOT a mammalian response to counteract decreasing body temperature?
    a. increase in voluntary muscle contraction
    b. increase in thyroxine release
    c. adrenaline-stimulated release and oxidation of glucose
    d. breakdown of fat stimulated by the autonomic nervous system
    e. dilation of blood vessels near the skin surface

12. Unlike a winter sleeper, a true hibernator:
    a. is aroused only when the environmental temperature is favorable.
    b. exhibits a marked decrease in metabolic rate.
    c. lowers its body temperature.
    d. lives on stored food reserves.
    e. cannot be awakened before a preprogrammed time period has passed.

13. A camel can lose water equal to more than _____ percent of its body weight and still function, whereas a human may become deaf or delirious with a loss of as little as _____ percent.
    a. 30; 5   d. 15; 10
    b. 25; 10   e. 10; 5
    c. 20; 5

14. The major mechanism of temperature regulation used by small desert animals is:
    a. sweating.
    b. panting.
    c. avoiding direct heat.
    d. countercurrent exchange.
    e. a light coat color that reflects the sun's rays.

15. Temperatures on land may exhibit an annual variation of as much as _____°C.
    a. 10 to 30   d. 75 to 85
    b. 40 to 50   e. almost 100
    c. 60 to 70

16. T or F   It is impossible for any poikilotherm to survive at temperatures below freezing.

17. T or F   Homeothermy made it possible for early mammals to invade environments that were thermally inhospitable to reptiles.

18. T or F   In order to sustain their higher metabolic rate, mammals eat more food and process it faster than do poikilotherms.

19. T or F   During arousal from hibernation, an animal warms up mainly by absorbing heat from the environment.

20. Identify the two primary sources of heat gain for living organisms.

21. Cite one example of an animal that maintains a constant body temperature even though it is an ectotherm.

22. In order to maintain themselves and to grow and reproduce, birds must eat almost constantly. Identify the aspects of a bird's lifestyle that are responsible for this phenomenon.

23. Identify two problems that might occur if an Arctic animal had feet as warm as its body core.

24. What is the major temperature-regulation problem facing fur seals and polar bears when they are on land? What behavioral strategies do they use to circumvent this problem?

PERFORMANCE ANALYSIS

1. c Within the temperature range at which enzymes function, the reaction rate of an enzymatic reaction doubles for every 10°C rise in temperature. (page 777)

2. a The temperature range within which most living organisms exist roughly corresponds to the temperature range of optimal enzyme function. (page 777)

3. d Transfer of heat between any two bodies occurs only if they are at different temperatures. Radiation is heat transfer by electromagnetic waves in the absence of direct contact. (pages 778, 779)

4. d The rate of heat loss is proportional to the surface-to-volume ratio of an object (or organism). The smaller the object, the higher the surface-to-volume ratio, and the faster the heat loss. (page 779)

5. e Terrestrial poikilotherms can maintain fairly stable body temperatures during their active hours by regulating the amount of exposure to solar radiation. (page 781)

6. a By definition, an animal that maintains a constant body temperature is a homeotherm. (pages 780, 781)

7. d The metabolic rate of a mammal is generally 10 times higher than that of a reptile of similar size and body temperature. (page 782)

8. e The hypothalamus of the brain integrates signals concerning body temperature and external temperature and initiates appropriate responses. (page 784)

9. b In a person with a fever who experiences chills, the thermostat in the hypothalamus has been reset to a higher level and body temperature has not yet reached that level. (The chills will cease when body temperature reaches the new thermostat setting.) (page 784)

10. a When air temperature is greater than body temperature, the human body unloads excess heat by sweating and by dilating peripheral blood vessels. (page 785)

11. e Mammals can counteract decreasing body temperature by several mechanisms: (1) increasing involuntary (shivering) and voluntary muscle contractions, (2) increasing thyroxine release (if exposure to cold temperature is prolonged), (3) release and oxidation of glucose, stimulated by adrenaline, (4) breakdown of fat, stimulated by the autonomic nervous system, and (5) constriction of peripheral blood vessels. (page 786)

12. b A true hibernator experiences a dramatic decrease in metabolic rate. An animal such as a black bear that spends much of the winter sleeping experiences a body temperature drop and some decrease in metabolic rate, but the reduction is not as pronounced as that of a true hibernator. (page 786)

13. b A camel is adapted to lose water equal to more than 25 percent of its body weight. If a human loses as little as 10 percent of body weight in water, he or she may become delirious, deaf, and insensitive to pain. (page 789)

14. c Most small desert animals are nocturnal—their major means of temperature regulation is the avoidance of direct heat. (page 789)

15. c In the course of a year, the temperature of a terrestrial environment may vary as much as 60 to 70° Celsius (which corresponds to a variation of 140 to 158° Fahrenheit). (page 781)

16. False Several poikilotherms, including the fishes and arthropods of polar regions, have antifreeze molecules in their body fluids or can enter a state of near-total dormancy, which enables them to survive extremely cold conditions. (page 778)

17. True Because the early mammals were homeotherms, they could function at peak efficiency under conditions at which reptiles were inactive, and thus could invade environments inhospitable to the reptiles. (page 782)

18. True Because of their higher metabolic rate, mammals must eat more food and process it more quickly than poikilotherms of the same size. This is possible because of the longer, more convoluted mammalian digestive system with its numerous villi and extensive microvilli. (page 782)

19. False On arousal from hibernation, an animal warms up mainly by generating metabolic heat. (page 786)

20. The two primary sources of heat gain for living organisms are exothermic chemical reactions within cells and the radiant energy of the sun. (page 778)

21. Fish that live deep in the ocean maintain constant body temperatures because the surrounding water temperature does not vary. (page 780)

22. Since birds must minimize body weight in order to fly, they cannot store large quantities of food and so must forage constantly. In searching for food, birds spend much of their time exposed to the environment, which renders them susceptible to heat loss. To meet the high-energy requirements of this lifestyle a bird may consume 30 percent of it body weight in high-protein foods per day. (page 782)

23. (1) If the feet of an Arctic animal were as warm as its body core, the snow might thaw and refreeze over its feet, trapping the animal. (2) Because of the great surface-to-volume ratio of extremities compared with the thorax and abdomen, an animal with extremities as warm as its body core would lose a tremendous amount of heat to the environment through its limbs. (page 787)

24. When fur seals and polar bears are on land, they need to unload excess heat. This can be accomplished by wetting their fur, which allows evaporative cooling, and by sleeping, which reduces heat production due to metabolism. (page 787)

Homeostasis III: The Immune Response

MAJOR CONCEPTS

The body has several nonspecific mechanisms for defense against foreign bodies and pathogenic microorganisms. The first line of defense is the skin and mucous membranes, which when intact are impermeable to penetration by pathogens. The most common route by which pathogens enter the body is by crossing a compromised mucous membrane.

The inflammatory response is the second line of defense. Chemicals released from damaged tissues cause blood vessels to dilate and attract white blood cells into the area. Two types of circulating white blood cells active in inflammation are granulocytes (neutrophils, basophils, and eosinophils), and monocytes, which are transformed into phagocytic macrophages in the tissues. Blood clots form to wall off the damaged tissue from surrounding healthy tissues. The cardinal signs of inflammation are heat, swelling, redness, and pain.

Interferons are chemicals that are released from cells infected with viruses and stimulate nearby cells to resist viral infection. The production of interferons by recombinant DNA techniques is enabling scientists to study their action and potential as therapeutic agents.

The third line of defense is the immune response, which is characterized by the production of chemicals and cells that act against a specific agent to which the body has been exposed. The organs involved in the immune response include all lymph nodes and lymphatic tissues (e.g., tonsils and Peyer's patches), bone marrow, thymus, and spleen. The principal cells of the immune response are B and T lymphocytes, but macrophages are also important participants.

Lymph nodes filter the lymph, removing foreign particles, dead cells, pathogens, and cancer cells. They contain high numbers of macrophages and lymphocytes and are one site of B lymphocyte activation and proliferation.

When presented with a foreign antigen possessing a structure complementary to its specific membrane antibodies, a B lymphocyte becomes activated and divides, producing plasma cells and memory cells. Plasma cells are short-lived antibody factories. Memory cells persist in the circulation much longer than do plasma cells (months or a life-time) and begin proliferating and producing antibodies quickly upon subsequent exposure to the same agent that triggered their production.

Antibodies (also known as immunoglobulins) are proteins that function in the immune response against specific agents, antigens. (An antigen is any particle that can stimulate an immune response.) Antibodies may function by (1) coating foreign particles, thereby facilitating their phagocytosis, (2) combining with a pathogen or toxin so that it is neutralized, or (3) acting with other components of the immune system to lyse foreign cells.

Vaccines protect against infection by stimulating antibody production against specific infectious agents. A vaccine may contain killed or weakened virus particles or bacteria, or may contain purified antigenic proteins produced by recombinant DNA technology.

The basic structure of an antibody consists of two heavy protein chains and two light protein chains. Each chain has constant and variable regions in its amino acid sequence. The variable regions determine the antigen-binding site. The five classes of antibodies, which are distinguished by the constant regions of their heavy chains, are functionally specialized.

Complement is a group of proteins, found in the blood, that function in the immune response. In various combinations, these proteins act as lytic enzymes; they also promote phagocytosis by coating foreign cells. In addition, complement mediates and enhances the inflammatory response.

B lymphocytes are produced in the bone marrow. During the initial differentiation of bone marrow stem cells into B lymphocytes, the gene sequences coding for surface antibodies are shuffled and assembled so that each B lymphocyte receives a unique type of surface antibody. Each B lymphocyte circulates in the body until it encounters the antigen corresponding to its surface antibodies, at which time it becomes activated.

T lymphocytes are the effectors of the cell-mediated immune response and are activated by interactions between surface receptors and complementary antigens. There are three major classes: helper T cells and suppressor T cells regulate the immune response; cytotoxic T cells destroy abnormal body cells. T cells undergo differentiation, selection, and maturation in the thymus, and the antigenic specificity of their surface receptors is determined in a manner analogous to that of B lymphocytes.

The appropriate function of T lymphocytes depends upon their ability to distinguish "self" from "not-self." The major histocompatibility complex (MHC) is a series of genes coding for a group of surface antigens found on nucleated cells. T lymphocytes learn what "self" is by exposure to MHC antigens during their development in the thymus.

A growing body of evidence indicates numerous interactions among the immune, endocrine, and nervous systems, in which the chemical mediators of each system influence the other systems.

One hypothesis about cancer that is gaining acceptance is that cancer represents a failure of the immune system, which is normally stimulated by the presence of a cancer cell. Cancer develops when the immune system fails to mount a successful response against the abnormal cells.

Transplanted organs and tissues are rejected primarily by the actions of cytotoxic T cells and "killer cells," which respond to the foreign MHC antigens. Patients receiving transplants have often been placed on immunosuppressive drugs that unfortunately have rendered them highly susceptible to infections. A fairly new drug, cyclosporin, acts only against those T cells involved in transplant rejection.

There are four major human blood groups (A, B, AB, and O) based on the presence or absence of A and B surface antigens on red blood cells. A separate surface antigen, Rh factor, may be present or absent as well. Antigens against Rh factor are responsible for hemolytic anemia of the newborn, which may affect an Rh-positive child born to an Rh-negative mother who has previously given birth to an Rh-positive child. Before any blood transfusion is performed, donor and recipient blood is typed and matched according to these and other factors.

Several types of disorders involve the immune system. Autoimmune diseases are those in which the immune system fails to distinguish between "self" and "not-self" and attacks body cells. Allergies occur when the immune system inappropriately responds to antigens that are not harmful. Acquired immune deficiency syndrome (AIDS) is caused by a retrovirus that destroys helper T cells, thereby decimating cell-mediated immunity. People with AIDS become susceptible to a number of opportunistic infections and other diseases.

HOW TO STUDY THE CHAPTER

Read the entire chapter through quickly, focusing on the major concepts.

Use the GUIDED STUDY OF THE CHAPTER to help you identify the important details as you **reread** the chapter. Writing out the answers to these questions will help fix them in your mind as well as provide you with a valuable study aid.

Answer the questions in TESTING YOUR UNDERSTANDING without the aid of your text. Check your answers against those in PERFORMANCE ANALYSIS. Analyzing your answers will give you valuable feedback on your level of understanding and preparedness for classroom testing.

GUIDED STUDY OF THE CHAPTER

I. Introduction *(page 791)*

> *Focus:* During the course of evolution, organisms have developed numerous nonspecific and specific mechanisms for defending themselves against invasion by pathogens and foreign substances.

II. Nonspecific Defenses *(pages 791–793)*

A. Anatomic Barriers *(page 791)*

1. a. Name the two components of the body's first line of defense against invaders.

   b. What characteristic of the skin contributes to its effectiveness as a barrier to invasion?

   c. Identify the feature common to mucous membranes that makes them effective barriers to infection even though the epithelium is fragile.

   d. Name three other features in various parts of the body that contribute to defense against invasion through mucous membranes.

2. What is the most common site of entry of microorganisms or toxins?

B. The Inflammatory Response *(pages 791–793)*

> *Focus:* **The inflammatory response is nonspecific and constitutes the second line of defense against invaders.**

3. Name two functions of histamine and the other chemicals that are released from cells when tissue damage occurs.

4. a. How do white blood cells "know" that tissue has been damaged?

   b. What do white blood cells do once they reach damaged tissue?

5. What role does blood clotting play?

6. Identify two consequences of the local temperature increase that accompanies inflammation.

> *Focus:* **The cardinal signs of inflammation are heat, redness, swelling, and pain.**

7. Where, and from what other cells, do white blood cells originate?

8. a. Name the three types of granulocytes.

   b. Granulocytes are identified based on what characteristic?

9. a. Which type of granulocyte is the most numerous and also the first to arrive at damaged tissue?

   b. Describe the process by which neutrophils leave the blood vessels and move toward injured tissue.

   c. When neutrophils arrive at the site of an infection, what do they do? (Include the term lysosomes in your answer.)

10. In what way do some microorganisms defend themselves against attack by neutrophils?

11. *Vocabulary:* Distinguish basophils from eosinophils on the basis of function.

12. What are mast cells?

13. *Vocabulary:* Differentiate between monocytes and macrophages.

14. a. How do monocytes locate the site of an infection?

    b. When do monocytes arrive at the site of infection, relative to neutrophils?

    c. What happens to monocytes once they arrive at infected tissue?

15. a. Name five locations of tissue macrophages.

    b. In addition to phagocytosis, macrophages perform what other defensive function?

16. Identify two possible systemic effects of inflammation.

C. **Interferons** *(page 793)*

17. Name two ways in which interferons differ from other defense mechanisms.

18. What observation led to the search for interferons?

> *Focus:* **Interferons are very small proteins that are active in minute quantities.**

19. Describe the mechanism by which interferons operate.

20. How do interferons affect white blood cells?

21. a. By what means are interferons being mass-produced?

    b. Identify two proposed clinical uses for interferons that are currently under investigation.

    c. What property of interferons initially led scientists to believe they might be potent anti-cancer agents?

II. The Immune System (pages 793–795)

22. How does the immune response differ from the other defense mechanisms of the body?

> *Focus:* **The immune response has two phases; one that occurs the first time an organism is exposed to an agent, and a more rapid response that occurs at any subsequent exposure to the same agent.**

23. a. Name the two types of cells involved in the immune response.

    b. In mammals, these cells differentiate and proliferate primarily in what two sites?

    c. List four other structures that participate in the immune response.

24. a. Describe the structure of a lymph node.

    b. What is the function of lymph nodes?

    c. Name five types of substances that may be filtered from the circulation by lymph nodes.

    d. What types of cells populate lymph nodes and are important to their function in the immune response?

25. Cite two reasons why the lymph nodes filtering lymph from an area affected by cancer are commonly removed during surgery.

26. a. Trace the flow of lymph through a lymph node.

    b. What prevents backflow of the lymph?

    c. Where in the lymph node do foreign particles and pathogens come into contact with immunologically active cells?

    d. *Vocabulary:* Define the following components of a lymph node. (Figure 39–5)

    Reticulum cells

    Medullary cords

    Mantle

> *Focus:* **Lymph nodes typically increase in size when they are actively fighting invaders. (Figure 39–5)**

27. Describe the role of the following lymphoid tissues in defense against invaders.

    Spleen

    Tonsils

    Peyer's patches

IV. B Lymphocytes and the Formation of Antibodies (pages 795–801)

A. General Remarks (page 795)

> *Focus:* **Antibodies, also known as immunoglobulins, are produced by B lymphocytes. Antibodies recognize and combine with particles that are foreign or "not-self."**

28. a. *Vocabulary:* What is an antigen?

   b. What types of substances may act as antigens?

Focus: The surface of a cell or virus may have multiple antigens, each of which stimulates a separate immune response.

B. **The B Lymphocyte: A Life History** *(pages 795–797)*

   29. a. Describe the range of circulating B lymphocytes through the body.

   b. What are the general characteristics of both circulating and sessile B lymphocytes?

   c. What structural feature of a B lymphocyte enables it to recognize an antigen?

   30. a. Describe the events that occur following the binding of an antigen to a B lymphocyte.

   b. Where do these events (B cell activation and proliferation) occur?

   c. Name the two types of cells resulting from B cell activation and identify the function of each type.

   31. a. How long does it take after recognition of an antigen for plasma cells to operate at maximum capacity?

   b. Before the use of antibiotics to treat infections, what change in the patient's condition often corresponded to the time when antibody production caught up with pathogen multiplication?

   c. How do antibiotics assist the immune system's fight against disease?

   32. a. What phenomenon is the basis of natural immunity to infectious diseases?

   b. This same phenomenon is also the basis for:

   33. a. Excluding recombinant DNA technology, vaccines can be prepared in what three ways?

   b. The use of recombinant DNA technology offers what two advantages to vaccine production?

   c. Name two diseases for which recombinant DNA vaccines have been developed.

   d. Identify three other diseases for which recombinant DNA may provide vaccines in the future.

Focus: The use of recombinant DNA techniques in vaccine production is revolutionizing this industry.

C. *Essay:* Death Certificate for Smallpox *(pages 796–797)*

34. In eighteenth century Europe, what percentage of (a) the whole population died of smallpox and (b) those who survived childhood experienced smallpox?

Focus: One of the earliest facts recognized about smallpox was that a person who had been infected once would never again be affected by the disease.

35. Which civilization first immunized individuals against smallpox and what technique was used?

36. *Vocabulary:* Define variolation.

37. a. Describe the *two* experiments performed by Edward Jenner that demonstrated the immunizing effects of cowpox virus against smallpox infection.

    b. Why was it possible to use cowpox inoculation to protect against smallpox?

38. What discovery made by Louis Pasteur provided the basis for developing modern vaccines?

39. Summarize the events leading to the World Health Organization recommendation that smallpox research laboratories be closed and the stores of virus destroyed.

D. **The Action of Antibodies** *(pages 797–798)*

40. Identify the three ways in which antibodies commonly act against invaders.

41. a. *Vocabulary:* What is complement?

    b. List three ways in which complement may function.

E. **The Structure of Antibodies** *(pages 798–799)*

42. a. What is multiple myeloma?

    b. What is its usefulness in immunology research?

43. Sketch a diagram of an antibody molecule identifying the following parts: constant regions, variable regions, light chains, heavy chains, and antigen-binding sites.

44. a. Discuss the constancy and variability of amino acid sequences in the antibody molecule.

    b. Describe the recognition/binding site of the molecule.

45. The C_H2 and C_H3 domains serve what functions?

46. a. What is the basis for classifying antibodies into five distinct classes?

  b. List the five distinct classes of immunoglobulins and characterize each class.

  c. Which class includes the principal circulating antibodies, on which the most research has been done?

F. The Clonal Selection Theory of Antibody Formation (*pages 799–801*)

  47. Summarize the clonal selection theory of antibody formation.

  48. a. What two predictions about the response of B lymphocytes were made by this model?

  b. What two experimental techniques have confirmed these predictions?

The Genetics of Antibody Formation (page 801)

  49. a. What apparent discrepancy faced the clonal selection model before the genetic mechanisms of antibody formation were elucidated?

  b. How are the genes that code for antibody chains assembled and at what stage in the life of a B lymphocyte does this assembly occur?

  c. How many combinations of the gene sequences are possible?

  d. In what two ways may additional genetic combinations be generated?

G. *Essay:* **Monoclonal Antibodies** (*pages 802–803*)

  50. Before the development of monoclonal antibodies, what problems faced researchers attempting to obtain pure fractions of a single antibody? (Include in your answer the underlying reasons for these problems.)

  51. a. Monoclonal antibodies can be produced by fusing what two cell types to form a hybrid cell that produces one type of antibody and can live in tissue culture?

  b. Separation of the hybrid cells from the two parent cell types was based on what two factors?

  c. How were the different types of hybrids separated?

  52. Name three scientific advancements made possible by monoclonal antibodies.

  53. a. Identify three current or potential clinical uses of monoclonal antibodies.

  b. Describe how monoclonal antibodies may be used in the future to treat certain cancers.

V. T Lymphocytes and Cell-Mediated Immunity (pages 801–808)

A. General Remarks (page 801)

54. *Vocabulary:* In general terms, what is cell-mediated immunity?

55. By what mechanism do cells involved in cell-mediated immunity recognize one another? (Figure 39–11)

56. Compare the "targets" of T lymphocytes with those of antibodies.

57. a. Name three functional types of T lymphocytes.

b. Which two types are the main regulators of the immune response?

58. a. Identify the targets of cytotoxic T cells.

b. How does a cytotoxic T cell recognize a target?

B. The T Lymphocyte: A Life History (pages 801–805)

> *Focus:* Like B lymphocytes, T lymphocytes are the offspring of stem cells in the bone marrow.

59. a. Name the organ in which immature T cells develop.

b. What event, important to a T cell's life history, occurs in the 8-week-old human fetus?

> *Focus:* During the differentiation of T cells, at least three different types of membrane glycoproteins become displayed on the surface of the cell. These glycoproteins determine the function and antigenic specificity of the cell.

60. a. For each type of T cell, indicate whether it contains T4 (CD4) or T8 (CD8) membrane glycoprotein.

Helper T cell

Suppressor T cell

Cytotoxic T cell

b. What is the usual ratio of T4 to T8 cells?

61. How are T cell receptors (by which the cell recognizes foreign antigens displayed on body cells) similar to and different from antibodies?

62. Characterize the following portions of the T-cell receptor. (Figure 39–12)

Alpha chains

Beta chains

Helical portions of constant regions

> *Focus:* The functions of five other membrane proteins (known collectively as T3 or CD3) are poorly understood but they are thought to mediate T-cell activation.

63. What two types of cells are discarded during T-cell selection?

64. In what two ways does the fate of a T lymphocyte resemble that of a B lymphocyte?

65. *Review Question:* Distinguish between the three stages in a T cell's life history: differentiation, selection, and maturation.

The Major Histocompatibility Complex (page 805)

66. a. *Vocabulary:* What is the major histocompatibility complex and what function does it serve?

b. What types of cells display MHC antigens?

c. What type of studies led to the discovery of the MHC?

d. How many (i) different genes and (ii) alleles for those genes are involved in the MHC?

Focus: **Each person's major histocompatibility complex is unique to that individual.**

---

67. Summarize the mechanism by which thymic selection of T cells occurs.

68. Distinguish between Class I and Class II MHC antigens on the basis of location and function.

69. What structural feature of a Class I MHC antigen has recently been elucidated by x-ray crystallography studies?

C. *Essay:* Children of the *Desaparecidos:* An Application of MHC-Antigen Testing *(page 806)*

Focus: **Each human inherits two sets of alleles for a Class I MHC antigen and for a Class II MHC antigen, one set from each parent.**

---

70. Describe how MHC antigens can be used to determine the probable parentage of a person.

71. *Vocabulary:* What is a haplotype?

72. How could the haplotype of Paula Eva's maternal grandfather be deduced?

D. **The Functions of the T Lymphocytes** *(pages 806–808)*

73. a. What do the surface receptors of a cytotoxic T cell recognize?

b. What happens when a cytotoxic T cell encounters a cell with these recognition sites?

74. What three types of substances can be secreted by cytotoxic T cells?

75. *Vocabulary:* What are lymphokines? (Include their source and function in your answer.)

76. What are "killer cells"?

77. a. What do the surface receptors of a helper T cell recognize?

b. What two cell types trigger the activation of helper T cells?

c. Describe the actions of an activated helper T cell.

78. *Vocabulary:* What are interleukins? (Include their source and function in your answer.)

79. What role do helper T cells play in antibody production? (See also Figure 39–17.)

80. Identify two important roles of macrophages in the immune response. (Figure 39–17)

> *Focus:* **The mechanism by which the immune response is normally suppressed is poorly understood, but it may involve lymphocytes bearing the T8 antigen, cytotoxic T cells, and/or helper T cells.**

81. What is known about suppressor T cells?

82. Summarize the interactions between the immune system and the endocrine and nervous systems.

VI. Cancer and the Immune Response *(pages 808–809)*

> *Focus:* **Many cancer researchers believe that cancer cells routinely stimulate an immune response and that actual cases of cancer represent failures of the immune system to control the abnormal cells.**

83. What are two selective pressures that may have favored the evolution of cell-mediated immunity?

VII. Tissue Transplants *(pages 810–812)*

A. Organ Transplants *(page 810)*

84. a. If skin is grafted from one individual to another (not an identical twin) and the recipient is not treated with immunosuppressive drugs, how long after the transplantation can an immune response to the graft be demonstrated?

    b. How long after the surgery does it take for the graft to die if no measures are taken to suppress the recipient's immune system?

c. What types of cells are responsible for transplant rejection?

85. What development has made possible more careful matching of tissue types before transplant operations?

86. a. Name the major disadvantage to using immunosuppressive drugs in transplant patients.

    b. Identify the source of cyclosporin and describe how it works.

87. a. Describe one alternative to immunosuppressive drugs that is under investigation as a way to reduce transplant rejection.

    b. This approach is based on what characteristic of long-term grafts?

> *Focus:* **The ultimate acceptance of a transplant depends upon changes that occur in the graft itself rather than changes in the recipient's immune system.**

B. Blood Transfusions *(pages 810–812)*

88. Outline the simple experiment by Karl Landsteiner that led him to the conclusion that there are different categories of blood.

89. a. List the four major human blood groups.

    b. What two characteristics define these groups?

90. a. Identify all possible genotypes for persons with each blood type.

    A

    B

    AB

    O

    b. Indicate which antibodies are present in the plasma of a person with each blood type.

    A

    B

    AB

    O

91. a. Specifically, what happens if a person receives blood from a person of a different blood type? (Figure 39–19)

    b. Type O blood was once considered to be a "universal donor" type that could be administered fairly safely to persons of other blood types in an emergency. However, type O blood contains antibodies to A and B surface antigens. Explain why such transfusions were possible. (Figure 39–19)

92. What information about parentage can and cannot be determined by blood typing?

The Rh Factor (pages 811–812)

93. a. Summarize the events leading to hemolytic disease of a newborn.

    b. Which infants are at risk for developing this condition?

    c. How can this syndrome now be prevented?

VIII. Disorders of the Immune System *(pages 812–818)*

A. Autoimmune Diseases *(page 812)*

94. a. *Vocabulary:* What is an autoimmune disease?

    b. Cite three examples of autoimmune disorders.

    c. What other disorders are likely caused by autoimmune responses?

B. Allergies *(pages 812–813)*

95. a. Which type of immunoglobulin mediates allergic responses?

    b. Where are these immunoglobulin molecules located when the allergic response is initiated?

    c. What four types of cells are actively involved in an allergic response?

    d. What event triggers an allergic response and what reactions typically result?

96. a. *Vocabulary:* What is anaphylactic shock?

   b. Name two physiological events that occur during anaphylactic shock and have serious consequences.

97. a. How do (i) antihistamines and (ii) steroid hormones relieve the symptoms of allergies?

   b. What is questionable about the use of allergy preparations that contain both antihistamines and decongestants?

C. **Acquired Immune Deficiency Syndrome (AIDS)** *(pages 813–818)*

> *Focus:* **Throughout human history, epidemics of fatal infectious diseases have been commonplace occurrences. With the development of antibiotics, vaccines, and other antimicrobial agents, the impact of infectious diseases in economically advanced societies decreased dramatically. Just when we thought serious infectious diseases were history, AIDS appeared on the scene.**

98. a. Name three rare diseases that led to the discovery of the AIDS virus.

   b. What was the common denominator between these three diseases?

99. Most of the early victims of the AIDS virus belonged to what four groups of people?

The AIDS Virus and Its Effects (pages 814–816)

100. a. To what virus class does the AIDS virus belong?

   b. The formal name of the AIDS virus is:

101. a. Describe the structure of an AIDS virus particle.

   b. What feature of the virus structure is responsible for its deadly effects?

102. Macrophages are thought to perform what two roles in an AIDS virus infection?

103. a. Summarize the events that occur when an AIDS virus particle binds to a helper T cell.

   b. What is the end result of this infection?

104. Of what value are the antibodies produced early in the infection in terms of neutralizing the virus?

105. What changes occur in the lymph nodes of a person with AIDS? (See also Figure 39–24.)

106. a. Identify five secondary diseases commonly affecting AIDS patients.

   b. AIDS patients most often die from either _____ or _____ .

107. a. Describe the range of symptoms characteristic of an AIDS patient with nervous system involvement.

   b. What two types of pathogens may be causing these symptoms?

   c. How is the AIDS virus thought to enter the nervous system?

108. a. Account for the time lag between infection with the virus and the appearance of symptoms.

b. What event apparently triggers the activation of latent AIDS provirus?

Transmission of the AIDS Virus (page 817)

109. Name the virus from which the AIDS virus is thought to have descended and identify its natural host. (Figure 39–25)

110. a. The AIDS virus may be transmitted by what three means?

b. High levels of the virus occur in what two tissues or body fluids?

c. By what route(s) does the virus gain entry into a human?

d. How well does the virus persist in the environment without the nutritional support of a correct medium?

e. What are the risks for people living with and caring for people with AIDS?

111. a. Which four groups are at the highest risk for contracting the AIDS virus in North America?

b. Which group is most at risk in equatorial Africa?

Focus: **Once infected with the AIDS virus, a person is apparently infected for life and may transmit the virus even if he or she is not showing signs of illness.**

The Prospects (pages 817–818)

Focus: **As of November 1988, 2 million people in the United States are thought to carry the virus; over 78,000 AIDS cases have been documented, and of these, about 44,000 people have died from AIDS-related causes.**

112. Current therapy for AIDS patients is directed at what two targets?

113. What features of AIDS infections may make the development of an effective vaccine difficult?

114. Name one practical use of techniques to detect antibodies to AIDS.

115. Account for the decreasing rate at which the AIDS virus is spreading through the homosexual population.

116. The greatest increase in new AIDS cases is currently occurring among what groups?

Note: Educating the public about the transmission of the AIDS virus and the means to prevent transmission is essential to controlling the spread of the disease.

IX. **Summary** *(pages 818–819):* Read the summary. If you are familiar with the essential features of the material presented there, you are ready to complete the section TESTING YOUR UNDERSTANDING.

TESTING YOUR UNDERSTANDING

After you have completed this examination, compare your answers with those in the section that follows.

1. Which event is NOT part of the inflammatory response?
   a. Chemicals that attract white blood cells are released from injured tissues.
   b. The temperature of the inflamed tissue rises.
   c. Blood flow into the inflamed area increases.
   d. The permeability of capillaries in the area increases.
   e. All of these events are part of the inflammatory response.

2. What characteristic is used to classify the three types of granulocytes?
   a. staining property
   b. shape of the nucleus
   c. size relative to the size of red blood cells
   d. contents of the granules
   e. ratio of nuclear volume to cytoplasmic volume

3. Mast cells are specialized, noncirculating forms of what cell type?
   a. monocytes
   b. lymphocytes
   c. neutrophils
   d. basophils
   e. eosinophils

4. Which of the following is NOT a site where macrophages may lodge to perform their phagocytic function?
   a. lymph nodes
   b. liver
   c. lungs
   d. spleen
   e. All of these are sites where macrophages may lodge.

5. Because interferons inhibit cell proliferation, scientists initially hoped they would be highly effective in treating:
   a. a wide range of cancers.
   b. acquired immune deficiency syndrome (AIDS).
   c. allergic reactions.
   d. organ transplant patients.
   e. a variety of autoimmune disorders.

6. The follicles in which lymphocytes proliferate are located in the _____ of the lymph node.
   a. peripheral cortical areas
   b. peripheral lymphatic sinuses
   c. medullary cords
   d. central cistern
   e. mantle

7. The spleen traps foreign materials carried in the _____, whereas the tonsils trap potential pathogens carried in the _____.
   a. lymph; air
   b. blood; lymph
   c. lymph; food
   d. blood; air
   e. blood; food

8. What types of substances may act as antigens?
   a. any foreign substance
   b. only foreign proteins
   c. only foreign polysaccharides
   d. only foreign glycoproteins
   e. virtually all foreign proteins and most foreign polysaccharides

9. How many B lymphocytes are on alert at any given time in the human body?
   a. 2 million
   b. 200 million
   c. 2 billion
   d. 200 billion
   e. 2 trillion

10. Plasma cells are:
    a. inactive B lymphocytes circulating in the blood.
    b. inactive T lymphocytes circulating in the blood.
    c. inactive B and T lymphocytes circulating in the blood.
    d. descendants of activated B lymphocytes that manufacture antibodies.
    e. descendants of activated T lymphocytes that are being transported in the blood to their target tissue.

11. In Europe by the eighteenth century, _____ persons died of smallpox and _____ percent of those surviving childhood had experienced the disease.
    a. 1 in 5; 90
    b. 1 in 10; 50
    c. 1 in 10; 95
    d. 1 in 50; 50
    e. 1 in 100; 95

12. Who discovered that a virus or bacterium grown in a host other than its natural host often loses its virulence while retaining its capacity to stimulate antibody production?
    a. Edward Jenner
    b. Louis Pasteur
    c. Lady Mary Wortley Montagu
    d. Gerald Edelman
    e. Karl Landsteiner

13. Which statement is NOT true of complement?
    a. Certain combinations of complement proteins function as lytic enzymes to destroy bacterial cells.
    b. Complement may promote the phagocytosis of foreign cells by coating them.
    c. Complement enhances the inflammatory response.
    d. Complement acts only in conjunction with antibodies.
    e. Complement is a group of blood-borne proteins, some of which function in the immune response.

14. The various domains of an antibody molecule have specific activities. Interactions with complement involve the _____ domain, whereas the binding of phagocytic cells to an agglutinated mass of antibodies and foreign particles is made possible by the _____ domain.
    a. C_H1; C_H2
    b. C_H2; C_H3
    c. C_H3; C_L
    d. C_L; C_H1
    e. C_H2; C_H1

15. The five classes of antibodies are distinguished by the:
    a. constant regions of their heavy chains.
    b. constant regions of their light chains.
    c. variable regions of their heavy chains.
    d. variable regions of their light chains.
    e. constant regions of both their heavy and light chains.

16. Which type of antibody is involved in allergic reactions and the expulsion of parasites?
    a. IgA   d. IgG
    b. IgD   e. IgM
    c. IgE

17. The DNA sequences coding for the variable regions of human antibody molecules, which are rearranged during B lymphocyte differentiation, may occur in _____ possible combinations.
    a. 100,000
    b. 10 million
    c. 100 million
    d. 10 billion
    e. 18 billion

18. T lymphocytes were discovered as a consequence of studies of:
    a. new stains for use in light microscopy.
    b. lymph node anatomy and physiology.
    c. the role of the spleen in the immune response.
    d. the function of the thymus.
    e. B lymphocyte maturation and differentiation.

19. Which statement is NOT true of the membrane receptors by which T lymphocytes recognize foreign antigens?
    a. The genes coding for the two chains of the receptor are shuffled and assembled during differentiation.
    b. The receptor is an antibody produced by the T lymphocyte.
    c. The genes coding for T-cell receptors are distinct from those coding for antibodies.
    d. The receptor alpha chain is characterized by acidic amino acids.
    e. The receptor beta chain is characterized by basic amino acids.

20. Helper T cells are characterized by the _____ membrane glycoprotein.
    a. T2
    b. T3
    c. T4
    d. T6
    e. T8

21. Which statement is NOT true of Class I MHC antigens?
    a. They are found on molecules throughout the body.
    b. The three-dimensional structure of a Class I antigen has been determined.
    c. Both chains are encoded by MHC genes.
    d. They are essential to the action of cytotoxic T cells.
    e. They are structurally distinct from Class II MHC antigens.

22. Interleukins are:
    a. hormones, secreted by helper T cells, that stimulate the differentiation of both B and T lymphocytes.
    b. cytotoxic chemicals secreted by cytotoxic T cells.
    c. chemicals, secreted by helper T cells, that attract macrophages.
    d. a group of interferons secreted by cytotoxic T cells.
    e. regulatory enzymes secreted by helper T cells to influence the activity of cytotoxic T cells.

23. Which statement is NOT true of red blood cell surface antigens?
    a. Red blood cells do not have MHC antigens on their surface.
    b. Antibodies against the antigens displayed on an individual's own red blood cells are present in the plasma.
    c. The alleles for the A and B antigens are codominant.
    d. The allele for the O antigen is recessive.
    e. Antibodies against the "not-self" red blood cell surface antigens are always present in the plasma.

24. The cells responsible for transplant rejection are:
    a. macrophages and suppressor T cells.
    b. suppressor T cells and cytotoxic T cells.
    c. "killer cells" and macrophages.
    d. cytotoxic T cells and "killer cells."
    e. "killer cells" and suppressor T cells.

25. The immune system mounts a cell-mediated response against cancer cells because they:
    a. secrete lymphokines that activate macrophages.
    b. secrete interleukins that activate helper T cells.
    c. bear antigens on their surfaces that identify them as foreign.
    d. kill adjacent cells, thereby initiating an inflammatory response and activating macrophages.
    e. secrete chemicals that attract neutrophils to them.

26. Which statement is NOT true of allergic reactions?
    a. They are mediated by IgE molecules displayed on mast cells and noncirculating basophils.
    b. Histamine and other chemicals released from mast cells and basophils trigger an inflammatory response.
    c. The reaction typically involves an epithelial cell surface.
    d. A systemic reaction may occur if the chemicals released from mast cells and basophils enter the circulation.
    e. The drugs of choice for treating allergic reactions are decongestants.

27. What epidemiological clue led to the discovery of acquired immune deficiency syndrome (AIDS) and its causative agent?
    a. a high incidence of a rare cancer and fatal infections caused by ordinarily innocuous protists in young men
    b. a high incidence of fatal bacterial pneumonia in urban prostitutes
    c. clusters of Kaposi's sarcoma among intravenous drug users

d. a high incidence of fatal intestinal infections in recipients of whole blood or blood products

e. a high incidence of *Pneumocystis carinii* pneumonia among immigrants from Africa

28. Which person is NOT at a high risk of contracting the AIDS virus?
   a. an intravenous drug user who shares needles
   b. a child whose classmate has AIDS
   c. a homosexual man
   d. the wife of a bisexual man
   e. a child born to a mother who has AIDS

29. One of the earliest signs of an AIDS virus infection is:
   a. bacterial pneumonia.
   b. severe weight loss.
   c. swollen lymph nodes.
   d. chronic diarrhea.
   e. heart failure.

30. The longest documented time period from initial infection with the AIDS virus to onset of symptoms is:
   a. 18 months.
   b. 3 years.
   c. 5 years.
   d. 7 years.
   e. greater than 10 years.

31. How many people are currently thought to be infected with the AIDS virus in the United States?
   a. 2 million
   b. 1 million
   c. 500,000
   d. 178,000
   e. 78,000

32. **T or F** Since most of the studies of immunity have involved mammals, scientists know a great deal about mammalian defenses against invading organisms but very little about the defense mechanisms of other vertebrates.

33. **T or F** Plasma cells and memory cells are produced from two different types of B lymphocytes.

34. **T or F** The eventual acceptance of transplanted tissue depends on changes in the recipient's immune system so that the donor tissue is recognized as "self."

35. **T or F** Antibodies present in the donor's blood usually cause no problems for a recipient because they are diluted by the recipient's blood.

36. **T or F** Because of the lipid bilayer around its nucleic acid core, the AIDS virus usually enters a target cell by diffusing directly through the cell membrane.

37. **T or F** To date, there has been no report of a person becoming infected with the AIDS virus while nursing an affected family member or friend.

38. Identify the two components of the first line of defense against invading microorganisms. Name three features in various parts of the body that facilitate defense against invasion through mucous membranes.

39. What are two advantages of using synthetic purified antigens produced by recombinant DNA technology in vaccine production? Name four diseases for which vaccines have been or are being developed using this technique.

40. What two predictions did the clonal selection model make? These predictions have been confirmed by what two areas of research?

41. Summarize the inheritance of Class I and Class II MHC antigens. To what extent are the alleles for these antigens expressed?

42. How may hemolytic disease of the newborn be prevented?

43. Define an autoimmune disease and cite three examples.

44. Cite two facts indicating that the development of a truly protective vaccine against the AIDS virus may be extraordinarily difficult.

PERFORMANCE ANALYSIS

1. **e** In the inflammatory response, chemicals that attract white blood cells to the site are released from injured cells and sometimes from invading cells. Other chemicals cause local blood vessels to dilate and the permeability of capillaries to increase. As vessels dilate, blood flow to the injured tissues increases, thereby raising the local temperature. (page 792)

2. **a** The three types of granulocytes are classified based on the staining characteristics of their cytoplasmic granules. (page 792)

3. **d** Mast cells are specialized, noncirculating basophils found in connective tissue. During an inflammatory response, mast cells release histamine. (page 792)

4. **e** Macrophages may lodge and conduct their phagocytic function in the lymph nodes, lungs, spleen, and liver. (page 792)

5. **a** When the ability of interferons to inhibit cell proliferation was first discovered, scientists hoped they would be effective in treating a wide range of cancers. Unfortunately, this hope has not been fulfilled. (page 793)

6. **a** The lymphoid follicles, in which lymphocytes proliferate, are located in the peripheral cortical areas of lymph nodes. (page 794)

7. **d** The spleen traps foreign materials carried in the blood, the tonsils trap airborne foreign particles, and the lymph nodes filter the lymph. (pages 794, 795)

8. **e** Most foreign polysaccharides and virtually all foreign proteins can act as antigens. (page 795)

9. **é** Approximately 2 trillion B lymphocytes are on alert at any given time in the human body. (page 795)

10. **d** Plasma cells are descendants of activated B lymphocytes and function as antibody factories. (page 795)

11. **c** In Europe by the eighteenth century, 1 in 10 persons died of smallpox and 95 percent of those surviving childhood had experienced the disease. (essay, page 796)

12. **b** Louis Pasteur discovered that a virus or bacterium grown in a host other than its natural host loses its virulence while retaining its capacity to stimulate antigen production. This fact has historically been the basis for vaccine production. (essay, page 797)

13. **d** Complement is a group of plasma proteins that function in the inflammatory and immune responses. By coating foreign particles, complement may promote their phagocytosis. Certain combinations of complement proteins function as lytic enzymes to destroy bacterial cells. Complement often, but not always, acts in conjunction with antibodies. (page 798)

14. **b** On an antibody molecule, the C_H2 domain mediates interactions with complement and the C_H3 domain enables phagocytic cells to bind to an agglutinated mass of antibodies and foreign particles. (Figure 39–8)

15. **a** The constant regions of the heavy chains are identical for all members of a particular antibody (or immunoglobulin) class. For example, all IgG molecules have the same constant regions of their heavy chains, and all IgM molecules have the same constant regions in their heavy chains, but they differ from those of the IgG molecules. (page 799)

16. **c** IgE molecules are involved in allergic reactions and the expulsion of parasites. (page 799)

17. **e** In the human genome, there are 18 billion possible combinations of the DNA sequences coding for the variable regions of antibody molecules. Each human is capable of making antibodies to 100 million different antigens. (page 801)

18. **d** T lymphocytes were discovered during studies to determine the function of the thymus gland. In the eighth week of fetal life, primitive blood cells migrate into the thymus. Here, the T-lymphocyte precursors undergo differentiation, selection, and maturation. (pages 801, 804)

19. **b** The membrane receptors by which T lymphocytes recognize foreign antigens are not antibodies. However, the genes coding for these receptors undergo rearrangement during T-cell differentiation in a manner analogous to that of the genes coding for antibodies. (page 804, Figure 39–12)

20. **c** Helper T cells have T4 (CD4) membrane glycoproteins whereas cytotoxic and suppressor T cells have T8 (CD8) membrane glycoproteins. (page 804)

21. **c** The gene coding for the alpha chain of Class I MHC antigens is part of the major histocompatibility complex. However, the gene that codes for the other protein chain is not part of the MHC. (page 805, Figure 39–13)

22. **a** Interleukins are hormones that are secreted by helper T cells and stimulate the differentiation of both B and T lymphocytes. (page 808)

23. **b** An individual has plasma antibodies against those red blood cell antigens that are "not-self," even if he or she has never had a blood transfusion. For example, a person with type B blood has antibodies against type A red cell surface antigens. (pages 810, 811)

24. **d** The cells responsible for a transplant rejection are cytotoxic T cells and "killer cells." (page 810)

25. **c** Even though cancer cells are body cells, and might therefore be expected to be identified as "self," they can be identified by the immune system as foreign because they display foreign antigens on their surfaces. (pages 808, 809)

26. **e** The drugs of choice for treating allergic reactions are antihistamines and/or cortisone steroids, depending upon the severity of the reaction. Most decongestants stimulate histamine release, which would only complicate matters since histamine is responsible for many of the symptoms of an allergic reaction. (pages 812, 813)

27. **a** The evidence that first clued scientists to the existence of the AIDS virus was an unusually high incidence of two disorders in young men: Kaposi's sarcoma and fatal infections caused by generally harmless opportunistic protists. (page 814)

28. **b** Homosexual and bisexual men, intravenous drug users who share needles, sexual partners of these individuals, and children born to mothers infected with the AIDS virus are at a very high risk of contracting the virus. Individuals who have only casual contact with an AIDS victim, such as the classmate of a child with AIDS, are not at risk of contracting the virus. (page 817)

29. **c** As is true of many infections, one of the first signs of an AIDS virus infection is swollen lymph nodes. (page 816)

30. **e** Based on studies performed on serum collected from homosexual men in the 1970s, the longest documented time period from initial infection with the AIDS virus to onset of symptoms is greater than 10 years. (page 816)

31. **a** It has been estimated that 2 million people in the United States are currently infected with the AIDS virus. (page 817)

32. **True** Since most of the research in immunology has been performed on mammals, very little is known about the immune systems of other vertebrates. (page 791)

33. **False** A plasma cell and a memory cell are produced following the activation of a single B lymphocyte by interaction with its corresponding antigen. (page 795)

34. **False** The eventual acceptance of a transplant depends upon changes in the tissue itself that transform it into tissue recognized as "self" by the host's immune system. (page 810)

35. **True** Historically, persons with type O blood were referred to as universal donors since their red blood cells have no surface antigens to react with circulating anti-A and anti-B antibodies. The antibodies contained in the

type O blood cause no problem for a recipient since the blood is diluted by the recipient's blood upon transfusion. (The transfusion of blood that does not match the recipient's blood type is no longer commonly practiced.) (Figure 39–19)

36. **False** The lipid bilayer of the AIDS virus is studded with glycoproteins that are a perfect match for the T4 receptors on helper T cells. The AIDS virus enters these cells by receptor-mediated endocytosis. (page 815)

37. **True** All the evidence accumulated thus far indicates that the AIDS virus, although highly virulent, is extremely fragile and not highly infectious and that a person is unlikely to contract the virus from casual contact with an infected person or from objects used by an infected person. No person caring for a friend or family member infected with the AIDS virus has yet contracted the disease. (page 817)

38. The first line of defense against invading microorganisms consists of the skin and mucous membranes. Examples of features that facilitate the protective function of mucous membranes are constant flushing with fluids that contain antimicrobial substances (mucus, tears, and saliva), the cilia of the respiratory tract, the extreme pH of the stomach, and the resident bacteria of the large intestine. (page 791)

39. (1) The possibility of a disease-causing infection is eliminated. (2) Contaminants that might produce an adverse reaction are not introduced into the vaccine. Examples of diseases for which recombinant vaccines have been developed (or are being developed) include hoof-and-mouth disease of livestock, hepatitis B virus, malaria, schistosomiasis, and other parasitic diseases. (pages 796, 797)

40. (1) Only a very small number of lymphocytes will respond to any given antigen. (2) A single plasma cell will produce only one antibody. These predictions have been confirmed by studies involving multiple myeloma cells and during the development of monoclonal antibodies. (page 800)

41. There are at least 20 different genes involved in the code for the two classes of MHC antigens. Each human inherits two copies (one from each parent) of the loci coding for Class I MHC antigens and two copies of the loci coding for Class II MHC antigens. All of the MHC alleles are expressed in an individual. (essay, page 806)

42. If an Rh-negative woman gives birth to an Rh-positive child, she can be injected with antibodies against the fetal Rh antigen within 72 hours of delivery. This precaution will prevent her from manufacturing her own antibodies against the Rh antigen, which would endanger a subsequent Rh-positive child she might have. (page 812)

43. Autoimmune diseases are those in which the body's ability to distinguish "self" from "not-self" malfunctions and the immune system begins producing antibodies against the body's own tissues. Examples include myasthenia gravis, lupus erythematosus, and several types of anemia. (page 812)

44. (1) The antibodies formed against the AIDS virus during an infection are not effective in controlling the virus. (2) The genes coding for key portions of the AIDS virus protein coat have a high rate of mutation. (page 818)

REVIEW TEST 11

Section 6: Chapters 37–39

This review test is *not* designed to cover all of the important information in these chapters. However, it does touch on the major topic categories in each chapter. It will also give you valuable practice in taking this type of test. When you are finished, check your answers with those provided in the following PERFORMANCE ANALYSIS section.

1. Uric acid is NOT the major nitrogenous waste product in:
   a. insects.
   b. chickens.
   c. snakes.
   d. humans.
   e. eagles.

2. The high concentration of salts and urea in the interstitial fluid bathing the loop of Henle results primarily from the properties of the:
   a. proximal convoluted tubule and descending loop of Henle.
   b. ascending loop of Henle and distal convoluted tubule.
   c. distal convoluted tubule and collecting duct.
   d. collecting duct and ascending loop of Henle.
   e. ascending and descending loops of Henle.

3. Why are tropical fish unable to maintain a body temperature lower than that of their aquatic environment?
   a. They produce too much heat by the muscular movements associated with swimming.
   b. They have no means of unloading excess heat.
   c. They do not have sufficient fat to insulate themselves.
   d. Their metabolic rate is too high.
   e. Heat transfer through the gills is unidirectional, from outside to inside the body.

4. Which of these homeotherms would have the greatest difficulty remaining cool if placed in the open sun on a hot, humid day?
   a. a person wearing thin summer clothing
   b. a fox with long ears
   c. a black Labrador retriever
   d. a white Leghorn chicken
   e. a coyote with a tan-colored coat

5. Before the genetics of antibody formation were elucidated, the clonal selection model was in trouble because:
   a. numerous potential antigens were discovered for which no corresponding B lymphocyte existed.
   b. the genes coding for the production of antibodies are regulated by mechanisms that differ from those regulating the expression of other genes.
   c. the number of different antigens against which antibodies can be produced is greater than the number of structural genes in the human genome.
   d. studies of monoclonal antibodies initially seemed to contradict the model.
   e. the existence of movable genetic elements had not yet been proposed.

6. Which statement is NOT true of cytotoxic T cells?
   a. They are regulated by macrophages attracted to their site of activity.
   b. Once activated, they proliferate into active cells and memory cells.
   c. They secrete lymphokines.
   d. Some secrete cytotoxins.
   e. Some secrete interferons.

7. Receptors that detect changes in _____ are found in the heart, aorta, and carotid arteries.
   a. blood volume
   b. ADH levels
   c. total plasma solute concentration
   d. dissolved oxygen
   e. urea levels

8. During an active AIDS virus infection, the body can only launch an effective immune response against:
   a. cells harboring the AIDS virus.
   b. the virus itself.
   c. other invading microorganisms.
   d. cancer cells that may develop.
   e. The premise is false; during an active AIDS virus infection, the body is unable to mount an effective response against any foreign antigens.

9. T or F   The immune response consists of two phases: the first phase is a generalized, nonspecific reaction to any foreign invader; the second phase is a rapid, specific response to the particular pathogenic agent.

10. T or F   In settling paternity cases, blood types can only prove that a person is not the father of a particular child.

11. T or F   As long as they are dry, fur and feathers promote heat conservation by trapping air close to the animal body.

12. T or F   Since their core body temperatures are so much higher than the surrounding water temperature, marine

mammals lose a great deal of body heat to the water by conduction.

13. What are the advantages of hibernation to an animal?

14. List three areas of research that have advanced because of the use of monoclonal antibodies.

15. Identify the function of each of the three types of membrane glycoproteins synthesized during T lymphocyte differentiation.

16. Describe the proposed physiological function of the kidney in early freshwater fish.

PERFORMANCE ANALYSIS

1. **d** Uric acid is the form in which nitrogenous wastes are excreted from birds, terrestrial reptiles, and insects. In humans, nitrogenous wastes are converted into urea by the liver and the urea is removed from the blood by the kidneys. (page 766)

2. **d** The lower portion of the collecting duct is permeable to urea, which diffuses from the collecting duct into the surrounding interstitial fluid. Sodium and chloride ions are actively pumped out of the renal tubule in the upper portion of the ascending branch of the loop of Henle, which is not permeable to water. These two factors contribute to the high ion and urea concentration present in the medulla, through which the loops of Henle and the collecting ducts pass. (page 773)

3. **b** Most fish are unable to maintain a body temperature that is either significantly greater than or lower than that of the water in which they live. They lose a great deal of heat as blood flows through the gills. They have no mechanisms for either unloading excess heat or conserving heat. (page 780)

4. **c** Dark objects absorb solar energy more readily than do light objects. In the open sun on a hot, humid day, a black dog will have the most trouble regulating body temperature. The black coat will absorb the energy of the sun and the humid conditions will reduce the effectiveness of evaporative cooling by panting. (page 781)

5. **c** Each human is apparently capable of making antibodies against 100 million different antigens. Since this number is greater than the number of structural genes in the human genome, the clonal selection model of antibody formation was in difficulty until the genetics of antibody production were elucidated. (page 801)

6. **a** Activated cytotoxic cells proliferate into active cells and memory cells and secrete lymphokines. In addition, some secrete cytotoxins and some secrete interferons. As is true for all lymphocytes, their activity is regulated by helper T cells and suppressor T cells. (page 807)

7. **a** Kidney function is affected by blood pressure and blood volume. Receptors that detect changes in blood volume are located in the heart, aorta, and carotid arteries. (page 774)

8. **e** Helper T cells and suppressor T cells are the major regulators of the immune response, including the activities of B lymphocytes. Since the activated AIDS virus destroys the body's helper T cells, AIDS patients are unable to mount an effective immune response to any foreign invaders. (pages 801, 814)

9. **False** The primary response of the immune system to an invader involves the production of plasma cells and memory cells the first time a particular antigen is recognized by the immune system. Subsequent responses to the same antigen are mediated by the memory cells and are consequently more rapid. (page 793)

10. **True** In paternity cases, blood types can only prove that a person is *not* the father of a particular child. Since there are only four blood types, blood typing cannot prove with certainty that a man is the father of a child. However, MHC antigen analysis is a very reliable way to determine biological relationships within a family. (page 811)

11. **True** Dry fur and feathers insulate the body by trapping air close to the body. If they are wet, the fur and feathers lie flat against the body, obliterating the insulating air space. (pages 779, 786)

12. **False** Marine mammals have a very thick (compared with terrestrial mammals) layer of subcutaneous fat that insulates their body core and minimizes heat loss. Their skin temperature may be only a few degrees higher than that of the surrounding water. (page 787)

13. During hibernation, heart rate, respiratory rate, oxygen consumption, and metabolic rate slow down dramatically. There is evidence that aging processes stop. This minimizes energy expenditures during times when food supplies are scarce and environmental conditions are unfavorable. (page 786)

14. Three areas of research that have benefited from the development of monoclonal antibodies are studies of antibody structure and function, studies of T lymphocyte structure and function, and purification of naturally occurring substances present in very small quantities. (essay, page 802)

15. The first type of membrane glycoprotein is correlated with T lymphocyte function. Helper T cells bear the T4 molecule, whereas cytotoxic and suppressor T cells bear the T8 molecule. The second type is the receptor by which a T cell recognizes cells of the body itself and foreign antigens that are displayed on these cells. The third type is thought to function in the interactions among the different types of T lymphocytes. (page 804)

16. It is generally believed that fish evolved in fresh water and later moved to marine environments. In a freshwater environment, the original function of the kidney was probably to pump out excess water while conserving salt and other desirable solutes. Like modern freshwater fish, these fish would have excreted a hypotonic urine. (page 767)

Integration and Control I: The Endocrine System

MAJOR CONCEPTS

The two systems responsible for the integration and control of body functions are the nervous system and the endocrine system. The increasing discovery of interactions between these two components leads some authorities to consider them as two facets of one neuroendocrine system. The way in which messages are transmitted by these systems differs. Signals sent by the nervous system travel very quickly and are short-lived. Messages sent by the release of hormones take longer to reach their targets and have greater durations of action.

Glands may be composed of specialized glandular epithelial tissue or neurosecretory cells. There are two basic types of glands: exocrine glands secrete their products into ducts, whereas endocrine glands release their products into the bloodstream.

The hypothalamus of the brain is linked to the endocrine system by its direct influence on the pituitary gland. Under hypothalamic influence (through releasing or inhibitory hormones) the pituitary releases growth hormone, prolactin, and tropic hormones that affect other glands, including the thyroid, adrenal cortex, and gonads. The hypothalamus also produces two hormones (ADH and oxytocin) that travel through a short portal system to the posterior lobe of the pituitary, where they are stored until they are released.

The synthesis and release of a specific hormone is controlled by intricate negative (and sometimes positive) feedback mechanisms, which may involve several hormones.

The thyroid secretes thyroxine, which stimulates and maintains metabolic processes. It also secretes calcitonin, which lowers blood calcium levels by stimulating the deposition of calcium in bone. Located adjacent to or embedded in the thyroid are the parathyroid glands. Parathyroid hormone increases blood calcium levels by stimulating the release of calcium from bone, by increasing intestinal calcium absorption by activation of vitamin D, and by inhibiting calcium excretion.

The adrenal glands have two separate secretory tissues. The outer cortex produces two classes of compounds. Glucocorticoids (e.g., cortisol) function in the metabolism of carbohydrates, proteins, and lipids and suppress the immune system. Mineralocorticoids (e.g., aldosterone) are important in regulating ion and water balance. The inner medulla is composed of neurosecretory cells that produce adrenaline and noradrenaline, hormonal mediators of the "fight-or-flight" response of the sympathetic nervous system.

The pancreas secretes insulin, glucagon, and somatostatin, which are involved in the regulation of glucose metabolism.

The pineal gland secretes melatonin, which apparently regulates circadian rhythms in some animals. The pineal glands of some lower vertebrates have light-sensitive cells.

Prostaglandins are fatty acid derivatives that are produced by cell membranes. Their target tissues are usually the tissues in which they are produced or the tissues of another individual. They are active at concentrations much lower than those of most hormones. There are several different types of prostaglandins and their effects vary—the activity of some classes directly opposes the effects of other classes. Prostaglandins influence smooth muscles of various organs, platelet aggregation, and the inflammatory response. The potent anti-inflammatory effect of aspirin is due to its inhibition of prostaglandin production.

There are two basic mechanisms by which hormones exert their effects. Steroids and thyroid hormones pass through cell membranes and bind to intracellular receptors. Hormones derived from proteins or amino acids bind to receptors in the target cell membrane, stimulating the release of a second messenger inside the cell. Cyclic AMP is the second messenger for many hormones.

Hormones that have traditionally been thought of as "mammalian" have been discovered in many other organisms, including protists and fungi.

HOW TO STUDY THE CHAPTER

Read the entire chapter through quickly, focusing on the major concepts.

Use the GUIDED STUDY OF THE CHAPTER to help you identify the important details as you **reread** the chapter. Writing out the answers to these questions will help fix them in your mind as well as provide you with a valuable study aid.

Answer the questions in TESTING YOUR UNDERSTANDING without the aid of your text. Check your answers against those in PERFORMANCE ANALYSIS. Analyzing your answers will give you valuable feedback on your level of understanding and preparedness for classroom testing.

GUIDED STUDY OF THE CHAPTER

I. Introduction *(pages 821–822)*

> *Focus:* Chemical signals are the means by which homeostasis and the integration and control of body functions are accomplished in living systems.

1. a. What two types of receptors receive chemical signals from outside the cell?

   b. In what two ways can cells respond to such signals?

   c. List three ways in which chemical signals can be transmitted to other cells.

2. Fill in the following table contrasting the endocrine and nervous systems. (See also Figure 40–1.)

Table for Question 2

	ENDOCRINE SYSTEM	NERVOUS SYSTEM
Structural components of system		
Time interval between initiation of signal and arrival at target		
Duration of action at target		

3. How are hormones and neurotransmitters similar? (Figure 40–1)

4. What evidence supports consideration of the nervous and endocrine systems as two components of the neuroendocrine system?

5. *Vocabulary:* Distinguish between neurotransmitters and neurosecretory cells.

6. Name the two cell types capable of endocrine function.

> *Focus:* The basic principles of chemical communication among cells apply to both endocrine and nervous functions.

II. Glands and Their Products: An Overview (*pages 823–824*)

7. a. *Vocabulary:* Define a hormone.

   b. Name four types of cells that secrete hormones.

8. a. *Vocabulary:* Distinguish between endocrine and exocrine glands, naming three examples of each type. (See also Figure 40–3.)

   b. Name one gland that has both endocrine and exocrine function. (Figure 40–2)

9. Identify three general types of hormones (based on chemical composition) and indicate the site in the body where each type is degraded.

Focus: **Hormones are active in very small amounts, and their concentrations are tightly regulated.**

---

10. Name two factors important in the regulation of hormone activity.

Focus: **Study Table 40–1. Appreciate the number and variety of body activities under hormonal control.**

III. The Pituitary Gland (*pages 825–826*)

A. General Remarks (*page 825*)

11. Hormones from the pituitary influence which three organs or organ systems?

12. Secretion of pituitary hormones is influenced by what structure?

13. Where is the pituitary located?

B. The Anterior Lobe (*page 825*)

14. Provide the following information about growth hormone (somatotropin).

    Four functions:

    Result of deficit in children:

    Result of excess in children:

    Result of excess in adults:

15. Identify the function of prolactin and describe how its production is controlled.

16. a. *Vocabulary:* What is a tropic hormone?

    b. List four tropic hormones secreted by the anterior pituitary and identify the target and/or action of each one.

C. The Intermediate and Posterior Lobes (*page 826*)

17. What hormone is produced by the intermediate lobe in many vertebrates and what is its function?

18. Name the function of the posterior pituitary.

IV. The Hypothalamus (*pages 826–828*)

A. The Pituitary-Hypothalamic Axis (*pages 826–827*)

Focus: **The hypothalamus is the major link between the nervous and endocrine systems. At least nine hormones produced in the hypothalamus directly stimulate or inhibit the release of pituitary hormones.**

---

19. a. The hormones of the hypothalamus are unusual in what two ways?

b. What is the role of the portal system between the hypothalamus and anterior pituitary? (See also Figure 40–5.)

c. How do oxytocin and ADH reach the posterior pituitary, where they are stored? (Figure 40–5)

20. a. Distinguish between releasing hormones and inhibitory hormones.

b. Identify the action of (a) thyrotropin-releasing hormone (TRH), (b) gonadotropin-releasing hormone (GnRH), and (c) somatostatin.

21. What is the role of the hypothalamus in feedback control?

B. **Other Hypothalamic Hormones** (*page 828*)

22. a. Name three activities influenced by oxytocin.

b. What controls the release of oxytocin?

23. Identify two functions of antidiuretic hormone (ADH), one of which is not very pronounced in humans.

24. Account for the fact that ADH has some oxytocin effect and oxytocin has some ADH effect.

V. **The Thyroid Gland** (*pages 828–829*)

25. Construct a flow diagram to summarize the negative feedback system by which the secretion of thyroxin from the thyroid gland is controlled.

26. a. Thyroxine is made up of what constituents?

b. Describe two effects of thyroxine.

c. What is triiodothyronine?

27. Describe the consequences of the following conditions:

Hyperthyroidism

Hypothyroidism in infants

Hypothyroidism in adults

28. Identify the effect of calcitonin.

VI. **The Parathyroid Glands** (*page 829*)

29. a. Where are the parathyroid glands located?

b. Parathyroid hormone functions in:

c. Name three body functions that may be severely affected by minor changes in blood calcium concentration.

30. a. Identify the specific effect of parathyroid hormone.

b. Name three ways in which it exerts this effect.

31. What factor influences the production of calcitonin and parathyroid hormone?

32. Summarize the consequences of hyperparathyroidism.

33. What happens if the parathyroid glands are removed and hormone replacement is not instituted?

VII. *Essay:* **The Regulation of Bone Density** *(pages 830–831)*

34. *Vocabulary:* Distinguish between osteoblasts and osteoclasts.

35. a. *Vocabulary:* What is osteoporosis?

    b. Name two factors contributing to the development of osteoporosis. (You must read the entire essay to answer this question.)

    c. Osteoporosis resembles what hormonal disorder?

    d. Identify two effects of osteoporosis.

36. Where do colony-stimulating factors originate and what are their effects?

37. a. Name four molecules that apparently stimulate osteoclast activity.

    b. How may osteoclast activity be reduced therapeutically?

38. a. Identify two compounds that influence the absorption of calcium from the intestine.

    b. Why are postmenopausal women particularly susceptible to osteoporosis?

Focus: Current evidence suggests that calcium supplementation without appropriate hormone and anti-inflammatory therapy has little effect on the progression of osteoporosis.

39. What is thought to be a good preventive measure that children and young adults can take against osteoporosis in later years?

VIII. **Adrenal Cortex** *(pages 829–832)*

A. **General Remarks** *(page 829)*

40. a. *Vocabulary:* What are corticosteroids?

    b. Name the two groups of steroid hormones produced in the adrenal cortex.

B. **Glucocorticoids** *(pages 830–831)*

41. a. Identify two general roles of cortisol in the body.

    b. Under what conditions do cortisol levels increase?

    c. Glucocorticoids work in concert with which branch of the nervous system?

42. a. Why do periods of stress increase a person's susceptibility to illness?

    b. Glucocorticoids may be used to treat what two types of disorders?

    c. Name three adverse effects of glucocorticoid therapy.

43. a. What stimulates the release of cortisols?

b. How is the secretion of cortisols controlled? (Include in your answer the role of an interleukin.)

Focus: **Corticosteroids may normally function as part of the mechanism regulating the inflammatory and immune responses.**

C. **Mineralocorticoids** *(page 832)*

44. What is the function of (a) mineralocorticoids in general and (b) aldosterone in particular?

45. Describe the result of mineralocorticoid deficiency.

46. What other hormones are produced by the adrenal cortex?

IX. **Adrenal Medulla** *(page 832)*

47. The adrenal medulla is composed of what type of cells?

48. a. Name two compounds secreted by the adrenal medulla.

b. These compounds, which are synthesized from modified amino acids (Figure 40–11), are known as:

c. These two compounds are synthesized from what amino acid? (Figure 40–11)

d. Identify five effects of adrenaline and noradrenaline.

49. What stimulates the adrenal medulla?

Focus: **Adrenaline is also known as epinephrine, and noradrenaline as norepinephrine.**

X. **The Pancreas** *(pages 832–833)*

50. Fill in the following table.

Table for Question 50

	INSULIN	GLUCAGON
Site of production		
Stimulus for release		
Effect on blood glucose levels		
Mechanism of action		

---

51. a. Describe what happens if there is an insulin deficiency.

b. Name one cause of death in an untreated diabetic and describe how this condition arises.

52. a. Cite two locations from which somatostatin has been isolated.

b. Identify two ways in which somatostatin influences glucose metabolism.

53. Summarize the role of each hormone in the regulation of blood glucose levels. (See also Figure 40–12.)

Adrenaline

Cortisol

Glucagon

Growth hormone

Insulin

Noradrenaline

Somatostatin

54. Explain why maintenance of proper blood glucose levels is so important.

55. Under what circumstances do brain cells use fatty acids as their energy source?

56. How can insulin overdose lead to death?

XI. The Pineal Gland *(page 833)*

A. General Remarks *(page 833)*

57. Where is the human pineal gland located?

Focus: **In lower vertebrates, the pineal gland contains light-sensitive cells.**

---

58. a. What hormone is secreted by the pineal gland?

b. Describe the influence of light and dark on melatonin production in humans.

59. How does melatonin influence the skin pigment of amphibian larvae?

60. Name three effects of melatonin in birds.

61. What evidence suggests that the pineal gland may be involved in sexual maturation of humans?

Focus: **Current evidence suggests that the pineal gland functions as a biological timekeeper and that, in at least some species, light influences its function.**

B. *Essay:* **Circadian Rhythms** *(page 834)*

62. Summarize the findings in fruit flies that indicate a genetic basis for circadian rhythms.

63. Name nine phenomena that exhibit circadian rhythms in humans.

64. a. Identify four consequences of "jet lag."

b. For what length of time after a person changes time zones may he or she experience the effects of jet lag?

c. What two situations may have "set the stage" for the accident at Three Mile Island nuclear power plant?

65. How may circadian fluctuations in physiological parameters influence diagnosis and therapy? Cite at least two examples.

XII. Prostaglandins *(pages 833–836)*

A. General Remarks *(pages 833–835)*

66. a. In which body fluid do prostaglandins occur in high levels?

b. What is the origin of the prostaglandins found in semen?

67. List four ways in which prostaglandins differ from hormones.

68. a. Prostaglandins are synthesized in cell membranes from what precursor? (Figure 40–13)

b. Name four substances that may stimulate the production of this precursor. (Figure 40–13)

69. *Vocabulary:* What are leukotrienes and in what body functions do they play a role? (Figure 40–13)

70. Identify three types of substances that may be produced from endoperoxide intermediates and indicate one function for each. (Figure 40–13)

71. Aspirin is an anti-inflammatory agent. Describe its influence on arachidonic acid metabolism. (Figure 40–13)

Focus: **Prostaglandins, which are very potent chemicals, are released in minute quantities and broken down rapidly.**

B. **Stimulation of Smooth Muscle** *(page 835)*

72. a. How are prostaglandins thought to function in human reproduction?

b. What evidence in males and females supports this proposed role?

73. a. What has been postulated as the role of prostaglandins in menstruation and labor?

b. *Vocabulary:* Define dysmenorrhea.

c. Identify three means by which prostaglandins contribute to dysmenorrhea.

d. What strategy is currently employed in the treatment of dysmenorrhea?

C. **Other Prostaglandin Effects** *(pages 835–836)*

74. Name five effects of prostaglandins not previously mentioned.

75. a. Identify three cell types that produce leukotrienes.

b. Increased prostaglandin levels have been implicated as factors in what types of disorders?

XIII. **Mechanisms of Action of Hormones** *(pages 836–838)*

A. **General Remarks** *(page 836)*

76. a. How does the means by which hormones transmit messages differ from that of neurons?

b. What three factors determine whether or not a hormone exerts an effect?

Focus: **Hormones exert specific effects because only cells that have receptors for a particular hormone can receive its message.**

77. a. Name two possible locations for hormone receptors.

b. Name two groups of hormones that interact with each type of receptor.

B. **Intracellular Receptors** *(page 836)*

78. a. Steroid hormones are available to intracellular receptors because of what characteristics?

b. Where in a target cell are steroid hormone receptors located?

79. a. Summarize the events that occur after a steroid hormone binds to its specific receptor.

b. Name three possible functions of the proteins whose synthesis is stimulated by a hormone. (See also Figure 40–14.)

80. a. How is thyroid hormone thought to enter cells?

b. Where is its receptor located?

C. Membrane Receptors *(pages 837–838)*

81. a. How does adult-onset diabetes differ from juvenile-onset diabetes?

b. What is the most effective means for treating (i) adult-onset and (ii) juvenile-onset diabetes?

82. Once a hormone binds to a membrane receptor, either of two events may occur. What are these two events?

83. a. Name the compound that is the second messenger for many hormones.

b. How is this compound formed? (Include the name of the enzyme.)

84. Identify the role of (a) protein kinase and (b) phosphorylase *a* in the amplification cascade triggered when adrenaline binds to liver cells.

Focus: In the second messenger mechanism, the effect of one hormone molecule is amplified by the creation of many molecules of cyclic AMP, each of which participates in the next step.

---

85. Indicate one function of cyclic AMP in (a) cellular slime molds and (b) *Escherichia coli.*

Focus: The discovery of "mammalian" hormones in unicellular organisms is compelling evidence of the evolutionary continuity of life.

---

86. A. *Review Question:* Match each hormone listed with its site of production. Answers may be used only once. A blank may have more than one answer, as indicated in parentheses.

1. adrenaline
2. adrenocorticotropic hormone (ACTH)
3. aldosterone
4. antidiuretic hormone (ADH)
5. calcitonin
6. cortisol
7. glucagon
8. growth hormone
9. insulin
10. melatonin
11. noradrenaline
12. oxytocin
13. parathyroid hormone
14. prolactin
15. thyroid-stimulating hormone (TSH)
16. thyroxine

a. _____ adrenal cortex (2 answers)

b. _____ adrenal medulla (2 answers)

c. _____ hypothalamus (2 answers)

d. _____ pancreas (2 answers)

e. _____ parathyroid

f. _____ pineal

g. _____ pituitary, anterior lobe (4 answers)

h. _____ thyroid (2 answers)

B. *Review Question:* Match each hormone listed in (A) with the correct description of its function(s). Use each answer only once. A blank may have more than one answer, as indicated in parentheses.

a. _____ stimulates metabolic activities

b. _____ stimulates the adrenal cortex

c. _____ regulates circadian rhythms

d. _____ increase blood glucose, increase heart rate, dilate or constrict specific blood vessels (2 answers)

e. _____ affects carbohydrate, protein, and lipid metabolism; suppresses the immune system

f. _____ stimulates milk production and secretion in a "primed" gland

g. _____ inhibits the release of calcium from bone

h. _____ influences ion and water balance

i. _____ decreases blood glucose, increases glycogen production

j. _____ controls water excretion

k. _____ stimulates the thyroid gland

l. _____ stimulates the release of calcium from bone; stimulates conversion of vitamin D to the active form; inhibits calcium excretion

m. _____ stimulates breakdown of glycogen to glucose in the liver

n. _____ stimulates uterine contractions and milk ejection

o. _____ stimulates growth of bone; inhibits glucose oxidation; promotes fatty acid catabolism

XIV. Summary *(page 839)*: Read the summary. If you are familiar with the essential features of the material presented there, you are ready to complete the section TESTING YOUR UNDERSTANDING.

TESTING YOUR UNDERSTANDING

After you have completed this examination, compare your answers with those in the section that follows.

1. Hormones that are derivatives of amino acids are broken down by enzymes in the:
   a. spleen.
   b. liver.
   c. blood.
   d. stomach.
   e. gallbladder.

2. During childhood, an excess of the hormone somatotropin results in the production of a:
   a. goiter.
   b. midget.
   c. hermaphrodite.
   d. giant.
   e. diabetic.

3. A tropic hormone is one that:
   a. stimulates the thyroid gland.
   b. acts on exocrine as opposed to endocrine glands.
   c. causes the organism to move toward light.
   d. acts on another endocrine gland.
   e. inhibits production of sex hormones by the gonads.

4. In many vertebrates, the intermediate lobe of the pituitary gland is the source of:
   a. adrenocorticotropic hormone.
   b. melanocyte-stimulating hormone.
   c. thyroid-stimulating hormone.
   d. gonadotropin.
   e. growth hormone.

5. Somatostatin is a hypothalamic hormone that:
   a. stimulates bone growth.
   b. causes lactation in mammals.
   c. prevents the release of LH and FSH.
   d. stimulates protein synthesis in the body cells.
   e. inhibits the release of growth hormone.

6. Oxytocin release can be initiated by mechanical stimulation of the:
   a. vagina.   d. pituitary.
   b. uterus.   e. gonads.
   c. hypothalamus.

7. ADH is sometimes called _____ because it _____.
   a. cardiac peptide; increases the strength of the heart contractions
   b. pseudooxytocin; produces so many of the same effects as true oxytocin
   c. antidiabetic hormone; increases glucose uptake by cells
   d. somatostatin; inhibits the release of somatotropin
   e. vasopressin; increases blood pressure in some vertebrates

8. Excitability, nervousness, excessive sweating, high blood pressure, and weight loss are symptoms of:
   a. excessive thyroxine.
   b. inadequate thyroxine.
   c. excessive oxytocin.
   d. inadequate oxytocin.
   e. inadequate growth hormone.

9. Dwarfism and mental deficiency may be caused by too little _____ during infancy.
   a. oxytocin   d. thyroxine
   b. calcitonin   e. adrenaline
   c. prolactin

10. Calcitonin and parathyroid hormone regulate the blood concentration of _____ ions.
    a. bicarbonate and hydrogen
    b. calcium and bicarbonate
    c. sodium and potassium
    d. sulfate and calcium
    e. calcium and phosphate

11. The symptoms of osteoporosis are similar to those of:
    a. hyperthyroidism.
    b. hyperparathyroidism.
    c. hypothyroidism.
    d. hypoparathyroidism.
    e. iodine deficiency.

12. Monocytes and macrophages secrete _____, which stimulates the release of ACTH by the pituitary gland.
   a. an interleukin
   b. an interferon
   c. histamine
   d. a leukotriene
   e. vasopressin

13. Blood levels of sodium and potassium are regulated by the kidneys under the influence of:
   a. cortisol.
   b. oxytocin.
   c. aldosterone.
   d. thyroxine.
   e. ADH.

14. A bearded lady is most likely the victim of a(n) _____ tumor.
   a. pituitary
   b. adrenal cortical
   c. adrenal medullary
   d. thyroid
   e. pineal

15. Which event is NOT an effect of adrenaline and noradrenaline?
   a. dilation of respiratory passages
   b. increase in heart rate
   c. decrease in respiratory rate
   d. increase in blood glucose
   e. increase in contraction strength of heart muscle

16. The cells of which organ generally use only glucose as an energy source?
   a. brain
   b. heart
   c. skeletal muscle
   d. liver
   e. spleen

17. Which structure may function as a biological timekeeper?
   a. pineal gland
   b. pituitary gland
   c. hypothalamus
   d. cerebellum
   e. thymus gland

18. The prostaglandins found in semen are synthesized in the:
   a. prostate gland.
   b. seminal vesicles.
   c. testes.
   d. thyroid gland.
   e. pineal gland.

19. Most prostaglandins are formed by:
   a. oxygenation of arachidonic acid.
   b. iodination of an amino acid.
   c. breakdown of a nucleic acid.
   d. a limited number of tissues.
   e. the prostate gland.

20. The leukotrienes released by macrophages and mast cells during an inflammatory response are:
   a. amines.
   b. protein hormones.
   c. prostaglandins.
   d. small peptides.
   e. steroid hormones.

21. Which trait is NOT associated with circadian rhythms in humans?
   a. body temperature
   b. alcohol tolerance
   c. secretion of hormones
   d. urinary excretion of ions
   e. All of these traits are associated with circadian rhythms in humans.

22. In addition to prolactin, several hormones are necessary for milk production by mammary glands. Which of the following is NOT one of these hormones?
   a. growth hormone
   b. estrogen
   c. calcitonin
   d. progesterone
   e. adrenal steroids

23. The receptor molecule for thyroxine resides:
   a. in the cytoplasm.
   b. in the mitochondria.
   c. on the cell surface.
   d. in the endoplasmic reticulum.
   e. in the nucleus.

24. Which of these common diseases is most effectively treated with dietary management?
   a. rheumatoid arthritis
   b. asthma
   c. juvenile-onset diabetes
   d. adult-onset diabetes
   e. dwarfism

25. T or F   Chemical stimuli are the mediators of integration and control throughout the living world.

26. T or F   There are two distinctly different, independently functioning systems involved in integration and control in large animals—one based on hormones and the other based on nerve impulses.

27. T or F   The mineralocorticoids produced by the adrenal cortex suppress the inflammatory and immune responses and are thought to be part of the body's regulation of its defense system.

28. T or F   In diabetes mellitus, the blood glucose level is so high that the kidney cannot reabsorb all of the glucose entering the glomerular filtrate.

29. T or F   The hormone melatonin causes darkening of the skin in larval amphibians.

30. Match each hormone with its site of production. Answers may be used more than once. Each blank will have only one answer.

   1. anterior pituitary
   2. hypothalamus
   3. thyroid
   4. adrenal cortex

5. adrenal medulla
6. pancreas
7. pineal

a. _____ calcitonin

b. _____ aldosterone

c. _____ insulin

d. _____ thyroxine

e. _____ oxytocin

f. _____ prolactin

g. _____ cortisol

h. _____ melatonin

i. _____ somatotropin

j. _____ ADH

k. _____ adrenaline

l. _____ glucagon

m. _____ ACTH

n. _____ TSH

31. Identify two important aspects of the tight regulatory control that is maintained over hormones.

32. Describe the mechanism by which milk production is maintained in response to suckling.

33. Identify two actions of prostaglandins that contribute to the pain of dysmenorrhea.

34. There are two general mechanisms by which hormones interact with their receptor molecules. Briefly describe these two mechanisms and indicate which type of hormone operates by each mechanism.

PERFORMANCE ANALYSIS

1. **c** Amines, hormones that are derivatives of amino acids, are broken down by enzymes in the blood. Steroid, protein, and peptide hormones are broken down by the liver. (page 823)

2. **d** During childhood, excess somatotropin (growth hormone) results in a giant. If excess growth hormone is produced in an adult, the result is acromegaly, a condition in which the bones of the hands, feet, and jaw enlarge. (page 825)

3. **d** A tropic hormone is one that acts on another endocrine gland, regulating the secretion of its hormones. (page 825)

4. **b** In many vertebrates, the intermediate lobe of the pituitary produces melanocyte-stimulating hormone. (page 826)

5. **e** Somatostatin, which is produced by the hypothalamus, inhibits pituitary release of growth hormone (somatotropin). (page 827)

6. **b** The release of oxytocin, which causes uterine con-

tractions during labor and after birth, can be stimulated by movements of the fetus or by increasing pressure within the uterine wall. (page 828)

7. **e** ADH (antidiuretic hormone) is sometimes referred to as vasopressin because it increases blood pressure in some vertebrates. (page 828)

8. **a** Symptoms of hyperthyroidism (excessive thyroxine production) include excitability, nervousness, excessive sweating, high blood pressure, and weight loss. (page 829)

9. **d** Inadequate production of thyroxine during infancy may result in dwarfism and mental deficiency. (page 829)

10. **e** Calcium and phosphate ions exist in the blood in a reciprocal relationship. The levels of these two ions are influenced by calcitonin and parathyroid hormone. (page 829)

11. **b** The symptoms of hyperparathyroidism and osteoporosis include soft, fragile bones due to calcium loss. In addition, there may be a loss of height, which results from shrinking of the vertebrae. (page 829; essay, page 830)

12. **a** In a positive feedback loop influencing the production of cortisol, an interleukin secreted by activated monocytes and macrophages stimulates the release of ACTH from the pituitary gland. (page 831)

13. **c** Aldosterone promotes the retention of sodium by the kidneys and the concomitant excretion of potassium. (page 832)

14. **b** In a woman, excessive production of male sex hormones by an adrenal cortical tumor can stimulate the development of secondary male sex characteristics, including the growth of a beard. (page 832)

15. **c** Adrenaline and noradrenaline are chemical mediators of the sympathetic nervous system, which prepares the body to fight or flee. The responses include dilation of respiratory passages, increased respiratory rate and heart rate, increased blood glucose concentration, and increased strength of heart muscle contraction. (page 832)

16. **a** Brain cells generally use only glucose as an energy source and are therefore quite sensitive to the effects of low blood sugar. However, after several days of fasting, the brain cells begin to use fatty acids for energy. (page 833)

17. **a** The pineal gland secretes the hormone melatonin, the production of which fluctuates on a daily basis and is influenced by exposure to light in several animal species. The function of the pineal has not been conclusively determined but there is evidence suggesting it may be a biological timekeeper. (page 833)

18. **b** The prostaglandins found in semen are manufactured in the seminal vesicles. (page 833)

19. **a** Most prostaglandins are formed by the oxygenation of arachidonic acid. They are produced by cell membranes in most organs of the body. (page 833)

20. **c** Leukotrienes, which are chemical mediators of the inflammatory response, are actually prostaglandins. (page 836)

21. **e** Several functions in humans, including body temperature, hormone production, urinary excretion of ions, alcohol tolerance, and respiratory and heart rates, follow circadian rhythms. (essay, page 834)

22. **c** The production and secretion of milk involve the interaction of several hormones, including growth hormone, estrogen, progesterone, adrenal steroids, prolactin, and oxytocin. (page 836)

23. **e** The receptor molecule for thyroxine is located in the nucleus. Receptors for steroid hormones are typically located in the cytoplasm of target cells. The receptors for protein hormones are in cell membranes. (page 836)

24. **d** Adult-onset diabetes results from a decrease in the number of insulin receptors on target cells, not from inadequate insulin. These patients are most effectively treated with dietary modifications. (page 837)

25. **True** Molecules similar or identical to mammalian neurotransmitters and hormones have been found in a wide variety of organisms, including prokaryotes and protists. (page 821)

26. **False** The effects of the nervous and endocrine systems are so interrelated that some scientists prefer to refer to them as one system, the neuroendocrine system. (page 821)

27. **False** The *glucocorticoids* produced by the adrenal cortex suppress the inflammatory and immune responses. Mineralocorticoids function in regulating ions, especially potassium and sodium. (pages 830, 832)

28. **True** In diabetes mellitus, the level of glucose in the blood exceeds the reabsorptive capacity of the kidney, so glucose is excreted in the urine. The presence of glucose in the urine is the basis for the diagnosis of diabetes. (page 832)

29. **False** In larval amphibians, melatonin causes the pigment granules in skin cells to aggregate near the center of the cells, resulting in *blanching* of the skin. (page 833)

30. The answers are **a3, b4, c6, d3, e2, f1, g4, h7, i1, j2, k5, l6, m1,** and **n1.** (Table 40–1)

31. (1) Negative feedback mechanisms regulate the production of hormones. (2) Hormones are rapidly degraded by enzymes in the liver or in the blood and therefore do not persist in the circulation. (page 823)

32. Prolactin, which is produced by the anterior pituitary, stimulates milk secretion in mammals. An inhibitory hormone synthesized in the hypothalamus prevents its production. As long as an infant is suckling, the nerve impulses prevent the synthesis of the inhibitory hormone, so prolactin is produced and milk secretion continues. When nursing stops, the inhibitory hormone is again produced and milk production ceases. (page 825)

33. The high prostaglandin levels in the menstrual fluid of women with dysmenorrhea cause strong, rapid contractions of the uterine walls (causing cramps) and a reduced blood supply to the tissue. It has been proposed that prostaglandins also stimulate the rapid firing of pain nerve endings. (In some women, the prostaglandins also stimulate contractions of intestinal smooth muscle, causing diarrhea and nausea.) The most widely available drugs that control the symptoms of dysmenorrhea are those that inhibit prostaglandin production. (page 835)

34. Steroid hormones and thyroxine diffuse through cell membranes into cells. The receptors for steroid hormones are in the cytoplasm and those for thyroxine are in the nucleus. The binding of these hormones to the receptor stimulates a response by the cell. Protein hormones bind to receptors in cell membranes, stimulating the release of a second messenger, which triggers the appropriate events in the cell. (You should be familiar with the details of both mechanisms.) (pages 836, 837)

Integration and Control II: The Nervous System

MAJOR CONCEPTS

A diversity of nervous systems parallels the diversity present in the animal kingdom. The simplest nervous system that coordinates activities at the level of the whole organism is present in cnidarians and consists of neurons in a diffuse network. The next level of complexity is represented by the two nerve cords with anterior ganglia, characteristic of planarians. Earthworms and arthropods exemplify further advancements, but the nervous systems of vertebrates are by far the most complex.

The vertebrate nervous system can be divided into different parts based on anatomical and functional distinctions. The central nervous system consists of the brain and spinal cord. The peripheral nervous system comprises all nerves carrying impulses between the central nervous system and organs and tissues of the body. The peripheral nervous system has nerves that carry either sensory or motor impulses. Motor nerves are either somatic (innervating skeletal muscle) or autonomic (innervating smooth and cardiac muscle, glands, and organs).

The autonomic nervous system has two divisions, sympathetic and parasympathetic, which often have antagonistic actions. The sympathetic division prepares the body to deal with stressful situations (i.e., to fight or flee). The parasympathetic division stimulates restorative activities, such as digestion.

The functional unit of the vertebrate nervous system is the neuron, which consists of a cell body and processes known as dendrites and axons. Dendrites receive nerve impulses from sensory organs or from the axons of other neurons and transmit them toward the cell body. Axons transmit nerve impulses away from the cell body to other neurons or to effector organs.

The initiation and propagation of a nerve impulse depend on concentration gradients of sodium and potassium ions that are maintained by sodium-potassium pumps. The concentration of potassium ions is higher inside the neuron than outside, but the concentration of sodium ions is higher in the extracellular fluid than in the cytoplasm. At rest, the inside of the neuron is slightly negatively charged with respect to the outside (the resting potential). When a neuron is stimulated, the permeability of the membrane to sodium ions increases and sodium ions rush in, reversing the polarity so that the inside of the membrane is positive. (This is an action potential.) Potassium ions then flow out of the cell, restoring the resting potential. The sodium-potassium pump, which is an integral membrane protein, restores the concentration gradients to their previous status. Nerve impulses are self-propagating and can travel over considerable distances with undiminished strength.

Ions flow through gated ion channels, which are integral membrane proteins; conformational changes in these proteins open and close the ion channels. Depending upon the specific ion channel, the conformational changes associated with opening and closing may be regulated by the voltage across the membrane or by neurotransmitters.

Many nerves are coated with a myelin sheath that is formed when specialized glial cells (Schwann cells or neuroglia) wrap around the nerves. The sheath, which is interrupted at regular intervals called nodes, acts as an insulator. The movement of sodium and potassium ions associated with action potentials occurs only at the nodes of myelinated fibers. The effect is that a nerve impulse "jumps" from node to node, greatly increasing its velocity of travel along the nerve. Another consequence of myelination is that much less energy is expended by the sodium-potassium pump to restore the resting potential than in unmyelinated fibers.

Nerve impulses are transmitted from one axon to another across synapses. Synapses may be either electrical, in which the impulse is transmitted between the cells directly through gap junctions, or chemical. In chemical synapses, chemical neurotransmitters carry the signal across the synapse. Two common neurotransmitters in the peripheral nervous system are acetylcholine and norepinephrine. A variety of neurotransmitters are present in the vertebrate brain. Several naturally occurring compounds that resemble opiates structurally and functionally have been identified in the brain.

Neurotransmitters are present in vesicles in the presynaptic neuron. When an action potential arrives at a synapse, calcium ion channels open and calcium ions flow into the axon. This causes synaptic vesicles to fuse with the axon membrane, emptying neurotransmitter molecules into the synaptic cleft. These molecules diffuse across the cleft and bind to receptors on the postsynaptic membrane, stimulating a series of events that may or may not result in propagation of the nerve impulse. Following their release, neurotransmitters are rapidly removed or destroyed.

There are two principal mechanisms by which neurotransmitters act upon postsynaptic cells. In one mechanism, the binding of the neurotransmitter to its receptor causes a conformational change in an integral membrane protein that functions as a gated ion channel. The altered flow of ions results in a change in the degree of polarization of the postsynaptic cell membrane. In the second mechanism, the binding of the neurotransmitter to its receptor triggers the release of a second messenger, usually cyclic AMP or cyclic GMP, which stimulates a complex set of events. Like the first mechanism, the end result is a change in membrane polarization; however, the change in polarization occurs more slowly when a second messenger is involved. Some neurotransmitters function by only one of these mechanisms; others bind to two different types of receptors, activating the two different mechanisms.

Signals may be excitatory or inhibitory in nature. The axon hillock of the neuron is the region in which all incoming signals are processed. A nerve impulse originates from the axon hillock if the net effect of all incoming signals is stimulatory.

The presence in prokaryotes, protists, and fungi of chemicals involved in vertebrate cell-cell communication is compelling evidence of a common evolutionary origin for these communication mechanisms.

HOW TO STUDY THE CHAPTER

Read the entire chapter through quickly, focusing on the major concepts.

Use the GUIDED STUDY OF THE CHAPTER to help you identify the important details as you **reread** the chapter. Writing out the answers to these questions will help fix them in your mind as well as provide you with a valuable study aid.

Answer the questions in TESTING YOUR UNDERSTANDING without the aid of your text. Check your answers against those in PERFORMANCE ANALYSIS. Analyzing your answers will give you valuable feedback on your level of understanding and preparedness for classroom testing.

GUIDED STUDY OF THE CHAPTER

I. Introduction *(page 841)*

1. What facts provide strong evidence that endocrine and nervous systems evolved from primitive cell-cell communication mechanisms?

2. A nervous system is distinguished by what specialized feature?

3. Neurons transmit signals to what two types of target?

4. *Vocabulary:* What is a synapse?

> *Focus:* The active life style of animals is possible because of the rapid integration of incoming information and response to that information made possible by the nervous system.

II. Evolution of Nervous Systems *(pages 841–842)*

5. a. Outline the increasing structural and functional complexity of the nervous systems of *Hydra*, planarians, earthworms, and arthropods.

   b. What features of the vertebrate nervous system represent advances over those described in (a)?

c. What has been the major trend in the evolution of vertebrate nervous systems?

III. Organization of the Vertebrate Nervous System *(pages 842–847)*

A. General Remarks *(pages 842–843)*

6. Arrange the following components of the vertebrate nervous system in a diagram illustrating their correct relationships: autonomic system, central nervous system, parasympathetic system, peripheral nervous system, somatic system, and sympathetic system.

7. a. The functional unit of the vertebrate nervous system is the _____.

   b. Name the three parts of a neuron and identify the function(s) of each part. (See also page 708.)

8. *Vocabulary:* Distinguish between the members of the following pairs of terms: neuroglia/Schwann cells; ganglia/nuclei; tracts/nerves.

9. a. Describe the process by which a myelin sheath is formed. (Figure 41–3)

   b. What component of myelin sheaths gives them their white color?

   c. The function of the myelin sheath is to:

B. The Central Nervous System *(pages 843–844)*

10. Identify the two components of the central nervous system.

11. a. Distinguish between the composition of the gray matter and the white matter of the spinal cord.

    b. Describe the functional and anatomical differences in the connection of the sensory and motor roots of spinal nerves to the spinal cord. (Figure 41–4)

12. Name three major functional components of the brainstem.

C. The Peripheral Nervous System *(pages 844–847)*

13. *Vocabulary:* Distinguish between efferent and afferent neurons.

14. *Vocabulary:* How do cranial nerves differ from spinal nerves?

> *Focus:* **Cranial nerves and spinal nerves carry both sensory *and* motor information.**

15. Identify four possible fates of a signal carried by a sensory neuron.

16. The cell bodies of motor neurons are located _____, and may receive impulses from what three types of neurons?

17. Label and then color the diagram on the following page of a reflex arc. Draw arrows to indicate the flow of nerve impulses through the reflex arc. Labels: cell body of motor neuron, cell body of sensory neuron, gray matter, motor neuron, nerve endings in skin, nerve endings on muscle fibers, sensory neuron, white matter.

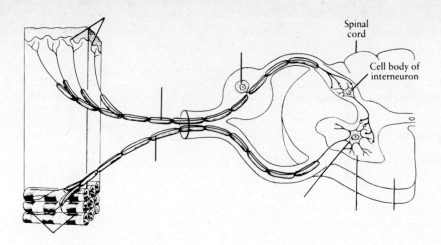

Spinal cord

Cell body of interneuron

18. *Vocabulary:* Distinguish between polysynaptic and monosynaptic reflex arcs.

19. What is the function of relay neurons?

Divisions of the Peripheral Nervous System: Somatic and Autonomic (page 845)

20. Identify the effectors innervated by the (a) autonomic and (b) somatic nervous systems.

21. Why are the terms "voluntary" and "involuntary" not precise when applied to motor function of the nervous system? Cite examples to illustrate your answer.

22. Fill in the following table summarizing differences between motor fibers of the somatic and autonomic nervous systems.

Table for Question 22

	SOMATIC SYSTEM	AUTONOMIC SYSTEM
Anatomical arrangement		
Functions: stimulatory vs. inhibitory		
Source(s) of sensory input		
Conscious awareness of reflex arcs (yes or no)		

23. *Vocabulary:* Distinguish between preganglionic and postganglionic neuron fibers.

Divisions of the Autonomic Nervous System: Sympathetic and Parasympathetic (pages 845–847)

24. Fill in the following table outlining the differences between the sympathetic and parasympathetic divisions of the autonomic nervous system.

Table for Question 24

	SYMPATHETIC	PARASYMPATHETIC
Origin of fibers		
Location of ganglia		
Neurotransmitter at preganglionic nerve endings		
Neurotransmitter at postganglionic nerve endings		
General type of body activities each system controls		
Specific examples of activities promoted by each system		

Focus: **Most internal organs are innervated by nerves from both the sympathetic and parasympathetic divisions.**

25. Describe (specifically) how the sympathetic division prepares the body for "fight or flight."

IV. The Nerve Impulse *(pages 847–851)*

A. General Remarks *(pages 847–848)*

26. *Vocabulary:* Distinguish between conductors and insulators.

27. a. What is an electric potential?

    b. How is an electric potential measured?

28. Identify three lines of evidence demonstrating that a nerve impulse is not an electric current.

29. Which animal has been used extensively in nerve conduction studies?

30. Describe how the electric potential across an axon membrane is measured. (See also Figure 41–11.)

31. a. *Vocabulary:* Distinguish between the resting potential and the action potential of a neuron.

    b. What is a nerve impulse?

Focus: Neurons have an "all or none" response: all the action potentials for a particular neuron are of the same magnitude. The intensity of a stimulus is communicated by the *frequency* with which nerve impulses are transmitted.

B. The Ionic Basis of the Action Potential *(pages 848–850)*

32. a. In general, what creates the electrical potential of the axon?

    b. Name the two ions responsible for creating the resting potential and describe the relative concentrations of these ions inside and outside the axon.

c. Identify three factors influencing the distribution of these ions.

d. Describe three aspects of axon membrane permeability that are important in creating the resting potential.

33. a. *Vocabulary:* What is meant by gated ion channels?

    b. The ion gates may be opened or closed in response to what two factors?

34. Describe in detail how sodium and potassium ions establish the resting potential.

35. a. What change occurs in the neuron membrane when it is stimulated?

    b. Identify two factors that influence sodium ions to rush into the axon when it is stimulated.

    c. Reversal of polarity across the membrane produces the:

    d. The influx of sodium ions is counteracted by what phenomenon, which restores the resting potential?

    e. Only very small changes in ion concentrations occur in generation of the action potential. Why is this advantageous?

C. Propagation of the Impulse *(pages 850–851)*

36. a. What phenomenon is responsible for the self-propagating nature of the action potential?

    b. Indicate the effect of axon depolarization on membrane permeability. (Explain the term voltage-gated in your answer.)

c. What prevents the action potential from traveling in a reverse direction?

Focus: **The nature of the action potential allows nerve impulses to travel over considerable distances with no reduction in their intensity.**

The Role of the Myelin Sheath (page 851)

37. a. Describe how the presence of a myelin sheath permits an increased speed of nerve impulse conduction compared with unmyelinated nerves.

   b. *Vocabulary:* Define saltatory conduction.

   c. How fast can an impulse travel along some large, myelinated nerve fibers?

   d. In addition to increased velocity, what other advantage does saltatory conduction have?

V. The Synapse *(pages 851–856)*

A. General Remarks *(pages 851–853)*

38. a. Describe how an electrical synapse functions.

   b. Where have electrical synapses been identified?

39. How does a chemical synapse differ from an electrical synapse?

40. a. *Vocabulary:* What is the synaptic cleft?

   b. What feature of the synaptic cleft anchors the two cell membranes together? (Figure 41–17)

41. *Vocabulary:* Distinguish between the presynaptic and postsynaptic cells of a synapse.

42. How does the signal transmitted across a synapse differ from the action potential that is propagated along a nerve?

43. a. Where are neurotransmitters synthesized and how do they reach the synapse?

   b. Describe what happens when an action potential arrives at a synapse.

   c. Name three potential fates of neurotransmitters that are released at a synapse.

Focus: **The membranes of neurotransmitter vesicles are re-formed after releasing their contents and are transported back into the cytoplasm to be recycled into new vesicles.**

---

44. New synaptic vesicles are derived from what intracellular structure?

45. What synaptic events occur by (a) exocytosis and (b) endocytosis?

46. List four areas of research being studied by scientists interested in neurotransmitters.

B. Neurotransmitters *(pages 853–855)*

47. a. Name the two major neurotransmitters of the peripheral nervous system.

   b. Identify one major neurotransmitter in the brain and cite its apparent function in the hypothalamus.

48. Severe depression may be related to what phenomenon in the brain?

49. Name and identify the activity of three neurotransmitters of the central nervous system, all of which are amino acid derivatives.

50. What are the neurotransmitter deficiencies in (a) Parkinson's and (b) Huntington's diseases?

51. Identify the effect of and describe the mechanism by which each compound (or group of compounds) exerts its mood-altering effects.

Caffeine, nicotine, amphetamines

Chlorpromazine

LSD

52. a. Summarize the two major mechanisms by which neurotransmitters exert their effects.

b. Both mechanisms have the same general consequence. What is it?

c. How does this consequence differ between the two mechanisms?

Focus: A neurotransmitter may have some receptors that function by one mechanism and other receptors that operate by the other mechanism.

53. a. *Vocabulary:* What are neuromodulators?

b. How does their function resemble and differ from that of neurotransmitters?

c. Name five types of neuromodulators.

C. *Essay:* Internal Opiates: The Endorphins *(pages 854–855)*

Focus: The opiates are the most potent pain-killers known and are highly addictive.

54. What property of opiates greatly enhances their pain-killing effect?

55. Name three locations of opiate receptors in the body.

56. Describe the action of opiates when they bind to neurons bearing their receptors.

57. *Vocabulary:* What is an endorphin?

58. a. Where are enkephalins located in the body?

b. What is the structure of these substances?

59. a. Endorphins other than enkephalins are produced primarily by what organ?

b. The most common of these, _____, is related in what way to ACTH?

60. Name two proposed functions of endorphins.

61. What is the significance of the fact that some macrophages bear endorphin receptors?

62. How are exogenous opiates thought to affect the production of endorphins?

D. **The Integration of Information** *(pages 855–856)*

63. a. What happens to the electrical charge on the interior of a cell when the cell receives an impulse that is (i) excitatory and (ii) inhibitory?

    b. *Vocabulary:* Distinguish between depolarization and hyperpolarization.

    c. Give an example of each type of molecule: excitatory and inhibitory.

64. a. What is the function of the axon hillock?

    b. How is an action potential initiated?

    c. Name four factors that influence the processing of information within a neuron.

65. a. Where are inhibitory or modulatory synapses located on a neuron? (Figure 41–20)

    b. How do these synapses influence the amount of neurotransmitter released? (Figure 41–20)

Focus: Whether or not an action potential is initiated depends on the *sum* of all signals reaching the axon hillock during a given time period.

VI. **Summary** *(pages 856–857)*: Read the summary. If you are familiar with the essential features of the material presented there, you are ready to complete the section TESTING YOUR UNDERSTANDING.

TESTING YOUR UNDERSTANDING

After you have completed this examination, compare your answers with those in the section that follows.

1. The functional unit of the nervous system is the:
   a. neuron.   d. glial cell.
   b. synapse.   e. Schwann cell.
   c. ganglion.

2. The two major anatomical subdivisions of the vertebrate nervous system are _____ (brain and spinal cord) and _____ (sensory and motor pathways).
   a. peripheral; central
   b. somatic; autonomic
   c. autonomic; somatic
   d. central; peripheral
   e. sympathetic; parasympathetic

3. Within the central nervous system, clusters of nerve cell bodies are known as:
   a. neuroglia.   d. nuclei.
   b. ganglia.   e. tracts.
   c. Schwann cells.

4. Clusters of axons that lie outside the central nervous system are called:
   a. neuroglia.   d. nuclei.
   b. ganglia.   e. nerves.
   c. tracts.

5. Motor neurons are _____, and carry signals _____ the central nervous system. Sensory neurons are _____, and carry signals _____ the central nervous system.
   a. afferent; into; efferent; out of
   b. efferent; into; afferent; out of
   c. afferent; out of; efferent; into
   d. efferent; out of; afferent; into
   e. either afferent or efferent; either into or out of; afferent; into

6. Identify the correct order through which action potentials travel in a complete polysynaptic reflex arc involving skeletal muscle.
   a. motor neurons, sensory neurons, interneurons, relay neurons
   b. sensory neurons, relay neurons, motor neurons
   c. sensory neurons, interneurons and relay neurons, motor neurons

d. relay neurons, interneurons, motor neurons

e. relay neurons, interneurons, sensory neurons

7. There are several differences between the autonomic and somatic nervous systems. Which of these is NOT such a difference?

a. The axons of the somatic system travel from the central nervous system to the effector organs without synapsing.

b. The somatic system can only stimulate or not stimulate an effector.

c. The autonomic system can either stimulate or inhibit an effector.

d. A person is usually aware of somatic reflex arcs but unaware of autonomic reflexes.

e. The somatic system uses only motor neurons, whereas the autonomic system uses only sensory neurons.

8. Axons of the sympathetic nervous system originate in the _____ regions of the spinal cord.

a. cranial and sacral

b. thoracic and sacral

c. cranial and lumbar

d. thoracic and lumbar

e. sacral and lumbar

9. The ganglia of the sympathetic nervous system are located:

a. within the brainstem.

b. within the spinal cord.

c. near the central nervous system.

d. within the effector organs.

e. adjacent to the effector organs.

10. Which activity is NOT accomplished by the parasympathetic nervous system?

a. increased intestinal movements

b. increased salivation

c. increased respiratory rate

d. recovery following orgasm

e. decreased heart rate

11. Across an axon membrane, a difference in the amount of charge of 70 millivolts, with the interior negatively charged with respect to the exterior, corresponds to a(n):

a. resting potential.

b. nerve impulse.

c. electrostatic impulse.

d. hyperpolarized state.

e. action potential.

12. On the inside of a neuron membrane at the resting potential, the concentration of _____ times greater than on the outside.

a. potassium is 10

b. potassium is 20

c. potassium is 30

d. sodium is 10

e. sodium is 20

13. The opening and closing of the gates in the ion channels of neuron cell membranes may be regulated by:

a. the voltage across the membrane.

b. the concentration gradient of the ion in question.

c. the internal hydrogen ion concentration.

d. hormones from the hypothalamus.

e. the ratio of sodium ions to potassium ions on the inside versus the outside of the membrane.

14. As potassium ions leak out of a resting neuron, a negative charge builds up on the inside because:

a. chloride ions leak into the cell at the same rate.

b. the membrane is not permeable to negative ions.

c. negative ions are pumped into the cell.

d. sodium ions are pumped out of the cell.

e. potassium ions are not pumped back into the cell.

15. When an action potential reaches a synapse, an influx of _____ ions causes fusion of synaptic vesicles with the cell membrane, resulting in the release of neurotransmitter.

a. potassium

b. calcium

c. sodium

d. chloride

e. magnesium

16. The synaptic cleft of a chemical synapse is approximately _____ wide.

a. 10 micrometers

b. 20 micrometers

c. 30 micrometers

d. 10 nanometers

e. 20 nanometers

17. Membrane for the formation of new synaptic vesicles is formed by the:

a. mitochondria.

b. smooth endoplasmic reticulum.

c. rough endoplasmic reticulum.

d. axon terminals.

e. recycling of phagocytic vesicles.

18. Caffeine exerts its effect on a person by:

a. inhibiting dopamine production.

b. substituting for excitatory neurotransmitters.

c. inhibiting serotonin production.

d. blocking postsynaptic receptors for inhibitory transmitters.

e. inhibiting GABA production.

19. Small peptides released from the presynaptic membrane, which may diffuse over a great distance and affect numerous cells in a local area of the central nervous system, are called:

a. neuromodulators.

b. neurotransmitters.

c. neurosecretors.

d. neuroinhibitors.

e. neuromodifiers.

20. T or F   All reflex arcs, somatic and autonomic, send messages along relay neurons to the conscious centers of the brain.

21. T or F   Both the preganglionic and postganglionic axons of the parasympathetic nervous system release acetylcholine.

22. T or F   Luigi Galvani was the first scientist to note that Schwann cells insulate a nerve fiber.

23. T or F   One mechanism by which neurotransmitters

function is that the binding of the neurotransmitter to its receptor either opens or closes ion channel gates.

24. **T or F** Because of the large concentration gradients of sodium (10:1; outside to inside) and potassium (30:1; inside to outside), restoration of the resting potential involves the active transport of many sodium and potassium ions across the neuron cell membrane.

25. **T or F** In a myelinated axon, sodium and potassium ions traverse the cell membrane only at the nodes.

26. **T or F** The major neurotransmitters of the central nervous system are amino acid derivatives.

27. **T or F** In all responses of a postsynaptic membrane to a neurotransmitter, either cyclic AMP or cyclic GMP acts as a second messenger.

28. **T or F** The enkephalins are abundant in the central nervous system and the adrenal medulla whereas the other endorphins are produced largely by the pituitary gland.

29. Match each organism with the correct description of its nervous system. Use each answer only once. Each blank will have only one answer.

   1. vertebrate
   2. earthworm
   3. *Hydra*
   4. arthropod
   5. planarian

   a. _____ diffuse neuronal network

   b. _____ two ventral nerve chords and two ganglia

   c. _____ fused double ventral nerve cord and ganglia in each segment

   d. _____ double ventral nerve cord and brain

   e. _____ dorsal nerve cord and brain

30. Match each neurotransmitter with its associated condition. Use each answer only once. A blank may have only one answer.

   1. serotonin
   2. noradrenaline
   3. dopamine
   4. GABA

   a. _____ deficiency produces severe depression

   b. _____ loss of synapses in Huntington's disease

   c. _____ decreased production occurs in Parkinson's disease

   d. _____ increasing levels cause induction of sleep

31. Name the three main components of the gray matter of the spinal cord.

32. Distinguish between the somatic and autonomic nervous systems with respect to voluntary or involuntary nature and the type(s) of tissue(s) or organ(s) innervated.

33. Identify three factors that govern the distribution of ions on the two sides of a neuron cell membrane.

PERFORMANCE ANALYSIS

1. **a** The neuron is the functional unit of the nervous system. Each neuron consists of a cell body and one or more axons and dendrites. (pages 841, 842)

2. **d** The two major anatomical subdivisions of the vertebrate nervous system are the central nervous system (brain and spinal cord) and the peripheral nervous system (sensory and motor pathways leading to and from the central nervous system). (page 842)

3. **d** Within the central nervous system, clusters of nerve cell bodies are called nuclei. Outside the central nervous system, such clusters are called ganglia. (page 842)

4. **e** Within the central nervous system, clusters of axons are called tracts. Outside the central nervous system, they are called nerves. (page 842)

5. **d** Afferent means to carry toward or into. Efferent means to carry away from or out of. Motor neurons are efferent neurons that carry impulses *from* the central nervous system to skeletal muscles. Sensory neurons are afferent neurons that carry impulses from sensory receptors *to* the central nervous system. (page 844)

6. **c** In a polysynaptic reflex arc, the action potential travels from sensory neurons to interneurons and relay neurons in the spinal cord. From the interneurons the nerve impulses travel to motor neurons, which stimulate the appropriate muscles. The relay neurons carry the signals to the brain, where conscious recognition of the reflex activity occurs. (page 844)

7. **e** The somatic nervous system sends only motor neurons to its effector muscles. The autonomic nervous system has both sensory and motor neurons. (page 845)

8. **d** Axons of the sympathetic division of the nervous system originate from the thoracic and lumbar regions of the spinal cord. (page 847)

9. **c** The ganglia of the sympathetic nervous system are located near the central nervous system. The ganglia of the parasympathetic division are located within or near the effector organs. (page 847)

10. **c** The parasympathetic nervous system promotes restorative functions of the body. Specific activities include stimulation of the digestive system, increased salivation, slowing of the heart and respiratory rates, and recovery following orgasm. (page 847)

11. **a** The resting potential of a neuron is a 70 millivolt difference between the inside and the outside of the cell membrane—the inside of the membrane is negatively charged relative to the outside. (page 848)

12. **c** At the resting potential, the concentration of potassium ions inside the cell is 30 times greater than that of the extracellular fluid. Conversely, the concentration in sodium ions outside the neuron is 10 times the cytoplasmic concentration. (page 849)

13. **a** In a neuron cell membrane, the opening and closing of the ion channel gates may be regulated by the voltage across the membrane or by the action of neurotransmitters. (page 850)

14. **b** A negative charge develops on the inside of a neuron cell membrane because the membrane is slightly permeable to potassium ions, which leak out at a slow rate. It is not permeable to negative ions, which remain inside the cell. (page 850)

15. **b** When an action potential arrives at an axon terminal, calcium ion channels open and calcium ions flow into the axon. In response to this influx of calcium ions, synaptic vesicles fuse with the cell membrane, releasing neurotransmitter. (page 852)

16. **e** The width of the synaptic cleft at a chemical synapse (i.e., the distance between the presynaptic and postsynaptic neurons) is approximately 20 nanometers. (page 852)

17. **b** The smooth endoplasmic reticulum manufactures new synaptic vesicles in which neurotransmitters are packaged. The membrane from used synaptic vesicles, which have fused with the neuron cell membrane, can be returned to the cytoplasm and recycled into new synaptic vesicles. (page 853)

18. **b** Caffeine exerts its stimulatory effects by substituting for excitatory neurotransmitters at synapses. (page 853)

19. **a** Neuromodulators are small peptides that may be released from axon terminals along with the principal neurotransmitter or from other cells. They may move directly to the postsynaptic membrane or they may diffuse over great distances, affecting many cells in a local area of the central nervous system. They bind to specific receptors and modulate the effects of neurotransmitters. (pages 854, 855)

20. **False** The reflexes of the autonomic nervous system generally do not send signals to the conscious centers of the brain. (page 845)

21. **True** Acetylcholine is the neurotransmitter of both preganglionic and postganglionic axons of the parasympathetic nervous system. In the sympathetic nervous system, most postganglionic neurons release noradrenaline, and the preganglionic neurons release acetylcholine. (page 847)

22. **False** Luigi Galvani noted that the passage of an electric current along the neuron of a frog leg made the muscle twitch. (page 847)

23. **True** When a neurotransmitter binds to its receptor, the conformation of membrane proteins functioning as ion channel gates changes. The channel may open or close, depending upon the neurotransmitter. If enough stimulatory signals are received that the membrane polarity reverses, an action potential may be initiated. (pages 853, 854)

24. **False** Even though the ion concentration gradients are steep, only a few sodium and potassium ions cross the membrane during an action potential. The resting potential is quickly restored when these few ions are transported by the sodium-potassium pump. (page 850)

25. **True** Two advantages of this saltatory conduction in myelinated fibers are that impulse transmission is faster and that less energy is expended by the sodium-potassium pump to restore the resting potential. (page 851)

26. **True** The major neurotransmitters of the central nervous system are amino acid derivatives, including noradrenaline, dopamine, serotonin, and GABA. (page 853)

27. **False** There are two mechanisms by which neurotransmitters stimulate action potentials in postsynaptic membranes. One mechanism is described in the answer to question 23. In the other mechanism, the binding of a neurotransmitter to its receptor on the postsynaptic membrane stimulates the release of a second messenger, which is generally cyclic AMP or cyclic GMP. (pages 853, 854)

28. **True** The enkephalins are abundant in the central nervous system and in the adrenal medulla. The other endorphins are produced mainly by the pituitary but may be produced by other organs as well. (essay, page 854)

29. The answers are: **a3, b5, c2, d4,** and **e1.** (pages 841, 842)

30. The answers are: **a2, b4, c3,** and **d1.** (page 853)

31. The three main components of spinal cord gray matter are interneurons, cell bodies of motor neurons, and neuroglia. (page 843)

32. The somatic nervous system innervates skeletal muscles and its activities are largely voluntary (except for reflexes). The autonomic nervous system innervates body organs and its activities are not subject to voluntary control (except in individuals who have had biofeedback or similar training). The sympathetic division of the autonomic system prepares the body for stressful situations. The parasympathetic division promotes restorative activities. (page 845)

33. Three factors that determine the distribution of ions across a neuron cell membrane are the ion concentration gradients, the attraction of oppositely charged particles and repulsion of similarly charged particles, and the properties of the membrane itself. (page 849)

Integration and Control III: Sensory Perception and Motor Response

CHAPTER ORGANIZATION

MAJOR CONCEPTS

Different sensory receptors respond to specific types of stimuli by initiating action potentials. As in the initiation and propagation of nerve impulses, the mechanism involves a change in the membrane permeability of either the sensory receptor or an associated neuron. The intensity of a stimulus is communicated by the frequency of action potentials produced—the more intense the stimulus, the higher the frequency. The distinction among different stimuli lies in the perception of the signals in the brain, not in the individual sensory receptors or in the mechanism by which the signals are sent.

Sensory receptors may be classified according to the type of stimulus to which they respond. The basic types in mammals are mechanoreceptors, chemoreceptors, photoreceptors, temperature receptors, and pain receptors. Some animals other than humans may also have electroreceptors and/or magnetoreceptors.

Sensory receptors may also be classified based on the type of information they provide. Interoceptors monitor internal body conditions such as blood pressure, temperature, and blood ion concentrations. Proprioceptors provide information about the location of body parts relative to one another. Exteroceptors collect information about the external environment.

The senses of taste and smell are accomplished by chemoreceptors. Taste is the detection of liquid-borne chemicals; smell is the detection of airborne chemicals. Both senses require that the chemical stimulus be dissolved in a fluid film in order to be detected. According to current evidence, the binding of an odor molecule to its receptor involves cyclic AMP as a second messenger. The ensuing series of events triggers the opening of sodium ion channels, resulting in a change in the olfactory cell membrane potential.

Hearing and balance are accomplished by separate organs in the mammalian ear. The receptor for both senses is the hair cell. When the cilia of a hair cell are stimulated by the movement of fluid in the organs, an action potential is initiated.

The sense of balance is the role of three bony semicircular canals positioned at right angles to each other. In response to movements of the head, fluid in the canals moves, triggering action potentials in the hair cells.

In the sound receptor of the mammalian ear, sound waves set a membrane and a series of bones in motion, which causes vibration in another membrane, which in turn initiates fluid waves in the cochlea. In the organ of Corti, hair cells are arranged along a membrane that vibrates in response to the cochlear fluid waves. The pitch of sound is distinguishable because different portions of the membrane vibrate at different frequencies.

The vertebrate eye has three basic layers, an external connective tissue layer (the sclera), a middle vascular layer (the choroid), and an inner layer of light-sensitive cells and neurons (the retina). A transparent lens focuses the visual image on the retina. The retina contains two types of light-sensitive cells: cones are responsible for high resolution and color vision; rods are responsible for black-and-white vision and for vision in dim light.

The initiation of action potentials in the retina depends upon photoactive pigments in the rods and cones. When light of certain wavelengths strikes pigments in the rods, the pigment molecules change shape, triggering a cascade of reactions resulting in altered membrane permeability of the rod or cone. In contrast to other sensory receptors, the rod membrane becomes *hyperpolarized*. This alters the signal transmission of the ganglion cells, the axons of which combine to form the optic nerves. The brain interprets the signals carried by the optic nerves and constructs a visual image.

There are three different types of cones, each of which contains a different photoactive pigment. These pigments respond to wavelengths of light in the red, blue, or green portions of the spectrum. Color vision is a function of differing combinations of stimulated cones.

Skeletal muscles are the effectors of the somatic nervous system. A muscle fiber is a single, multinucleated cell surrounded by a membrane capable of propagating an action potential. Each muscle fiber contains tightly packed myofibrils running the length of the cell. A specialized endoplasmic reticulum, the sarcoplasmic reticulum, encases each myofibril. A system of transverse tubules, formed by invaginations of the cell membrane, lies perpendicular to the myofibrils.

Myofibrils are subdivided into sections known as sarcomeres, the contractile units of skeletal muscle. Sarcomeres are composed of thick (myosin) and thin (actin) filaments that slide past one another during muscle contraction. In response to an action potential, calcium ions are released from sacs in the sarcoplasmic reticulum. The calcium ions combine with troponin molecules located on the tropomyosin chains that lie along the actin molecules of the thin filament. The binding of calcium to troponin results in shifting of the tropomyosin chains and exposure of the cross-bridge binding sites of the contractile machinery. The energy for muscle contraction is provided by ATP.

Neuromuscular junctions resemble synapses between nerves in that chemical neurotransmitters carry the signal across the synapse. They differ in that the signals at neuromuscular junctions are all stimulatory. Many drugs and chemicals affect the neuromuscular junction.

A motor unit is a motor neuron and all the muscle fibers it innervates. In a single muscle, fibers from different motor units are intermingled. The number of muscle fibers in a motor unit determines the fineness of control. The strength of a muscle contraction depends on the number of motor units activated and the frequency of stimulation.

HOW TO STUDY THE CHAPTER

Read the entire chapter through quickly, focusing on the major concepts.

Use the GUIDED STUDY OF THE CHAPTER to help you identify the important details as you **reread** the chapter. Writing out the answers to these questions will help fix them in your mind as well as provide you with a valuable study aid.

Answer the questions in TESTING YOUR UNDERSTANDING without the aid of your text. Check your answers against those in PERFORMANCE ANALYSIS. Analyzing your answers will give you valuable feedback on your level of understanding and preparedness for classroom testing.

GUIDED STUDY OF THE CHAPTER

I. Introduction *(page 858)*

1. Name three functions of sensory perception in animals.

Focus: **The nervous system of an animal receives sensory input from the environment and initiates an appropriate response by stimulating skeletal muscles.**

II. Sensory Receptors and the Initiation of Nerve Impulses *(pages 858–872)*

A. General Remarks *(pages 858–859)*

2. The initiation of a nerve impulse in a sensory neuron depends upon what phenomenon?

3. List five different types of stimuli.

4. a. The mechanism by which a sensory receptor responds to a stimulus has what feature in common with the initiation of a nerve impulse?

   b. By what means does the nervous system perceive the intensity of a stimulus?

5. How does an organism distinguish among different types of sensory information?

B. Types of Sensory Receptors (pages 859–860)

6. a. List five different types of sensory receptors present in most animals.

   b. Name two other types of sensory receptors possessed by some animals.

> **Focus:** "If a tree falls in a forest but no animal is present to hear it fall, does it make a sound?" "Sound" is not a property intrinsic to an object but rather is a function of sensory receptors to detect air waves of a certain frequency range. (A similar statement is true of visual stimuli.)

7. a. *Vocabulary:* Distinguish among interoceptors, proprioceptors, and exteroceptors with respect to the general type of information detected. Cite examples of each receptor type.

   b. Identify the stimuli to which each receptor in the skin responds.

   Free nerve ending

   Free nerve ending + hair follicle

   Meissner's corpuscle

   Merkel cell

   Pacinian corpuscle

   c. Where are Meissner's corpuscles and Merkel cells particularly abundant.

   d. Describe how Pacinian corpuscles function.

   e. What is one possible mechanism by which free nerve endings detect painful stimuli?

C. Chemoreception: Taste and Smell (pages 860–862)

8. What aspect of human evolution may have been responsible for the reduced sensitivity of humans to olfactory stimuli compared with other animals?

Taste (page 860)

9. *Vocabulary:* Define taste.

10. a. What clue does the carp brain provide concerning the importance of taste in this species?

    b. Taste receptors are located in what two regions of a catfish?

11. a. Where are taste receptors located in vertebrates?

    b. What is the vomeronasal organ of reptiles and amphibians? (Figure 42–4)

12. a. Name the four primary tastes.

    b. Distinguish between the response of *taste receptors* and *taste buds* to each of the four primary tastes.

> **Focus:** Although most animals appear to have similar abilities to detect taste stimuli, different animals have distinct preferences.

Smell (pages 861–862)

> **Focus:** Taste and smell are accomplished by similar mechanisms, but the two types of input are received by separate receptors and processed in separate areas of the brain.

13. a. *Vocabulary:* Define smell.

b. What must happen to airborne materials before they can be "smelled"?

c. Where are the chemoreceptors for smell located in humans?

14. Describe the functions of these components of olfactory epithelium. (Include in your answer the roles of microvilli and cilia.) (Figure 42-5)

Supporting cells

Basal cells

Olfactory cells

15. Summarize the mechanism by which olfactory stimuli are received and signals are sent to the brain.

Focus: **The change in membrane potential across the axon hillock resulting from the action of stimulatory, inhibitory, and modulatory molecules determines whether or not an action potential is initiated in olfactory epithelium.**

---

16. Outline the hypothesis of scent discrimination presented by John Amoore. Include a list of the proposed seven primary odors.

D. *Essay:* **Chemical Communication in Mammals** *(page 863)*

17. Cite four examples of chemical communication among mammals.

18. a. In humans, what role in communication has been proposed for the chemicals that cause "body odor" and for pubic and axillary hair?

b. Summarize the evidence that pheromones function in communication among humans.

E. **Mechanoreception: Balance and Hearing** *(pages 863–865)*

19. a. Name the two mechanoreceptors present in the mammalian ear and indicate the type of stimuli detected by each one.

b. What type of sensory receptor functions in both of these organs?

20. Describe the mechanism by which the semicircular canals operate. (Figure 42-6)

21. a. Trace the flow of sound through the mammalian ear. Indicate where sound waves are converted to (i) mechanical motion, (ii) fluid waves, and (iii) action potentials.

b. In the human ear, how is the original stimulus amplified before reaching the sensory receptor cells of the cochlea?

c. What structure prevents pressure build-up in the cochlear canals and how does it operate?

22. a. The middle ear is the region that houses the hammer, anvil, and stirrup. What structure enables air pressure in the middle ear to be equalized with atmospheric pressure? (Figure 42-6)

b. Account for the ease with which middle ear infections may develop. (Figure 42–6)

23. a. Describe the anatomical relationship between the organ of Corti and the fluid through which the pressure waves created by sound waves travel.

b. Summarize the mechanism by which fluid waves in the cochlea are transformed into action potentials.

c. The auditory nerve links the ear to what region of the brain?

d. How does the brain distinguish between sounds of different frequencies?

24. a. Humans can generally detect sound waves in the frequency range of _____ to _____ cycles per second.

b. How does frequency detection vary with age in humans?

25. What is the usefulness of the human inability to hear sounds of frequencies below 12 cycles per second?

26. In addition to sound waves traveling in the air, what other source of sound can humans detect?

Focus: **Although the basic structure of the ear is the same among mammals, species vary with respect to the frequency range they can detect.**

F. **Photoreception: Vision** *(pages 866–872)*

27. Name three of the most highly developed photoreceptors in modern animals.

28. a. *Vocabulary:* Distinguish among the sclera, choroid, and retina. (Figure 42–9)

b. Describe the modifications of the anterior sclera to form the cornea. (Figure 42–9)

c. Distinguish between the iris and the pupil. (Figure 42–9)

d. Which region of the retina is responsible for the greatest visual acuity? (Figure 42–9)

29. Why is the vertebrate eye called a "camera" eye?

30. a. How is the mammalian mechanism for focusing the visual image different from that in fish and amphibians?

b. Distinguish between the conditions responsible for nearsightedness and farsightedness. (Figure 42–10)

31. a. *Vocabulary:* Define stereoscopic vision.

b. Upon what factor does stereoscopic vision depend?

c. *Vocabulary:* What is parallax?

d. How does the human brain determine the distance of (i) a near object and (ii) a far object?

32. a. Name two types of animals in which stereoscopic vision is of great adaptive value.

b. Identify one type of animal that does not have stereoscopic vision and explain why it is of a lesser advantage to this group.

The Retina (pages 868–869)

33. a. Name and distinguish between the two types of photoreceptor cells in the vertebrate eye, based on degree of resolution and light-sensitivity.

b. How does the retina of nocturnal animals differ from that of diurnal animals?

34. a. What fraction of the light that strikes the cornea reaches the retina?

b. What are the two possible fates of light that falls on the retina?

c. Some nocturnal vertebrates have what adaptation that maximizes the amount of light striking the retina?

35. a. Trace the pathway traveled by action potentials from the photoreceptor cells to the optic nerve.

b. The blind spot is formed by:

Information Processing in the Retina (pages 869–870)

36. a. In the retina as a whole, for every ganglion cell axon there are _____ photoreceptors in the retina.

b. Describe the anatomical relationships responsible for this ratio.

c. Name the five types of cells that participate in the processing of visual information.

37. a. *Vocabulary:* Define the fovea. (Include the type of photoreceptor cell present.)

b. How does the ratio of photoreceptor to ganglion cells in the fovea differ from that in other parts of the retina?

c. What two anatomical features of the bird retina are responsible for a visual acuity that is eight times greater than that of the human eye?

d. In primates, the evolution of a single, central fovea is thought to be related to what activity?

G. *Essay:* What the Frog's Eye Tells the Frog's Brain
(page 870)

38. Describe the similarities and differences between the frog's eye and the human eye.

39. a. Jerome Lettvin discovered what information about the function of the frog retina?

b. How does this discovery correlate with the fact that frogs in captivity will starve if they are not given live prey?

Visual Pigments and the Capture of Light (pages 871–872)

40. Identify the functions of the following parts of a rod. (Figure 42–16)

Lamellae

Foot

Inner segment

41. a. Name the two components common to the light-sensitive pigments in the vertebrate eye.

b. Which component varies in structure from one pigment to another?

c. Arthropods, mollusks, and vertebrates have almost identical:

d. Explain the importance of correct diet to normal vision.

e. Similar light-sensitive compounds have been located in what unicellular organisms?

42. a. *Vocabulary:* What is rhodopsin?

b. Outline the sequence of events that occurs when rhodopsin absorbs light.

c. What compound acts as a second messenger?

d. What compound supplies energy to restore rhodopsin to its original state?

e. When most sensory receptors are stimulated, depolarization of the cell membrane occurs. How does the response of the rod cell differ, and what event results?

43. a. The perception of colors depends upon what factors?

b. When was this hypothesis first proposed and what type of evidence supported it.

c. Describe in general terms the mechanism by which different colors are perceived.

d. Where does the processing of information about color occur?

e. Where are the genes located that code for the pigments sensitive to red and green?

f. Comparing the nucleotide sequences of the three pigment genes led to what conclusion about their evolutionary origin?

44. a. Human photoreceptors are sensitive to what portion of the electromagnetic spectrum?

b. Why do humans generally not perceive ultraviolet light?

c. Under what circumstances may a human perceive ultraviolet light?

45. Cite two examples in which color perception is important to vertebrate behavior.

III. The Response to Sensory Information: Muscle Contraction *(pages 873–879)*

A. General Remarks

> *Focus:* Skeletal muscle is the effector of the somatic nervous system. Muscle is the principal tissue in the vertebrate body.

46. What are the effectors of the autonomic nervous system?

B. The Structure of Skeletal Muscle (pages 873–874)

47. a. *Vocabulary*: Distinguish between muscle fiber and myofibril.

    b. *Vocabulary*: Distinguish among sarcolemma, sarcoplasmic reticulum, and the T system.

48. Where are the nuclei of a muscle fiber located?

Focus: Because of the structure of the T system, only a lipid bilayer separates each myofibril from the extracellular fluid.

---

49. a. *Vocabulary*: What is a sarcomere?

    b. Name the two contractile proteins that compose the (i) thick and (ii) thin fibers of a sarcomere.

    c. Describe the arrangement of the thick and thin filaments when viewed in (i) longitudinal and (ii) transverse section.

    d. Identify the composition of the following sarcomere regions. (See also Figure 42–19.)

       M line

       Z line

       I band

       A band

       H zone

C. The Contractile Machinery (pages 874–876)

50. a. Describe the composition and structure of thin and thick sarcomere filaments.

    b. Name two functions of the heads of myosin molecules.

51. a. Outline the mechanism by which thick and thin filaments slide past each other.

    b. Cite two functions of ATP in muscle contraction.

    c. What phenomenon causes rigor mortis?

52. How does the contraction of (a) cardiac and (b) smooth muscle compare with that of skeletal muscle?

The Regulation of Contraction (page 876)

53. a. The regulation of skeletal muscle contraction depends upon what three factors?

    b. Describe the structure of the two protein factors.

    c. Summarize the interaction of these three factors in the control of muscle contraction.

D. *Essay:* Twitch Now, Pay Later (page 878)

54. Fill in the following table on vertebrate muscle fiber types.

Table for Question 54

	RED	WHITE
Degree of blood supply		
Number of mitochondria		
Amount of myoglobin		
Fuel source		
Process that supplies energy		
Type of activity for which fiber is best adapted		

55. a. What determines the proportion of red to white fibers in a person's body?

    b. How may information regarding muscle fiber type be used to direct potential athletes to sports for which their muscles are most suited?

c. To what extent can red and white muscle fibers be modified with training?

E. **The Neuromuscular Junction** *(pages 876–878)*

56. a. *Vocabulary:* What is a neuromuscular junction?

b. Describe the mechanism by which a signal crosses a neuromuscular junction.

c. Synaptic transmission at a neuromuscular junction differs from that between neurons in what manner?

d. Describe the events that follow combination of acetylcholine with the sarcolemmal receptors.

e. How does the stimulation of (i) smooth muscle and (ii) cardiac muscle differ from that of a skeletal muscle cell?

57. Name two chemicals that interrupt the function of the neuromuscular junction and state the specific effect of each one.

F. **The Motor Unit** *(pages 878–879)*

58. a. *Vocabulary:* Define a motor unit.

b. What characteristic of a motor unit determines the fineness of control? (Compare eye movements with biceps contraction.)

c. Describe the arrangement of muscle fibers from various motor units in a muscle.

d. Name two factors that determine the strength of a muscle contraction.

Focus: **Complex activities involve well-ordered contractions of different groups of fibers in antagonistic muscles in different regions of the body.**

IV. **Summary** *(pages 879–880):* Read the summary. If you are familiar with the essential features of the material presented there, you are ready to complete the section TESTING YOUR UNDERSTANDING.

TESTING YOUR UNDERSTANDING

After you have completed this examination, compare your answers with those in the section that follows.

1. Which of the following is NOT a function of sensory perception?
   a. identifying predators
   b. finding food
   c. sensing changes in the weather
   d. providing tools for learning
   e. All of these activities are accomplished through sensory perception.

2. The perception of the intensity of a stimulus depends on the:
   a. length of the sensory neuron axons.
   b. frequency of action potentials.
   c. type of receptor stimulated.
   d. number of receptors stimulated.
   e. region of the brain that is stimulated.

3. Which of these stimuli is NOT detected by exteroceptors?
   a. light   d. air temperature
   b. touch   e. blood pressure
   c. sound

4. The sensory receptors associated with hair follicles are:
   a. Meissner's corpuscles.   d. Pacinian corpuscles.
   b. free nerve endings.   e. proprioceptors.
   c. Merkel cells.

5. Which statement is true of Meissner's corpuscles, Merkel cells, and Pacinian corpuscles?
   a. They are stimulated by light touch.
   b. They are stimulated by pressure and vibrations.
   c. They are stimulated by pain.

d. They all consist of one or more free nerve endings with one or more outer layers of connective tissue.

e. They are located in the superficial layers of the skin.

6. What characteristic is thought to be responsible for a primate's less sensitive sense of smell relative to other mammals?

a. tree-dwelling lifestyle of ancestors

b. stereoscopic vision

c. well-developed sense of touch

d. high manual dexterity

e. keen intelligence

7. In humans, the olfactory epithelium of each nasal passage contains _____ receptor cells.

a. 500,000   d. 50 million

b. 1 million   e. 5 billion

c. 5 million

8. Which event is NOT part of the mechanism of olfaction?

a. An odor molecule binds to its complementary receptor molecule on an olfactory cell cilium.

b. Cyclic AMP binds to and opens a sodium-ion membrane channel.

c. An enzyme catalyzes the phosphorylation of GDP to GTP.

d. Adenylate cyclase catalyzes the formation of cyclic AMP from ATP.

e. The sudden influx of potassium ions into the olfactory cell initiates an action potential.

9. The cochlea and the semicircular canals of the ears have what feature in common?

a. The energy of moving fluid is transduced into action potentials.

b. They both function in sound perception.

c. Both structures are involved in maintaining balance.

d. The primary sensory receptors are free nerve endings.

e. Interconnected, movable bones are the transducers.

10. In the human ear, the energy of sound waves is transduced into the energy of fluid waves by the action of:

a. two membranes and three bones.

b. three membranes and three bones.

c. two membranes and two bones.

d. two membranes, three bones, and ciliated cells.

e. five membranes, three bones, and ciliated cells.

11. An action potential is initiated in an auditory sensory neuron when:

a. vibration of a hair cell opens ion channels in its membrane, causing depolarization.

b. adequate stimulation occurs at its synapse with hair cells.

c. sodium ion channels open in the tectorial membrane, reversing its polarity.

d. calcium ion channels open in the basilar membrane, reversing its polarity.

e. neurotransmitter molecules are released from any hair cell.

12. Different frequencies of sound can be distinguished because:

a. the basilar membrane does not vibrate uniformly along its length.

b. the cilia of different hair cells vary in length and respond to distinct frequencies.

c. the various regions of the tectorial membrane vibrate at different frequencies.

d. the frequency with which the oval window vibrates varies with the frequency of the sound.

e. some frequencies of sound are conducted through the bones of the skull.

13. Humans can hear sounds ranging from about _____ cycles per second.

a. 10 to 10,000   d. 20 to 40,000

b. 16 to 20,000   e. 250 to 80,000

c. 20 to 25,000

14. The layer of the eye that contains blood vessels is the:

a. sclera.   d. cornea.

b. retina.   e. choroid.

c. fovea.

15. The structure regulating the amount of light that enters the eye is the:

a. pupil.   d. fovea.

b. ciliary body.   e. cornea.

c. iris.

16. In mammals, the viewed image is focused on the retina by the action of the:

a. iris.   d. ciliary muscles.

b. sclera.   e. cornea.

c. fovea.

17. What feature of the eyes of some nocturnal vertebrates maximizes their vision in dim light?

a. Their pupils dilate to a much greater degree than those of diurnal vertebrates.

b. The retina is much closer to the cornea, so more incoming light reaches the photoreceptors.

c. They have a reflective layer behind the photoreceptors.

d. Their photoreceptors point toward the front of the eye.

e. The iris is reduced in size, allowing more light to enter the eye.

18. In the human retina as a whole, the ratio of photoreceptors to ganglion cell axons is:

a. 10 to 1.   d. 125 to 1.

b. 50 to 1.   e. 225 to 1.

c. 75 to 1.

19. The central fovea of primates is apparently correlated with:

a. an arboreal lifestyle.

b. the need to recognize predators at a distance.

c. the development of the hands for fine motor coordination.

d. the ability to judge the distance to the next branch.

e. an extended period of parental care with its associated learning.

20. Light-sensitive compounds similar to those found in the vertebrate eye have NOT been identified in:

a. halobacteria.

b. *Chlamydomonas.*

c. arthropods.

d. mollusks.

e. Light-sensitive compounds similar to those of the vertebrate eye have been identified in all of these organisms.

21. In a human rod cell, rhodopsin molecules are located in (on) the:

a. inner segment.

b. foot.

c. lamellae.

d. nucleus.

e. cristae of the mitochondria.

22. Which event is NOT part of the mechanism by which light energy stimulates a response in rod cells?

a. The retinal and opsin portions of pigment molecules change shape.

b. Membrane channels for sodium ions open in the rod cell membrane.

c. The rod cell membrane is hyperpolarized.

d. The second messenger in the reaction is cyclic GMP.

e. Energy supplied by GTP returns rhodopsin to its original shape.

23. Which humans are able to read by ultraviolet light?

a. all humans

b. all persons with red-green colorblindness

c. anyone who has had surgery to remove cataracts

d. only persons homozygous for the gene coding for the pigment stimulated by ultraviolet light

e. persons who are either homozygous or heterozygous for the gene coding for the pigment stimulated by ultraviolet light

24. The sarcolemma is the:

a. lipid bilayer enclosing each myofibril.

b. contractile unit of a muscle fiber.

c. membrane forming the sarcoplasmic reticulum.

d. cell membrane surrounding a muscle fiber.

e. specialized synapse between a motor nerve and a muscle fiber.

25. Which statement is NOT true of the sarcoplasmic reticulum?

a. It is an invagination of the sarcolemma.

b. The sacs of the sarcoplasmic reticulum contain calcium ions.

c. Each myofibril is encased by sarcoplasmic reticulum.

d. It is a specialized endoplasmic reticulum.

e. The T system traverses the sarcoplasmic reticulum perpendicular to the myofibrils.

26. Which statement is NOT true of sarcomeres?

a. During muscle contraction, the thin filaments slide between the thick filaments.

b. Their repetitive arrangement gives skeletal muscle its striated appearance.

c. They are the contractile units of skeletal, cardiac, and smooth muscle.

d. Each sarcomere contains actin and myosin filaments in a parallel arrangement.

e. Each myosin filament is surrounded by six actin filaments.

27. In skeletal muscle, the region where the thin filaments of adjacent sarcomeres interweave is the:

a. I band. d. Z line.

b. A band. e. M line.

c. H zone.

28. Which statement is NOT true of muscle stimulation at a neuromuscular junction in skeletal muscle?

a. The movement of sodium and potassium ions during the action potential activates the contractile machinery.

b. Only excitatory messages are transmitted at the synapse.

c. The neurotransmitter is acetylcholine.

d. The action potential travels along the sarcolemma and the T system.

e. Calcium ions are released from the sarcoplasmic reticulum as long as the fiber is stimulated.

29. Curare causes paralysis by:

a. preventing the release of acetylcholine from nerve endings.

b. preventing the breakdown of acetylcholine at the neuromuscular junction.

c. binding to acetylcholine that has been released into the synaptic cleft.

d. destroying acetylcholine receptors on the sarcolemma.

e. binding to and blocking acetylcholine receptors on the sarcolemma.

30. T or F The various senses are perceived as different because each sensory receptor sends signals to the brain in a form that is unique for that sensation.

31. T or F Smell is the chemoreception of substances borne on water vapor droplets.

32. T or F The molecules associated with each primary taste stimulate different types of taste receptors.

33. T or F Farsightedness is the inability to focus on a near object because the eyeball is too short; it may be corrected by a convex lens.

34. T or F Rods, which are responsible for black-and-white vision, are more sensitive to light than cones.

35. T or F Because of the multiple foveas and the tighter packing of photoreceptors in the retina, the visual acuity of birds is about three times that of humans.

36. **T or F** Because the bug detector of the frog eye is only stimulated by a small moving object, a frog will starve if it is not fed living prey.

37. **T or F** Smooth and cardiac muscles are effectors of the somatic nervous system, whereas skeletal muscles are effectors of the autonomic nervous system.

38. In humans, what role has been proposed for the chemicals that cause body odor, and for pubic and axillary hair, all of which begin to develop at puberty?

39. Describe the mechanism by which sound waves are amplified in the ear.

40. Cite two examples illustrating the importance of color in animal behavior.

41. Distinguish between red and white muscle fibers with respect to speed of contraction, size, blood supply, energy source(s), and type of activity for which each is best adapted.

42. What two factors determine the strength of contraction of a muscle as a whole?

PERFORMANCE ANALYSIS

1. **e** Sensory perception provides the means by which an animal learns about and interacts with its environment. All of the activities listed are accomplished by sensory perception. (page 858)

2. **b** The frequency with which action potentials are initiated determines the perceived intensity of the stimulus. (page 859)

3. **e** By definition, exteroceptors detect stimuli such as light, touch, air temperature, and air pressure waves (sound) from the external environment. Interoceptors detect stimuli from the internal environment, such as changes in blood pressure. (page 859)

4. **b** The sensory receptors associated with hair follicles are free nerve endings. When a hair is touched or bent, a nerve impulse is initiated. (page 859)

5. **d** Meissner's corpuscles, Merkel cells, and Pacinian corpuscles all consist of one or more free nerve endings surrounded by one or more layers of connective tissue. (page 860)

6. **a** For tree-dwelling primates, vision was far more important for survival than olfaction, a fact that is thought to account for the inferior sense of smell in primates compared with other mammals. (page 860)

7. **d** The olfactory epithelium of each human nasal passage contains approximately 50 million receptor cells. (page 861)

8. **e** An influx of *sodium* ions into an olfactory cell initiates an action potential. (page 861)

9. **a** The cochlea and the semicircular canals of the ears both contain fluid, the movement of which triggers action potentials in hair cells. (pages 863, 864)

10. **b** The **hammer, anvil,** and **stirrup** of the middle ear transduce the air pressure waves striking the **tympanic membrane** into fluid waves when the stirrup vibrates against the **oval window membrane**. The **round window membrane** moves in response to the pressure waves initiated by the oval window membrane, thereby preventing pressure buildup in the cochlear fluid. (page 864)

11. **b** When adequate stimulation occurs at its synapse with hair cells, an action potential is initiated in an auditory sensory neuron. (page 864)

12. **a** The basilar membrane varies in thickness and flexibility along its length. Consequently, different sections of the basilar membrane are stimulated by different frequencies of sound. (page 865)

13. **b** The frequency range that can be detected by most humans is 16 to 20,000 cycles per second. (page 865)

14. **e** The vascular layer of the eye is the choroid, the front portion of which is modified to form the iris, ciliary body, and suspensory ligaments. (Figure 42–9)

15. **c** The muscles of the iris contract in response to the amount of light striking the retina, adjusting the size of the pupil and thereby regulating the amount of light entering the eye. (Note: There are two sets of muscles in the iris. When the sphincter muscle contracts, the pupil becomes smaller. Contraction of the radial muscles, which are arranged like the rays of a starburst, causes the pupil to increase in size.) (Figure 42–9)

16. **d** In mammals, the ciliary muscles change the shape of the lens, which focuses an image on the retina. (page 867)

17. **c** Some nocturnal vertebrates have a reflective layer (the tapetum) behind the retina. This layer reflects light back to the retina, maximizing vision in dim light. (page 868)

18. **d** Except for the fovea, the ratio of photoreceptors to ganglion cell axons is 125 to 1. In the fovea, there is a one-to-one relationship between cones and ganglion cells. (page 869)

19. **c** The single, central fovea characteristic of primate retinas is apparently correlated with the development of fine motor coordination of the hands. (page 870)

20. **e** In another example of universality among living organisms, light-sensitive compounds similar to those found in the vertebrate eye have been identified in widely diverse organisms, including halobacteria, *Chlamydomonas*, arthropods, and mollusks. (page 871)

21. **c** The light-sensitive rhodopsin molecules are located on the lamellae of the rod cells in the human eye. (Figure 42–16)

22. **b** When rhodopsin absorbs light, the retinal changes shape, which triggers a change in the shape of the opsin. A series of reactions ensues, in which cyclic GMP acts as a second messenger. Energy supplied by GTP returns rhodopsin molecules to their original shape. The end result of the series of reactions is the hyperpolarization of the

rod cell membrane, which alters the release of neuro-transmitter to the bipolar cells. (pages 871, 872)

23. **c** Human photoreceptors are sensitive to ultraviolet (UV) wavelengths of light, but a yellow pigment in the lens prevents these rays from reaching the retina in most people. Those individuals who have had surgery to remove a cataractous lens can see by ultraviolet light. (A cataract is a condition in which the lens becomes opaque.) Since ultraviolet light may be damaging to the retina, people who have undergone cataract surgery are often advised to wear UV-filtering sunglasses in bright sunlight. (page 872)

24. **d** The sarcolemma is the cell membrane surrounding a muscle fiber. (Remember that each muscle fiber is composed of one muscle cell.) Invaginations of the sarcolemma constitute the transverse tubules of the T system. (page 873)

25. **a** The sarcoplasmic reticulum is a specialized endoplasmic reticulum that surrounds myofibrils in muscle cells. The sacs of the sarcoplasmic reticulum contain calcium ions that are released when an action potential reaches the cell. The T system traverses the sarcoplasmic reticulum perpendicular to the myofibrils. (page 874)

26. **c** Sarcomeres are the contractile units of skeletal and cardiac muscle. The structure of the contractile unit of smooth muscle is still under investigation; it is not a sarcomere. (pages 874, 875, 876)

27. **d** As observed in an electron micrograph of a sarcomere, the region where the thin filaments of adjacent sarcomeres interweave is the Z line. (page 874, Figure 42–19)

28. **a** The contractile machinery of skeletal muscle is activated by calcium ions, which are released from the sarcoplasmic reticulum as long as a muscle fiber is stimulated. (page 876)

29. **e** Curare causes paralysis by binding to (and blocking) acetylcholine receptors on the sarcolemma. Consequently, the action potential is not initiated in the sarcolemma and muscle contraction does not occur. (page 878)

30. **False** The signals sent to the brain from the various sensory receptors are identical. The perception of these signals as different forms of stimuli occurs because the signals are sent to different sections of the brain. For example, all signals received by the visual cortex (located at the back of the brain) are interpreted by the brain as vision, which is why you may "see stars" if you receive a sharp blow to the back of your head. (page 859)

31. **False** Smell is the chemoreception of airborne particles. However, these particles must be dissolved in the layer of watery mucus covering the olfactory epithelium before they can be perceived. (page 861)

32. **True** The molecules associated with each primary taste stimulate different types of taste receptors. However, each taste bud may respond to (and send signals corresponding to) more than one category of molecule. (page 860)

33. **True** A person who is farsighted is unable to focus on a near object because the eyeball is too short (or because the lens is no longer able to accommodate adequately, as often occurs as a person ages). Farsightedness may be corrected by a convex lens. (Figure 42–10)

34. **True** Rods are more sensitive to light than are cones. Nocturnal animals typically have retinas composed entirely (or nearly so) of rods. (page 868)

35. **False** The visual acuity of birds is approximately eight times that of humans. (page 870)

36. **True** The frog retina contains ganglion cells that are only stimulated by bug-sized moving objects. Since a frog will strike only at a small, moving object, a captive frog will starve if it is not offered living prey. (essay, page 870)

37. **False** Skeletal muscles are the effectors of the somatic nervous system. Smooth and cardiac muscles are innervated by the sympathetic and parasympathetic divisions of the autonomic nervous system. (page 873)

38. It has been proposed that the chemicals causing human body odor, the production of which begins at puberty, may have served to attract males to females at one time in evolutionary history. The pubic and axillary hair may have functioned to retain and concentrate these chemicals. (essay, page 863)

39. Sound waves are amplified in the ear because the oval window membrane is smaller than the tympanic membrane and therefore is subject to more force per unit area. This greater force per unit area is translated into greater pressure in the cochlear fluid waves. (page 864)

40. (1) The underbelly of a male stickleback fish turns red during the breeding season. This red pigmentation elicits fighting behavior among males. Captive males will assume an aggressive posture toward any red object, whether or not it is shaped like a fish. (2) The bright plumage of many species of birds is important in attracting potential mates and in courtship ceremonies. (page 872)

41. Red muscle fibers are relatively small, slow-contracting, and have a good blood supply. They have abundant mitochondria and myoglobin; oxidative phosphorylation of fatty acids provides most of their energy. They are adapted for continuous use that requires endurance. White muscle fibers are larger, and fast-contracting. They have a poorer blood supply, fewer mitochondria, and less myoglobin. They depend upon glucose or the anaerobic glycolysis of glycogen for energy. They are adapted for quick bursts of power. (essay, page 878)

42. Two factors that determine the strength of whole muscle contraction are the number of motor units activated and the frequency with which they are stimulated. (pages 878, 879)

Integration and Control IV: The Vertebrate Brain

MAJOR CONCEPTS

Centralization of the integration and control of body activities in the brain has been a major evolutionary trend in vertebrates. Brain structure and function are simplest in the fishes, and complexity increases in reptiles, birds, and mammals. The most complex brains are possessed by primates, particularly humans.

The vertebrate brain evolved as a series of three bulges at the anterior end of the dorsal nerve tube. In birds and mammals, these bulges increased in size and folded over on one another.

The hindbrain and midbrain of birds and mammals together form the brainstem and cerebellum. The brainstem contains centers that control respiration, heartbeat, and other life-sustaining activities. The cerebellum is responsible for fine-tuning complex muscular movements. In lower vertebrates, the midbrain contains the optic lobes.

The two major regions of the forebrain are the diencephalon and the telencephalon. The diencephalon contains the thalamus and hypothalamus and is a major coordinating center. The telencephalon has changed the most in the course of vertebrate evolution. In fishes, it is concerned primarily with olfaction. The telencephalon is greatly enlarged in mammals; the largest portion is the cerebrum.

The cerebral cortex is the most recent evolutionary advancement and is present in reptiles, birds, and mammals, but not fish. In reptiles and birds, the cerebral cortex is only rudimentary. The mammalian cortex varies from a simple, smooth structure in primitive mammals to a highly convoluted structure in primates. The cerebral cortex is the location of the brain regions that process visual, auditory, sensory, and motor information.

There are two major integrating networks in the brain that involve several regions of the brain. The reticular activating system functions in arousal and consciousness and filters important incoming information from unimportant background information. The limbic system links the hypothalamus with the cerebral cortex and is thought to be the mechanism by which drives and emotions are translated into actions.

The cerebrum is divided into right and left hemispheres which are connected by fiber tracts of the corpus callosum. The left hemisphere generally contains the centers responsible for speech and the right hemisphere is the usual location of spatial perception and musical abilities.

Evidence indicates that the functional organization of the brain is more vertical than horizontal. Communication between the sensory and motor cortices occurs largely through the thalamus, a "lower" brain center.

The unmapped regions of the cortex (i.e., those that have not been associated with specific functions) constitute the intrinsic processing areas. These regions receive and process information from different areas of the brain and relay it to other regions.

There are two basic types of memory: short-term and long-term. According to available information, the establishment of long-term memory involves alterations in the synapses by which neurons communicate with one another. Several different regions of the brain function in the consolidation and storage of memory, including the hippocampus, amygdala, thalamus, mammillary body, basal forebrain, and prefrontal cortex. Current evidence indicates that REM sleep is important to information processing and memory consolidation. There are a number of memory disorders which vary with the region of the brain injured and the type of injury suffered.

HOW TO STUDY THE CHAPTER

Read the entire chapter through quickly, focusing on the major concepts.

Use the GUIDED STUDY OF THE CHAPTER to help you identify the important details as you **reread** the chapter. Writing out the answers to these questions will help fix them in your mind as well as provide you with a valuable study aid.

Answer the questions in TESTING YOUR UNDERSTANDING without the aid of your text. Check your answers against those in PERFORMANCE ANALYSIS. Analyzing your answers will give you valuable feedback on your level of understanding and preparedness for classroom testing.

GUIDED STUDY OF THE CHAPTER

I. Introduction *(page 881)*

1. What trend characterizes the evolution of vertebrate integration and control mechanisms?

2. Identify two general functions of the vertebrate brain.

3. a. Describe the composition of gray and white matter in the brain.

   b. The gray matter of the brain contains about _____ neurons. In some areas, there are _____ cell bodies per cubic centimeter.

   c. Each neuron of the brain may synapse with as many as _____ other neurons.

II. The Structural Organization of the Brain: An Evolutionary Perspective *(pages 881–884)*

A. **General Remarks** *(pages 881–882)*

4. a. In what form did the vertebrate brain emerge in the course of evolution?

   b. *Vocabulary:* What are the ventricles?

   c. The fluid in the ventricles also fills what other cavity?

5. How does the arrangement of the three bulges of the brain in lower vertebrates differ from their arrangement in birds and mammals?

B. **Hindbrain and Midbrain** *(pages 882–883)*

6. a. What two major brain regions constitute the hindbrain and midbrain?

   b. The brainstem contains nuclei involved in what types of functions?

   c. *Vocabulary:* What is the medulla and what body functions are controlled there?

   d. In addition to medullary functions, what else occurs in the brainstem?

Focus: **Many of the fiber tracts that connect the spinal cord and the brain cross over in the brainstem.**

---

7. a. Identify the function of the cerebellum.

   b. How does the relative size of the cerebellum differ in reptiles, mammals, and birds?

8. Name two activities that occur in the pons.

9. How does the midbrain of mammals differ from that of lower vertebrates?

C. **Forebrain** *(pages 883–884)*

10. Name the two major regions of the forebrain.

11. a. What is the general function of the diencephalon?

   b. Distinguish between the thalamus and hypothalamus on the basis of location and functions.

> *Focus:* The telencephalon is the brain region that has changed the most during vertebrate evolution.

12. How does the telencephalon differ in fishes, birds, and mammals with respect to function and the major component?

13. a. *Vocabulary:* Distinguish between cerebellum, cerebrum, and cerebral cortex.

   b. What proportion of the total human brain volume is occupied by the cerebrum?

   c. *Vocabulary:* What is the corpus callosum?

   d. Describe the composition of the gray matter of the cerebral cortex and white matter of the corpus callosum and brainstem. (Figure 43–4)

III. **Brain Circuits** *(pages 884–885)*

A. **General Remarks** *(page 884)*

> *Focus:* Two major neuron networks that facilitate communication among the different regions of the brain are the reticular activating system and the limbic system.

B. **The Reticular Activating System** *(page 885)*

14. a. Name the two structural components of the reticular activating system.

   b. Identify two activities performed by this system.

   c. Stimulation of the system results in what activity?

   d. Cite two pieces of evidence supporting the existence of a filtering system.

C. **The Limbic System** *(page 885)*

15. The limbic system links what major two regions with other areas of the brain?

16. Name two activities of the limbic system.

17. The structures of the limbic system correspond to what part of the reptile brain?

IV. **The Cerebral Cortex** *(pages 885–893)*

A. **General Remarks** *(page 885)*

> *Focus:* The cerebral cortex is a thin layer covering the cerebral hemispheres and represents the most recent evolutionary development in the vertebrate brain.

18. What percentage of the brain's neurons are in the cerebral cortex?

19. Contrast the complexity of the cerebral cortex in fish and amphibians, reptiles and birds, primitive mammals, and primates.

20. In the following diagram, label the two principal fissures and distinguish the four major lobes of the cerebral cortex by coloring and labeling them.

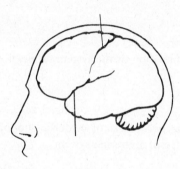

B. Motor and Sensory Cortices *(pages 885–888)*

21. List four methods by which scientists have localized the control of different activities to specific regions of the cerebral cortex.

22. What type of neuron is absent from the brain such that touching the brain does not cause pain?

23. Indicate the location of the motor, sensory, auditory, and visual cortices on the diagram accompanying question 20.

24. a. What is the role of the motor cortex?

b. How does the amount of motor cortex devoted to a specific set of muscles correlate with the function of those muscles? (Give an example.)

25. a. What is the function of the sensory cortex?

b. Name the five regions of the human body where sensory receptors are most abundant.

26. How do neurons in the auditory cortex relate to hair cells in the cochlea?

27. a. Contrast the proportion of the retina occupied by the fovea with the proportion of visual cortex allotted to process information from the fovea.

b. What two factors provide evidence of the importance of eye-hand coordination in primates?

The Perception of Form (pages 887–888)

28. Describe the functional relationship between retinal ganglion cells and cells of the visual cortex.

29. Name four specific visual images that trigger specific neurons in the visual cortex of a cat, as elucidated in the studies of Hubel and Wiesel.

30. Identify the location of the lateral geniculate nucleus and summarize its roles.

31. a. *Vocabulary:* What is the optic chiasm? (Figure 43–11)

b. What occurs in the optic chiasm and how does this influence the perception of visual images? (Figure 43–11)

C. Left Brain/Right Brain (pages 888–891)

32. *Vocabulary:* Define aphasia.

33. a. What percentage of (i) right-handed people and (ii) left-handed people have speech centers in the left cerebral cortex?

   b. Name the two areas of the left cerebral hemisphere associated with speech and indicate the consequences of an injury that is localized to each area.

34. Why has the left hemisphere often been regarded as the dominant side of the brain?

35. a. Identify two abilities that are centered in the right cerebral hemisphere.

   b. What is the evidence for this?

36. a. What is an evolutionary advantage of the specialization of right and left cerebral hemispheres for different tasks?

   b. Cite evidence (in humans and in birds) supporting the proposal that specialization of right and left hemispheres is in part a consequence of development.

37. Describe how the song nuclei of songbirds may vary in size according to sex and season of the year.

Split Brain (pages 889–891)

> *Focus:* If the corpus callosum is severed, the two hemispheres of the cerebral cortex function independently, with the right brain responding to stimuli perceived by the left side of the body and the left brain responding to stimuli perceived by the right side of the body.

38. a. What experimental subjects were used by Sperry and associates?

   b. Describe two examples illustrating the independent function of the two brain hemispheres.

D. Essay: Electrical Activity of the Brain (pages 890–891)

39. a. *Vocabulary:* What is an electroencephalogram?

   b. How can the electrical activity of the brain be measured?

   c. Name four diagnostic uses of electroencephalography.

40. Distinguish between alpha waves and beta waves with respect to (a) body state during which they are produced and (b) type of thought process with which they are associated. (See also the caption to the illustration.)

41. a. *Vocabulary:* Define paradoxical sleep.

   b. Name three ways in which REM sleep differs from the other sleep stages.

c. What is the relationship between REM sleep and dreams?

d. Identify one proposed function of REM sleep and cite evidence supporting this proposal.

42. a. The use of the SQUID is based upon what principle?

b. Identify two uses of the SQUID.

E. **Intrinsic Processing Areas** *(pages 892–893)*

43. What idea was behind use of the term "silent" cortex?

44. What evidence indicates that the organization of the brain is more vertical than horizontal?

45. According to current evidence, where does communication between the sensory and motor cortices occur?

46. Where are the intrinsic processing areas and what occurs there?

47. a. Name one technique used by neurobiologists to map the intrinsic processing areas.

b. Discuss the role of the following brain regions in processing visual information.

Posterior parietal lobe

Lower temporal lobe

48. a. How does the proportion of the cerebral cortex devoted to intrinsic processing vary among humans, other primates, and other mammals?

b. What does this suggest about possible roles of the intrinsic processing areas?

c. Half of the total brain area involved in intrinsic processing is in which part of the brain?

d. List three proposed functions of the intrinsic processing areas upon which scientists currently agree.

V. **Learning and Memory** *(pages 893–898)*

A. **General Remarks** *(pages 893–894)*

49. a. *Vocabulary:* Define learning.

b. What does this definition imply concerning the physiological basis of memory?

50. What conclusion did Karl Lashley draw concerning the nature of memory and what evidence led him to this conclusion?

51. a. *Vocabulary:* Distinguish between short-term and long-term memory.

b. The establishment of long-term memory is currently thought to involve what physiological change?

c. If a person experiences a head injury and "loses memory" of all prior events, but later recovers completely, what did this person actually lose? Explain your answer.

d. What kind of memory deficit occurs with bilateral destruction of the hippocampus?

B. Anatomical Pathways of Memory (pages 894–895)

52. a. Name six regions of the brain involved in memory.

    b. For each of the following conditions, indicate which portion (or portions) of the brain is typically affected. (Figure 43–16)

    Stroke

    Korsakoff's syndrome

    Alzheimer's disease

    Inflammation or temporary interruption of oxygen supply to the brain

53. a. Trace the pathway along which information is thought to travel through the brain in the consolidation of long-term memory.

> **Focus:** A positive feedback circuit that parallels this path is important in information processing.

    b. Which structure is a major producer of acetylcholine for this pathway and what is the importance of this neurotransmitter?

54. a. How does the amygdala function in memory?

    b. Name four brain regions with which the amygdala forms connections.

55. The emotional content of memories is due to connections between what two structures?

56. The recognition of sensory input depends upon what aspect of memory?

57. Stimulation of what region of the brain has evoked vivid memories in some patients?

58. *Vocabulary:* Distinguish "habit" memory from "recognition" memory in terms of the type of activities stored and the part of the brain involved.

C. *Essay:* Alzheimer's Disease (pages 896–897)

59. a. What major symptom characterizes Alzheimer's disease?

    b. Describe the progressive nature and ultimate outcome of the disease.

60. a. Name three abnormalities consistently found in the brain during autopsies of Alzheimer's patients.

    b. These abnormalities are most apparent in which regions of the brain?

    c. What lesion often affects the basal forebrain? What consequences does this lesion have for other regions?

61. What doubts remain concerning the causative factors of Alzheimer's disease?

62. Cite evidence supporting the involvement of genetic factors in the development of Alzheimer's disease.

63. a. What external factors may participate in the development of Alzheimer's disease?

    b. Describe one example supporting the role of external factors.

D. Synaptic Modification (pages 895–898)

64. Name two reasons why invertebrates are useful in studying the mechanisms by which the nervous system functions.

65. Distinguish between habituation and sensitization in terms of events at the synapse. Use an example in your discussion.

> *Focus:* Evidence from several sources indicates that alterations in the strength of synaptic transmission are critical to memory and learning.

66. Name three of the sources referred to in the Focus statement.

67. a. The alterations of synaptic transmission are thought to depend on changes in which cells?

    b. Name three factors that may be involved in the mechanism by which the strength of synaptic transmissions are altered.

> *Focus:* Some neurobiologists believe that a fuller understanding of learning and memory will come from research on a simple model organism.

68. Some scientists disagree with the belief expressed in the Focus statement. Why?

VI. **Summary** *(pages 898–899):* Read the summary. If you are familiar with the essential features of the material presented there, you are ready to complete the section TESTING YOUR UNDERSTANDING.

TESTING YOUR UNDERSTANDING

After you have completed this examination, compare your answers with those in the section that follows.

1. The white matter of the brain is composed of:
   a. neuron cell bodies.
   b. myelinated fibers.
   c. neuroglia.
   d. unmyelinated fibers.
   e. Schwann cells.

2. Which statement is NOT true of the brainstem?
   a. It contains clusters of nerve cell bodies called nuclei.
   b. It includes the medulla.
   c. It contains sensory and motor neurons that serve the skin, muscles, and other structures of the head.
   d. It has undergone much change in the course of evolution from fish to humans.
   e. Many fiber tracts cross over in the brainstem.

3. The execution and fine-tuning of complex patterns of muscular movements is a function of the:
   a. medulla.
   b. pons.
   c. midbrain.
   d. cerebellum.
   e. thalamus.

4. The pons is a ventral enlargement of the brainstem that:
   a. controls respiration and heartbeat.
   b. is the seat of human emotions.
   c. provides communication between the left and right hemispheres of the cerebellum.
   d. processes visual information in mammals.
   e. serves as a relay-and-reflex center.

5. In mammals, visual information is processed in the _____, whereas in lower vertebrates the optic lobes are located in the _____.
   a. forebrain; midbrain
   b. cerebellum; midbrain
   c. pons; forebrain
   d. medulla; pons
   e. hindbrain; cerebellum

6. The major relay center between the brainstem and the higher brain centers is the _____, which consists of two egg-shaped masses of gray matter tucked within the _____.
   a. pons; cerebellum
   b. cerebellum; cerebrum
   c. thalamus; cerebrum
   d. hypothalamus; cerebellum
   e. medulla; midbrain

7. Which statement is NOT characteristic of the hypothalamus?
   a. It contains the mammalian thermostat.
   b. It coordinates the activities associated with thirst.
   c. It produces ADH and oxytocin.
   d. It integrates the activities of the nervous and endocrine systems.
   e. It is the main relay center between the brainstem and higher brain centers.

8. Which major brain region has changed the most during vertebrate evolution?
   a. diencephalon
   b. telencephalon
   c. medulla
   d. cerebellum
   e. pons

9. The function of the corpus callosum is to connect the:
   a. two cerebral hemispheres.
   b. two optic lobes.
   c. two halves of the cerebellum.
   d. hypothalamus and thalamus.
   e. thalamus to higher brain centers.

0. A sleeping person whose EEG shows a period of rapid, low-amplitude waves similar to those observed in an alert person:
   a. has just gone to sleep.
   b. is in a deep sleep.
   c. has entered REM sleep.
   d. is awakening from sleep.
   e. is sleeping restlessly.

11. A cerebral cortex is ABSENT in:
    a. reptiles only.
    b. fish only.
    c. amphibians only.
    d. both reptiles and amphibians.
    e. both fish and amphibians.

12. Which two regions of the cerebral cortex are separated by the lateral sulcus?
    a. frontal and temporal lobes
    b. temporal and occipital lobes
    c. occipital and parietal lobes
    d. parietal and frontal lobes
    e. temporal and parietal lobes

13. Which two investigators discovered that different groups of cells in the visual cortex respond to different types of visual stimuli?
    a. Watson and Crick
    b. Broca and Wernicke
    c. Hershey and Chase
    d. Nottebohm and Sperry
    e. Hubel and Wiesel

14. The neurons that sort signals representing different components of the visual image are located in the:
    a. lateral geniculate nucleus.
    b. reticular activating system.
    c. hypothalamus.
    d. visual cortex.
    e. limbic system.

15. A person with damage to Wernicke's area of the brain may experience difficulty in:
    a. formulating a meaningful visual image.
    b. hand-eye coordination.
    c. comprehending the speech of others.
    d. regulating respiratory rate and heart rate.
    e. moving the mouth and lips to produce speech.

16. Which description is NEVER characteristic of a person with a dominant left hemisphere following damage to the right side of the brain?
    a. difficulty following directions in an unfamiliar city
    b. inability to recognize the voices of friends
    c. difficulty recognizing faces
    d. difficulty in speaking
    e. loss of musical ability

17. The processing of visual information into a mental image occurs in the:
    a. thalamus.
    b. lower portion of the temporal lobe.
    c. visual cortex.
    d. posterior portion of the parietal lobe.
    e. cerebellum.

18. The intrinsic processing areas are NOT concerned with:
    a. integration of sensory information with emotions.
    b. organization of ideas.
    c. long-range planning and intention.
    d. retention of integrated sensory information in memory.
    e. integration of activities performed by the skeletal muscles.

19. Amnesia from which a person completely recovers actually involves the:
    a. loss of short-term memory.
    b. loss of long-term memory.
    c. inability to establish short-term memories.
    d. inability to establish long-term memories.
    e. inability to retrieve memories.

20. Which structure is NOT thought to be involved in the consolidation and storage of memory?
    a. hippocampus
    b. amygdala
    c. hypothalamus
    d. thalamus
    e. basal forebrain

21. Identify the hypothesized order of structures through which information is transferred during memory consolidation.

    1. mammillary body
    2. thalamus
    3. hippocampus
    4. basal forebrain
    5. amygdala
    6. prefrontal cortex

    a. 2 and 3; 4 and 5; 1 and 6
    b. 2 and 5; 1 and 3; 4 and 6
    c. 3 and 4; 2 and 5; 1 and 6
    d. 3 and 5; 1 and 2; 4 and 6
    e. 1 and 2; 3 and 4; 5 and 6

22. Acetylcholine released by the _____ is vital to the processes that occur in the amygdala and _____.
    a. prefrontal cortex; hippocampus
    b. basal forebrain; mammillary body
    c. thalamus; mammillary body
    d. basal forebrain; hippocampus
    e. prefrontal cortex; thalamus

23. The emotional content of memories apparently results from a linkage between the:
    a. thalamus and prefrontal cortex.
    b. amygdala and hypothalamus.
    c. mammillary body and thalamus.
    d. amygdala and hippocampus.
    e. thalamus and basal forebrain.

24. Habit or procedural memory apparently depends on a distinct set of pathways involving the:
    a. corpus striatum.
    b. amygdala.
    c. thalamus.
    d. mammillary body.
    e. corpus callosum.

25. **T or F** The terms telencephalon, diencephalon, and hindbrain are used to describe the principal regions of all vertebrate brains.

26. **T or F** All sensory systems throughout the body send signals to the reticular activating system.

27. **T or F** In the motor cortex, the size of the area servicing a particular part of the body is roughly proportional to the complexity of movement of that body part.

28. **T or F** The techniques utilized in mapping the cerebral cortex are painful because pain receptors are stimulated when the brain is touched.

29. **T or F** In 90 percent of all humans, the speech centers are located in the left cerebral cortex.

30. **T or F** In many songbirds, the size of the song center varies from season to season.

31. **T or F** Habituation of an animal to a stimulus results in an increase in the amount of neurotransmitter released at the synapses of the sensory neurons.

32. **T or F** It is hypothesized that alterations in the strength of synaptic transmissions are involved in memory and learning.

33. What two facts regarding the allocation of functional space in the cerebral cortex provide evidence for the importance of hand-eye coordination in primate evolution?

34. Why can a split-brain patient verbally identify an object if it is felt with the right hand, but not if it is felt with the left hand?

35. Distinguish between short-term and long-term memory with respect to mechanism and part(s) of the brain involved.

36. Cite two reasons why invertebrates are useful in studying the cellular and molecular mechanisms involved in memory.

PERFORMANCE ANALYSIS

1. **b** The white matter of the brain is composed of myelinated nerve fibers. The gray matter contains neuron cell bodies and glial cells. (page 881)

2. **d** The brainstem has undergone little change over the course of vertebrate evolution. It contains nuclei involved in reflexes, centers that control vital functions (including respiration and heartbeat), and sensory and motor neurons that serve the head. Many fiber tracts coming from the body cross over in the brainstem so that information from the left side of the body is processed by the right side of the brain. (page 882)

3. **d** The cerebellum is responsible for the execution and fine-tuning of complex patterns of muscular movement. It reaches its greatest relative size in birds. (page 883)

4. **c** The pons provides communication between the left and right cerebellar hemispheres. (page 883)

5. **a** Visual information is processed in the forebrain of mammals and involves structures in both the diencephalon and telencephalon. By contrast, the optic lobes of lower vertebrates are located in the midbrain. (page 883)

6. **c** The thalamus is the major relay center between the brainstem and the higher centers of the brain. It consists of two egg-shaped masses of gray matter within the cerebrum. (page 883)

7. **e** The thalamus, not the hypothalamus, is the major relay center between the brainstem and higher brain centers. The hypothalamus contains the mammalian thermostat, produces ADH and oxytocin, and coordinates activities associated with sex, hunger, thirst, pleasure, pain, and anger. Through its influence on the pituitary, it provides the major integrating link between the nervous and endocrine systems. (page 883)

8. **b** Over the course of vertebrate evolution, the telencephalon has changed the most. In fish it is concerned primarily with processing olfactory information; in mammals it consists mainly of the cerebrum. (page 883)

9. **a** The corpus callosum connects the left and right cerebral hemispheres. Severing the corpus callosum results in a split-brain patient. (page 884)

10. **c** REM (rapid-eye movement) sleep, which is also known as paradoxical sleep, is characterized by an EEG pattern of rapid, low-amplitude waves similar to those in an alert person. (essay, page 890)

11. **e** Neither fish nor amphibians possess a cerebral cortex. Only a rudimentary cerebral cortex is present in reptiles and birds. Mammals have cerebral cortices of varying levels of complexity. (page 885)

12. **a** The frontal and temporal lobes of the cerebral cortex are separated by the lateral sulcus. The central sulcus divides the parietal and frontal lobes. (Figure 43–7)

13. **e** In studies of the cat, David H. Hubel and Torsten N. Wiesel discovered that different groups of cells in the visual cortex respond to different types of visual stimuli. (page 887)

14. **a** The lateral geniculate nucleus, which resides in the thalamus, contains neurons that sort the signals representing different components of the visual image. It then relays these signals to the visual cortex. (page 888)

15. **c** Damage to Wernicke's area of the brain results in impaired comprehension of both spoken and written words. The person may be capable of fluent but meaningless speech. If a person with damage to Broca's area is able to speak, the speech is slow and labored, but comprehension is unimpaired. (page 888)

16. **d** The left hemisphere is considered to be dominant because it contains the speech centers and controls the

dominant hand in most people. In such a person, spatial perception and musical ability reside in the right hemisphere. (page 888, 889)

17. **b**   Several brain regions are involved in processing visual information, including the lateral geniculate nucleus of the thalamus, the visual cortex, the posterior region of the parietal lobe, and the lower portion of the temporal lobe, which processes information into a mental image of the whole subject. (page 892)

18. **e**   The intrinsic processing areas are concerned with the organization of ideas, long-range planning and intention, the integration of sensory information with emotions, and memory. Integration of activities performed by skeletal muscles is the function of the motor cortex and the cerebellum. (page 893)

19. **e**   A loss of long-term memory from which a person eventually recovers actually involves the temporary inability to retrieve memories. (page 893)

20. **c**   The consolidation and storage of memory involves several structures, including the hippocampus, amygdala, thalamus, basal forebrain, mammillary body, and prefrontal cortex. (page 894)

21. **d**   According to the current hypotheses of memory consolidation and storage, information is transmitted from sensory cortical areas to the hippocampus and amygdala. From these centers, signals travel over independent pathways to the thalamus and mammillary body, which relay the information to the basal forebrain and prefrontal cortex. (page 894)

22. **d**   The basal forebrain is a major source of acetylcholine, which is apparently vital for processes occurring in other parts of the memory circuit, particularly the amygdala and the hippocampus. (page 894)

23. **b**   Connections between the amygdala and the hypothalamus are apparently responsible for the emotional content of memories. (page 894)

24. **a**   Habit or procedural memory is thought to depend on pathways that include the corpus striatum and are distinct from the pathways involved in recognition memory. (page 895)

25. **False**   The major regions of the vertebrate brain are the forebrain, which includes the telencephalon and diencephalon, and the hindbrain and midbrain, which together comprise the brainstem and cerebellum. (page 882 and Table 43–1)

26. **True**   The reticular activating system receives signals from all sensory systems in the body. (page 885)

27. **True**   The various portions of the body are represented in the motor cortex according to the complexity of the movement. For example, the region of the motor cortex controlling the human hand is larger in relation to brain size than the region of the motor cortex controlling a cat's paw. (page 886)

28. **False**   The brain itself has no pain receptors, so touching the brain or stimulating certain brain centers is not intrinsically painful. (page 886)

29. **False**   The speech centers are located in the left cerebral cortex in 90 percent of all right-handed people and 65 percent of all left-handed people. (page 888)

30. **True**   In many songbirds, the size of the song center increases during certain seasons of the year (particularly those in which the birds mate and defend territories) and decreases during other seasons of the year. (page 889)

31. **False**   In habituation of an animal to a stimulus, a decrease of neurotransmitter released at the synapses of sensory neurons results in a reduced response to the stimulus. By contrast, sensitization results in an increase in the amount of neurotransmitter released, with a concomitant increased response of the animal to the stimulus. (page 896)

32. **True**   Evidence gained through studies of the mollusks *Aplysia* and *Hermissenda* and of tissue preparations from the hippocampus of various mammals supports the hypothesis that changes in the strength of synaptic transmissions are important to memory and learning. (page 897)

33. (1) The fovea provides high-resolution color vision, particularly of objects within arm's length. Although the fovea represents only about 1 percent of the area of the retina, it projects information to nearly 50 percent of the visual cortex. (2) Substantial portions of the motor and sensory cortices are devoted to the hands. (page 887)

34. Information from the right hand is processed by the left brain, where the speech centers reside in most humans. Conversely, information from the left hand is processed by the right brain, which is concerned with spatial perceptions and musical ability in most people. In split-brain patients, the right brain "knows" what the left hand is doing, but is unable to articulate the activity. (page 890)

35. Short-term memory involves the hippocampus; an example is remembering an unfamiliar phone number long enough to dial it. The transfer of short-term memory to long-term memory involves pathways that traverse the hippocampus, amygdala, thalamus, mammillary body, basal forebrain, and prefrontal cortex. Current evidence indicates that the establishment of long-term memory involves repetitions of an experience. (page 893)

36. (1) The neurons of invertebrates are large and unmyelinated. (2) Invertebrate nervous systems have far fewer neurons than those of vertebrates and their circuits are correspondingly less complicated. (page 895)

REVIEW TEST 12

Section 6: Chapters 40–43

This review test is *not* designed to cover all of the important information in these chapters. However, it does touch on the major topic categories in each chapter. It will also give you valuable practice in taking this type of test. When you are finished, check your answers with those provided in the following PERFORMANCE ANALYSIS section.

1. All of these are endocrine glands EXCEPT for the _____, which has an exocrine function.
   a. liver
   b. thyroid
   c. adrenal cortex
   d. ovary
   e. hypothalamus

2. Parasympathetic stimulation _____ heart rate and _____ movements of intestinal smooth muscle. Sympathetic stimulation _____ heart rate and _____ movements of intestinal smooth muscle.
   a. increases; decreases; decreases; increases
   b. decreases; increases; decreases; increases
   c. decreases; increases; increases; decreases
   d. increases; decreases; increases; decreases
   e. decreases; increases; increases; increases

3. Which of the following is NOT a possible fate of a neurotransmitter that has been released into a synaptic cleft?
   a. It may be rapidly destroyed by a specific enzyme.
   b. It may diffuse away.
   c. It may be taken up again by the axon terminal.
   d. It may bind to a receptor on the postsynaptic cell.
   e. All of these are possible fates of a neurotransmitter that has been released into a synaptic cleft.

4. Which of the following is NOT one of the "primary" odors proposed by John Amoore?
   a. camphoric
   b. oily
   c. musky
   d. floral
   e. putrid

5. Build-up of pressure in the cochlear fluid is prevented by the:
   a. round window membrane.
   b. oval window membrane.
   c. tympanic membrane.
   d. auditory tube.
   e. semicircular canals.

6. Which situation would NOT be handled primarily by the limbic system?
   a. a hungry person searches for food
   b. information is memorized by a student studying for final exams
   c. a female dog in season seeks a mate
   d. a person awakens to the sound of a nearby siren
   e. a thirsty person seeks a water fountain

7. Which activity or sensation is NOT associated with function of the hypothalamus?
   a. sex drives
   b. temperature regulation
   c. pain
   d. thirst
   e. vision

8. Overactive parathyroid glands can cause:
   a. violent muscular contractions.
   b. tumors within bone tissue.
   c. decreased excretion of calcium in the urine.
   d. softening of the bones due to calcium loss.
   e. an increase in height due to thickening of the vertebrae.

9. What molecule has been implicated in the aggregation of slime mold amoebas, regulation of the *lac* operon, and mediation of the response to many protein hormones?
   a. glucose
   b. lactose
   c. mRNA
   d. cortisol
   e. cyclic AMP

10. Animals that have a ganglion cluster large enough to be considered a brain include BOTH:
   a. earthworms and arthropods.
   b. vertebrates and arthropods.
   c. hydrozoans and vertebrates.
   d. arthropods and planarians.
   e. planarians and vertebrates.

11. Which statement is NOT true of the mechanism of skeletal muscle contraction?
   a. ATP provides the energy for muscle contraction.
   b. The somatic nervous system innervates skeletal muscles.
   c. Globular troponin molecules act as enzymes to split ATP.
   d. Calcium ions released from the sarcoplasmic reticulum turn the contractile machinery on.
   e. The globular heads of the myosin molecules are the cross-bridge binding sites at which force is exerted on actin molecules.

12. T or F   A split-brain patient is unable to identify objects felt only with the left hand.

13. T or F   The intrinsic processing areas neither receive information directly from sense organs nor transmit signals to neurons outside the brain.

14. T or F   Signals from the hypothalamus stimulate the production of some pituitary hormones and inhibit the production of others.

15. T or F   Some prostaglandins are being investigated as possible therapeutic agents to halt premature labor in women.

16. T or F   The withdrawal of your hand from a hot stove is pure reflex, even though you may say, "Ouch!"

17. T or F   Alzheimer's disease is associated with the degeneration of neurons in the hippocampus and amygdala.

18. Match each functional cortex with the lobe of the cerebral cortex in which it is located.

   1. sensory cortex   3. motor cortex
   2. visual cortex   4. auditory cortex

   a. _____ frontal lobe
   b. _____ parietal lobe
   c. _____ temporal lobe
   d. _____ occipital lobe

19. Name and state the role of each of the seven hormones involved in regulating blood glucose concentration.

20. Cite three lines of evidence that refuted the electric current model of nerve impulse transmission.

21. Identify two advantages of impulse transmission along myelinated axons relative to impulse transmission along unmyelinated neurons.

22. What is the adaptive value of stereoscopic vision to tree-dwelling animals and to predators? What type of vision is most valuable to prey species?

23. Identify two roles of ATP in muscle contraction. What causes rigor mortis (the stiffened muscles of a corpse)?

PERFORMANCE ANALYSIS

1. **a**   Exocrine glands secrete their products into ducts. The liver produces bile, which travels to the gallbladder in ducts and is stored there. From the gallbladder, bile travels through a duct to the duodenum. The thyroid, adrenal cortex, hypothalamus, and ovaries all function as endocrine glands, which secrete their products into interstitial tissues, from which they move into the bloodstream. (page 823, Figure 40–3)

2. **c**   The parasympathetic nervous system promotes restorative functions of the body—it decreases heart rate and increases activity of the digestive system. The sympathetic nervous system prepares the body for "fight or flight"—it increases heart rate and decreases activity of the digestive system. (page 847)

3. **e**   Once a neurotransmitter has been released into a synaptic cleft, it may diffuse away, be taken up again by the axon terminal, bind to a receptor on the postsynaptic membrane, or be broken down by specific enzymes. (page 853)

4. **b**   The seven primary odors proposed by John Amoore are camphoric, musky, floral, putrid, pepperminty, etherlike, and pungent. (page 861)

5. **a**   Each time a pressure wave is initiated in the cochlear fluid, the round window membrane bows outward into the middle ear, preventing the increase of pressure in the cochlea. The auditory tube enables the pressure in the middle ear to be equalized with atmospheric pressure. (page 864)

6. **d**   The limbic system is thought to be the circuit by which drives and emotions (e.g., hunger, thirst, need for a mate) are translated into actions. It is also a major circuit in memory consolidation. The awakening of a person to a siren involves arousal, which is a function of the reticular activating system. (page 885)

7. **e**   The hypothalamus contains nuclei that coordinate activities related to sex, hunger, thirst, pleasure, pain, and anger. It also contains the thermostat by which body temperature is maintained at the appropriate level. The brain centers associated with vision include the visual cortex and the lateral geniculate nucleus of the thalamus. (page 883)

8. **d**   Parathyroid hormone elevates blood calcium levels by stimulating the removal of calcium from bone. In hyperparathyroidism, the presence of excessive parathyroid hormone eventually results in softening of the bones due to calcium loss. (page 829)

9. **e**   Cyclic AMP is one of the most common second messengers in the body and can be activated by numerous stimuli. It functions in the aggregation of slime mold

amoebas and in regulation of the *lac* operon, and is the second messenger for many protein hormones. (page 838)

10. **b** In vertebrates and arthropods, the anterior cluster of ganglia is large enough to be considered a brain. (page 842)

11. **c** The globular heads of the myosin molecules act as enzymes that split ATP, thus providing energy for muscle contraction. (page 875)

12. **False** A split-brain patient can identify objects felt only with the left hand, but cannot *verbalize* the identity of the object. For instance, if the person feels the number 2 in the left hand, he or she may hold up two fingers with the right hand. (page 890)

13. **True** The intrinsic processing areas of the brain receive information from and relay information to other brain centers. They do not communicate directly with structures outside the brain. (page 892)

14. **True** The hypothalamus releases some hormones that stimulate the secretion of pituitary hormones and others that inhibit the secretion of pituitary hormones. (page 826)

15. **False** Since prostaglandins stimulate uterine muscle contractions during labor, compounds that *inhibit* prostaglandin synthesis are being investigated for their potential to halt premature labor. (page 835)

16. **True** The withdrawal response is a pure reflex in that the hand (or foot) is pulled away from the painful stimulus before the brain perceives the injury. The painful stimulus is consciously recognized because the sensory neurons synapse with relay neurons in the spinal cord that carry this information to the brain. (page 845)

17. **True** The degeneration of neurons in the hippocampus and amygdala is characteristic of Alzheimer's disease, but it is not known whether this is a cause or a consequence of the disease. There is strong evidence supporting the role of genetic factors in the development of early-onset Alzheimer's disease. (essay, page 896)

18. The answers are: **a3, b1, c4,** and **d2.** (Figures 43–7 and 43–8)

19. **Growth hormone** (somatotropin) inhibits the uptake and oxidation of glucose by some types of cells and stimulates the breakdown of fatty acids, a function that conserves glucose. **Cortisol** promotes the formation of glucose from protein and fat and decreases the utilization of glucose by most cells in the body except those of the brain and heart. **Adrenaline** and **noradrenaline** promote the breakdown of glycogen into glucose, thereby increasing blood glucose concentration. **Insulin** promotes the uptake and utilization of glucose by cells of the body. **Glucagon** stimulates the breakdown of glycogen to glucose in the liver and the breakdown of fats and proteins. The inhibitory effects of **somatostatin** on the digestive tract serve to regulate the rate at which glucose and other nutrients are absorbed into the bloodstream. Evidence suggests that somatostatin also influences the synthesis of insulin and glucagon. (pages 825, 830, 832, 833)

20. (1) The axon is a poor conductor of an electric current. (2) The nerve impulse travels at a much slower rate than an electric current. (3) The strength of a nerve impulse does not diminish as it is transmitted, whereas the strength of an electric current decreases as the distance from the source increases. (page 848)

21. (1) Conduction of a nerve impulse along a myelinated neuron (saltatory conduction) is more rapid than impulse conduction along an unmyelinated neuron. (2) Since ion flow across the neuron membrane (and thus the need for sodium-potassium pump activity) occurs only at the nodes, impulse conduction requires far less energy in a myelinated axon than in an unmyelinated axon. (page 851)

22. Stereoscopic vision enables animals to determine the distance to a nearby object. For tree-dwelling animals, stereoscopic vision facilitates movement from limb to limb in the treetops. Predators use stereoscopic vision to judge the distance to a prey item. The most valuable type of vision for prey species is provided by eyes placed on opposite sides of the head. This wider field of vision allows the animal to detect the approach of predators from any direction. (page 867)

23. (1) Hydrolysis of ATP by myosin provides the energy needed for muscle contraction. (2) The combination of a new ATP molecule with myosin releases myosin from actin. Rigor mortis occurs because ATP is not available to release myosin from actin and the actin-myosin crossbridges become locked. (page 875)

The Continuity of Life I: Reproduction

MAJOR CONCEPTS

Most vertebrates and all mammals reproduce sexually. Gametes are produced in the gonads: testes in males and ovaries in females. Male gametes are adapted for motility, whereas female gametes store nutrients. Fertilization is external in most fish and amphibians but is internal in reptiles, birds, and mammals.

Once sexual maturity is reached, sperm are produced continuously in the seminiferous tubules of the testes. Spermatogenic cells of the tubules undergo meiosis to produce spermatids, which differentiate into sperm. Sertoli cells of the tubules provide nutrients to the meiotic cells and developing sperm.

Sperm travel from the seminiferous tubules to the epididymis, where they gain motility. They then pass into the vas deferens, where they are stored. The paired vasa differentia empty into the urethra, through which sperm are expelled from the body.

The semen consists of the sperm plus secretions from the seminal vesicles and the prostate gland. Seminal vesicle fluid nourishes the sperm and contains prostaglandins that stimulate muscular contractions in the female reproductive tract. Fluid from the prostate gland neutralizes the acidity of the vagina.

Sperm production is under hormonal control, primarily by testosterone produced by interstitial cells of the testis and FSH produced by the anterior pituitary. The production of FSH is stimulated by GnRH from the hypothalamus. Hormone production is regulated by negative feedback mechanisms. Testosterone is also responsible for the secondary sex characteristics that accompany the onset of puberty in the male.

The penis is specialized for delivery of sperm to the female. In humans, it is composed primarily of spongy erectile tissue. The penis becomes turgid following increased blood flow into the erectile tissues with associated constriction of the veins draining the penis. During erection, the bulbourethral glands discharge a fluid that lubricates the urethra and aids penetration of the penis into the vagina. Male orgasm is characterized by muscular contractions in the reproductive organs and ejaculation of the sperm. Erection and ejaculation are mediated by reflex arcs in the lower spinal cord.

The major female reproductive structures are the ovaries, uterus, cervix, vagina, and vulva. The uterus houses and nourishes the developing embryo. The cervix is a muscular sphincter that separates the uterus from the vagina and seals closed during pregnancy. The vagina is a muscular tube that receives the penis and through which the infant passes at birth. The external genitalia are collectively known as the vulva and consist of the clitoris and labia.

In the human female, primary oocytes form during fetal development and reach prophase of meiosis I by birth. They remain in this stage until just prior to ovulation, when the first meiotic division is completed. The second meiotic division occurs shortly following fertilization.

Sexual maturity in the human female is marked by the first menstrual cycle, which is the cyclic development and shedding of the uterine endometrium. These events are under hormonal control.

The hypothalamic hormone GnRH stimulates the release of FSH and LH. Oocytes develop within ovarian follicles under the influence of FSH. A rise in estrogens produced by follicular cells stimulates a rise in LH, which triggers ovulation. Follicular estrogens also stimulate development of the uterine endometrium in preparation for embryonic implantation. After ovulation, the follicular cells form the corpus luteum, which secretes estrogens and progesterones. The released oocyte travels down the oviduct, where it may be fertilized if sperm are present.

The egg or embryo arrives in the uterus about three days after ovulation. After floating free in the uterus for a few days, an embryo implants into the endometrium and begins to grow. During early pregnancy, progesterone produced by the corpus luteum maintains the uterine lining in a state conducive to growth of the embryo. If fertilization and implantation do not occur, the corpus luteum disintegrates and the uterine lining is shed (menstruation). Orgasm occurs in the female but is not necessary for conception.

Estrus is the fertile period during which oocytes are released. Females of nearly all mammalian species except humans will mate only during estrus. Depending upon the species, estrus may occur only once a year (deer and wolves) or every few days (rats and mice).

Humans have devised several techniques for preventing conception, which vary widely in their effectiveness. The most effective technique is surgical sterilization, which is permanent. Estrogens and/or progesterone taken by women inhibit FSH and LH production and therefore maturation of oocytes. These preparations have a high success rate. Intrauterine devices apparently prevent conception by interfering with implantation of an embryo. Various chemical and physical barriers have been designed for use by males and females and their effectiveness depends upon the diligence and care with which they are used.

HOW TO STUDY THE CHAPTER

Read the entire chapter through quickly, focusing on the major concepts.

Use the GUIDED STUDY OF THE CHAPTER to help you identify the important details as you **reread** the chapter. Writing out the answers to these questions will help fix them in your mind as well as provide you with a valuable study aid.

Answer the questions in TESTING YOUR UNDERSTANDING without the aid of your text. Check your answers against those in PERFORMANCE ANALYSIS. Analyzing your answers will give you valuable feedback on your level of understanding and preparedness for classroom testing.

GUIDED STUDY OF THE CHAPTER

I. Introduction (pages 900–901)

> *Focus:* Sexual reproduction, which characterizes most vertebrates and all mammals, involves the processes of meiosis and fertilization.

1. In vertebrates, the male and female gametes are specialized for what functions?

2. a. *Vocabulary:* Define nonoverlapping generations.

   b. What types of animals have nonoverlapping generations?

   c. How does vertebrate reproduction differ from this system?

> *Focus:* Parental care of the young is highly developed in birds and mammals.

3. a. Among the vertebrates, identify the groups that exhibit (i) internal and (ii) external fertilization.

   b. In which two vertebrate groups do the young always develop inside the mother's body?

II. The Male Reproductive System (pages 901–907)

A. General Remarks (page 901)

4. a. Where are sperm cells produced in mammals?

   b. Where do the testes develop in the embryo?

   c. What is the function of the scrotum?

   d. By what mechanism does the scrotum perform this function?

B. Spermatogenesis (pages 901–903)

5. *Vocabulary:* Distinguish between seminiferous tubules and interstitial cells.

> *Focus:* In a sexually mature human male, sperm are produced continuously.

6. Name the two cell types in the seminiferous tubules and identify their functions.

7. A cross section of a seminiferous tubule may not reveal all of the stages of spermatogenesis. Why?

8. a. Characterize spermatogonia with respect to location and chromosome number.

   b. Name two possible fates of a spermatogonium.

9. a. Distinguish primary and secondary spermatocytes based on chromosome number.

   b. Characterize the chromosomes of secondary spermatocytes.

10. a. How are spermatids formed?

    b. How are spermatozoa formed?

    c. In humans, the development of mature sperm cells from spermatogonia takes _____ weeks.

Differentiation of Spermatids (pages 902–903)

> *Focus:* Differentiation is the process by which a cell undergoes changes that suit it to a specific function (or functions).

11. a. What is the first visible sign of spermatid differentiation?

    b. Describe the acrosome and identify its function in fertilization.

    c. Where is the acrosomal vesicle located in a differentiating spermatid?

12. a. Describe centriole movements in the differentiating spermatid.

    b. Identify the two fates of the centrioles.

> *Focus:* The axial filament of a sperm flagellum has the 9 + 2 microtubule arrangement characteristic of eukaryotic cilia and flagella.

13. Indicate the function of the spiral of mitochondria present at the basal end of a flagellum.

14. a. Describe the structure of the fibrous sheath that covers most of the axial filament.

    b. What is the likely function of the fibers forming in the sheath?

15. What other changes take place in the nucleus and cytoplasm as a spermatid differentiates?

16. The mature sperm cell consists of what five components?

> *Focus:* The entire sperm cell, including the flagellum, is contained within the cell membrane.

17. The primary function of the sperm cell is:

C. **Pathway of the Sperm** *(pages 903–905)*

> *Focus:* The sperm leave the seminiferous tubules to enter the epididymis and then the vas deferens. Each vas deferens empties into the urethra, through which the sperm are ejaculated from the body.

18. a. Fluids from what two sources are mixed with sperm as they travel through the reproductive tract? (Figure 44–5)

    b. *Vocabulary:* Define semen. (Figure 44–5)

19. a. Where in the male reproductive tract do the sperm gain motility?

    b. This process requires how much time?

    c. Where do sperm gain *maximum* motility?

20. a. By what mechanism are sperm propelled through the vasa deferentia?

    b. *Vocabulary:* Name the components of a spermatic cord.

21. a. What procedure is performed in a vasectomy?

    b. Summarize the effects of a vasectomy on hormone levels and sexual potency. (Figure 44–6)

    c. Identify the major disadvantage of a vasectomy. (Figure 44–6)

22. Trace the path of the vas deferens from the epididymis to the urethra.

23. Identify the two substances that leave the body through the urethra.

Erection of the Penis and Orgasm in the Male (pages 904–905)

24. What is the function of the penis?

25. a. Name five groups of animals in which some members possess a penis.

    b. Which animals have a cloaca, and what is its function?

    c. The penis characterizes *all* males of which group?

26. a. Describe the erectile tissues of the human penis.

    b. Distinguish between the glans penis and the bulb of the penis.

    c. What does circumcision involve?

27. Summarize the mechanism by which penile tissues become erect.

28. a. Identify the function of fluid from the bulbourethral glands.

    b. When is this fluid released?

29. a. In the reflex arcs preceding ejaculation, the nerve impulses travel from mechanoreceptors in the _____ and _____ to the _____ and then to motor neurons innervating muscles of the reproductive system.

    b. Two of the first sets of muscles to contract are located:

30. a. Name two components of seminal vesicle fluid and identify the function of each component.

b. What is accomplished by the fluid secreted from the prostate gland?

31. Ejaculation is accomplished by the action of what muscles?

32. a. What is the volume of semen in an average human ejaculate?

b. How many sperm are present in an average human ejaculate?

c. What fraction of the ejaculated sperm actually reach the oviducts?

d. A human male is generally sterile if the concentration of sperm in his semen is less than _____ sperm per milliliter.

D. The Role of Hormones (pages 905–907)

33. a. *Vocabulary:* What are androgens?

b. Name the major androgen.

c. Testosterone is produced by the _____ and is essential for _____.

d. Where are other androgens produced?

Focus: During fetal development, a fetus that is genetically programmed to be male develops into a male because of the influence of fetal androgens.

---

34. What event marks the beginning of puberty in a male?

35. a. Name six secondary sex characteristics that result from the influence of testosterone in human males.

b. Identify four examples of secondary sex characteristics resulting from testosterone influence that occur in other animals.

c. Cite three examples of behavior patterns influenced by testosterone.

Focus: For as long as animals have been domesticated, males have been castrated to improve their meat and make them easier to handle.

---

36. Why were the effects of testosterone the first hormonal effects to be noted and studied?

Regulation of Hormone Production (pages 906–907)

37. Name the hormone(s) performing each function and identify its (their) source. (See also Figure 44–10.)

Stimulates FSH and LH production

Stimulates testosterone production

Inhibits FSH production

Inhibits LH production

Inhibit GnRH production (two hormones)

Stimulate sperm production (two hormones)

38. a. Name three environmental factors that may influence testosterone production in some animals.

b. Cite two examples in which testosterone production may be influenced by social circumstances surrounding the male.

c. How may emotional state affect testosterone production in human males?

39. a. What are anabolic steroids?

b. Why were they first developed?

40. a. How do anabolic steroids affect the reproductive system of adult males?

b. What other effects do they have?

c. How do these drugs affect adolescent males?

E. *Essay:* Sex and the Brain *(pages 908–909)*

41. a. Summarize the influence of testosterone on male songbird brains.

b. What is the influence of administered testosterone on female songbird brains?

c. What conclusion can be drawn from these studies of songbirds?

42. a. Identify the hormonal basis for estrus behavior in female rats.

b. *Vocabulary:* What is lordosis?

43. Summarize the influence of testosterone on mating behavior in normal and castrated male rats.

44. a. What structure of the male rat brain contains an area sensitive to testosterone?

b. What were the effects of (i) electrical stimulation and (ii) injection of estrogen in this region of the brain?

Focus: **The differentiation of male and female brains occurs during embryonic development and is mediated by sex hormones, particularly testosterone.**

45. a. Identify the two developmental stages in which sex hormones affect rat brains.

b. Describe the experiments that revealed this information.

46. How did experimentally elevated testosterone levels created in pregnant rhesus monkeys during the critical period in gestation affect the behavior of their female offspring?

47. At what stage in human embryonic development does the critical surge in testosterone production occur?

III. The Female Reproduction System (pages 907-915)

A. General Remarks (pages 907-909)

48. Female gametes are produced in the:

49. Cite one anatomical reason why childbirth is more difficult for humans than for other mammals. (Figure 44-11)

> *Focus:* **The uterus is the organ in which the embryo develops in placental mammals.**

50. a. *Vocabulary:* What is the endometrium?

    b. Describe the two layers of the endometrium.

51. Identify four functions of uterine wall contractions.

52. a. *Vocabulary:* Distinguish between the cervix and the vagina.

    b. Account for the normal mild acidity of the vagina.

53. a. *Vocabulary:* Define the vulva.

    b. How does the structure of the clitoris resemble that of the penis?

    c. Distinguish between the labia majora and the labia minora.

B. Oogenesis (pages 910-911)

54. a. In human females, primary oocytes are in what meiotic phase at birth?

    b. When is the first meiotic division completed?

    c. What are the products of this division?

55. a. At what age does menopause typically occur in women?

    b. What time period may elapse between the beginning and end of the first meiotic division of an oocyte?

56. *Vocabulary:* Define ovulation.

57. a. Name two processes involved in oocyte maturation.

    b. What does the increase in size reflect?

58. a. By what process is the cytoplasm distributed unevenly in the second meiotic division?

    b. When does the second meiotic division occur?

    c. What are the products of this division?

    d. What is the eventual fate of the polar bodies?

59. a. What constitutes an ovarian follicle?

    b. Name two functions of the follicular cells.

    c. Cite two functions of estrogens secreted by the follicle.

Focus: **In humans, several follicles typically develop during a single cycle but usually only one releases its oocyte.**

C. **Pathway of the Oocyte** *(pages 912–913)*

60. a. By what mechanism does the oocyte move from the follicle to the oviduct?

    b. How effective is this mechanism?

61. a. What forces propel the oocyte through the oviduct?

    b. List the correct time interval for each situation:

    Journey from ovary to uterus

    Life span of unfertilized oocyte after ovulation

    Interval during which fertilization must occur after ovulation

    Time of implantation relative to fertilization

    Time of implantation relative to arrival in the uterus

62. Where does fertilization occur?

63. What happens if the oocyte is not fertilized?

64. Name and describe the process by which women may be surgically sterilized.

65. a. If a woman has blocked oviducts and desires to conceive, how may this be accomplished? (Name and describe the technique.)

    b. State the success rate of in vitro fertilization.

Orgasm in the Female (pages 912–913)

66. How does sexual arousal in the female resemble and differ from that in the male?

67. Female orgasm is characterized by what events?

68. a. Under their own power, it has been calculated that the journey of the sperm to the oviducts would require how much time?

    b. What may speed their movements along?

Focus: **Although the muscular contractions associated with female orgasm may speed the movement of sperm, orgasm is not required for conception.**

D. **Hormonal Regulation in Females** *(pages 913–915)*

The Menstrual Cycle (pages 913–915)

69. a. *Vocabulary:* What is the menstrual cycle?

    b. Control of the menstrual cycle is centered in the

    _____.

Focus: **Sex hormones are steroids. Although they are structurally similar, they have specific and distinct functions.**

70. a. Name five hormones involved in the menstrual cycle.

b. Identify the hormone or hormones characterized below.

Low concentrations inhibit production of GnRH and FSH

High concentrations inhibit secretion of GnRH (two hormones together)

High concentrations increase the sensitivity of the pituitary to GnRH

High concentrations stimulate production of GnRH

Stimulate follicle maturation (two hormones)

Stimulates endometrial growth

c. What sequence of hormonal events stimulates ovulation?

71. a. Identify the origin of the corpus luteum. (See also Figure 44-13.)

b. Name two hormones produced by the corpus luteum and indicate their functions.

c. What happens to the corpus luteum when fertilization does not occur?

d. What happens to the endometrium if the corpus luteum does not persist?

72. a. How long is an average menstrual cycle?

b. Why is the rhythm method unreliable as a contraceptive technique?

c. How may external factors influence the menstrual cycle?

73. a. What event marks the beginning of puberty in human females?

b. What secondary sex characteristics accompany puberty in the human female?

Estrus (page 915)

74. a. *Vocabulary:* Define estrus.

b. Summarize the variation in the frequency of estrus in mammals.

> *Focus:* Some mammals, including cats, rabbits, and mink, ovulate only under the stimulus of copulation.

75. Human females are receptive to mating during infertile periods. How is this behavior pattern thought to be related to the development of human social structure?

IV. Contraceptive Techniques *(pages 915–917)*

> *Focus:* Eighty percent of the women of childbearing age who regularly engage in sexual intercourse without using contraceptives conceive within a year.

76. In Table 44–1, two figures are given for the effectiveness of each technique. What does each number represent?

77. By what mechanism do contraceptive pills containing estrogens and progesterone work?

78. What factor is contributing to the increasing popularity of the condom?

Tidbit: Use of latex condoms with spermicidal jellies is an effective contraceptive technique, and is the most effective means to prevent transmission of venereal diseases, including AIDS, during sexual intercourse. A diaphragm with spermicide does not provide protection against AIDS and other venereal diseases.

79. Considering the Focus statement at the bottom of page 543, how effective are the following contraceptive techniques? (Use the figures in Table 44–1 for effectiveness that reflect *actual* experience.)

Rhythm

Douche

Withdrawal

Diaphragm with spermicidal jelly

Condom

"The pill" (estrogens and progesterone)

Surgical sterilization

V. Summary *(pages 917–918):* Read the summary. If you are familiar with the essential features of the material presented there, you are ready to complete the section TESTING YOUR UNDERSTANDING.

TESTING YOUR UNDERSTANDING

After you have completed this examination, compare your answers with those in the section that follows.

1. Fertilization is NOT internal in:
  a. marsupials.
  b. amphibians.
  c. reptiles.
  d. monotreme mammals.
  e. birds.

2. In a mature human male, the total length of the seminiferous tubules in both testes is _____ meters.
  a. 8   d. 200
  b. 25   e. 500
  c. 80

3. Mature sperm do not gain full motility until they reach the:
  a. epididymis.
  b. vas deferens.
  c. prostate gland.
  d. urethra.
  e. female reproductive tract.

4. What is the purpose of the thin, milky, alkaline fluid secreted by the prostate gland?
  a. It serves as a lubricant, facilitating the movement of sperm.
  b. It nourishes the sperm cells.
  c. It stimulates contractions in the musculature of the uterus and oviducts.
  d. It helps neutralize the normally acidic pH of the female reproductive tract.
  e. It stimulates mechanoreceptors in the penis and scrotum.

5. Indicate the correct order of structures through which sperm pass in the male reproductive system.
  1. urethra   3. seminiferous tubules
  2. epididymis   4. vas deferens
  a. 3, 2, 4, 1   d. 3, 4, 1, 2
  b. 1, 3, 4, 2   e. 3, 4, 2, 1
  c. 4, 3, 2, 1

6. The first muscles to contract when mechanoreceptors of the penis and scrotum are stimulated are those:
  a. surrounding the prostate gland.
  b. in the scrotum.
  c. surrounding the urethra.
  d. in the seminal vesicles.
  e. surrounding the seminiferous tubules.

7. Of the 300 to 400 million human sperm cells deposited in the female vagina, only _____ reach the oviducts, where fertilization occurs.
  a. 1 to 2 million   d. 1,000 to 2,000
  b. 100,000 to 200,000 e. a few hundred
  c. 10,000 to 20,000

8. The actual site of sperm production is the _____ of each testis.
   a. seminiferous tubules
   b. interstitial cells
   c. vas deferens
   d. Sertoli cells
   e. epididymis

9. Which cell type undergoes the first meiotic division in human males?
   a. secondary spermatocyte
   b. spermatid
   c. spermatogenic cell
   d. primary spermatocyte
   e. spermatogonium

10. The first visible sign of spermatid differentiation is the:
    a. formation of the flagellum by a centriole.
    b. migration of centrioles to the cell membrane.
    c. appearance of vesicles containing dark granules in the Golgi complex.
    d. formation of the neck of the sperm.
    e. aggregation of mitochondria into a sheath.

11. One of the centrioles in a differentiating sperm forms part of a connecting piece that links the flagellum to the:
    a. nucleus.
    b. acrosome.
    c. Golgi complex.
    d. mitochondrial sheath.
    e. cell membrane.

12. Which of the following is NOT a secondary sex characteristic in human males?
    a. growth of the larynx with deepening of the voice
    b. increase in skeletal size
    c. increase in body hair
    d. production of sperm
    e. overactive sebaceous glands, resulting in acne

13. The loss of testosterone due to castration of domestic food animals has several advantageous consequences. Which of the following is NOT one of these consequences?
    a. The meat is less tough.
    b. They possess increased body fat.
    c. They are easier to manage.
    d. They may be less aggressive toward other males.
    e. All of these are advantageous consequences of castrating domestic animals.

14. Which hormone is of theoretical interest as a potential male contraceptive?
    a. FSH
    b. LH
    c. inhibin
    d. estrogen
    e. testosterone

15. Which female structure(s) is (are) homologous with the scrotal sac of the male?
    a. clitoris
    b. labia majora
    c. labia minora
    d. vaginal wall
    e. cervix

16. What is the importance of female orgasm to human conception?
    a. It is necessary for conception.
    b. It may increase the possibility of conception.
    c. It actually interferes with conception.
    d. It has no role other than sexual gratification.
    e. So far, no role has been proposed for female orgasm.

17. An oocyte and the specialized cells surrounding it are known as a(n):
    a. polar body.
    b. oogonium.
    c. ovarian follicle.
    d. corpus luteum.
    e. ovum.

18. Women who are infertile because of blocked oviducts may be able to bear children by employing which technique?
    a. tubal ligation
    b. superovulation
    c. in vitro fertilization
    d. the rhythm method
    e. dilation and curettement

19. In humans, ovulation is triggered by a sudden surge of:
    a. LH.
    b. FSH.
    c. progesterone.
    d. testosterone.
    e. oxytocin.

20. T or F The division of labor between males and females appears to have arisen from the development of family units in the course of human evolution.

21. T or F The penis is a male reproductive structure found only in vertebrates and found in all vertebrates.

22. T or F The purpose of the scrotum is to keep the sperm-producing structures at body temperature, which is required for normal sperm development.

23. T or F The smooth muscles of the uterine wall contract not only during menstruation and labor, but all the time.

24. T or F Prior to fertilization, a human oocyte undergoes both the first and second meiotic divisions.

25. T or F If implantation does not occur, the blood levels of estrogens and progesterone fall, and menstruation ensues.

26. T or F The human is one of the few mammals in which the female reproductive cycle does not respond to environmental influences.

27. T or F In human females, enlargement of the breasts and hips in response to increasing levels of female sex hormones always follows the actual onset of puberty.

28. Label these structures in the diagram on the following page of a mature sperm: acrosome, mitochondrial sheath, nucleus, and tail. Identify the function of each structure.

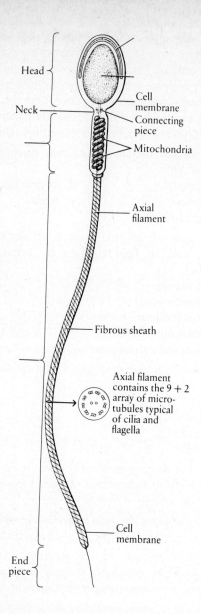

Head

Neck

Cell membrane

Connecting piece

Mitochondria

Axial filament

Fibrous sheath

Axial filament contains the 9 + 2 array of micro-tubules typical of cilia and flagella

Cell membrane

End piece

29. Match each birth-control technique with its mode of action. Use each answer only once. A blank may have more than one answer, as indicated in parentheses.

1. rhythm
2. vaginal foam
3. intrauterine device
4. tubal ligation
5. morning-after pill
6. condom

a. _____ surgical obstruction of the oviducts
b. _____ abstinence from intercourse when conception is likely to occur
c. _____ prevents sperm from entering the female
d. _____ kills sperm before they enter the uterus
e. _____ prevents implantation of a zygote (2 answers)

30. Describe the two-stage response of the brain of male rats to testosterone. Indicate when these two stages occur in the life of the rat.

31. Cite two examples in which the testosterone level in a male mammal is influenced by environmental or social circumstances.

32. Why must the fertilization of a human oocyte occur in an oviduct?

33. Identify the hormonal stimulus for each event in the menstrual cycle: ovulation, follicle development, development of the endometrium, onset of the menstrual flow.

PERFORMANCE ANALYSIS

1. **b** Fertilization is internal in all reptiles, mammals, and birds. Fertilization is typically external in fishes and amphibians. (page 900)

2. **e** The total length of the seminiferous tubules in both testes of a mature human male is approximately 500 meters. (page 901)

3. **e** Sperm gain partial motility in the epididymis. They do not gain full motility until they reach the female reproductive tract. (page 903)

4. **d** The fluid secreted by the prostate gland helps neutralize the normally acidic pH of the female reproductive tract, which is spermicidal. (page 905)

5. **a** In the male reproductive tract, sperm develop in the seminiferous tubules and pass through the epididymis and into the vas deferens, where they are stored. During ejaculation, the sperm are propelled from the vas deferens through the urethra. (pages 903, 904)

6. **b** When the mechanoreceptors of the penis and scrotum are stimulated, the first muscles to contract are those of the scrotum, which pull the testes closer to the body, and those surrounding the epididymis and vasa deferentia, which propel the sperm into the urethra. (page 905)

7. **e** Only a few hundred of the sperm cells deposited in the female vagina reach the oviducts. (page 905)

8. **a** Sperm cells are produced in the seminiferous tubules of the testes. (page 901)

9. **d** The first meiotic division occurs in primary spermatocytes and produces secondary spermatocytes. The second meiotic division occurs in secondary spermatocytes, resulting in spermatids. (page 902)

10. **c** The first visible sign of spermatid differentiation is the appearance within the Golgi complex of vesicles containing dark granules. (page 902)

11. **a** During spermatid differentiation, one of the centrioles forms part of a structure connecting the flagellum to the nucleus. (page 902)

12. **d** Secondary sex characteristics are traits associated with sex hormones but not directly involved in reproduction. (pages 905, 906)

13. **e** All of the consequences listed are reasons why male domestic animals (especially food animals) are commonly castrated. Reduced aggression toward other males varies among different species and may be influenced by the age at which the surgery is performed. For example, castrating a mature stallion nearly always results in behavioral modi-

fication, whereas castrating a mature male dog will affect behavior in some cases but not in others. (page 906)

14. c The hormone inhibin, which is produced by the Sertoli cells and inhibits FSH production, is of theoretical interest as a potential male contraceptive. (page 906)

15. b The labia majora of the female are homologous to the scrotal sac of the male. (page 909)

16. b On orgasm, the cervix dips into the upper vagina where semen collects. Also, orgasm may produce contraction of the oviducts that could speed the movement of sperm. These effects may increase the chance of conception, but orgasm is not by any means required for conception. (page 913)

17. c An ovarian follicle consists of the oocyte and the specialized cells of the ovary that surround and support the oocyte. (page 911)

18. c A woman who has blocked oviducts may bear children if she utilizes in vitro fertilization. In this technique, her mature oocyte is harvested surgically, fertilized by her husband's sperm in a Petri dish with nutrient solution, and the zygote is placed in her uterus. The success rate of this technique is 15 to 20 percent. (page 912)

19. a In humans (and some other mammals in which ovulation is spontaneous) ovulation is triggered by a sudden surge of LH. (page 913)

20. True A division of labor in which the male hunted and protected the family and the female cared for the young seems to have been a natural outcome of the development of family units in human evolution. (page 915)

21. False The penis has evolved independently in some species of insects and other invertebrates. It is present in all mammalian species and in some species of reptiles and birds. (page 904)

22. False The function of the scrotum is to keep the sperm at a lower temperature than body temperature, which is too warm for sperm development. (page 901)

23. True The smooth muscles in the wall of the uterus are constantly contracting in rhythmic waves. The contractions are strongest during menstruation and labor. (page 908)

24. False In humans, the second meiotic division occurs *after* fertilization. (page 910)

25. True Menstruation is apparently triggered by falling blood levels of estrogens and progesterone, which occurs unless an embryo implants into the endometrium. (page 914)

26. False The menstrual cycle of human females can be influenced by emotional trauma and by chemical stimuli (e.g., the underarm secretions of other females). (page 914)

27. False In humans, enlargement of the breasts and hips in response to increasing hormone levels may occur prior to the actual onset of puberty, i.e., menstruation. (page 915)

28. The **acrosome** contains enzymes that help the sperm penetrate the outer layers of an unfertilized secondary oocyte. The **mitochondrial sheath** contains the mitochondria that provide the energy for sperm motility. The **tail** is a specialized flagellum by which the sperm moves. The **nucleus** carries the hereditary information of the male. (pages 902, 903, Figure 44–4)

29. The answers are **a4, b1, c6, d2, e3,** and **e5.** (Table 44–1)

30. The brain of the male rat is influenced by testosterone during an early period, in which brain cells are differentiating, and after sexual maturity, when the differentiated brain cells respond to the hormones. The brain must be exposed to testosterone during the early period if the response of the mature rat to testosterone is to be normal. The first stage is from birth to 10 days of age and the second stage is after sexual maturity is reached. (essay, page 909)

31. When a bull sees a cow, his blood testosterone level increases. Men in an army training camp or in combat have lower testosterone levels than those in noncombat positions. (page 907)

32. The human oocyte must be fertilized in an oviduct because it takes approximately three days for the oocyte to reach the uterus. Although an unfertilized oocyte lives for 72 hours after ovulation, it is susceptible to fertilization for less than half this time. (page 912)

33. Ovulation is triggered by a surge of LH. Follicle development is stimulated by FSH released by the pituitary in response to GnRH. The endometrium develops under the influence of estrogens produced mainly by the ovarian follicle. The onset of menstruation follows decreasing levels of estrogens and progesterone. (pages 913, 914)

The Continuity of Life II: Development

MAJOR CONCEPTS

The development of a fertilized egg into a complete organism involves the processes of growth (increase in size), differentiation (specialization of cells, tissues, and organs), and morphogenesis (shaping of the adult body form).

Although the details vary from one group to the next, all coelomates pass through the same basic developmental stages. Three species in which development has been studied extensively are the sea urchin, amphibian, and chick.

The early embryo is called a morula. The first stage of development is cleavage, or early cell division, which results in the formation of the blastula. The blastula, a fluid-filled sphere of a single layer of cells, has the same total volume as the fertilized egg. The fluid-filled cavity is the blastocoel. Two important ratios increase as a consequence of cleavage: the surface-to-volume ratio and the ratio of nuclear volume to cytoplasmic volume.

Next, the cell migrations of gastrulation result in the formation of the primitive gut and the blastopore, and the establishment of the three primary tissue layers (ectoderm, mesoderm, and endoderm). The primary tissue layers begin to differentiate, the coelom forms, and organ systems begin to develop.

The sea urchin is a favorite subject for developmental studies because fertilization and development are external, the egg and embryo are nearly transparent, and the process is rapid.

Sea urchin eggs are surrounded by two layers outside the cell membrane: the vitelline envelope, which contains species-specific protein receptors, and the jelly layers. Fertilization occurs when acrosomal enzymes dissolve the jelly layers, sperm cells bind to the vitelline receptors, the cell membranes of sperm and egg fuse, and the sperm nucleus enters and fuses with the egg nucleus.

There are at least four consequences of fertilization. (1) The vitelline membrane is transformed into the fertilization membrane, and the entry of additional sperm cells is prevented. (2) The egg is activated metabolically. (3) The genotype of the new individual is established. (4) Mitosis begins.

In some species, egg activation can be triggered in unfertilized eggs by certain stimuli, including exposure to a hypertonic solution, pricking with a glass needle, or mild electric shock. The nucleus is *not* essential to egg activation, which is mediated by previously transcribed maternal mRNA residing in the cytoplasm.

In sea urchins, the development of organ systems is determined by cytoplasmic determinants distributed in two gradients, animal and vegetal. The cytoplasmic determinants are thought to be messenger RNA or protein molecules. In addition to cytoplasmic determination, the interactions among individual cells influence development of the organism.

Amphibians also undergo cleavage and gastrulation, but the details differ from those of sea urchin development. Amphibian eggs have a much greater amount of yolk and the animal and vegetal portions differ markedly in appearance. A major reorganization of the cytoplasm occurs immediately following fertilization, resulting in gray crescent formation. The gray crescent becomes the dorsal lip of the blastopore, which is the organizer of developmental processes in amphibians.

During gastrulation, the primary embryonic tissues develop and are rearranged in the embryo. Mesoderm gives rise to lateral plate mesoderm and to chordamesoderm, which develops into the notochord. Ectoderm gives rise to neural ectoderm, which develops into the brain and spinal cord, and to epidermal ectoderm, which gives rise to the epidermis of the skin. By the end of gastrulation, the notochord and neural plate have formed.

Embryonic induction is the process by which one tissue contacts and influences the differentiation of an adjacent tissue. Current evidence indicates that this process is chemically mediated. The "master" inducer in amphibians is the dorsal lip of the blastopore, the organizer. Secondary induction involves the induction of one tissue by an adjacent tissue that was itself induced.

Chick development differs from that of amphibians in several ways. The egg cell is surrounded by membranes and a shell that permit its survival in a terrestrial environment, and it contains a large amount of yolk. The major consequence of this large yolk content is that cleavage involves only part of the egg mass. Cleavage occurs in a small region of the cytoplasm on top of the yolk in which the nucleus resides. The resulting blastula is called a blastodisc. In chick gastrulation, a primitive streak forms instead of a blastopore.

The extraembryonic membranes of a chick originate as extensions of the blastodisc, and each membrane is composed of two primary tissue types. The yolk sac gradually surrounds the yolk and supplies nutrients to the embryo. The amnion encloses the fluid-filled amniotic cavity, which protects the embryo from mechanical and thermal shocks. The allantois encloses excreted nitrogenous wastes. The chorioallantoic membrane, which develops fairly late from fusion of the allantois and the chorion, conducts gas exchange between the embryo and the environment.

Organogenesis is the formation of organ systems. It begins with the inductive interaction of the chordamesoderm and the overlying ectoderm. Ectoderm develops into the nervous system, the skin, and epithelium. Mesoderm gives rise to the circulatory, reproductive, and excretory systems, and the notochord, muscles, and skeleton. Endoderm develops into glands and the linings of the respiratory and digestive tracts. These are only generalities; each organ and organ system contains components originating from all three tissue layers.

Morphogenesis is accomplished by only a few basic processes. These include the deposition of extracellular material, changes in the rates of cell growth and division, altered adhesion among neighboring cells, and changes in cell shape due to extension or contraction.

Pattern formation is the general process that results in the differences characteristic of homologous structures in different species. One example is limb development among vertebrates. Two factors determining the developmental fate of limb cells are the length of time spent near the apical ectodermal ridge and the influence of chemical mediators, particularly retinoic acid.

Human embryonic development differs from that of amphibians and chicks in several ways. The blastula, known as a blastocyst, consists of a sphere formed from two layers of cells (the trophoblast) enclosing an inner cell mass that is attached to the sphere. The embryo develops from the inner cell mass and the trophoblast develops into the chorion. The trophoblast releases the hormone chorionic gonadotropin, which maintains early pregnancy by stimulating the corpus luteum to continue estrogen and progesterone production.

When the human embryo implants into the lining of the uterus, extraembryonic membranes develop, but there are several differences between mammalian and chick membranes. Human yolk sacs do not contain yolk. The allantois becomes a primitive urinary bladder and the allantoic mesoderm develops into the blood vessels of the embryonic placenta. The chorion develops villi that project into the endometrium and form a major component of the placenta. As in the chick, the amniotic cavity is a fluid-filled shock absorber.

The placenta is formed from chorionic villi of the embryo and the maternal endometrium. It is the structure through which the exchange of nutrients, gases, and wastes between mother and embryo is accomplished. The embryo is attached to the placenta by the umbilical cord. As the pregnancy progresses, the placenta becomes the major site of progesterone and estrogen production and the corpus luteum deteriorates.

During the first three months of human embryonic development, all the major organ systems are laid down, the germ cells migrate to the developing gonads, and sexual development begins. The human embryo is usually referred to as a fetus after two months of development.

The embryo is most sensitive to the influence of external factors during the first two months. Developmental defects may be caused by exposure to certain drugs, x-rays, infectious agents, or unhealthy habits of the mother such as heavy alcohol consumption and smoking. The specific defect depends upon which organ systems were developing at the time of the insult.

The second three months of human development are characterized mainly by increasing development of organ systems and some increase in size.

The major increase in size occurs during the final three months. The fetal nervous system undergoes major development during this time. If maternal protein intake is inadequate, the fetal nervous system may be adversely affected. As the physiology of the fetus grows to resemble that of the adult, the fetus is prone to acquire any chemical dependencies of the mother. During the last month before birth, the fetus obtains antibodies from the mother, which cross the placenta by active transport.

The most important factor in infant mortality is weight at birth. Two-thirds of all infants who die are underweight (less than 2.5 kilograms or 5.5 pounds). The major cause of low birth weight is maternal malnutrition.

Labor is divided into three stages: dilation (of the cervix), expulsion (of the fetus), and the placental stage, all of which are accomplished by contractions of the uterus.

HOW TO STUDY THE CHAPTER

Read the entire chapter through quickly, focusing on the major concepts.

Use the GUIDED STUDY OF THE CHAPTER to help you identify the important details as you **reread** the chapter. Writing out the answers to these questions will help fix them in your mind as well as provide you with a valuable study aid.

Answer the questions in TESTING YOUR UNDERSTANDING without the aid of your text. Check your answers against those in PERFORMANCE ANALYSIS. Analyzing your answers will give you valuable feedback on your level of understanding and preparedness for classroom testing.

GUIDED STUDY OF THE CHAPTER

I. **Introduction** (*page 919*)

1. Name and define three processes that constitute development.

2. Our understanding of developmental processes is largely due to studies of what three organisms?

II. **Development of the Sea Urchin** (*pages 920–926*)

A. **General Remarks** (*page 920*)

3. a. Which group of invertebrate organisms are the most useful models for the study of early vertebrate development? Why?

b. Cite four reasons why the sea urchin has long been a favorite study subject of embryologists.

4. *Vocabulary:* What is a pluteus?

B. **Fertilization and Activation of the Egg** (*pages 920–921*)

5. Development begins with what event?

6. How does the sea urchin sperm cell resemble the human sperm cell?

7. List in order, from most superficial to most internal, the three layers surrounding an unfertilized egg. (See also Figure 45–3.)

8. a. *Vocabulary:* What is the vitelline envelope?

b. What feature of this structure facilitates fertilization?

9. a. Describe the process by which a sperm cell reaches and penetrates the vitelline envelope.

b. What event constitutes fertilization?

10. a. Where are cortical granules located and what do they contain? (Figure 45–3)

b. What event triggers the release of cortical granule contents? (Figure 45–3)

c. Identify the role of each substance released from cortical granules. (Figure 45–3)

11. a. After one sperm cell penetrates the egg cell membrane, additional sperm are prevented from entering the egg. Summarize the mechanism by which this is accomplished.

b. What is the fertilization membrane and from what structure is it formed?

12. List (briefly) four consequences of fertilization.

13. What evidence indicates that the egg has been metabolically activated by fertilization?

14. What structures take part in formation of the mitotic spindle of the zygote?

Focus: Under certain laboratory conditions, an unfertilized egg may be metabolically activated.

15. Name three events or situations that may trigger egg activation.

16. a. What evidence indicates that the nucleus is *not* essential for the activation of some eggs?

b. The burst of biosynthetic activity that follows fertilization is accomplished because of the presence of what substance stored in the egg cytoplasm?

C. From Zygote to Pluteus (pages 921–924)

17. a. *Vocabulary:* What is a morula?

b. What developmental process characterizes a morula?

18. a. *Vocabulary:* Distinguish among blastocoel, blastula, and blastomere.

b. How is a blastocoel created?

19. Although overall volume stays the same, cleavage changes what two characteristics of a zygote?

Focus: The end result of cleavage is a blastula, a single layer of cells enclosing a cavity, the blastocoel.

20. a. What two events are accomplished during gastrulation?

b. The first step in gastrulation is:

21. Identify the origin of the cells forming the primary mesenchyme.

22. a. Name the cavity formed during gastrulation.

b. This cavity eventually develops into what structure?

c. What is the fate of the blastopore?

d. Diagram the spatial relationship between the archenteron and the blastocoel.

e. What is the fate of the blastocoel? (Figure 45–7)

23. a. Name and identify the locations of the three embryonic tissue layers formed during gastrulation.

b. Mesoderm originates from what group of cells?

Focus: By the end of gastrulation, the archenteron has formed, three embryonic tissue layers are present, and the anterior-posterior axis of the embryo is obvious.

---

24. Cite three examples of cell differentiation that occur once gastrulation is complete.

D. The Influence of the Cytoplasm *(pages 925–926)*

25. a. Distinguish between the vegetal and animal halves of a sea urchin embryo.

b. Describe the first three cleavage divisions in terms of the relationship of each to the vegetal and animal poles.

c. What happens if a developing embryo is divided along

(i) either of the first two cleavage planes?

(ii) the third cleavage plane?

Focus: The phenomena described in question 25 occur because of cytoplasmic determinants localized in various regions of the embryo.

---

26. a. What substances are thought to act as cytoplasmic determinants?

b. How are they localized in the cytoplasm and in different regions of the developing embryo?

27. What evidence indicates that interactions among blastomeres are important in development?

E. *Essay:* The Cytoplasmic Determination of Germ Cells *(pages 924–925)*

28. a. *Vocabulary:* What are germ cells?

b. Why might the early determination of germ cells be of adaptive value to an organism?

c. Why is this early determination of particular importance in organisms that undergo complete metamorphosis?

29. a. How does cleavage in *Drosophila* differ from that in sea urchins?

b. Describe the *Drosophila* blastula.

30. a. What are pole cells?

b. How is their fate determined?

c. Summarize the experiment that demonstrated the role of polar plasm.

d. Polar plasm granules, which are present in the egg cytoplasm *before* fertilization, contain what substance?

III. Development of the Amphibian *(pages 926–931)*

A. General Remarks *(pages 926–927)*

31. Identify two ways in which amphibian eggs differ from sea urchin eggs.

32. a. Describe the cytoplasmic reorganization that occurs immediately after formation of amphibian eggs.

b. Of what importance is the gray crescent to subsequent embryonic development?

c. What experimental evidence supports the role of the gray crescent?

d. Who discovered this?

e. Compare the influence of the gray crescent on amphibian development with that of the animal and vegetal poles on sea urchin development. (See Figures 45–8 and 45–10.)

B. Cleavage and Blastula Formation *(page 927)*

33. a. What feature accounts for the major differences between modes of cleavage in sea urchins and amphibians?

b. Describe cleavage in an amphibian egg, emphasizing how this process differs from cleavage in sea urchins.

C. Gastrulation and Neural Tube Formation *(pages 927–929)*

34. a. In general terms, how does gastrulation differ in amphibians and sea urchins?

b. Describe gastrulation in amphibians, emphasizing how the process differs from sea urchin development.

c. How is the anterior-posterior axis of the animal determined?

d. From what cells is the yolk plug formed?

e. Characterize the cells lining the archenteron.

35. Distinguish between chordamesoderm and lateral plate mesoderm.

36. Distinguish between neural ectoderm and epidermal ectoderm.

37. List three signs of differentiation apparent by the end of gastrulation.

38. How is the neural tube formed?

39. *Vocabulary:* What are somites and from what tissue are they formed?

40. The coelom develops in what specific tissue?

41. Which primary embryonic tissue contains the yolk?

42. Embryologists have developed fate maps using what technique? (See also Figure 45–15a.)

43. Cite examples of organs or tissues that develop from the following structures. (Figure 45–15b)

Ectoderm

Neural ectoderm

Endoderm

Mesoderm

Somites

Lateral plate mesoderm

D. The Role of Tissue Interactions *(pages 930–931)*

Focus: Differentiation results from the selective activation and inactivation of specific genes in a cell. Some cells differentiate fairly early in embryonic life and their fate becomes fixed; differentiation is irreversible in these cells. Other cells may differentiate but retain full developmental potential.

44. Which tissue is the first to differentiate in the amphibian?

The Organizer (page 930)

45. a. Define the developmental relationship between the gray crescent and the dorsal lip of the blastopore.

b. Describe the different experimental results obtained by Spemann from (i) dividing a developing blastula and (ii) dividing the embryo when gastrulation had started.

46. Outline the experiments of Hilde Mangold in which the organizing role of the dorsal lip of the blastopore was elucidated.

Focus: A transplanted dorsal lip will induce development of the *recipient* cells in the same manner in which it would have organized the cells of the embryo from which it was removed.

47. What tissue serves as the primary organizer in all chordate embryos that have been studied?

Induction and Inducers (pages 930–931)

48. a. *Vocabulary:* Define embryonic induction.

b. What experimental results indicate a chemical basis for this process?

Focus: The capacity of the organizer to induce neural plate formation is *not* species-specific.

49. Cite evidence indicating that an inducing chemical activates a latent potential rather than endowing a cell with the capacity to perform an activity.

IV. Development of the Chick *(pages 931–940)*

A. General Remarks *(pages 931–932)*

50. Name three ways in which a hen's egg differs from the eggs previously described.

> *Focus:* **The chick is more developmentally advanced on emergence than are the larvae of the sea urchin and frog.**

51. *Vocabulary:* What is the chalaza? (Of what is it composed and what is its function?) (Figure 45–17)

52. a. How does cleavage in a hen's egg differ from that in amphibians?

    b. What is the structural basis for this difference?

53. Define and describe the relationship among the following: blastodisc, epiblast, hypoblast, blastocoel.

54. The primitive streak of a bird embryo corresponds to what structure in a developing amphibian?

55. a. Describe gastrulation in the bird egg, emphasizing how it differs from the same process in amphibians.

    b. The hypoblast cells develop into what structures?

c. From what embryonic tissue does the notochord develop?

56. Summarize the similarities between frog and bird embryos at the end of gastrulation.

B. Extraembryonic Membranes of the Chick *(pages 932–934)*

> *Focus:* **Each extraembryonic membrane forms from a combination of two of the three primary tissue types.**

57. Identify the origin of each of the three primary tissues that develop into the extraembryonic membranes.

    Endoderm

    Ectoderm

    Mesoderm

58. How do the two components of the yolk sac contribute to its function?

59. a. Describe how the amniotic cavity is formed.

    b. Of what protective value is the amniotic fluid?

    c. Identify the two primary tissues from which the amnion is formed.

    d. Distinguish between the amnion and the chorion.

60. a. From what primary tissues is the allantois constructed?

b. What is the function of the allantois and what does it contain?

61. a. How is the chorioallantoic membrane formed?

b. Identify the function of this membrane.

c. By late embryonic development, the chorioallantoic membrane encloses what structures? (Figure 45–21c)

C. **Organogenesis: The Formation of Organ Systems** *(pages 935–937)*

62. a. *Vocabulary:* Define organogenesis.

b. What is the first step of organogenesis?

Differentiation of the Ectoderm (pages 935–936)

63. Characterize the early development of motor neurons.

64. Identify four fates of neural crest cells.

65. How does the vertebral column form?

66. Development of the brain begins with what step?

67. Describe the appearance of the primordia for ears, nostrils, and eyes. (Figure 45–23)

68. a. Identify the two components that interact to develop eyes. (See also Figure 45–24.)

b. Indicate the origin of each structure of the eye:

Retina

Pigment layer

Pupil

Lens

Cornea

Optic nerve

Focus: **The retina is a differentiated extension of the brain.**

69. a. *Vocabulary:* Define secondary induction.

b. Development of an eye involves how many levels of tissue induction?

Focus: **Processes similar to the development of the eye result in the organs of olfaction and hearing.**

Differentiation of the Mesoderm (page 936)

70. Indicate three fates of somites.

71. How does the segmental pattern formed by somites differ in aquatic and terrestrial vertebrates?

72. Excretory and reproductive structures form from the:

73. Identify the fates of the two sheets formed in the lateral plate mesoderm during coelom formation.

Differentiation of the Endoderm (page 937)

74. Endoderm gives rise to what types of organs?

75. a. How does development of respiratory structures differ in aquatic and terrestrial vertebrates?

    b. The endodermal pouches that give rise to the gills in fish give rise to what structures in terrestrial vertebrates?

    c. Identify the developmental origin of lungs.

76. How do the accessory digestive organs originate?

> *Focus:* **All organs consist of tissues derived from more than one primary tissue.**

77. Illustrate this Focus statement with an example.

D. **Morphogenesis: The Shaping of Body Form** *(pages 938–940)*

> *Focus:* **There appear to be only a few basic processes that become modified in forming the various shapes of body structures.**

78. Name the four basic processes involved in morphogenesis.

79. Describe the mechanism by which neural plate cells change shape in formation of the neural tube. Include the roles of microtubules and actin microfilaments. (See also Figure 45–28.)

80. What type of differences arise from pattern formation?

81. Identify the first visible sign of wing development.

82. a. Name and indicate the function of the two major groups of cells in a developing wing bud.

    b. What factor determines the fate of an individual wing bud cell?

    b. Correlate the amount of time a cell spends near the apical ectodermal ridge with the location of the wing tissue into which it differentiates.

83. a. What substance influences the differentiation of wing structures from anterior to posterior?

    b. By what mechanism is this substance thought to work?

> *Focus:* **The integration and control of development and of other physiological processes are evidently similar.**

E. *Essay:* **Genetic Control of Development: The Homeobox** *(pages 942–943)*

84. State the general mechanism by which cytoplasmic determinants and other substances involved in differentiation ultimately operate.

85. a. Distinguish between segmentation genes and homeotic genes in *Drosophila*.

b. Cite two examples illustrating the developmental disruptions that may result from mutations of homeotic genes.

86. a. Name two possible functions of the proteins coded for by homeotic genes.

b. Describe the evidence for one of these functions.

87. a. *Vocabulary:* What is the homeobox?

b. Identify one proposed function of the homeobox and cite evidence supporting this proposed function.

Focus: Development in animals may be controlled by master genes that are universal in the animal kingdom.

V. Development of the Human Embryo *(pages 940–950)*

A. General Remarks *(pages 940–941)*

Focus: The basic patterns of development, which are similar throughout the animal kingdom, constitute evidence of the fundamental unity of life.

---

88. Where is the fertilized egg when the first divisions occur?

89. a. In humans, how long after fertilization do the divisions occur that produce two, four, and eight cells?

b. During these early cleavage divisions, the embryo has what requirements for survival?

90. a. How does the mammalian blastocyst differ structurally from the blastulas studied previously?

b. *Vocabulary:* Describe the trophoblast and identify its function.

c. The trophoblast is the precursor of what embryonic structure?

91. Identify the source and function of chorionic gonadotropin.

Focus: Most pregnancy tests depend upon the detection of chorionic gonadotropin in either blood or urine.

---

92. a. Describe the process of implantation.

b. What role does the trophoblast play in implantation?

B. Extraembryonic Membranes *(page 944)*

93. a. How do the eggs of monotremes, marsupials, and placental mammals differ with respect to yolk content?

b. In placental mammals, the absence of yolk has what effect on the processes of cleavage and gastrulation?

94. a. What is the fate of the following structures in mammals?

Allantois

Allantoic mesoderm

Allantoic stalk

b. How are waste disposal and nutrient acquisition accomplished in the mammalian embryo?

95. How does the amniotic cavity of mammals resemble that of chicks?

96. a. The chorion develops from what components?

   b. What event marks the beginning of the mature placenta?

97. Discuss amniocentesis and chorionic villus biopsy. Include the reasons for performing these procedures and their relative advantages and disadvantages.

C. The Placenta (pages 944–945)

98. Describe the structure and extent of the placenta three weeks after conception.

> *Focus:* The placenta is formed from maternal endometrium and from extraembryonic chorion, but the embryonic and maternal blood supplies are not in direct contact.

99. a. Name the blood vessel that supplies the maternal blood space. (Figure 45–34)

   b. Describe the exchange of materials between embryonic and maternal circulatory systems.

100. Summarize the three functions of the placenta. (Figure 45–34)

101. Name three changes that occur in the mother in response to her pregnancy.

102. The embryo is attached to the placenta by the _____.

103. a. Since the chorionic component of the placenta is a foreign material, one might expect the mother's immune system to attack it. What prevents this expected rejection?

   b. If a cell-mediated immune response does occur, what is the consequence?

104. a. Name and identify the source of the hormones that maintain pregnancy by the end of the third month.

   b. What happens to the corpus luteum?

   c. What is one proposed cause of miscarriages that occur during the third month of pregnancy?

D. The First Trimester (pages 945–948)

> *Focus:* During embryonic development, the embryo is measured from crown to rump.

105. Summarize the state of embryonic development at each time interval following conception. Focus particularly on the development of the heart, gonads, and other major organ systems. Note the length of the embryo where possible.

3 weeks

4 weeks

5 weeks

106. a. Summarize how the gene mentioned in the Focus statement is thought to influence sexual development.

    b. Identify two other factors that influence sexual development.

107. a. What developmental changes occur during the second month?

    b. What is the approximate weight and length of the embryo by the end of the second month?

    c. At what stage is a human embryo referred to as a fetus?

    d. Name the major blood-forming organ of the fetus at this time.

108. a. At what stage is a human embryo most susceptible to the influence of external factors?

    b. *Vocabulary:* What are teratogens?

    c. Name five factors that may cause abnormal fetal development if the fetus is exposed to them at this time.

    d. Identify four organs that may be damaged if a pregnant woman has a rubella infection at this stage.

109. Summarize developmental progress during the third month.

110. What is the approximate weight and length of the fetus by the end of the third month?

E. **The Second Trimester** *(page 948)*

111. At what stage do fetal movements become obvious to the mother?

112. Describe fetal development during the following time periods. Note the weight and length of the fetus by the end of each period.

    Fourth month

    Fifth month

    Sixth month

113. The youngest fetus to survive outside the uterus was how old?

114. *Vocabulary:* Distinguish between lanugo and meconium.

F. **The Final Trimester** *(pages 948–949)*

115. What is the major change that occurs during the last trimester?

116. What major organ system undergoes significant development during this period?

117. How may inadequate maternal nutrition during the third trimester adversely affect the fetus?

118. What is the consequence of the fact noted in the Focus statement for a fetus whose mother participates in substance abuse?

119. What events occur in the last month of pregnancy?

120. a. How do maternal antibodies cross the placenta to enter the fetus?

    b. At what age does the infant begin replacing maternal antibodies with its own?

121. a. Identify the major factor in infant mortality.

    b. State the birth weights consistent with the following risk categories.

       High risk of death or severe brain damage

       Low weight (mortality within one month of birth 40 times greater than for heavier infants)

    c. What fraction of all infants who die have a low birth weight?

    d. How does the United States rank in infant mortality statistics?

    e. Name the major cause of low birth weight.

    f. How does the cost of improved maternal nutrition compare with the cost of maintaining a low weight infant in the hospital?

    g. What is the difference in infant mortality rates for black and white Americans?

    h. What simple measures could improve these statistics?

G. Birth (page 950)

122. How is the projected date of birth calculated relative to the day of conception and to the beginning of the last menstrual period?

123. a. Name the three stages of labor and summarize the major events of each stage.

    b. During what stage does the amniotic sac usually rupture?

    c. What is accomplished by the minor uterine contractions that continue after delivery?

124. What is the chemical basis for the alertness of the infant immediately after birth? (Figure 45–42)

125. How are the functions of vital organs (particularly the heart, brain, and lungs) protected from oxygen deprivation during delivery? (Figure 45–42)

VI. Epilogue (page 951)

VII. Summary (pages 951–953): Read the summary. If you are familiar with the essential features of the material presented there, you are ready to complete the section TESTING YOUR UNDERSTANDING.

TESTING YOUR UNDERSTANDING

After you have completed this examination, compare your answers with those in the section that follows.

1. Which of the following is NOT a reason that scientists find sea urchins to be suitable organisms for the study of embryonic development?
   a. The zygote develops into a free-swimming larva within 48 hours.
   b. The eggs and early embryo are almost transparent, allowing observation of early developmental events.
   c. There are many eggs produced at one time and fertilization is external.
   d. The egg can readily be seen with the unaided eye.
   e. Sea urchins are abundant in pleasant places where scientists enjoy spending the summer.

2. In a sea urchin, sperm receptors are located in (on) the:
   a. nucleus.   d. cell membrane.
   b. cytoplasm.   e. vitelline envelope.
   c. jelly layers.

3. In the formation of the fertilization membrane of sea urchin eggs, the vitelline envelope lifts off the cell membrane in response to an increasing concentration of free _____ ions, which are released into the cytoplasm from the endoplasmic reticulum.
   a. calcium
   b. chloride
   c. sodium
   d. potassium
   e. phosphate

4. The sea urchin embryo becomes a:
   a. blastula by taking up fluids osmotically.
   b. planula by undergoing cleavage.
   c. morula by undergoing differentiation.
   d. pluteus by undergoing cleavage.
   e. gastrula by taking up fluids osmotically.

5. In a sea urchin embryo, mesoderm is derived from:
   a. ectoderm.
   b. endoderm.
   c. primary mesenchyme.
   d. the roof of the archenteron.
   e. the blastodisc.

6. The active factor within the polar plasm of a *Drosophila* embryo that induces cells to become future gametes seems to be _____ of maternal origin.
   a. DNA
   b. RNA
   c. structural proteins
   d. enzymes
   e. carbohydrates

7. In the frog egg, a massive reorganization of cytoplasm immediately following fertilization involves the:
   a. development of a gray, crescent-shaped cap over the entire vegetal pole.
   b. elevation of the fertilization membrane.
   c. movement of calcium ions into the polar regions of the cell.
   d. rotation of the pigment cap toward the point of sperm entry.
   e. movement of the yolk mass into the vegetal pole.

8. Which event is NOT part of gastrulation in a frog embryo?
   a. formation of the archenteron
   b. formation of a yolk plug
   c. migration of cells from the surface to the interior of the embryo
   d. formation of the neural tube
   e. establishment of the three primary tissue layers

9. The first visible sign of cellular differentiation in the frog gastrula is the:
   a. closure of the neural tube.
   b. formation of the blastopore.
   c. formation of the notochord.
   d. segmentation of the somites.
   e. creation of the coelom.

10. At the time that a fertilized chicken egg is laid, the blastodisc is about _____ millimeters in diameter and contains about _____ cells.
    a. 2; 100,000
    b. 4; 200,000
    c. 6; 300,000
    d. 8; 400,000
    e. 10; 500,000

11. The two extraembryonic membranes of the chick that are formed from ectoderm and mesoderm are the:
    a. yolk sac and amnion.
    b. amnion and allantois.
    c. allantois and yolk sac.
    d. chorion and allantois.
    e. amnion and chorion.

12. In the chick embryo, the function of the chorioallantoic membrane, which forms late in embryonic development from _____, is _____.
    a. ectoderm, mesoderm, and endoderm; storage of metabolic wastes and their elimination to the environment
    b. ectoderm, mesoderm, and endoderm; exchange of gases between the embryo and the external environment
    c. ectoderm and mesoderm; protection from mechanical and thermal shocks
    d. endoderm and mesoderm; storage of metabolic wastes
    e. ectoderm and endoderm; nutrient supply to the embryo

13. Which of the following is NOT a cell type that develops from neural crest cells?
    a. sensory neurons   d. vertebral column cells
    b. melanocytes   e. adrenal medulla cells
    c. Schwann cells

14. Which structure does NOT develop from epidermal ectoderm?
    a. ear primordia
    b. nostril primordia
    c. lens primordia
    d. skin
    e. optic vesicle

15. The eyes develop from interactions between what two entities?
    a. epidermal ectoderm and optic vesicles
    b. optic vesicles and neural tube cells
    c. epidermal ectoderm and lens primordium
    d. forebrain and optic vesicle
    e. optic cup and optic vesicle

16. Which structure is NOT of endodermal origin?
    a. gallbladder
    b. secretory cells of the liver
    c. ciliated lining of the lungs
    d. pancreas
    e. hepatic vein

17. Which statement is NOT part of the mechanism by which the neural tube is formed from the neural plate?
    a. Microtubules in neural plate cells are aligned parallel to the dorsal-ventral axis of the embryo.
    b. A band of actin filaments, which run parallel to the left-right axis of the embryo, contracts.
    c. The rate of cell division in cells at the periphery of the neural plate increases relative to that for cells in the center.
    d. Neural plate cells elongate.
    e. Neural plate cells become wedge-shaped.

18. The developmental process that results in the differences between homologous structures of different species is called:
    a. organogenesis.
    b. pattern formation.
    c. morphogenesis.
    d. embryonic induction.
    e. evolutionary progression.

19. In the developing chick wing, the region of actively dividing mesodermal cells is known as the:
    a. progress zone.
    b. apical ectodermal ridge.
    c. meristematic zone.
    d. pattern formation.
    e. mesodermal ridge.

20. The four-cell stage of the human embryo is created _____ hours after fertilization.
    a. 24
    b. 36
    c. 48
    d. 60
    e. 72

21. In the human, pregnancy is protected when _____ secreted by the _____ maintain(s) the endometrium and prevent(s) menstruation.
    a. gonadotropin-releasing hormone; trophoblast
    b. chorionic gonadotropin; chorion
    c. estrogens and progesterone; corpus luteum
    d. chorionic gonadotropin; corpus luteum
    e. estrogens and progesterone; trophoblast

22. In the human embryo, nitrogenous wastes:
    a. are transported to the maternal bloodstream as urea and ammonia.
    b. are transported as uric acid to the maternal bloodstream, where they are converted to urea and ammonia.
    c. are stored in the primitive bladder formed from the allantois.
    d. diffuse into the amniotic cavity, where they contribute to the osmolality of amniotic fluid.
    e. are recycled in the fetal liver to make amino acids.

23. The umbilical cord develops from the:
    a. chorion.
    b. amnion.
    c. allantois.
    d. yolk sac.
    e. yolk sac and amnion.

24. Which event does NOT usually occur in the human mother as the placenta develops?
    a. increase of blood volume
    b. production of antibodies against fetal cells
    c. increase of appetite
    d. increase in absorption of certain nutrients
    e. decrease in production of chorionic gonadotropin

25. In humans, most of the major organ systems begin to form during the _____ of pregnancy.
    a. second week
    b. third week
    c. fourth week
    d. second month
    e. third month

26. Which embryonic structure is NOT among those commonly affected if a woman contracts rubella (German measles) during the fourth to the twelfth weeks of pregnancy?
    a. heart
    b. eye
    c. liver
    d. ear
    e. brain

27. Reflexes first appear during the _____ month of human development.
    a. second
    b. third
    c. fourth
    d. fifth
    e. sixth

28. Movements of the fetus become quite obvious to the mother during the _____ month of pregnancy.
    a. third
    b. fourth
    c. fifth
    d. sixth
    e. seventh

29. The soft body hair called _____ develops on a fetus by the end of the _____ month.
    a. lanugo; fifth
    b. lanugo; fourth
    c. lanugo; third
    d. meconium; fifth
    e. meconium; fourth

30. During the last two months of pregnancy, the fetus increases in size by _____ percent.
   a. 50　　　　　　　　d. 150
   b. 75　　　　　　　　e. 200
   c. 100

31. A human fetus obtains maternal antibodies by _____ across the placenta.
   a. simple diffusion
   b. facilitated diffusion
   c. active transport
   d. bulk flow
   e. The premise is false; human infants obtain maternal antibodies only by nursing.

32. In humans, the date of birth is estimated to be _____ days after the onset of the last menstrual period.
   a. 245　　　　　　　　d. 280
   b. 258　　　　　　　　e. 295
   c. 266

33. The alertness of a newborn human infant is due to a surge of:
   a. dopamine and epinephrine.
   b. serotonin and dopamine.
   c. parasympathetic activity.
   d. acetylcholine and norepinephrine.
   e. adrenaline and noradrenaline.

34. T or F   If an eight-celled sea urchin embryo is divided such that the animal and vegetal poles reside in different halves, each half will continue development to produce an entire pluteus larva.

35. T or F   If sea urchin cells that have been separated by agitation in calcium-free seawater are not reunited, they will soon die, even if they are returned to normal sea water.

36. T or F   Differentiation results from the selective activation and inactivation of specific genes in the nucleus.

37. T or F   When any portion of an amphibian embryo is transplanted to another embryo of similar age, it develops strictly according to the site to which it is transplanted.

38. T or F   The chemical that mediates embryonic induction endows the target tissue with the capacity to differentiate into a specific tissue type.

39. T or F   The segmental pattern created by somites in the embryo is retained in a nearly unchanged form in all adult vertebrates.

40. T or F   In wing development, the specific sequence of events triggered in a cell by the action of retinoic acid apparently depends on the number of receptors activated, and therefore on the location of cells within the retinoic acid concentration gradient.

41. T or F   One advantage of chorionic villus biopsy over amniocentesis is that the biopsy can be taken earlier in (the eighth or ninth week of) the pregnancy.

42. T or F   In most pregnancies, a selective immunosuppression occurs so that the maternal immune system does not reject the chorionic component of the placenta.

43. T or F   The external genitalia of a human embryo become masculinized under the influence of a DNA-binding regulatory protein encoded by a gene on the Y chromosome.

44. T or F   The most significant factor in the neonatal survival of humans is weight at birth.

45. List the four major consequences of the fertilization of an egg.

46. Identify four features common to chick embryos and frog embryos that have completed gastrulation.

47. In *Drosophila* development, what types of functions might be performed by the proteins encoded by the homeotic genes?

48. Name the two tissues from which the placenta develops. Identify three functions of the placenta.

PERFORMANCE ANALYSIS

1. **d**   Sea urchin eggs are produced in large numbers and fertilization is external. The eggs and early embryos are transparent, but they must be viewed under a microscope. (page 920)

2. **e**   Sperm receptors are located in the vitelline envelope of a sea urchin egg. (page 920)

3. **a**   The vitelline envelope becomes the fertilization membrane when a sea urchin egg is fertilized. When a sperm contacts the egg, the cell membrane is depolarized by an influx of sodium ions into the cytoplasm. (This polarity reversal renders the egg unresponsive to the advances of other sperm.) In response to the sodium ion influx, calcium ions sequestered in the endoplasmic reticulum are released into the cytoplasm. This increase in free calcium ion concentration causes the vitelline membrane to lift off the surface of the egg. The vitelline membrane then develops into the fertilization membrane. (page 921)

4. **a**   As cells of the sea urchin morula divide, sodium ions are pumped into the extracellular spaces. Water follows osmotically, and a cavity, the blastocoel, is created. (page 921)

5. **c**   The mesoderm of a sea urchin embryo develops from the primary mesenchyme. (page 924)

6. **b**   Current evidence indicates that the cytoplasmic determinants within the polar plasm of a *Drosophila* embryo are RNA of maternal origin. (essay, page 925)

7. **d**   The reorganization of the frog egg cytoplasm that occurs immediately following fertilization involves the rotation of the pigment cap toward the point of sperm entry and the appearance of the gray crescent on the side opposite the point of sperm entry. (pages 926, 927)

8. **d** Gastrulation in the frog embryo involves the migration of cells from the surface to the interior of the embryo, forming the the archenteron and the three primary tissue layers. The yolk plug is also formed during gastrulation. (pages 927, 928, 929)

9. **c** The formation of the notochord is the first visible sign of cellular differentiation in the frog gastrula. (page 928)

10. **a** When a fertilized chicken egg is laid, the blastodisc is approximately 2 millimeters in diameter and contains about 100,000 cells. (pages 931, 932)

11. **e** In the chick embryo, each extraembryonic membrane is formed from two primary tissue layers. The amnion and chorion are both formed from ectoderm and mesoderm. The yolk sac and allantois develop from endoderm and mesoderm. (page 934)

12. **b** The chorioallantoic membrane, which develops from the chorion and the allantois (and consequently from all three primary tissue layers), is the respiratory organ of the chick embryo in the advanced stages of development. (page 934)

13. **d** All neurons, including the modified neurons of the adrenal medulla, develop from neural crest cells. Melanocytes and Schwann cells are also derived from neural crest cells. The cells of the vertebral column, which is first cartilage and later becomes bone, develop from mesoderm. (page 935)

14. **e** The optic vesicles develop from neural ectoderm that projects outward from the brain. They eventually develop into the retinas of the eyes, meaning that the retinas are (embryonically) extensions of the brain. (page 936, Figures 45–23, 45–24)

15. **a** The vertebrate eye develops during a series of inductions that begin with interactions between the epidermal ectoderm and the optic vesicles, which protrude from the neural ectoderm. (page 936, Figure 45–24)

16. **e** Endoderm gives rise to the inner linings of the digestive and respiratory tracts and to glands, including the liver, pancreas, and gallbladder. Blood vessels are of mesodermal origin. (page 937)

17. **c** The neural tube is formed from the neural plate as the cells at the periphery of the plate change shape. A change in the rate of cell division is not involved. (page 938, Figure 45–28)

18. **b** Pattern formation is the developmental process whereby the homologous structures of different species develop. That is, the forelimb bud of a horse embryo develops into a leg and the forelimb bud of a chick embryo develops into a wing. (page 938)

19. **a** The cells in the progress zone of a developing chick wing are actively dividing mesodermal cells. (page 939)

20. **d** In human development, the two-cell stage is reached 36 hours after fertilization, the four-cell stage 60 hours after fertilization, and the eight-cell stage by the third day. (page 940)

21. **c** The early human pregnancy is protected when the trophoblast arrives in the uterus and begins producing chorionic gonadotropin. This hormone stimulates the production of estrogens and progesterone by the corpus luteum, which prevents menstruation. (page 941)

22. **a** In mammalian embryos, nitrogenous wastes are transported to the maternal blood stream as urea and ammonia. The mother's kidneys then remove these wastes from the blood. (page 944)

23. **c** The umbilical cord characteristic of placental mammals develops from the allantois. (page 944)

24. **b** Even though the implanted chorion of the fetus is actually a graft of foreign tissue, a selective suppression of the mother's immune response normally prevents rejection of the placenta and consequently of the fetus. (page 944)

25. **b** In humans, most of the major organ systems begin to form during the third week of development and are laid down by the end of the first trimester. (page 945)

26. **c** If a woman who has not previously been exposed to rubella contracts the disease during the fourth to twelfth weeks of pregnancy, the developing embryo can be adversely affected. Organs commonly damaged include the heart, lens of the eye, inner ear, and brain. (page 947)

27. **b** Motor reflexes appear in a developing fetus by the end of the third month of development. (page 947)

28. **b** The fetus begins to move during the third month, but the mother may or may not be aware of these movements. Strong movements become obvious during the fourth month. (page 948)

29. **a** The soft body hair of the fetus is called lanugo and develops by the end of the fifth month. (page 948)

30. **c** The fetus doubles in size (an increase of 100 percent) during the last two months of pregnancy. (page 948)

31. **c** Maternal antibodies cross the placenta by selective active transport during the last month of pregnancy. A nursing infant also obtains antibodies in the milk. (page 948)

32. **d** In humans, the birth date is calculated to be 266 days after conception or 280 days after the onset of the last menstrual period. (page 950)

33. **e** The alertness of the newborn human infant, which is thought to be important in the formation of the infant-mother bond, is due to a surge of adrenaline and noradrenaline that occurs during birth. (Figure 45–42)

34. **False** In order for the normal development of two embryos to occur, each half of a separated eight-cell sea urchin embryo must contain cells of the animal pole and cells of the vegetal pole. (page 925)

35. **True**   If the cells of a sea urchin embryo have been separated by agitation in calcium-free seawater, they must be stirred gently to promote reaggregation when they are replaced in normal seawater if they are to survive and resume normal development. (page 926)

36. **True**   The process of differentiation involves the selective activation and deactivation of specific genes in cells. (page 930)

37. **False**   If cells from the dorsal lip of the blastopore of an amphibian embryo are transplanted into another embryo, the transplanted cells develop into the neural tube of a second embryo and organize the recipient cells to become the other tissues of a second embryo, which develops as a Siamese twin on the recipient. If cells of any other type are transplanted, they develop according to the site to which they are grafted. (page 930)

38. **False**   Every cell has the genetic potential to develop into its predestined tissue. The chemical mediators of embryonic induction merely activate a specific pattern of gene activity, which results in the differentiation of the cell. (page 931)

39. **False**   Although the segmental pattern created by somites is readily apparent in the embryos of all vertebrates, only the aquatic vertebrates retain this distinct segmental pattern in adulthood. (page 936)

40. **True**   During wing development, retinoic acid is released from the posterior wing margin and migrates toward the anterior wing margin, creating a concentration gradient. The development of the wing cells is influenced by the number of retinoic receptors activated, which is related to the location of the cells within the concentration gradient. (page 940)

41. **True**   Chorionic villus biopsy can be performed during the eighth or ninth week of pregnancy whereas amniocentesis cannot be successfully performed until the sixteenth week of pregnancy. If the test reveals a fetal condition that warrants an abortion for medical reasons, the procedure is much safer for the mother earlier in pregnancy than after four months of fetal development. (page 944)

42. **True**   Even though the fetal chorion is technically a graft of foreign tissue, the mother's immune system is selectively suppressed so that an immune reaction is not mounted against the fetus. (page 945)

43. **False**   Masculinization of the external genitalia is stimulated by testosterone produced by the developing testes. The development of the embryonic gonads into testes is directed by a gene on the Y chromosome, which codes for a DNA-binding regulatory protein. (page 946)

44. **True**   Weight at birth is the single most important factor in human infant mortality. Birth weight is greatly influenced by the nutritional level of the mother during pregnancy. (page 949)

45. (1) When the fertilizing sperm makes contact with the egg, changes occur in the cell membrane that render it unresponsive to other sperm. A series of reactions is initiated that results in the formation of the fertilization membrane from the vitelline membrane. (2) The nucleus of the sperm enters the egg and fuses with the egg nucleus, forming the diploid nucleus of the zygote. (3) The egg is activated metabolically. (4) The zygote begins to divide by mitosis. (pages 920, 921)

46. (1) Concentric circles of ectoderm, mesoderm, and endoderm surround the archenteron. (2) The notochord has differentiated. (3) The somites have formed. (4) The edges of the neural plate are elevated, the first step in the formation of the neural tube. (page 932)

47. In *Drosophila*, homeotic genes control the identity of the body segments. The proteins encoded by these genes probably regulate the expression of other genes or are involved in the regulation of basic cellular activities. (essay, page 942)

48. The placenta develops from the chorion of the embryo and the endometrium of the uterus. Its three functions are gas exchange between mother and fetus, transfer of fetal wastes to the maternal bloodstream, and supply of nutrients to the fetus. (pages 944, 945)

Section 6: Chapters 44–45

This review test is *not* designed to cover all of the important information in these chapters. However, it does touch on the major topic categories in each chapter. It will also give you valuable practice in taking this type of test. When you are finished, check your answers with those provided in the following PERFORMANCE ANALYSIS section.

1. Nutrients are supplied to the developing cells during spermatogenesis by:
   a. the prostate gland.
   b. interstitial cells.
   c. the bulbourethral gland.
   d. Sertoli cells.
   e. spermatogonia.

2. The vagina of the human female has a pH of _____ due to the presence of _____.
   a. 2 to 3; acetic acid
   b. 4 to 5; lactic acid
   c. 5 to 6; carbonic acid
   d. 6 to 7; an efficient buffer system
   e. 7 to 8; sodium bicarbonate

3. From the most internal to the most external, the structures of a sea urchin egg are:
   a. jelly layers; vitelline envelope; cytoplasm; cell membrane
   b. cell membrane; cytoplasm; jelly layers; vitelline envelope
   c. cytoplasm; jelly layers; cell membrane; vitelline envelope
   d. cytoplasm; cell membrane; vitelline envelope; jelly layers
   e. jelly layers; cytoplasm; vitelline envelope; cell membrane

4. "One wall of the blastula moves inward and forms a tubular extension toward the other wall." This statement describes gastrulation in the:
   a. human.
   b. fruit fly.
   c. chicken.
   d. frog.
   e. sea urchin.

5. The function of the amnion, which is composed of _____, is _____.
   a. mesoderm and endoderm; nutrient supply to the embryo
   b. endoderm and ectoderm; storage of metabolic wastes
   c. ectoderm and mesoderm; protection from mechanical and thermal shocks
   d. ectoderm and mesoderm; nutrient supply to the embryo
   e. mesoderm and endoderm; storage of metabolic wastes

6. Immediately after completion of meiosis I, _____ are present within a human ovarian follicle.
   a. a primary oocyte and one polar body
   b. a secondary oocyte and one polar body
   c. a secondary oocyte and two polar bodies
   d. an egg and three polar bodies
   e. Nothing is present within the follicle because ovulation occurs before meiosis I is completed.

7. Which statement is NOT true of estrogens?
   a. They stimulate growth of the endometrium.
   b. Low concentrations inhibit FSH synthesis.
   c. They stimulate the development of ovarian follicles.
   d. High concentrations stimulate GnRH secretion.
   e. High concentrations of progesterone in the presence of estrogens inhibit GnRH secretion.

8. The birth control pill that consists of progesterone and estrogen prevents conception by:
   a. inhibiting the production of LH and FSH by the pituitary.
   b. preventing implantation of a fertilized ovum in the uterus.
   c. creating a spermicidal environment in the female reproductive tract.
   d. making the endometrium an inhospitable environment for the young embryo.
   e. accelerating menstruation, thus "washing out" the new embryo if conception has occurred.

9. In the human embryo, the trophoblast is the precursor of the:
   a. chorion.
   b. amnion.
   c. allantois.
   d. yolk sac.
   e. amnion and allantois.

10. To date, the youngest fetus to survive outside the uterus had completed approximately _____ weeks of development.
    a. 20
    b. 23
    c. 26
    d. 29
    e. 31

11. T or F   If they are separated, each cell of a frog embryo at the two-cell stage may develop into a normal embryo only if each one receives cytoplasm containing part of the gray crescent.

12. T or F   Once a cell has become differentiated, its nucleus cannot support normal development if it is transplanted to an enucleated egg.

13. T or F   The first role of androgens in a human male is to bring about puberty beginning at approximately 10 years of age.

14. T or F   Embryonic gonads develop into ovaries in the absence of a gene coding for a DNA-binding regulatory protein, which is located on the Y chromosome.

15. "No organ system is derived from only one type of tissue." Cite at least one example that illustrates this statement.

16. Describe the mechanism by which the human penis becomes hard and enlarged during sexual arousal.

PERFORMANCE ANALYSIS

1. **d**   During spermatogenesis, the Sertoli cells of the testes supply nutrients to the developing spermatogenic cells. (page 902)

2. **b**   The bacteria that normally inhabit the vagina produce lactic acid, which keeps the pH of the vagina between 4 and 5. (page 909)

3. **d**   During fertilization, the enzymes of the sperm acrosome digest a path through the jelly layers, which are external to the vitelline envelope. The vitelline envelope, which surrounds and is attached to the cell membrane, contains species-specific receptors to which the fertilizing sperm binds. The sperm cell penetrates the vitelline envelope, the cell membranes of sperm and egg fuse, and the sperm nucleus enters the egg's cytoplasm. (page 920)

4. **e**   Gastrulation in the sea urchin embryo is analogous to pushing one wall of a partially inflated beach ball in with a fist. In this analogy, the fist ends up in the archenteron. (page 923)

5. **c**   The amnion is composed of ectoderm and mesoderm and encloses the amniotic cavity, a fluid-filled cavity that protects the developing embryo from thermal and mechanical shocks. (page 934)

6. **b**   Meiosis I is completed a few hours before ovulation, and produces the secondary oocyte and one polar body. (page 910)

7. **c**   Follicle development is stimulated by FSH, which is released from the pituitary under the influence of GnRH from the hypothalamus. (page 913)

8. **a**   The birth control pill containing estrogens and progesterone inhibits the production of LH and FSH, thereby preventing ovarian follicle maturation and ovulation. (page 915)

9. **a**   The trophoblast is the precursor of the chorion of the mammalian embryo. In the early embryo, it produces chorionic gonadotropin, which stimulates the corpus luteum to continue producing estrogens and progesterone, thus maintaining the pregnancy. (page 940)

10. **b**   The youngest surviving fetus on record had completed 23 weeks of development and required multiple life-support measures. (page 948)

11. **True**   The gray crescent of the frog embryo contains the material that develops into the organizer of the dorsal lip of the blastopore. This material is essential to the normal development of an embryo. (page 927)

12. **False**   Differentiation involves the selective activation and deactivation of genes. Depending on the type of cell and the stage of differentiation, the nucleus of a differentiated cell may direct normal embryonic development when transplanted into an enucleated egg. (page 930)

13. **False**   In early embryonic development, androgens produced by the developing testes stimulate development of the male external genitalia. (page 905)

14. **True**   A gene carried on the Y chromosome is essential for the development of embryonic gonads into testes. In the absence of this gene, the gonads develop into ovaries. (page 946)

15. In the intestine, the lining is derived from endoderm, the blood vessels and muscles in the wall are derived from mesoderm, and the nerves are derived from ectoderm. All organ systems are made up of two or all three embryonic tissues. (page 937)

16. The human penis contains three masses of spongy erectile tissue. When stimulated, the flow of blood into these tissues increases and they become distended. This distention constricts the veins, keeping the blood in the engorged tissues. (page 904)

PART **3**

Biology of Populations

SECTION **7**

Evolution

Evolution: Theory and Evidence

MAJOR CONCEPTS

Although he was not the first or the only scientist to propose a theory of evolution, Charles Darwin is given credit as the creator of the modern concept of evolution for two reasons. His text, *The Origin of Species*, provided copious, convincing evidence in favor of evolution, and he accurately deduced the mechanism by which evolution occurs, namely, natural selection.

Darwin's concept of evolution was based on five premises. (1) Organisms reproduce themselves accurately. (2) For most species, the number of individuals born far exceeds the number surviving to reproductive age. (3) Chance variations occur among individuals in a population and some of these variations are inheritable. (4) The survival and subsequent reproduction of an individual depend upon the interaction between the individual and the environment. Individuals that are better adapted to their environments will leave more surviving offspring than poorly adapted individuals. (5) Over time, natural selection produces an accumulation of changes in a population.

Microevolutionary events are small-scale evolutionary phenomena in which the process of natural selection can be observed. Examples include the peppered moth in industrial England, the resistance of insects to insecticides, and antibiotic resistance in bacteria.

Macroevolution refers to large-scale changes over time or "descent with modification." There are five categories of evidence for macroevolution: (1) the number and diversity of species, (2) the distribution of plants and animals across the globe (biogeography), (3) the fossil record, (4) homology of anatomical structures and biochemical pathways, and (5) the imperfection of adaptations, which are only as good as they must be.

Contemporary evolutionary theory is a synthesis of Darwin's concept of evolution by natural selection and the genetic principles discovered by Mendel and de Vries, and is expanded upon daily by evolutionary biologists around the world.

HOW TO STUDY THE CHAPTER

Read the entire chapter through quickly, focusing on the major concepts.

Use the GUIDED STUDY OF THE CHAPTER to help you identify the important details as you **reread** the chapter. Writing out the answers to these questions will help fix them in your mind as well as provide you with a valuable study aid.

Answer the questions in TESTING YOUR UNDERSTANDING without the aid of your text. Check your answers against those in PERFORMANCE ANALYSIS. Analyzing your answers will give you valuable feedback on your level of understanding and preparedness for classroom testing.

GUIDED STUDY OF THE CHAPTER

I. Introduction (page 961)

> *Focus:* Since the time of Darwin, Wallace, and Bates, evidence has been steadily accumulating that the earth has a long history and all living organisms have descended from more primitive life forms.

II. Darwin's Theory (pages 961–962)

1. Cite two reasons why Darwin has rightfully received credit for the theory of evolution.

2. What aspects of Darwin's nature contributed to the "intellectual avalanche" created by his evolutionary theory?

3. List the five premises upon which Darwin's concept of evolution was based.

4. *Vocabulary:* Define natural selection.

III. Evidence for Microevolution (pages 962–965)

A. General Remarks (page 962)

5. *Vocabulary:* What is microevolution?

6. What is one origin of selective pressures that have produced microevolutionary changes?

B. The Peppered Moth (pages 962–963)

7. a. Describe what happened to the peppered moth population in England as industrialization advanced.

   b. The phenomenon you just described is called _____.

   c. What was the origin of the black peppered moth?

8. a. State the hypothesis put forth by H. B. D. Kettlewell to explain the changes in the peppered moth population.

   b. How did he test his hypothesis (and prove it to be correct)?

> *Focus:* Neither color variation in peppered moths is absolutely superior to the other. The superiority of an alternative characteristic is relative to the circumstances, i.e., the selective pressures, involved.

9. What is happening to the peppered moth population in Britain now that industrial pollution is decreasing?

C. Insecticide Resistance *(page 964)*

10. Name two reasons why insecticides are no longer perceived to be miracle chemicals.

Focus: **An extreme example of adaptation and insecticide resistance is provided by a species that can detoxify DDT and use the resulting molecules as food.**

---

11. a. Scale insects that are resistant to the effects of hydrocyanic gas fumigation possess what capability relative to susceptible insects?

    b. How many genes are involved in the inheritance of this resistance?

D. Drug Resistance in Bacteria *(pages 964–965)*

12. a. Summarize the procedure by which the Lederbergs isolated drug-resistant bacteria.

    b. How did these drug-resistant bacteria arise in their cultures?

    c. Where are the genes for drug resistance carried in bacteria and what does this imply concerning the spread of drug resistance through a bacterial population?

IV. Evidence for Macroevolution *(pages 965–972)*

A. General Remarks *(page 965)*

13. *Vocabulary:* Distinguish macroevolution from microevolution.

14. List the five broad categories into which the evidence for microevolution can be divided.

B. The Number of Species *(page 965)*

Focus: **The observations of Darwin, Wallace, and Bates in their journeys revealed that species are not nearly as distinct as had previously been believed.**

---

15. a. What information presented the first serious challenge to the concept of special creation?

    b. Cite two arguments creationists used in trying to make this information fit their doctrine.

C. Biogeography *(page 966)*

16. *Vocabulary:* What is biogeography?

17. a. How did the creationists explain the adaptation of an organism to the environment in which it is found?

b. Cite four biogeographical observations that prompted Darwin to question this explanation of the creationists.

> *Focus:* Although biogeography cannot directly disprove special creation, it lends strong evidence for the historical connectedness of living organisms.

D. The Fossil Record *(pages 966–968)*

18. a. What features made the coasts of South America particularly interesting to Darwin?

    b. How did Darwin's geologic studies of the South American coast support his observations of living organisms in the area? (Cite one specific example.)

19. With respect to Darwin's concept of evolution, what information (a) was supplied and (b) was *not* supplied by the fossil record?

20. What impressive evidence in support of evolution was provided by O. C. Marsh?

21. Summarize T. H. Huxley's use of Marsh's evidence to support Darwin's theory.

> *Focus:* In recent decades, many groups of organisms have been identified for which the fossil record reveals a gradual change in anatomical characteristics and even indicates pathways of divergence from common ancestors.

E. *Essay:* The Record in the Rocks *(pages 970–971)*

22. By what processes are geologic strata formed, hidden, and revealed?

23. Name four ways in which strata may be distinguished from one another.

> *Focus:* The fossil record is built from evidence gained from identifiable strata in many different locations.

24. How did many geologic periods receive their names?

25. The early attempts to date geologic strata were performed on what basis?

26. a. The first scientific estimate of the age of the earth was made by:

b. On what did he base his calculation?

c. What information was he lacking that made his estimate incorrect?

27. a. What technique is most commonly used now to date geologic strata?

   b. The age of the earth is currently estimated to be

   _____.

   c. Dating strata with radioactive isotopes is based on what property of these isotopes?

   d. *Vocabulary:* Define the half-life of a radioactive isotope.

   e. When does a radiometric clock start to "tick"?

   Focus: **By measuring the proportion of ^{238}U to ^{206}Pb, the age of a rock can be estimated fairly accurately.**

   ---

   f. How many different isotopes are now used to date rocks?

28. Radiometric clocks are important to modern students of evolution for what two reasons?

F. **Homology** (*page 969*)

29. a. How does homology of structure and function provide evidence supporting evolution?

   b. Cite three examples of structural homology among vertebrates.

   c. Identify several examples of homology at the subcellular or biochemical level.

   d. Which molecular homology provides the strongest evidence of the evolutionary relatedness of living organisms?

G. **The Imperfection of Adaptation** (*pages 970–972*)

30. *Vocabulary:* List three biological definitions of the term adaptation.

   Focus: **The apparent perfection of adaptations provided the creationists with strong evidence against evolution (or so they believed).**

   ---

31. In actuality, adaptations are far from perfect. How does this fact provide evidence in favor of evolution?

32. What did Darwin deduce from (a) his studies of the ways in which flowers are pollinated and (b) his observations on defensive mimicry? (See also Figure 46–13.)

V. The Theory Today (pages 972–973)

> *Focus:* **All of modern biology affirms the relatedness of living organisms and the gradual divergence of the species from one another over the course of time.**

33. In current discussions of evolution, the central questions relate to what issue?

34. What three questions was Darwin unable to answer, that were solved as the science of genetics developed?

35. a. What does the synthetic theory of evolution comprise?

b. Name four scientists that contributed to the synthetic theory of evolution and identify the major scientific specialty of each scientist. (Figure 46–14)

36. a. Cite two reasons why some aspects of the synthetic theory of evolution have been challenged recently.

b. The current controversies revolve around what two features of evolutionary theory?

VI. Summary *(page 973):* Read the summary. If you are familiar with the essential features of the material presented there, you are ready to complete the section TESTING YOUR UNDERSTANDING.

TESTING YOUR UNDERSTANDING

After you have completed this examination, compare your answers with those in the section that follows.

1. Which statement does NOT represent a premise upon which the evolutionary concepts of Darwin and Wallace rested?
   a. Interactions between chance variations and the environment determine which individuals in a population will reproduce.
   b. For most species, the number of individuals that reproduce is small compared with the number produced in a generation.
   c. There is stability in the process of reproduction from one generation to the next.
   d. The interactions between organisms and their environment produce inheritable changes that accumulate over time.
   e. The process by which favorable variations survive and accumulate in populations is natural selection.

2. A major weakness in Darwin's evolutionary theory was that:
   a. plants were not considered during its development.
   b. it was a synthetic, or artificial, explanation of the evidence.
   c. the fossil record was so incomplete in his time.
   d. it did not account for the many marine invertebrates.
   e. it did not include the mechanism by which traits are inherited.

3. Hidden cameras placed by H. B. D. Kettlewell revealed:
   a. entomologists capturing his experimental moths for their collections.
   b. ornithologists capturing his experimental moths to use as bird food.
   c. birds eating primarily dark-colored moths.
   d. birds eating primarily light-colored moths.
   e. birds eating both light and dark moths in varying proportions, depending upon the location.

4. Which statement is NOT true of industrial melanism?
   a. It is a clear-cut case of natural selection at work.
   b. It is a rare and isolated phenomenon.
   c. Predation is the main selective force in the phenomenon.
   d. It illustrates the impact of human activities on other species.
   e. It demonstrates the speed at which evolution may occur.

5. Which of these statements does NOT apply to the red scale insects of southern California that are sensitive to hydrocyanic gas?
   a. High levels of gas do not kill them, but do render them sterile.
   b. Sensitivity is an inherited characteristic.
   c. The mutation rendering some insects resistant occurred in a single gene.
   d. Their spiracles can remain closed for only one minute.
   e. Mortality approaches 100 percent at gas concentrations that leave significant numbers of resistant insects alive.

6. In bacteria, genes for resistance to drugs are generally carried on:
   a. the host chromosome.
   b. viroids.
   c. oncogenes.
   d. plasmids.
   e. transposons.

7. "Descent with modification" is how Darwin described:
   a. homology.
   b. adaptation.
   c. natural selection.
   d. macroevolution.
   e. microevolution.

8. "Why are there placental rabbits in England, but no marsupial counterparts? Conversely, why are there marsupials resembling hares in Australia, but no native placental counterparts?" Which of the following does NOT provide clues to answer these questions?
   a. the fossil record
   b. homology
   c. biogeography
   d. the conservative nature of evolution
   e. the number and diversity of species

9. The coastline of South America was ideal for Darwin's geologic studies because:
   a. many different strata were exposed.
   b. fossilized horses were abundant.
   c. fossilized giant armadillos were abundant.
   d. he was particularly interested in fossils of marine invertebrates.
   e. South America had many fossilized marsupials similar to those he found in Australia.

10. The study of fossilized horses that strongly supported Darwin's theory of evolution was performed by:
    a. Darwin himself.
    b. Thomas Huxley.
    c. Othniel Marsh.
    d. William Smith.
    e. Alfred Wallace.

11. Periods within geologic eras are often named after:
    a. the scientists who discovered them.
    b. localities where they were first studied extensively.
    c. a series of Greek deities.
    d. the predominant type of rock in strata formed during the period.
    e. the type of fossils most abundant in deposits formed during the period.

12. Which situation is an example of physiological adaptation?
    a. legs that just reach the ground
    b. absence of eyes in cave-dwelling crickets
    c. increased numbers of red blood cells produced in response to life at high altitudes
    d. development of gills in a fish
    e. high visual acuity of the eyes of birds of prey

13. According to Darwin, most adaptations are:
    a. perfected structures or behaviors.
    b. random occurrences.
    c. only as good as they must be.
    d. not necessary for the survival of the organism.
    e. the result of sudden genetic changes.

14. T or F If Charles Darwin had never lived, someone else eventually would have convinced the scientific community of the reality of evolution.

15. T or F Darwin believed that, in some instances, the evolutionary process could be documented within a human lifetime.

16. T or F As far as Darwin was concerned, the fossil record provided copious evidence about the mechanisms by which evolution occurs.

17. T or F The cryptic structure of a walking-stick insect is a good example of an adaptation.

18. Identify the two controversies currently surrounding the synthetic theory of evolution.

19. Cite two reasons why the heavy use of insecticides has fallen into disfavor as a primary means of pest control.

20. In the experiments of Joshua and Esther Lederberg, why was it important that they kept the original culture containing colonies that had never been exposed to penicillin?

21. How did proponents of special creation explain the distribution of organisms?

PERFORMANCE ANALYSIS

1. **d** The evolutionary concepts of Darwin and Wallace were based on several premises. (1) There is stability in the process of reproduction. (2) In most species, the number of organisms surviving to reproduce is much less than the number initially produced. (3) There are chance variations among individuals in a population, some of which may be inherited. (4) Natural selection operates on populations. (5) Over time, natural selection leads to the accumulation of changes that differentiate groups of organisms from one another. (page 962)

2. **e** Darwin (and the rest of the scientific community at that time) was not aware of Mendel's principles of heredity or of de Vries's concept of mutations. The absence of a genetic mechanism for evolution was the major weakness in Darwin's theory. (page 972)

3. **e** The hidden cameras placed at Kettlewell's test sites revealed that birds do in fact prey on peppered moths and that cryptic coloration and associated behavior of the moths are critical to avoiding predation by birds. (page 963)

4. **b** Industrial melanism has been documented in many moth species both in England and in the United States. (page 963)

5. **a** Red scale insects are either killed by hydrocyanic gas (sensitive form) or survive apparently unharmed (resistant form). There is no evidence that the gas affects the ability of survivors to reproduce. (page 964)

6. **d** The genes for resistance to drugs are generally carried on plasmids. In some cases, such plasmids may be transferred among bacteria of different species. (page 965)

7. **d** Darwin described the process that is now called macroevolution as "descent with modification." (page 965)

8. **e** The issue of the number and diversity of species does not really address the distribution of placental rabbits in England and their marsupial counterparts in Australia. This phenomenon can be studied in terms of biogeography, the fossil record, homology, and the conservative nature of evolution, by which the evolution of organisms is limited to building on the materials available. (pages 966–972)

9. **a** The exposure of many different geologic strata along the coastline of South America made it an ideal site for Darwin's geologic studies. (pages 966, 967)

10. **c** Othniel C. Marsh studied the fossils of equines and published a genealogy of the horse, which strongly supported Darwin's theory of evolution. (page 968)

11. **b** In many cases, the periods within geologic eras are named after the locations in which they were initially studied most intensely. (essay, page 970)

12. **c** A physiological adaptation is an alteration of the physiology of an organism in response to a new environment. An example is the production of increased numbers of red blood cells when an individual changes residence to live at a higher elevation. See also the answer to question 17. (page 970)

13. **c** According to Darwin and many of his followers, adaptations are not perfect, but are only as good as they need to be to contribute to the survival and reproductive success of an organism. (page 971)

14. **True** There were enough people working on the theory of evolution that eventually it would have been recognized by the scientific community as a whole. However, it may not have been widely accepted until after the genetic studies of Mendel and de Vries became known. It was Darwin's vast personal knowledge, dogged intellectual pursuit of the subject, and well-versed argument that convinced his contemporaries of the validity of his theory. (page 961)

15. **False** Darwin did not think in terms of microevolutionary processes and believed that evolution only occurred over long periods of time. (page 962)

16. **False** Darwin reported that, for him, the fossil record revealed very little about the mechanisms by which evolution occurs. (page 967)

17. **True** There are three meanings for the term adaptation in biology. (1) Adaptation can refer to a process of becoming better suited to the environment. This can occur within the individual as a physiological adaptation. Conversely, evolutionary adaptation occurs within a population over many generations. (2) Adaptation can refer to the state of being adjusted to the environment. (3) Adaptation can refer to a particular characteristic that makes the organism better suited to the environment, such as the camouflaged structure of a walking-stick insect. (page 972)

18. The rate and tempo of macroevolutionary change and the role of chance in determining the course of evolution are the two current controversies surrounding the synthetic theory of evolution. The general concept of evolution and the basic tenets of the synthetic theory are widely accepted by modern scientists and are not under debate. (page 973)

19. (1) Many insecticides persist in the environment and harm or kill organisms that were not the intended target. (2) Persistent use of the same type of chemical selects for resistant insect varieties, which are difficult to control once they become the predominant pests. (page 964)

20. The original colonies had never been exposed to penicillin, but one of those colonies was in fact resistant to penicillin. This demonstrated that penicillin-resistance was due to a genetic mutation that occurred by chance rather than one that was induced by exposure to the drug. (pages 964, 965)

21. The champions of special creation accounted for the distribution of organisms by contending that each organism was created for a particular way of life and was placed in the environment for which it was best suited. (page 965)

The Genetic Basis of Evolution

CHAPTER ORGANIZATION

MAJOR CONCEPTS

Population genetics is the branch of biology in which Darwin's theory of evolution was finally reconciled with Mendel's principles of heredity. The focus of population genetics is on the frequencies of alleles within populations. A population geneticist defines evolution in terms of changes in allele frequencies within a population over time and perceives an individual organism as containing a fraction of the gene pool. In population genetics, fitness is defined in terms of the relative number of surviving offspring left by an individual.

In general, natural populations exhibit a great degree of genetic variation. Variations arise by mutation and are the raw material upon which natural selection acts. Genetic variability has long been demonstrated by animal breeders and horticulturists using artificial selection. More recently, genetic variation has been quantified by studying the structural differences of one or more specific proteins present in a population. These differences in structure, as revealed by electrophoresis, represent differences in the genes coding for the proteins. With the advances in molecular genetics, genetic variation is being studied in terms of differences in DNA structure. The importance of minute changes in DNA structure to the fitness of an organism is currently being debated.

Working independently, Hardy and Weinberg demonstrated that, in a ideal population in which certain restrictions apply, the frequency of alleles and genotypes remains constant from one generation to the next. The five restrictions are: (1) no mutations occur; (2) there is no net movement of individuals into or out of the population; (3) the population is large enough for the laws of probability to apply; (4) mating is random; and (5) all alleles are equally viable. The Hardy-Weinberg equation that mathematically describes this equilibrium for a gene with only two alleles is $p^2 + 2pq + q^2 = 1$, where p is the frequency of the dominant allele and q is the frequency of the recessive allele. Although no natural population conforms to the restrictions of Hardy-Weinberg equilibrium, the concept is useful in tracking changes in allele frequencies over several generations and provides a technique for quantifying evolutionary changes.

Five major forces change gene frequencies in populations. Natural selection is the major factor. Mutations produce new variations, thereby providing raw material upon which natural selection can act. Gene flow refers to the movement of organisms and/or alleles into and out of populations. Genetic drift occurs when a population fragments or rapidly declines in size and the gene pool of the fragment (or survivors) is different from that of the parent population. Several mechanisms exist (particularly in animals) that contribute to nonrandom mating.

Genetic variability is preserved in populations by sexual reproduction, mechanisms that promote outbreeding, diploidy, and heterozygote superiority. Duplicate genes play a crucial role in allowing the accumulation of mutations and thus evolutionary change.

HOW TO STUDY THE CHAPTER

Read the entire chapter through quickly, focusing on the major concepts.

Use the GUIDED STUDY OF THE CHAPTER to help you identify the important details as you **reread** the chapter. Writing out the answers to these questions will help fix them in your mind as well as provide you with a valuable study aid.

Answer the questions in TESTING YOUR UNDERSTANDING without the aid of your text. Check your answers against those in PERFORMANCE ANALYSIS. Analyzing your answers will give you valuable feedback on your level of understanding and preparedness for classroom testing.

GUIDED STUDY OF THE CHAPTER

I. Introduction (page 974)

> *Focus:* The early geneticists did not accept Darwin's theory of evolution by gradual change partly because they were working with specimens that did not exhibit the great range of variations that Darwin observed during his travels.

1. What did the early geneticists consider to be the chief agent of evolution?

2. What change in attitudes led to the reconciliation of Mendelian genetics with Darwinian evolution?

3. a. *Vocabulary:* What is population genetics?

   b. *Vocabulary:* Cite a population geneticist's definition of a population.

4. *Vocabulary:* What is a population's gene pool?

5. a. How does a population geneticist perceive an individual organism?

   b. How would a population geneticist define evolution?

   c. State the definition of fitness that applies to population genetics.

II. *Essay:* Survival of the Fittest (page 975)

6. a. What was "social Darwinism" and how did this philosophy originate?

   b. To what degree does this philosophy accurately reflect evolution?

III. The Extent of Variation (pages 975–978)

A. General Remarks (page 975)

> *Focus:* The continuity of the reproductive process results from the accurate replication of DNA and its transfer to daughter cells during meiosis and mitosis.

7. Identify the raw materials upon which evolutionary forces act.

8. a. How did creationists perceive variations?

b. How did Darwin's view of variations differ from that of the creationists?

Focus: In modern population genetics, emphasis has been placed on determining the extent of variation in populations and the mechanisms by which these variations are preserved in gene pools.

B. Breeding Experiments (pages 976–977)

9. a. Vocabulary: Define artificial selection.

b. How does artificial selection parallel natural selection?

10. What did the pigeon breeders of Darwin's time demonstrate about the extent of variation?

Bristle Number in Drosophila (pages 976–977)

11. Summarize the initial results (obtained after about 25 generations) of the breeding experiments in which Drosophila were selected for high and low bristle number.

Focus: The bristle-number breeding experiments demonstrated that wide genetic variability existed in the population at the beginning of the experiment.

12. a. What happened to the low-bristle-number line?

b. How was the high-bristle-number line handled differently and what were the results?

c. Account for the results obtained in (b).

13. a. What is a major problem in virtually all cases of intense inbreeding?

b. What is the genetic explanation for this problem?

Focus: During natural selection, the entire phenotype is selected, not just one or a few specific characteristics.

C. Quantifying Variability (page 978)

14. a. Studying proteins can provide insights into the degree of genetic variability within a population because:

b. Describe the technique by which proteins are separated during electrophoresis. (Figure 47–6)

15. a. What did Hubby and Lewontin discover in their studies of Drosophila enzymes?

b. Summarize their conclusions.

c. State the estimated degree of genetic variation in the *Drosophila* population at large and the degree of heterozygosity within individual fruit flies.

16. a. Studies in humans have revealed what level of heterozygosity within individuals?

b. What fraction of genes in a given human population have two or more alleles?

17. Why is DNA sequencing expected to reveal a greater extent of genetic variation than that demonstrated by protein electrophoresis?

D. **Explaining the Extent of Variation** *(page 978)*

18. Why did some geneticists expect that populations would be genetically almost uniform?

19. a. Summarize the role of variation in fitness as viewed by the selectionists.

b. How does the view of the neutralists differ?

c. What evidence supports the neutralists' perspective?

IV. **A Steady State: The Hardy-Weinberg Equilibrium** *(pages 979–981)*

A. **General Remarks** *(page 979)*

20. In the early 1900s, what question arose about the existence of dominant and recessive alleles in a population?

> *Focus:* Hardy and Weinberg demonstrated that the genetic recombination that occurs in diploid organisms as a consequence of sexual reproduction does not by itself change the overall composition of the gene pool.

21. a. List the five conditions of an idealized Hardy-Weinberg population.

b. Under the conditions listed in (a), what will happen to the allele frequencies and the genotype frequencies of a specific trait over several generations?

22. a. Define each element in the Hardy-Weinberg equation:

p^2

$2pq$

q^2

b. *p* plus *q* must always equal _____.

B. Derivation of the Hardy-Weinberg Equation (pages 979–980)

> **Focus:** The allele frequency is the proportion of one allele relative to its complementary allele in a population.

23. For a gene with two alleles, if the frequency of one allele is known, how may the frequency of the other allele be calculated?

24. In using the Hardy-Weinberg equation to estimate allele frequencies in offspring, what two additional assumptions are being made? (Figure 47–8)

25. a. In a certain population, the particular gene of interest has two alleles (*B* and *b*). The frequency of allele *b* is 0.3. Use the Hardy-Weinberg equation to calculate these frequencies: allele *B* and genotypes *BB*, *Bb*, and *bb*.

    b. Assuming that the population meets the Hardy-Weinberg restrictions, calculate these frequencies in the next generation: allele *B*, allele *b*, and genotypes *BB*, *Bb*, and *bb*.

    c. Would a population geneticist say that evolution has occurred? Explain your answer.

> **Focus:** In a population that meets the five restrictions of the Hardy-Weinberg equilibrium, neither allele frequencies nor genotype frequencies change from one generation to the next.

C. The Effect of Multiple Alleles (page 980)

26. Write the Hardy-Weinberg equation for a gene that has three alleles.

D. The Significance of the Hardy-Weinberg Equilibrium (page 981)

27. Since no natural populations meet the restrictions of Hardy-Weinberg equilibrium, of what value is the Hardy-Weinberg equation?

28. In order to determine genotype frequencies in a population, what information is required?

An Application of the Hardy-Weinberg Equation (page 981)

29. Describe how the Hardy-Weinberg equation can be used to trace changing gene frequencies in a population.

V. The Agents of Change (pages 981–984)

A. General Remarks (page 981)

30. a. List five agents that cause changes in gene frequency within a population.

    b. Which is the major force acting in natural populations?

B. Mutations (pages 981-982)

31. a. *Vocabulary:* State the definition of mutations according to population genetics.

b. Identify four mechanisms by which mutations may occur.

c. In what types of genes can mutations occur?

d. Name four agents known to cause mutations.

e. How do most mutations arise?

32. "Mutations are generally said to occur at random, or by chance." Explain this statement.

Focus: **Although environment may influence the *rate* at which mutations occur, the particular mutation produced is not affected by the environment.**

---

33. a. What is the "hopeful monster" theory of mutations? (Figure 47-10)

b. Why has this concept been abandoned? (Figure 47-10)

Focus: **Different genes, and different alleles of a particular gene, have different rates of mutation.**

---

34. a. Identify the range for the spontaneous mutation rate producing mutations detectable in the phenotype.

b. Why do the rates of mutation vary among genes and alleles?

c. Each human is estimated to carry _____ new mutations.

35. What is the role of mutations in evolution?

C. Gene Flow (page 982)

36. a. *Vocabulary:* Define gene flow.

b. State two mechanisms by which gene flow may occur.

c. Identify two results of gene flow.

37. What is the overall effect of gene flow and how does this effect differ from that of natural selection?

38. How can gene flow be prevented?

D. Genetic Drift (pages 982-983)

39. The Hardy-Weinberg equation only holds true if the population is large. Explain why this is true.

40. *Vocabulary:* Define genetic drift.

The Founder Effect (page 983)

41. a. *Vocabulary:* What is the founder effect?

b. Describe one example.

Population Bottleneck (page 983)

42. a. *Vocabulary:* What is a population bottleneck?

    b. What types of events may cause a population bottleneck? (See if you can think of some not listed in the text.)

43. Cite the evidence indicating that a loss of genetic variability occurred in elephant seals off the coast of California.

44. What two effects can a population bottleneck have on alleles in a gene pool?

E. **Nonrandom Mating** *(page 984)*

45. Cite examples of mechanisms by which nonrandom mating may be accomplished in animals and plants.

46. *Vocabulary:* Define polymorphism.

Focus: **Nonrandom mating may result in changes in genotype frequencies *without* accompanying changes in allele frequencies.**

---

47. How could nonrandom mating cause changes in allele frequencies?

VI. **Preservation and Promotion of Variability** *(pages 984–988)*

A. **Sexual Reproduction** *(page 984)*

48. Name three means by which sexual reproduction produces new genetic combinations.

49. Explain why asexual reproduction fails to produce new genetic combinations.

50. What is the only biological advantage of sexual reproduction?

B. *Essay:* **Why Sex?** *(pages 986–987)*

51. Identify three disadvantages of sexual reproduction.

52. a. How does each of the following hypotheses account for the advantages of sexual reproduction?

    "Best man" hypothesis

    "Tangled bank" hypothesis

    "Red Queen" hypothesis

    b. What evidence indicates that the best man hypothesis is probably incorrect, but is consistent with the other two hypotheses?

C. **Mechanisms That Promote Outbreeding** *(pages 984–985)*

53. What is meant by outbreeding?

54. a. Name four mechanisms in plants that promote outbreeding.

b. Describe how self-sterility alleles in plants promote outbreeding. (Include how the frequencies of these alleles fluctuate.)

c. How does the self-sterility system promote variability in a population?

55. Hermaphrodites rarely fertilize their own eggs. What, then, is the advantage of hermaphroditism? (See also Figure 47-15.)

56. a. Cite one example in which a behavioral strategy promotes outbreeding among animals.

b. List three possible reasons for the taboos against incest that are prevalent in human societies.

D. **Diploidy** (*pages 986–987*)

57. How does diploidy contribute to the maintenance of genetic variability?

Focus: **As the frequency of a recessive allele decreases in a population, the proportion of the allele existing in the homozygous state decreases and removal of the allele by natural selection slows down.**

58. a. What is eugenics?

b. Setting aside any ethical considerations, why is the elimination of an "undesirable" recessive allele from a population an almost insurmountable task?

Note: In a planned breeding program, such as those involving domestic animals and plants, the efficient removal of an undesirable recessive allele from the population depends on identification of heterozygotes and their removal from the breeding population.

E. **Heterozygote Superiority** (*pages 987–988*)

59. *Vocabulary:* Define heterozygote superiority.

Sickle Cell Anemia (*page 988*)

60. a. What percentage of some African peoples are heterozygotes for the sickling allele?

b. What two factors favor preservation of the sickling allele?

Heterosis, or Hybrid Vigor (*page 988*)

61. a. *Vocabulary:* What is heterosis?

b. Heterosis is a mechanism by which _____ are maintained in populations.

c. Hybrid vigor apparently results from what phenomenon?

d. State two possible reasons for the superiority of hybrid organisms.

VII. **The Origin of Genetic Variation** (*pages 988–989*)

62. Summarize the mechanism by which genetic variation is thought to have originated.

VIII. **Summary** (*pages 989–990*): Read the summary. If you are familiar with the essential features of the material presented there, you are ready to complete the section TESTING YOUR UNDERSTANDING.

After you have completed this examination, compare your answers with those in the section that follows.

1. There were several reasons why some of the early geneticists did not accept the concept of gradual change proposed by Darwin. Which of the following is NOT such a reason?
   a. There was not much variation in the laboratory organisms they studied.
   b. They believed most mutations to be harmful.
   c. They believed mutation, which resulted in rapid change, to be the agent of evolution.
   d. They believed in special creation.
   e. Occasionally, useful mutations would rapidly change the evolutionary course of a population.

2. A major problem with intense artificial selection, such as that in the experiments involving abdominal bristle number in *Drosophila*, is:
   a. a decrease in body size.
   b. loss of the selected trait.
   c. increased rate of mutation.
   d. increased mortality of juveniles.
   e. reduced fertility.

3. Which of the following is NOT a technique by which the latent genetic variability of a population may be assessed?
   a. breeding studies involving intense selection
   b. protein electrophoresis
   c. amino acid sequencing of proteins
   d. nucleotide sequencing of DNA
   e. chromosome mapping

4. In human populations, it is estimated that individuals are heterozygous for at least _____ percent of their genes.
   a. 7
   b. 12
   c. 18
   d. 25
   e. 30

5. Why is DNA sequencing expected to result in a higher estimate of genetic variability than protein electrophoresis?
   a. Not every amino acid difference represents a functional change in a protein molecule.
   b. Some amino acid variations are not detectable by electrophoresis.
   c. DNA sequencing is an older, more established technique.
   d. More scientists know how to do DNA sequencing than electrophoresis.
   e. DNA is more easily damaged than are amino acids.

6. Which terms used in the Hardy-Weinberg equation represent the allele frequencies?
   a. p^2 and q^2
   b. $2pq$
   c. p and q
   d. p^2 and $2pq$
   e. $2pq$ and q^2

7. In a population that is in Hardy-Weinberg equilibrium, if the frequency of the A allele is 0.75 and the frequency of the a allele is 0.60, then the sum of p and q will be:
   a. 1.35
   b. 0.65
   c. 0.40
   d. 0.25
   e. These frequencies cannot exist in the population described.

8. In a population in Hardy-Weinberg equilibrium, if the genotype frequencies are 81 percent AA, 18 percent Aa, and 1 percent aa, what are the frequencies of the A and a alleles, respectively?
   a. 0.81; 0.01
   b. 0.18; 0.18
   c. 0.9; 0.1
   d. 0.09; 0.01
   e. 0.01; 0.09

9. Which of these expressions is a form of the Hardy-Weinberg equation that accounts for the presence of three alleles for a single gene?
   a. $(p + q)^2 = 1$
   b. $(p + q)^3 = 1$
   c. $(p + q + r)^2 = 1$
   d. $(p + q + r)^3 = 1$
   e. $p^3 + 3pq + q^3 = 1$

10. Tay-Sachs disease is inherited as a Mendelian recessive and occurs in Ashkenazic Jews at a frequency of 1 per 3,600 births. Estimate the percentage of this population that are heterozygote carriers of the Tay-Sachs allele.
    a. 3.3 percent
    b. 1.7 percent
    c. 0.983 percent
    d. 0.06 percent
    e. 0.033 percent

11. It is estimated that each human carries about _____ new mutations.
    a. 2
    b. 4
    c. 6
    d. 8
    e. 10

12. The overall effect of gene flow is to:
    a. create population bottlenecks.
    b. increase the efficiency of natural selection.
    c. create sexual barriers between populations.
    d. decrease the differences between populations.
    e. accelerate the process of speciation.

13. Genetic drift has the largest effect on allele frequencies when:
    a. mating is random.
    b. populations are small.
    c. gene flow is prevented.
    d. allele frequencies are equal.
    e. natural selection is prevented.

14. Which phenomenon may produce changes in the genotype frequencies without necessarily affecting the allele frequencies of a population?
a. mutation
b. gene flow
c. genetic drift
d. population bottleneck
e. nonrandom mating

15. Suppose that the squirrels in the park near the Grand Canyon experienced an epidemic disease that killed 90 percent of them. What is a probable consequence of this event?
a. The survivors would exhibit an increase in resistance to all diseases.
b. The genetic variability in the population would decrease.
c. The mutation rate in genes for disease resistance would increase.
d. The preepidemic gene frequencies would be reestablished by genetic drift.
e. The survivors would migrate to a new, disease-free environment.

16. The phenomenon in which the heterozygous genotype is favored over either homozygote is known as:
a. heterosis.
b. polymorphism.
c. hybrid vigor.
d. heterozygote superiority.
e. heterogamy.

17. T or F   Pigeon fanciers, who produced a number of exotic breeds by means of artificial selection for specific traits, demonstrated a large amount of variability in the gene pool of the ancestral wild pigeons.

18. T or F   Variability at some gene loci is so great that it is possible to find five or more alleles for some enzymes in a natural population.

19. T or F   Because of natural selection, many more harmful mutations are produced than beneficial mutations.

20. T or F   From an evolutionary viewpoint, one of the greatest problems with asexual reproduction is that genetic variability is limited.

21. T or F   Among African hunting dogs, outbreeding is ensured when the young males leave their families of origin and mate with females from other family groups.

22. T or F   As the frequency of a recessive allele increases in a population, the proportion of the allele found in heterozygotes also increases.

23. How would a population geneticist define a population of organisms? Cite one example that fits this definition.

24. Summarize the five conditions that must characterize an idealized population in order for Hardy-Weinberg equilibrium to exist.

25. Briefly describe two mechanisms by which genetic drift influences allele frequencies in populations. Cite one example of each mechanism.

26. List three possible reasons for the strong prohibitions against incest present in most human cultures.

27. Compare and contrast the "best man" and "tangled bank" hypotheses as explanations for the persistence of sexual reproduction.

PERFORMANCE ANALYSIS

1. d   Early geneticists did not see a great deal of variation in the laboratory organisms they studied. They saw beneficial mutations causing rapid, rather than gradual, change and believed that mutations were the major force of evolution. They also believed that most mutations were harmful. (page 974)

2. e   Intense artificial selection (especially if that selection involves intense inbreeding) often results in a decrease of fertility or overall reproductive success, such as that experienced in the studies selecting for high and low bristle number in *Drosophila*. (page 977)

3. e   Chromosome mapping merely determines the location of genes on chromosomes, but it does not directly provide information about the degree of genetic variability in a population. (pages 977, 978)

4. a   Studies of human tissues have led to the estimate that individuals are heterozygous for at least 7 percent of their genes. Further, 25 percent of the genes in a human population are represented by two or more alleles. (page 978)

5. b   DNA sequencing is expected to demonstrate higher levels of genetic variability because some changes in the DNA sequence do not alter the amino acid sequence and because some amino acid variations are not detectable by electrophoresis. (page 978)

6. c   In the Hardy-Weinberg equation, the frequency of the dominant allele is represented by p and the frequency of the recessive allele by q. (page 979)

7. e   In any population, the frequencies of all alleles for a single gene *must* add up to 1.0. The situation described cannot exist. (page 979)

8. c   The frequency of the dominant genotype (p^2) is 0.81. Therefore, $p = 0.9$ and $q = 0.1$. (page 980)

9. c   The form of the Hardy-Weinberg equation that takes into account three alleles for a gene is $(p + q + r)^2 = 1$, where r represents the frequency of the third allele. (page 980)

10. a   The frequency of Tay-Sachs disease (q^2) is $1/3{,}600$ or 0.0002778. Therefore, $q = 0.0167$, $p = 0.09833$, and $2pq$ (the frequency of heterozygote carriers) = 0.0328 or 3.3 percent. (page 980)

11. a   It has been estimated that each human carries two new mutations in his or her gametes. (page 982)

12. **d** Gene flow is the movement of genes between two populations. The overall effect is to reduce the differences in the gene pools of the two populations. (page 982)

13. **b** Genetic drift is a change in the gene pool that results from chance. Its impact is greatest when populations are small. In such cases, allele frequencies may change dramatically by the addition or loss of one or a few individuals. (page 983)

14. **e** Nonrandom mating may result in changes in genotype frequencies without affecting allele frequencies in a population. An example is the preferential mating of blue snow geese with other blue snow geese and of white snow geese with other white snow geese. (page 984)

15. **b** Any time a population experiences a "crash" (a sudden, dramatic loss of a large proportion of individuals), a population bottleneck occurs and the result is a loss of genetic variability. (page 983)

16. **d** Heterozygote superiority is the phenomenon in which the survival and reproductive success of the heterozygote is greater than in either homozygote. (page 987)

17. **True** The exotic breeds of pigeons, all of which were developed from one species of wild pigeon, were produced by artificial selection for different characteristics. This demonstrated a vast latent genetic variation in the original breeding stock. (The different breeds of domestic dogs were produced in a similar fashion.) (page 976)

18. **True** In natural populations, five or more alleles may be present at a single gene locus. (page 978)

19. **False** Mutations occur at random and arise independently of their resulting effects. Most mutations are probably harmless and undetectable except by DNA sequencing, protein electrophoresis, or amino acid sequencing techniques. (page 982)

20. **True** Asexual reproduction is more efficient than sexual reproduction in terms of time and resources expended. However, the potential for new genetic combinations is extremely limited compared with the potential realized in sexual reproduction. (page 984)

21. **False** Among African hunting dogs, the young females leave the family of origin. Young males leave the family of origin in lions, gorillas, baboons, and many other species. (page 985)

22. **False** As the frequency of a recessive allele *decreases* in a population, the proportion of the allele found in heterozygotes increases. This is why the attempted elimination of recessive traits from a population (or breeding program) is so difficult; the recessive allele can "hide" in the heterozygous state. (page 986, Table 47–1)

23. A population geneticist defines a population as an interbreeding group of organisms of the same species, which is localized in space and time. An example would be all the fish of one species in a pond. (page 974)

24. (1) There must be no mutations. (2) There is no net movement of individuals into or out of the population. (3) The population is large enough that chance alone is not likely to alter allele frequencies. (4) Random mating occurs. (5) All alleles are equally viable and the offspring of all possible matings are equally likely to survive to reproduce. (page 979)

25. The **founder effect** can result in altered allele frequencies if the allele frequencies in the founding population differ from those in the parent population. An example is the Old Order Amish of Lancaster, Pennsylvania, which experience a very high incidence of an allele for dwarfism and polydactylism that is extremely rare in the rest of the human population. In a **population bottleneck,** the population size is dramatically reduced in a fairly short period of time, and the allele frequencies in the survivors are different from those of the original population. The population that becomes established from the survivors will probably have less genetic variability than the original population. An example is the northern elephant seal population along the coast of California and the Baja peninsula, which was hunted nearly to extinction before being protected by law. (page 983)

26. (1) Strong psychological taboos may prohibit incestuous behavior in some cultures. (2) Early human societies may have observed the adverse effects that result when inbreeding brings together harmful recessive alleles. (3) Avoidance of mating with relatives may be a genetically determined behavioral mechanism similar to those observed in other mammals. (page 985)

27. The **"best man" hypothesis** argues that the environment is very changeable and offspring are likely to live in an environment that is very different from that of the parents. Sexual reproduction provides a way of producing offspring different from their parents and therefore more likely to survive the changing conditions. The **"tangled bank" hypothesis** emphasizes the diversity of the environment, which provides a variety of different but limited opportunities to be exploited. Sexual reproduction produces offspring capable of utilizing different sets of resources, thereby reducing competition among closely related individuals. The "best man" hypothesis is almost certainly incorrect, for reasons established by Graham Bell of McGill University. (essay, pages 986, 987)

CHAPTER 48

Natural Selection

CHAPTER ORGANIZATION

MAJOR CONCEPTS

A population geneticist defines natural selection as the differential rate of reproduction of different genotypes in a population. This results from interactions between individuals and their environment. Natural selection maintains genetic variability in populations; there are numerous different selective forces, each of which may favor the survival of a slightly different genotype.

Balanced polymorphism is one example of the genetic variability maintained by natural selection. It is the coexistence within a population of more than one phenotype in fairly stable proportions. Examples include human blood types (the selective value of which is under debate) and shell color and banding patterns in *Cepaea* land snails.

Natural selection acts upon the entire phenotype of an organism. Most phenotypic characteristics result from the expression of several different genes. Coadaptive gene complexes are groups of genes that work together to produce a coordinated phenotypic characteristic. For a particular phenotypic characteristic, there may be more than one genotypic route.

There are three types of natural selection based on the effect of selection on the distribution of characteristics within a population. Stabilizing selection tends to increase uniformity in a population. Disruptive selection maintains two or more distinct genotypes in a population. Directional selection results in the gradual replacement of one phenotype by another. None of these three types is affected by the frequency at which the various phenotypes occur in a population.

Two other types of natural selection are frequency-dependent selection and sexual selection. In frequency-dependent selection, the least common phenotypes are favored and gradually become more common. As their proportion increases, these formerly rare phenotypes consequently become less favored. Sexual selection occurs in two forms. Intrasexual selection is the competition between members of the same sex for opportunities to mate. In intersexual selection, members of one sex exert strong selective pressures on the opposite sex through their criteria for mate selection. In some species, one consequence of sexual selection is sexual dimorphism, in which males and females have markedly different phenotypic characteristics.

Natural selection results in adaptations that can be correlated with selective pressures of the environment, including those pressures imposed by other organisms. A cline is a graded variation in a trait (such as body size) over a wide geographic distribution. It may be correlated with such environmental factors as temperature and humidity. Ecotypes are distinct phenotypes of a species that result from the occupation of different habitats. At least some of the phenotypic variations among ecotypes are due to genetic variations.

Coevolution occurs when two or more populations interact and exert strong selective pressures on each other. Examples include flowers and their pollinators, milkweed plants and monarch butterflies, and mimicry among insects.

There are two basic types of mimicry among insects. In Müllerian mimicry, each individual possesses a noxious characteristic (bad taste to predators, stingers, etc.) and all individuals have a similar warning coloration. Examples are bees, wasps, and hornets. Batesian mimics have coloration similar to species that possess a noxious characteristic, but are themselves harmless. An example is monarch butterflies (the noxious model) and viceroy butterflies (the innocuous mimic). Müllerian mimicry benefits all individuals, since the experience of a predator with any individual results in association of the color pattern with the noxious stimulus. Batesian mimicry benefits only the mimic. The model is doubly at a disadvantage because it experiences attacks by inexperienced predators and by predators that were first exposed to the innocuous mimic.

Evolutionary progress builds on the past and is limited by structural and developmental constraints within a population. Thus, complex processes (i.e., metabolic pathways and patterns of embryonic development) are accomplished by well-established and ancient mechanisms. This conservatism of evolution is evidence of the relatedness of living organisms.

Evolutionary progress is not always measured in increasing complexity and may result in the loss of certain capacities. Examples are wingless insects and cave-dwelling arthropods that lack eyes. Further, natural selection can only act on the phenotypic characteristics already present in a population. Thus, adaptations are rarely perfect but are only as good as they must be to promote survival and reproductive success.

In addition to coevolution, there are two other patterns of evolution produced by natural selection. Convergent evolution occurs when organisms that are not closely related look similar because they have evolved under similar selection processes. Divergent evolution occurs when a fraction of a population becomes separated and is subjected to selection pressures different from those acting on the parent population. Ecotypes and/or new species may arise in this fashion.

HOW TO STUDY THE CHAPTER

Read the entire chapter through quickly, focusing on the major concepts.

Use the GUIDED STUDY OF THE CHAPTER to help you identify the important details as you **reread** the chapter. Writing out the answers to these questions will help fix them in your mind as well as provide you with a valuable study aid.

Answer the questions in TESTING YOUR UNDERSTANDING without the aid of your text. Check your answers against those in PERFORMANCE ANALYSIS. Analyzing your answers will give you valuable feedback on your level of understanding and preparedness for classroom testing.

GUIDED STUDY OF THE CHAPTER

I. Introduction *(page 991)*

1. a. Darwin developed his concept of natural selection after reading what publication? (See also pages 7–8.)

   b. How did Darwin define natural selection?

2. a. *Vocabulary:* Redefine natural selection in terms of population genetics.

   b. Differential reproductive success is the result of what phenomenon?

   c. Differential reproductive success results in:

II. Natural Selection and the Maintenance of Variability *(pages 991–993)*

A. General Remarks *(pages 991–992)*

Focus: Early in the development of the synthetic theory, some biologists argued that natural selection would reduce genetic variation in a population. Modern population genetics has proved this position to be incorrect.

B. Balanced Polymorphism: Color and Banding in Snail Shells *(pages 992–993)*

3. *Vocabulary:* Distinguish balanced polymorphism from transient polymorphism and cite one example of each.

4. According to the fossil record, balanced polymorphism has existed in snails for how long?

5. In studies of *Cepaea* snails, how did researchers determine what types of snails were preyed upon in different habitats?

6. Characterize the predation of snails (banded, unbanded, yellow, dark brown) in environments with the following backgrounds.

Uniform

Mottled

Green

Dark

7. What is surprising about the findings of both banded and unbanded snails living in a uniform habitat?

8. What factors other than predation apparently contribute to the maintenance of genetic variation in *Cepaea* snail populations?

Focus: As occurs in most natural systems, several selection pressures operate to maintain shell color variations in *Cepaea* snails.

C. *Essay:* Human Blood Groups: A Puzzle *(page 996)*

Focus: Some biologists believe that the various human blood types are maintained by natural selection. Others argue that they have no selective value.

---

9. a. Summarize the correlation between blood type and life expectancy in humans.

b. List the health risks associated with blood types A and O.

c. Do these conditions act as selective forces? Explain your answer.

10. What proposal, which is currently unsubstantiated, has been made concerning blood types and the possible selective forces that may, at least in the past, have maintained this polymorphism?

11. What factors may account for the geographic distribution of blood types?

III. **What Is Selected?** *(pages 993–994)*

Focus: During natural selection, the entire phenotype of an organism is selected.

---

12. Define phenotype in terms of evolutionary theory.

Focus: Since most phenotypic characteristics result from the interaction of several genes, a particular phenotype may be obtained by a number of different genotypic routes.

---

13. How is this Focus statement illustrated by the selection experiments for bristle number in *Drosophila*?

14. In addition to coloration, name three characteristics important to peppered moths in avoiding predation.

15. *Vocabulary:* Distinguish between coadaptive gene complexes and supergenes.

16. How may supergenes be protected in *Drosophila*?

17. a. Identify two factors that determine an organism's phenotype.

b. How does the IQ test illustrate the interaction between genotype and environment in determining the phenotype?

c. What does an IQ test predict?

> *Focus:* **The phenotype and genotype of sexually reproducing organisms are transitory. Only the individual genes survive through the generations.**

IV. Types of Selection *(pages 994–1001)*

A. General Remarks *(pages 994–995)*

18. a. List the three types of natural selection based on the effect of selection on the distribution of characteristics within a population.

b. Identify two other types of natural selection.

B. Stabilizing Selection *(page 995)*

19. a. *Vocabulary:* Define stabilizing selection and cite one example.

b. What is the optimal clutch size, i.e., that giving rise to the greatest number of offspring, in Swiss starlings?

c. Name three factors involved in determining clutch size in birds.

C. Disruptive Selection *(pages 996–997)*

20. *Vocabulary:* Distinguish disruptive selection from stabilizing selection.

21. Describe two examples of disruptive selection in nature, focusing on the selective pressures that maintain the preferred phenotypes.

22. Disruptive selection may lead to the formation of:

23. In the coho salmon, what two other types of natural selection are also at work?

D. Directional Selection *(page 997)*

24. What effect does directional selection have on a population?

25. Cite three examples of directional selection.

> *Focus:* **Stabilizing selection increases uniformity within a population. Directional selection and disruptive selection produce adaptive change.**

E. Frequency-Dependent Selection *(pages 997–998)*

26. *Vocabulary:* Define frequency-dependent selection.

27. Describe how frequency-dependent selection occurs in predator-prey interactions. Note how polymorphism is maintained.

28. Cite two other examples of frequency-dependent selection.

F. **Sexual Selection** *(pages 998–1001)*

29. a. *Vocabulary:* What is sexual selection?

    b. *Vocabulary:* Distinguish between and cite examples of intrasexual selection and intersexual selection.

30. Name three different components of sexual reproduction for which energy must be allocated.

Focus: **Energy allocation in sexual reproduction directly affects the nature of the mate selection process.**

---

31. *Vocabulary:* Distinguish among polygynous, monogamous, and polyandrous species based on the strategy of each gender and the relative investment of each gender in parental care.

Focus: **Polygyny, monogamy, and polyandry are points along a continuum describing the relative energy investment of each gender in the basic components of sexual reproduction.**

---

32. What circumstances can affect the type of mating system used by a particular species? Identify examples.

33. a. *Vocabulary:* Define sexual dimorphism and identify the presumed chief cause of this phenomenon.

    b. Sexual dimorphism is most apparent among species exhibiting which mating strategy?

    c. What is the adaptive advantage of the relatively dull coloration of the female of many species?

    d. What could be considered a maladaptive aspect of males with brilliant plumage?

34. How do the differences in appearance of males and females vary in (a) monogamous and (b) polyandrous species?

35. a. Why did Darwin categorize sexual selection as a force separate from natural selection?

    b. Why is sexual selection now considered by many scientists to be one of the mechanisms by which natural selection is accomplished?

G. *Essay:* **Male Ornamentation: The Role of Female Preference** *(pages 1000–1001)*

36. a. State two conflicting hypotheses concerning the role of male ornamentation in natural selection.

b. The evidence of what study supports Darwin's hypothesis?

37. How is mate selection accomplished in the long-tailed widowbirds of Kenya?

38. a. Describe Andersson's experimental technique to test the role of female preference in the evolution of male ornamentation.

   b. An experimental control was created to ensure that the techniques used (in this case, cutting and gluing tail feathers) did not influence the test results. What was the control group for this study?

39. a. Summarize the results of this study.

   b. What conclusions can be drawn?

Note: The tail feathers of the male widowbirds would be restored to the pre-test length in the course of the normal molting process.

V. The Result of Natural Selection: Adaptation
(pages 1001–1005)

A. General Remarks *(pages 1001–1002)*

Focus: Natural selection results in adaptations that often can be correlated with selective pressures imposed by the environment or other organisms.

B. Adaptation to the Physical Environment: Clines and Ecotypes *(pages 1002–1003)*

40. a. *Vocabulary:* What is a cline?

   b. List three characteristics that may exhibit clinal variation, and use an example to describe the variations for one of these.

41. *Vocabulary:* Distinguish an ecotype from a cline.

42. a. What were Clausen, Keck, and Hiesey attempting to determine by their experiments on *Potentilla glandulosa*?

   b. Summarize their methods, results, and conclusions.
   Methods

   Results

   Conclusions

Focus: The emergence of ecotypes is often a first step in the formation of new species.

C. Adaptation to the Biological Environment: Coevolution *(pages 1003–1005)*

43. *Vocabulary:* Define coevolution.

44. What is one of the most important examples of coevolution?

Milkweed, Monarchs, and Mimics (pages 1003–1005)

45. a. What feature of milkweed plants discourages predation?

b. How have monarch butterflies circumvented this defense mechanism?

c. What is the origin of the noxious substance present in monarch butterflies that discourages predators from eating them?

d. How does the presence of noxious chemicals in the monarch butterfly benefit the viceroy butterfly? (Figure 48–13)

46. Animal species that use poisonous or distasteful chemicals in their defensive strategies often have what feature in common?

47. *Vocabulary:* Fill in the following table to distinguish Batesian mimicry from Müllerian mimicry. (See also Figure 48–15.)

Table for Question 47

	MÜLLERIAN	BATESIAN
Model has a noxious trait (yes or no)		
Mimic has a noxious trait (yes or no)		
Who benefits? (model or mimic or both)		

48. Outline the experiments of Jane Brower that confirmed the selective value of Batesian mimicry.

49. a. In what way does the model suffer from the success of its Batesian mimic?

b. Name two conditions under which the advantages of Batesian mimicry are maximized.

VI. Evolution and the Idea of Progress *(pages 1005–1006)*

A. General Remarks *(pages 1005–1006)*

> *Focus:* Natural selection tends to push populations toward better solutions to the problems of survival, but it does not necessarily result in *individuals* that are better adapted to their *immediate* environment.

50. Explain this statement: "... populations ... are often a generation behind."

B. Developmental and Structural Constraints *(page 1006)*

> *Focus:* Natural selection does not create adaptations de novo, it must work with the genotypes and phenotypes already present in a population.

51. Name two human discomforts that are consequences of using a skeleton designed for quadrupeds as the basis for an animal with an upright posture.

52. What is Stephen Jay Gould's attitude toward imperfect adaptations?

53. How does embryonic development exemplify the conservative nature of evolution?

C. Eyeless Arthropods and Other Degenerates *(page 1006)*

> *Focus:* Evolutionary progress is not defined by a continual increase in complexity and may result in the loss of capacities.

54. a. Cite three examples in which organisms have lost a characteristic or ability.

   b. By what mechanism does this loss occur?

   c. In terms of evolutionary progress, are such organisms more or less advanced than their counterparts that have retained these capacities? Explain your answer.

VII. Patterns of Evolution *(pages 1007–1008)*

A. General Remarks *(page 1007)*

> *Focus:* **Three broad patterns of evolution (coevolution, convergent evolution, and divergent evolution) may be produced by natural selection.**

B. Convergent Evolution *(page 1007)*

55. a. Some organisms that are only distantly related may appear quite similar. Explain this phenomenon, including the conditions under which it occurs.

   b. This phenomenon is called:

   c. Cite two examples.

C. Divergent Evolution *(page 1008)*

56. a. How does divergent evolution occur?

   b. Identify one example.

   c. Name two possible results of divergent evolution.

VIII. Summary *(pages 1008–1009)*: Read the summary. If you are familiar with the essential features of the material presented there, you are ready to complete the section TESTING YOUR UNDERSTANDING.

TESTING YOUR UNDERSTANDING

After you have completed this examination, compare your answers with those in the section that follows.

1. Which statement is NOT true of studies of the forces maintaining genetic diversity in *Cepaea* land snails?
   a. The proportions of shell types found at thrush anvils were compared with those of nearby snail colonies.
   b. Unbanded snails are more common in habitats with uniform backgrounds.
   c. The highest proportion of yellow shells exists in the greenest environments.
   d. Physiological factors correlated with shell patterns maintain highly visible phenotypes in some populations.
   e. Banded snails dominate the population in rocky habitats.

2. Which statement is NOT true of the human A, B, AB, and O blood groups?
   a. The presence of these blood groups in other primates indicates that they are part of the human evolutionary legacy.
   b. The life expectancy of Caucasian males is greatest for those with type O blood.
   c. The B allele is not found in Native Americans and Australian aborigines who have not mixed with Europeans.
   d. Persons with type A blood are at a higher risk of developing pernicious anemia.
   e. The selective value of these blood groups is illustrated by the correlation between susceptibility to certain infectious diseases and specific blood types.

3. A coadaptive gene complex is a group of genes:
   a. on a single chromosome that promote the survival of an organism.
   b. coding for a characteristic that favors the reproductive success of an organism.
   c. coding for traits that impair the survival ability of an organism.
   d. that work together to produce coordinated phenotypic traits.
   e. coding for the production of hormones that enhance reproductive success.

4. Which of the following is NOT an example of sexual dimorphism?
   a. brilliant plumage of peacocks
   b. antlers of male deer and elk
   c. two distinct sizes of male salmon

d. larger body size and canines of male baboons
e. long tails of male long-tailed widowbirds

5. Which statement is NOT true of a cline?
   a. It is a graded variation in a trait or group of traits exhibited by members of one species located in different geographical regions.
   b. North-south clines are common in animal species.
   c. The size range of extremities within some species of North American mammals exemplifies a cline.
   d. It is a distinct phenotype characteristic of a species living in a particular habitat.
   e. Clines may be established fairly rapidly in terms of evolutionary time.

6. Which pair does NOT constitute an example of coevolution?
   a. milkweed plants and mimics of the monarch butterfly
   b. monarch butterflies and their mimics
   c. flowers and their pollinators
   d. lynx and their favorite prey, snowshoe hares
   e. milkweed plants and monarch butterflies

7. The conservative nature of evolution is MOST readily apparent in the:
   a. embryonic development of chordates.
   b. forelimb variations of vertebrates.
   c. evolution of eyes in the animal kingdom.
   d. evolution of sound receptors in the animal kingdom.
   e. chemical nature of light-sensitive pigments characteristic of plants and animals.

8. When natural selection produces very similar phenotypes in distantly related organisms, _____ is said to have occurred.
   a. coevolution
   b. convergent evolution
   c. divergent evolution
   d. disruptive selection
   e. stabilizing selection

9. Divergent evolution occurs because two populations:
   a. experience similar selection pressures.
   b. become separated and follow different evolutionary routes.
   c. developed from different ancestral populations.
   d. exert significant selective pressures on each other.
   e. occupied the same habitat and their adaptations diverged, thereby reducing competition.

10. T or F   According to Darwin, organisms that survive the process of natural selection are innately superior to other members of their species.

11. T or F   By eliminating "less fit" individuals from a population, natural selection reduces genetic variability.

12. T or F   In the course of evolution, the phenotype is a transient entity; only the individual genes survive from one generation to the next.

13. T or F   In Swiss starlings, as clutch size increases beyond the optimum number, the number of surviving young decreases apparently because of lack of space in the nest; the smaller nestlings are pushed out of the nest by larger siblings.

14. T or F   In a population, directional selection is likely to result in the replacement of one allele (or group of alleles) with another allele (or group of alleles).

15. T or F   In the predation of corixid bugs by fish, the most common color form experiences the greatest predation, regardless of its degree of camouflage relative to the other forms.

16. T or F   The maintenance of variation in a plant population by means of self-sterility alleles is a form of sexual selection.

17. T or F   Polygynous species are those in which the females attempt to maximize the number of mates while the males are quite choosy about the females with which they mate.

18. T or F   The formation of ecotypes is one mechanism by which new species may be established.

19. T or F   Müllerian mimicry is adaptive for all individuals involved.

20. T or F   All Batesian mimics have a noxious characteristic.

21. T or F   Natural selection results in populations that are better adapted to the immediate environment.

22. T or F   The loss of certain abilities by some organisms reflects a regression to a lower evolutionary state.

23. Match each type of selection with the correct example. An answer may be used more than once. A blank may have more than one answer, as indicated in parentheses.

   1. directional   4. sexual
   2. disruptive   5. stabilizing
   3. frequency-dependent

   a. _____ Male coho salmon that fertilize eggs are typically small two-year-olds or large three-year-olds. (3 answers)

   b. _____ Self-sterility alleles in plants maintain genetic variation. (2 answers)

   c. _____ A breeder of miniature horses selects the smallest male he can find to breed to his mares.

   d. _____ A breeder of beagles selects only those individuals between 13 and 15 inches tall, although larger and smaller individuals exist.

   e. _____ Two male elephant seals fight for control of a harem.

24. Define and distinguish between transient polymorphism and balanced polymorphism. Cite one example of each phenomenon.

25. In the study of coho salmon conducted by Mart R. Gross, what factors promoted the persistence in the population of small two-year-old males and large three-year-old males?

26. What hypothesis was Malte Andersson testing by studying long-tailed widowbirds? What features of the experimental design enabled Andersson to obtain conclusive results?

27. Identify the two conditions under which Batesian mimics derive the greatest advantage. In this type of mimicry, the model suffers in what two ways?

PERFORMANCE ANALYSIS

1. **e** *Cepaea* snails are preyed upon by song thrushes, which select the most highly visible snails. Snails with banded shells are most visible in uniform habitats, such as rocks, and would be found in lower numbers in such habitats than snails with unbanded shells. (page 992)

2. **e** It has been suggested, but *not* substantiated, that the selective value of the human blood groups is correlated with susceptibility or resistance to certain infectious diseases. (essay, page 996)

3. **d** By definition, a coadaptive gene complex is a group of genes that work together to produce coordinated phenotypic traits. If such groups exist on a single chromosome, they are called supergenes. (page 993)

4. **c** Sexual dimorphism refers to the differences between males and females that are related to obtaining a mate rather than to the reproductive process itself. Such differences result largely from sexual selection. Although the two distinct sizes of male salmon also result from sexual selection, they do not exemplify sexual dimorphism. (Differences between male and female salmon that are related to obtaining a mate would represent sexual dimorphism.) (pages 997, 1000)

5. **d** Ecotypes are distinct phenotypes characteristic of a species living in a variety of different habitats. Examples are the ecotypes of *Potentilla glandulosa* that live at different altitudes. (page 1002)

6. **a** Coevolution occurs when two populations exert strong selective pressures on each other over the course of evolution. Examples include predators and their prey (including herbivores and the plants they eat), flowers and their pollinators, hosts and their pathogens (or parasites), and mimics and their models. Two specific examples are milkweed plants and monarch butterflies, which eat the plants, and monarch butterflies and their mimics, which benefit from a predator's previous experience eating a monarch butterfly. Since the mimics of monarchs and the milkweed plants do not exert selective pressure directly on each other, they do not exemplify coevolution. (pages 1003, 1004)

7. **a** Numerous examples abound concerning the conservative nature of evolution. It is most readily apparent in the embryonic development of chordates, which follows predetermined pathways. (page 1006)

8. **b** Convergent evolution is said to have occurred when unrelated or distantly related organisms possess similar phenotypic characteristics because they have evolved under similar selective pressures. (page 1007)

9. **b** Divergent evolution occurs when one population becomes separated from all other populations of the species, is subjected to different selective pressures, and follows a separate evolutionary route. It is one mechanism by which new species may arise. (page 1008)

10. **False** Darwin did not claim that the individuals that survived the process of natural selection were innately superior to other members of their species. Rather, he contended that they were better adapted to their environment—or more accurately, to the environment of their parents. (page 991)

11. **False** Natural selection acts to maintain and enhance genetic variability in populations. (pages 991, 992)

12. **True** The individual organism, and its phenotype, are but transient vessels carrying the genes that persist through the generations. (page 994)

13. **False** In Swiss starling clutches larger than the ideal size for the species, the number of surviving young is lower than that for a clutch of optimum size largely because the parents can adequately nourish only a limited number of young. (page 995)

14. **True** Directional selection is the selection for one extreme of a range of traits. Over time, the allele(s) for the extreme being favored replaces the allele(s) for the other phenotypic extreme. (page 997)

15. **True** Corixid bugs, which are preyed upon by fish, exist in three different color variations that represent varying degrees of camouflage. When the proportions of the three forms are unequal, the most common color variation is preyed upon most, regardless of its cryptic value. This is an example of frequency-dependent selection. (Figure 48–8)

16. **True** One aspect of sexual selection is mate selection. The statement is true in the sense that self-sterility alleles determine pollination and fertilization (i.e., mate selection) in plants. (page 998)

17. **False** Polygynous species are those in which a few males sire most of the young and the females invest more heavily in care of the young. Polyandrous species are those in which a female has more than one mate, the males typically have only one mate, and the males perform most or all of the parental care. (page 999)

18. **True** The formation of ecotypes, which is an example of divergent evolution, is one mechanism by which new species may be established. (page 1003)

19. **True** In Müllerian mimicry, all the mimics have a noxious characteristic and each individual acts as a model for all species of similar appearance. Models and mimics alike benefit from the mimicry. Conversely, Batesian mimics do not have a noxious characteristic. Such mimics benefit from the experience of predators with the model species. However, the model suffers from inexperienced predators and from predators whose first experience was with an innocuous mimic. (page 1004)

20. **False** See the answer to question 19. (page 1004)

21. **False** In changing environmental conditions, natural populations are often one generation behind. Because of natural selection, young are produced by the organisms best adapted to the environment at the time of mating and conception. Therefore, the young of each generation are adapted to the environment of their parents, which may or may not be their immediate environment. (page 1006)

22. **False** Many highly evolved animals are characterized by the loss of an ability to perform a task. Examples include flightless birds and blind cave fish. (page 1006)

23. The answers are **a2, a3, a4, b3, b4, c1, d5,** and **e4.** (pages 995–998)

24. In transient polymorphism, one phenotype typically replaces the other phenotype in the population. An example is industrial melanism in peppered moths. Balanced polymorphism refers to the persistence of different phenotypes in a population over time. The shell color variations of *Cepaea* land snails is an example. (page 992)

25. The small two-year-old males can hide among rocks and debris in stream bottoms, or swim through shallow areas to get close enough to fertilize a nest. The smaller individuals are less likely to be discovered by larger males. The large three-year-old males are at an advantage in fighting other males for the right to fertilize a nest of eggs. (page 997)

26. Andersson was testing Darwin's hypothesis that female choice exerts selective pressure favoring male ornamentation. Several features of the experimental design enabled conclusive results to be obtained. (1) The study was conducted in a natural environment. (2) The territories of the males contained similar numbers of females before the study began. (3) Controls were created—the tails of half the birds with tails of normal length were cut and glued. (essay, pages 1000, 1001)

27. A Batesian mimic derives the greatest advantage if it appears in the environment after the model and if it is less numerous than the model. The model suffers from inexperienced predators that have not learned to avoid potential prey items with its appearance and from predators whose first experience was with the harmless mimic. (pages 1004, 1005)

REVIEW TEST 14

Section 7: Chapters 46–48

This review test is *not* designed to cover all of the important information in these chapters. However, it does touch on the major topic categories in each chapter. It will also give you valuable practice in taking this type of test. When you are finished, check your answers with those provided in the following PERFORMANCE ANALYSIS section.

1. What was the origin of the dark form of the peppered moth, *Biston betularia*?
   a. It arose as a spontaneous mutation whose survival was favored in the polluted environment.
   b. It was introduced from Pittsburgh, Pennsylvania.
   c. It was blown into London from the western coast of England.
   d. It was blown into London from the southeastern coast of England.
   e. It was due to a mutation induced by mutagenic air pollutants.

2. Darwin found evidence of a gradual change over time in one line of organisms, which he believed was the pattern of evolution, in the fossil record of:
   a. horses.
   b. giant armadillos.
   c. shells of marine invertebrates.
   d. Australian marsupials.
   e. Darwin did not find fossil evidence of a gradual change over time in a single lineage of organisms.

3. Most mutations:
   a. occur at the same rate, about one in 1 million gametes.
   b. occur randomly and are independent of their subsequent effects.
   c. are induced by ultraviolet light from the sun.
   d. are caused by harmful environmental chemicals.
   e. occur as a result of exposure to medical or dental x-rays.

4. Which statement is NOT true of the phenotype of an organism?
   a. The phenotype is determined in part by the genotype.
   b. The phenotype is determined in part by the interaction of the genotype with the environment.

   c. Natural selection acts on the whole phenotype.
   d. For any given phenotype, there may be several corresponding genotypes.
   e. All of these statements are true of the phenotype.

5. Which of the following is probably NOT a factor involved in determining clutch size in birds?
   a. availability of calcium for the egg shells
   b. size of the nest built by the parents
   c. length of time the female will mate
   d. synthesis of proteins for the yolk and albumen
   e. genetic predisposition to lay a certain number of eggs

6. The combination of Darwin's concept of evolution with Mendel's principles of genetics gave rise to which branch of biology?
   a. ecology
   b. ethology
   c. population genetics
   d. social Darwinism
   e. evolutionary genetics

7. In a population at Hardy-Weinberg equilibrium, if 16 percent of the population has blue eyes (*bb*), what percentage of the population would be expected to be heterozygous (and thus have brown eyes)?
   a. 1 percent
   b. 4 percent
   c. 6 percent
   d. 24 percent
   e. 48 percent

8. Which organism does NOT exemplify a highly advanced state of evolution?
   a. a cave-dwelling fish that has no eyes
   b. an ostrich, which is a flightless bird
   c. a digestive tract parasite that has no internal digestive system
   d. human immunodeficiency virus, which causes a disease that is invariably fatal
   e. a flea, which has no wings

9. T or F The result of natural selection is that individual organisms become better adapted to the environment.

10. **T** or **F** Although Darwin was unable to answer the question of why there are so many species, the modern synthetic theory of evolution has provided an adequate explanation.

11. **T** or **F** Within a plant population, rare self-sterility alleles are more likely to be involved in flower pollination than are common ones.

12. Match the scientists with the correct area of investigation. Use each answer only once. Each blank will have only one answer.

   1. H. B. D. Kettlewell
   2. Joshua and Esther Lederberg
   3. Lord Kelvin
   4. Othniel C. Marsh
   5. Charles Darwin

   a. _____ drug resistance in bacteria
   b. _____ first widely accepted theory of evolution
   c. _____ natural selection in peppered moths
   d. _____ fossil record of equine evolution
   e. _____ age of the earth

13. List at least five molecular or biochemical characteristics that make clear the unity of all living organisms.

14. Compare and contrast the views of Darwin and Linnaeus with respect to the role of variation within a species.

PERFORMANCE ANALYSIS

1. **a** The dark variety of the peppered moth originated as a spontaneous mutation whose survival was favored in the polluted environments of industrial England. (page 963)

2. **e** Although evidence of gradual change over time in one line of organisms has been found by modern paleontologists, Darwin did not find this evidence. (page 967)

3. **b** Many agents are known to increase the rate of mutations, but most mutations occur randomly and their occurrence is independent of their subsequent effects. (page 982)

4. **e** The phenotype of an organism is determined by the genotype and by the interaction of the genotype with the environment. Natural selection acts on the entire phenotype, not just on one or a few characteristics, as occurs in artificial selection. As exemplified in the *Drosophila* breeding experiments selecting for high bristle number, a particular phenotype may have more than one corresponding genotype. (page 993)

5. **b** Clutch size in birds is influenced by several factors, including the length of time the female will mate, the synthesis of proteins for the yolk and albumen, the availabil-

ity of calcium for the egg shells, and the genetic predisposition to lay a certain number of eggs. (page 995)

6. **c** The field of population genetics grew out of a merging of Darwin's theory of evolution and Mendelian principles of genetics. (page 974)

7. **e** The frequency of blue eyes (the homozygous recessive phenotype represented by the term q^2) is 0.16. Therefore, $q = 0.4$, $p = 0.6$, and the frequency of the heterozygous phenotype ($2pq$) is 0.48, or 48 percent. (page 979)

8. **d** The loss of a superfluous ability over the course of evolution illustrates an advanced evolutionary state. In the coevolution of hosts and parasites (including pathogens), a parasite often becomes less virulent to its host. In highly evolved host-parasite relationships, the two organisms coexist and the host may be only mildly affected by the presence of the parasite if it is affected at all. A parasite that is highly virulent typically indicates that the history of the parasite-host relationship is of a fairly short duration. (page 1006)

9. **False** The overall effect of natural selection is that *populations* become better adapted to the environment. However, this improvement may be one generation behind because environmental conditions may change from one generation to the next. In constantly changing environments, each generation is adapted to the environment in which its parents lived. (page 1006)

10. **False** The reason for the existence of so many species is a question that remains unanswered by the synthetic theory of evolution. (page 965)

11. **True** In a plant population, the rare self-sterility alleles are more likely to encounter flowers and gametes containing different self-sterility alleles than are the more common self-sterility alleles. Therefore, pollen grains containing the rare alleles are more likely to pollinate flowers than are pollen grains with the more common alleles. (page 985)

12. The answers are **a2, b5, c1, d4,** and **e3.** (pages 961, 963, 964, 968, and essay, page 970)

13. Molecular or biochemical characteristics that provide evidence for the unity of all living organisms include the two-ply nature of cell membranes, the internal organization of eukaryotic cilia and flagella, glycolysis, the role of ribosomes in protein synthesis, the electron transport chain, the many roles of ATP, and the genetic code. (page 969)

14. Darwin perceived the variations within a population of organisms to be the raw material upon which evolutionary forces act. To Darwin, there was no ideal type for a given species. Linnaeus and other creationists believed that an ideal type exists for each species and that variations from this ideal represent imperfections. (page 975)

C H A P T E R **49**

On the Origin of Species

MAJOR CONCEPTS

Speciation has occurred when two diverging populations become reproductively isolated from one another and cannot interbreed to produce fertile offspring. Two basic mechanisms of speciation are allopatric and sympatric. Allopatric speciation occurs when two populations are separated by spatial boundaries that prevent contact between members of the populations. Sympatric speciation is the evolution of a new species in the presence of the parent population. One mechanism by which sympatric speciation occurs is the appearance of polyploid plants. Sympatric speciation resulting from disruptive selection is a controversial issue because it is difficult to prove.

Once a species has evolved, several mechanisms maintain genetic isolation from all other species. Premating isolating mechanisms prevent the mating of members of two species and may be behavioral, chemical, or temporal in nature. Postmating isolating mechanisms generally prevent either fertilization or the development of a fertilized egg. An exception is the production of sterile offspring, which are a reproductive dead end.

The finches of the Galapagos Islands provide an exquisite example of speciation in response to the pressures of natural selection. The types and availability of food items are major selective forces influencing body size and beak shape and size in both tree and ground finch species.

There are two basic patterns of evolution. Phyletic change is the gradual change within one lineage due to directional selection. Cladogenesis is the splitting of lineages as various groups branch off from the ancestral line. Adaptive radiation, which is the rapid diversification of an ancestral group in many directions when a new habitat is opened, occurs as a result of both phyletic change and cladogenesis. The extinction of species has occurred on a regular basis throughout evolutionary history and influences evolution by vacating habitats as species become extinct.

Analysis of the fossil record has revealed a steady background extinction rate of about 180 to 300 species every one million years. Every 26 million years, a period of mass extinctions occurred. The mass extinction of dinosaurs and other large terrestrial animals at the end of the Cretaceous period corresponds with the appearance of unusually high levels of iridium in clays at the Cretaceous-Tertiary boundary of sediments. One hypothesis suggests that the high iridium levels resulted from the collision of an asteroid (asteroids are rich in iridium) with the earth, and the extinctions occurred when a cloud of debris produced by the impact circled the earth, interrupting global climatic patterns and photosynthesis. There are flaws in this hypothesis, but it is the only explanation for the iridium anomaly that has withstood extensive challenge.

The evolution of the modern horse has been considered a classic example of phyletic change. However, with the discovery of more fossils, it is apparent that cladogenesis also occurred, but that the other branches that evolved died out.

Darwin viewed the fossil record as an incomplete record of evolutionary history since there are few examples that demonstrate the gradual change in a line of evolving organisms, which he believed to be the tempo of evolution. The contemporary model of punctuated equilibrium contends that the fossil record accurately reflects evolutionary history and that evolution actually occurs by periods of rapid change alternating with periods during which little or no change occurs. Further, the model contends that cladogenesis is the major mechanism of evolutionary change and that natural selection occurs among species as well as among individuals.

HOW TO STUDY THE CHAPTER

Read the entire chapter through quickly, focusing on the major concepts.

Use the GUIDED STUDY OF THE CHAPTER to help you identify the important details as you **reread** the chapter. Writing out the answers to these questions will help fix them in your mind as well as provide you with a valuable study aid.

Answer the questions in TESTING YOUR UNDERSTANDING without the aid of your text. Check your answers against those in PERFORMANCE ANALYSIS. Analyzing your answers will give you valuable feedback on your level of understanding and preparedness for classroom testing.

GUIDED STUDY OF THE CHAPTER

I. **Introduction** *(page 1010)*

> *Focus:* **Speciation is the link between microevolution and macroevolution.**

II. **Modes of Speciation** *(pages 1010–1011)*

A. **General Remarks** *(page 1010)*

1. What is the key feature of the modern definition of a species?

2. Define species in terms of population genetics.

3. What two questions must be answered in explaining the mode of speciation?

4. *Vocabulary:* Distinguish between allopatric and sympatric speciation.

B. **Allopatric Speciation** *(pages 1010–1011)*

5. *Vocabulary:* Distinguish between races and subspecies.

6. What factors contribute to the likelihood of speciation?

7. a. List six types of geographic barriers that may separate populations and facilitate speciation.

   b. Cite examples that illustrate two of the geographic barriers listed in (a).

8. a. Identify the roles of the founder effect and gene flow in allopatric speciation.

   b. What is the most likely fate of a small, isolated population?

C. *Essay:* **The Breakup of Pangaea** *(pages 1012–1013)*

9. a. Who first proposed the theory of continental drift?

   b. Why was this theory initially rejected?

10. a. Summarize the new theory of plate tectonics.

    b. What happens where two plates:

    Collide?

    Separate?

    Move in the same direction at different speeds?

    c. In addition to those in (b), what other plate motion may characterize the meeting of two plates?

11. a. When did Pangaea begin to break apart?

    b. What land masses were initially created when Pangaea broke apart and what modern continents were those land masses composed of?

12. Name the geologic formation created when India collided with Asia.

13. a. When did the major land masses finally separate into their modern configuration?

    b. When and how did North and South America become joined by the Isthmus of Panama?

14. Cite two pieces of fossil evidence supporting the concepts of Pangaea and of continental drift.

D. Sympatric Speciation *(pages 1012–1015)*

Polyploidy (pages 1012–1014)

15. a. *Vocabulary:* Define polyploidy.

    b. Identify two mechanisms by which polyploidy may occur.

    c. How may polyploidy be induced in the laboratory?

    d. *Vocabulary:* Distinguish between autopolyploidy and allopolyploidy.

16. *Vocabulary:* What is a hybrid organism?

> *Focus:* **Some hybrids are better adapted to the immediate environment than either parent.**

17. Illustrate the Focus statement with an example.

18. How do asexually reproducing hybrid plants conform to and deviate from the definition of a species?

19. a. Why are hybrids often sterile?

b. A sterile hybrid plant may become a new species by what mechanism?

Focus: **Sympatric speciation through polyploidy is an important mechanism of speciation among natural and agricultural plants.**

---

20. Cite one example of speciation by polyploidy in the animal kingdom.

Disruptive Selection (pages 1014–1015)

21. Name two reasons why it is difficult to prove that disruptive selection is a mechanism of sympatric speciation.

22. a. What evidence supports the notion that the races of *Rhagoletis pomonella* are an example of sympatric speciation?

b. Why do some biologists contend that this is actually an example of allopatric speciation?

III. Maintaining Genetic Isolation *(pages 1015–1017)*

A. General Remarks *(page 1015)*

23. a. Identify and distinguish between two mechanisms that maintain genetic isolation among species.

b. Which mechanism usually functions to prevent interbreeding?

c. Why is the mechanism named in (b) "an expected consequence of sexual selection"?

B. Premating Isolating Mechanisms *(page 1016)*

24. List and cite examples of five different premating isolating mechanisms.

C. Postmating Isolating Mechanisms *(page 1016)*

25. List six postmating isolating mechanisms.

26. How do postmating isolating mechanisms reinforce premating isolating mechanisms?

D. *Essay:* Creating Sexual Chaos *(page 1017)*

27. a. Certain members of the genus *Heliothis* cause damage to what agricultural products?

b. What is a disadvantage of using insecticides to kill these pests?

28. a. Describe the discovery made by scientists during a test of synthetic pheromones as potential attractants to draw male budworms to traps.

b. What natural phenomenon may limit the effectiveness of this chemical as an insect control technique?

IV. An Example: Darwin's Finches *(pages 1017–1021)*

A. General Remarks *(pages 1017–1020)*

29. What is the proposed origin of the Galapagos finches?

30. Briefly describe the diversity of habitats available to finches on the Galapagos islands.

> *Focus:* The variety of habitats on the Galapagos Islands exerted varying selection pressures on different finch groups, resulting in the development of the different species.

31. How did the geographic arrangement of the islands facilitate finch speciation?

> *Focus:* The fact that finches are poor long-distance fliers limited gene flow among the populations of the different islands.

32. Speciation resulting from geographic isolation, such as occurred with Darwin's finches, takes at least _____ years.

> *Focus:* Several species of Darwin's finches were probably evolving at the same time on different islands.

33. Describe the ancestral finch from which all other Galapagos finches evolved.

34. The major characteristic distinguishing the six ground finch species from one another is _____. (Figure 49–9)

35. Describe beak variations among the tree finches.

36. Identify the features of the warbler finch that are characteristic of (a) finches and (b) warblers.

> *Focus:* The Galapagos finches have been classified into three genera.

37. a. Within each finch genus, what two factors maintain reproductive isolation among the species?

    b. Describe how these two factors operate to keep the species separate.

B. The Effect of Natural Selection on Body and Beak Size *(pages 1020–1021)*

38. a. What data did Peter Grant and his colleagues collect during their studies of *Geospiza fortis*?

    b. Summarize the selection pressures experienced by the *G. fortis* population during the study period.

    c. Describe how the finch population changed in response to changing selection pressures.

V. The Evidence of the Fossil Record *(pages 1021–1025)*

A. General Remarks *(page 1021)*

B. Phyletic Change *(page 1021)*

39. a. *Vocabulary:* Define phyletic change.

b. What mechanism of selection produces phyletic change?

40. Phyletic change is the large-scale version of what observable events?

C. Cladogenesis *(pages 1021–1022)*

41. *Vocabulary:* Distinguish cladogenesis from phyletic change.

42. a. What is Ernst Mayr's opinion concerning the importance of cladogenesis in evolution?

b. Why can favorable genetic combinations grow rapidly in number and frequency in small, isolated populations?

D. Adaptive Radiation *(pages 1022–1023)*

43. a. *Vocabulary:* What is adaptive radiation?

b. What event is necessary for adaptive radiation to occur?

44. George Gaylord Simpson ascribed what significance to the role of adaptive radiation in evolution?

45. Identify three examples of adaptive radiation.

E. Extinction *(pages 1024–1025)*

46. a. Outline the findings of J. John Sepkoski, Jr., and David M. Raup in their statistical analysis of marine animal extinction.

b. When did the greatest extinction occur?

c. What fraction of living species became extinct at that time?

47. During the great extinction that occurred at the end of the Cretaceous period, which groups of organisms were the hardest hit and which groups were relatively unaffected?

48. a. What discovery, initially made by Walter Alvarez and colleagues, prompted the formulation of a new hypothesis concerning the extinctions that occurred at the end of the Cretaceous period?

b. Summarize this new hypothesis.

c. What evidence (i) conflicts with and (ii) supports this hypothesis?

49. Name four areas of study that have been initiated or revitalized since the discovery of the iridium anomaly.

> **Focus:** Every mass extinction has been followed by great evolutionary diversification of the survivors, who filled the habitats vacated by the extinct species.

VI. *Equus:* **A Case Study** *(page 1025–1027)*

50. a. Describe the appearance and eating habits of Eohippus.

b. Describe how each of the following characteristics changed during the course of equine evolution, and correlate these changes with changing environmental conditions.

Body size

Dentition and diet

Placement of the eye

Bone structure of legs and feet

51. What evolutionary adaptation gave the larger horses the ability to escape from predators?

52. The evolution of *Equus* is the result of what pattern(s) of evolution?

VII. Punctuated Equilibria *(pages 1028–1029)*

53. How did Charles Darwin describe the fossil record?

54. Prior to the model of punctuated equilibrium, how was the paucity of fossil evidence for slow phyletic change explained?

55. a. How does the perception of the fossil record held by Niles Eldredge and Stephen Jay Gould differ from that held by Darwin?

b. Trace the logic of Eldredge and Gould in arriving at this position.

Focus: The model of punctuated equilibria contends that evolution does not result from gradual, steady change but rather from periods of rapid change alternating with fairly long periods during which little or no change occurs.

56. a. As the model of punctuated equilibria has been modified, what two major concepts have been added?

b. How does horse evolution support these two concepts?

c. What two basic mechanisms of evolution does the revised model of punctuated equilibria suggest?

57. What have been the effects of the punctuated equilibrium model on the work of evolutionary biologists?

VIII. **Summary** *(pages 1029–1030):* Read the summary. If you are familiar with the essential features of the material presented there, you are ready to complete the section TESTING YOUR UNDERSTANDING.

TESTING YOUR UNDERSTANDING

After you have completed this examination, compare your answers with those in the section that follows.

1. The key feature defining a species is:
   a. geographic isolation.
   b. reproductive isolation.
   c. a common ancestry.
   d. its interactions with other species.
   e. its role in the environment.

2. Which statement is NOT true of the Isthmus of Panama?
   a. Speciation occurred in the Atlantic and Pacific Oceans when the Isthmus was submerged.
   b. The Isthmus has submerged and reemerged repeatedly throughout geologic history.
   c. When emerged, the Isthmus separated the Atlantic and Pacific Oceans.
   d. When the Isthmus was submerged, the two continents became "islands."
   e. The Atlantic and Pacific Oceans became "islands" when the Isthmus reemerged.

3. A small, peripheral population that becomes isolated is likely to differ from the parent population because of:
   a. natural selection.
   b. the founder effect.
   c. mutation.
   d. nonrandom mating.
   e. migration.

4. The distribution of the snaggle-toothed reptile *Mesosaurus* in _____ was strong evidence favoring the continental drift theory.
   a. Africa and Australia
   b. England and Greenland
   c. Antarctica and Australia
   d. India and Asia
   e. South America and Africa

5. Speciation that results from the doubling of chromosome number within a species is known as:
   a. allopolyploidy.
   b. autopolyploidy.
   c. homopolyploidy.
   d. heteropolyploidy.
   e. synpolyploidy.

6. Which of the following is NOT a postmating isolating mechanism?
   a. nonviability of sperm in the female reproductive tract
   b. inability of sperm to fuse with the ovum
   c. reproductive cycles on different schedules

d. production of hardy but sterile offspring

e. structural differences that prevent insemination

7. The studies of Peter Grant and his coworkers on Daphne Major have shown that:
   a. the finch species on that island belong to three different genera.
   b. finches may quickly abandon seed-eating in favor of fruit-eating if seeds are in short supply.
   c. an influx of birds from other islands keeps the gene pool in a state of disequilibrium.
   d. selective forces can have a dramatic effect on the phenotype in a relatively short time.
   e. the average size of the finches increases in response to the warm winds and longer growing season associated with an El Niño.

8. Evolutionary change within a single lineage of organisms that results in the formation of new species is known as:
   a. cladogenesis.
   b. adaptive radiation.
   c. extinction.
   d. stabilizing evolution.
   e. phyletic change.

9. Adaptive radiation is generally considered to result from a combination of:
   a. genetic drift and migration.
   b. migration and directional selection.
   c. directional selection and disruptive selection.
   d. cladogenesis and phyletic change.
   e. phyletic change and genetic drift.

10. According to the study of J. John Sepkoski, Jr., and David M. Raup, during the past 250 million years there has been a background extinction rate among marine organisms of _____ species extinctions every one million years.
    a. 20 to 60
    b. 75 to 150
    c. 180 to 300
    d. 480 to 600
    e. 750 to 1,000

11. Dinosaurs became extinct approximately _____ million years ago.
    a. 35
    b. 50
    c. 65
    d. 80
    e. 95

12. Which scientific activity is NOT among those initiated following the discovery of the iridium anomaly?
    a. reexamination of the fossil record
    b. geochemical studies of boundary layers
    c. a search for terrestrial sources to account for high iridium levels
    d. reanalysis of geologic phenomena known to have occurred in periods of mass extinction
    e. a search for subterranian craters created by large meteors or asteroids

13. The punctuated equilibria model was proposed as a general mechanism of evolution by:
    a. Niles Eldredge and Stephen Jay Gould.
    b. Steven M. Stanley and Ernst Mayr.
    c. Alfred Wegener and Guy Bush.
    d. J. John Sepkoski, Jr., and David M. Raup.
    e. Peter Grant and George Gaylord Simpson.

14. Steven M. Stanley and the proponents of the more radical version of punctuated equilibria emphasized:
    a. the selection of individuals as the mechanism of evolution.
    b. the tempo of evolution.
    c. gradual change in the fossil record.
    d. cladogenesis as the pattern of evolution.
    e. the role of Mendelian genetics in evolution.

15. T or F   Most hybrid organisms (plant or animal) are sterile because the chromosomes separate unequally during meiosis, producing nonviable gametes.

16. T or F   As a consequence of sexual selection, premating isolating mechanisms are generally sufficient to prevent interbreeding between two species.

17. T or F   Although the warbler finch is similar to the other Galapagos finches, its external appearance and behavior clearly place it with the warblers taxonomically.

18. T or F   In cladogenesis, favorable genetic combinations that occur in small populations can increase rapidly in number and frequency, resulting in the spurts of rapid evolutionary change that seem to be documented in the fossil record.

19. T or F   At each mass extinction, the course of evolution has been dramatically altered, with some branches of the evolutionary tree permanently eliminated and others undergoing adaptive radiation.

20. T or F   The evolution of the horse is an excellent example of phyletic change because there has been only one equine species living at any one time during history.

21. Cite two reasons why sympatric speciation by disruptive selection is difficult to prove.

22. Briefly identify and describe the two premating isolating mechanisms that are most important in maintaining reproductive isolation among Darwin's finches.

23. Explain how the model of allopatric speciation provides support for the punctuated equilibria model of evolution.

PERFORMANCE ANALYSIS

1. **b**   The essential feature defining a species is its reproductive isolation from all other species. (page 1010)

2. **a**   The Isthmus of Panama has submerged and reemerged throughout geologic time. When it was submerged, organisms from the Atlantic and Pacific Oceans could mix. At these times, the North and South American continents were separated by a body of water, blocking the

movement of land animals. When the Isthmus reemerged, the land animals could mingle but the marine life of the two oceans was separated into "islands." (page 1011)

3. **b** The founder effect is a type of genetic drift in which the gene pool of a population fragment differs from that of the main population from which it separated. (page 1011)

4. **e** The discovery of *Mesosaurus* fossils in South America and Africa, but nowhere else in the world, was strong evidence supporting the concept of continental drift. (essay, page 1013)

5. **b** Autopolyploidy refers to a doubling of chromosome number within a single species. Allopolyploidy refers to the doubling of chromosome number in a hybrid created from two species. Both processes can result in the formation of a new species. (page 1012)

6. **c** Postmating isolating mechanisms are those that function if copulation occurs between members of two different species. The separation of species because their reproductive cycles are on different schedules is a *premating* isolating mechanism, since it would prevent copulation. (page 1016)

7. **d** Peter Grant's team, which studied Galapagos finches on Daphne Major for 15 years, demonstrated that selective forces can have a dramatic effect on the phenotype of a species within a relatively short time period. (page 1020)

8. **e** Phyletic change refers to evolutionary change within a single lineage of organisms over time, with the gradual development of new species and their replacement of old species. Cladogenesis refers to a branching pattern of evolution in which several related lines of organisms evolved at the same time from a common ancestor or ancestral group. (page 1021)

9. **d** Adaptive radiation is generally considered to result from both cladogenesis and phyletic change. (page 1022)

10. **c** Sepkoski and Raup reviewed the pattern of extinction of marine organisms revealed in the fossil record and discovered a background rate of 180 to 300 extinctions every 1 million years, with massive extinctions occurring approximately every 26 million years. (page 1024)

11. **c** Dinosaurs became extinct approximately 65 million years ago, at the end of the Cretaceous period. (page 1024)

12. **e** The discovery of the iridium anomaly at the Cretaceous-Tertiary boundary in many sediments has stimulated a reexamination of the fossil record, geochemical studies of boundary layers, a search for terrestrial sources of high iridium, and a reanalysis of the geologic phenomena known to have occurred at the times of mass extinctions. (page 1025)

13. **a** Niles Eldredge and Stephen Jay Gould proposed the punctuated equilibrium model of evolution. A more radi-

cal revision was later proposed by Steven M. Stanley. (page 1028)

14. **d** Steven M. Stanley and the champions of his more radical version of punctuated equilibria emphasize cladogenesis as the overall pattern of evolution. (page 1028)

15. **False** Most hybrid organisms are sterile because their chromosomes are unable to pair during meiosis. (page 1014)

16. **True** Premating isolating mechanisms are generally adequate to prevent matings among members of different species. This is generally considered to be one consequence of sexual selection. (page 1015)

17. **False** The warbler finch of the Galapagos Islands would be classified as a warbler if only its external appearance and behavior were considered. However, its internal anatomy clearly places it among the finches. (page 1020)

18. **True** Cladogenesis is the pattern of evolution that results from the branching off from parent populations of groups of organisms, which then undergo separate evolutionary courses. (page 1022)

19. **True** Each mass extinction has involved the elimination of some branches of organisms, but has allowed the subsequent adaptive radiation of other groups into the vacated environments. (page 1025)

20. **False** When it was first studied, horse evolution appeared to be a classic example of phyletic change. However, as more fossils have been discovered, it is apparent that the evolution of this group had more branches than was initially realized and that cladogenesis more accurately describes the pattern of horse evolution. (page 1027)

21. (1) It is difficult to prove that two similar, coexisting species were never geographically separated (i.e., allopatric speciation cannot be ruled out.) (2) Two forms of one species may be in the process of diverging, but it is impossible to know if they will ever become separate species. (page 1014)

22. The two premating isolating mechanisms that are most important in maintaining genetic isolation among Darwin's finches are species-specific songs and beak size and shape. The songs, which are sung by males only, serve as long-range identity cues, whereas beak size and shape serve as short-range cues. (page 1020)

23. Allopatric speciation involves the geographic separation of a group from the main population. If a group fragmented from the main population, became geographically isolated, and experienced different selection pressures, it would have undergone a period of rapid change. If the new species then expanded and replaced the original parent population, the result would be the fossil pattern commonly observed. (page 1028)

C H A P T E R 50

The Evolution of the Hominids

MAJOR CONCEPTS

The major trends in primate evolution appear to be adaptations to arboreal life. Compared with other vertebrates, the primate forelimb (hand and arm) is relatively unspecialized. The hand, with its opposing thumb, is adapted for grasping objects and for dexterity. The arm can be widely rotated in its socket. Primates have stereoscopic vision, color discrimination, and greater visual acuity than most vertebrates (except for birds). As mammals, primates nurse their young and have fairly extended periods of parental care, which enables the young to learn from the adults. The upright posture of primates is apparently an adaptation to life in the trees.

There are two major groups of primates: the prosimians and the anthropoids. Prosimians are exemplified by lemurs, tar-siers, bush babies, and lorises. These animals resemble the presumed primate ancestor and most are small and arboreal. Many are nocturnal. The anthropoids include the monkeys, apes, and humans.

There are two groups of monkeys: platyrrhines (New World monkeys) and catarrhines (Old World monkeys). The platyrrhines (flat-nosed) are native to Central and South America, are strictly arboreal, and have prehensile tails. The catarrhines (downward-nosed) are native to Africa, include both terrestrial and arboreal species, and use their tails for balance, but not for hanging.

The hominoids, which are descendants of catarrhines, include apes and humans. There are four genera of modern apes: gibbons, orangutans, chimpanzees, and gorillas. Apes have relatively long arms and short legs and all are capable of brachiation. There is a wide range of size in the apes, from the gibbons at about 6 kilograms to the gorillas at 180 kilograms. Social behavior of ape species varies greatly and ranges from the solitary orangutan to troops of gorillas and chimpanzees.

The hominids are members of the family Hominidae and include modern humans and several extinct species. Members of the genus *Australopithecus* were the first recognized fossil hominids. Anthropologists disagree on the number of species that existed, but at least three species are widely recognized. These species are *A. africanus*, a more robust form called *A. robustus*, and an even more robust form, *A. boisei*. Two other proposed species are *A. afarensis* and a specimen designated WT 17000, whose classification status is under debate. The exact relationships among these varieties and their relationship to members of the genus *Homo* and to modern humans is currently a subject of intense discussion and debate.

Three long-standing concepts regarding hominid evolution have been laid to rest by recent discoveries. The supposition that hominid evolution occurred by a straight phyletic progression has been disproven by the discovery of many branches in the hominid evolutionary tree. The premise that the superior intelligence of hominids led to that elusive quality of humanness was laid to rest by overwhelming evidence that hominids were bipedal and walked erect long before there was a significant increase in brain size. Historically, it has been assumed that the selective pressure favoring bipedalism was the freeing of the hands for tool use. Three alternative selective pressures have been proposed: (1) bipedalism enabled males to transport food back to females and young; (2) bipedalism was triggered by the spreading grasslands that occurred at the end of the Miocene; and (3) in a female cooperative, bipedalism favored those who could carry young, gather food, and transport it for sharing.

Homo habilis, the first member of the genus *Homo*, was so classified because its brain size was larger than that of the australopithecines and it apparently used tools. Its discovery and classification were controversial because it provided the first evidence that two different hominid species lived concurrently.

The most distinctive features of *Homo erectus* are the use of the hand ax and of fire. The hand axes that have been found throughout the range of *H. erectus* closely resemble one another, which suggests that a cultural tradition had emerged in which information was passed between groups and from one generation to the next. The brain capacity of *H. erectus* overlapped that of modern humans.

The archaic *Homo sapiens* fossils represent the oldest members of the species that includes modern humans. Specimens so classified are dated between 400,000 and 200,000 years ago. They had larger brains, smaller teeth, less prominent brow ridges, and more bulging foreheads than *H. erectus*. Most archaic *H. sapiens* specimens have been found in Europe, with a few being identified in Asia and Africa.

Homo sapiens neanderthalensis was abundant from 150,000 to 35,000 years ago. Most specimens have been found in Europe, but some have also been identified in the Near East and Central Asia. The tools used by Neanderthals were more varied and more sophisticated than those employed by *H. erectus*. Neanderthals are thought to have worn clothing made of animal skins. They buried their dead in such a way that anthropologists suspect they believed in life after death. They disappeared abruptly about 30,000 years ago.

All hominid specimens of the last 30,000 years have been classified as *Homo sapiens sapiens*, the variety to which modern humans belong. The tools of the Cro-Magnons were far more advanced than those of the Neanderthals—Cro-Magnons used tools to make other tools. Cro-Magnons were responsible for the cave art that has been found in several locations in Spain and France.

There are two major hypotheses concerning the evolution of modern humans. The candelabra hypothesis contends that *H. erectus* migrated from Africa and established populations in other regions of the world. These populations then evolved separately into the different modern races. The opposing Noah's Ark model contends that a small group of *Homo sapiens sapiens* started in one place and colonized the entire world.

Results of a study involving the use of mitochondrial DNA as a molecular clock have supported the Noah's Ark hypothesis. Five important conclusions came from this study. (1) There was a single common female ancestor or ancestral population (only one set of mitochondrial DNA was passed on). (2) The female lived about 200,000 year ago. (3) She lived in Africa. (4) The original founding population left Africa a little more than 100,000 years ago. (5) There is no evidence that new mitochondrial DNA was introduced into the original population or into any of the colonizing populations. These conclusions must be supported by other data before they gain complete acceptance by the scientific community. If they are supported by other data, there are two major implications of these conclusions: humans evolved fairly recently from a common source, and colonies of *Homo sapiens sapiens* replaced populations of *Homo erectus* and Neanderthals as they expanded their range.

HOW TO STUDY THE CHAPTER

Read the entire chapter through quickly, focusing on the major concepts.

Use the GUIDED STUDY OF THE CHAPTER to help you identify the important details as you **reread** the chapter. Writing out the answers to these questions will help fix them in your mind as well as provide you with a valuable study aid.

Answer the questions in TESTING YOUR UNDERSTANDING without the aid of your text. Check your answers against those in PERFORMANCE ANALYSIS. Analyzing your answers will give you valuable feedback on your level of understanding and preparedness for classroom testing.

GUIDED STUDY OF THE CHAPTER

I. Introduction *(page 1031)*

> *Focus:* The first mammals appeared approximately 200 million years ago, at about the time of the first dinosaurs.

1. a. From what ancestors did the first mammals evolve?

   b. Describe the first mammals.

   c. Identify the three principal lineages of mammals.

2. What event allowed the rapid evolution and diversification of the mammals?

II. Trends in Primate Evolution *(pages 1031–1033)*

A. General Remarks *(page 1031)*

3. Most trends in primate evolution reflect adaptations to what lifestyle?

B. The Primate Hand and Arm *(pages 1032–1033)*

4. "The extremities of primates are relatively unspecialized." Explain this statement, using examples.

5. What feature of the primate hand greatly increases gripping power and manual dexterity?

6. What characteristic of the forearm enables a primate to rotate its hand through a semicircle without moving the elbow or the upper arm?

7. How does the shoulder joint of some primates differ from that of other mammals?

8. a. What is the adaptive value of fingernails as opposed to claws?

   b. How does this complement the feature mentioned in question 5?

C. Visual Acuity *(page 1033)*

9. a. How did the habitat of early primates influence the evolution of vision and olfaction?

   b. Which other group of animals experienced similar evolutionary pressures?

10. What anatomical arrangement of the eyes is required for stereoscopic vision?

11. Name two features of the primate retina associated with increased visual acuity.

D. Care of the Young *(page 1033)*

12. What characteristic unique to mammals contributes to the strong mother-young relationships, and to long periods in which the young depend upon the parents?

13. Identify the selection pressure that favored single births among many primates. (Figure 50–4)

E. Uprightness *(page 1033)*

14. Identify one consequence of the upright posture of primates.

15. Upright posture was a necessary precondition for what evolutionary development?

III. Major Lines of Primate Evolution *(pages 1034–1038)*

A. Prosimians *(page 1034)*

16. Name the two major groups of primates and identify examples of each group.

17. What environments were occupied by prosimians during the Paleocene and Eocene?

18. a. Characterize modern prosimians.

    b. What features of the tarsier make it well-adapted for its arboreal, nocturnal lifestyle? (Figure 50–6)

B. **Monkeys** *(pages 1034–1035)*

19. Name six ways in which monkeys differ from prosimians.

20. a. Describe the social roles of males and females in many monkey troops.

    b. What techniques might monkeys employ to drive off a predator?

21. During what period did monkeys evolve and what animals were their probable ancestors?

22. *Vocabulary:* Use the following table to distinguish between platyrrhines and catarrhines.

Table for Question 22

	PLATYRRHINES	CATARRHINES
Common name		
Distinguishing facial feature		
Site of evolution		
Lifestyle(s) (arboreal or terrestrial)		
Use of the tail		
Examples		

23. What event physically separated the populations ancestral to platyrrhines and catarrhines?

24. Describe the means of locomotion in ground-dwelling Old World monkeys.

C. **Apes** *(pages 1035–1038)*

25. a. Name the two groups of primates that are classified as hominoids.

    b. Hominoids are considered to be cousins of which type of monkey?

> *Focus:* The apes that inhabited Kenya and Uganda during the Miocene have been assigned to the genus *Proconsul.*

26. a. Who discovered *Proconsul africanus*?

    b. Characterize *P. africanus* with respect to size and lifestyle.

    c. Where and when did *Proconsul* and related apes live?

27. Different body parts of *P. africanus* resembled three different types of present-day primates. Name these three primates and the corresponding body parts.

28. Identify the four major genera of modern apes and indicate which primates belong to each genus.

29. Name three ways in which apes differ from monkeys.

30. a. *Vocabulary:* What is brachiation?

b. The upright suspension of brachiation is believed to have influenced primate evolution in what manner?

31. What anatomical feature of apes lends them a partially erect posture when they are on all fours?

> *Focus:* The size range of modern apes is broad, extending from 6-kilogram gibbons to 180-kilogram gorillas. In some species, there is a marked size difference between males and females.

32. Name three features that characterize the head of a gorilla.

Social Behavior (pages 1036–1038)

33. Characterize the social structure of (a) gibbons and (b) orangutans.

34. a. Characterize the social structure of a gorilla troop.

b. How is the mate of a female gorilla determined?

c. For what time period may a female gorilla nurse one infant?

35. a. Characterize the social structure of a chimpanzee troop.

b. Describe the diet of a chimpanzee. What degree of cooperation may occur during food-gathering?

c. How is the mate of a female chimpanzee determined?

d. To what extent do related individuals maintain bonds within the group?

36. To what degree are chimpanzee groups territorial?

> *Focus:* Chimpanzee groups are not exclusive and members come and go freely.

IV. **The Emergence of the Hominids** *(pages 1038–1050)*

A. **The First Hominids** *(page 1038)*

37. a. Name four features of the skull examined by Raymond Dart in 1924 that distinguished it from both modern apes and their ancestors.

b. What did he name this new fossil?

c. Where was this fossil found?

d. Cite three reasons why Dart's report and conclusions were ignored by anthropologists.

> *Focus:* Although the australopithecines are now widely accepted as hominids, the number of species that may have existed and their evolutionary relationship to modern humans are the subjects of intense debate.

38. Identify four reasons behind the debates concerning the place of australopithecines in hominid evolution.

B. *Essay:* **The Footprints of Laetoli** *(page 1039)*

> *Focus:* **The footprints made 3.6 million years ago in a volcanic ash deposit indicate conclusively that hominids with a fully erect, bipedal human gait existed at that time.**

39. How were these footprints discovered?

C. Current Status of the Australopithecines *(pages 1038–1042)*

40. Over what time period did autralopithecines inhabit the earth?

41. Fill in the following table comparing the two most widely accepted species of the genus *Australopithecus.*

Table for Question 41

	A. AFRICANUS	A. ROBUSTUS
Size		
Body type		
Brain capacity		
Skull and dental features		
Posture		

42. What do we know about the diet of *A. robustus?*

43. How did *A. boisei* differ from *A. robustus?*

44. How was the age of Zinjanthropus estimated?

45. a. What was the original conclusion regarding the evolutionary relationship among *A. africanus, A. robustus,* and *A. boisei?*

b. What discovery invalidated this conclusion?

c. What is currently thought to be the relationship among these three species and WT 17000?

46. a. Why do some anthropologists believe that the "First Family" should be classified separately from the other australopithecine species? (See also Figure 50–15.)

b. What species name was given to the "First Family" by their discoverer?

c. Characterize the members of the "First Family."

d. What evidence suggests that hominids walked fully erect before there was a significant increase in brain size? (Figures 50–15, 50–16)

47. Outline the debate among paleoanthropologists about classification of the Hadar and Laetoli fossils.

D. *Homo habilis* (page 1042)

    48. a. On what basis did Louis Leakey assign the name *Homo habilis* to the fossil found in 1962 at Olduvai?

        b. Why was this proposal not readily accepted?

        c. What later discovery supported this classification?

    49. How did the fossil OH 62 compare with skull 1470 (*H. habilis*) and Lucy (*A. afarensis*)?

    50. Describe the tools, and their possible uses, that were first found at the Olduvai site.

    51. a. What are the three possible origins of *H. habilis*?

        b. Cite the evidence that leads some anthropologists to argue that *Homo habilis* should be classified as an australopithecine.

> **Focus:** The classification of *Homo habilis* and its evolutionary relationship to the australopithecines is being actively debated.

E. New Concepts in Hominid Evolution (pages 1043–1044)

    52. Summarize the evidence that discredited each of the following concepts.

        a. The evolution of hominids occurred by a direct phyletic progression.

        b. The superior intelligence of humans was the characteristic that led to the evolution of "humanness."

        c. The selective pressures favoring bipedalism included freeing the hands for tool use.

    53. a. While it was accepted as authentic, what influence did the Piltdown forgery have on the concept of hominid evolution?

        b. Why was this archaeological "find" so readily accepted at the time of its "discovery"?

    54. a. What traits, believed by Dart, Ardrey, and many of their contemporaries to be characteristic of early hominids, were thought to have been part of the selective pressures favoring bipedalism?

        b. How do the thoughts of (i) Owen Lovejoy and (ii) Matt Cartmill on the evolution of bipedalism differ from those of the scientists mentioned in (a)?

c. What other two situations have been proposed as the source of selective pressures favoring bipedalism?

F. *Homo erectus* (pages 1044–1045)

55. How did *Homo erectus* resemble and differ from modern man?

56. a. Fossils of *Homo erectus* have been found in what regions of the world?

    b. On what continent have fossils of *Homo erectus* NOT been found?

    c. What evidence indicates that *Homo erectus* may have inhabited Europe?

57. a. What tool was characteristic of *Homo erectus*?

    b. What does the distribution of this tool suggest about *Homo erectus* societies?

Focus: **Homo erectus used fire and inhabited caves.**

---

58. How would the use of fire have affected the diet and dwelling places of *H. erectus*?

59. a. How many years separate the youngest example of *H. habilis* from the oldest example of *H. erectus*?

    b. Name two evolutionary events that might account for this relatively short time period.

G. *Homo sapiens* (pages 1045–1047)

60. What criteria are used to classify a fossil as *Homo sapiens*?

61. List the three varieties of *Homo sapiens* that are currently recognized.

Archaic Homo sapiens (page 1046)

62. How did archaic *H. sapiens* differ from *H. erectus*?

63. Where have most of the archaic *H. sapiens* fossils been found?

Homo sapiens neanderthalensis (pages 1046–1047)

64. Where have most of the Neanderthal fossils been found?

65. How did Neanderthal anatomy differ from that of modern humans?

66. Describe tool use by the Neanderthals.

67. What evidence suggests that the Neanderthals believed in life after death?

68. Describe the distribution of Neanderthals at the time of their extinction 30,000 years ago.

Homo sapiens sapiens (page 1047)

> *Focus:* **All hominid fossils since the extinction of the Neanderthals are anatomically identical to modern humans.**

69. a. What other name is commonly used to identify fossils of this type?

    b. Where were the *first* fossils of this type found?

70. How did the tools of Cro-Magnons differ from those of earlier hominids with respect to variety, method of construction, and use?

H. *Essay:* **The Art of the Caves** (page 1048)

> *Focus:* **The cave paintings of western Spain and southern France were painted by Cro-Magnons.**

71. Modern archaeologists have speculated about the possible meaning of these cave paintings. List several suggestions concerning their significance.

I. **The Origin of Modern Humans** (*pages 1047–1050*)

72. Summarize the candelabra hypothesis of modern human evolution and cite the supporting evidence.

73. a. How does the Noah's Ark model of modern human evolution differ from the candelabra model?

    b. What fossil evidence supports this model?

74. Cite two reasons why mitochondrial DNA is useful for studying evolutionary relationships of recent times (times measured in thousands, rather than millions, of years).

75. a. List four conclusions drawn from Rebecca Cann's study of mitochondrial DNA from five different geographic regions. Cite the evidence supporting each conclusion.

    b. If these conclusions are correct, what does this indicate about the evolution of modern humans?

    c. Identify one possible explanation for the sudden disappearance of all hominid species other than *Homo sapiens sapiens*.

are familiar with the essential features of the material presented there, you are ready to complete the section TESTING YOUR UNDERSTANDING.

TESTING YOUR UNDERSTANDING

After you have completed this examination, compare your answers with those in the section that follows.

1. Based on various forms of fossil evidence, which of the following is NOT a trait believed to have been characteristic of the first mammals?
   a. nocturnal vision   d. erect posture
   b. homeothermy   e. small size
   c. carnivorous diet

2. Humans are members of which mammalian order?
   a. Carnivora   d. Omnivora
   b. Edentata   e. Primates
   c. Insectivora

3. Most trends in primate evolution are apparently related to various adaptations to what type of lifestyle?
   a. arboreal   d. agricultural
   b. terrestrial   e. hunting-gathering
   c. aquatic

4. Which of these mammals can rotate its forelimbs freely in their shoulder sockets?
   a. horse   d. cow
   b. dog   e. gorilla
   c. pig

5. If primates had claws instead of nails, their fingers (or toes) would not be as useful for:
   a. climbing trees.
   b. swinging from branches.
   c. manipulating objects.
   d. walking on all fours.
   e. digging in the ground.

6. Which structure(s) of the primate eye is (are) responsible for high visual acuity?
   a. rods
   b. the fovea
   c. the cornea
   d. the lens
   e. It is not the eye that is responsible for visual acuity but the intricate processing of visual stimuli by the complex brain.

7. Our fellow primates look "human" to us because of their:
   a. tendency to look straight ahead when assuming a vertical posture.
   b. opposable thumb and the presence of nails instead of claws on their digits.
   c. similar skeletal structures, particularly of the torso.
   d. use of tools and similar ways of solving problems.
   e. family-oriented interactions with their young.

8. Which description does NOT apply to most modern prosimians?
   a. small or medium-sized
   b. furry
   c. arboreal
   d. herbivorous or insectivorous
   e. All of these describe most modern prosimians.

9. Anthropoids are:
   a. mostly small.
   b. insectivorous.
   c. the higher primates.
   d. nocturnal.
   e. more primitive than prosimians.

10. In the arboreal Old World monkeys, the tail is used for:
    a. climbing.   d. prehension.
    b. swinging.   e. combat.
    c. balance

11. The hominoids comprise:
    a. all races of humans, including extinct ones.
    b. anthropoids and prosimians.
    c. monkeys, apes, and humans.
    d. humans and apes.
    e. New World and Old World monkeys.

12. *Proconsul africanus* was first discovered by _____. It seems to have been a tree-dwelling, fruit-eating hominoid about the size of a _____.
    a. Mary Leakey; baboon
    b. David Pilbeam; chimpanzee
    c. Richard Leakey; gibbon
    d. Raymond Dart; macaque
    e. Louis Leakey; gorilla

13. Which statement is NOT true of gorillas?
    a. They walk on all fours.
    b. They are primarily terrestrial.
    c. A male may challenge an adversary by beating his chest.
    d. Males are typically solitary and the females live in small groups with their young.
    e. Females may nurse their young for two to four years.

14. Receptive females of which primates typically mate with more than one male?
    a. gibbons
    b. chimpanzees
    c. gorillas
    d. gorillas and chimpanzees
    e. gibbons and gorillas

15. The fossil find that sparked a great surge of interest in research into human origins was the 1959 discovery of _____ by Mary Leakey.
    a. *Australopithecus africanus*
    b. *Proconsul*
    c. *Australopithecus afarensis*
    d. Zinjanthropus
    e. *Homo habilis*

16. The footprints at Laetoli demonstrated that hominids roamed the Rift Valley some _____ million years ago.
    a. 4.0
    b. 3.6
    c. 2.8
    d. 2.4
    e. 1.8

17. In 1962, Louis Leakey gave a newly discovered Olduvai specimen the name *Homo habilis* because:
    a. it would have weighed about 100 kilograms.
    b. its skull did not have the prominent crest characteristic of australopithecines.
    c. numerous hand axes were located in the same stratum.
    d. it was much more robust than the australopithecines.
    e. its brain was larger than that of the australopithecines and it was associated with stone tools.

18. Which statement does NOT reflect one of the now widely accepted concepts that emerged from the chaos created by the more recent discoveries of hominid fossils?
    a. Hominid evolution followed a branching course similar to that of *Equus*, and most of those branches led to extinction.
    b. Increasing brain size and intelligence were the adaptations that set hominids on the evolutionary path to humanness.
    c. The australopithecines and *Homo habilis* evolved in Africa over a period of 3 million years.
    d. In early hominid evolution, multiple species of hominids coexisted.
    e. The bipedal gait of early hominids set the evolutionary course of the human race.

19. Which of these fossil remains proved to be a forgery?
    a. the nutcracker man
    b. the Taung child
    c. the Piltdown man
    d. Zinjanthropus
    e. *Proconsul*

20. The skeleton of *Homo erectus* differs from that of *Homo sapiens* primarily in the:
    a. shape of the skull.
    b. length of the long bones.
    c. evidence for knuckle-walking.
    d. position of the eye sockets.
    e. length of the toes.

21. The tool that was distinctive to *Homo erectus* was the:
    a. spear.
    b. slingshot.
    c. hand ax.
    d. flint scraper.
    e. grinding stone.

22. Which statement does NOT apply to Neanderthals?
    a. Their fossil remains have been found in Europe, the Near East, and Central Asia.
    b. Their skulls had low foreheads and heavy brow ridges.
    c. Their tools were made of stone.
    d. Their dead were often buried with food, weapons, or flowers.
    e. They appear to have been a subspecies of *Homo erectus*.

23. Neanderthals disappeared abruptly some _____ years ago.
    a. 30,000
    b. 60,000
    c. 90,000
    d. 120,000
    e. 150,000

24. According to current evidence, modern humans appeared in Europe about _____ years ago.
    a. 20,000
    b. 40,000
    c. 60,000
    d. 80,000
    e. 100,000

25. T or F   Ground-dwelling Old World monkeys walk on all fours.

26. T or F   One of the compelling reasons for studying the behavior of apes is that human behavior can be interpreted in terms of the behaviors exhibited by other primates.

27. T or F   Despite the fact that more than one million years separated Lucy and specimen OH 62, they were both about the same size and had similar limb structures.

28. T or F   The roots and lineage of modern human evolution became clear upon the recent discovery of a specimen that is intermediate between *Homo sapiens neanderthalensis* and *Homo sapiens sapiens*.

29. Identify three of the four concepts that have been proposed regarding the origin and classification of *Homo habilis*.

30. Briefly summarize the advantages conferred on *Homo erectus* by the use of fire.

31. Explain the basis for the concept of a mitochondrial Eve.

PERFORMANCE ANALYSIS

1. **d**   The first mammals are thought to have been mouse-sized, homeothermic, and basically carnivorous, and most likely were nocturnal. (page 1031)

2. **e**   Humans are members of the order Primates. (page 1031)

3. **a**   Most of the trends in primate evolution, including visual acuity, adaptations of the hand and arm, parental care, and upright posture, appear to be related to the arboreal lifestyle of early primates. (page 1031)

4. **e**   Primates, including gorillas, have shoulder joints that enable them to rotate their forelimbs freely. (pages 1032, 1033)

5. **c** The presence of nails instead of claws on the primate hand enabled the end of the fingers to become very sensitive, a feature important in manipulating objects. (page 1033)

6. **b** The fovea is a concentration of cones that is particularly well developed in the eyes of primates (and of birds) and is responsible for the great visual acuity characteristic of these animals. (page 1033)

7. **a** The tendency of primates to look directly ahead when assuming a vertical posture is what gives them a "human" appearance. (page 1033)

8. **e** Most modern prosimians are small to medium-sized, furry, arboreal, and either herbivorous or insectivorous. Many are nocturnal. (page 1034)

9. **c** Anthropoids are the higher primates, including the monkeys, apes, and humans. The prosimians are the lower primates. (page 1034)

10. **c** Old World monkeys use their tails for balance, whereas New World monkeys use their tails as prehensile organs. (page 1035)

11. **d** Hominoids include the humans and apes. Members of the family Hominidae, including modern humans and fossil specimens, are called hominids. (page 1035)

12. **a** Mary Leakey discovered *Proconsul africanus*, a hominoid that lived in the woodlands of Africa during the Miocene. It was about the size of a baboon and is thought to have been an arboreal fruit-eater. (page 1035)

13. **d** Gorillas are highly social animals and live in troops, each of which is led by a large, mature male. (page 1036)

14. **b** Receptive female chimpanzees may mate with more than one male during estrus. (page 1036)

15. **d** The discovery of the nutcracker man, Zinjanthropus, stimulated a great surge of interest in the evolutionary origins of modern humans. (page 1040)

16. **b** The hominid footprints in the Rift Valley at Laetoli were made some 3.6 million years ago. (essay, page 1039)

17. **e** Louis Leakey named a specimen found in Olduvai *Homo habilis* because it had a larger brain size than that of the australopithecines (the genus *Homo*) and because it was apparently associated with stone tools (*habilis*, handy). (page 1042)

18. **b** It was bipedalism, not superior intelligence, that set the early hominids on the evolutionary course toward humanness. (pages 1043, 1044, 1047)

19. **c** The Piltdown man was a "fossil find" that was fabricated from the skull of a modern human and the modified jaw of an orangutan. When first "discovered" it was widely accepted by the scientific community because it was exactly what the scientists had been seeking. (page 1044)

20. **a** The skeleton of *Homo erectus* differed from that of *H. sapiens* mainly in the shape of the skull; the skull of the former was much heavier and had a lower forehead. (page 1045)

21. **c** The hand ax was the tool distinctive to *Homo erectus*. Its widespread distribution suggested communication among groups of hominids and the passing of information from one generation to the next. (page 1045)

22. **e** Neanderthals are classified as a subspecies of *Homo sapiens*. (page 1046)

23. **a** The Neanderthals disappeared suddenly about 30,000 years ago. (page 1047)

24. **b** Modern humans of the species *Homo sapiens* first appeared in Europe about 40,000 years ago. (page 1047)

25. **True** Old World monkeys include arboreal and terrestrial forms. The terrestrial varieties walk on all fours. (page 1035)

26. **False** The social behavior patterns of primates are so diverse that studying them for the purpose of interpreting human behavior patterns is not valid. (page 1038)

27. **True** Lucy is a member of the australopithecine First Family. Although Lucy and the hominid specimen known as OH 62 were about the same size and had a similar limb structure, the skull and dentition of OH 62 are more similar to specimens classified as *H. habilis*. (page 1042)

28. **False** To date, no specimen intermediate between the Neanderthals and modern humans has been discovered. (page 1047)

29. *Homo habilis* may have descended from either *Australopithecus afarensis* or *A. africanus*. Some scientists contend that it coexisted with other hominid species. Still others argue that *H. habilis* should be classified as an australopithecine. (page 1042)

30. Fire enabled members of *Homo erectus* to extend their diet to include foods that were inedible raw. Fires also discouraged other potential cave dwellers from inhabiting their caves and provided a center for social activities. (page 1045)

31. When an egg is fertilized, the DNA in the zygote nucleus is a combination of that of the sperm and egg. However, *all* the mitochondria, which have their own DNA, are of maternal origin. The studies of mitochondrial DNA performed by Rebecca Cann suggest that there was one common ancestral population from which all the races of modern humans have descended. (pages 1049, 1050)

C H A P T E R **51**

Animal Behavior and Its Evolution

MAJOR CONCEPTS

The behavioral characteristics of an organism are as much a consequence of natural selection as are its physical and bio-logical features, and the evolution of all these traits is governed by the same factors. Behavioral traits may be determined by the action of a single gene, which may be pleiotropic in some cases, or by the interaction of multiple genes.

For any behavior, there is a proximate and an ultimate cause. The proximate cause is the sequence of physiological events leading immediately to a particular behavior. The ultimate cause refers to the adaptive value of a behavior pattern—the reason it evolved.

Fixed action patterns are highly stereotyped, rigid, predictable behavior patterns that are triggered by specific stimuli, known as sign stimuli. Releasers are sign stimuli that are communication signals exchanged between members of a species. No learning is required for the performance of a fixed action pattern and the behavior appears in essentially complete form the first time it is elicited.

Learning is a process in which the responses of an organism to stimuli are modified through experience. The capacity to learn is roughly correlated with the length of the life span and with the size and complexity of the brain.

Three major types of learning are associative learning, imprinting, and imitative learning. There are two types of associative learning. Classical conditioning occurs during animal training, when an animal is taught to perform a specific action in response to a command, for which it receives a reward. Operant conditioning is trial-and-error learning. Imprinting is the means by which the young of many species learn to recognize members of their own species or species-specific behaviors. The young must be exposed to the appropriate stimulus during a critical period if adult behavior is to be normal. Imitative learning occurs when animals learn by mimicking the actions of others.

Within a society, communications among individuals maintain societal structure. Four basic types of behavior constitute social behaviors: selfish, cooperative, altruistic, and spiteful. These behaviors are defined according to the effect (in terms of increasing or decreasing fitness) on the donor and the recipient. The first three types are well documented in natural animal societies. Spiteful behavior has only been documented among humans.

Insects may be classified as solitary, subsocial, or eusocial based on degree of parental care, persistence of the nest site, division of labor among individuals, and overlap of generations. Eusocial insects, such as honey bees, have complex societies characterized by rigid caste systems and division of labor.

Vertebrate societies are often characterized by dominance hierarchies and territoriality. In a dominance hierarchy, the dominant individuals have priority access to food, shelter, and reproductive opportunities. Dominance hierarchies are established by actual or ritualized conflict behavior. Once a dominance hierarchy is established, overt conflict ceases until some outside factor upsets the hierarchy. The hierarchy is maintained by ritualized behaviors in which subordinate members signal their lower rank to dominant members. Subordinate individuals typically wait for openings to occur in higher ranks rather than challenge dominant individuals directly.

Territories may be held by an individual or by a group. Territories may be used for courtship and breeding, feeding, and/or rearing young. Territorial boundaries are typically established and maintained by ritualized behavior patterns such as vocalizations, displays, and scent marking rather than by open conflict. A territory owner has a "home team advantage" and is nearly always successful at defending the territory against invaders. Territories are vigorously defended against competitors of the same species. Territory defense against members of others species generally only applies to predators and to animals competing for a limited resource within the territory.

Kin selection refers to the differential reproductive success of groups of related individuals. The key to kin selection is the effect of a nonreproducing individual on the reproductive success of its relatives. Kin selection is operating when a youngster remains with his parents to help raise younger siblings. In these situations, the helper often contributes more of his genes to the next generation by helping raise siblings than if it left the family to raise its own young.

The evolutionary advantage of kin selection can be demonstrated by calculating degrees of relatedness among individuals. In a diploid species that reproduces sexually, the degree of relatedness between a parent and its offspring is 0.5, and the degree of relatedness between full siblings is also 0.5. In haplodiploid honey bees, the behavior of each caste in the society correlates with its degree of relatedness to the next generation.

The essence of the selfish gene concept is that an organism is merely a gene's way of making more genes. The focus is on the persistence of genes and the transience of individual organisms. Although an oversimplification, the selfish gene concept sheds light on some behavior patterns, especially those in which conflicts of interest among related individuals arise. Such conflicts of interest occur between mammalian mother and nursing young, between males and females of the same species, and when the males of some species take over a group of females with young. It also explains why it is advantageous for a subordinate member of a society or one without a territory to wait for an opportunity to mate rather than to fight for the opportunity and be killed.

Reciprocal altruism is exemplified by the statement, "you scratch my back and I'll scratch yours." In order for it to be a successful strategy in a society, individuals must meet more than once and be able to recognize one another. Each individual must also cooperate on the first encounter and reciprocate on the next encounter. If these conditions apply, cooperation pays off and cheating is punished. Reciprocal altruism is thought to have originated in groups of relatives in which the principle of kin selection would apply. Evidence is accumulating that individuals of many different species can distinguish relatives from nonrelated members of the same species.

The role of inheritance in human behavior is a matter of debate—some scientists argue that human behavior is fundamentally governed by genes, and other scientists contend that environmental and cultural influences are far more important.

HOW TO STUDY THE CHAPTER

Read the entire chapter through quickly, focusing on the major concepts.

Use the GUIDED STUDY OF THE CHAPTER to help you identify the important details as you **reread** the chapter. Writing out the answers to these questions will help fix them in your mind as well as provide you with a valuable study aid.

Answer the questions in TESTING YOUR UNDERSTANDING without the aid of your text. Check your answers against those in PERFORMANCE ANALYSIS. Analyzing your answers will give you valuable feedback on your level of understanding and preparedness for classroom testing.

GUIDED STUDY OF THE CHAPTER

I. **Introduction** *(page 1052)*

　　1. List four activities that constitute behavior.

Focus:　**The behavioral characteristics of an individual are products of natural selection.**

---

　　2. Identify four factors that determine the evolution of behavioral characteristics.

3. a. *Vocabulary:* Define ethology.

b. Name three scientists who pioneered the field of ethology.

4. List eight types of scientists that study behavior.

5. a. What three stimuli induce a young bird to open its mouth for food? (Figure 51–1)

b. What stimulus may induce a parent bird to feed only its own young?

II. The Genetic Basis of Behavior (pages 1052–1053)

6. Outline the minimal steps involved in translating the genetic components of behavior into actions.

7. a. Distinguish between hygienic and unhygienic strains of honey bees.

b. Summarize the results of Walter Rothenbuhler's breeding experiments on hygienic and unhygienic honey bees.

c. What conclusions did Rothenbuhler draw concerning the inheritance of hygienic behavior in bees?

> *Focus:* Behavior patterns may be due to single gene effects, including pleiotropic genes, or, more commonly, to the interactions of multiple genes.

8. Fruit flies carrying the *Hk* allele have what characteristics?

III. Proximate and Ultimate Causation (pages 1053–1054))

9. *Vocabulary:* Distinguish between a proximate cause and an ultimate cause.

10. a. Identify the proximate cause and the ultimate cause of the startle behavior exhibited by many fish.

b. Describe the structure of the Mauthner cells of fish and their function in the startle response.

IV. Fixed Action Patterns (pages 1054–1055)

11. *Vocabulary:* What are fixed action patterns?

> *Focus:* Fixed action patterns are specific and constant for a species and appear in nearly complete form the first time the organism is exposed to the appropriate stimulus.

12. *Vocabulary:* Distinguish between sign stimuli and releasers.

13. Define innate releasing mechanisms and describe their role in fixed action patterns.

14. Cite one example of a fixed action pattern.

> *Focus:* **In general, once a fixed action pattern has been initiated, it continues until the behavior has been completed.**

15. What is important for the correct outcome of many fixed action patterns?

16. For sticklebacks, identify the specific releaser for female attraction and male-male aggression.

V. Learning *(pages 1055–1058)*

A. General Remarks *(page 1055)*

17. Name two factors upon which all behavior patterns depend.

18. a. *Vocabulary:* Define learning.

b. The capacity to learn is loosely correlated with what two characteristics?

c. In general terms, describe animals whose behavior involves (i) fixed action patterns and (ii) learning.

19. a. *Vocabulary:* What is habituation?

b. Cite three examples of habituation.

B. Associative Learning *(pages 1055–1056)*

20. a. *Vocabulary:* What is associative learning?

b. Distinguish between classical and operant conditioning and cite one example of each.

> *Focus:* **Conditioned behavior is learned by the process of association.**

21. Use examples from Pavlov's experiments to define the following terms.

Unconditional stimulus

Unconditional response

Conditional stimulus

Conditional response

C. Imprinting *(pages 1056–1057)*

22. a. *Vocabulary:* What is imprinting?

b. How is imprinting of adaptive value to an individual?

> *Focus:* **Imprinting typically must occur within a critical period after birth.**

23. *Vocabulary:* What are precocial birds? (See also Figure 51-7.)

The Song of the White-Crowned Sparrow (pages 1056–1057)

24. a. Describe song development in a juvenile male white-crowned sparrow living in the wild.

b. A sparrow's song imparts what two pieces of information?

25. According to the results of Peter Marler's laboratory experiments, describe the ability of male white-crowned sparrows to sing under the following conditions.

a. Hand-reared bird never exposed to adult song:

b. Bird exposed to adult song during critical period then kept in sound isolation:

c. Bird exposed to adult song during critical period then deafened:

d. Bird exposed to songs of other sparrow species during the critical period (but not to songs of its own species) then kept in sound isolation:

26. How does the critical period for exposure to the adult song compare with the time at which a young male starts to sing?

27. What three requirements must be met for normal song development in male white-crowned sparrows?

D. Imitative Learning *(pages 1057–1058)*

28. Describe one example of imitative learning in birds other than the white-crowned song sparrows.

29. a. Cite two examples of imitative learning in the macaques of Koshima Island, Japan.

b. New behaviors spread through the macaques according to what pattern?

VI. Social Behavior: An Introduction *(pages 1058–1059)*

30. a. *Vocabulary:* Define a society.

b. What force holds a society together?

31. *Vocabulary:* What is altruism?

32. a. List the four types of social behavior proposed by W. D. Hamilton.

b. Which types increase the fitness of the donor?

c. Which types increase the fitness of the recipient?

d. Which types decrease the fitness of the donor?

e. Which types decrease the fitness of the recipient?

f. Which three types of social behavior are well-documented in natural, nonhuman animal societies?

33. The evolution of which two types of behavior initially presented a puzzle to ethologists? Explain the basis for this puzzle.

VII. Insect Societies (pages 1059–1061)

A. General Remarks (page 1059)

> *Focus:* Insect societies are among the most complex in the animal kingdom.

B. Stages of Socialization (pages 1059–1060)

> *Focus:* Social insects evolved from solitary ancestors.

34. Fill in the following table by indicating which traits apply to each insect lifestyle. (You will need to read the next section on honey bees to complete the table.)

Table for Question 34

	SOLITARY	SUBSOCIAL	EUSOCIAL
Care of young (describe)			
Persistence of nest/colony			
Overlap between generations			
Division of labor			
Example(s)			

C. Honey Bees (pages 1060–1061)

35. Fill in the following table.

Table for Question 35

	QUEEN	WORKERS	DRONES
Sex			
Ploidy			
Origin			

36. Summarize the development of workers from fertilized egg to emergence from the pupa, including the time spent in each stage (if provided in the text).

37. a. Identify the three separate occupations of a worker and describe the duties that accompany each occupation.

    b. What is the average life span of a worker?

38. a. Name and describe the structures of a worker honey bee that are involved in collecting and carrying pollen. (Figure 51–11)

    b. When does transfer of pollen from the collecting structures to the storage compartment occur? (Figure 51–11)

39. How does the visual spectrum perceived by honey bees differ from that perceived by humans? (Figure 51–11)

The Queen (pages 1060–1061)

40. How does the larval development of a queen differ from that of a worker?

41. a. By what general mechanism does the queen influence activity of the other colony inhabitants? (See also Figure 51–12.)

    b. What prevents ovarian development in workers?

42. What happens when a hive loses its queen?

43. From what substance is honey made? (Figure 51–12)

The Annual Cycle (page 1061)

44. How do honey bees maintain their temperature while overwintering in the hive?

45. a. Describe the process by which a new bee colony forms.

   b. By what mechanism is the departure of the old queen coordinated with the emergence of a new queen?

46. Identify two sources (one major and one minor) of the unfertilized eggs that develop into drones.

47. a. How many nuptial flights will a queen make in her lifetime?

   b. What is the origin of the drones with which a queen mates?

> *Focus:* A queen honey bee makes one nuptial flight in her life, during which she may mate with more than one drone to obtain sperm.

48. Where are sperm stored in the queen?

49. During what season does a queen typically lay unfertilized eggs?

50. Summarize the role of drones in the life of a colony.

VIII. **Vertebrate Societies** *(pages 1062–1065)*

A. **General Remarks** *(page 1062)*

> *Focus:* Although only one vertebrate society is known to have a rigid caste system, many do have a highly structured social order.

51. How does the caste system of naked mole rats resemble that of eusocial insects? (Figure 51–13)

B. **Dominance Hierarchies** *(pages 1062–1063)*

52. a. In general, how are dominance hierarchies maintained?

   b. What two factors are determined by a dominance hierarchy?

53. a. Describe the pecking order in domestic chickens.

   b. High-ranking hens have what privileges?

> *Focus:* Once a pecking order is established, fighting does not occur unless members are added to or removed from the flock.

54. How does a pecking order influence reproduction in a flock?

> *Focus:* Dominance hierarchies minimize conflict (and the energy wasted on fighting) within a group.

55. How do the subordinate members of a wolf pack contribute to the reproductive success of the dominant members? (Figure 51–14)

56. Indicate how (a) a submissive wolf and (b) a submissive baboon acknowledge their subordinate rank to a dominant member of the society. (Figure 51–14)

57. What three messages may the act of presenting communicate in a baboon society? (Figure 51-14)

Tidbit: Domestic dogs have adapted so well to life with humans because the social structure of the wild canid society was replaced with that of a human "pack" during the process of domestication. When a dog licks its owner's face, it is acknowledging the human as "top dog."

C. **Territories and Territoriality** *(pages 1063-1065)*

58. *Vocabulary:* Define a territory and territoriality.

Territoriality in Birds (page 1063)

59. Who first recognized territoriality in birds?

60. a. In birds, members of which sex generally establish a territory?

b. Name four activities that may occur within a territory.

c. Describe three types of territories based on the activities that take place within them.

61. What is the fundamental advantage of territoriality to birds?

62. In some species, failure of a male bird to secure a territory may have what consequence?

Territorial Defense (pages 1064-1065)

Focus: Territories are vigorously defended against all rivals of the same species that enter the territorial boundaries.

63. In defending his territory, how does a male bird react to animals of other species that enter his territory?

64. a. How successful is a territory owner in defending his or her territory against invaders?

b. Describe five types of behavior used in territorial defense.

65. Briefly describe the territorial activity of a male cichlid.

Focus: Territorial defense is usually accomplished by ritualized behaviors rather than by actual violence.

66. What else besides territorial defense may be accomplished by these ritualized behaviors?

D. *Essay: A Tale of Two Territories* *(pages 1066-1067)*

67. Summarize the conflict between the older swans and the younger pair during the summers of 1986 and 1987.

68. a. What strategy was adopted by the younger swan pair in 1988 that resulted in the successful rearing of young?

b. What degree of violence characterized this territorial dispute?

IX. Kin Selection (pages 1065–1070)

A. General Remarks (pages 1065–1069)

69. a. Summarize the concept of group selection—now considered to be invalid by most biologists—that was proposed by Wynne-Edwards.

    b. On what basis was the group selection hypothesis rejected?

    c. What was the usefulness of Wynne-Edwards's idea?

70. Darwin suggested what general mechanism might maintain sterile castes within a population?

71. a. Define kin selection and summarize Hamilton's hypothesis of kin selection, including the principle upon which it is based.

    b. What is the critical factor in kin selection?

72. For mammals, indicate the likelihood that the members of each of the following pairs share an allele and the degree of relatedness between them. Show the calculations and reasoning by which you arrive at your answers.

    Father—offspring

    Mother—offspring

    Two siblings

73. a. Define haplodiploidy.

    b. In honey bees, indicate the likelihood that the members of each of the following pairs share an allele and the degree of relatedness between them. Show your work.

    Two workers

    Queen—worker

    Drone—worker

    c. Correlate these figures with the behavior of workers and drones in the colony.

74. Identify two criticisms of Hamilton's use of honey bees as an example of kin selection.

75. How does the concept of inclusive fitness differ from Darwin's definition of fitness?

B. Tests of the Hypothesis (pages 1069–1070)

76. Describe how silver-backed jackal societies support Hamilton's concept of inclusive fitness and kin selection.

77. a. How does the genetic contribution of helper Florida scrub jays compare with that of helper silver-backed jackals?

b. In spite of the data in Table 51–4, why is it advantageous for young jays to remain with their parents as helpers?

c. In Florida scrub jays, how does acquisition of a territory correlate with helping behavior?

X. The Selfish Gene *(pages 1070–1074)*

A. General Remarks *(pages 1070–1071)*

> *Focus:* According to the selfish gene concept, an organism is merely a gene's way of making more genes.

78. Who articulated the selfish gene concept?

B. Conflicts of Interest *(pages 1071–1074)*

> *Focus:* In focusing on the survival of genes, it became apparent to biologists that conflicts of interest may exist between closely related individuals.

79. Outline the conflict of interest between a mammalian mother and nursing young.

80. a. Cite two examples of infanticide by males who take over a group of females with young. (See also Figure 51–20.)

b. Who first reported such infanticide?

c. What makes this an evolutionarily successful strategy?

Male vs. Female (pages 1071–1074)

> *Focus:* In most animal species, the female has a much greater investment in producing and rearing young than the male.

81. Summarize the relative investment of males and females in producing the next generation. (For now, consider only the most common situation.)

82. What behaviors might males exhibit in demonstrating their worth to potential mates?

83. a. Selection for a reproductive strategy in which the male participates equally in care of the young depends on what three factors?

b. In this situation, how does prolonged courtship demonstrate that the interests of both the male and the female will be protected?

> *Focus:* In systems where both parents rear the young, the female who cheats loses nothing in terms of passing her genes on to the next generation. The male, however, loses completely in terms of reproductive success if his mate cheats.

84. What evidence is there of "cheating" by the females of some bird species?

C. *Essay:* Arts and Crafts of Bowerbirds *(page 1073)*

85. List several objects that male bowerbirds may use to decorate their bowers.

86. What is the relationship between the degree to which a bower is decorated and the brightness of a male's plumage?

87. The bower is probably an evolutionary derivative of what structure?

88. Identify the advantages of this system to male and female bowerbirds.

D. The Advantage of Waiting *(page 1074)*

89. How is waiting sometimes the best way of ensuring reproductive success?

90. Cite two examples illustrating "the advantage of waiting."

XI. Reciprocal Altruism *(pages 1074–1075)*

A. General Remarks *(pages 1074–1075)*

91. a. *Vocabulary:* Define reciprocal altruism.

b. Cite three general examples of reciprocal altruism.

c. Describe one specific example of reciprocal altruism.

92. What two conditions are necessary if reciprocal altruism is to succeed as a strategy?

93. Reciprocal altruism is believed to have originated in what fashion?

> *Focus:* Accumulating data indicate that a wide variety of animals can distinguish relatives from nonrelated members of the same species.

B. *Essay:* Kin Recognition in Tadpoles *(pages 1076–1077)*

94. a. Outline the experiments of Blaustein and O'Hara.

b. The tadpoles identified relatives by what mechanism?

c. Summarize the two hypotheses proposed to explain this phenomenon.

> *Focus:* In most species in which kin recognition has been identified, the ability to recognize kin appears early in development.

95. Name three animal behaviors that involve kin recognition.

XII. The Biology of Human Behavior *(page 1075)*

96. Outline the opposing viewpoints concerning the role of heredity in human behavior.

97. What consequences can follow from the idea that human behavior is biologically determined?

XIII. Summary *(pages 1075–1079):* Read the summary. If you are familiar with the essential features of the material presented there, you are ready to complete the section TESTING YOUR UNDERSTANDING.

TESTING YOUR UNDERSTANDING

After you have completed this examination, compare your answers with those in the section that follows.

1. Ethology is the study of:
   a. behavior patterns and their evolutionary origins.
   b. the interactions among populations within a community.
   c. the coevolution of predators and their prey.
   d. cultural customs characteristic of various human societies.
   e. the adaptive value to a species of specific physical and behavioral characteristics.

2. Which activity is NOT part of the mechanism by which the information coding for a genetically determined behavior pattern is converted into an action by the animal?
   a. Specific molecules are synthesized.
   b. Molecules that are synthesized are organized into structures capable of responding to stimuli.
   c. Multiple repetitions of the stimulus are required before the animal responds appropriately.
   d. Pathways develop that transmit information within the organism.
   e. Structures and pathways are modified as an organism interacts with its environment.

3. Which statement is NOT true of hygienic and unhygienic behavior in honey-bee colonies?
   a. The testcross performed in Rothenbuhler's study revealed the 9:3:3:1 phenotypic ratio typical of characteristics governed by two independently assorting genes.
   b. There are two separate behaviors involved in hygienic behavior.
   c. Unhygienic behavior is inherited as a Mendelian dominant trait.
   d. A testcross between hygienic bees and hybrid bees revealed that two genes are involved in this behavior pattern.
   e. Bees that have a dominant allele for either of the two genes involved are functionally unhygienic.

4. The proximate cause of the startle behavior characteristic of many fish and larval amphibians is the:
   a. activation of the Malpighian system.
   b. stimulation of Mauthner cells.
   c. operation of a monosynaptic reflex arc.
   d. operation of a polysynaptic reflex arc.
   e. resulting escape from predators.

5. Which of the following does NOT represent a fixed action pattern and its sign stimulus or releaser?
   a. aggressive behavior in a male stickleback; a red object
   b. egg-rolling behavior of greylag geese; an egg outside the nest
   c. prey-capture behavior of a frog; a bug-sized moving object
   d. alarm calls by ground squirrels; a potential predator
   e. the gaping of a nestling bird's mouth; a light touch on the nest

6. Which statement is NOT true of learning?
   a. A common example of learning is the gaping of a nestling bird's mouth when the parent lands on the nest.
   b. The capacity for learning is apparently correlated with the length of the life span.
   c. The size and complexity of an animal's brain is related to its capacity to learn.
   d. Behavior patterns in animals with short life spans tend to be in the form of fixed action patterns.

   e. Habituation is one of the simplest forms of learning.

7. In the classical conditioning experiments of Ivan Pavlov, the salivation of the dog when it heard a bell ring was a(n):
   a. unconditional stimulus.
   b. unconditional response.
   c. trial-and-error response.
   d. conditional stimulus.
   e. conditional response.

8. Which situation is an example of imprinting?
   a. Male white-crowned sparrows must hear the song of a mature male when they are between 10 and 50 days of age if they are to sing correctly themselves.
   b. A dog learns by accident that a sliding glass door will open if it jumps and hits the handle at an angle with its paws.
   c. Eleven species of birds learned to tear open the paper caps of milk bottles—a behavior originated by titmice in their neighborhood.
   d. A dog learns to sit in response to a verbal command because it receives food when it does so.
   e. Ground squirrels stop responding to distress calls if they are made frequently in the absence of predators.

9. Which activity is NOT among those acquired by macaques of Koshima Island, Japan, after scientists started feeding them on the beaches?
   a. washing sweet potatoes to remove sand
   b. swimming and diving for seaweed
   c. washing wheat kernels to separate them from sand
   d. dipping sweet potatoes in the ocean to pick up salt
   e. throwing wheat kernels in the air so the wind can separate them from the sand

10. Who first proposed a classification of social behaviors based on the effect of the behavior on the fitness of the donor and recipient?
    a. Konrad Lorenz
    b. Peter Marler
    c. W. D. Hamilton
    d. Niko Tinbergen
    e. V. C. Wynne-Edwards

11. In cooperative behavior, the fitness of the donor _____, and the fitness of the recipient _____.
    a. increases; decreases
    b. increases; increases
    c. increases; remains the same
    d. decreases; increases
    e. decreases; decreases

12. When the evolution of social behaviors was first studied, scientists expected which type of behavior would be produced by natural selection?
    a. spiteful
    b. selfish
    c. cooperative
    d. altruistic
    e. The existence of all of these behaviors was predicted.

13. Which statement characterizes solitary insects?
    a. The female returns to feed her larvae.
    b. The young often lay their eggs in the same nest in which they hatched.
    c. The generations do not overlap.
    d. The nest site persists for several generations.
    e. The males cooperate with the females in caring for the young.

14. Which activity is NOT performed by a houseworking bee?
    a. enlarging the comb
    b. cleaning empty cells in the comb
    c. guarding the hive entrance
    d. removing sick or dead individuals from the hive
    e. feeding the larvae, queen, and drones

15. The eyes of honey bees do not detect _____ light, but can detect _____ light.
    a. red; violet
    b. violet; red
    c. ultraviolet; red
    d. red; ultraviolet
    e. blue; infrared

16. A new colony of honey bees forms when:
    a. a group of workers and drones leaves a hive and migrates to a suitable site.
    b. the established queen of one hive leaves, taking half the workers with her.
    c. a newly hatched queen goes on her maiden flight and takes half the workers with her to a new site.
    d. a group of drones kidnap a queen during her maiden flight and half the workers from the queen's hive follow her.
    e. the scouts of an established hive kidnap their queen and take her to a new site.

17. Which vertebrate species has a caste system similar to that of the eusocial insects?
    a. timber wolf
    b. baboon
    c. prairie dog
    d. African mole
    e. naked mole rat

18. Which statement is NOT true of the pecking order characteristic of hens?
    a. High-ranking hens typically have a sleek, well-groomed appearance.
    b. Low-ranking hens usually do not reproduce.
    c. If a new hen is added to the group, she automatically assumes the lowest rank.
    d. Once established, the pecking order is maintained by ritualized behaviors.
    e. In terms of reproductive success, the outcome is equivalent to what it would be if there was no social order.

19. Which statement is NOT true of territories?
    a. They may be defended by an individual or by a group.
    b. They may be defended against members of the same or of different species.
    c. They may be used only for courtship and mating.
    d. Defense of the territory is performed only by males.
    e. A male bird without a territory may not mature sexually.

20. Who first proposed the concept of group selection, which hypothesized that natural selection favors groups in which individuals, for the benefit of the entire group, do not reproduce?
    a. V. C. Wynne-Edwards
    b. W. D. Hamilton
    c. Eliot Howard
    d. Karl von Frisch
    e. Richard Dawkins

21. What is the critical factor in kin selection?
    a. the inclusive fitness of the family as a whole
    b. the reproductive success of one family relative to that of another family
    c. the degree of altruistic behavior exhibited by members of one family relative to that exhibited by members of other families
    d. the degree of relatedness among members of the family group
    e. the effect of a nonreproducing family member on the reproductive success of its relatives

22. The likelihood that you and your full sibling share a particular allele obtained from your mother is:
    a. 1.0.
    b. 0.75.
    c. 0.50.
    d. 0.25.
    e. impossible to determine from the information given.

23. In haplodiploid honey bees, the degree of relatedness between a worker and her father is:
    a. 1.0.
    b. 0.75.
    c. 0.50.
    d. 0.25.
    e. impossible to determine from the information given.

24. Sarah Blaffer Hrdy was the scientist who:
    a. observed the cheating behavior of female herons.
    b. reported infanticide by male Hanuman langurs when they took over a group of females.
    c. studied helping behavior of silver-backed jackals.
    d. proposed the adaptive value of waiting for an opportunity to reproduce rather than fighting to obtain the opportunity.
    e. has studied the territorial behavior of swans on Long Island, New York, for several years.

25. Which statement does NOT reflect an advantage of the long courtship characteristic of some bird species?
    a. It ensures that the offspring belong to the male.

b. It demonstrates the intent of the male to remain with the female.

c. Nest-building may be part of the courtship ritual.

d. It provides time for the partners to acquire the feeding territory needed to raise the young.

e. The female has the opportunity to assess the male's qualities.

26. In which situation is waiting NOT an advantageous strategy?

a. A hen ranks 26 in a flock of 30 hens.

b. A male Florida scrub jay lives in an area with no available territories.

c. A young male lion is being driven from his pride.

d. The young of a female Hanuman langur has just been killed by a new male.

e. A pair of swans live in an area where all the territories are occupied.

27. How do members of a school of Cascades frog tadpoles learn of a predator in their midst?

a. An injured tadpole releases a chemical that warns other school members of danger.

b. A captured tadpole releases a distress call.

c. When one member sees the predator, it darts away and the other members follow.

d. They detect subtle movements of the water created by the predator.

e. All of these are mechanisms by which the presence of a predator may be communicated.

28. T or F The factors that govern the evolution of behavioral characteristics are the same as those governing the evolution of other characteristics.

29. T or F Habituation is a form of learning in which an animal repeats an activity so many times that the activity becomes a habit and does not require conscious thought.

30. T or F The novel food-washing behavior that spread through the macaques of Koshima Island, Japan, was initiated by the dominant male in the group.

31. T or F When British titmice started tearing the foil caps off milk bottles and drinking the cream, at least 11 other species acquired the behavior by imitation.

32. T or F For territorial species of birds, the close proximity of a rival is sufficient to elicit an aggressive response by the territory owner.

33. T or F The reproductive success of a male is always maximized by inseminating as many females as possible. Thus for males, it does not really matter if the mate cheats, since he has offspring by other females.

34. T or F The competition between rival male bowerbirds may include raids to damage another bird's bower and steal decorations.

35. T or F The ability to recognize related individuals is present in a wide variety of animals.

36. Cite one example of a behavior pattern inherited by a pleiotropic gene and describe the characteristic effects of the gene.

37. Distinguish between classical and operant conditioning and cite one example of each type of learning.

38. In Florida scrub jays, young males who remain with their parents as helpers make less of a contribution to the next generation than if they left and reared their own young. Cite two reasons why it is advantageous for them to remain as helpers. Under what conditions do they leave to rear their own young?

39. Cite three examples illustrating the conflicts of interest that exist between members of a nuclear family or between actual or potential mates.

40. Identify two conditions that must be met if reciprocal altruism is to be a successful strategy. How is this type of behavior thought to originate in a society?

PERFORMANCE ANALYSIS

1. a Ethology is the comparative study of patterns of behavior and the construction of hypotheses concerning their evolutionary origins. (page 1052)

2. c Genetically determined behavior patterns are typically rigid, stereotyped, and predictable. They appear in nearly complete form the first time they are elicited by the appropriate stimulus. (page 1053)

3. a The testcross performed by Rothenbuhler revealed the 1:1:1:1 phenotypic ratio characteristic of two independently assorting genes. (page 1053)

4. b The proximate cause of a behavior is the physiological mechanism by which the behavior is accomplished. The proximate cause for startle behavior of many fish and larval amphibians is stimulation of two Mauthner cells, which are giant neurons whose cell bodies lie within the medulla and whose axons stimulate motor neurons in every segment of each side of the body. (page 1054)

5. d Fixed action patterns are innate behavior patterns that are rigid, highly stereotyped, and appear in nearly complete form the first time the behavior is elicited. They are initiated by external stimuli known as sign stimuli. Releasers are sign stimuli that are communication signals exchanged between members of a species. The alarm calls given by ground squirrels are not performed by every animal who sees the predator; therefore, they are not the result of a fixed action pattern. (page 1054)

6. a The gaping of a nestling bird's mouth when the parent lands on the nest is an example of a fixed action pattern. (page 1055)

7. e In his experiments, Pavlov conditioned a dog by presenting food to the dog when he rang a bell. The dog

learned to associate food with the ringing bell and began to salivate at the sound of the bell. Initially, the food was the unconditional stimulus that triggered the unconditional response, salivation. When the ringing bell (a conditional stimulus) stimulated salivation in the absence of food, salivation was a conditional response. (page 1055)

8. **a** Imprinting is the type of learning in which young learn to distinguish members of their own species from members of other species. They may also acquire species-specific behavior patterns through imprinting. Imprinting requires the exposure of the young to the appropriate stimulus during a critical period in the life of the young. Examples include the following behavior of precocial birds and song development in white-crowned sparrows. (page 1056)

9. **e** Several new behaviors began in the macaques of Koshima Island, Japan, and spread through the troop by imitation. A single female named Imo initiated two food-handling behaviors—washing the sand off sweet potatoes by dipping them in the ocean, and throwing wheat into the water to separate it from the sand. The macaques also started swimming and diving as a consequence of being fed on the beaches. (pages 1057, 1058)

10. **c** W. D. Hamilton classified social behaviors based on the effect of the behavior on the fitness of the donor and the recipient. His four classifications were selfish, cooperative, altruistic, and spiteful. (page 1059, Table 51-1)

11. **b** In cooperative behavior, the fitness of both the donor and the recipient increases as a result of the cooperation. (Table 51-1)

12. **b** In early discussions on the evolution of behavior patterns, scientists expected that selfish behaviors would be favored by natural selection. The evolution of cooperative and altruistic behavior patterns was a puzzle until the advent of the selfish gene hypothesis and the concepts of inclusive fitness and kin selection, which provided valuable insights that enabled biologists to form hypotheses concerning the evolutionary origins of these types of behavior. (page 1059)

13. **c** Females of solitary insect species generally lay their eggs, stock the nest with food, leave, and die before the eggs hatch. (page 1059)

14. **e** Houseworking honey bees may enlarge the comb, clean empty cells, guard the hive entrance, and remove dead or sick bees from the hive. Nurse workers feed the larvae, queen, and drones. (page 1060)

15. **d** The eyes of honey bees detect ultraviolet light but not light in the red wavelengths. (Figure 51-11)

16. **b** A new colony of honey bees may be established when the queen from an established hive leaves, taking about half the workers with her. Her departure is coordinated with the emergence of a new queen by sound signals transmitted through the hive. (page 1061)

17. **e** Naked mole rats of Africa have a caste system similar to that of the eusocial insects in which there is a single reproducing female. Both male and female members of the colony work by foraging for food and by maintaining the tunnels in which they live. (Figure 51-13)

18. **c** If a new member is added to a flock of chickens, the entire pecking order will have to be reestablished, a process that involves considerable conflict. (page 1062)

19. **d** Although males are the primary territory defenders in many species, females may also participate in territory defense. (pages 1063, 1064)

20. **a** V. C. Wynne-Edwards proposed the concept of group selection, an idea that did not last very long among mainstream behaviorists, but did stimulate much discussion. This debate led to the development of several important concepts concerning the origin of altruistic behaviors. (page 1065)

21. **e** The critical factor in kin selection is the effect of a nonreproducing family member on the reproductive success of its relatives. (page 1067)

22. **d** The likelihood that you obtained a particular allele from your mother is 0.5 (you received one-half of your genes from her). The same is true of a full sibling. The likelihood that you and your sibling both possess a particular allele belonging to your mother is governed by the product rule of probability, which states that the probability of two events occurring together equals the probability of one event occurring alone multiplied by the probability of the other event occurring alone. $0.5 \times 0.5 = 0.25$. The degree of relatedness for full siblings takes into account both parents and is calculated by adding the chance that you both share an allele from your father to the chance that you both share an allele from your mother. $0.25 + 0.25 = 0.5$. (pages 1067, 1068)

23. **c** A worker receives half her genes from her haploid father, who has only one set, so the degree of relatedness to her father is 0.5. (For a more complete discussion on coefficients of relatedness and their calculations in haplodiploid species, the hopelessly confused and extremely curious are referred to *An Introduction to Behavioural Ecology*, by J. R. Krebs and N. B. Davies, Sinauer Associates, 1981, pages 19 and 198–200.) (page 1068)

24. **b** Sarah Blaffer Hrdy was the first scientist to report infanticide by males at the time of taking over a group of females. She observed this behavior in Hanuman langurs. (page 1071)

25. **d** In territorial bird species, the male generally acquires the territory *before* the courtship begins, and the quality of the territory is one criterion evaluated by potential mates. (page 1072)

26. **d** If a female Hanuman langur has just lost her young to infanticide, she can best maximize her reproductive success by producing more young as soon as possible by mating with the new male. (page 1074)

27. **a** Tadpoles of Cascades frogs swim in schools. When one member of the school is captured and injured by a predator, it releases a chemical warning into the water and the other tadpoles swim away. Of course, this does not benefit the victim, but it does benefit the other school members, who are relatives of the victim. (essay, page 1076)

28. **True** All factors that are determined by genes are subject to the forces of natural selection. (page 1052)

29. **False** In habituation, an animal learns to ignore a repeated but harmless stimulus to which it initially responded. An example is the habituation of amoeba to a light from which it cannot escape. (A classic fairy tale example is the story of the boy who cried wolf; his villagers underwent habituation.) (page 1055)

30. **False** The novel food-washing behaviors exhibited by the macaques of Koshima Island were initiated by a female, called Imo by the researchers studying the group. (page 1057)

31. **True** This is a classic example of imitative learning. (page 1057)

32. **False** In territorial bird species, it is not the close proximity of a rival to the territory owner that initiates defensive behavior but rather entry into the territory. As long as a rival remains outside the boundaries, the territory owner will not initiate defensive behavior. (page 1064)

33. **False** In many species, the male pair bonds with only one female and will not mate with another female, thus he loses completely (in terms of reproductive success) if she cheats. The female does not lose since all her offspring have half her genes, no matter who fathers them. (page 1074)

34. **True** The competition between male bowerbirds may be quite intense and may include raids by one male on the bower of another, during which the raided bower is damaged and decorations may be stolen. (essay, page 1073)

35. **True** Biologists have discovered the ability to recognize relatives in many types of organisms, including insects, amphibians, birds, and mammals. (page 1075)

36. Fruit flies that possess the *Hk* allele are more active than those lacking it, jump violently when a shadow passes overhead, exhibit rapid leg movements when anesthetized, and demonstrate abnormal mating behaviors. (page 1053)

37. Both classical and operant conditioning are types of associative learning. As a consequence of classical conditioning, an animal displays a particular behavior pattern in response to a specific stimulus. An example is a dog that has been trained to go to its owner (and perhaps receive a treat) when the owner says, "Come!" Operant conditioning occurs when an animal learns a behavior by trial and error. An example is pecking behavior of young chicks; both accuracy and efficiency (fewer pecks per food item obtained) increase with experience. (pages 1055, 1056)

38. (1) All the available territories may be occupied. By remaining and helping rear younger siblings, the number of surviving young is greater than the number that could be raised by his parents alone. Thus, the young male increases his parents' (and indirectly, his) contribution to the gene pool. (2) A young male often obtains a territory by inheriting part of the territory belonging to his parents. Young males leave the family of origin as soon as a territory becomes available. (page 1070)

39. (1) In mammals, it is in the best interest of the young to nurse for as long as possible. It is in the best interest of the mother to wean the young at an appropriate time and begin a new reproductive investment. (2) In most animal species, it is in the best interest of the male to inseminate as many females as possible. The female's interests are best served by selecting the best male to sire her offspring. (3) In some species, including lions and Hanuman langurs, when a new male takes over a group of females, he may kill the young. The females are then available to mate with the new male. (page 1071)

40. (1) The individuals in the society must meet frequently and be able to recognize one another. (2) Both parties must cooperate on the first encounter and on subsequent encounters. Reciprocally altruistic behaviors are thought to originate among related individuals in a society. (page 1075)

REVIEW TEST 15

Section 7: Chapters 49–51

This review test is *not* designed to cover all of the important information in these chapters. However, it does touch on the major topic categories in each chapter. It will also give you valuable practice in taking this type of test. When you are finished, check your answers with those provided in the following PERFORMANCE ANALYSIS section.

1. Populations that can interbreed freely but differ from one another somewhat are referred to as different:
   a. genera.
   b. subspecies.
   c. species.
   d. races.
   e. breeds.

2. Dinosaurs had disappeared from the earth by the end of the _____ period, about _____ million years ago.
   a. Jurassic; 65
   b. Permian; 65
   c. Cretaceous; 65
   d. Tertiary; 130
   e. Triassic; 130

3. Which statement reflects an unusual feature of the tools used by the Cro-Magnons?
   a. They were simple and made of stone.
   b. They included tools made by using another tool.
   c. Many of them were flakes shaped in a large variety of ways.
   d. They included tools used to shape bone and ivory.
   e. They included needles and fishing hooks.

4. Which statement is NOT characteristic of a fixed action pattern?
   a. Several encounters with the sign stimulus are required before the behavior pattern is completed correctly.
   b. The behavior pattern is usually elicited by a very specific stimulus.
   c. In most cases, once the behavior pattern is initiated, it must be carried through to completion.
   d. Such a behavior pattern is highly stereotyped, rigid, and predictable.
   e. The behavior pattern involves innate releasing mechanisms in the brain.

5. Which statement is NOT true of the selfish gene concept?
   a. When it was proposed, biologists realized that conflicts of interest occur among closely related animals.
   b. It has shed light on the adaptive value of courtship and mating behaviors in birds.
   c. It provides insights into the infanticide that occurs when a new male lion takes over a group of females with young.
   d. It provides support for the kin selection hypothesis.
   e. It provides insight into the mechanism underlying group selection.

6. Who proposed the concept of reciprocal altruism?
   a. Gerald Wilkinson
   b. Robert Trivers
   c. Richard Dawkins
   d. Sarah Blaffer Hrdy
   e. V. C. Wynne-Edwards

7. T or F   One plausible explanation for the origin of the Galapagos finches is that the ancestor (or ancestral group) was blown there from the South American mainland during a storm.

8. T or F   Some authorities estimate that as few as 1/1,000 of 1 percent of all species that have ever lived are living today.

9. T or F   The upright suspension of brachiation is thought to have played a role in the transition from crouching to erect posture in primate evolution.

10. T or F   Bipedalism apparently developed to allow early male hominids to hold weapons and defend their mates and young.

11. T or F   In general, far more examples of gradual change have been found in the fossil record than anyone expected, substantiating Darwin's concept of the tempo of evolution.

12. T or F   In a honey-bee society, the drones, which develop from unfertilized eggs, guard the hive from potential invaders.

13. T or F   The group selection hypothesis has been rejected by nearly all biologists because restraint-of-breeding alleles, if they existed, could not be perpetuated in a population.

14. Match each type of learning with the correct example. Use each answer only once. Each blank will have only one answer.
   1. habituation
   2. classical conditioning
   3. operant conditioning
   4. imprinting
   5. imitation

a. _____ A gosling will follow the first moving object it sees after hatching.

b. _____ A trained dog sits when given the verbal command, "Sit!"

c. _____ An amoeba ceases to respond to a bright light from which it cannot escape.

d. _____ Macaques on a Japanese Island learned to wash their food by watching one member of the group do so.

e. _____ The pecking accuracy of a growing chick improves with practice.

15. Summarize the advantages that accrue to an individual or group that holds a territory.

16. Cite at least two examples of premating isolating mechanisms that involve either sound, smell, or vision.

17. A great deal of controversy surrounds the australopithecine fossils, including debates about the actual number of species and which, if any, were ancestors of the genus *Homo*. Cite three of the four reasons for this controversy.

PERFORMANCE ANALYSIS

1. **d** Races are defined as populations of a species that differ somewhat but can still interbreed. When races are distinct enough to warrant a unique Latin name, they are called subspecies. (page 1010)

2. **c** By the end of the Cretaceous period, which occurred 65 million years ago, the dinosaurs had disappeared from the earth. (pages 1024, 1031)

3. **a** The tools of earlier hominids, including *Homo habilis* and *Homo erectus*, were made of stone. However, the tools of the Cro-Magnons were numerous and quite complex by comparison, and this group regularly used tools to make other tools. (page 1047)

4. **a** A fixed action pattern appears in a nearly complete form the first time the behavior is elicited. (page 1054)

5. **e** The group selection hypothesis has been rejected by nearly all biologists. The selfish gene concept does provide support for the kin selection hypothesis. (pages 1070, 1071)

6. **b** Robert Trivers proposed the concept of reciprocal altruism. Gerald Wilkinson reported reciprocal altruism among related and unrelated individuals in a vampire bat society. (page 1074)

7. **True** It has been proposed that the ancestor (or ancestral group) of the Galapagos finches came from the South American mainland and may have been blown there during a storm. (page 1017)

8. **True** The vast majority of all species that have evolved have subsequently become extinct. Some scientists believe that as few as 1/1,000 of 1 percent of all species may still be in existence today, and the figure is certainly less than 1/10 of 1 percent. (page 1024)

9. **True** The upright suspension of brachiation is thought to have been a factor in the transition of body structures associated with crouching to those associated with an erect posture. (page 1033)

10. **False** The freeing of the hands for tool (or weapon) use was apparently *not* a selective pressure involved in the evolution of bipedalism. A more likely selective pressure was the ability of males to procure food at a distance and carry it back to females and young. Or a female collective may have existed and favored those individuals that could carry children and collect and transport food. Other scientists contend that the transition to bipedalism was triggered by the change in climate at the end of the Miocene, which caused an increase in grasslands at the expense of forests. (page 1044)

11. **False** If Darwin's concept of evolution as gradual change over time was true, more examples of this type of change would have been expected in the fossil record. Darwin thought the incompleteness of the fossil record was the reason for the paucity of examples supporting gradual change. Niles Eldredge and Stephen Jay Gould have proposed that the fossil record is in fact an accurate reflection of the evolutionary process and that evolution is characterized by periods of rapid change interspersed with periods of relative stability during which organisms undergo little change. This is the punctuated equilibrium model of evolution. (page 1028)

12. **False** Honey bee drones contribute nothing to the protection of the hive. Their only role in the society is to mate with queens, usually from neighboring colonies. (page 1061)

13. **True** Perhaps the greatest contribution of the group selection hypothesis was that it stimulated thought (and debate), that eventually led to the kin selection hypothesis. (page 1065)

14. The answers are **a4, b2, c1, d5,** and **e3.** (pages 1055–1057)

15. An individual that holds a territory is likely to have the opportunity to reproduce. (Individuals who are unable to secure territories generally do not reproduce.) A territory may provide nesting sites and food supplies to which the territory holder may have exclusive access. (page 1063)

16. Premating isolating mechanisms include the courtship calls of birds and insects, pheromones released by females of a species to attract males, structural differences (e.g., the shape of a beak or ornamentation), and courtship rituals, all of which are species-specific. (page 1016)

17. (1) The number of hominid fossils is small compared with that of some other organisms. (2) The hominid fossils that are available consist primarily of fragmented parts of skeletons. (3) The australopithecines appear to have been a heterogeneous group of individuals. (4) The search for human ancestors has become intensely competitive, and there is a great temptation to call each new discovery by a new name. (page 1038)

CHAPTER 52

Population Dynamics: The Numbers of Organisms

CHAPTER ORGANIZATION

MAJOR CONCEPTS

In order of increasing complexity, the levels of biological organization are subatomic particle, atom, molecule, cell, tissue, organ, organ system, organism, *population*, community, ecosystem, and biosphere.

Ecology is the discipline in which the interactions of organisms with their environment and with other organisms are studied. The focus of ecology is on populations, communities, ecosystems, and the biosphere rather than on the individual organism.

Several properties characterize populations (but not individuals), including growth and mortality, age structure, density, and dispersion. Population growth may be exponential or

logistic. Exponential growth occurs when population size increases at a constant rate. In logistic growth, the rate of increase is high when the population is small and decreases as population size approaches the carrying capacity of the environment. In addition to its importance in ecological studies, an understanding of population dynamics has practical applications, especially in cases where natural populations are harvested regularly.

Three basic mortality patterns are high mortality of young with decreasing mortality as age increases, constant mortality throughout the life span, and low mortality early in life with increasing mortality as individuals near the end of the biological life span. Natural populations often exhibit a combination of these three patterns.

The age structure of a population consists of the proportion of individuals in each age category. The age structure is influenced by the mortality pattern and in turn affects population growth patterns.

Density is the number of individuals per unit of area or volume. Dispersion refers to the distribution of individuals (or populations) in a specified space. The three basic dispersion patterns are regular, random, and clumped. Dispersion patterns may be influenced by distribution of essential resources and by behavior patterns.

Population size is often regulated by limiting factors, which include the physiological tolerance of an organism to environmental conditions, available shelter, and food supply. Limiting factors may be density-dependent or density-independent. In some populations, population size fluctuates on a regular cycle that is apparently independent of limiting factors. The causes of population cycles are as yet undetermined and are being actively investigated.

Life-history patterns are groups of coadapted traits that influence reproductive performance. For convenience, patterns at the extremes of a continuum have been labeled as prodigal (*r*-selected) and prudent (*K*-selected). Prodigal species typically produce many, small young in one reproductive effort, provide little or no parental care, and mature rapidly. Prudent species have fewer, large young that mature slowly, provide intense parental care, and reproduce many times during the life span. Prodigal species are opportunistic and adept at colonizing new environments. Prudent species are best adapted to stable, established environments. Prodigal species typically recover rapidly following a population crash. Prudent species require a longer period to recover from a population crash and are much more susceptible to extinction when numbers are low than are prodigal species. A species may possess characteristics of both life-history patterns. Also, there is variation of patterns among individuals within a species and among populations within a species.

Many species reproduce asexually, which allows faster population growth than does sexual reproduction. Plants that reproduce asexually by runners and similar structures have the added advantage of providing nutritional support for the new plant. Parthenogenesis is one form of asexual reproduction that is considerably more efficient than sexual reproduction. Many organisms are capable of both asexual and sexual reproduction, and typically reproduce asexually when resources are abundant and switch to sexual reproduction as resources become scarce.

HOW TO STUDY THE CHAPTER

Read the entire chapter through quickly, focusing on the major concepts.

Use the GUIDED STUDY OF THE CHAPTER to help you identify the important details as you **reread** the chapter. Writing out the answers to these questions will help fix them in your mind as well as provide you with a valuable study aid.

Answer the questions in TESTING YOUR UNDERSTANDING without the aid of your text. Check your answers against those in PERFORMANCE ANALYSIS. Analyzing your answers will give you valuable feedback on your level of understanding and preparedness for classroom testing.

GUIDED STUDY OF THE CHAPTER

I. Introduction (page 1087)

1. a. *Vocabulary:* Define ecology.

   b. The science of ecology seeks answers to what questions?

2. Why can ecology be considered to be both the oldest and the youngest discipline of the biological sciences?

3. Cite three practical applications of population dynamics.

A. *Essay:* **An Example of Ecological Modeling: Transmission of the AIDS Virus** *(pages 1090–1091)*

4. a. Cite two disadvantages of using models for ecological studies.

   b. In what way are models useful in studying ecological problems?

5. a. Models of HIV transmission were designed to predict what two phenomena?

   b. What is the practical usefulness of these predictions?

6. Identify three factors considered by May and Anderson in the development of their models.

7. a. *Vocabulary:* Define the average number of secondary infections and indicate its abbreviation in the May-Anderson model.

b. For a sexually transmitted disease, the average number of secondary infections depends on what three factors?

c. Which of these factors can be altered, and how?

8. Exponential growth in numbers of newly HIV-infected individuals is occurring in which population?

9. How can the information gained from using this model benefit health professionals?

II. Properties of Populations (pages 1087–1094)

A. General Remarks (pages 1087–1088)

> *Focus:* In the levels of biological organization, the population is the next level up beyond the individual organism. The individual has a transitory existence; the population persists.

10. Name five properties of populations that are not features of an individual organism.

B. Patterns of Population Growth (pages 1088–1091)

11. *Vocabulary:* Define the rate of increase of a population.

12. a. Write an equation for a population's per capita rate of increase.

b. What two conditions apply when using this formula?

13. a. *Vocabulary:* What is exponential growth?

b. Define these terms from the equation for exponential population growth.

dN/dt

r

N

14. a. Short-term exponential growth characterizes what types of populations? (See also Figure 52–2.)

b. Cite three examples of populations growing exponentially.

15. a. Name seven factors that generally limit exponential population growth.

b. Short-term exponential growth characterizes what type of species?

16. a. In general, what induces a population crash?

b. Name two specific factors that may result in a population crash.

The Effect of the Carrying Capacity (pages 1089–1091)

17. a. *Vocabulary:* Define carrying capacity.

b. What factor(s) may determine carrying capacity for (i) an animal population and (ii) a plant population?

18. Cite two examples of the fact that limiting factors may vary seasonally.

19. a. In the following equation, what does K represent?

$$\frac{dN}{dt} = rN\left(\frac{K-N}{K}\right)$$

b. Characterize the rate of population growth when population size is very small.

c. What happens to the rate of population growth as population size approaches the carrying capacity?

d. What happens to the growth rate when $N = K$?

20. On the following axes, draw the curves characteristic of (a) exponential growth and (b) logistic growth. Draw in and label the carrying capacity on graph (b).

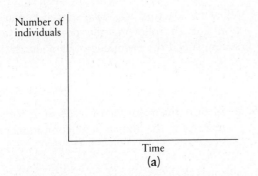

Number of individuals

Time
(a)

Number of individuals

Time
(b)

21. Cite two practical applications of the logistic growth model.

C. Mortality Patterns *(page 1092)*

22. a. Describe two extreme mortality patterns. (See also Figure 52–5.)

b. How does the mortality pattern characteristic of birds represent a combination of these extreme forms?

D. Age Structure *(page 1092)*

23. What characteristic of a population is affected by its mortality pattern?

24. *Vocabulary:* What is the age structure of a population?

25. a. How may knowledge of a population's age structure be used?

b. Explain why the population of India would continue to grow even if individuals currently alive only replace themselves.

26. a. Cite two reasons why the United States population has continued to increase despite a reduced reproductive rate.

b. Account for the population boom of the 1980s. (Figure 52–7)

c. *Vocabulary:* Define a cohort. (Figure 52–7)

Focus: A population that is not increasing has a stable age structure.

E. **Density and Dispersion** *(pages 1093–1094)*

27. *Vocabulary:* Distinguish between population density and dispersion.

28. Name, diagram, and describe the three basic dispersion patterns.

Focus: Dispersion patterns may be applied to individuals of one species, to populations of a species, or to populations of different species. (Figure 52–8)

---

29. Identify two factors that may influence the dispersion pattern of a population and cite one example of each factor.

Focus: A population's dispersion pattern may vary at different stages of the life cycle or during seasonal fluctuations of an essential resource.

---

30. Cite two examples illustrating the Focus statement.

Focus: For a particular population, the dispersion pattern depends on the total amount of space included in the observations.

F. *Essay:* **The Human Population Explosion** *(pages 1096–1097)*

31. Identify the factors responsible for the first two periods of rapid growth in the human population and indicate when these periods of rapid growth began.

32. What is the present (1989) population of the earth, and what is its overall rate of growth?

33. a. What factor is thought to have limited population growth of humans when they lived as hunter-gatherers?

   b. Account for this in terms of physiological events.

   c. Identify one social pressure that would encourage a long interval between children.

34. What two factors may have slowed human population growth following the population surge that accompanied the beginning of agriculture?

35. a. Identify the major factor responsible for the rapid increase in the human population in recent times.

   b. Cite three factors contributing to the rapid drop in death rate that has occurred in many tropical countries since 1940. (Figure (b), page 1097)

36. Some experts argue that increasing the standard of living and decreasing the death rate would slow population growth. What arguments suggest this may not be true?

37. What is the relationship between birth rate and economic development?

38. Account for the continued high birth rate in India in spite of that country's problems in feeding its population.

39. What complex issues are posed by the current global trends in population growth and food supply?

III. The Regulation of Population Size *(pages 1095–1099)*

A. General Remarks *(page 1095)*

> *Focus:* **For any given population, the total size fluctuates over time and this fluctuation affects other interacting populations.**

40. Identify two examples of extreme fluctuations in population size or density. (See also Figure 52–10.)

B. Limiting Factors *(pages 1095–1098)*

41. Name six factors for which an organism's range of tolerance is significant to population growth.

42. In the following graph of a tolerance range, label the optimum range, zones of physiological stress, and zones of intolerance, and indicate the zones in which organisms will be abundant, infrequent, and absent.

43. Outline one example in which limiting factors influence population growth.

Density-Dependent and Density-Independent Factors (page 1098)

44. a. *Vocabulary:* Distinguish between density-dependent and density-independent factors affecting population growth.

b. Name three factors that are density-dependent.

c. Cite one reasonably clear example of a density-independent factor.

45. Use an example to illustrate the difficulty in categorizing some factors as density-dependent versus density-independent.

C. Population Cycles *(pages 1098–1099)*

> *Focus:* **The populations of many organisms fluctuate in regular cycles, a phenomenon that is still a mystery to biologists.**

46. Cite one example of animals that undergo population fluctuations.

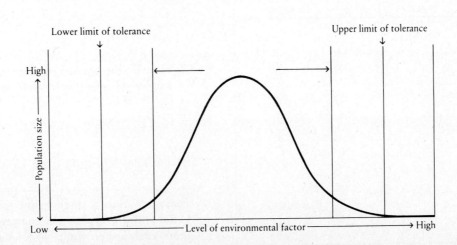

47. a. Summarize four hypotheses regarding density-dependent factors that have been proposed as possible causes of the cyclic population fluctuations seen in voles.

b. Name two density-independent factors that have been proposed as possible causes of vole population cycles.

IV. Life-History Patterns (pages 1099–1104)

A. General Remarks (page 1099)

48. *Vocabulary:* What are life-history patterns?

49. What two extremes of life-history strategies are represented by the oyster and the elephant?

Focus: Life-history patterns vary among individuals within a population and among populations and are therefore subject to natural selection.

B. The Alternatives (pages 1099–1102)

Focus: The characteristics associated with alternative life-history patterns represent extremes of a continuum.

---

50. Fill in the following table.

Table for Question 50

	PRODIGAL	PRUDENT
Synonyms (two each)		
Number of young		
Size of young		
Rate of maturation		
Degree of parental care		
Number of reproductive efforts in life		
Situation to which pattern is best adapted		

51. Name three scientists who have characterized life-history patterns.

52. In what way is the *r*-selected and *K*-selected distinction an oversimplification?

53. Which group of animals provides the longest period of parental care relative to total life span? (Figure 52–13)

Early or Late Reproduction (page 1100)

54. Compare the advantages of early and late reproduction.

55. Identify an example of organisms whose life histories entail (a) early reproduction and (b) late reproduction, following an extended growth period.

56. Describe the influence of early versus late reproduction on rate of population growth.

57. How do the age and experience of the mother affect survival of offspring in some mammals?

Small or Large Offspring (page 1102)

Focus: Within a single plant species, seed size varies with the type of environment in which the individuals are growing.

---

58. a. What was the relationship between seed size and seed number in the study of goldenrods in old field and prairie environments?

b. How does this fit the predictions of MacArthur and Wilson?

C. The Asexual Advantage *(pages 1102–1103)*

> *Focus:* **A population that reproduces asexually can grow more rapidly than if it reproduces sexually.**

59. Name a major advantage of reproduction by runners.

Parthenogenesis (page 1103)

60. *Vocabulary:* Define parthenogenesis.

61. Account for the greater efficiency of parthenogenesis relative to sexual reproduction.

62. Cite two advantages of parthenogenesis in plants.

63. a. What evidence suggests that parthenogenetic dandelions evolved from sexual forms?

b. In the dandelions studied by Otto Solbrig, what characteristics of genotypes D and A contributed to their persistence in the environment?

64. Name three types of animals in which some species may reproduce only asexually.

> *Focus:* **Many organisms reproduce sexually and asexually, depending on environmental conditions.**

65. In freshwater *Daphnia*, indicate the conditions under which (a) sexual and (b) asexual reproduction occurs.

66. a. Describe parthenogenetic reproduction in the gall midge.

b. Under what conditions does this mode of reproduction occur?

D. Some Consequences of Life-History Patterns *(pages 1103–1104)*

67. Discuss the consequences of a population crash (and the relative risk of extinction) for (a) an *r*-selected species and (b) a *K*-selected species.

68. Name two examples of animals with *K*-selected strategies that are close to extinction.

V. Summary *(pages 1104–1105):* Read the summary. If you are familiar with the essential features of the material presented there, you are ready to complete the section TESTING YOUR UNDERSTANDING.

TESTING YOUR UNDERSTANDING

After you have completed this examination, compare your answers with those in the section that follows.

1. Which trait is more a property of individuals than of populations?
   a. growth pattern
   b. age structure
   c. density
   d. mortality
   e. dispersion

2. At the present time, AIDS is spreading exponentially among:
   a. homosexual men.
   b. intravenous drug users.
   c. heterosexual hemophiliacs.
   d. infants of heterosexual women.
   e. Exponential spread of the disease is occurring in all of these groups.

3. Which situation does NOT illustrate seasonal changes in the carrying capacity of an environment?
   a. Lion cubs in the Serengeti plain often die after the migrating wildebeests have left the area.
   b. An extremely abundant crop of acorns enables more deer than usual to survive over the winter.
   c. An extremely dry summer reduces the number of water holes by half.
   d. Gypsy moth caterpillars starve to death after consuming their entire food supply.
   e. All of these situations illustrate seasonal changes in carrying capacity.

4. Once a saltmarsh song sparrow survives its first, risk-laden year, its chance of dying _____ each year for the rest of its life.
   a. decreases
   b. remains constant
   c. increases
   d. either increases or decreases, depending on environmental conditions,
   e. is entirely unpredictable

5. Which factor is NOT among those that may influence the dispersion pattern of a population?
   a. distribution of essential resources
   b. social interactions among individuals
   c. mortality pattern of the population
   d. stage of the life cycle
   e. season of the year

6. Which factor is most likely to have a density-independent influence on population growth?
   a. water phosphorus concentration
   b. availability of shelter
   c. weather conditions
   d. infectious diseases
   e. abundance of prey

7. Which statement is true of life-history patterns?
   a. They are subject to the pressures of natural selection.
   b. They are usually learned.
   c. They have no genetic component.
   d. They exist only in organisms that exhibit significant parental care.
   e. They may be grouped into one of two mutually exclusive categories.

8. Life-history patterns were dubbed prudent and prodigal by:
   a. E. O. Wilson.
   b. Robert MacArthur.
   c. Otto Solbrig.
   d. Charles Elton.
   e. G. E. Hutchinson.

9. Which statement is NOT true of reproduction by parthenogenesis?
   a. It requires males only for fertilizing the eggs, then they die.
   b. It is a form of asexual reproduction.
   c. It usually produces large numbers of offspring at one time.
   d. It may allow for wider dispersal than sexual reproduction.
   e. In some species, it alternates with sexual reproduction.

10. T or F   If a population is growing exponentially, a graph of the equation describing its growth is a straight line, the slope of which is equal to the per capita rate of increase.

11. T or F   The sigmoid shape of the logistic growth curve is due to a decline in the rate of increase as population size increases.

12. T or F   Under the appropriate conditions of population age structure, it is possible for a population to increase in size rapidly even though reproductive individuals are only replacing themselves.

13. T or F   Most natural populations eventually reach an equilibrium stage, at which point population size remains fairly constant.

14. T or F   In *Daphnia*, asexual reproduction occurs under conditions of environmental stress.

15. Distinguish between the rate of increase of a population and the per capita rate of increase of a population.

16. Explain how it is possible for a given species to exhibit two different dispersion patterns simultaneously.

17. What four hypotheses have been proposed to explain the periodic fluctuation in population size experienced by voles of the genus *Microtus*?

18. Cite one example of an organism that exhibits both *r*-selected and *K*-selected characteristics.

19. Contrast the life-history pattern of a long-lived oak tree with that of a short-lived annual flower and identify the circumstances that make both patterns evolutionarily successful.

PERFORMANCE ANALYSIS

1. d Mortality refers to the death of individuals. The mortality pattern—how many and which individuals in a population die at what ages—is a property of a population. (page 1088)

2. b Human immunodeficiency virus is currently spreading at an exponential rate among intravenous drug users and their sexual partners. Its spread among homosexual men has slowed dramatically in recent years. (essay, page 1091)

3. d The starvation of gypsy moths during a population boom is not due to seasonal changes in the environment. (page 1089)

4. b The mortality pattern for many bird species is typically very high during the first year of life. After one year of age, the mortality rate is fairly constant over the rest of the life span. (page 1092)

5. c The dispersion pattern of a population may be influenced by several factors, including the distribution of essential resources, social interactions among individuals, the stage of the life cycle, and the season of the year. (page 1094)

6. c Weather-related factors that influence birth rates and death rates *may* be density-independent. Nutrient supply, infectious diseases, and the availability of shelter are all density-dependent factors. (page 1098)

7. a As are all the other inheritable characteristics of a species, life-history patterns are subject to the forces of natural selection. (page 1099)

8. e G. E. Hutchinson termed life-history patterns prudent and prodigal. These patterns were later named *r*-selected (prodigal) and *K*-selected (prudent) by Robert MacArthur and E. O. Wilson. (page 1099)

9. a Parthenogenesis is a form of asexual reproduction in which offspring develop from unfertilized eggs. (page 1103)

10. False The graph of exponential growth is a curve, the slope of which increases steadily over time. Although the per capita rate of increase remains the same, the growth rate of the population increases because the number of reproducing individuals is constantly increasing. (page 1088)

11. True In logistic growth, the growth rate of the population decreases as the population approaches the carrying capacity. This results in the sigmoid shape of the curve. (page 1090)

12. True This situation is characteristic of populations whose age structure is characterized by a large number of individuals at or below reproductive age. (page 1092)

13. False The population size of natural populations that have reached an equilibrium state may fluctuate around the carrying capacity. Also, regular population cycles may occur once equilibrium is reached. (pages 1089, 1098; Figure 52–4)

14. False In organisms that can reproduce both asexually and sexually, asexual reproduction typically occurs when conditions for rapid growth are favorable or when environmental conditions are fairly constant. Sexual reproduction occurs when resources begin to decline or when environmental conditions are unstable. (page 1103)

15. The rate of increase of a population is the increase in the number of individuals in a given unit of time per individual present. This value includes immigration and emigration as well as reproduction. The per capita rate of increase considers only population increase due to reproduction. (page 1088)

16. The dispersion pattern of a species depends upon the scale that is being considered. For example, at the local level, a territorial bird species may exhibit a regular dispersion pattern. On a regional or continental level, the distribution may be clumped because the birds only live in certain types of habitats. (page 1094)

17. (1) At peak density, the voles exhaust the food supply. (2) When vole densities become very high, hormonal imbalances occur that inhibit reproduction. (3) As the number of voles increase, the number of predators also increases; increased predation pressures cause the vole population to crash. (4) When population density is very high, certain members of the population disperse to less-densely populated regions. (pages 1098, 1099)

18. Some species of intertidal starfish produce large numbers of eggs (*r*-selected) but have long lives (*K*-selected). (page 1099)

19. The short life span of an annual plant provides a selection pressure favoring the production of a large number of seeds as soon as possible. The much longer life span of the oak tree, and the fact that it grows in a forest, provides a selection pressure favoring years (or decades) of growth to a large size before it reproduces. Once it reaches a certain size and age, its chances of continued survival are quite high. Selection favors production of a moderate number of acorns every year for the rest of its life. (page 1100)

CHAPTER 53

Interactions in Communities

CHAPTER ORGANIZATION

MAJOR CONCEPTS

An ecological community is composed of all the populations inhabiting a common environment and interacting with one another. The interactions among populations are major forces of natural selection, influencing population sizes and species composition in the community. Types of interactions commonly observed within communities are competition, predation, and symbioses.

The best approach to studying the interactions among populations in communities is under debate. Some ecologists argue that careful observation alone is sufficient; others maintain that manipulating natural study areas and establishing experimental controls are necessary to obtain valid information.

Competition for a limited resource may occur between individuals of the same species (intraspecific) or between members of different species (interspecific). Resources for which individuals may compete include food, water, light, and living space.

The principle of competitive exclusion contends that if two populations compete for the same limited resource, one population will eventually be eliminated from the community. In natural systems, similar species occupying the same habitat frequently partition available resources, thereby minimizing competition. Animals may partition resources by dividing the habitat into feeding zones, with each species foraging in slightly different areas or on different types of food. Plants sharing the same habitat may occupy slightly different microenvironments. However, if a superior competitor is introduced into a community and resources are not partitioned, a resident species may be eliminated.

The ecological niche may be described as the occupation of a species within a community. A description of an organism's niche includes its physical and biological requirements and behavioral characteristics. The fundamental niche of a species is determined by its physiological limitations. The realized niche is that portion of the fundamental niche occupied by the species in a natural community. It may be much smaller than the fundamental niche because of interactions with other populations. A corollary to the principle of competitive exclusion is the idea that no two populations can occupy exactly the same realized niche at the same time in the same habitat.

When similar species occupy the same habitat and have overlapping characteristics, these characteristics tend to diverge. This phenomenon is known as character displacement. Beak size in Darwin's finches is often cited as an example of character displacement. This characteristic may represent actual divergence due to the selective pressures of interspecific competition. However, some scientists argue that the beak variations may be present because the species evolved in different environments (under differing selective pressures) and later migrated to occupy the same environment. In order to establish conclusive evidence of character displacement, the divergence of two coexisting species must be documented as it happens.

Predation involves an arms race between predators and their prey. In prey species, natural selection favors techniques for escaping predation. At the same time, selection pressures favor predators who employ the most successful means of capturing prey. In both groups, physical and behavioral traits are important.

Methods of escaping predation include cryptic coloration, which is often accompanied by behavior patterns that enhance the camouflage, secure hiding places, and temporal strategies.

Prey population size is an important component of the carrying capacity of an environment for a predator species. In some cases predation may limit prey population size to a level below the environment's carrying capacity for the prey species. However, this does not always occur. Predators typically prey upon more than one species, with the prey species that is most abundant at a particular time often experiencing the greatest predation. By preying upon species that are competitively superior, predation can maintain a greater species diversity in a community than if no predators were present.

There are three basic types of symbiosis based on who benefits and who is harmed by the relationship. In mutualism, both species benefit from the relationship. In commensalism, one species benefits and the other species neither benefits nor is harmed. Parasitism can be considered as a special case of predation in which the predator (parasite) is much smaller than the prey (host).

According to the island biogeography model, the number of species on an island will remain the same over time but the species composition will continuously change. The total number of species and the population sizes are influenced by the size of the island. Another important factor influencing species diversity is the distance of the island from a source of new colonizing species.

The intermediate disturbance hypothesis contends that the degree of species diversity in a community is a function of the time interval between disturbances. Up to a point, as the interval between disturbances increases, so does species diversity. Beyond this point, as the interval between disturbances increases, species diversity declines.

Ecological succession is the process whereby an undisturbed community undergoes changes in composition over time. The replacement of photosynthetic organisms by new species usually precedes changes in the animal life. The facilitation hypothesis states that each successional stage "prepares the way" for the following stage. The inhibition hypothesis con-

tends that the early species initially prevent colonization by later arrivals but are subsequently replaced because later arrivals are better competitors. The tolerance hypothesis suggests that early species have no effect on subsequent colonization and the community composition represents the organisms best adapted to current conditions. Evidence exists (in different communities) supporting the first two hypotheses.

HOW TO STUDY THE CHAPTER

Read the entire chapter through quickly, focusing on the major concepts.

Use the GUIDED STUDY OF THE CHAPTER to help you identify the important details as you **reread** the chapter. Writing out the answers to these questions will help fix them in your mind as well as provide you with a valuable study aid.

Answer the questions in TESTING YOUR UNDERSTANDING without the aid of your text. Check your answers against those in PERFORMANCE ANALYSIS. Analyzing your answers will give you valuable feedback on your level of understanding and preparedness for classroom testing.

GUIDED STUDY OF THE CHAPTER

I. Introduction *(page 1106)*

    1. *Vocabulary*: State the ecological definition of a community.

    2. a. Interactions among organisms influence a community in what three ways?

       b. Name the three categories of interaction within communities.

II. Competition *(pages 1106–1114)*

A. General Remarks *(page 1106)*

    3. *Vocabulary*: Define competition, distinguishing between intraspecific and interspecific competition.

4. For individuals, what is one consequence of competition?

5. Identify four resources for which living organisms may compete.

6. *Vocabulary*: Distinguish between interference competition and exploitative competition.

7. Under what conditions is competition most intense?

B. The Current Debate *(pages 1106–1107)*

> *Focus:* Historically, competition has been considered to be a major force determining community composition and structure. This concept is currently under debate.

> *Focus:* Some ecologists believe that careful observations of organisms in their natural environments are adequate to identify the factors influencing their distribution.

8. How did Jared Diamond explain the distribution of the thrush *Turdus poliocephalus* on 10 islands in the southwestern Pacific?

> *Focus:* Other ecologists maintain that controlled experiments conducted in natural settings are required to identify factors influencing the structure of a particular community.

9. In addition to competition, name three other factors that may be influencing the distribution of *T. poliocephalus*.

C. The Principle of Competitive Exclusion *(pages 1108–1109)*

10. a. Who formulated the principle of competitive exclusion?

b. State this principle.

11. a. Outline the experiment in which Gause substantiated this principle.

b. The *Paramecium* species with what characteristic eliminated its competitor?

12. a. In experiments with two duckweed species of the genus *Lemna*, how did the two species differ with respect to rate of growth?

b. What characteristic did the "winner" possess that was essential to its success?

> *Focus:* In studies of competition in *Paramecium* and *Lemna,* varying the conditions may produce different results, but one species is always eliminated in a laboratory setting.

The Ecological Niche (page 1109)

13. *Vocabulary*: Define an ecological niche and distinguish niche from habitat.

14. The description of a niche includes what types of information?

D. Resource Partitioning (pages 1109–1111)

Focus: When two similar species are present in a community, resources are frequently partitioned.

15. Describe the resource partitioning exhibited by the following organisms. (Be sure to include the resources that are being partitioned.)

    Woodland warblers

    Bog mosses

    African ungulates: browsers

    African ungulates: grazers

The Role of Past Competition in Resource Partitioning (pages 1110–1111)

16. What factor has long been thought to be the cause of resource partitioning?

17. a. *Vocabulary:* Define character displacement.

    b. Describe one frequently cited example of character displacement.

18. a. How is niche overlap thought by some ecologists to cause character displacement?

    b. State one opposing explanation for the different beak sizes exemplified by Darwin's finches.

Focus: Some ecologists contend that species must be shown to be diverging in the here and now in order to provide conclusive evidence for character displacement.

19. Ecologists agree on what aspect of beak differences in Darwin's finches?

E. Experimental Approaches to the Study of Competition (pages 1112–1114)

20. Name three factors that determine the intensity of competition within a community.

Focus: In order to demonstrate conclusively that competition is occurring, controlled experiments in natural settings are required.

21. a. In field experiments of competition, how are the suspected competitors manipulated?

    b. What results indicate that competition is occurring?

22. Identify three difficulties involved in conducting controlled experiments in natural settings.

Barnacles in Scotland (pages 1112–1113)

23. What characteristics of barnacles make them good subjects for ecological studies of competition?

24. a. How do the environmental conditions of the high intertidal zone inhabited by *Chthamalus* differ from those in the lower zone occupied by *Semibalanus*?

b. Identify the major factor restricting each barnacle species to the habitat in which it is found naturally.

25. *Vocabulary*: Distinguish between fundamental and realized niches, including the factors that determine each type of niche.

26. Relate the breadth of an organism's fundamental niche to its competitive ability.

The Prevalence of Competition (page 1113)

27. State the prevalence of competition reported in the separate reviews conducted by Connell and Schoener.

F. **Winner Takes All** *(pages 1113–1114)*

Focus: **Competition between species can lead to the elimination of a species from a community.**

28. Cite three examples illustrating the Focus statement.

29. a. Why is it difficult to conclude that competition is responsible for elimination of a species?

b. What other factors may be involved?

III. **Predation** *(pages 1114–1119)*

A. **General Remarks** *(page 1114)*

30. a. *Vocabulary*: Define predation.

b. Identify the general predator-prey relationships that exist among plants, animals, and fungi.

c. What are foraging strategies?

31. How does predation influence the evolution of predators and prey?

32. Name two aspects of community structure that are affected by predation.

B. **The Arms Race** *(pages 1114–1116)*

33. a. Summarize the arms race occurring between *Eleodes longicollis* and the grasshopper mouse.

b. How has *Megasida obliterata* benefited from this arms race? (Figure 53–12)

34. a. Cite three mechanisms that have evolved in legumes for avoiding bruchid weevil predation.

b. Indicate which defense mechanism has been circumvented by the weevils and describe how this is accomplished.

Bats, Moths, and Biosonar (pages 1115–1116)

35. a. What did Spallanzani conclude about the navigation of bats in the dark?

   b. How did he reach this conclusion?

36. a. Who discovered the actual mechanism by which bats navigate at night?

   b. Summarize this mechanism.

37. Describe how the Doppler effect provides a bat with information about the direction of movement of an object relative to its flight path.

38. a. Name three scientists who have studied sound detection by moths.

   b. Characterize the frequency range to which a moth's ears are sensitive.

39. What defense mechanisms help moths avoid predation by bats?

C. Escape from Predation *(page 1116)*

40. Name three ways of escaping predation that do not involve direct contact between predator and prey.

41. The use of camouflage to avoid predation involves what corollary behavioral adaptation? (Figure 53–14)

42. Describe the adaptive value of the synchronized calving exhibited by wildebeests.

D. Predation and Population Dynamics *(pages 1116–1119)*

> *Focus:* Predation may influence the age structure of a prey population but it is not clear that predation necessarily reduces the prey population to below the carrying capacity of the environment.

43. How can predation affect life-history patterns?

44. a. In a population, prey animals with what characteristic experience the heaviest predation pressure?

   b. Cite one piece of evidence for this.

45. Cite one clear example in which a prey species was limited by a predator.

> *Focus:* Most predators will prey on more than one species, with the most abundant prey typically experiencing the heaviest predation.

46. How does the availability of prey influence predator populations?

47. a. Describe the population cycles of the lynx and the snowshoe hare as revealed by pelt data of the Hudson Bay Company.

b. Summarize three possible explanations for these data.

c. What two pieces of recent evidence suggest the presence of a regular cycling of the hare population?

E. **Predation and Species Diversity** *(page 1119)*

Focus: **Predation is one factor important in maintaining species diversity in a community.**

48. a. How does the starfish *Pisaster* influence the structure of the community in which it lives?

b. Who elucidated this role of *Pisaster*?

c. Describe the experimental techniques used to elucidate *Pisaster*'s influence on community structure.

49. How does the herbivorous marine snail *Littorina littorea* influence the growth of algal species in intertidal pools of the New England coast?

50. a. Describe the role of rabbits in maintaining species diversity in the English countryside.

b. How was this role demonstrated?

IV. **Symbiosis** *(pages 1119–1122)*

A. **General Remarks** *(pages 1119–1120)*

Focus: **Symbioses that persist for many generations can profoundly influence the evolution of both members of the relationship.**

51. Fill in the table.

Table for Question 51

	PARASITISM	MUTUALISM	COMMENSALISM
Who benefits?			
Who is harmed?			
Example(s)			

B. **Parasitism** *(page 1120)*

52. *Vocabulary*: Define parasitism in the context of predation.

53. Which individuals in a population are most susceptible to parasitism?

54. Explain why it is disadvantageous for a parasite to cause severe disease and death.

55. Outline the coevolution of European rabbits and myxoma virus in Australia.

C. Mutualism (pages 1120–1122)

56. Cite three examples of mutualism involving the roots of plants.

57. Indicate how each member benefits from the symbiotic association. (Figure 53–19)

Sea anemone and hermit crab

Cleaner fish and larger "dirty" fish

Aphids and ants

Oxpeckers and their hosts

Ants and Acacias (pages 1121–1122)

58. a. In the ant-acacia relationships characteristic of Mexico and Central America, indicate what benefits accrue to each partner in the symbiosis.

Ant

Acacia

b. What are Beltian bodies?

c. How did Janzen demonstrate the beneficial effects for the acacia?

59. a. Other than the main participants, name three species that benefit from the ant-acacia relationship.

b. What is the secret to their success?

V. Community Composition and the Question of Stability (pages 1123–1129)

A. General Remarks (page 1123)

> *Focus:* Two issues of concern to ecologists are the factors determining the number of species in a community and the factors responsible for changes in community composition.

B. The Island Biogeography Model (pages 1123–1124)

60. a. State the basic premise of the equilibrium hypothesis of island biogeography.

b. Who developed this hypothesis?

61. a. Describe the experimental technique used by Wilson and Simberloff to test the equilibrium hypothesis of island biogeography.

b. What were the results of this study?

62. a. Identify two reasons why immigration rate declines as more species reach an island. (Figure 53–21)

b. Why does extinction rate increase more rapidly as species number increases? (Figure 53–21)

> *Focus:* According to the equilibrium model of island biogeography, once the equilibrium number of species is reached, species number will remain the same although species composition will continue to change.

63. Identify the two most important factors influencing the species diversity of an island.

64. Rates of immigration to islands are a function of what factor?

65. How does island size influence the equilibrium number of species and the extinction rate?

C. *Essay:* **Conservation Biology and the Island Biogeography Model** *(pages 1126–1127)*

66. Cite four human activities that fragment natural environments.

67. a. What implications does the equilibrium hypothesis have for conservation biology?

    b. Describe two situations in which the predictions of the island biogeography model have been upheld following the disturbance of an environment by human activities.

    c. Describe one similar situation that involved natural events.

68. a. What technique has been proposed (and tested) to reduce the isolation of ecological communities?

    b. What were the results of the test?

69. Identify two physical changes and three biological changes that occur in tropical forests that are fragmented.

70. a. What important prediction of the island biogeography model has been upheld by research?

    b. What two factors will determine the number of species lost in the establishment of a nature reserve?

D. **The Intermediate Disturbance Hypothesis** *(pages 1124–1125)*

Focus: **In some communities, species diversity is apparently a function of the frequency and magnitude of disturbances.**

71. a. Name two types of communities that are particularly rich in diversity and number of species.

    b. Describe the disturbances that can occur in these environments.

72. If disturbances are frequent, what types of organisms will inhabit the community?

73. What does the intermediate disturbance hypothesis predict regarding species diversity in a community?

74. What is thought to be the major factor in declining species diversity as the interval between disturbances increases?

75. Describe the observations of Wayne Sousa and explain how they support the intermediate disturbance hypothesis.

E. Ecological Succession *(pages 1125–1129)*

76. *Vocabulary:* What is ecological succession?

77. Outline ecological succession as it might occur in an abandoned field in North America.

78. a. Name and distinguish between the three hypotheses regarding the mechanism of ecological succession.

b. How do the studies of Wayne Sousa support the inhibition hypothesis?

c. In the inhibition hypothesis, how do later species manage to become established?

d. How do the studies of Teresa Turner support the facilitation hypothesis?

> *Focus:* Each community is unique, and patterns of species interaction that occur in one community may not apply to other communities.

79. *Review Question:* Match each scientist with the correct study or concept. An answer may be use more than once. A blank may have more than one answer, as indicated in parentheses.

1. F. E. Clements
2. Joseph H. Connell
3. Jared Diamond
4. M. Brock Fenton
5. James H. Fullard
6. G. F. Gause
7. Donald Griffin
8. Daniel Janzen
9. Jane Lubchenco
10. Robert MacArthur
11. R. T. Paine
12. Kenneth Roeder
13. Daniel Simberloff
14. Lazzara Spallanzani
15. Wayne Sousa
16. Teresa Turner
17. E. O. Wilson

a. _____ Tested the island model of biogeography. (2 answers)

b. _____ Studied algae on intertidal boulders; the results supported the intermediate disturbance hypothesis.

c. _____ Developed the facilitation hypothesis of ecological succession.

d. _____ Studied algae on intertidal boulders; the results supported the inhibition hypothesis of ecological succession.

e. _____ Studied surfgrass colonization; the results supported the facilitation hypothesis of ecological succession.

f. _____ Proposed that competition limits the distribution of a thrush species on 10 islands in the southwestern Pacific.

g. _____ Formulated the principle of competitive exclusion.

h. _____ Documented resource partitioning by five species of New England warblers.

i. _____ He is considered by many to be the founder of modern experimental ecology.

j. _____ Concluded that bats use the sense of hearing to navigate in the dark.

k. _____ Discovered that bats navigate by means of echolocation.

l. _____ Revealed that more than 95 percent of the moth species in Ontario have functioning ears. (2 answers)

m. _____ Discovered that some moths detect bat cries at distances of 40 meters.

n. _____ Studied the influence of *Pisaster* starfish on species diversity.

o. _____ Studied the influence of a marine snail on species diversity.

p. _____ Performed a detailed analysis of the ant-acacia relationship.

q. _____ Formulated the island model of biogeography. (2 answers)

r. _____ Documented competitive exclusion among barnacles.

VI. Summary *(pages 1129–1130):* Read the summary. If you are familiar with the essential features of the material presented there, you are ready to complete the section TESTING YOUR UNDERSTANDING.

TESTING YOUR UNDERSTANDING

After you have completed this examination, compare your answers with those in the section that follows.

1. When one species, which has been competing with another species, is eliminated from an environment, _____ has occurred.
   a. interference competition
   b. exploitative competition
   c. competitive exclusion
   d. overpredation by the competitors, resulting in exhaustion of their prey,
   e. overpredation of the competitors by predators in the community

2. By analogy, the niche is an organism's _____, whereas the habitat is its _____.
   a. diet; address
   b. life-history strategy; geographical distribution
   c. population; community
   d. profession; address
   e. occupation; community

3. Five closely related species of warblers can coexist in New England forests because they:
   a. partition resources.
   b. occupy different habitats.
   c. subdivide their realized niches.
   d. are highly territorial.
   e. eat different types of foods.

4. What do ecologists accept as evidence of competition between species in an experimentally manipulated community?
   a. a change in the population size of the species NOT being manipulated
   b. a change in the distribution of the species being manipulated
   c. a change in the population size or distribution of the species NOT being manipulated
   d. a change in the population size or distribution of the species being manipulated

   e. It is not possible to determine the existence of competition by experimental manipulation.

5. The founder of modern experimental ecology is generally acknowledged to be:
   a. Joseph Connell.
   b. Charles Darwin.
   c. Thomas Schoener.
   d. G. F. Gause.
   e. Robert MacArthur.

6. Which species is an example of the winner in winner-take-all competition between two species?
   a. oak tree
   b. herring gull
   c. bluebird
   d. gerbil
   e. house mouse

7. Of the moth species studied in Ontario by Fenton and Fullard, more than _____ percent can detect the cries of their bat predators. (Select the highest correct number.)
   a. 35
   b. 50
   c. 65
   d. 80
   e. 95

8. Which pair best exemplifies the evolutionary arms race between predator and prey?
   a. herring gulls and ring-billed gulls
   b. voles and house mice
   c. *Paramecium aurelia* and *Paramecium caudatum*
   d. *Semibalanus* barnacles and herring gulls
   e. *Eleodes longicollis* (beetles) and grasshopper mice

9. In a wildebeest herd, 80 percent of the young are born within a three-week period. This is advantageous because:
   a. predators are rarely courageous enough to attack a large herd of wildebeests.
   b. this interval coincides with the time during which predators are busy bearing their own young.
   c. the older members of the herd form a protective ring around the young until they can defend themselves.
   d. the chance that a specific wildebeest calf will survive is greatly increased.
   e. at this time of the year, most of the predators have migrated to other regions.

10. In a complex food web involving several predator and prey species, when one type of prey becomes less abundant, the predators:
    a. migrate to other areas.
    b. prey on other species with increased intensity.
    c. experience a population crash because of the reduced food supply.
    d. undergo character displacement, which reduces competition.
    e. switch from a carnivorous to an omnivorous or herbivorous diet.

11. When the marine snail *Littorina littorea* preys on the green alga *Enteromorpha* in intertidal pools of New England:
    a. several other algal species grow in the same pools.
    b. *Enteromorpha* is eliminated from the pools.
    c. the number of chitons and limpets increases.
    d. other species that prey on *Enteromorpha* are competitively excluded from the pools.
    e. the pools become blanketed with mussels.

12. *Pinnixa* crabs living in the burrows of *Chaetopterus* worms is an example of:
    a. mutualism.
    b. competitive exclusion.
    c. parasitism.
    d. character displacement.
    e. commensalism.

13. Which statement does NOT apply to parasites?
    a. They often affect the very old or very young.
    b. They usually do not kill their hosts.
    c. Well-adapted parasites cause severe disease in their hosts.
    d. They predispose the host to other diseases.
    e. Their hosts are more susceptible to other predators and adverse environmental factors.

14. Which association is NOT an example of mutualism?
    a. crabs and sea anemones
    b. myxoma virus and European rabbits
    c. aphids and ants
    d. oxpeckers and zebras
    e. cleaner fish and their hosts

15. The number of species that repopulate an island after mass extinction depends:
    a. only on the size of the island.
    b. only on the distance from the island to the mainland.
    c. only on the number of species initially present.
    d. primarily on the size of the island and its distance from the mainland.
    e. largely on the size of the island and the number of species originally present.

16. Which situation does NOT represent the subdivision of an ecological community?
    a. construction of a dam across a river
    b. clearing a field in a forest
    c. selectively removing one species of tree from a mixed forest
    d. erecting a fence between two pastures
    e. building a roadway through a marsh

17. Progressive changes in the photosynthetic components of a community, which are accompanied by changes in the animal life, are called:
    a. ecological succession.
    b. character displacement.
    c. competitive exclusion.
    d. genetic drift.
    e. intermediate disturbances.

18. T or F   The more ecologically similar two species are, the more intense the competition between them is likely to be.

19. T or F   In laboratory studies of competition between *Paramecium aurelia* and *Paramecium caudatum*, *P. aurelia* always outcompeted *P. caudatum*, no matter what culture conditions were used.

20. T or F   Ecologists agree that Darwin's finches are the best example of character displacement resulting from competition in the evolutionary past.

21. T or F   The realized niche of the barnacle *Semibalanus*, which inhabits the lower intertidal zone, is nearly identical to its fundamental niche.

22. T or F   Predation affects the evolution of both predator and prey.

23. T or F   Predation is the primary force reducing a prey population below the carrying capacity of its environment.

24. T or F   According to the island biogeography model of MacArthur and Wilson, the number of species and the species composition on a given island remain constant.

25. T or F   In some communities, ecological succession supports the inhibition hypothesis, whereas succession in other communities supports the facilitation hypothesis.

26. Distinguish between interference competition and exploitative competition.

27. Describe how these grazing ungulates of the African plains are able to coexist: zebras, gazelles, and wildebeests.

28. Describe the evolutionary adaptations in the "arms race" between legumes and the bruchid weevils that feed on their seeds.

29. Cite two recent discoveries suggesting that the snowshoe hare and lynx population cycles may be caused by some factor other than predator-prey interactions.

30. Summarize the intermediate disturbance hypothesis and cite one experiment that provided supporting evidence.

PERFORMANCE ANALYSIS

1. **c**   According to the principle of competitive exclusion, if two species are directly competing for the same limited resources, one species will eventually eliminate the other from the community or from part of the community, as revealed by Gause's experiments with *Paramecium* and Joseph Connell's studies of barnacles in Scotland. (page 1108)

2. **d**   The description of a species' ecological niche includes physical tolerance factors, requirements for food and shelter, and behavior patterns. It is the species' way of life

or role in the community. The habitat is the actual environment in which members of a species live. (page 1109)

3. **a** Closely related species may coexist in an environment by partitioning resources. No two species may occupy exactly the same niche for an indefinite period of time—one species will eventually eliminate the other. (page 1109)

4. **c** In studies of competition in natural communities where scientists have manipulated one population under study, a change in the population size or distribution of the population that was *not* manipulated is regarded as evidence that competition was occurring. (page 1112)

5. **a** Joseph Connell is often regarded as the founder of modern experimental ecology. (page 1112)

6. **b** The herring gull, which is extremely efficient at obtaining food in garbage dumps, exemplifies a species that is competitively superior and excludes other would-be competitors—in this case, the ring-billed gull and the laughing gull—from the community. (page 1114)

7. **e** Ninety-five percent of the moth species in Ontario have functional ears that can detect the cries of the bats who prey on them. (page 1115)

8. **e** The beetle *Eleodes longicollis,* when threatened, sprays a noxious chemical from a gland at the tip of its abdomen. Grasshopper mice grab these beetles, jam their abdomens into the ground, and eat them head first. (page 1114)

9. **d** The production of large numbers of young at the same time increases the chance that a particular calf will survive. The predators will feed until sated, but the seasonal nature of the food supply limits the predator population size. (page 1116)

10. **b** In most communities, when one prey species decreases in numbers, the predators will prey on other species with increasing frequency. (page 1118)

11. **a** By preying on the competitively superior green alga *Enteromorpha,* the marine snail *Littorina littorea* maintains species diversity in intertidal pools. (page 1119)

12. **e** Commensalism is a type of symbiosis from which one species benefits but the other species is neither harmed nor benefited. (page 1120)

13. **c** Well-adapted parasites cause little or no disease in their hosts, a situation that guarantees a prolonged relationship. (page 1120)

14. **b** The myxoma virus is a parasite that is usually deadly to European rabbits. It was introduced into Australia to control the population of imported European rabbits. Initially, the plan worked; however, resistant rabbits and less virulent virus strains were rapidly selected and spread through the land. All of the other examples are mutualistic relationships. (pages 1120, 1121, Figure 53–19)

15. **d** The size of the island and its distance from a source of colonizers (usually the mainland) are the two most important factors determining the equilibrium number of species on an island. (page 1124)

16. **c** Ecological communities can be subdivided in many ways, creating ecological islands. The selective harvesting of one species of tree from a mixed forest would disturb the community structure but would not subdivide it. (essay, page 1126)

17. **a** By definition, ecological succession is the progressive change in the photosynthetic components of a community, which is accompanied by corresponding changes in the animal life. (page 1125)

18. **True** The greater the degree of niche overlap, the more intense is the competition between two species. Many ecologists would also argue that niche overlap provides the selective pressure favoring character displacement. (page 1106)

19. **False** Although one species always eliminated the other species, the outcome (i.e., the winner) could be altered by changing the culture conditions. (page 1108)

20. **False** Although ecologists generally agree that Darwin's finches are a classic example of character displacement, many argue that the role of competition in character displacement can only be proved if two competing populations are shown to be diverging *in the present;* speculating on the role of past competition is just that, speculation. (page 1111)

21. **True** Joseph Connell's studies of barnacles illustrate the concepts of fundamental and realized niches. The fundamental niche is defined by the limits of tolerance of the organism, whereas the realized niche is that which is actually occupied. *Semibalanus* is restricted to the lower zone because it cannot survive in the harsher conditions of the higher intertidal zone inhabited by *Chthamalus.* Therefore, its fundamental niche is nearly identical to its realized niche. (page 1113)

22. **True** Predation exerts evolutionary pressure on prey species by favoring those species that avoid predators. At the same time, predators with adaptations that render them efficient at catching prey are favored. (page 1114)

23. **False** Although availability of prey is a major component of the carrying capacity of a predator population, predators do not generally reduce a prey population below its carrying capacity. (page 1116)

24. **False** According to the equilibrium model of island biogeography, once an island reaches equilibrium, the number of species will remain the same but species composition will continuously change. (page 1123)

25. **True** No one hypothesis has yet been formulated to adequately explain the mechanisms of ecological succession. It is likely that different mechanisms operate in different types of communities. (page 1128)

26. Interference competition involves direct interactions between competitors. In exploitative competition, competitors do not interact directly but rather, by using resources themselves, leave less for the competing species. (page 1106)

27. The grazing ungulates of the African plains coexist because they partition resources. The zebras eat older, tougher grass stems. Wildebeests eat younger stems and blades near the ground. Gazelles eat the smallest green shoots of grass and small dicots. (page 1110)

28. Bruchid weevils lay eggs on developing legume pods. When the larvae hatch, each one burrows into a seed, consuming it as it grows. Several adaptations have evolved in legumes to avoid this predation. Some have individual seeds too small to nourish a larva. Some produce a chemical that inhibits the protein-digesting enzymes of insects. (However, bruchids have a metabolic pathway that circumvents the effects of these enzymes.) Soybean seeds contain a chemical that inhibits bruchid larva development. (page 1114)

29. (1) When overbrowsed, certain types of plants produce new shoots and leaves containing chemicals toxic to the hares. (2) The hare population undergoes similar population cycles on an island where there are no lynx. (page 1119)

30. The intermediate disturbance hypothesis contends that species diversity in a community is related to the frequency of major disturbances. As the interval between disturbances increases, species diversity increases to a maximum level. However, if the interval between disturbances increases further, species diversity declines, possibly because of interspecific competition. Studies conducted by Wayne Sousa on coastal boulders revealed that the largest boulders, which are rarely disturbed, and the smallest boulders, which are frequently disturbed, have the lowest number of algal species. The medium-sized boulders, which experience an intermediate frequency of disturbances, have the greatest species diversity. (page 1124)

Ecosystems

MAJOR CONCEPTS

An ecosystem is a combination of biotic and abiotic components through which energy flows and materials cycle. Ecosystems may vary widely in size, depending upon the scale at which processes are being studied.

With few exceptions, the energy to support life on earth is supplied by the sun. Only about 50 percent of the solar radiation that strikes the earth's atmosphere reaches the earth's surface. The remainder is either reflected back into space or absorbed by the atmosphere.

There are four layers of the earth's atmosphere, which are distinguished by temperature differences. The troposphere is the layer extending approximately 10 kilometers above the surface of the earth and containing 75 percent of all the molecules in the atmosphere. Nearly all of the phenomena known

as weather occur within this layer. The next layer out is the stratosphere, which contains the ozone layer. Ozone molecules absorb most of the ultraviolet rays, high levels of which damage organic molecules. The mesosphere and the thermosphere are the next two layers. Temperature decreases in the mesosphere only to increase again in the thermosphere.

Water vapor and carbon dioxide in the atmosphere retain the infrared rays reflected from the earth's surface, thereby maintaining planetary temperatures compatible with life. Changes in the earth's reflectivity or disturbances in the atmosphere, such as increased cloud cover and altered carbon dioxide levels, could upset the energy balance of the entire planet.

The distribution of life on earth is mainly a consequence of the uneven distribution of the sun's energy over the surface of the planet. The resulting temperature variations, coupled with the earth's rotation, are responsible for the major patterns in wind, weather, and climate observed on the planet.

Air masses move because warm air is less dense than cool air and consequently rises. As an air mass rises and cools, its capacity to hold moisture decreases and precipitation occurs. This phenomenon is responsible for the rainy climate characteristic of the equator and for the creation of a rain shadow on the leeward side of a mountain range.

Energy flows through an ecosystem from one trophic level to the next. The first trophic level is composed of primary producers, which are always autotrophs. Primary consumers (herbivores and omnivores in terrestrial ecosystems) feed on the primary producers. Secondary consumers (carnivores and omnivores) prey on primary consumers. The most common number of trophic levels in a food chain is four, although there may occasionally be five or six levels. In most ecosystems, producers far outweigh consumers in terms of total biomass. Detritivores (scavengers and decomposers) obtain energy from both primary producers and consumers.

Gross productivity is a measure of the rate at which energy is assimilated by a group of organisms, such as those in a particular trophic level of an ecosystem. Net productivity is the gross productivity minus the cost of all the metabolic activiies of the organisms being studied. Net primary production is the net productivity of the producers in an ecosystem over a given time period, often one growing season.

Only 1 to 3 percent of the solar energy striking photosynthetic organisms is captured and stored in the bonds of organic molecules. As energy is transferred from one trophic level to the next, only about 10 percent of the energy stored at the lower trophic level is converted to biomass of the consumer. (Actual efficiencies of energy transfer may vary from 2 to 20 percent, depending upon the organisms involved.)

The transfer of energy between trophic levels determines the ecosystem structure, which can be depicted graphically in pyramids of numbers and pyramids of biomass. Pyramids of numbers may be upright, with fewer individuals at each higher trophic level, or inverted, with fewer producers than primary consumers. Pyramids of biomass are nearly always upright; exceptions occur when the primary producers have extremely high reproduction rates.

Although most life on earth depends on solar energy, marine chemosynthetic ecosystems that are powered by geothermal energy have been discovered. The critical reaction occurring in these ecosystems is the reduction of sulfate in sea water to hydrogen sulfide, which occurs at the high temperature and pressure of deep-sea rifts. Chemosynthetic bacteria—the primary producers—oxidize the sulfide to obtain the energy needed to extract carbon from carbon dioxide dissolved in the water. In some cases, chemosynthetic bacteria are symbiotic with other organisms in the ecosystem. Consumers in these systems include tube worms, bivalves, crustaceans, and octopods.

Inorganic substances move through ecosystems in biogeochemical cycles. These cycles have three geological components: the atmosphere, the lithosphere, and the hydrosphere. The biological components are the producers, consumers, and detritivores. Substances that cycle through ecosystems include nitrogen, carbon, water, ions, and minerals. Producers obtain inorganic substances from the environment and make them available to consumers. Decomposers break down organic molecules contained in the bodies of producers, consumers, and scavengers and release inorganic products back into the environment.

One biogeochemical cycle that is important to living organisms is the nitrogen cycle, which consists of three basic stages. In ammonification, nitrogenous compounds in organic matter are decomposed principally by soil-dwelling bacteria and fungi, which use proteins and amino acids and release excess nitrogen in the form of ammonia or ammonium ion. During nitrification, one group of soil bacteria oxidizes ammonia or ammonium ion to nitrite, and another group oxidizes nitrite to nitrate; energy is released at both steps. (An additional source of nitrates is nitrogen fixation by symbiotic microorganisms.) Nitrate is the form in which most nitrogen moves into plant roots. Inside the cells of plants, nitrate is reduced back to ammonium during assimilation, which requires energy.

In forest ecosystems, the plant life is critical to the retention of nutrients within the ecosystem. In tropical forests, nearly all the minerals in the ecosystem are present in the vegetation.

Substances that are not eliminated from the bodies of consumers accumulate and increase in concentration as they are passed along the food chain. Examples are the pesticide DDT, strontium-90 from radioactive fallout produced during nuclear testing in the 1950s, and cesium-137 from radioactive fallout produced during the Chernobyl nuclear power plant accident in 1986.

HOW TO STUDY THE CHAPTER

Read the entire chapter through quickly, focusing on the major concepts.

Use the GUIDED STUDY OF THE CHAPTER to help you identify the important details as you **reread** the chapter. Writing out the answers to these questions will help fix them in your mind as well as provide you with a valuable study aid.

Answer the questions in TESTING YOUR UNDERSTANDING without the aid of your text. Check your answers against those in PERFORMANCE ANALYSIS. Analyzing your answers will give you valuable feedback on your level of understanding and preparedness for classroom testing.

GUIDED STUDY OF THE CHAPTER

I. **Introduction** (pages 1131–1132)

1. Name two consequences of the interactions among populations in a community.

2. The cycling of material between living organisms and the abiotic environment depends upon what type of organism?

3. a. Define an ecosystem.

   b. Cite three examples of natural or artificial ecosystems.

II. Solar Energy *(pages 1132–1137)*

A. General Remarks *(page 1132)*

4. What factor determines global climatic and weather patterns? (Figure 54–1)

5. a. At what rate does solar energy reach the surface of the earth's atmosphere?

   b. *Vocabulary:* What is the solar constant?

   c. Why is only a fraction of the solar constant available to living organisms?

B. The Influence of the Atmosphere *(pages 1132–1133)*

6. What factor distinguishes the four layers of the atmosphere?

7. List the four layers of the atmosphere in order from the earth out into space and indicate the distance of each layer's outer boundary from the earth. (See also Figure 54–2.)

8. Of all the molecules in the earth's atmosphere, _____ percent lie within the troposphere.

9. Nearly all weather phenomena occur within which layer?

10. What event marks the boundary between troposphere and stratosphere?

11. Account for the temperature characteristics of the stratosphere.

12. a. How is ozone formed?

   b. Why is the presence of ozone important?

13. What fraction of the atmosphere is contained within the outer boundary of the stratosphere?

14. Account for the temperature characteristics of the mesosphere and thermosphere.

15. Indicate the fraction of incoming solar energy that:

   a. is reflected back by clouds.

   _____

   b. is absorbed by the atmosphere.

   _____

   (i) by water vapor, dust, and water droplets.

   _____

   (ii) by oxygen and ozone in the stratosphere and mesosphere. _____

   c. reaches the earth's surface.

   _____

16. Much of the energy absorbed by the atmosphere is stored in what form?

17. What is the importance of absorption of ultraviolet light by the ozone in the stratosphere?

18. Indicate the consequences of solar energy absorption by (a) the oceans and (b) the land.

19. Alteration of what factors could change the global energy balance of the earth?

C. *Essay:* The Threatened Ozone Layer *(pages 1134–1135)*

20. a. What fraction of the ultraviolet radiation reaching the stratosphere is converted into heat?

   b. By what mechanism does this conversion occur?

21. Cite two pieces of evidence that the ozone layer is being damaged.

22. a. Summarize the mechanism by which chlorofluro-carbons damage the ozone layer.

    b. How much damage can each chlorine atom do?

    c. Where are such chemicals used?

23. a. What is the direct consequence of a reduced ozone layer?

    b. How do increased levels of ultraviolet radiation affect (i) phytoplankton, (ii) crop plants, and (iii) nitrogen-fixing bacteria?

    c. What are two effects of increased ultraviolet radiation on humans?

24. According to current projections, the ozone layer will be reduced by what fraction by the year 2000?

    By the year 2050?

25. a. What steps are being taken to halt the destruction of the ozone layer?

    b. How effective do you think these steps will be? Substantiate your answer.

D. **Climate, Wind, and Weather** *(pages 1136–1137)*

26. Identify the major factor determining the distribution of life on earth.

27. Why do the equatorial regions receive more energy per unit area than the polar regions?

28. Which factor is responsible for the seasonal variations in temperature that occur in the Northern and Southern Hemispheres? (See also Figure 54–3.)

29. What two factors determine the major patterns of air circulation and rainfall?

Focus: **Warm air, which is less dense than cold air, rises and cools as it gains altitude.**

Focus: **Air masses associated with precipitation are rising and cooling, which reduces their moisture-holding capacity, resulting in rain or snow.**

30. Account for the following environmental conditions in terms of air mass movement.

    Tropical rains at the equator

    Deserts at 30° north and south latitude

    The polar front

Focus: **Zones of descending air masses are associated with regions on the planet's surface that have little or no rainfall.**

31. What factors are responsible for the major wind patterns?

32. Describe the conditions under which a "rain shadow" forms.

III. **The Flow of Energy** *(pages 1137–1145)*

A. **General Remarks** *(page 1137)*

33. Name the most important factor influencing the organization of ecosystems.

34. What fraction of the solar energy reaching the earth's surface is captured by living systems?

35. On a global basis, how much organic matter is produced annually by photosynthetic organisms?

B. **Trophic Levels** *(pages 1137–1143)*

36. *Vocabulary:* Distinguish among food chain, trophic level, and food web.

Focus: **The position of a species in the food web is an important component of its ecological niche.**

Producers (pages 1138–1139)

37. a. Characterize the primary producers of the first trophic level.

b. How does the total mass of producers compare with that of consumers?

38. a. *Vocabulary:* Distinguish between gross productivity and net productivity.

b. How is net productivity usually expressed?

c. What does net productivity measure?

39. *Vocabulary:* Define biomass and net primary production.

40. a. Identify three key factors influencing productivity in terrestrial ecosystems.

b. What factor often limits productivity in aquatic ecosystems?

Consumers (pages 1139–1141)

41. How does the energy stored in producers enter the animal world?

42. What happens to most of the chemical energy obtained by a consumer from digested food?

43. The increase in animal biomass is the sum of what two components?

Focus: **Animal biomass increase represents energy available to the next trophic level.**

---

44. Characterize secondary consumers.

45. *Vocabulary:* What are top predators?

46. a. What was the *maximum* number of trophic levels documented in one study of top predators?

b. What was the *most common* number of trophic levels in this study?

Focus: **With each higher trophic level, the amount of energy (stored in biomass) available to the consumers decreases.**

Detritivores (pages 1142–1143)

47. a. *Vocabulary:* What are detritivores?

    b. From what sources do detritivores obtain nutrients and energy?

Focus: Detritivores include scavengers and decomposers.

48. Name two sources of energy exploited by decomposers that are not usable by animals.

49. What fraction of a forest's primary productivity is eventually consumed by detritivores?

50. Under what conditions will energy stored in organic matter remain unutilized?

C. *Essay:* Chemosynthetic Ecosystems *(pages 1140–1141)*

51. What is the energy source for chemosynthetic ecosystems?

52. Describe the deep-sea oasis discovered by the crew of the Alvin.

53. Which chemical reaction is crucial to the existence of the chemosynthetic ecosystems that have been identified?

Focus: The specific composition of chemosynthetic ecosystems varies but the primary producers in all known instances are chemosynthetic bacteria.

54. Name six nonbacterial organisms found in these ecosystems.

55. How do tube worms in the Garden of Eden oasis obtain energy?

56. Identify three other environments (besides deep-sea environments) in which chemosynthetic ecosystems have been found.

57. What is the source of food for the black abalone found off the Palos Verde peninsula?

58. Name three environments that have not yet been studied that are likely prospects for the existence of chemosynthetic ecosystems.

D. **Efficiency of Energy Transfer** *(page 1143)*

59. How is the shortness of food chains generally explained?

60. a. Approximately what fraction of the energy stored at one trophic level is converted to biomass at the next trophic level?

    b. What is the actual range of variation in energy transfer efficiencies?

61. Carnivores expend more energy than herbivores in what activity?

62. Summarize the basis for the position in favor of humans living lower on the food chain.

E. *Essay:* **Energy Costs of Foodgathering** *(page 1144)*

63. For each calorie invested in food gathering, what is the return on the investment in each situation?

    Organisms in natural populations

    Human hunter-gatherers

    Shifting agriculture (no fertilizers)

    "Modern" agriculture in 1900

    Modern agriculture today

64. Technology allows us to use what sources of energy that are unavailable to other animals?

F. **Energy Transfer and Ecosystem Structure** *(pages 1144–1145)*

> *Focus:* **Energy transfer within an ecosystem determines the structure of the ecosystem in terms of the numbers of organisms and amount of biomass present.**

65. Identify the similarities and differences between the pyramids of numbers for grassland and forest ecosystems.

66. a. What shape characterizes pyramids of biomass for most ecosystems?

    b. Under what conditions may this pyramid be inverted? Cite one example.

> *Focus:* **Pyramids of numbers and of biomass indicate only the amount of organic matter present at a specific time.**

IV. **Biogeochemical Cycles** *(pages 1145–1152)*

A. **General Remarks** *(pages 1145–1146)*

67. Name at least six substances that cycle through ecosystems.

68. a. *Vocabulary:* Define a biogeochemical cycle.

    b. Name and describe the three geological components of biogeochemical cycles.

    c. Name the three biological components of biogeochemical cycles.

> *Focus:* **Decomposers release inorganic substances from the organic compounds of dead organisms. These inorganic materials are then available for assimilation by primary producers.**

69. For any reaction in which _____ kcal/mole are made available, there is some organism that can use the energy.

    Note: Study Figure 54–14. Focus on the organisms through which phosphorus cycles in terrestrial and aquatic ecosystems and the phosphorus-containing compounds at each stage of the cycle.

B. **The Nitrogen Cycle** *(pages 1146–1149)*

70. a. What is the chief reservoir of nitrogen?

    b. Plants depend on what source of nitrogen?

71. a. Identify the three stages of the nitrogen cycle.

b. Fill in the following table.

Table for Question 71

	AMMONIFICATION	NITRIFICATION	ASSIMILATION
Starting compound(s)			
Product(s)			
Energy required or released?			
Participating organism(s)			

72. Name four sources of nitrates. (Figure 54–15)

73. a. Most nitrogen enters plant roots in what form?

   b. In plants, nitrogen in what form is actually incorporated into organic molecules?

74. Identify four factors responsible for the loss of nitrates from the soil.

75. a. *Vocabulary:* What is denitrification?

   b. Under what conditions does denitrification occur?

76. What organisms convert gaseous nitrogen into nitrogenous compounds?

C. *Essay:* **Agricultural Ecosystems and a Hungry World** *(pages 1148–1149)*

   77. a. According to current information, agriculture originated in what region of the world?

   b. What natural features of this area facilitated agricultural development?

78. Outline the spread of agriculture and its independent origination in other parts of the world.

79. a. Identify four characteristics of agricultural ecosystems.

   b. Which of these characteristics was most important in early agriculture?

   c. Identify five consequences of the exponential growth of the human population that started with the beginnings of agriculture.

80. What portion of the world human population is malnourished or undernourished?

81. a. What is the major emphasis of the Green Revolution?

   b. Describe the effects of the Green Revolution on agricultural yields in Mexico.

82. Why has the Green Revolution come under criticism? (4 reasons)

Focus: Although food production continues to rise, the problems of food supply will not be solved while population continues to rise.

Focus: The problem of world-wide hunger will not be solved by technology alone, but rather by social and political actions.

D. Recycling in a Forest Ecosystem *(pages 1149–1150)*

Focus: The plant life of a community plays a major role in nutrient retention.

83. a. Outline the approach used in the study conducted in Hubbard Brook Experimental Forest.

b. Summarize the influence of deforestation on the following activities or agents.

Water runoff

Calcium

Potassium

The nitrogen cycle

Streams draining the test area

Focus: In tropical rain forests, nearly all the mineral nutrients in the ecosystem exist in living organisms.

84. What are the consequences of this Focus statement for attempts at agriculture in areas cleared of tropical forests?

E. Concentration of Elements *(pages 1150–1152)*

85. a. What two factors are responsible for the concentration of elements in the bodies of living organisms?

b. In what way is this concentrating effect valuable?

86. Organisms at what trophic level(s) are most affected by the accumulation of toxic substances such as DDT?

87. What effect do toxic levels of DDT have in birds of prey? (Figure 54–17)

88. Trace strontium-90 from the fallout of nuclear tests in the 1950s through the food chains to its accumulation in (a) North Americans and (b) Eskimos.

89. What is the half-life of strontium-90?

Focus: The exposure to radiation from strontium-90 in bones continues. The long-term effects of this exposure are unknown.

The Lessons of Chernobyl (pages 1151–1152)

90. a. Approximately how many people in how many European countries were affected by fallout from the accident at the Chernobyl nuclear power plant?

  b. What determined the dispersal of the radioactive cloud?

91. a. What radioactive element was the main component of the Chernobyl fallout?

  b. What is the half-life of this element?

  c. Trace the route of this element through the food chain from fallout to humans.

92. What are the consequences of this situation for the Sami people of Norway?

93. Identify three lessons that were reinforced by the Chernobyl incident.

V. Summary *(pages 1152–1153):* Read the summary. If you are familiar with the essential features of the material presented there, you are ready to complete the section TESTING YOUR UNDERSTANDING.

TESTING YOUR UNDERSTANDING

After you have completed this examination, compare your answers with those in the section that follows.

1. Identify the value of the solar constant.
   a. 1.94 calories per square meter per minute
   b. 1.3 calories per square centimeter per minute
   c. 1.94×10^{24} calories per square centimeter per year
   d. 1.3×10^{24} calories per year
   e. 1.94×10^{24} calories per year

2. The troposphere extends about _____ kilometers above the earth's surface and contains about _____ percent of the molecules in the atmosphere.
   a. 25; 90
   b. 10; 75
   c. 8; 60
   d. 5; 45
   e. 3; 30

3. Which event would have little or no effect on the amount of solar energy available to organisms on earth?
   a. reducing the ozone layer
   b. increasing the earth's reflectivity
   c. increasing the atmospheric carbon dioxide content
   d. decreasing the earth's cloud cover
   e. Any of these events could have a dramatic effect on the amount of energy available to living organisms.

4. A low-pressure area known as the polar front occurs at:
   a. 60° north and south latitudes.
   b. 30° north and south latitudes.
   c. 60° north and 30° south latitudes.
   d. 30° south and 60° north latitudes.
   e. the north and south poles.

5. The Sierra Nevada mountains lie along the western coast of the United States. Why are the eastern slopes of these mountains dry and desertlike?
   a. Hot, dry air off the deserts is rising and picking up moisture.
   b. They are in the rain shadow of the prevailing westerly winds.
   c. The air currents that carry moisture-laden air are blocked by the mountains.
   d. The soil is so poor that little vegetation grows there to release moisture into the air.
   e. There are no large bodies of water in this region to release water into the air.

6. Of the solar energy that reaches the earth's surface, only a very small fraction, _____ percent, is diverted into living systems.
   a. 0.1
   b. 0.2
   c. 0.3
   d. 0.4
   e. 0.5

7. All heterotrophs combined account for only _____ percent of the organic matter on earth.
   a. 1
   b. 3
   c. 5
   d. 7
   e. 9

8. In many aquatic ecosystems, the principal factor affecting productivity is the:
   a. wavelengths of light that penetrate the water.
   b. intensity of the sunlight.
   c. availability of essential minerals.
   d. availability of dissolved carbon dioxide.
   e. temperature of the water.

9. In the chemosynthetic bacteria of deep-sea rift ecosystems, the energy required to incorporate carbon from carbon dioxide into organic molecules is derived from the metabolism of:
   a. dissolved oxygen.
   b. hydrogen sulfide.
   c. hydrochloric acid.
   d. sulfuric acid.
   e. sodium sulfate.

10. The most common number of trophic levels in a food chain is:
    a. two.
    b. three.
    c. four.
    d. five.
    e. six.

11. If a given square meter of land surface receives 1,200 kilocalories of light energy per day, how much energy is available to the first level of carnivores?
    a. 120 to 360 kilocalories
    b. 12 to 36 kilocalories
    c. 1.2 to 3.6 kilocalories
    d. 0.12 to 0.36 kilocalorie
    e. 0.012 to 0.036 kilocalorie

12. The process in which microorganisms release excess nitrogen in the form of ammonia or ammonium ions is known as:
    a. nitrogenation.
    b. nitrification.
    c. ammonification.
    d. deamination.
    e. denitrification.

13 Poorly drained soils have a low oxygen content. Consequently, nitrates may be reduced to elemental nitrogen in a process called:
    a. ammonification.
    b. assimilation.
    c. denitrogenation.
    d. denitrification.
    e. nitrogen fixation.

14. Based on current data, it has been estimated that by the year 2000 the ozone layer will be reduced by a factor of _____ percent.
    a. 1
    b. 3
    c. 5
    d. 10
    e. 30

15. T or F   Since cool air holds less moisture, as cool air descends toward the earth, it loses its moisture as rain or snow.

16. T or F   In a complex food web, predators typically take more than one type of prey.

17. T or F   Of the organic material consumed by herbivores, much is eliminated undigested.

18. T or F   The energy available to a given trophic level is the sum total of all the energy in the trophic levels below it.

19. T or F   A pyramid of numbers may be inverted if the primary producer is large with respect to the size of the primary consumers.

20. T or F   In the case of the nitrogen cycle, each decomposer can perform multiple functions involved in the movement of nitrogen through the cycle.

21. Explain why most forms of terrestrial life could not exist in the absence of the ozone layer surrounding the earth.

22. Distinguish between the two types of detritivores—scavengers and decomposers.

23. Identify (a) the three geological and (b) the three biological components of biogeochemical cycles.

24. List four characteristics of agricultural ecosystems. These traits are characteristic of what type of ecosystems?

25. What is the importance of plants to nutrient retention and cycling in forest ecosystems?

26. What are the consequences of the Chernobyl accident for the Sami peoples of Norway?

PERFORMANCE ANALYSIS

1. **d**   The solar constant is the amount of energy reaching the upper surface of the earth's atmosphere. It equals 1.3×10^{24} calories per year, which is equivalent to an average of 1.94 calories per square centimeter per minute. (page 1132)

2. **b**   The troposphere is the atmospheric layer closest to the earth's surface. It extends approximately 10 kilometers above the surface and contains 75 percent of the molecules in the atmosphere. (pages 1132, 1133)

3. **e**   All of the events listed could affect the energy balance of the entire planet. The specific effects and the degree of change required to initiate such effects are currently under debate. (page 1133)

4. **a** Low-pressure areas occur in regions of rising warm air, which occur at the equator and at 60° north and south latitudes. The low-pressure areas at 60° north and south latitudes are known as polar fronts. (page 1137)

5. **b** The prevailing winds over that part of the country are westerlies that carry moisture in from the ocean. As air masses encounter the mountains, they rise, cool, and lose their moisture on the western slopes. This creates a rain shadow on the eastern slopes. (page 1137)

6. **a** Only about 0.1 percent of the solar energy reaching the earth's surface is incorporated into living systems. (page 1137)

7. **a** Plants and algae account for 99 percent of all the organic matter on earth; heterotrophs constitute the remaining 1 percent. (page 1138)

8. **c** The major factor affecting productivity in aquatic ecosystems is often the availability of essential minerals. The major factors influencing the productivity of terrestrial ecosystems are the duration and intensity of sunlight, temperature, and precipitation. (page 1139)

9. **b** The energy source powering the chemosynthetic ecosystems of deep-sea rifts is hydrogen sulfide, which is produced when heat and high pressure in the rift supply energy for the reduction of sulfate in sea water. (essay, page 1140)

10. **c** Although an occasional food chain will have five or six trophic levels, the most common number is four—primary producer, primary consumer, secondary consumer, and tertiary consumer (top predator). (page 1141)

11. **d** Only 1 to 3 percent of the light energy falling on vegetation is incorporated into energy in the bonds of organic molecules. At each transfer of energy from one trophic level to the next, only 10 percent of the energy available at one level is assimilated into biomass at the next higher level. Therefore, of the original 1,200 calories, the primary producers retain 12 to 36 kilocalorie, the herbivores retain 1.2 to 3.6 kilocalories, and the first level of carnivores retain 0.12 to 0.36 kilocalorie. (page 1143)

12. **c** Ammonification is the process in which microorganisms release excess nitrogen as ammonia or ammonium ions. (page 1147)

13. **d** When oxygen is in short supply, as occurs in poorly drained soils, microorganisms reduce nitrates to elemental nitrogen. (page 1148)

14. **c** It has been estimated that the ozone layer will be reduced by 5 percent by the year 2000. A reduction of only 1 percent will result in an increased incidence of human skin cancer and may damage the human immune system. (essay, page 1134)

15. **False** Falling cool air is typically quite dry; the major deserts of the world are located in regions characterized by descending cool air masses. Precipitation occurs as warm air masses rise, expand, and cool, thereby reducing their moisture-holding capacity. (page 1137)

16. **True** in most cases, predators will feed on more than one type of prey. (page 1138)

17. **True** Much of the organic matter ingested by herbivores is eliminated undigested since these animals have a limited capacity to utilize the structural polysaccharides (i.e., cellulose and lignin) characteristic of primary producers. (page 1139)

18. **False** Only 10 percent of the energy contained in a trophic level is incorporated into new biomass at the next highest trophic level. (page 1141)

19. **True** An example of this situation would be a large tree supporting numerous insects. (page 1145)

20. **False** In the nitrogen cycle, each decomposer is highly specialized and performs only one step in the movement of nitrogen through the cycle. (page 1147)

21. Ultraviolet and other high-energy radiations damage organic molecules and would be lethal to most forms of terrestrial life if they reached the earth's surface in large quantities. By absorbing the energy of these radiations, the ozone layer acts as a protective shield over the planet. (page 1133; essay, page 1134)

22. Detritivores can be thought of as highly specialized consumers. Scavengers live on dead, rather than living, prey. Decomposers have adaptations that enable them to use sources of chemical energy—such as cellulose and nitrogenous wastes—that cannot be used by animals. (page 1142)

23. The three geological components of biogeochemical cycles are the atmosphere, the lithosphere, and the hydrosphere. The three biological components are producers, consumers, and detritivores. (page 1145)

24. Four traits of agricultural ecosystems are a small number of species, a relatively low total biomass, a high net productivity relative to biomass, and a limited capacity to retain nutrients. These traits are characteristic of ecosystems subjected to frequent, large-scale disturbances. (essay, page 1148)

25. Studies in experimental forests and in tropical forests have revealed that plants are critically important to nutrient retention and cycling in these ecosystems. Most of the minerals present in tropical rain forests are contained within the vegetation. Further, natural forests are extremely efficient in conserving minerals. (pages 1149, 1150)

26. The Sami peoples of Norway have traditionally depended upon reindeer milk and meat for food and for income. Because reindeer eat primarily lichens and the lichens have accumulated cesium-137 from the Chernobyl fallout, it will be 30 years before the Sami peoples can use the meat and milk from their herds. This is an entire generation, time for a cultural tradition centuries old to be lost forever. (page 1152)

The Biosphere

MAJOR CONCEPTS

The biosphere is the thin zone on the earth's surface in which life can exist. It extends from the depths of the oceans to the upper limit at which the atmosphere can support life. It also includes that portion of the soil inhabited by living organisms.

Freshwater environments may be classified as running water or as standing water. In rivers and streams, the nature of the inhabitants is determined largely by the swiftness of the current. As a stream flows away from its source, it typically becomes wider, its speed decreases, and it begins to take on characteristics of standing water.

Lakes and ponds may be divided into three distinct zones. The littoral zone, which is the most richly inhabited, is the region at the edge of the water. The limnetic zone is the zone of open water; it extends to the limit of light penetration. This zone is innabited mainly by phytoplankton and fish. The profundal zone lies deep to the limnetic zone and has no photosynthetic inhabitants. Detritivores are its major inhabitants.

Marine environments can be categorized as oceans, including the continental shelves, and seashores. The oceans cover nearly 75 percent of the earth's surface. Life in the open oceans can be classified as pelagic (free-floating) or benthic (bottom-dwelling). Although life extends to the depths of the oceans, photosynthesis is limited to the upper several meters because of the absorption of light by seawater. The continental shelves, which receive a constant supply of nutrients from the land, are richly populated by a great variety of organisms.

The major ocean currents influence life in the oceans and along the coastal regions of continents. The ocean currents are determined by the combined effects of the major wind patterns and the earth's rotation. Disturbances of the normal ocean currents, such as those occurring in an El Niño, can have devastating results on marine and terrestrial organisms.

There are three general types of seashores: rocky, sandy, and muddy. Rocky seashores typically are the most densely populated. Rocky seashores have three distinct zones, the inhabitants of which are determined by gradients of light, temperature, and wave action, by competitive interactions, and by predation pressures. The shifting bottoms of sandy seashores have fewer inhabitants than either rocky or muddy seashores. Mud flats, salt marshes, and estuaries receive a constant flow of nutrients from the land and are important spawning grounds and nurseries for many forms of marine life.

Patterns of temperature and precipitation largely determine the life forms characteristic of terrestrial environments. These patterns are influenced by the angle of the earth's axis relative to the sun, the earth's rotation, and surface features of the continents such as mountain ranges. The mineral content of soils also influences the type of plant life that can grow in a region, and consequently the type of animal life that can be supported.

With respect to mean atmospheric temperature, a gain in elevation of about 100 meters is analogous to an increase in latitude of one degree. In spite of the similarities between high-altitude and high-latitude environments, there are three major differences. (1) At high altitudes, the atmosphere is less dense and solar radiation more intense. (2) Most of the water vapor in the atmosphere, which has important temperature-modulating effects, occurs within 2000 meters above sea level. (3) At high latitudes, there is marked seasonal variation in day length that does not characterize all high-altitude environments.

Terrestrial environments can be classified into several categories, or biomes, based on the dominant types of vegetation. Each biome has characteristic temperature and precipitation patterns. Although not closely related phylogenetically, the plants and animals of the same biome on different continents are often quite similar in appearance, reflecting evolutionary adaptations to similar selection pressures.

Temperate deciduous forests occur in regions with a mild growing season at least four months long and characterized by moderate precipitation, followed by a colder period during which plant growth is minimal or arrested. Because of the expense of growing and shedding leaves every year, the soils supporting these forests must be rich and hold nutrients.

Deciduous forests typically have four layers: tree layer, shrub layer, field layer, and ground layer. These forests support abundant animal life.

Coniferous forests exist where summers are too short for deciduous forests to survive. The dominant plants are evergreen trees with needle-like leaves. Deciduous trees may grow along stream banks. Unlike deciduous forests, the ground layer of coniferous forests has little vegetation. Coniferous forest biomes include the taiga, alpine forests, and mixed west-coast forests of California and Chile.

The tundra is a type of grassland that exists where winters are too long and severe to support even conifers. Tundra is found at high altitudes as well as at high latitudes. This biome is characterized by permafrost, a permanently frozen subsoil. Perennial herbaceous plants, mosses, and lichens support a variety of animal life.

Temperate grasslands are generally located in the interior of continents and are characterized by periodic droughts, hot-cold seasons, and rolling to flat terrain. They are transitional between temperate forests and deserts and are maintained by grazing and by periodic fires. Sod-forming grasses with legumes and annuals are the major plant forms.

Savannas are tropical grasslands with scattered clumps of trees. The seasons are wet-dry rather than hot-cold. The balance between trees and grasses depends upon water supply: more water favors the growth of trees; less water favors the growth of grasses. Overgrazing can increase water supply in the soil to the extent that trees increase in number and shade out the grasses. The savannas of Africa support large, diverse populations of herbivores.

Mediterranean scrub biomes are found in regions with mild winters and long, dry summers. Small trees or spiny shrubs with leaves adapted for water conservation are the dominant forms of vegetation.

The large deserts of the world are located at latitudes of 30° north and south, where cold air masses are falling and warming. Deserts are characterized by less than 25 centimeters of rainfall per year. The paucity of atmospheric water vapor in deserts is responsible for wide temperature fluctuations between day and night. Perennial plants have multiple adaptations for water conservation. The reproductive cycles of annual plants go quickly from seed to flower to seed after a rain. The animal inhabitants also have adaptations for water conservation.

Most of the tropical forests are located near the equator where the mean daily temperature is constant throughout the year and day length varies by less than one hour. Rainfall may be seasonal, as in tropical mixed forests, or it may be abundant all year, as in tropical rain forests. In tropical rain forests, the limiting factor for which plants compete is light. Dead organisms are decomposed rapidly and frequent heavy rains quickly leach nutrients from the soil. Consequently, most of the nutrients present are contained in living organisms and the soils are typically very poor.

As a result of the pressures of growing human populations, the tropical forests are being destroyed at an alarming rate. Accompanying their destruction are mass extinctions of plant and animal species. Scientists predict that continued destruction of these forests will affect the atmosphere of the entire planet.

HOW TO STUDY THE CHAPTER

Read the entire chapter through quickly, focusing on the major concepts.

Use the GUIDED STUDY OF THE CHAPTER to help you identify the important details as you **reread** the chapter. Writing out the answers to these questions will help fix them in your mind as well as provide you with a valuable study aid.

Answer the questions in TESTING YOUR UNDERSTANDING without the aid of your text. Check your answers against those in PERFORMANCE ANALYSIS. Analyzing your answers will give you valuable feedback on your level of understanding and preparedness for classroom testing.

GUIDED STUDY OF THE CHAPTER

I. Introduction *(page 1154)*

　　1. *Vocabulary:* Define the biosphere and describe its extent.

II. Life in the Waters (pages 1154–1160)

A. General Remarks (page 1154)

> *Focus:* **The biosphere is mainly composed of aquatic environments and their inhabitants.**

2. Name the two general categories of freshwater environments.

3. What are the two basic categories of marine environments?

B. Rivers and Streams (pages 1154–1155)

> *Focus:* **The water in rivers and streams is continuously moving.**

4. Identify three sources of rivers and streams.

5. a. What factor determines the character of life in a stream?

   b. How does the swiftness of the current change as the stream moves away from its source?

6. a. Where do most organisms live in swift streams?

   b. Characterize life in the pools of streams.

C. Lakes and Ponds (page 1155)

7. Fill in the following table on the three zones of a lake or pond.

Table for Question 7

	LITTORAL	LIMNETIC	PROFUNDAL
Location (or boundaries)			
Photosynthetic organisms (identify)			
Other inhabitants			

8. Which zone is the most richly inhabited?

9. What are marshes?

10. How do nutrients released by decomposers in the profundal zone become distributed throughout the lake or pond?

D. The Oceans (pages 1155–1157)

> *Focus:* **The oceans cover nearly 75 percent of the earth's surface.**

11. What is the average depth of the oceans?

12. What types of organisms inhabit most of the oceans?

13. a. In clear water, what fraction of the sunlight that strikes the surface penetrates to a depth of:
    1 meter?
    50 meters?

    b. What wavelengths of light are absorbed first as light travels through water?

    c. What wavelengths of light are available to photosynthetic organisms that grow at the limits of light penetration?

14. *Vocabulary:* Use the following table to distinguish between the pelagic and benthic divisions of the ocean.

Table for Question 14

	PELAGIC	BENTHIC
Location		
Major food source(s)		
Examples of inhabitants		

15. Describe the components of plankton.

16. a. Contrast the productivity of the open ocean with that of terrestrial environments.

   b. What factor is presumed to be responsible for the low productivity of the open ocean?

Focus: **The major ocean currents influence ocean life and the climates of coastal regions.**

17. a. Describe the major patterns of water circulation in the Northern and Southern Hemispheres.

   b. Cite three specific examples illustrating the influence of ocean currents on coastal climates. (See also Figure 55–4.)

18. a. *Vocabulary:* What is upwelling and what factors are responsible for this phenomenon?

   b. Characterize ocean life in regions of upwelling.

E. *Essay:* **El Niño** *(page 1158)*

Focus: **An El Niño is a disturbance of the normal patterns of water circulation in the Pacific Ocean.**

19. At what intervals do major and minor El Niño events occur?

20. a. Describe the air pressure systems over the Pacific Ocean that normally determine weather patterns.

   b. Describe the water currents that result from the wind patterns.

21. a. Outline the air pressure, wind, and current changes that occur during an El Niño.

   b. What effect does an El Niño have on the food chains along the western coasts of North and South America?

22. Identify four consequences of the El Niño of 1982–1983.

23. What events are thought to be associated with the initiation of an El Niño?

F. **The Seashore** *(pages 1157–1160)*

24. a. *Vocabulary:* What are continental shelves?

   b. How does the primary productivity of the continental shelf environments of temperate zones compare with the productivity of other environments?

25. a. What are the major primary producers of the continental shelf environments in temperate zones?

b. List some of the other inhabitants of these shallow coastal waters.

26. List three types of temperate zone seashore environments.

27. Animal and plant inhabitants of rocky coasts typically have adaptations for what activity? (Describe several examples.)

28. a. *Vocabulary:* Complete the following table to distinguish between the three zones characteristic of a rocky coast.

Table for Question 28

	SUPRATIDAL	INTERTIDAL	SUBTIDAL
Exposure to seawater			
Life forms (examples)			

b. What three factors are responsible for this zonation?

29. a. Why do sandy beaches typically have fewer bottom inhabitants than rocky coasts?

b. Name several sand-dwelling animals and describe their diets.

c. What inhabitants of sandy beaches are important in stabilizing sand dunes?

30. a. Mud flats, salt marshes, and estuaries are rich in animal life for what reason?

b. These environments serve what important role in the life of ocean inhabitants?

c. What communities serve this role in the tidal areas of tropical and subtropical regions?

Focus: The draining and filling of mud flats, salt marshes and estuaries, destroys spawning grounds and nurseries for marine organisms and may have severe consequences for ocean life.

III. Life on the Land *(pages 1160–1177)*

A. General Remarks *(pages 1160–1161)*

31. a. Name two physical factors that most immediately influence terrestrial organisms.

b. These two factors are influenced by what three phenomena?

32. The continents are composed largely of what type of rock?

Focus: The organisms present on the earth's surface are greatly influenced by the mineral content of the soil and the patterns of rainfall.

33. Discuss the effect of latitude and elevation on mean atmospheric temperature.

34. Identify three important differences between high-latitude habitats and high-altitude habitats.

B. The Concept of the Biome *(pages 1161–1163)*

35. *Vocabulary:* What is a biome? (Explain as fully as possible.)

> *Focus:* **Because they are influenced by similar environmental conditions, organisms in the same biome in different regions of the world resemble one another even though they are not closely related.**

36. Organisms in biomes often exemplify what pattern of evolution?

C. Temperate Forests *(pages 1164–1166)*

37. Characterize the climate in which temperate deciduous forests are located.

38. a. The shedding of leaves evolved as an adaptation to what stress?

    b. How long must the growing season be in order for this strategy to be of adaptive value?

    c. What other environmental factor is critical to the success of deciduous plants?

39. List and characterize the four layers of plant growth in a temperate deciduous forest.

> *Focus:* **The dominant trees in temperate deciduous forests vary in different regions.**

40. What factor is largely responsible for the variation in dominant tree species in temperate deciduous forests?

41. Name at least five trees that may be dominant species in the temperate deciduous forests of North America.

42. Discuss the animal life supported by a temperate deciduous forest.

43. a. Characterize the soil of a temperate deciduous forest.

    b. What features of this soil make it good potential farmland?

D. Coniferous Forests *(pages 1166–1167)*

44. Why are conifers out-competed by deciduous trees in temperate regions with adequate rainfall and rich soil?

45. Identify three coniferous forest biomes.

The Taiga (page 1166)

46. Characterize the taiga with respect to prevailing weather conditions, major plant inhabitants, type of ground cover, abundance and location of deciduous trees, and quantity of annual plants.

47. Cite examples of the principal large and small animals living in the taiga.

48. In what two ways does the ground layer of the taiga differ from that of the deciduous forest?

The Pacific Northwest (page 1166)

49. a. To what environmental conditions have the forests of the Pacific Northwest adapted?

   b. What factors place deciduous trees at a disadvantage in this environment?

   c. Describe three adaptations of the resident trees to this environment.

E. **The Tundra** *(pages 1168–1169)*

50. a. Characterize the climate of the tundra.

   b. In what two locations may tundra biomes be found?

51. Identify and describe the most characteristic feature of the tundra.

52. a. Identify two factors that limit plant growth in the tundra.

   b. What types of plants dominate tundra vegetation?

Focus: **All the angiosperms of the tundra are perennials.**

53. Identify the major herbivores and carnivores of tundra biomes.

54. What features of the tundra biome attract migratory birds?

Focus: **In many areas of the tundra, the growing season is less than two months.**

F. **Temperate Grasslands** *(pages 1169–1170)*

55. a. Grasslands are considered to be transitional environments between what two other biome types?

   b. Where are they usually located on a continent?

56. List four distinguishing features of temperate grasslands.

57. State the different names used to describe grasslands on four different continents.

58. What type of vegetation is typical of temperate grasslands?

59. Why are the grasslands of North America drier at their western border than they are in the east?

60. Cite examples of herbivores and carnivores that may be found in temperate grasslands.

61. Identify two factors that maintain the grasslands and prevent their conversion to forests.

G. Tropical Grasslands: Savannas *(page 1170)*

62. *Vocabulary:* What are savannas?

63. What factors determine the transition from open forest with grassy undergrowth to savanna?

64. a. Name the key factor in maintaining the balance between grasses and woody plants.

    b. What happens to this balance if rainfall increases or if the grasses are overgrazed?

    c. What happens to this balance if rainfall decreases?

65. Describe the exquisite adaptation of savanna grasses to their climate.

Focus: **The African savannas are inhabited by the greatest numbers and varieties of large herbivores on the earth.**

H. Mediterranean Scrub *(page 1171)*

66. a. What climate is typical of Mediterranean scrub biomes?

    b. Characterize the vegetation of this biome.

    c. What name does this biome have in the United States?
    Mediterranean regions?
    Chile?

67. What types of animals live in this environment?

I. The Desert *(pages 1172–1173)*

68. In terms of global air-mass movements, account for the location of the great deserts of the earth at latitudes of 30° north and south.

69. What factor is contributing to the southward extension of the Sahara desert?

70. a. Deserts may be defined as regions where rainfall is less than _____ centimeters per year.

    b. Describe the adaptations of plants and animals that enable them to survive in this extremely dry environment.

71. Account for the dramatic daily temperature fluctuations that may occur in the desert compared with the more constant temperatures characteristic of more humid regions at the same latitude and elevation.

J. Tropical Forests *(pages 1174–1177)*

Focus: **Tropical forests are located in the equatorial zone, where the mean daily temperature and day length are fairly constant throughout the year.**

72. Identify three types of tropical environments based on annual rainfall patterns.

Tropical Rain Forest (pages 1174–1176)

73. What rainfall pattern characterizes tropical rain forests?

74. a. Contrast the diversity of plant species in a tropical rain forest with that of a temperate deciduous forest.

    b. What percentage of the plant species in a tropical rain forest are trees?

75. Name the limiting factor for which plants compete in a tropical rain forest.

76. Characterize the vegetation at each level of the tropical rain forest.

    Upper tree story

    Lower tree story (canopy)

    Forest floor

77. a. *Vocabulary:* What are epiphytes and where do they live?

    b. What are some adaptations of epiphytes to this lifestyle?

78. At what level do many of the animals of the tropical rain forest live?

79. a. Account for the infertility of the soils in tropical rain forests.

    b. Name the soil type characteristic of many tropical rain forests.

    c. Where are most of the nutrients located in this biome?

Tropical Forests, Mass Extinction, and Human Responsibility (pages 1176–1177)

80. a. Describe the major factors contributing to the rapid demise of tropical rain forests.

    b. If the present rate of forest destruction continues, what will be the situation by the turn of the century?

81. a. Identify three factors contributing to increasing levels of atmospheric carbon dioxide.

    b. Explain why scientists fear that increasing levels of carbon dioxide in the atmosphere will result in a significant increase in global temperature.

c. Name four consequences of this predicted increase in temperature.

82. List several factors that must be considered in finding a solution to the mass destruction of the tropical rain forests.

IV. **Summary** *(page 1177):* Read the summary. If you are familiar with the essential features of the material presented there, you are ready to complete the section TESTING YOUR UNDERSTANDING.

TESTING YOUR UNDERSTANDING

After you have completed this examination, compare your answers with those in the section that follows.

1. The character of life in a stream is determined largely by the:
   a. swiftness of the current.
   b. depth of the riffles.
   c. number of pools per mile.
   d. vegetation growing along the banks.
   e. nutrient composition of the stream bottom.

2. Which statement characterizes the littoral zone of a lake?
   a. It is the zone of open water.
   b. It extends down to the limits of light penetration.
   c. It is the most richly inhabited zone of a lake.
   d. Its principal occupants are detritivores.
   e. All of these statements are true of the littoral zone.

3. Oceans cover approximately what portion of the earth's surface?
   a. three-fourths
   b. two-thirds
   c. five-eighths
   d. one-half
   e. one-quarter

4. Which of the following are NOT an example of pelagic organisms?
   a. adult fish
   b. algae
   c. shrimp
   d. clams
   e. zooplankton

5. Productivity in the open ocean is limited primarily by the:
   a. number of hours of daylight.
   b. cold-water currents from the poles.
   c. availability of mineral nutrients.
   d. heavy grazing of phytoplankton by herbivores.
   e. depth to which light rays can penetrate.

6. Large numbers of fish may be found in waters in which:
   a. there is violent surf action.
   b. upwellings occur.
   c. warm currents flow.
   d. there is an abundance of coral skeletons.
   e. sponges populate the bottom in large numbers.

7. On the continental shelves of temperate latitudes, the large primary producers are:
   a. aquatic angiosperms.
   b. phytoplankton.
   c. green algae.
   d. corals.
   e. brown algae.

8. The patterns of life on land are determined largely by what two physical factors?
   a. precipitation and mineral content of the soil
   b. temperature and the angle of the earth's axis
   c. angle of the earth's axis and wind velocity
   d. temperature and precipitation
   e. angle of the earth's axis and mineral content of the soil

9. For every increase in altitude of 100 meters, the mean atmospheric temperature drops approximately _____ degree(s).
   a. 0.5
   b. 1.0
   c. 1.5
   d. 2.0
   e. 2.5

10. Within a particular biome, the organisms provide many examples of _____ evolution.
    a. divergent
    b. convergent
    c. dynamic
    d. unchanging
    e. gradual

11. The boundary between temperate mixed forests and northern coniferous forests occurs where:
    a. the growing season is too long for conifers to compete successfully.
    b. summers are too short and winters are too long for deciduous trees to grow well.
    c. the rainfall during the summer months drops dramatically.
    d. snow lies on the ground year round.
    e. temperatures drop below freezing nearly every day of the year.

12. Permafrost is the most distinctive feature of which biome?
    a. taiga
    b. tundra
    c. Pacific Northwest
    d. temperate grassland
    e. chaparral

13. Which statement is NOT characteristic of the tundra biome?
    a. There is a yearly freeze-thaw cycle.
    b. The dominant vegetation consists of small, stunted herbaceous plants.
    c. Caribou, reindeer, and lemmings inhabit the Arctic tundra.
    d. The tundra is devoid of insects and birds.
    e. The ground layer consists of mosses and lichens.

14. Which of these animals would you NOT expect to find in a temperate grassland?
    a. bison
    b. voles
    c. zebras
    d. gazelles
    e. caribou

15. Which statement does NOT reflect a reason why grasses are well-suited for growing in savannas?
    a. There are few native herbivores in this biome.
    b. Grasses grow well in fine, sandy soils.
    c. The aboveground portions of grass plants die during dry seasons.
    d. The dense root network of grasses extracts the maximum amount of water from the soil during the rainy seasons.
    e. The deep roots of the grasses can survive many months of drought.

16. Desert regions are characterized by less than _____ centimeters of rainfall per year.
    a. 5
    b. 10
    c. 15
    d. 20
    e. 25

17. Which statement does NOT reflect a characteristic of epiphytes?
    a. Many have spongy roots.
    b. A major problem facing epiphytes is obtaining adequate water.
    c. Many resemble desert succulents.
    d. Most have long vines by which they obtain nutrients from the soil.
    e. Some varieties trap organic debris in cup-shaped leaves.

18. For plants inhabiting a tropical rain forest, the critical competition is for:
    a. soil nutrients.
    b. water.
    c. space.
    d. carbon dioxide.
    e. light.

19. T or F The largest portion of the biosphere is occupied by terrestrial environments and their inhabitants.

20. T or F Biomes are defined by their characteristic temperature and precipitation patterns.

21. T or F The Pacific Northwest is characterized by wet winters and dry summers.

22. T or F Plants that populate Mediterranean scrub biomes are all phylogenetically related and closely resemble one another in growth patterns and appearance.

23. T or F Plants native to desert biomes often exhibit Crassulacean acid metabolism (CAM) or C_4 photosynthesis, both of which are adaptations that conserve water.

24. T or F More species of plants and animals live in tropical rain forests than in all the rest of the biomes combined.

25. What is an El Niño? What are the possible consequences of an El Niño to marine and terrestrial environments?

26. Mud flats, salt marshes, and estuaries occur in what locations? What is the importance of such areas to various life forms? What is happening to these regions?

27. Summarize the biological activities that occur on the floor of a deciduous forest and enrich the soil.

28. Based on your knowledge of nutrient cycling in tropical rain forests, predict the outcome of slash and burn agriculture and large-scale lumbering operations in these biomes.

PERFORMANCE ANALYSIS

1. **a** The swiftness of the current is the major factor influencing the character of life in a stream. (page 1155)

2. **c** The littoral zone of a lake, which is located along the lake margins, is the most richly populated region. The limnetic zone is the zone of open water extending down to the limits of light penetration. The profundal zone extends down from the limnetic zone. (page 1155)

3. **a** Nearly three-fourths of the earth's surface is covered by oceans. (page 1155)

4. **d** Pelagic organisms are those that are free-floating, such as algae, larval forms of many aquatic organisms, shrimp, many fish species, and plankton. Benthic organisms are bottom-dwellers and include clams, sponges, starfish, crustaceans, and some fish. (page 1156)

5. **c** The productivity of the open ocean is limited primarily by the availability of mineral nutrients. (page 1156)

6. **b** The world's richest fishing areas are located in regions where upwellings bring nutrients up from the ocean floor. (page 1157)

7. **e** The large primary producers of temperate continental shelves are brown algae. (page 1157)

8. **d** Temperature and precipitation largely determine the patterns of life on land. These factors are influenced by the angle of the earth's axis and by the earth's rotation. (page 1160)

9. **a** For each degree of decrease in latitude, mean atmospheric temperature decreases about 0.5°C. Increases in altitude produce a similar decrease in mean atmospheric temperature. An increase in altitude of 100 meters roughly corresponds to an increase in latitude of 1 degree. (page 1161)

10. **b** Biomes are defined in terms of the characteristic plant life. The plants (and animals) of a particular biome are subjected to similar selection pressures. As a consequence, similar adaptations to these conditions have evolved in unrelated organisms located on different continents. Thus, the biomes of different continents provide numerous examples of convergent evolution. (page 1161)

11. **b** The temperate mixed forests give way to northern coniferous forests where the summers are too short and the winters too long for deciduous trees to prosper. (page 1166)

12. **b** Permafrost, which is a permanently frozen subsoil, is a distinctive characteristic of the tundra. (page 1168)

13. **d** Many life forms, including insects and birds, populate the tundra. (pages 1168, 1169)

14. **e** Caribou are native to the tundra, where they live largely on lichens. (pages 1169, 1170)

15. **a** Grasses are well-suited for savannas. Grasslands are typically *maintained* by grazing herbivores and by periodic fires. *Overgrazing*, such as may occur if agricultural livestock are introduced, can lead to destruction of the grassland. (page 1170)

16. **e** Deserts have less than 25 centimeters of rainfall per year. (page 1173)

17. **d** Epiphytes typically germinate in the branches of trees high above the forest floor. They do not obtain nutrients from the soil. (page 1175)

18. **e** Competition for light is a critical survival factor among plants living in tropical rain forests. (page 1174)

19. **False** The biosphere is that part of the earth in which life exists; it includes the oceans and extends about 8 to 10 kilometers above sea level and a few meters into the soil. Nearly three-fourths of the earth's surface is covered by oceans. (page 1154)

20. **False** See the answer to question 10. (page 1161)

21. **True** The Pacific Northwest has wet winters and dry summers. (page 1166)

22. **False** Although they resemble one another superficially, plants that inhabit Mediterranean scrub biomes are not necessarily phylogenetically related. The inhabitants of the chaparral of the North American Southwest are not closely related to the inhabitants of similar biomes on other continents. (page 1171)

23. **True** CAM and C_4 photosynthesis are both adaptations that enable plants to conserve water. They are commonly found in plants that evolved in desert environments. (page 1173)

24. **True** The tropical rain forests have more species than all the rest of the biomes combined. (page 1174)

25. An El Niño is a disturbance of the normal patterns of water circulation in the Pacific Ocean. Changes in the location and strength of atmospheric pressure systems cause warm surface water of the western Pacific to flow toward North and South America. This traps the cold undercurrent along the Americas below the surface. The higher water temperatures block nutrient upwelling and food chains collapse. The changes in the atmospheric pressure systems also may result in changes in weather patterns, resulting in torrential rains in normally dry regions and severe droughts in other regions. (essay, page 1158)

26. Mud flats, salt marshes, and estuaries occur where fresh-water rivers flow into the oceans. They are rich in nutrients and are the spawning grounds and nurseries for many forms of marine life. These areas are rapidly being destroyed to make way for coastal resorts, recreation centers, and other developments. (pages 1159, 1160)

27. Detritivores rapidly break down the organic matter (plant and animal in origin) of the forest floor and return nutrients to the soil. Tree roots penetrate the soil deeply and add organic matter when they die. Insects and other arthropods live in the soil, burrowing through it and making passageways that turn the soil into a water-holding sponge. (page 1166)

28. Most of the nutrients of a tropical rain forest are contained in its vegetation. Decomposition occurs so rapidly that little organic matter is returned to the soil. Also, many soils of tropical rain forests are mostly clay; once the vegetation is removed from such soils, they either erode quickly or form thick crusts that cannot be cultivated after one or two seasons. The mineral content of tropical rain forest soils is very low and minerals are leached out very quickly by the frequent, heavy rainfalls. Deforestation of tropical rain forests quickly produces infertile soils that are unable to support the abundant plant life that was once present. (page 1176)

REVIEW TEST 16

Section 8: Chapters 52–55

This review test is *not* designed to cover all of the important information in these chapters. However, it does touch on the major topic categories in each chapter. It will also give you valuable practice in taking this type of test. When you are finished, check your answers with those provided in the following PERFORMANCE ANALYSIS section.

1. Which of these is a tool used by ecologists 40 years ago that is still used today?
   a. calculus
   b. statistics
   d. computer modeling
   d. observations of nature
   e. radio tracking devices

2. When two ecologically similar species persist in the same community:
   a. they occupy different niches.
   b. they occupy different habitats.
   c. their territories do not overlap.
   d. one is usually nocturnal and the other is diurnal.
   e. they respond differently to temperature, light, and humidity.

3. Which of these is a poor example of an ecosystem?
   a. a pond
   b. a meadow
   c. the earth's atmosphere
   d. a self-sufficient terrarium
   e. the entire surface of the earth

4. Which of the following animals do NOT feed on insects in the littoral zone of lakes?
   a. ducks
   b. frogs
   c. salamanders
   d. water turtles
   e. All of these animals feed on insects in the littoral zone of lakes.

5. The nature of grasslands is maintained by grazing animals and:
   a. sporadic rainfall.
   b. cold/hot seasons.
   c. wet/dry seasons.
   d. poor soils.
   e. periodic fires.

6. The gases of the atmosphere are transparent to visible light, but _____ are NOT transparent to infrared rays reflected from the earth's surface, a phenomenon responsible for the greenhouse effect.
   a. nitrogen and water vapor
   b. carbon dioxide and nitrogen
   c. carbon dioxide and water vapor
   d. ozone and water vapor
   e. carbon dioxide and ozone

7. The major cause of the reduction in the ozone layer is:
   a. fallout from the Chernobyl nuclear power plant accident.
   b. chlorofluorocarbons released into the air.
   c. increasing atmospheric carbon dioxide levels.
   d. sulfates released into the atmosphere from coal-burning power plants.
   e. lead released during the combustion of leaded fossil fuels.

8. T or F   Energy trapped in detritus can reenter the food web when consumers prey on detritivores.

9. T or F   An opportunistic species is one that, after invading a new environment, grows exponentially until it uses up the local resources; then it either enters a dormant state or moves on to a new location.

10. T or F   The particular life-history pattern used by a species is independent of the age structure and mortality pattern of the population.

11. T or F   In character displacement, when two different species compete for the same limited resource, one species will eventually be displaced or eliminated.

12. T or F   Predation may affect both the number of individuals in a prey population and the diversity of species within the community.

13. **T** or **F** Food production in modern technological societies yields 0.1 calorie for every calorie expended. This can occur because energy stored in fossil fuels is utilized in food production.

14. **T** or **F** Extinction rates are higher on smaller islands than on larger islands, a fact that is critically important when nature reserves are established.

15. Determining the presence or absence of competition in a community by experimental means is often difficult. Cite three reasons for this.

16. Distinguish between the terms symbiosis, parasitism, mutualism, and commensalism.

17. Identify and discuss two factors that cause certain elements to become concentrated in the bodies of living organisms. Indicate the trophic level(s) at which the most concentration occurs.

18. Name three types of seashores and distinguish between them on the basis of inhabitants and water activity.

19. Use India as an example to discuss the economic and social factors that contribute to high birth rates in some developing nations.

20. Distinguish between density-dependent and density-independent factors that influence the growth of a population.

21. Contrast northern coniferous forests and temperate deciduous forests with respect to composition and relative rates of decomposition of the ground layers.

22. Distinguish between tropical mixed forests and tropical rain forests with respect to the seasonality of rainfall.

PERFORMANCE ANALYSIS

1. **d** The early ecologists were sensitive observers of nature. Modern ecologists must also be diligent observers, but they also use statistics, calculus, computer modeling, and numerous technical devices in their work. (page 1087)

2. **a** If two ecologically similar species persist in the same community, they must occupy different niches. (page 1109)

3. **c** An ecosystem is a combination of biotic and abiotic components through which energy flows and materials cycle. Examples include a pond, a meadow, a self-sufficient terrarium or aquarium, and the entire surface of the earth. (page 1131)

4. **e** The littoral zone of lakes is at the edges and is the most richly inhabited zone. (page 1155)

5. **e** Grassland biomes are maintained by grazing animals and by periodic fires. (page 1170)

6. **c** Carbon dioxide and water vapor are not transparent to infrared rays radiated from the earth's surface. Such rays are trapped in the atmosphere and create the warm temperatures necessary for life to exist. Increasing carbon dioxide levels are causing warming of the earth's atmosphere and thus the earth's surface. (page 1133)

7. **b** The major cause of the reduction in the ozone layer is the release of chlorofluorocarbons into the atmosphere. These compounds are commonly found in refrigerators, freezers, air conditioners, and home fire extinguishers. (essay, page 1134)

8. **True** Detritivores consume dead prey (scavengers) or break down materials that cannot be utilized by other organisms (decomposers). When these organisms are themselves consumed, the energy they have ingested reenters food webs. (page 1143)

9. **True** Opportunistic species, which often conduct early, rapid, and copious reproduction (i.e., grow exponentially) are well-adapted for colonizing new environments. They may be eliminated from a community if they exhaust the local resources or if slower-growing species are superior competitors for the same resources. (page 1089)

10. **False** The life-history pattern that is most successful for a species depends largely on other properties of the population, including age structure and mortality pattern. (pages 1098, 1099)

11. **False** Character displacement refers to the divergence of overlapping characteristics possessed by two populations in the same environment. The result is a reduction of competition between the populations. (page 1110)

12. **True** Predation may affect the size and age structure of a prey population and may increase species diversity within a community. (page 1114)

13. **True** Food production in technologically advanced societies can be very wasteful of energy because the agricultural techniques use fossil fuels to plant, fertilize, and raise crops. (essay, page 1144)

14. **True** Establishing a nature reserve and then clearing the forest around the reserve creates an island. The equilibrium model of island biogeography predicts that the number of species within the island (reserve) will decrease under such circumstances. (page 1124; essay, page 1126)

15. (1) To conduct a controlled experiment, both experimental and control areas are required and the conditions in both areas should be as close to identical as possible. (2) The study must be repeated at different times of the year and in subsequent years to rule out the possibility that the results were due to seasonal variations. (3) The manipulations required to control the experimental conditions might interfere with the natural behavior of the organisms, in which case the results would be meaningless. (page 1112)

16. **Symbiosis** is a close and long-term association between organisms of two different species. **Parasitism** is a symbiosis in which one partner benefits at the expense of the other. In **mutualism**, both partners benefit from the asso-

ciation. In **commensalism**, one partner benefits, but the other partner neither benefits nor is harmed. (pages 1119, 1120)

17. (1) The selective uptake of materials by living cells enables organisms to accumulate substances in their bodies at a higher concentration than exists in the environment. (2) The movement of nonmetabolizable materials through food chains results in the amplification of those materials at higher trophic levels. (For example, if each fish in a particular region contains 10 units of DDT in its body, a bird who eats five fish will have 50 units of DDT in its body.) Consumers, and especially top predators (including humans), are the organisms in which concentration of materials is the greatest. (page 1150)

18. Inhabitants of **rocky coasts** are adapted to rising and falling tides and waves pounding upon rocks. Many have adaptations for clinging to rocks. Typical inhabitants include a variety of algae, mussels, oysters, barnacles, starfish, and other invertebrates. Three zones that characterize rocky coasts are supratidal, intertidal, and subtidal. **Sandy beaches** have fewer bottom dwellers than do rocky coasts because the shifting sands are unstable. Some animals live below the sand, including clams, ghost crabs, and lugworms. **Mud flats, salt marshes, and estuaries** occur where freshwater rivers and streams empty into the ocean. They are richly inhabited by plants and animals and serve as spawning grounds and nurseries for many types of marine life. (pages 1157–1160)

19. In India, the desire for large families has deep roots in the culture. Further, children (meaning sons in this case) tra-

ditionally care for their elderly parents, providing a kind of "social security." The mortality rate among children is such that a couple must have an average of five children (remember that half of them will be girls) if there is to be a 95 percent chance that one son will survive to the father's sixty-fifth birthday. (essay, page 1097)

20. Factors that influence the birth rate or the mortality rate of a population as population density changes are referred to as density-dependent. Examples are disease, limited food supply, and predation. Factors that influence birth rate or mortality rate regardless of population density are said to be density-independent. Such factors are often weather-related. (page 1098)

21. Temperate deciduous forests typically grow in rich soils. The ground layer vegetation consists of mosses and liverworts and the ground is often covered with leaf litter. Decomposition is fairly rapid and the return of nutrients to the soil by detritivores contributes to soil fertility. Northern coniferous forests have a thick layer of needles and dead twigs on the ground, matted together by fungal mycelia. The ground layer is less richly populated than that of deciduous forests and decomposition of the accumulated litter is slower. (pages 1164, 1166)

22. Tropical mixed forests and monsoon forests occur in areas where there are distinct wet and dry seasons. Tropical rain forests are found in areas where rainfall is abundant year-round. (page 1174)